SOIL COLLOIDS

PROPERTIES AND ION BINDING

SURFACTANT SCIENCE SERIES

FOUNDING EDITOR
MARTIN J. SCHICK
1918–1998

SERIES EDITOR
ARTHUR T. HUBBARD
Santa Barbara Science Project
Santa Barbara, California

1. Nonionic Surfactants, *edited by Martin J. Schick* (see also Volumes 19, 23, and 60)
2. Solvent Properties of Surfactant Solutions, *edited by Kozo Shinoda* (see Volume 55)
3. Surfactant Biodegradation, *R. D. Swisher* (see Volume 18)
4. Cationic Surfactants, *edited by Eric Jungermann* (see also Volumes 34, 37, and 53)
5. Detergency: Theory and Test Methods (in three parts), *edited by W. G. Cutler and R. C. Davis* (see also Volume 20)
6. Emulsions and Emulsion Technology (in three parts), *edited by Kenneth J. Lissant*
7. Anionic Surfactants (in two parts), *edited by Warner M. Linfield* (see Volume 56)
8. Anionic Surfactants: Chemical Analysis, *edited by John Cross*
9. Stabilization of Colloidal Dispersions by Polymer Adsorption, *Tatsuo Sato and Richard Ruch*
10. Anionic Surfactants: Biochemistry, Toxicology, Dermatology, *edited by Christian Gloxhuber* (see Volume 43)
11. Anionic Surfactants: Physical Chemistry of Surfactant Action, *edited by E. H. Lucassen-Reynders*
12. Amphoteric Surfactants, *edited by B. R. Bluestein and Clifford L. Hilton* (see Volume 59)
13. Demulsification: Industrial Applications, *Kenneth J. Lissant*
14. Surfactants in Textile Processing, *Arved Datyner*
15. Electrical Phenomena at Interfaces: Fundamentals, Measurements, and Applications, *edited by Ayao Kitahara and Akira Watanabe*
16. Surfactants in Cosmetics, *edited by Martin M. Rieger* (see Volume 68)
17. Interfacial Phenomena: Equilibrium and Dynamic Effects, *Clarence A. Miller and P. Neogi*
18. Surfactant Biodegradation: Second Edition, Revised and Expanded, *R. D. Swisher*
19. Nonionic Surfactants: Chemical Analysis, *edited by John Cross*
20. Detergency: Theory and Technology, *edited by W. Gale Cutler and Erik Kissa*
21. Interfacial Phenomena in Apolar Media, *edited by Hans-Friedrich Eicke and Geoffrey D. Parfitt*
22. Surfactant Solutions: New Methods of Investigation, *edited by Raoul Zana*
23. Nonionic Surfactants: Physical Chemistry, *edited by Martin J. Schick*
24. Microemulsion Systems, *edited by Henri L. Rosano and Marc Clausse*
25. Biosurfactants and Biotechnology, *edited by Naim Kosaric, W. L. Cairns, and Neil C. C. Gray*
26. Surfactants in Emerging Technologies, *edited by Milton J. Rosen*

27. Reagents in Mineral Technology, *edited by P. Somasundaran and Brij M. Moudgil*
28. Surfactants in Chemical/Process Engineering, *edited by Darsh T. Wasan, Martin E. Ginn, and Dinesh O. Shah*
29. Thin Liquid Films, *edited by I. B. Ivanov*
30. Microemulsions and Related Systems: Formulation, Solvency, and Physical Properties, *edited by Maurice Bourrel and Robert S. Schechter*
31. Crystallization and Polymorphism of Fats and Fatty Acids, *edited by Nissim Garti and Kiyotaka Sato*
32. Interfacial Phenomena in Coal Technology, *edited by Gregory D. Botsaris and Yuli M. Glazman*
33. Surfactant-Based Separation Processes, *edited by John F. Scamehorn and Jeffrey H. Harwell*
34. Cationic Surfactants: Organic Chemistry, *edited by James M. Richmond*
35. Alkylene Oxides and Their Polymers, *F. E. Bailey, Jr., and Joseph V. Koleske*
36. Interfacial Phenomena in Petroleum Recovery, *edited by Norman R. Morrow*
37. Cationic Surfactants: Physical Chemistry, *edited by Donn N. Rubingh and Paul M. Holland*
38. Kinetics and Catalysis in Microheterogeneous Systems, *edited by M. Grätzel and K. Kalyanasundaram*
39. Interfacial Phenomena in Biological Systems, *edited by Max Bender*
40. Analysis of Surfactants, *Thomas M. Schmitt* (see Volume 96)
41. Light Scattering by Liquid Surfaces and Complementary Techniques, *edited by Dominique Langevin*
42. Polymeric Surfactants, *Irja Piirma*
43. Anionic Surfactants: Biochemistry, Toxicology, Dermatology, Second Edition, Revised and Expanded, *edited by Christian Gloxhuber and Klaus Künstler*
44. Organized Solutions: Surfactants in Science and Technology, *edited by Stig E. Friberg and Björn Lindman*
45. Defoaming: Theory and Industrial Applications, *edited by P. R. Garrett*
46. Mixed Surfactant Systems, *edited by Keizo Ogino and Masahiko Abe*
47. Coagulation and Flocculation: Theory and Applications, *edited by Bohuslav Dobiás*
48. Biosurfactants: Production Properties Applications, *edited by Naim Kosaric*
49. Wettability, *edited by John C. Berg*
50. Fluorinated Surfactants: Synthesis Properties Applications, *Erik Kissa*
51. Surface and Colloid Chemistry in Advanced Ceramics Processing, *edited by Robert J. Pugh and Lennart Bergström*
52. Technological Applications of Dispersions, *edited by Robert B. McKay*
53. Cationic Surfactants: Analytical and Biological Evaluation, *edited by John Cross and Edward J. Singer*
54. Surfactants in Agrochemicals, *Tharwat F. Tadros*
55. Solubilization in Surfactant Aggregates, *edited by Sherril D. Christian and John F. Scamehorn*
56. Anionic Surfactants: Organic Chemistry, *edited by Helmut W. Stache*
57. Foams: Theory, Measurements, and Applications, *edited by Robert K. Prud'homme and Saad A. Khan*

58. The Preparation of Dispersions in Liquids, *H. N. Stein*
59. Amphoteric Surfactants: Second Edition, *edited by Eric G. Lomax*
60. Nonionic Surfactants: Polyoxyalkylene Block Copolymers, *edited by Vaughn M. Nace*
61. Emulsions and Emulsion Stability, *edited by Johan Sjöblom*
62. Vesicles, *edited by Morton Rosoff*
63. Applied Surface Thermodynamics, *edited by A. W. Neumann and Jan K. Spelt*
64. Surfactants in Solution, *edited by Arun K. Chattopadhyay and K. L. Mittal*
65. Detergents in the Environment, *edited by Milan Johann Schwuger*
66. Industrial Applications of Microemulsions, *edited by Conxita Solans and Hironobu Kunieda*
67. Liquid Detergents, *edited by Kuo-Yann Lai*
68. Surfactants in Cosmetics: Second Edition, Revised and Expanded, *edited by Martin M. Rieger and Linda D. Rhein*
69. Enzymes in Detergency, *edited by Jan H. van Ee, Onno Misset, and Erik J. Baas*
70. Structure-Performance Relationships in Surfactants, *edited by Kunio Esumi and Minoru Ueno*
71. Powdered Detergents, *edited by Michael S. Showell*
72. Nonionic Surfactants: Organic Chemistry, *edited by Nico M. van Os*
73. Anionic Surfactants: Analytical Chemistry, Second Edition, Revised and Expanded, *edited by John Cross*
74. Novel Surfactants: Preparation, Applications, and Biodegradability, *edited by Krister Holmberg*
75. Biopolymers at Interfaces, *edited by Martin Malmsten*
76. Electrical Phenomena at Interfaces: Fundamentals, Measurements, and Applications, Second Edition, Revised and Expanded, *edited by Hiroyuki Ohshima and Kunio Furusawa*
77. Polymer-Surfactant Systems, *edited by Jan C. T. Kwak*
78. Surfaces of Nanoparticles and Porous Materials, *edited by James A. Schwarz and Cristian I. Contescu*
79. Surface Chemistry and Electrochemistry of Membranes, *edited by Torben Smith Sørensen*
80. Interfacial Phenomena in Chromatography, *edited by Emile Pefferkorn*
81. Solid–Liquid Dispersions, *Bohuslav Dobiás, Xueping Qiu, and Wolfgang von Rybinski*
82. Handbook of Detergents, editor in chief: Uri Zoller Part A: Properties, *edited by Guy Broze*
83. Modern Characterization Methods of Surfactant Systems, *edited by Bernard P. Binks*
84. Dispersions: Characterization, Testing, and Measurement, *Erik Kissa*
85. Interfacial Forces and Fields: Theory and Applications, *edited by Jyh-Ping Hsu*
86. Silicone Surfactants, *edited by Randal M. Hill*

87. Surface Characterization Methods: Principles, Techniques, and Applications, *edited by Andrew J. Milling*
88. Interfacial Dynamics, *edited by Nikola Kallay*
89. Computational Methods in Surface and Colloid Science, *edited by Malgorzata Borówko*
90. Adsorption on Silica Surfaces, *edited by Eugène Papirer*
91. Nonionic Surfactants: Alkyl Polyglucosides, *edited by Dieter Balzer and Harald Lüders*
92. Fine Particles: Synthesis, Characterization, and Mechanisms of Growth, *edited by Tadao Sugimoto*
93. Thermal Behavior of Dispersed Systems, *edited by Nissim Garti*
94. Surface Characteristics of Fibers and Textiles, *edited by Christopher M. Pastore and Paul Kiekens*
95. Liquid Interfaces in Chemical, Biological, and Pharmaceutical Applications, *edited by Alexander G. Volkov*
96. Analysis of Surfactants: Second Edition, Revised and Expanded, *Thomas M. Schmitt*
97. Fluorinated Surfactants and Repellents: Second Edition, Revised and Expanded, *Erik Kissa*
98. Detergency of Specialty Surfactants, *edited by Floyd E. Friedli*
99. Physical Chemistry of Polyelectrolytes, *edited by Tsetska Radeva*
100. Reactions and Synthesis in Surfactant Systems, *edited by John Texter*
101. Protein-Based Surfactants: Synthesis, Physicochemical Properties, and Applications, *edited by Ifendu A. Nnanna and Jiding Xia*
102. Chemical Properties of Material Surfaces, *Marek Kosmulski*
103. Oxide Surfaces, *edited by James A. Wingrave*
104. Polymers in Particulate Systems: Properties and Applications, *edited by Vincent A. Hackley, P. Somasundaran, and Jennifer A. Lewis*
105. Colloid and Surface Properties of Clays and Related Minerals, *Rossman F. Giese and Carel J. van Oss*
106. Interfacial Electrokinetics and Electrophoresis, *edited by Ángel V. Delgado*
107. Adsorption: Theory, Modeling, and Analysis, *edited by József Tóth*
108. Interfacial Applications in Environmental Engineering, *edited by Mark A. Keane*
109. Adsorption and Aggregation of Surfactants in Solution, *edited by K. L. Mittal and Dinesh O. Shah*
110. Biopolymers at Interfaces: Second Edition, Revised and Expanded, *edited by Martin Malmsten*
111. Biomolecular Films: Design, Function, and Applications, *edited by James F. Rusling*
112. Structure–Performance Relationships in Surfactants: Second Edition, Revised and Expanded, *edited by Kunio Esumi and Minoru Ueno*
113. Liquid Interfacial Systems: Oscillations and Instability, *Rudolph V. Birikh, Vladimir A. Briskman, Manuel G. Velarde, and Jean-Claude Legros*
114. Novel Surfactants: Preparation, Applications, and Biodegradability: Second Edition, Revised and Expanded, *edited by Krister Holmberg*

115. Colloidal Polymers: Synthesis and Characterization, *edited by Abdelhamid Elaissari*
116. Colloidal Biomolecules, Biomaterials, and Biomedical Applications, *edited by Abdelhamid Elaissari*
117. Gemini Surfactants: Synthesis, Interfacial and Solution-Phase Behavior, and Applications, *edited by Raoul Zana and Jiding Xia*
118. Colloidal Science of Flotation, *Anh V. Nguyen and Hans Joachim Schulze*
119. Surface and Interfacial Tension: Measurement, Theory, and Applications, *edited by Stanley Hartland*
120. Microporous Media: Synthesis, Properties, and Modeling, *Freddy Romm*
121. Handbook of Detergents, editor in chief: Uri Zoller, Part B: Environmental Impact, *edited by Uri Zoller*
122. Luminous Chemical Vapor Deposition and Interface Engineering, *Hirotsugu Yasuda*
123. Handbook of Detergents, editor in chief: Uri Zoller, Part C: Analysis, *edited by Heinrich Waldhoff and Rüdiger Spilker*
124. Mixed Surfactant Systems: Second Edition, Revised and Expanded, *edited by Masahiko Abe and John F. Scamehorn*
125. Dynamics of Surfactant Self-Assemblies: Micelles, Microemulsions, Vesicles and Lyotropic Phases, *edited by Raoul Zana*
126. Coagulation and Flocculation: Second Edition, *edited by Hansjoachim Stechemesser and Bohulav Dobiás*
127. Bicontinuous Liquid Crystals, *edited by Matthew L. Lynch and Patrick T. Spicer*
128. Handbook of Detergents, editor in chief: Uri Zoller, Part D: Formulation, *edited by Michael S. Showell*
129. Liquid Detergents: Second Edition, *edited by Kuo-Yann Lai*
130. Finely Dispersed Particles: Micro-, Nano-, and Atto-Engineering, *edited by Aleksandar M. Spasic and Jyh-Ping Hsu*
131. Colloidal Silica: Fundamentals and Applications, *edited by Horacio E. Bergna and William O. Roberts*
132. Emulsions and Emulsion Stability, Second Edition, *edited by Johan Sjöblom*
133. Micellar Catalysis, *Mohammad Niyaz Khan*
134. Molecular and Colloidal Electro-Optics, *Stoyl P. Stoylov and Maria V. Stoimenova*
135. Surfactants in Personal Care Products and Decorative Cosmetics, Third Edition, *edited by Linda D. Rhein, Mitchell Schlossman, Anthony O'Lenick, and P. Somasundaran*
136. Rheology of Particulate Dispersions and Composites, *Rajinder Pal*
137. Powders and Fibers: Interfacial Science and Applications, *edited by Michel Nardin and Eugène Papirer*
138. Wetting and Spreading Dynamics, *edited by Victor Starov, Manuel G. Velarde, and Clayton Radke*
139. Interfacial Phenomena: Equilibrium and Dynamic Effects, Second Edition, *edited by Clarence A. Miller and P. Neogi*

140. Giant Micelles: Properties and Applications, *edited by Raoul Zana and Eric W. Kaler*
141. Handbook of Detergents, editor in chief: Uri Zoller, Part E: Applications, *edited by Uri Zoller*
142. Handbook of Detergents, editor in chief: Uri Zoller, Part F: Production, *edited by Uri Zoller and co-edited by Paul Sosis*
143. Sugar-Based Surfactants: Fundamentals and Applications, *edited by Cristóbal Carnero Ruiz*
144. Microemulsions: Properties and Applications, *edited by Monzer Fanun*
145. Surface Charging and Points of Zero Charge, *Marek Kosmulski*
146. Structure and Functional Properties of Colloidal Systems, *edited by Roque Hidalgo-Álvarez*
147. Nanoscience: Colloidal and Interfacial Aspects, *edited by Victor M. Starov*
148. Interfacial Chemistry of Rocks and Soils, *Noémi M. Nagy and József Kónya*
149. Electrocatalysis: Computational, Experimental, and Industrial Aspects, *edited by Carlos Fernando Zinola*
150. Colloids in Drug Delivery, *edited by Monzer Fanun*
151. Applied Surface Thermodynamics: Second Edition, *edited by A. W. Neumann, Robert David, and Yi Y. Zuo*
152. Colloids in Biotechnology, *edited by Monzer Fanun*
153. Electrokinetic Particle Transport in Micro/Nano-fluidics: Direct Numerical Simulation Analysis, *Shizhi Qian and Ye Ai*
154. Nuclear Magnetic Resonance Studies of Interfacial Phenomena, *Vladimir M. Gun'ko and Vladimir V. Turov*
155. The Science of Defoaming: Theory, Experiment and Applications, *Peter R. Garrett*
156. Soil Colloids: Properties and Ion Binding, *Fernando V. Molina*

SOIL COLLOIDS

PROPERTIES AND ION BINDING

Fernando V. Molina
University of Bueno Aires
Buenos Aires, Argentina

CRC Press
Taylor & Francis Group
Boca Raton London New York

CRC Press is an imprint of the
Taylor & Francis Group, an **informa** business

CRC Press
Taylor & Francis Group
6000 Broken Sound Parkway NW, Suite 300
Boca Raton, FL 33487-2742

Version Date: 20130220

International Standard Book Number-13: 978-1-4398-5114-2 (Hardback)

Library of Congress Cataloging-in-Publication Data

Molina, Fernando V.
Soil colloids : properties and ion binding / author, Fernando V. Molina.
pages cm. -- (Surfactant science series)
Includes bibliographical references and index.
ISBN 978-1-4398-5114-2 (alk. paper)
1. Soil colloids. 2. Soils--Analysis. I. Title.

S593.3.M65 2013
578.75'7--dc23 2013005085

Visit the Taylor & Francis Web site at
http://www.taylorandfrancis.com

and the CRC Press Web site at
http://www.crcpress.com

To my wife, Estela.

Contents

PART II Soil Components

PART III Ion Binding to Soil Colloids

Preface

This book is devoted to the study of soil components and their interaction with ions from the point of view of colloid science. However, it is not directed toward colloid chemists, but an attempt has been made to bridge that field with the soil and environmental sciences. Thus, Parts I and II include many basic concepts on colloid and soil science; but one has to start somewhere, and so it is expected that the reader will know general chemistry and chemical thermodynamics, up to the concept of chemical potential. Also, understanding of electrostatics, and the consequently necessary mathematical skills, is assumed. Finally, a basic knowledge of crystallography and mineralogy is also expected.

I have made every effort to keep the difficulty of the text at a moderate level, especially regarding mathematical treatments. Sections 3.2.3, 3.2.4, and 3.4 have more advanced or specific materials.

I wish to express my gratitude to Professor Arthur Hubbard for inviting me to contribute to the Surfactant Science Series with this book, and also thank the people at CRC Press, Barbara Glunn and David Fausel, for their patience and willingness to collaborate. Special thanks to Dr. Marcelo Avena for his revision of the manuscript. Thanks also to my students and collaborators for their patience and to all the people who in one or another way contributed to this work.

List of Common Greek Symbols

γ_i^m	Activity coefficient (molality scale) of species *i*
γ_i^R	Activity coefficient (Raoult scale) of species *i*
γ_{AB}	Interfacial tension or interfacial Gibbs free energy, between phases A and B
Γ_i	Surface excess of species *i*
ε	Dielectric constant
ε_0	Vacuum permittivity
ε_r	Relative dielectric constant
ζ	Electrokinetic ("zeta") potential
η	Viscosity coefficient
θ_i	Degree of coverage of species *i* on a sorbent
ϑ	Specific particle charge (mol kg^{-1})
ϑ_d	Diffuse layer specific charge
ϑ_H	Net proton specific charge
ϑ_I	Net inner sphere sorption specific charge
ϑ_O	Net outer sphere sorption specific charge
κ	Debye inverse length
λ	Wavelength
$\tilde{\mu}_i$	Electrochemical potential of species *i*
μ_i	Chemical potential of species *i* / Electrophoretic mobility of ion *i*
ρ	Number density (molecules per unity area)
ρ_P	Polydispersity ratio
ρ_q	Charge density
σ	Charge density (C m^{-2})
σ_S, σ_0	Net surface charge density (C m^{-2})
σ_d	Diffuse layer surface charge density (C m^{-2})
σ_f	Fixed surface charge density (C m^{-2})
σ_H	Net proton surface charge density (C m^{-2})
σ_I	Net inner sphere sorption surface charge density (C m^{-2})
σ_O	Net outer sphere sorption surface charge density (C m^{-2})
ψ	Electrostatic potential
ψ_0	Surface electrostatic potential
ψ_1	Electrostatic potential at the inner Helmholtz plane
ψ_2	Electrostatic potential at the outer Helmholtz plane
Ψ	Dimensionless electrostatic potential

List of Common Symbols

A	Area
A_p	Particle area
A_S	Specific surface area
A_{ijk}	Hamaker constant for particles (or surfaces) i and j in a medium k
a_i	Activity of species i
B_F	Freundlich isotherm parameter
B_{FG}	Fowler–Guggenheim interaction parameter
C	Differential interfacial capacitance
C_d	Diffuse layer capacitance
C_i	Inner layer capacitance
C^{int}	Integral interfacial capacitance
c_i	Molar concentration of species i
c_i^{∞}	Concentration of ion i in the solution bulk
D	Diffusion coefficient
d	Distance between two particles or surfaces
E	Nernst potential; electric field
$\boldsymbol{e}$	Euler's number, base of natural logarithms
e	Elementary charge (electron charge absolute value)
$\boldsymbol{F}$	Faraday constant
F	Helmholtz free energy
$\tilde{F}$	Force
f_{AB}	Free energy of interaction between molecules A and B
$f(\log K)$	Affinity (or equilibrium constant) distribution function
G	Gibbs free energy
g_f	Gel fraction in humic electrostatic models
H	Enthalpy
h	Planck's constant
I	Ionic strength
I_i	Ionization potential of molecule i
K	Equilibrium constant
K'	Conditional equilibrium constant
$\tilde{K}$	Differential equilibrium function
K^{int}	Intrinsic sorption constant
K_{ex}	Thermodynamic exchange equilibrium constant or coefficient
$\boldsymbol{k}$	Boltzmann constant
M	Molecular weight
M_r	Molar mass
$\bar{M}_n$	Number-average molecular weight
$\bar{M}_w$	Weight-average molecular weight
$\bar{M}_z$	z-average molecular weight
$\bar{M}_v$	Viscosity-average molecular weight

m^0	Standard molality (= 1 molal)
m_i	Molality of species i
N_A	Avogadro's number
n	Sips distribution width parameter; heterogeneity parameter
n_i	Mole number of species i
P^0	Standard pressure (= 1 bar)
P_i	Pressure of gas i
pH_0	Point of zero net charge
$\mathrm{pH}_{H,0}$	Point of zero net proton charge
$\mathrm{pH}_{in,0}$	Point of zero net intrinsic charge
$\mathrm{pH}_{se,0}$	Point of zero salt effect
$\mathrm{pH}_{\mathrm{iep}}$	Isoelectric point
p_i	Dipolar moment of species i
Q_i	Specific amount sorbed of species i (moles per unity of mass)
Q_S	Specific amount sorbed in a monolayer
Q_{AE}	Anion exchange capacity
Q_{CE}	Cation exchange capacity
q_f	Fixed particle charge
q_d	Diffuse layer particle charge
q_H	Net proton particle charge
q_i	Charge of the ionic species i
q_I	Net inner sphere sorption particle charge
q_O	Net outer sphere sorption particle charge
$\boldsymbol{R}$	Gas constant
R_G	Radius of gyration of a macromolecule or colloidal particle
r	Distance between two ions or molecules; radial coordinate in the spherical coordinate system
T	Absolute temperature
U	Internal energy
U_{AB}	Potential energy of interaction between particles A and B
u_{AB}	Potential energy of interaction between molecules A and B
V	Volume
W	Work
x	Mole fraction
x_1	Position of the inner Helmholtz plane; position of "1" plane in surface complexation models
x_2	Position of the outer Helmholtz (Stern) plane; start of diffuse layer in surface complexation models
y	Equivalent fraction in exchanger phase
z_i	Numeric charge of ion i (in units of q_e)
z	The z coordinate. Charge of each ion in a symmetrical electrolyte.
Z	Coordination number

List of Common Acronyms

AAS	Atomic absorption spectrometry
AEC	Anion exchange capacity
AFM	Atomic force microscopy
BET	Brunauer–Emmet–Teller theory
BSM	Basic Stern model
CCC	Critical coagulation concentration
CCM	Constant capacitance model
CEC	Cation exchange capacity
DEF	Differential equilibrium function
DFT	Density functional theory
DL	Double layer
DLM	Diffuse layer model
DLVO	Derjaguin–Landau–Verwey–Overbeek theory
DSC	Differential scanning calorimetry
DTA	Differential thermal analysis
EELS	Electron energy-loss spectrometry
EEM	Excitation-emission matrix
EPN	Elastic polyelectrolyte network
EXAFS	Extended X-ray absorption fine structure
FA	Fulvic acid
FG	Fowler–Guggenheim
GC	Gouy–Chapman
GCSG	Gouy–Chapman–Stern–Grahame
GF-AAS	Graphite-furnace atomic absorption spectrometry
HA	Humic acid
HFO	Hydrous ferric oxide
HG-AAS	Hydride generation atomic absorption spectrometry
HS	Humic substances
ICP-AES	Inductively coupled plasma-atomic emission spectrometry
ICP-MS	Inductively coupled plasma-mass spectrometry
IEP	Isoelectric point
IHP	Inner Helmholtz Plane
IR	Infrared or infrared spectroscopy
ISM	Impenetrable sphere model
LF	Langmuir–Freundlich
LIA	Local isotherm approximation
MC	Monte Carlo
MPB	Modified Poisson–Boltzmann
NAA	Neutron activation analysis
NICA	Nonideal competitive adsorption
NIRS	Near-infrared spectroscopy

NMR	Nuclear magnetic resonance
NOM	Natural organic matter
OHP	Outer Helmholtz plane
PB	Poisson–Boltzmann
PDI	Potential determining ion
PM	Primitive model
PSM	Permeable sphere model
pzc	Point of zero net charge
pznpc	Point of zero net proton charge
pzse	Point of zero salt effect
RPM	Restricted primitive model
SCM	Surface complexation modeling
SEC	Size exclusion chromatography
SEM	Scanning electron microscopy (or microscope)
SOM	Soil organic matter
TEM	Transmission electron microscopy (or microscope)
TGA	Thermogravimetric analysis or thermogravimetry
TLM	Triple layer model
X-EDS	X-ray energy-dispersive spectroscopy
XAFS	X-ray absorption fine structure
XANES	X-ray absorption near-edge structure
XAS	X-ray absorption spectroscopy
XPS	X-ray photoelectron spectroscopy
XRD	X-ray diffraction
XRF	X-ray fluorescence

Part I

Foundations

1 Introduction

1.1 SOIL

Soil has been defined as a "thin but precious skin which covers the geological formations at the earth's surface with a unique, extremely fragile veneer" (Sumner 1999). Those are very adequate words. The soil, if we include subaquatic and submarine surfaces, sustains all life on earth by supporting nutrients for plants and microorganisms that are at the base of the food chain. On the earth's surface, however, not all parts are well suited for that task, as there are areas covered by rocks where plants can hardly survive. But rocks are subject to erosion and several transformations originated by geological, climatic, chemical, biological, and other factors; these processes, collectively known as weathering, transform rocks into particles of relatively small sizes, which constitute the basis of what we know as soil. The soil is thus essentially a porous structure, which is permeated by water, dissolving nutrients and other inorganic species—constituting the soil solution—and by air in the spaces not filled by the solution. This structure is the place where plant roots get their nutrients and water; some other living entities such as bacteria also get their food from it. The residuals and debris from plants and other living beings, including animals, also go to the ground, completing the deeply interrelated association of inorganic, organic, and biological matter that constitutes the soil. There we find essentially two types of chemical processes: reactions occurring in the soil solution, which are very similar to those in groundwaters, and the surface reactions taking place at the interface between the soil solution and the soil solid particles. These surface processes are of utmost importance in the soil properties and behavior.

Soils components can be classified into three major classes: namely, mineral, organic, and biological matter. Minerals include mainly oxides and clay minerals and also other materials of low crystallinity. Organic matter includes mainly humic substances; other organic materials of more recognizable nature, such as peptides and saccharides, are usually downgraded in a relatively short time. Biological matter, if considered separately from organics, mainly comprises bacteria. All these solids, either separately or (as actually they are) as a whole, are known to interact with ions in different forms. Most of them are of colloidal nature, that is, with sizes ranging from 1 nm to a few micrometers, and thus can be grouped under the term *soil colloids*.

1.2 SOIL CHEMISTRY, SOIL COLLOIDS, AND METAL IONS

Soil science is a broad interdisciplinary field of science, dealing with all physical, chemical, biological, agricultural, and mineralogical aspects related in some way to the soil, either as a subject of basic knowledge or as a very important resource. Within this broad field, soil chemistry embraces different chemical processes taking place, including, among others, mineral weathering, humification of organic plant residues, and the ionic reactions occurring in the soil solution and at the solid–solution interface on soil colloids. Ionic species play a significant role in soil, as most of the nutrients are present in that form. Metal ions are present in soil, both of natural origin (alkaline, alkaline earth metals, Fe[II/III], Al[III], etc.) and of foreign origin (i.e., anthropogenic), often pollutants (Cu[II], Cd[II], Pb[II], etc.). Pollutant metallic cations are commonly found in soils as a result of human activity, being sorbed (or bound) to soil materials under mechanisms that are still under study. The binding of ions to soil particles has a central role in defining the fate of foreign species, such as pollutants, contributing in great measure to contaminant fixation. On the other hand, the interaction between ionic nutrients and soil particles has a direct impact on nutrient availability. Thus, the knowledge of the interaction between ionic species and soil components is necessary for the understanding of all those problems.

This book is devoted to the subject of soil colloidal components and their interaction with ionic species. The fundamental problem of ion binding to natural colloids (both inorganic, such as clays and oxides, and organic, mainly humic substances) is a multidisciplinary research field that raises special challenges. These come mainly from the heterogeneous and (sometimes) ill-defined nature of natural colloids, especially humic substances. Soil scientists, on one side, and colloid chemists, on the other, tend to have different views of the problem. This book is intended as a contribution to fill the gap and also to bring the latest advances in this active research field.

As in many fields of science, progress in the subject of the interaction between ions and soil colloids goes from experimental studies and data collection to the development of models seeking to reproduce and predict the observed results, to finally the establishment of general theories, based on fundamental physicochemical principles explaining the behavior of the system. Modeling has different purposes, depending on the point of view: from a geochemical perspective, a working model, able to predict reasonably well the state (solid, complex, free, etc.) of different chemical species, is an important tool; from a more fundamental view, it is also a step in the way of revealing the physical nature of the systems and the interactions that define its properties. At present, much of the work is in the stage of formulating working models capable of predicting the state and fate of the chemical species present, either natural or foreign. An important part of this book including most of Part III and many of the fundamental concepts presented in Part I, is devoted to ion sorption modeling.

1.3 BOOK CONTENTS

The book content is arranged into three parts. Part I covers the fundamentals of colloid science. This is intended for soil scientists and other researchers not familiar with these principles. Because of that, and as there are many excellent books covering colloid and interface science fundamentals (Verwey and Overbeek 1948; Shaw 1992; Lyklema 1995; Birdi 1997; Hiemenz and Rajagopalan 1997; Hubbard 2002; Shchukin et al. 2002; Birdi 2009), this part discusses all the important concepts but without too much detail, such as extensive mathematical derivations. Part II deals with soil composition and its components, especially the main ones: clay and oxide minerals and humic substances. In a similar way to Part I, this is mainly intended for scientists and professionals outside the soil science field, as the topics are widely covered in many references (Sumner 1999; Sparks 2002; White 2005; Brady and Weil 2007; Sposito 2008; Tan 2009). It covers the composition and properties of these components, with a deeper coverage of their colloidal properties and the experimental facts of ion sorption on these colloids. Also, characterization methodology, which is an ever-evolving field, is treated with some detail. Part III is devoted to an in-depth coverage of ion binding (or complexation) to soil colloids, with a focus on modeling, including recent advances by different research groups. Chapter 11 comprises the fundamental principles, such as general concepts and the problem of the affinity spectrum (or distribution of complexation constants) arising from the heterogeneous nature of most natural colloids, especially organic ones. Even when the concept of affinity spectrum is usually considered in connection with humic substances, it can be equally applied to other substrates. Chapters 12 and 13 cover, respectively, ion binding to minerals and humics, including the different theoretical approaches. Finally, Chapter 14 deals with ion binding to several components together, or whole natural soils, and presents the main advancing lines in the modeling. Owing to the present state of knowledge, this chapter is more close to a review.

REFERENCES

Birdi, K. S. 1997. *Handbook of Surface and Colloid Chemistry*. 1st ed. Boca Raton, FL: CRC Press.

Birdi, K. S. 2009. *Surface and Colloid Chemistry: Principles and Applications*. 1st ed. Boca Raton, FL: CRC Press.

Brady, N. C., and R. R. Weil. 2007. *The Nature and Properties of Soils*. 14th ed. Upper Saddle River, NJ: Prentice Hall.

Hiemenz, P. C., and R. Rajagopalan. 1997. *Principles of Colloid and Surface Chemistry*. 3rd ed. New York: Marcel Dekker.

Hubbard, A. T. 2002. *Encyclopedia of Surface and Colloid Science*. 4 Volume Set. 1st ed. New York: Marcel Dekker.

Lyklema, J. 1995. *Fundamentals of Interface and Colloid Science: Solid-Liquid Interfaces*. London: Academic Press.

Shaw, D. 1992. *Introduction to Colloid and Surface Chemistry*. 4th ed. Oxford: Butterworth-Heinemann.

Shchukin, E. D., A. V. Pertsov, E. A. Amelina, and A. S. Zelenev. 2002. *Colloid and Surface Chemistry*, Vol. 12. 2nd ed. Amsterdam: Elsevier Science.

Sparks, D. L. 2002. *Environmental Soil Chemistry*. 2nd ed. San Diego, CA: Academic Press.

Sposito, G. 2008. *The Chemistry of Soils*. 2nd ed. New York: Oxford University Press

Sumner, M. E., ed. 1999. *Handbook of Soil Science*. 1st ed. Boca Raton, FL: CRC Press.

Tan, K. H. 2009. *Environmental Soil Science*. 3rd ed. Boca Raton, FL: CRC Press.

Verwey, E., and J. Th. G. Overbeek. 1948. *Theory of the Stability of Lyophobic Colloids*. 1st ed. Amsterdam: Elsevier.

White, R. E. 2005. *Principles and Practice of Soil Science: The Soil as a Natural Resource*. 4th ed. Malden, MA: Wiley-Blackwell.

2 Colloidal Particles and Colloidal Suspensions

2.1 COLLOIDAL PARTICLES

Particles are present all around us in great abundance. They include, for example, pollens, dusts, fibers, metals, bacteria, and soil components. Many soil particles, including the clay fraction of minerals (as it will be seen in Chapter 7) and organic matter, fall in the range of colloidal particles. The term "colloidal particle" refers to an entity large in comparison with atomic dimensions so as to have a supramolecular structure and properties, but small enough to remain in suspension for a finite amount of time (at least of the order of hours) when dispersed in a liquid medium such as water. This definition covers a broad range of sizes, from macromolecules and small crystalline nuclei of approximately 1 nm to macroscopic particles of the order of 1 μm (Verwey and Overbeek 1948; Lyklema 1995; Hiemenz and Rajagopalan 1997; Wilkinson and Lead 2007). In this range of sizes, solid materials are dominated by the surface properties rather than bulk ones. As a result, there are various effects and particular properties affecting the state of these particles and their interactions with small molecules and ions in the surrounding medium.

Colloidal particles are either found in nature or synthesized by humans for a great variety of purposes. Figures 2.1 through 2.5 show some examples of synthetic colloidal particles. In Figure 2.1a and b, examples of iron oxide particles synthesized to study colloid properties are shown; note the different geometries that are related to their crystalline structures and, to some extent, can be tuned through the synthetic procedure. Figure 2.2 shows typical spherical polymeric particles (Sevonkaev et al. 2010). These examples are of relatively large particles; Figure 2.3 shows thymine-functionalized gold nanoparticles of approximately 7 nm average diameter (Zhou et al. 2007). Figure 2.4 also shows gold nanoparticles, but with different sizes and shapes as a result of a different, biodirected synthesis procedure (Philip 2009). Finally, Figure 2.5 shows cobalt nanoparticles arranged in nanowires by covering with polystyrene and heating to further polymerize (Keng et al. 2009).

In the environment, many types of particles can be found, either natural or anthropogenic. Figure 2.6 shows examples of airborne particles collected in the city of Taiyuan, China, a highly polluted area, due to the many industries using coal as energy source (Xie et al. 2009). Figure 2.7 shows bacteria sampled in an atmospheric fallout at 15 cm over a forested soil located in Morvan, France; the bacteria bear solid granules (like the one signaled with an arrow) that contain significant amounts of Pb (Perdrial et al. 2008). Figure 2.8 shows an interesting example, albeit the size of these particles is beyond what is considered as colloidal (Vinod, Sashidhar, and Sukumar 2010); here, gum kondagogu (*Cochlospermum gossypium*),

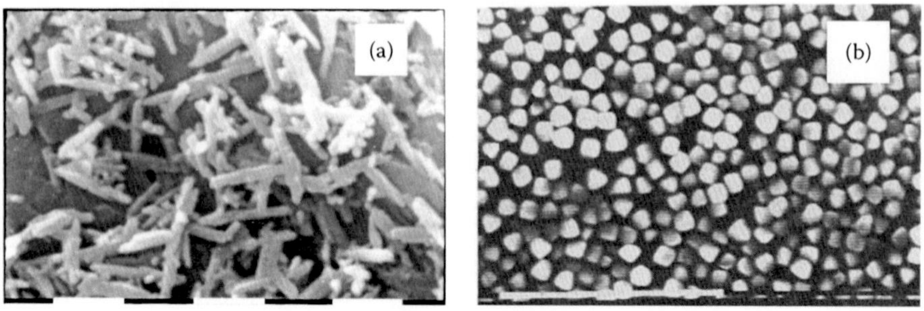

FIGURE 2.1 Examples of synthetic colloidal particles. (a) Scanning electron microscopy (SEM) image of goethite (FeOOH) particles; each scale bar is 1 μm. (b) SEM of hematite (α Fe_2O_3); each scale bar is 10 μm.

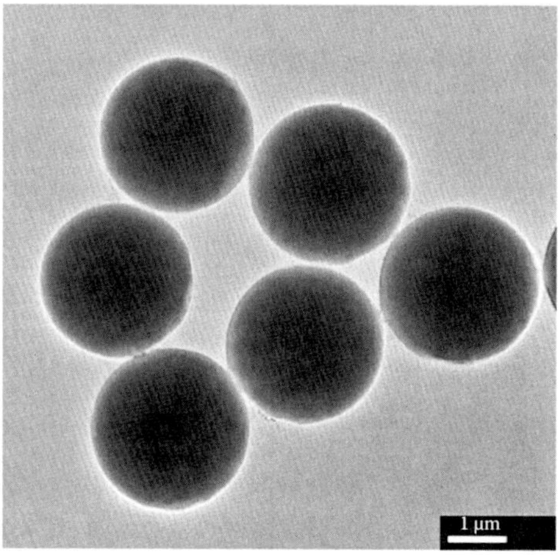

FIGURE 2.2 Transmission electron microscopy images of acrylate–methacrylate polymer particles. (Reprinted from *Colloids and Surfaces A: Physicochemical and Engineering Aspects*, 354, Sevonkaev, I. et al., Distribution of density in spherical colloidal particles by transmission electron microscopy, 16–21. Copyright 2010, with permission from Elsevier.)

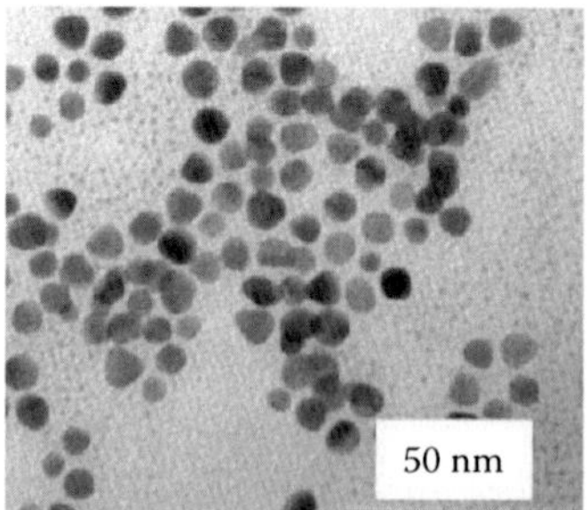

FIGURE 2.3 Transmission electron microscopy image of 7 nm average diameter gold particles. (Reprinted with permission from Zhou, J. et al., 2007, *Langmuir*, 23, no. 24, 12096–12103. Copyright 2007 American Chemical Society.)

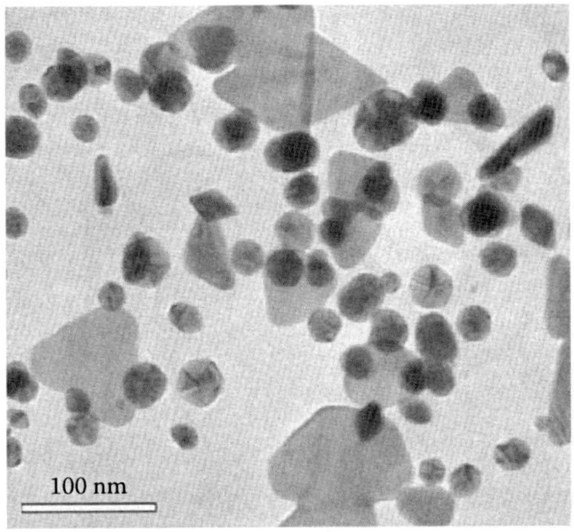

FIGURE 2.4 Transmission electron microscopy image of gold nanoparticles prepared by a different method, showing various sizes and shapes. (Reprinted from *Spectrochimica Acta Part A: Molecular and Biomolecular Spectroscopy*, 73, Philip, Honey mediated green synthesis of gold nanoparticles, 650–653. Copyright 2009, with permission from Elsevier.)

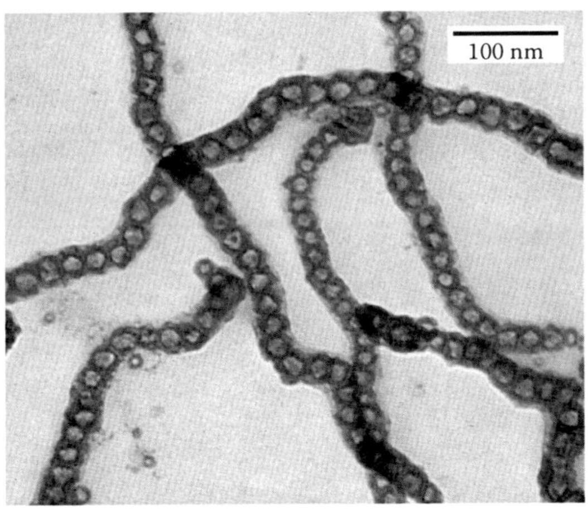

FIGURE 2.5 Transmission electron microscopy images of polystyrene-coated cobalt nanoparticles with particle diameter of 32 nm after oxidation for 1 week. (Reprinted with permission from Keng, P.Y. et al., 2009, *ACS Nano*, 3, no. 10, 3143–3157. Copyright 2009 American Chemical Society.)

a naturally occurring tree biopolymer, is tested as a biosorbent to remove metal ions from aqueous solutions; the change of morphology upon metal sorption is clearly visible. Figure 2.9 shows another biopolymer, glutinous rice starch, used in the food industry (Laovachirasuwan et al. 2010). Figure 2.10 presents another interesting example, showing silver nanoparticles grown in the plant *Brassica juncea* by adding silver salts to the feeding system.

FIGURE 2.6 Scanning electron microscopy images of airborne particles (sampled for 24 hours). Some overlapping or aggregation of particles is seen. (a) Soil particles; (b) a fly ash particle; (c) aggregated coal burning particles; (d) a sulfur-rich particle, possibly an ammonium sulfate particle; (e) an iron particle; (f) gypsum particles; (g) syngenite particles; (h) a natural quartz particle; (i) quartz particle agglomerate from coal combustion. (With kind permission from Springer Science + Business Media from *Air Quality, Atmosphere & Health*, Xie et al., Characterization of individual airborne particles in Taiyuan City, China, 2, 2009, 123–131.)

2.2 SOIL COLLOIDS

The term "soil colloid," according to Cameron (1915), was first coined by van Bemmelen around 1890, after observing parallelisms between the behavior of soils and known colloids. The study of colloidal properties of soil components has attracted the interest of scientists since the first decades of the twentieth century (Cameron 1915; Truog 1916; Gordon et al. 1921; Whitney 1921; Anderson and Mattson 1925; Fischer 1925). The topic of soil composition is detailed in Part II; at this point, we

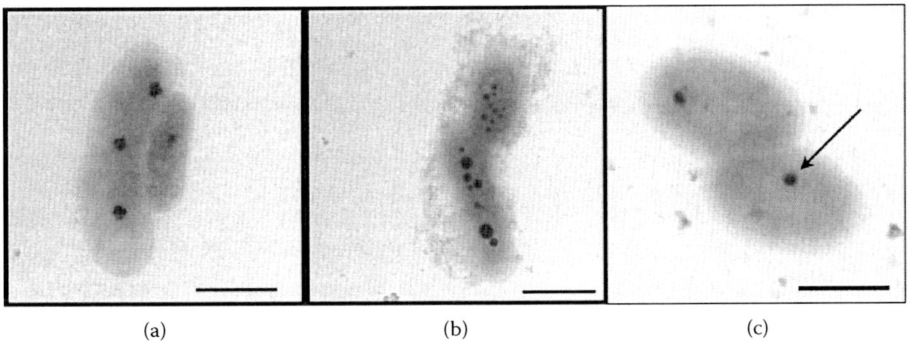

FIGURE 2.7 Transmission electron microscopy (TEM) microphotographs of coccobacillus-shaped bacteria sampled in the atmospheric fallout at 15 cm height above the forest ground. (a) Internal granules, opaque to TEM, are observed. In (b) a microcolony is found. (c) The granule indicated by an arrow was found by energy dispersive x-ray analysis to contain approximately 1% at Pb. Bars are 1 μm long. (Reprinted from *Chemical Geology*, 253, Perdrial et al., TEM evidence for intracellular accumulation of lead by bacteria in subsurface environments, 196–204. Copyright 2008, with permission from Elsevier.)

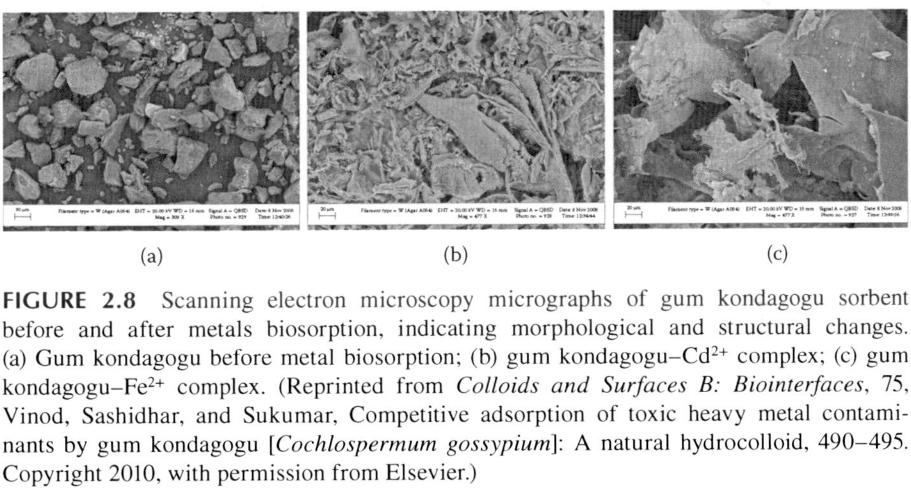

FIGURE 2.8 Scanning electron microscopy micrographs of gum kondagogu sorbent before and after metals biosorption, indicating morphological and structural changes. (a) Gum kondagogu before metal biosorption; (b) gum kondagogu–Cd^{2+} complex; (c) gum kondagogu–Fe^{2+} complex. (Reprinted from *Colloids and Surfaces B: Biointerfaces*, 75, Vinod, Sashidhar, and Sukumar, Competitive adsorption of toxic heavy metal contaminants by gum kondagogu [*Cochlospermum gossypium*]: A natural hydrocolloid, 490–495. Copyright 2010, with permission from Elsevier.)

can state that soil colloids mainly comprise the *clay fraction* of soil—that is, the fraction with particle size less than 2 μm—and also we can include the *fine silt fraction*, 2–10 μm (Skopp 1999). The main components are clay minerals, oxides, and organic matter (humic substances). As expected, there are a great variety of shapes and sizes, as it can be seen in the following examples. Figure 2.11 shows examples of colloids found in two French soils, relatively polluted, observed by transmission electron microscopy; the soil in Figure 2.11a shows organic matter in the form of amorphous domains around and between mineral particles, whereas organic matter in Figure 2.11b is almost transparent to electrons (even when the C contents is greater here). Inorganic particles are mainly clay minerals and some Fe oxides (Citeau et al. 2006). Figure 2.12 shows two samples of kaolinite, a typical clay mineral, displaying its platelike crystals coming from two North American soils (Zbik and Frost 2009).

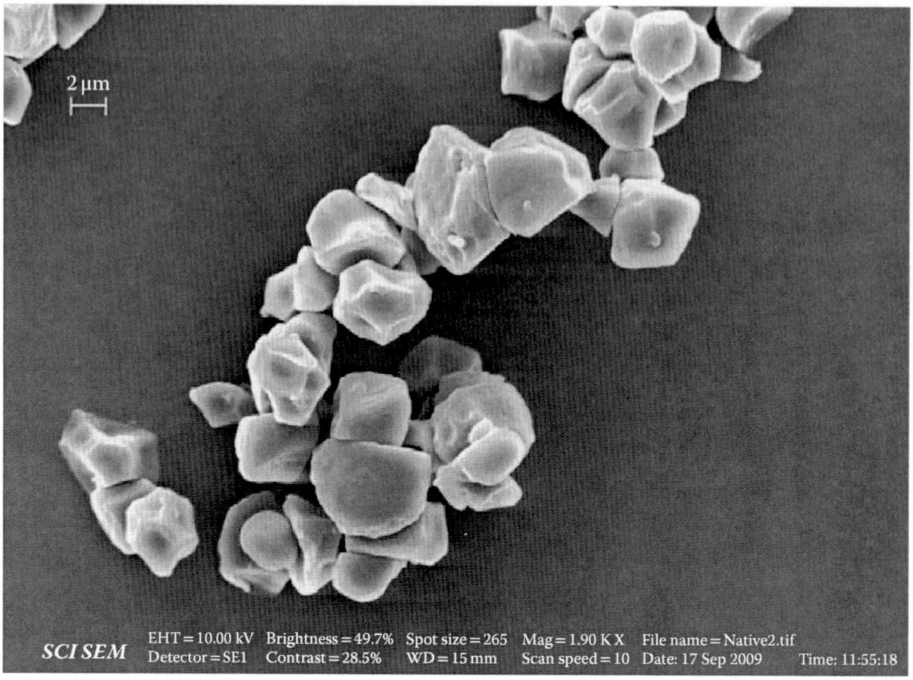

FIGURE 2.9 Scanning electron microscopy microphotograph of glutinous rice starch. (Reprinted from *Colloids and Surfaces B: Biointerfaces*, 78, Laovachirasuwan et al., The physicochemical properties of a spray dried glutinous rice starch biopolymer, 30–35. Copyright 2010, with permission from Elsevier.)

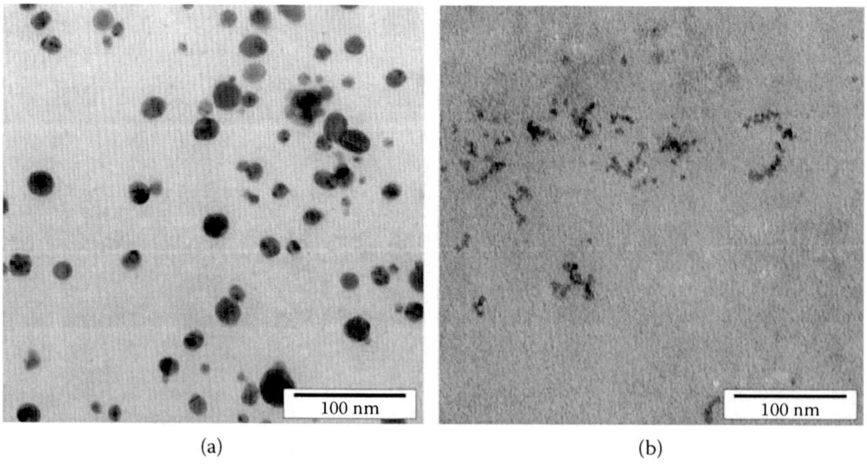

FIGURE 2.10 Transmission electron microscopy images of silver nanoparticles from plant samples irrigated with (a) 10 g/L Ag as $AgNO_3$ and (b) 10 g/L Ag as $Ag(NH_3)_2NO_3$. (With kind permission from Springer Science + Business Media from *Journal of Nanoparticle Research*, Haverkamp and Marshall. The mechanism of metal nanoparticle formation in plants: Limits on accumulation, 11, 2008, 1453–1463.)

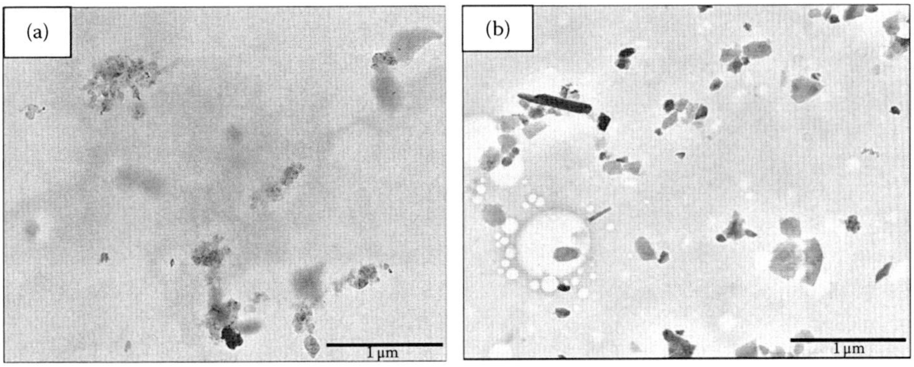

FIGURE 2.11 Transmission electron microscopy images of samples of two soils from France. (a) Organic matter in the form of amorphous domains around and between mineral particles, (b) organic matter is transparent to electrons. (Reprinted from *Colloids and Surfaces A: Physicochemical and Engineering Aspects*, 287, Citeau et al., Investigation of physico-chemical features of soil colloidal suspensions, 94–105. Copyright 2006, with permission from Elsevier.)

(a) (b)

FIGURE 2.12 High-resolution scanning electron microscopy images of (a) Birdwood soil kaolinite finest colloidal fraction and (b) Georgia soil kaolinite finest colloidal fraction. (Reprinted from *Journal of Colloid and Interface Science*, 339, Zbik and Frost, Microstructure differences in kaolinite suspensions, 110–116. Copyright 2009, with permission from Elsevier.)

Figure 2.13 shows another example of kaolinite, this time imaged with the aid of the atomic force microscope (AFM), showing more clearly the plate structure, which is distinctive of phyllosilicate minerals (Vaz, Herrmann, and Crestana 2002). Figure 2.14 shows a microphotograph of a commercial smectite, another class of clay minerals (Nègre et al. 2004). Figure 2.15 shows a shaded AFM image of humic particles sorbed onto the basal plane of cleaved muscovite (Plaschke et al. 1999); individual particles as well as aggregates are observed. More examples are presented in Part II, along with a discussion of the structure and properties of soil components. At present, we remark that soil chemical processes occur mainly at the surface of soil solids; that is, at the surface of soil colloidal particles. In the next section, we introduce the subject of surface properties, which are developed in the following chapters.

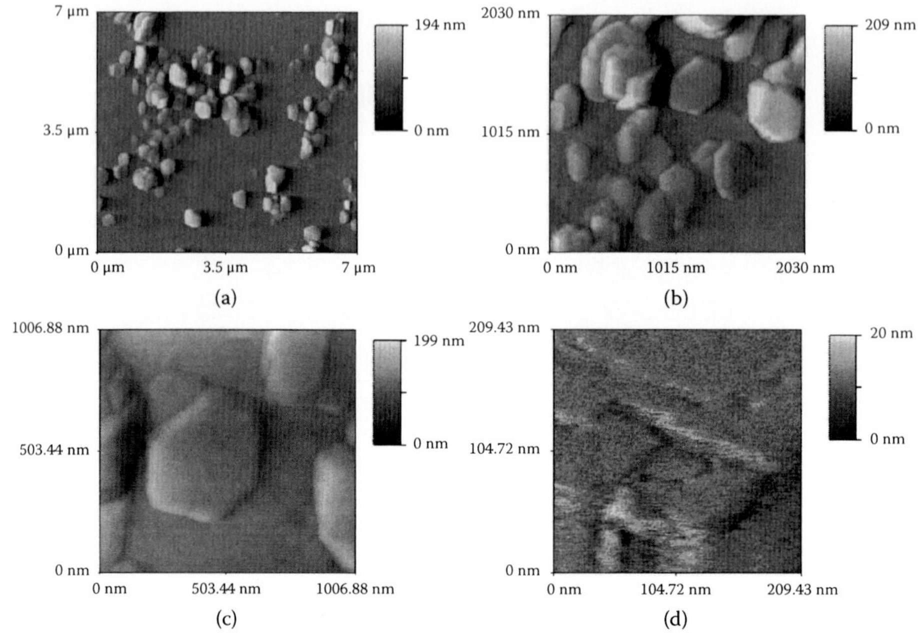

FIGURE 2.13 (a-d) Sequence of atomic force microscope images of kaolinite particles deposited on mica substrate, in four scan sizes (from 7 × 7 to 0.2 × 0.2 μm). (Reprinted from *Powder Technology*, 126, Vaz, Herrmann, and Crestana, Thickness and size distribution of clay-sized soil particles measured through atomic force microscopy, 51–58. Copyright 2002, with permission from Elsevier.)

FIGURE 2.14 Scanning electron microscopy micrograph of a synthetic smectite. (Reprinted from *Geoderma*, 121, Nègre et al., Characterization of model soil colloids by cryo-scanning electron microscopy, 1–16. Copyright 2004, with permission from Elsevier.)

2.3 COLLOIDAL SUSPENSIONS: GENERAL PROPERTIES AND CLASSIFICATION

The term "colloidal suspension" refers to a dispersion of colloidal particles in the bulk of another phase, usually in a different aggregation state; the most common use is for solid particles in a liquid medium. Regarding the definition of colloidal particles given before, it includes either macromolecules (e.g., proteins or polymers)

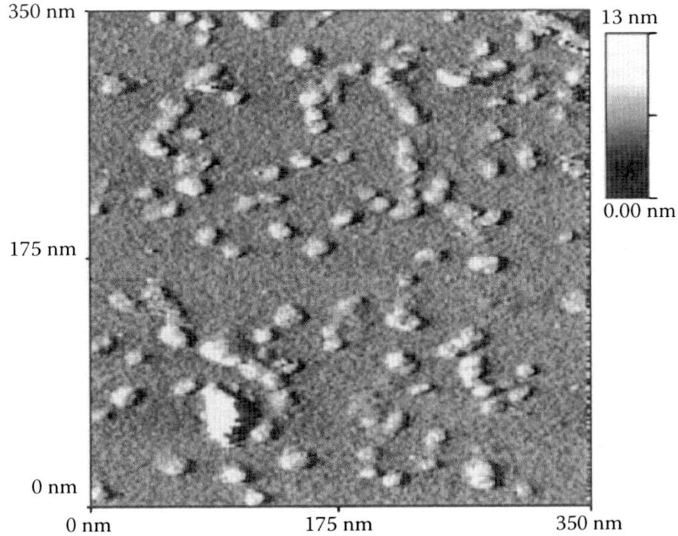

FIGURE 2.15 Atomic force microscope image (shaded) of humic acid colloids sorbed onto muscovite in an initial state of aggregation (pH 4): single particles, aggregates, and chain-like assemblies are seen. (Reprinted from *Colloids and Surfaces A: Physicochemical and Engineering Aspects*, 160, Plaschke et al., In situ AFM study of sorbed humic acid colloids at different pH, 269–279. Copyright 1999, with permission from Elsevier.)

with sizes of a few nanometer or small oxide particles of the same size. However, macromolecules may be soluble in the liquid medium (e.g., proteins in water), whereas oxide particles will not be soluble. Thus, in the first case, we will have a solution (there is no phase boundary in such case); whereas in the second case, a true suspension is obtained. There is a fundamental difference between them: a solution is thermodynamically stable, but a suspension is not. It can be shown that dispersed particles of the same material will have a tendency to *aggregate* (or *coagulate*) into clusters of individual particles bound by intermolecular forces, which will eventually sediment, or *condense* into larger particles (thus losing their former identity), which will also sediment. The fact that colloidal suspensions can remain stable for long times is a kinetic matter: there are in many cases energetic barriers (usually electrostatic) for these processes, making them slow, sometimes to the extent that suspensions remain stable for many years. However, this is a *kinetic* stability as opposed to thermodynamic stability. Kinetic stability can be broken by changing appropriate conditions, such as the addition of an indifferent electrolyte, whereas the former can be changed only by modifying thermodynamic variables, such as the temperature.

Colloids are broadly classified as *lyophobic* and *lyophilic,* depending on the affinity of the particles toward the dispersing medium; the former refers to particles with low affinity for the medium, whereas the latter conveys great solvent affinity particles (Verwey and Overbeek 1948). This classification, thus, should in fact be applied not to the particles themselves but to the suspension as a whole, because obviously the nature of the interactions will be dependent on both the particle and the dispersing medium. In the case of water, which is the main interest of this book, the more common terms *hydrophobic* and *hydrophilic* are usually employed. Suspensions of the first class are termed *sols*; examples of substances that particles can form sols are AgI, Au, clays, and crystalline oxides such as hematite and gibbsite. Suspensions have the dispersion of light as a distinctive feature, and in some cases the particles impart color to the suspension. This is the case, for example, of gold nanoparticles that, as it is well known, give colored suspensions as shown in Figure 2.16a, where the color is dependent on the refractive index (Underwood and Mulvaney 1994). Nanoparticles can be invisible to the naked eye, but can be revealed by the Tyndall effect (dispersion of light), as shown in Figure 2.16b (Li, Sheng, and Zhao 2005). Larger particles, such as those found in soil, usually give a milky appearance to the suspension, as seen in Figure 2.17 (Zbik and Frost 2009).

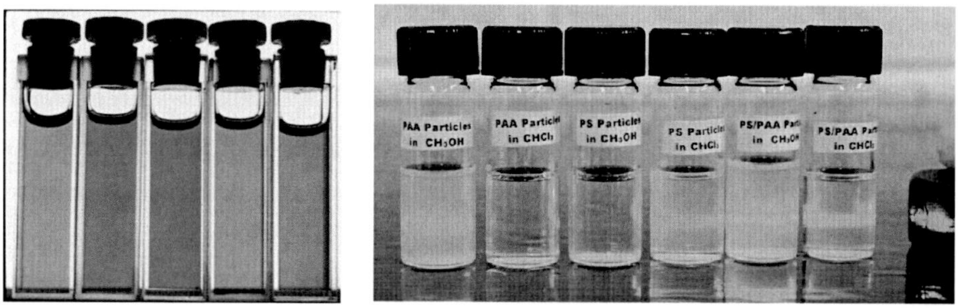

(a) (b)

FIGURE 2.16 (See color insert.) (a) Photograph of five sols of colloidal gold prepared in water and in mixtures of butyl acetate and carbon disulfide. Refractive indices of the solutions increase from left to right. (Reprinted with permission from Underwood, S., and P., Mulvaney, 1994, *Langmuir*, 10, no. 10, 3427–3430. Copyright 1994 American Chemical Society.) (b) Tyndall light-scattering experiment shows the difference between stable and unstable suspensions of polymer-coated silica particles in different solvents; those which disperse the light contain particles in suspension. (Reprinted with permission from Li, D. et al., 2005, *Journal of the American Chemical Society*, 127, no. 17, 6248–6256. Copyright 2005 American Chemical Society.)

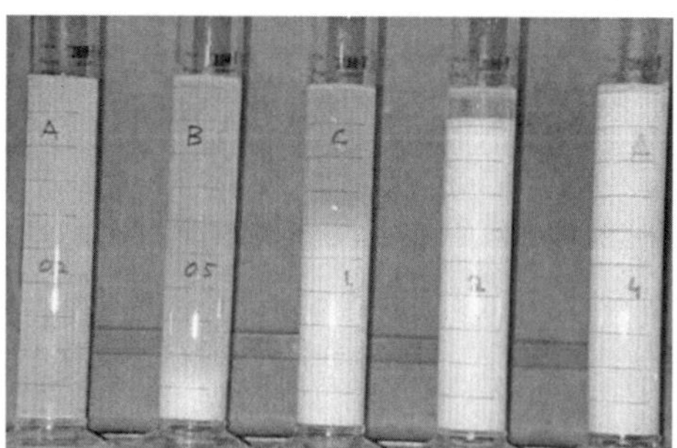

FIGURE 2.17 Suspensions of different concentrations of kaolinite (from Birdwood, South Australia) in water. (Reprinted from *Geoderma*, 121, Nègre et al., Characterization of model soil colloids by cryo-scanning electron microscopy, 1–16. Copyright 2004, with permission from Elsevier.)

Suspensions of hydrophilic particles are formed, for example, by hydrous (oxo) hydroxides such as $Fe(OH)_3$ and $Al(OH)_3$. There are some important differences between both types of suspensions. One of them is that hydrophobic colloids show, as discussed earlier in the text, the phenomenon of coagulation or *flocculation*; that is, the individual particles can aggregate in larger entities, with the addition of a relatively small amount of electrolyte to the dispersing water. In those aggregates, the individual particles are preserved, and can be redispersed (*peptized*) by replacing the electrolytic solution by water or a more diluted solution. This process can be observed through the Tyndall effect or microscopically, either directly or by the ultramicroscope. In a dispersed colloid, the individual particles can be observed subject to thermal (Brownian) motion, but upon coagulation they cluster to form larger aggregates that precipitate relatively quickly. Hydrophilic colloids in many cases do not show coagulation; thus, they may be at the borderline between suspensions and true solutions, as discussed earlier in the text.

2.3.1 Importance of Surface in Colloidal Systems

Colloid science is most often addressed in textbooks and courses on "surface and colloid (or colloid and surface) science." This should not be surprising, because surface is of paramount importance in colloidal systems. Consider, as shown in Figure 2.18, a cube of side $l = 1$ cm, made up of some solid substance; this cube has a volume $V = l^3 = 1$ cm^3 and a total area $A = 6l^2 = 6$ cm^2. If this cube is cut through the middle in each direction (Figure 2.18), one will have eight cubes of $l = 0.5$ cm each. The total volume will be, rather obviously, $V = 8l^3 = 1$ cm^3, so that the total mass is conserved, whereas the total area will be now $A = 8 \times 6l^2 = 12$ cm^2. If the resulting cubes are also cut in the same way repeatedly, the total area doubles each time. It can be seen easily that when 13 such cuts have been made, $l \approx 1$ μm and the area will have increased by a factor of $2^{13} = 8192$; and for dimensions of approximately 1 nm, 23 cuts should have been performed, giving a total area of $2^{23} \times 1$ cm$^2 \approx 8.4 \times 10^6$ cm^2, always with the same total volume and, consequently, the same mass. Thus, the smaller the particles, the more important their area is. It is, of course, at the surface where the particles interact with the surrounding medium and with themselves, hence the importance of the surface. The same can be said in the case of soils:

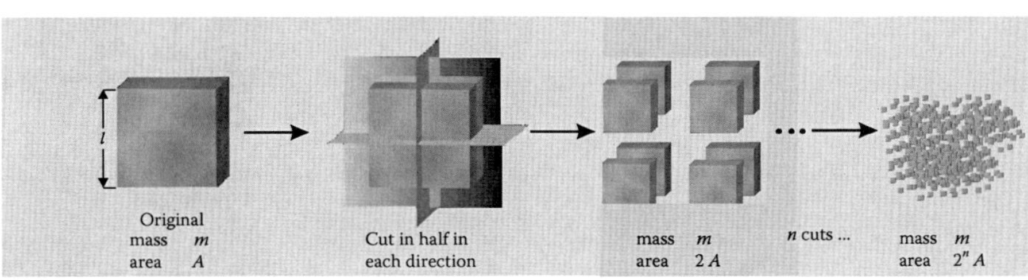

FIGURE 2.18 The importance of surface area as the particle size becomes smaller. Schematic drawing showing how the overall area increases as the particles are further divided. Each time the cube is cut in half along the three directions, the total area doubles, so that after n cuts, the area has increased by a factor of 2^n.

it is at the surface of soil colloids where many important chemical and physical processes take place, thus surface properties being determinant in many soil processes.

As a consequence, a relevant parameter in the characterization of particles is the *specific surface area*, or *specific area*—in short, A_S, which is the area per unit mass ($m^2 \cdot kg^{-1}$). It depends not only on particle size but also on surface roughness that clearly increases the actual area exposed to the surrounding medium. Surface roughness arises from the fact that actual particles, especially those found in natural environments, do not have perfectly flat surfaces (which is implied in the subdivision procedure given earlier in the text) but display defects and irregularities that increase the total area exposed to the surrounding phase.

Recalling the subdivision procedure outlined earlier, it should be remarked that in practice it cannot be continued indefinitely, because a material cannot be divided beyond a single molecule (in the case of molecular materials) or a small cluster of atoms in other cases. However, before that limit is reached, the properties of the material change as the particle size is reduced. This is primarily because a number of properties of a bulk material are a statistical consequence of the large (macroscopic) number of atoms/molecules involved; as the total number of molecules decreases, the bulk statistics no longer hold, and properties are different. Another cause of change, for crystalline particles such as oxides of Fe and Ti, is that the particle dimensions become comparables to quantum characteristic lengths, resulting in what is known as *quantum dots* (Ekimov and Onushchenko 1982; Ekimov, Efros, and Onushchenko 1985), showing interesting optical properties.

2.3.2 Surface Energy

Another fundamental concept in colloid science is that of the surface energy. The repeated cut process outlined in the previous section has an energy cost. Consider, in the bulk of a homogeneous condensed phase A (either liquid or solid), a molecule such as that depicted by the dark circle in Figure 2.19a. It is surrounded by an equal number of molecules that interact with the central one (except at very short distances where repulsive barriers arise) by van der Waals forces that keep the molecules bound in the phase. Assume, at first, that this condensed phase is in the presence of a vacuum. In vacuum, there are no interactions with the surroundings; thus, if our molecule is brought to the surface, as shown in Figure 2.19b, then some of these attractive interactions (in the direction normal to the surface) are lost. This means that work must be done to carry the molecule from the bulk to the surface; if another one is carried from surface to the bulk, obviously the net work is zero. But if the surface area is increased (e.g., changing the shape of the phase), more molecules must be brought to the surface; thus, a net work must be performed. The reversible work per unit area to increase the surface in these conditions is the *surface Gibbs free energy* or *surface tension* γ ($J \cdot m^{-2} = N \cdot m^{-1}$). The term "surface tension" arises from the interpretation that γ can be seen as the force that should be applied to "stretch" the surface linearly a unit length in any direction (Shaw 1992; Adamson and Gast 1997; Hiemenz and Rajagopalan 1997). If the phase depicted in Figure 2.19 is in contact with another solid or liquid phase B, then, to increase the surface of separation, again work should be done, which now is termed the *interfacial free energy* or

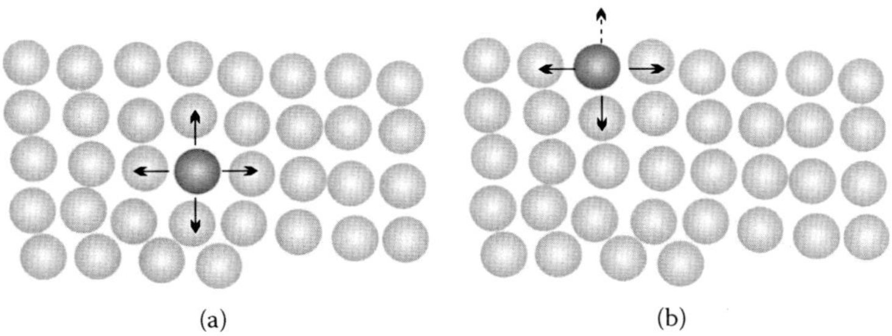

FIGURE 2.19 The concept of surface energy. (a) A molecule immersed in the bulk of a condensed phase is subject to attractive interactions in all directions (depicted as arrows). (b) When a molecule is brought to the surface in the presence of a dilute gas or vacuum, attractive forces in the direction normal to the surface (dotted arrow) are missing, so that a net work should be done.

interfacial tension γ_{AB}. Nevertheless, it is always the Gibbs free energy needed to expand the surface per unit area; in other words, the work, δW, involved in the reversible increase of the area of the A–B interface in an amount dA is given by

$$\delta W = \gamma_{AB} dA \tag{2.1}$$

Note that, for a thermodynamically stable interface, this should be a positive quantity; otherwise, the surface area will spontaneously increase without stopping until the two phases get mixed. If the second phase is a diluted gas such as air, the interfacial energy will be close to that in the presence of vacuum, so that customarily one speaks in this case of surface energy (or surface tension).

2.4 FUNDAMENTALS OF INTERFACIAL THERMODYNAMICS

2.4.1 Surface Excesses

The subject of the thermodynamic treatment of interfaces was solved for the first time by J. W. Gibbs in the 1870s (Gibbs 1961). We give here a short account emphasizing some fundamental concepts. Thorough treatments can be found in a number of sources (Adamson and Gast 1997; Hiemenz and Rajagopalan 1997; Levine 2008; Atkins and Paula 2009). We begin with the description given by Gibbs. Figure 2.20 schematizes the interface between two phases A and B; we assume here that A is an aqueous solution and B a solid or liquid phase, immiscible with A. The figure is a view normal to the interface. Assume that the vertical solid line is the physical separation between the two phases (more on this later). Each phase has its own values for physicochemical properties such as concentrations and chemical potentials of the different species, electrostatic potential, and entropy. Chemical potentials of species present in both phases should be equal in thermodynamic equilibrium (except for charged species, as we see in Chapter 3), but generally speaking the concentrations will be different. Other properties (e.g., electrostatic potential) will also have different values at each phase. However, they should not change abruptly at the interface

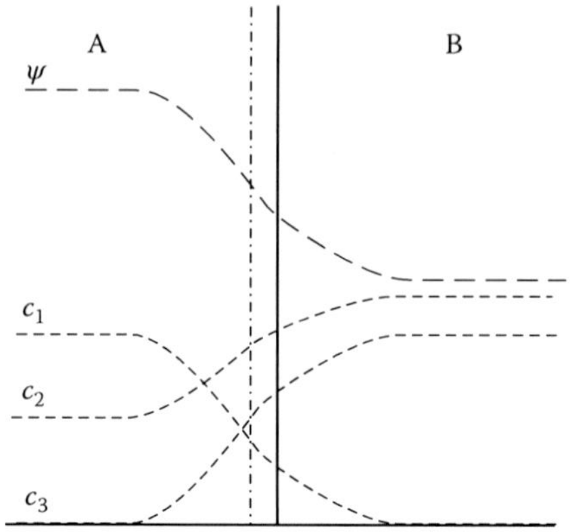

FIGURE 2.20 Gibbs concept of the interface structure. Phases A and B are assumed to be separated by the vertical solid line. The properties of each phase (dashed lines) vary smoothly across the interfacial region as schematically depicted; ψ is the electrostatic potential and the c_i are concentrations of different chemical species. The Gibbs plane of separation (dash-dot vertical line) can, in principle, be arbitrarily placed across the region.

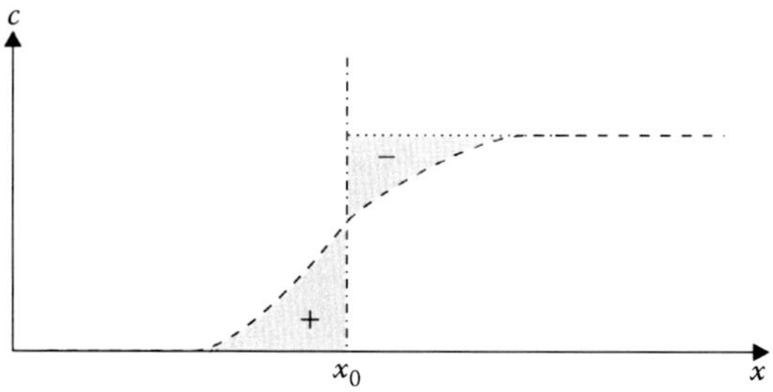

FIGURE 2.21 The surface excess: the vertical dash-dot line shows the Gibbs interfacial plane; the dashed line is the concentration profile c_i of some chemical species i. If that concentration varied abruptly at the interface, it will follow (from right to left) the horizontal portion of the dashed line, then the dotted line, the decrease to zero following the dash-dot line. The difference between that and the actual profile (dashed line) gives rise to the surface excess, given by the mole number contained in the "+" region (in excess) minus the mole number in the "–" region (in deficit), per unit area.

(i.e., they cannot follow a step function), because they, and in most cases their derivatives, should be defined everywhere. Thus, a relatively smooth change at the interface should occur, as schematized in Figure 2.20.

As seen in Figure 2.20, properties change across a small region around the interface, which is termed the *interfacial region*. If an arbitrary plane (the Gibbs plane) is placed somewhere within this region, such as at position x_0 indicated by the dash-dot line, there will be either accumulation or depletion of chemical species (as well as extensive properties such as entropy) at either side, relative to the bulk of each phase. This is detailed in Figure 2.21. The shaded areas represent regions where there is accumulation (that labeled with "+") and depletion (that marked with "–") relative to the hypothetical case of concentrations keeping their bulk value up to the separation plane. If the concentration difference in the shaded region is integrated along the interfacial region, the total number of moles accumulated of species i, n_{i+}, is obtained; likewise one can obtain the total mole number of depleted species, n_{i-} integrating the "–" region. Then the *surface excess* of species i can be defined as

$$\Gamma_i = \frac{n_{i+} - n_{i-}}{A} \tag{2.2}$$

where A is the interface area. Clearly, the surface excess can be negative, positive, or zero for different species. Furthermore, the same can happen for a given chemical species if x_0, the placement of the dividing plane, is changed along the x axis (Figure 2.21); thus, the surface excess of any species can be made null by assuming an appropriate placement of the Gibbs plane. This introduces an arbitrariness that, from the point of view of thermodynamics, is not relevant, but one should be careful when analyzing experimental results (Hiemenz and Rajagopalan 1997). When one of the phases is a solution, specially an aqueous solution, usually the solvent (set as species 1) is set to have $\Gamma_1 = 0$ in this way. The remaining species in general will have nonzero values of surface excess, defined by this particular placement of the dividing plane. The surface excesses defined in this way are termed *relative surface excesses.* Usually one is not interested in "absolute" surface excesses that have arbitrary values due to the arbitrary placement of the dividing plane. The relative surface excesses are usually noted Γ_i^j, where j is the component whose surface excess is zero. In the following, we consider only excesses relative to the solvent (Γ_i^1), as it is customarily done, but omitting the term "relative" and the superscript "1" for short.

It is important to remark that the dividing plane is not the physical division between phases. In fact, all the above treatments (and that following in the next sections) are carried out in terms of macroscopic thermodynamics, which considers only continuous variables and does not take care of molecules. Depending on the nature of the two phases, the thickness of the interfacial region schematized in Figures 2.20 and 2.21 may well be of molecular dimensions, so that the entire concentration change depicted in Figure 2.21 may span a few molecular diameters, especially when solid phases are involved.

2.4.2 Gibbs Equation

It can be shown, from the fundamental laws of thermodynamics, that for an infinitesimal change at the interface (Gibbs 1961; Shaw 1992; Adamson and Gast 1997; Hiemenz and Rajagopalan 1997; Levine 2008):

$$d\gamma = -\sum_i \Gamma_i d\mu_i \tag{2.3}$$

This is the fundamental equation for interfacial thermodynamics. In practice, however, it is more useful when dealing with interfaces not involving solid phases, because surface tension cannot be measured when dealing with solids. Equation 2.3 is valid regardless of the position of the dividing plane and consequently valid for either absolute or relative surface excesses. In the first case, the sum will run from

1 to the number of species present, n, whereas for solvent relative excesses, it will start at $i = 2$. If only one solute is present, Equation 2.3 becomes

$$d\gamma = -\Gamma_2 d\mu_2 \tag{2.4}$$

In this simpler form, the Gibbs equation gives the relationship between the surface excess and the surface energy of an interface. We recall that the chemical potential of the solute is given by

$$\mu_2 = \mu_2^0 + \boldsymbol{R}T \ln a_2 \tag{2.5}$$

where μ_2^0 is the standard chemical potential and a_2 is the solute activity; it is related to solute concentration by

$$a_2 = \frac{m_2 \gamma_2^m}{m^0} \tag{2.6}$$

with m_2 being the solute molality, γ_2^m the solute activity coefficient in molality scale, and m^0 the standard molality. In dilute solutions at room temperatures, both the activity coefficient and the solution density approach unity, so that under these conditions $a_2 \approx c_2$.

Introducing Equation 2.5 in Equation 2.4, we get

$$d\gamma = -\boldsymbol{R}T\Gamma_2 d\ln a_2 = -\frac{\boldsymbol{R}T\Gamma_2}{a_2} da_2 \tag{2.7}$$

so that

$$\Gamma_2 = -\frac{1}{\boldsymbol{R}T}\frac{\partial\gamma}{\partial\ln a_2}\bigg)_T = -\frac{a_2}{\boldsymbol{R}T}\frac{\partial\gamma}{\partial a_2}\bigg)_T \tag{2.8}$$

so that measuring the surface tension as a function of solute activity (or concentration, in dilute solutions) the surface excess—that is, the amount of solute adsorbed at the surface—can be obtained. As mentioned above, this is feasible for liquid–gas or liquid–liquid interfaces, not when solids are involved, so it is of limited use in soils. Nevertheless, it is instructive to examine some of their consequences.

Figure 2.22 shows typical variations of surface energy with solute concentration for different types of solutes. Curve 1 is the typical behavior for uncharged, poorly hydrated organic solutes in water; they show a tendency to accumulate at the surface due to polarity reasons, giving a positive surface excess and consequently a negative slope to γ. Curve 2 is found with electrolytes and highly hydrated organic compounds that, also due to polarity effects, dislike the surface and thus show negative Γ_2 (surface depletion or negative adsorption) and a positive slope. Curve 3 is for the special case of amphipathic species. These are molecules composed by a nonpolar part (usually a long aliphatic chain) and a polar part (either an ionizable group, such as sulfate or phosphate, or a highly hydrophilic group, such as polyethylene oxide chain). Figure 2.23 shows an example of experimental results for ether composed of a dodecyl hydrophobic chain and a hexaethylene oxide (Corkill, Goodman, and

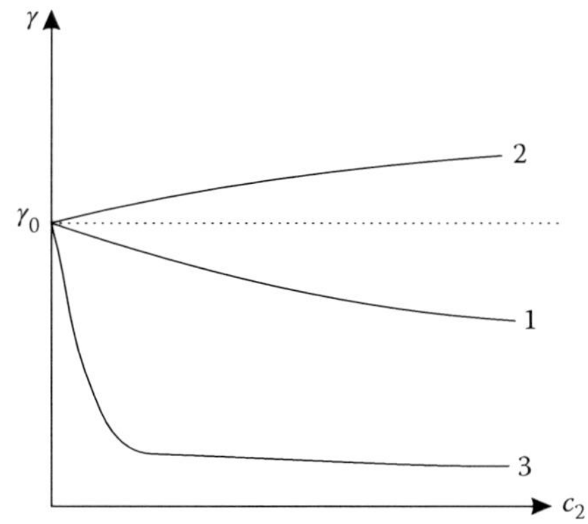

FIGURE 2.22 Schematic representation of the three types of surface energy variation with solute concentration. Curve 1 is for typical organic solutes of small molecules; curve 2 is for simple electrolytes; and curve 3 is for amphipathic solutes.

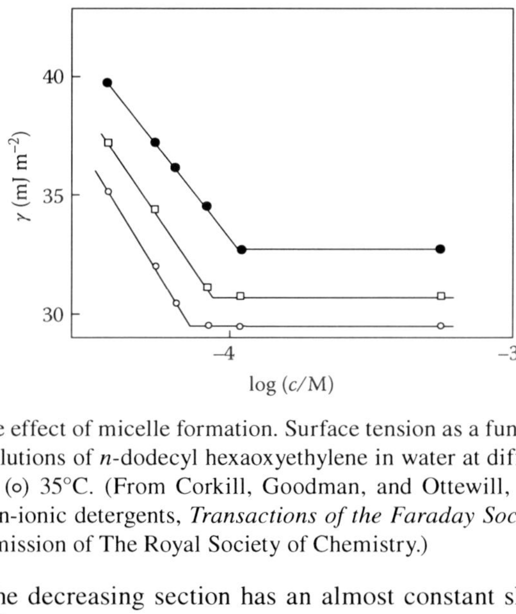

FIGURE 2.23 The effect of micelle formation. Surface tension as a function of logarithmic concentration for solutions of *n*-dodecyl hexaoxyethylene in water at different temperatures: (•) 15°C; (□) 25°C; (○) 35°C. (From Corkill, Goodman, and Ottewill, 1961, Micellization of homogeneous non-ionic detergents, *Transactions of the Faraday Society*, 57, 1627–1635. Reproduced by permission of The Royal Society of Chemistry.)

Ottewill 1961). The decreasing section has an almost constant slope equivalent to $\Gamma_2 = 2.93$ mol m^{-2}, indicating that the surface is saturated with this solute even at very low concentrations. After a particular, temperature-dependent, concentration value, there is a sharp change in slope to an almost zero value, indicating that there is almost no surface excess of this solute. This change is indicative of the formation of *micelles*, structures, or clusters where the amphipathic molecules group together with the nonpolar tails directed toward the interior of the micelle and the polar heads in the exterior part, in contact with water. Figure 2.24 shows an amphipathic molecule (the compound giving the results of Figure 2.23) and a simple example of a micelle made up of that compound. The concentration threshold for micelle formation (shown by the breaks in the lines in Figure 2.23) is the *critical micellar concentration*, a parameter dependent not only on the particular substance but also

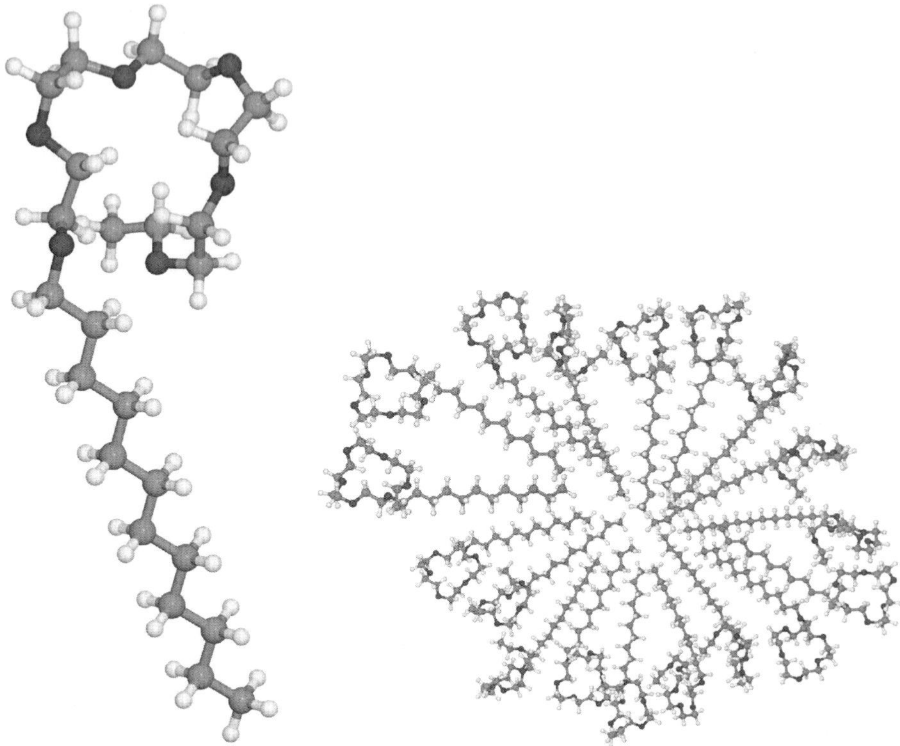
(a) (b)

FIGURE 2.24 (a) A molecular model of *n*-dodecyl hexaoxyethylene; the dark gray balls represent oxygen atoms; notice the separated hydrophilic "head" and hydrophobic "tail." (b) Schematic example of a micelle formed by molecules like that shown in (a) in water: the tails group together in the inner part and the heads are in contact with/immersed in the aqueous phase.

on other variables such as temperature and ionic strength (more on this is discussed in Chapter 6). Micelles are usually of colloidal size and display most colloidal properties. They can dissolve nonpolar compounds in the inner part and also can interact with ions and polar molecules at the surface. Amphipathic compounds and, eventually, micelles are usually not found in natural soils, but can be present as a consequence of human activity, such as detergent spillage.

2.5 SUMMARY

In this chapter, the fundamental concepts of colloid science have been introduced. The definition of colloidal particles, those with sizes (in all directions) ranging from ~1 nm to ~10 μm, has been presented, and their relevance in soil science is stated. The importance of surface properties was remarked, introducing several definitions. The specific surface area is the area per unit mass; the surface tension (or surface free energy) is defined as the Gibbs free energy per unit area. The surface excess of a given species is the amount (in moles per unit area) which is accumulated (or depleted) in the interfacial region relative to the bulk of the respective phase. The fundamental Gibbs relation (Equation 2.3) is, in principle, the basis for the interfacial thermodynamics. However, the concepts introduced so far are not directly applicable to charged species, because there is an electrostatic contribution to the free energy. Thus, Chapter 3 is devoted to the problem of the distribution of electrostatic potential and ion concentration in the interfacial region. Subsequently, Chapter 4 deals with sorption, having emphasis on soil components, and Chapter 5 is devoted to the composition of the charge along the interfacial region.

REFERENCES

Adamson, A. W., and A. P. Gast. 1997. *Physical Chemistry of Surfaces*. 6th ed. Hoboken, NJ: Wiley-Interscience.

Anderson, M. S., and S. E. Mattson. 1925. The relation between properties and chemical composition of soil colloids. *Science* 62, no. 1596. New Series: 114–115.

Atkins, P., and J. de Paula. 2009. *Physical Chemistry*. 9th ed. New York: W. H. Freeman.

Cameron, F. K. 1915. Soil colloids and the soil solution. *Journal of Physical Chemistry* 19, no. 1: 1–13.

Citeau, L., F. Gaboriaud, F. Elsass, F. Thomas, and I. Lamy. 2006. Investigation of physico-chemical features of soil colloidal suspensions. *Colloids and Surfaces A: Physicochemical and Engineering Aspects* 287, no. 1–3: 94–105. doi:10.1016/j.colsurfa.2006.03.040.

Corkill, J. M., J. F. Goodman, and R. H. Ottewill. 1961. Micellization of homogeneous non-ionic detergents. *Transactions of the Faraday Society* 57: 1627–1635. doi:10.1039/tf9615701627.

Ekimov, A. I., A. L. Efros, and A. A. Onushchenko. 1985. Quantum size effect in semiconductor microcrystals. *Solid State Communications* 56, no. 11: 921–924.

Ekimov, A. I., and A. A. Onushchenko. 1982. Quantum size effect in the optical spectra of semiconductor microcrystals. *Soviet Physics. Semiconductors* 16, no. 7: 775–778.

Fischer, E. A. 1925. Base exchange in relation to the swelling of soil colloids. *Transactions of the Faraday Society* 20, no. 4: 603–604.

Gibbs, J. W. 1961. *The Collected Works of J. Willard Gibbs*, vol. 1. New York: Dover publications.

Gordon, N. E., R. C. Wiley, E. B. Starky, A. L. Flenner, and D. C. Lichtenwalner. 1921. Adsorption by soil colloids. *Science* 54, no. 1406. New Series: 581–582.

Haverkamp, R. G., and A. T. Marshall. 2008. The mechanism of metal nanoparticle formation in plants: Limits on accumulation. *Journal of Nanoparticle Research* 11, no. 6 (October): 1453–1463. doi:10.1007/s11051-008-9533-6.

Hiemenz, P. C., and R. Rajagopalan. 1997. *Principles of Colloid and Surface Chemistry*. 3rd ed. New York: Marcel Dekker.

Keng, P. Y., B. Y. Kim, I. B. Shim, R. Sahoo, P. E. Veneman, N. R. Armstrong, H. Yoo et al. 2009. Colloidal polymerization of polymer-coated ferromagnetic nanoparticles into cobalt oxide nanowires. *ACS Nano* 3, no. 10: 3143–3157.

Laovachirasuwan, P., J. Peerapattana, V. Srijesdaruk, P. Chitropas, and M. Otsuka. 2010. The physicochemical properties of a spray dried glutinous rice starch biopolymer. *Colloids and Surfaces B: Biointerfaces* 78, no. 1: 30–35.

Levine, I. 2008. *Physical Chemistry*. 6th ed. New York: McGraw-Hill Science/Engineering/Math.

Li, D., X. Sheng, and B. Zhao. 2005. Environmentally responsive "Hairy" nanoparticles: Mixed homopolymer brushes on silica nanoparticles synthesized by living radical polymerization techniques. *Journal of the American Chemical Society* 127, no. 17: 6248–6256. doi:10.1021/ja0422561.

Lyklema, J. 1995. *Fundamentals of Interface and Colloid Science: Solid–Liquid Interfaces*. London: Academic Press.

Nègre, M., P. Leone, J. Trichet, C. Défarge, V. Boero, and M. Gennari. 2004. Characterization of model soil colloids by cryo-scanning electron microscopy. *Geoderma* 121, no. 1–2: 1–16. doi:10.1016/j.geoderma.2003.09.011.

Perdrial, N., N. Liewig, J. E. Delphin, and F. Elsass. 2008. TEM evidence for intracellular accumulation of lead by bacteria in subsurface environments. *Chemical Geology* 253, no. 3–4: 196–204.

Philip, D. 2009. Honey mediated green synthesis of gold nanoparticles. *Spectrochimica Acta Part A: Molecular and Biomolecular Spectroscopy* 73, no. 4: 650–653. doi:10.1016/j.saa.2009.03.007.

Plaschke, M., J. Römer, R. Klenze, and J. I. Kim. 1999. In situ AFM study of sorbed humic acid colloids at different pH. *Colloids and Surfaces A: Physicochemical and Engineering Aspects* 160, no. 3: 269–279. doi:10.1016/S0927-7757(99)00191-0.

Sevonkaev, I., I. Halaciuga, D. V. Goia, and E. Matijevic. 2010. Distribution of density in spherical colloidal particles by transmission electron microscopy. *Colloids and Surfaces A: Physicochemical and Engineering Aspects* 354, no. 1–3: 16–21.

Shaw, D. 1992. *Introduction to Colloid and Surface Chemistry*. 4th ed. Oxford: Butterworth-Heinemann.

Skopp, J. M. 1999. Physical properties of primary particles. In *Handbook of Soil Science*, ed. M. E. Sumner. 1st ed. Boca Raton, FL: CRC Press.

Truog, E. 1916. The cause and nature of soil acidity with special regard to colloids and adsorption. *Journal of Physical Chemistry* 20, no. 6: 457–484.

Underwood, S., and P. Mulvaney. 1994. Effect of the solution refractive index on the color of gold colloids. *Langmuir* 10, no. 10: 3427–3430.

Vaz, C. M. P., P. S. P. Herrmann, and S. Crestana. 2002. Thickness and size distribution of clay-sized soil particles measured through atomic force microscopy. *Powder Technology* 126, no. 1: 51–58.

Verwey, E., and J. Th. G. Overbeek. 1948. *Theory of the Stability of Lyophobic Colloids*. 1st ed. Amsterdam: Elsevier.

Vinod, V. T. P., R. B. Sashidhar, and A. A. Sukumar. 2010. Competitive adsorption of toxic heavy metal contaminants by gum kondagogu (*Cochlospermum gossypium*): A natural hydrocolloid. *Colloids and Surfaces B: Biointerfaces* 75, no. 2: 490–495. doi:10.1016/j.colsurfb.2009.09.023.

Whitney, M. 1921. The origin of soil colloids and the reason for the existence of this state of matters. *Science* 54, no. 1409. New Series: 653–656.

Wilkinson, K. J., and J. R. Lead. 2007. Environmental colloids and particles: Current knowledge and future developments. In *Environmental Colloids and Particles: Behaviour, Separation and Characterisation*, ed. K. J. Wilkinson and J. R. Lead, 10:1–116. 1st ed. IUPAC Series on Analytical and Physical Chemistry of Environmental Systems. Chichester, UK: Wiley.

Xie, R. K., H. M. Seip, L. Liu, and D. S. Zhang. 2009. Characterization of individual airborne particles in Taiyuan City, China. *Air Quality, Atmosphere & Health* 2, no. 3: 123–131. doi:10.1007/s11869-009-0039-x.

Zbik, M. S., and R. L. Frost. 2009. Micro-structure differences in kaolinite suspensions. *Journal of Colloid and Interface Science* 339, no. 1: 110–116.

Zhou, J., D. A. Beattie, J. Ralston, and R. Sedev. 2007. Colloid stability of thymine-functionalized gold nanoparticles. *Langmuir* 23, no. 24: 12096–12103. doi:10.1021/la7019878.

3 Electrified Interface

3.1 BASIC PRINCIPLES

As a general rule, particles suspended in a polar liquid medium such as water show an electrical charge. This can be verified by applying an electric field to such a suspension, which in most cases will cause a net displacement of the particles in a direction parallel to the field. The presence of a charge of any sign on the particle surface requires, for an electrically neutral system, a charge of opposite sign and the same magnitude on the liquid phase side. In other words, the surface charge on the particle is counterbalanced by ions on the liquid phase. Thus, in general, there is an electrostatic contribution to the energies associated with processes taking place at the particle surface. Although we will postpone the analysis of the origin and nature of the surface charge until Chapter 5, here we will consider as an experimental fact the presence of charge and will study the electrostatic potential profile and the ion distribution resulting from that charge. We will also restrict ourselves to aqueous media, as this is the most common case, and certainly the one present in soils.

A representation of the general case in two dimensions is depicted schematically in Figure 3.1. The particle (here represented as spherical) bears a positive surface charge, which causes an equivalent accumulation of negative ions near the surface, and a concomitant depletion of positive ones. Some electrolytes are considered to be present in the aqueous phase forming a solution, as this is the usual case. On the solution side, the negative charge (anion accumulation) is "smeared out"; this is due to the thermal agitation in the liquid phase, distributing the ions (which otherwise would attach to the particle surface) toward the bulk of the solution. As a result, the electrostatic potential, positive on the surface, decays asymptotically toward the bulk solution value (set here, as usual, as 0) at an infinite distance from the particle. Here, the particle is assumed to be impenetrable, so that we are not interested in the potential inside it; penetrable particles will be discussed in Part III.

Figure 3.1 shows the structure of an *interfacial region* between two distinct phases, the particle and the solution. As discussed in Section 2.4, it is strictly a region and not a plane, because the properties cannot vary abruptly, but must show some transition; for example, the electrical potential cannot "jump" at the hypothetical plane of the interface between two phases which, in the general case, will have different potential values, because a "jump" would imply an infinitely high electric field.

3.1.1 ELECTROCHEMICAL EQUILIBRIUM

This topic is extensively covered in many physical chemistry textbooks (McQuarrie and Simon 1997; Berry, Rice, and Ross 2000; Levine 2008; Atkins and Paula 2009), so it will be only summarized here. When dealing with neutral chemical species,

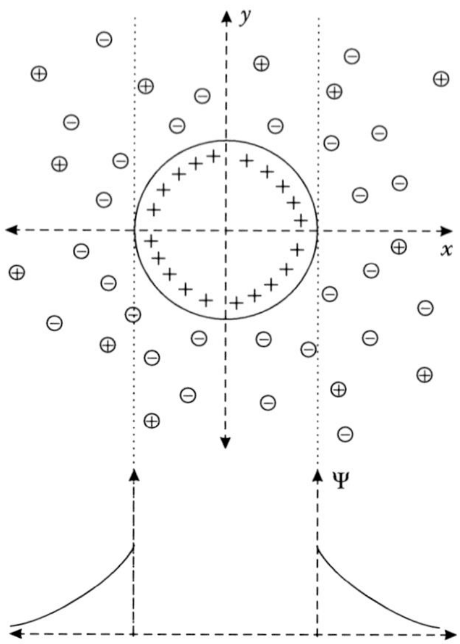

FIGURE 3.1 Schematic drawing of the diffuse layer around a spherical particle. The plot shows the approximate dependence of the electrostatic potential with distance.

the equilibrium condition for any chemical species across phase boundaries is the constancy of the chemical potential:

$$\mu_i^\alpha = \mu_i^\beta = \ldots \tag{3.1}$$

where (as detailed in Section 2.4)

$$\mu_i = \mu_i^\circ + \boldsymbol{R}T\ln\left(\frac{m_i\gamma_i^m}{m^\circ}\right) \approx \mu_i^\circ + \boldsymbol{R}T\ln\left(\frac{c_i\gamma_i^m}{c^\circ}\right) \tag{3.2}$$

is the chemical potential of species i and α, β, ... refer to different phases, μ_i° is the standard chemical potential, m_i is the molality of species i, m^0 is the standard molality, and γ_i^m is the activity coefficient in molality scale. Albeit strictly the chemical potential is defined in terms of molalities, in the following we will approximate by molar concentrations c_i as shown in the rightmost part of Equation 3.2; this approximation is valid for relatively diluted aqueous solutions. Equation 3.1 tells in fact that the Gibbs free energy of exchange of species i between any two phases in equilibrium is 0. However, when charged species (ions) are involved, an additional electrical work is involved in the transference of matter between phases, because in general each phase will have a different electrostatic potential. Thus, the electrochemical potential is introduced:

$$\tilde{\mu}_i^\alpha = \mu_i^\alpha + z_i\boldsymbol{F}\psi^\alpha \tag{3.3}$$

where z_i is the charge of ion i, $\boldsymbol{F}$ the Faraday constant, and ψ^α the electrostatic potential of phase α. For charged species, the equilibrium condition between different phases is the equality of the electrochemical potential. This is applied, for example, to redox processes between a metallic electrode and a solution. When considering microscopically an interfacial region, such as that shown in Figure 3.1, the equilibrium condition requires that the electrochemical potential should be constant across the entire region, thus resulting in accumulation or depletion of ions in the interfacial region, depending on the potential and ion charge signs.

3.1.2 Poisson and Poisson–Boltzmann Equations

3.1.2.1 Poisson Equation

The relationship between the electrostatic potential profile and the space charge distribution is given by the Poisson equation (Nussbaum 1965; Feynman, Leighton, and Sands 1970; Purcell 1984):

$$\nabla^2 \psi = -\frac{\rho_q}{\varepsilon} \tag{3.4}$$

This equation derives directly from the fundamental electromagnetism relations, the Maxwell equations, and relates the electrostatic potential, ψ, at any point in space with the charge density (i.e., electrical charge per unit volume), ρ_q, at the same point. In Equation 3.4, ρ_q is a *fixed* charge distribution, ε is the dielectric coefficient (or constant) of the medium, given by $\varepsilon = \varepsilon_r \varepsilon_0$, where ε_r is the relative dielectric constant, and ε_0 is the vacuum permittivity.

Equation 3.4 can be interpreted more easily considering a one-dimensional case:

$$\nabla^2 \psi = \frac{\partial^2 \psi}{\partial x^2} = -\frac{\rho_q}{\varepsilon} \tag{3.5}$$

Thus, the charge density defines the curvature of the potential profile; in an electrically neutral medium, the right-hand side of Equation 3.5 is 0 (Laplace equation), and consequently, the potential profile should be a straight line (i.e., a constant electric field). The presence of net charge at some region in space is required to find a curved potential profile; this is, unless special circumstances arise, the case when interfaces are present. On the other hand, in a bulk phase such an aqueous solution, only in conditions far from equilibrium, a charge imbalance can be found.

3.1.2.2 Poisson–Boltzmann Equation

Equation 3.4 applies to a fixed charge distribution. However, the charges in an electrolyte solution (the ions) are not fixed but mobile; even colloidal particles are moving, as they show Brownian motion. How can we relate this situation to the Poisson equation? The answer is that, in the absence of external forces, the motion of the charges in solution is at random, so that we can consider an *average* distribution. If the system is in thermal equilibrium, or not far away from it, the ions must follow the Boltzmann distribution, considering the potential energy of interaction of each charge with the electrostatic potential:

$$c_i(\mathbf{r}) = c_i^{\infty} e^{-\frac{z_i e \psi(\mathbf{r})}{kT}} \tag{3.6}$$

Here, $c_i(\mathbf{r})$ is the molar concentration of ion i at the point $\mathbf{r}$ in space ($\mathbf{r}$ is a 3D vector), $\psi(\mathbf{r})$ the potential at the same point, c_i^∞ is the concentration of the same ion in the solution bulk, z_i is the charge of that ion, e the elementary charge ($\boldsymbol{e}$ is Euler's number), $\boldsymbol{k}$ the Boltzmann constant, and T the absolute temperature. Equation 3.6 can be easily deduced from Equations 3.2 and 3.3, considering the constancy of the electrochemical potential and assuming that the activity coefficient does not change across the interfacial region. The charge density at any point in space will be thus given by

$$\rho_q(\mathbf{r}) = \sum_i z_i \boldsymbol{F} c_i(\mathbf{r}) \tag{3.7}$$

Introducing Equations 3.6 and 3.7 in Equation 3.4, we get the *Poisson–Boltzmann (PB) equation*:

$$\nabla^2\psi(\mathbf{r}) = -\frac{\boldsymbol{F}}{\varepsilon}\sum_i z_i c_i^\infty \boldsymbol{e}^{-\frac{z_i e\psi(\mathbf{r})}{\boldsymbol{k}T}} = -\frac{\boldsymbol{F}}{\varepsilon}\sum_i z_i c_i^\infty \boldsymbol{e}^{-\frac{z_i \boldsymbol{F}\psi(\mathbf{r})}{\boldsymbol{R}T}} \tag{3.8}$$

In the rightmost side, it has been fully written in molar units, $\boldsymbol{R}$ being the molar gas constant. The summation extends over all the ionic species present in the solution. Equation 3.8 is the fundamental expression used in the theoretical treatment of charged interfaces. In addition to the above considerations given to deduce Equation 3.7, when the PB equation is applied to a solution containing ions, important assumptions are involved: the solution is considered to be a continuous medium of constant ε, and the ions are considered point charges. However, these assumptions are strictly not valid for a number of reasons (Grahame 1947):

1. A liquid is composed of molecules with a certain structure and dipole moment, and furthermore, water is a strongly structured one due to hydrogen bonding.
2. The dielectric coefficient ε is in general not constant but can vary for several reasons; one of them is the restructuring of water due to a preferential molecular orientation in the presence of strong electric fields and surfaces; the latter is the case for water near a surface, where ε will have a value lower than that of the bulk water.
3. Ions have a finite size, not only on their own but also because they are hydrated in an aqueous solution. The main implications are that the treatment derived from Equation 3.8 will not be valid at high concentrations (Grahame 1947), and that ions cannot approach each other and the surface at infinitesimal distances.
4. The surface charge is "smeared out"—that is, considered uniformly distributed on the surface—and also the charge density in the solution side is taken on average, disregarding the discrete nature of ion charges.
5. From a more fundamental point of view, the PB equation is not consistent with Maxwell electrostatics (Spitzer 2003).

Even with the restrictions given, the PB equation is applicable to many cases, especially to environmental problems such as colloidal particles in soils and natural waters, as these systems, albeit never strictly in chemical equilibrium, are almost never far from equilibrium and do not have usually high electrolyte concentrations. The problems arising near the surface due to water orientation and ion approaching will be dealt in Sections 3.3.2 and 3.3.3. In Section 3.2, we develop and examine the predictions of the PB equation.

3.2 APPLICATION OF THE POISSON–BOLTZMANN EQUATION

The solution of a problem such as that depicted in Figure 3.1 involves solving Equation 3.8 with appropriate boundary conditions. Unfortunately, there is no general solution for the PB equation, and under most circumstances, no analytical solution can be found, so that one has to resort to approximate or numerical solutions. Here, we will examine some cases, mainly those of importance from the soil-science perspective.

3.2.1 Linearization of the PB Equation and the Debye–Hückel Theory

A basic approach to simplify the resolution of the PB equation is to linearize the exponential terms under the assumption of low-potential absolute values:

$$e^{-\frac{z_i F\psi}{RT}} \approx 1 - \frac{z_i F\psi}{RT} \tag{3.9}$$

Equation 3.9 is valid if $z_i F\psi/RT < 1$, or $\psi < 25$ mV, which is a rather low value. Applying Equation 3.9 to Equation 3.8, we get

$$\nabla^2\psi(\mathbf{r}) = -\frac{F}{\varepsilon}\sum_i z_i c_i^{\infty}\left(1 - \frac{z_i F\psi(\mathbf{r})}{RT}\right) = -\frac{F}{\varepsilon}\left(\sum_i z_i c_i^{\infty} - \sum_i \frac{z_i^2 c_i^{\infty} F\psi(\mathbf{r})}{RT}\right) \tag{3.10}$$

Note that due to the electroneutrality of the bulk solution, the first summation on the rightmost side of Equation 3.10 cancels out, so finally we obtain

$$\nabla^2\psi(\mathbf{r}) = \frac{F}{\varepsilon}\sum_i \frac{z_i^2 c_i^{\infty} F\psi(\mathbf{r})}{RT} \tag{3.11}$$

as the linearized form of the PB equation. It should be noted that the linearized PB equation, due to its linearity, is consistent with Maxwellian electrostatics.

The Debye–Hückel theory, originally developed to obtain expressions for activity coefficients, considers the charge distribution around a single ion of charge z and radius r_0 and assumes low-potential values, so that Equation 3.11 is solved in spherical coordinates with the boundary condition given by Gauss's law with spherical symmetry (Feynman, Leighton, and Sands 1970; Purcell 1984):

$$\frac{d\psi}{dr} = -\frac{ze}{4\pi\varepsilon r_0^2} \quad \text{at} \quad r = r_0 \tag{3.12}$$

and

$$\psi = 0 \quad \text{at} \quad r = \infty \tag{3.13}$$

with conditions 3.12 and 3.13, Equation 3.11 can be solved (Berry, Rice, and Ross 2000; Newman and Thomas-Alyea 2004) to give

$$\psi(r) = \frac{ze}{4\pi\varepsilon r}\frac{e^{\kappa(r_0 - r)}}{1+\kappa r_0} \tag{3.14}$$

where the parameter κ, known as Debye's inverse length, is given by

$$\kappa = \sqrt{\frac{2F^2 I}{\varepsilon RT}} \tag{3.15}$$

where

$$I = \frac{1}{2}\sum_i z_i^2 c_i^\infty \tag{3.16}$$

is the *ionic strength*, a property of electrolyte solutions. This is an important property, as it will determine the value of κ and, consequently, the spatial behavior of the potential and concentration profiles near the central ion. Figure 3.2a shows the potential profiles in the vicinity of an ion of charge $z = +1$, and Figure 3.2b shows the corresponding concentration profiles given by Equation 3.6. There is a region where potential and concentration values differ from those at the bulk phase but decay toward it asymptotically; this region is known as the *ion atmosphere* when around a single ion or the *diffuse layer* when around a particle (Verwey and Overbeek 1948; Delahay 1965; Hiemenz and Rajagopalan 1997; Bard and Faulkner 2000; Newman and Thomas-Alyea 2004), albeit it is essentially the same. Here, κ^{-1} is a characteristic length, known as the *Debye length* or *diffuse layer thickness*. As it is seen, the increase of I results in a faster decay of the profiles toward the bulk values; in other words, the double layer (DL) is thinner as the ionic strength increases.

From Equation 3.14, the potential at the surface, ψ_0, is given by

$$\psi_0 = \frac{ze}{4\pi\varepsilon r_0(1+\kappa r_0)} \tag{3.17}$$

It should be noted that for a singly charged ion, ψ_0 (r_0 assumed to be 0.2 nm from the center, considering the presence of an hydration layer) is about 92 mV, which is in fact outside the range of validity of the approximation made in Equation 3.9 ($\psi < 25$ mV). For sufficiently dilute solutions, the nearest ions would be far enough from the central one for the potential they sense to be sufficiently low, so that the results will be reasonable; this is the case of the Debye–Hückel limiting law, which gives accurate results only in dilute solutions ($I < 0.005$ M). In the case of spherical particles, where the above treatment can be in principle applied, the radius is greater,

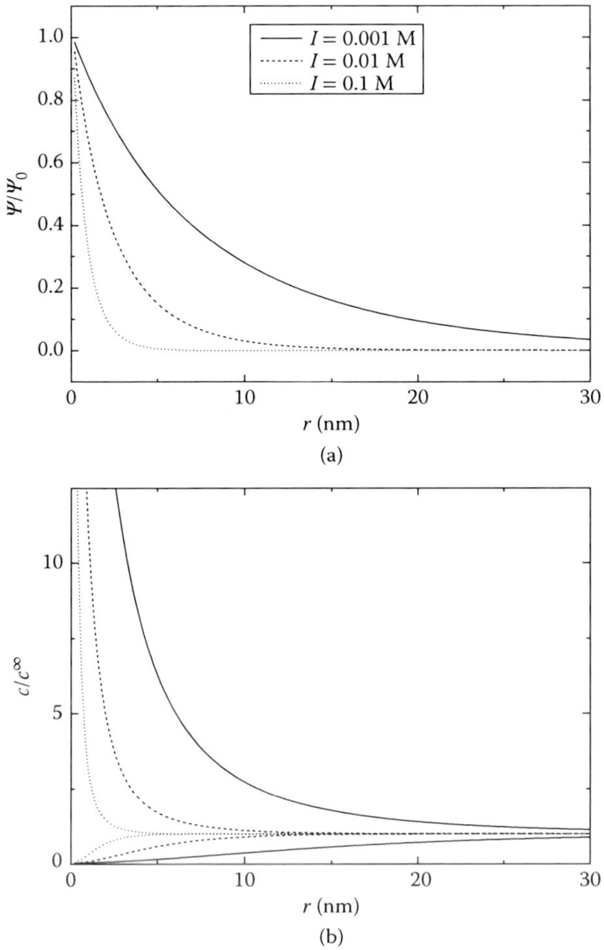

FIGURE 3.2 The interfacial region as predicted by the Debye–Hückel theory. (a) Electrostatic potential as a function of distance from the surface at different ionic strengths, for a positively charged surface; (b) ion concentrations as a function of distance for anions ($c_-/c^\infty > 1$) and cations ($c_+/c^\infty > 1$) for the same case as (a).

which would make the surface potential lower, but the total charge is several times that of a single ion, leading to a relatively high surface potential, and thus, the linearized PB equation is not generally valid for colloidal systems.

As we will be dealing with activities in the remainder of this book, it is worth noting that the Debye–Hückel theory leads to a limiting law for the activity coefficient γ_i given by (McQuarrie and Simon 1997; Berry, Rice, and Ross 2000; Atkins and Paula 2009)

$$\log \gamma_i^m = -A z_i^2 \sqrt{I} \tag{3.18}$$

where A is a constant depending on solvent and temperature, being $A = 0.509$ in water at 298.15 K, with I expressed in molar units. There are a number of extensions to Equation 3.18 (Lyklema 1991; McQuarrie and Simon 1997; Zuckerman, Fisher, and Bekiranov 2001; Wright 2007), from which we will mention, for practical reasons, only the semiempirical Davies equation (Davies 1962):

$$\log \gamma_i^m = -Az_i^2\left(\frac{\sqrt{I}}{1+\sqrt{I}} - BI\right) \tag{3.19}$$

where the constant B is empirical, having usually the value 0.3, but other values have also been used (Zachara and Westall 1998).

3.2.2 Gouy–Chapman Theory

Another treatment of the PB equation consists in considering a one-dimensional problem, that is, the diffuse layer in contact with a plane, infinite surface having a not zero *surface charge density* (charge per unit surface) σ_S. This approach, which was first introduced independently by G. Gouy and D. L. Chapman (Verwey and Overbeek 1948; Delahay 1965; Hiemenz and Rajagopalan 1997) for the case of the electrode–electrolyte interface, is in fact very similar to the Debye–Hückel theory, which was developed some 10 years after, and it has become known as the Gouy–Chapman (GC) theory. The situation is depicted in Figure 3.3. This type of interface is often called a *double layer*; this term, more commonly used in electrochemistry, refers to the two oppositely charged regions at both sides of the interface, which can also be regarded as a capacitor (we will see more on that in Section 3.3). The solid surface may be a metallic one in an electrochemical system (or buried in soil), a nonconducting macroscopic body, or even a particle; this case may be a plane surface from a crystalline solid such as a clay mineral, or a nonplanar surface, if the curvature radius is much greater than the ionic radii and the diffuse layer thickness, so that ions effectively "see" the surface as plane. In the case of conducting surfaces,

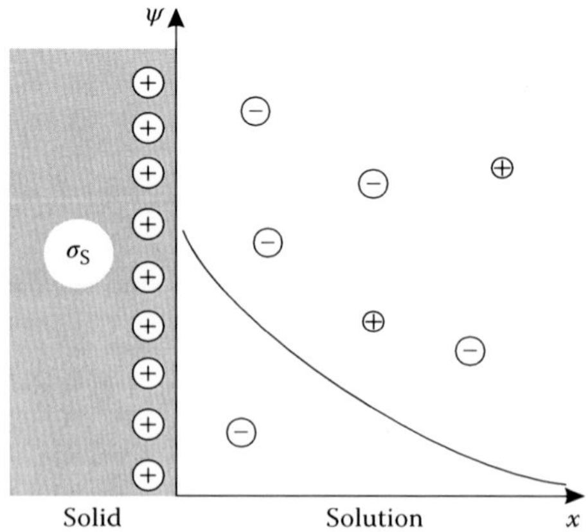

FIGURE 3.3 Schematic drawing of the Gouy–Chapman model of the interface. A plane surface bearing a surface charge σ_S (which in turn determines a surface potential ψ_0) is in contact with an electrolyte solution, where a diffuse layer is present.

the relevant electrical parameter may be the surface potential instead of the charge, whereas for nonconducting surfaces, it is always the charge, because it is necessarily fixed (from an electrostatic point of view). We begin the treatment by writing the PB equation in one dimension:

$$\frac{\partial^2 \psi(x)}{\partial x^2} = -\frac{F}{\varepsilon}\sum_i z_i c_i^{\infty} e^{-\frac{z_i F \psi(x)}{RT}} \tag{3.20}$$

which will be subject to the following boundary conditions:

$$\psi = 0 \quad \text{at} \quad x = \infty \tag{3.21}$$

and

$$\sigma_S = -\varepsilon \left.\frac{d\psi}{dx}\right)_{x=0} \tag{3.22}$$

Equation 3.22 is Gauss's law (Nussbaum 1965).

With the above conditions, we can deduce that (Delahay 1965; Singh and Uehara 1998; Ohshima 2006)

$$\left(\frac{\partial \psi}{\partial x}\right)^2 = \frac{2RT}{\varepsilon}\sum_i c_i^{\infty}\left(e^{-\frac{z_i F \psi}{RT}} - 1\right) \tag{3.23}$$

and

$$\sigma_S^2 = \left[2RT\varepsilon\sum_i c_i^{\infty}\left(e^{-\frac{z_i F \psi_0}{RT}} - 1\right)\right] \tag{3.24}$$

which gives a relationship between the charge and the potential at the surface. However, to progress further, integrating Equation 3.23, an additional simplification is needed. Guy and Chapman considered, as it is frequently used in experiments, a *symmetric (or z–z) electrolyte*; that is, equal charges (in absolute value) for anions and cations. For $z_+ = -z_- = z$ and $c_+^{\infty} = c_-^{\infty} = c^{\infty}$, we have

$$\sum_i c_i^{\infty}\left(e^{-\frac{z_i F \psi}{RT}} - 1\right) = 2c^{\infty}\left[\cosh\left(\frac{zF\psi}{2RT}\right) - 1\right] \tag{3.25}$$

Thus, Equation 3.23 reads

$$\left(\frac{\partial \psi}{\partial x}\right)^2 = \frac{4RTc^{\infty}}{\varepsilon}\left[\cosh\left(\frac{zF\psi}{RT}\right) - 1\right] \tag{3.26}$$

Equation 3.26 can be solved with the boundary conditions 3.21 and 3.22 (Singh and Uehara 1998) to give

$$\psi = \frac{2RT}{zF} \ln \left[\frac{\exp(\kappa x) + \tanh\left(\frac{zF\psi_0}{4RT}\right)}{\exp(\kappa x) - \tanh\left(\frac{zF\psi_0}{4RT}\right)} \right] \tag{3.27}$$

and

$$\sigma_S = -\sigma_d = \left(8c^{\infty}\varepsilon RT\right)^{1/2} \sinh\left(\frac{zF\psi_0}{2RT}\right) \tag{3.28}$$

where κ is given by Equation 3.15 for a symmetric electrolyte; that is, for $I = z^2c^{\infty}$; σ_d is the *diffuse layer charge density*, which compensates the surface charge. Figure 3.4 shows some examples of the predictions of the potential profile given by Equation 3.27 for an ionic strength $I = 0.01$ M, compared with the approximate relationship:

$$\psi = \psi_0 e^{-\kappa x} \tag{3.29}$$

The same general form is observed for different z values, albeit the decay is faster as z increases. Equation 3.29 appears to be a reasonable approximation for the case $z = 1$, and for other values shows qualitatively the same behavior. Ion concentrations following ψ by Equation 3.6 are shown in Figure 3.5. Again, the approximate potential given by Equation 3.29 predicts results close to the more exact Equation 3.27. It is observed that the GC theory predicts that the counterion concentration (the ions with charge sign opposite to that of the surface, here the anions) grows considerably as one approaches the surface. Indeed, close to the surface, c_- reaches very high values: one finds $c_-/c^{\infty} \approx 40$ for $z = 1$ and $\approx 10^5$ for $z = 3$! That leads to physically impossible concentration values. The answer lies, in part, in the assumption that ions are point

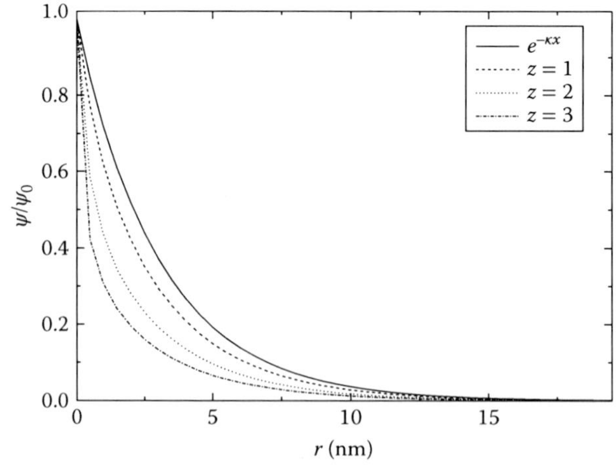

FIGURE 3.4 Electrostatic potential as a function of distance for a plane interface as predicted by the GC theory (Equation 3.27) for $I = 0.01$ M. The approximate case $\psi = \psi_0 e^{-\kappa x}$ (Equation 3.29) is also shown with a continuous line.

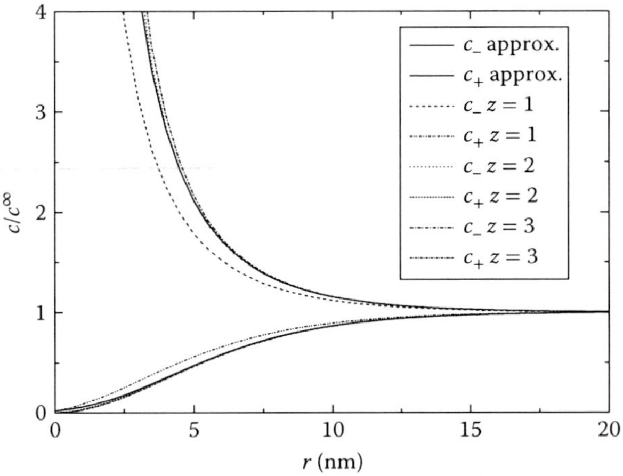

FIGURE 3.5 Ion concentrations as a function of distance for a plane interface as predicted by the GC theory, from potentials shown in Figure 3.4 in Equation 3.6, for different z values. The "approx" case corresponds to the approximate potential given by Equation 3.29.

charges that can approach the surface at infinitesimally short distances, and also in the assumption of a continuous medium of constant ε. For these and other reasons, the GC theory does not predict correctly the interfacial behavior (Grahame 1947; Delahay 1965; Henderson 1983; Baldelli 2008); however, it continues to be considered in the study of interfaces (Varela, García, and Mosquera 2003; Torres, van Roij, and Téllez 2006; Smagala and Fawcett 2007; Liu and Neretnieks 2008; Fawcett 2009; Tournassat et al. 2009; Guerrero-García, González-Tovar, and Olvera de la Cruz 2010; van Soestbergen, Biesheuvel, and Bazant 2010), because it gives analytical expressions for the electrostatic potential and, hence, the ion concentrations in the interfacial region. We will see that the Stern theory overcomes some shortcomings of the GC theory; but first we will look at cases other that planar interfaces and symmetric electrolytes.

3.2.3 Analytical Solutions for Other Cases*

A number of different solutions for the PB equation have been given in the literature. Here, we review some of them. Other cases can be found, for example, in a recent analysis of the problem for colloidal particles (Ohshima 2006).

3.2.3.1 Cylindrical Coordinates

Solution of the PB equation in cylindrical coordinates is relevant for the case of pores with charged walls filled with an electrolyte solution. This case has been treated analytically by several authors (Morrison and Osterle 1965; Rice and Whitehead 1965; Philip and Wooding 1970; Rice and Horne 1981; Sigal and Ginsburg 1981; Olivares and McQuarrie 1985; Rice 1985; Rice and Horne 1985). For the linearized PB Equation 3.11 applied to a pore filled with a uniform solution (Rice and Whitehead 1965), the potential is given by

* Section has more advanced or specific materials.

$$\psi(r) = \psi_0 \frac{I_0(\kappa r)}{I_0(\kappa a)} \tag{3.30}$$

where ψ_0 is the potential at the pore wall, a the pore radius, and I_0 the zero-order modified Bessel function of the first kind. The solution is independent of the x (longitudinal) direction due to the symmetry of the problem. Figure 3.6 shows the behavior of Equation 3.30; here, $r = 0$ lies at the center of the cylindrical pore, whereas at $r = a$ (the pore wall), there is a charged surface with potential ψ_0. The potential decays toward the pore center; for thin pores at relatively low ionic strength, $\psi(0)$ has a finite, nonzero value.

The case of a pore separating two different solutions was treated, among others, by Rice (1985). He solved the linearized PB equation by a finite Hankel transform. Although some of the required integrals could not be analytically evaluated in general, they were approximated with asymptotic expansions of the modified Bessel functions for large argument. The resulting expression for $\psi(x, r)$ (where x is the position along the x-axis and r the radial distance from that axis) can be written as

$$\psi(x,r) = -\frac{\sigma_S}{\varepsilon\kappa S^{x/2L}}\left[e^{-\tau(S^{x/L}-1)} + \frac{1}{S^{1/2}}e^{-\tau(S-S^{x/L})}\right] + \frac{2\sigma_S a}{\varepsilon}\sum_{n=0}^{\infty}\frac{1}{j_{1n}^2 + \kappa^2 a^2 S^{2x/L}}\frac{J_0(j_{1n} r/a)}{J_0(j_{1n})} \tag{3.31}$$

where σ_S is the surface charge density, S is the ratio of ionic strengths (higher I to lower I) between the two outer solutions, κ the Debye length of the lower I solution, L the pore length, a the pore radius, $\tau = \kappa L/\ln S$, J_0 is ordinary Bessel function of the first kind of order zero, and j_{1n} represents the nth zero of J_1. A more general solution, also in the form of a series, was given by Benham (1983). However, these expressions are quite cumbersome.

The opposite problem—that is, a charged rod (which can model either a linear macromolecule or a linear particle) immersed in an electrolyte—has been also addressed by a number of authors (Philip and Wooding 1970; Klein, Anderson, and Record 1981;

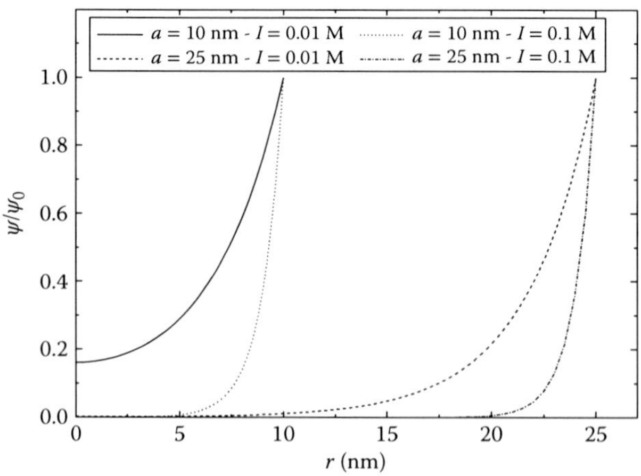

FIGURE 3.6 The electrostatic potential profile inside a cylindrical pore of radius a and infinite length for different a values and at different ionic strengths.

Van Keulen and Smit 1995; Shkel, Tsodikov, and Record 2002; Wang et al. 2005; Téllez and Trizac 2006).

3.2.3.2 Other Cases

A number of different approaches to obtain approximate analytical solutions for the PB equation have been given in the literature. Here, we will mention only a few. For a semi-infinite planar geometry, a general analytical solution for the nonlinear PB equation with any unmixed electrolyte was given by Chen and Singh (2002). It is an exact solution for symmetrical and z–$2z$/$2z$–z asymmetrical electrolytes and an approximate solution for other asymmetrical electrolytes (z–$3z$/$3z$–z or $2z$–$3z$/$3z$–$2z$). They obtained the following expression:

$$\psi(x) = -\frac{\boldsymbol{RT}}{z\boldsymbol{F}}\ln\left[\frac{3}{2}\tanh\left(\frac{\kappa x}{2}+\beta\right)^2 - \frac{1}{2}\right] \tag{3.32}$$

where

$$\beta = \operatorname{arctanh}\left[\sqrt{\frac{1+2e^{-\frac{zF\psi_0}{RT}}}{3}}\right] \tag{3.33}$$

where z is the charge of the lower charged ion. Figure 3.7 shows the potential profiles (in the form of reduced potential $\Psi = F\psi/\boldsymbol{RT}$) for different electrolyte types; it is seen that the charges of electrolyte ions affect the diffuse layer significantly. Figure 3.8 shows the potential profiles for different surface potentials; it is observed that the higher the surface potential is, the faster the diffuse layer shrinks, this phenomenon being more pronounced as the electrolyte is more asymmetrical. For nonplanar interfaces, Philip and Wooding (1970) proposed an approximate analytical solution for the case of a spherical particle. Tuinier (2003) gave approximate solutions of the PB equation for the electrostatic potentials around a charged sphere and charged cylinder. The expressions were obtained from combining various results in limits where exact solutions are available. The Tuinier equations were in agreement with numerical solutions to the PB equation for arbitrary values of the DL thickness. More recently, D'yachkov (2005a,b) gave accurate solutions for several geometries, in the form of power series expansions. The series are shown to converge quite accurately for about 10–20 terms, depending on the particular problem.

3.2.4 Numerical Solutions to the Poisson–Boltzmann Equation*

Beyond z–z electrolytes, or for nonplanar interfaces, analytical solutions to the PB equation are difficult or nonexistent. Thus, one often must resort to numerical solutions; fortunately, nowadays, with the power of digital computers and the availability of advanced numerical algorithms, that does not represent a very difficult task.

For a planar interface as in the GC theory but for unsymmetrical electrolytes, the numerical integration can be easily accomplished starting from Equation 3.23 because it is essentially a first-order differential equation. With the additional boundary condition

* Section has more advanced or specific materials.

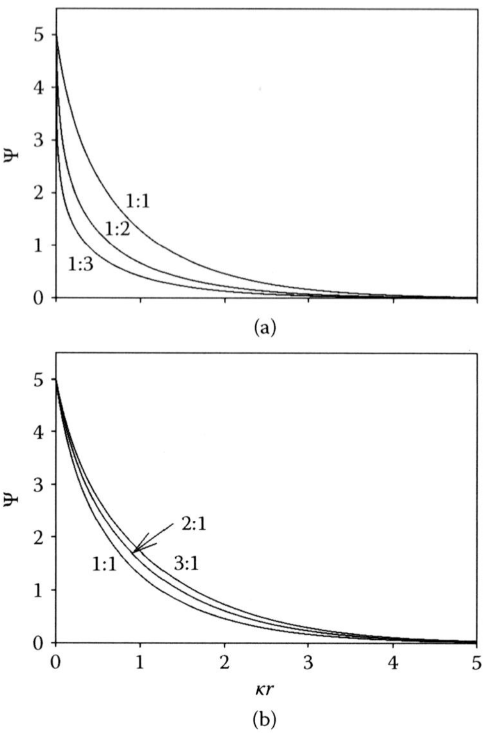

FIGURE 3.7 Plots of potential distributions near a charged planar surface according to Equation 3.32. (a) The charge of co-ions (ions with the same charge as the surface) is held constant; (b) the charge of counterions is held constant. (Reprinted from Chen, Z., and R.K. Singh., *J. Colloid. Interface. Sci.*, 245, 301–306, 2002. With permission.)

$$\psi = \psi_0 \quad \text{at} \quad x = 0 \tag{3.34}$$

the $\psi(x)$ profile can be obtained. Figure 3.9 shows some results for different cases, including a 1-1 electrolyte, all at the same ionic strength. As it can be seen, there are relatively small differences, all showing a similar nearly exponential decay.

Several numerical methods have been devised and applied to solve the PB equation for different geometries (Phillips 1995; Das et al. 1997; Dyshlovenko 2002; Lamm 2003; Bhuiyan, Outhwaite, and Henderson 2007; Lima, Tavares, and Biscaia 2007; Smagala and Fawcett 2007; Lu et al. 2008; Shi and Koehl 2008; Henderson and Boda 2009; Wang and Luo 2010). We will mention some accomplishments.

James and Williams (1985) employed a Galerkin finite element scheme to provide flexible numerical solutions to the PB equation in one and two dimensions, using a Newton sequence for the solution of the set of nonlinear equations arising from the finite-element discretization procedure; the procedure was shown to be applicable for different geometries. Dyshlovenko (2002) applied a finite-element solution of the PB equation with an adaptive mesh refinement procedure. The procedure allowed the gradual improvement of the solution and adjustment of the geometry of the problem. The approach was successfully applied to several problems

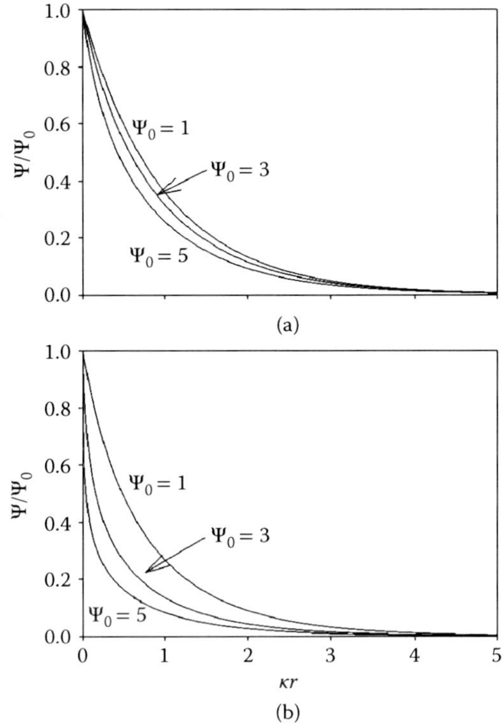

FIGURE 3.8 Plots of normalized potential distribution near a charged plane surface as given by Equation 3.32. (a) 1-1 Electrolyte; (b) 1-3 electrolyte. (Reprinted from Chen, Z., and R.K. Singh., *J. Colloid. Interface. Sci.*, 245, 301–306, 2002. With permission.)

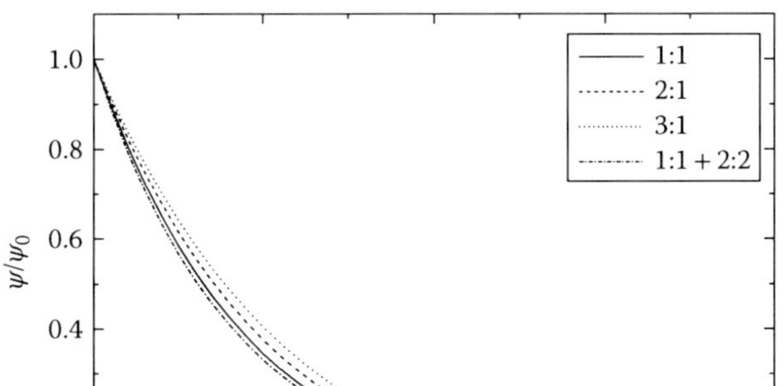

FIGURE 3.9 Results from the numerical integration of the Poisson–Boltzmann equation for different electrolyte types, in all cases at $I = 0.01$ M. In the last case, a mix of 1-1 and 2-2 electrolytes, both have the same concentration.

of colloidal interaction (see Chapter 6): two free identical particles, two identical particles confined in a charged cylindrical pore, and a particle near a charged plane. Sayyed-Ahmad, Tuncay, and Ortoleva (2004) reformulated the elliptic PB equation as the long-time solution of an advection–diffusion equation, using an efficient modified, memory-optimized, alternating direction implicit scheme to integrate the reformulated PB equation. The approach was demonstrated to give good results on protein composites. Lima, Tavares, and Biscaia (2007) solved the classical PB equation and a modified PB equation in bispherical coordinates, taking into account the van der Waals dispersion interactions between the ions and both surfaces. The finite volume method was used to solve the PB equation in bispherical coordinates; this method divides the calculation domain into a certain number of subdomains, where the physical law of conservation is valid and can be readily implemented, presenting low computational cost. The proposed method was validated by comparing the numerical results for the classical PB calculations with previous results reported in the literature. Shi and Koehl (2008) described a geometric approach for generating unstructured tetrahedral meshes and simplifications of these meshes that were reported to be well fitted for solving the PB equation using multigrid approaches. Recently, Wang and Luo (2010) carried out a comparative analysis of numerical methods, finding that modified incomplete Cholesky conjugate gradient and geometric multigrid are the most efficient in their test set; for these solvers, central processing unit (CPU) times increase approximately linearly with the numbers of grids. It was further found that geometric multigrid performs better in a large set of tested biomolecules. However, modified incomplete Cholesky conjugate gradient was found superior to geometric multigrid in molecular dynamics simulations of those tested molecules.

3.3 INTERFACIAL CAPACITANCE AND THE STERN AND GRAHAME MODELS

If the interface, as it is depicted, for example, in Figure 3.3, is considered from an electrical point of view, it should be realized that it looks similar to a capacitor: there are charges of opposite sign and equal magnitude at both sides of the interface. In a "normal" capacitor, there is a dielectric between the two charges, and the charges at both sides are localized in parallel planes (or curved surfaces). In the GC model, there is no separation, because ions are considered as point charges, which can approximate at a distance infinitely close to the surface, and the solution side charge is smeared out toward the bulk. The lack of separation means that the integral interfacial capacitance per unit area C^{int}, given, for example, for a parallel plate capacitor, by

$$C^{\mathrm{int}} = \frac{\varepsilon}{d} \tag{3.35}$$

where d is the distance of separation between plates—will go to infinity. It should be noted that the total capacitance of a plane capacitor is proportional to the area.

$$C_{\mathrm{TOT}} = \frac{A\varepsilon}{d} \tag{3.36}$$

However, the relevant magnitude from a fundamental point of view is the capacitance per unit area, and in the following, unless otherwise noted, we will refer always to this last magnitude.

Of course, the capacitance is an electrical property, which shows up in the context of some electric circuit. In the case of nonconducting materials, as with soil colloids, there are no direct ways to measure or otherwise detect its presence; however, the charge separation exists, and the concept of capacitance is present in some treatments and models of adsorption to soil colloids. Furthermore, it is at the origin of the important Stern modification of the GC theory, thus it is relevant to consider here.

3.3.1 Interfacial Capacitance

The *differential interfacial capacitance*, or simply the interfacial capacitance, is defined as (Grahame 1947; Delahay 1965; Singh and Uehara 1998)

$$C = \left.\frac{\partial \sigma_S}{\partial \Delta\psi}\right)_{T,P,\mu_i} = \left.\frac{\partial \sigma_S}{\partial \psi_0}\right)_{T,P,\mu_i} \tag{3.37}$$

where $\Delta\psi$ is the potential difference across the interface; in the right-hand side, it has been considered that $\psi = 0$ in the solution bulk. It is more relevant than the *integral capacitance* (Grahame 1947; Delahay 1965) defined, as for a common capacitor, by

$$C^{\text{int}} = \frac{\sigma_S}{\psi_0} \tag{3.38}$$

The differential capacitance, working in electrochemistry with metallic electrodes, is a magnitude experimentally accessible, measured with relative ease (Grahame 1947; Delahay 1965; Bard and Faulkner 2000). From Equation 3.24, it can be found that the capacitance, as given by the GC theory, is

$$C = \varepsilon \boldsymbol{F} \frac{\sum_i z_i c_i^\infty e^{-(z_i \boldsymbol{F}\psi_0/\boldsymbol{R}T)}}{\left[2\boldsymbol{R}T\varepsilon \sum_i c_i^\infty \left(e^{-(z_i \boldsymbol{F}\psi_0/\boldsymbol{R}T)} - 1\right)\right]^{1/2}} \tag{3.39}$$

or, for a z–z electrolyte, by

$$C = z\boldsymbol{F}\left(\frac{2c^\infty \varepsilon}{\boldsymbol{R}T}\right)^{1/2} \cosh\left(\frac{z\boldsymbol{F}\psi_0}{2\boldsymbol{R}T}\right) = \varepsilon\kappa \cosh\left(\frac{z\boldsymbol{F}\psi_0}{2\boldsymbol{R}T}\right) \tag{3.40}$$

Figure 3.10 shows (dash-dot line) the prediction of Equation 3.40, a hyperbolic cosine function that grows exponentially as the surface potential ψ_0 departs from zero. The circles and continuous line show experimental data from the classical Grahame measurements (Grahame 1954) for a mercury-dropping electrode immersed in a 0.001 M NaF solution. In comparing both curves, it has been taken into account that $\psi_0 = 0$ at the minimum of the experimental curve (more on this later), because the potentials are measured against a normal calomel electrode as reference (Grahame 1954; Delahay 1965; Singh and Uehara 1998; Newman and Thomas-Alyea 2004).

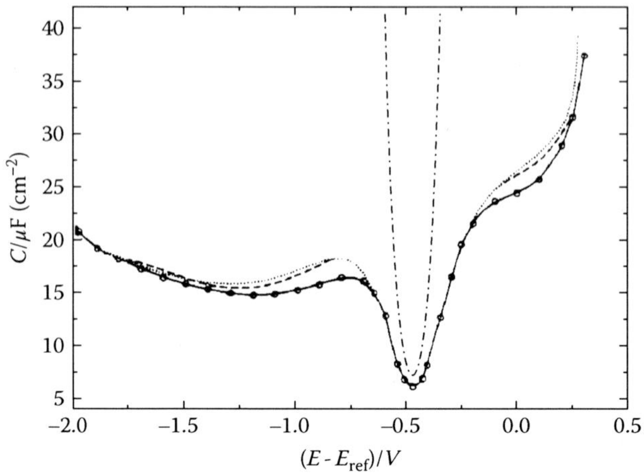

FIGURE 3.10 Differential capacitance of the interface Hg/0.001 M NaF. The potentials are measured against a normal calomel electrode. (—O—), Experimental results; (····), calculated by method I; (— —), calculated by method II; and (—— – —— – ——), Gouy–Chapman prediction. (Adapted from Grahame, D. C., *J. Am. Chem. Soc.*, 76, 4819–4823, 1954. With permission.)

Clearly, only the region near the minimum is reasonably predicted by the GC theory, but for the remainder of the potential range, the discrepancies are very high. At higher electrolyte concentrations, the discrepancies are even higher.

3.3.2 Stern Modification

The conclusion is that the GC theory is not satisfactory, mainly due to the assumption of the ions being point charges. It was Stern (1924) who clearly recognized this fact and proposed a modification to overcome that limitation. The modified model is depicted in Figure 3.11; an additional plane, the Helmholtz plane, Stern plane, or "2" plane is defined, as the distance of closest approach to the surface. Ions cannot get closer to the surface, and the space between the plane and the surface, termed the *inner layer*, is filled by solvent, which behaves as a dielectric. This model is interpreted, from an electrical point of view, as two capacitors in series: one is the "inner layer" or Stern capacitance, C_i, and the other is the diffuse layer capacitance, C_d. The series connection means that the inverse of the capacitance is additive:

$$\frac{1}{C} = \frac{1}{C_i} + \frac{1}{C_d} \tag{3.41}$$

where C_d is given by the GC theory. Now, however, the electrostatic potential determining the diffuse layer is no longer the surface potential ψ_0, but the potential at the plane of closest approach ψ_2 (we give this plane the name "2," anticipating the next section); thus, Equations 3.27 through 3.29 become

$$\psi = \frac{2\boldsymbol{RT}}{z\boldsymbol{F}} \ln\left[\frac{\exp[\kappa(x - x_2)] + \tanh(z\boldsymbol{F}\psi_2/4\boldsymbol{RT})}{\exp[\kappa(x - x_2)] - \tanh(z\boldsymbol{F}\psi_2/4\boldsymbol{RT})}\right] \tag{3.42}$$

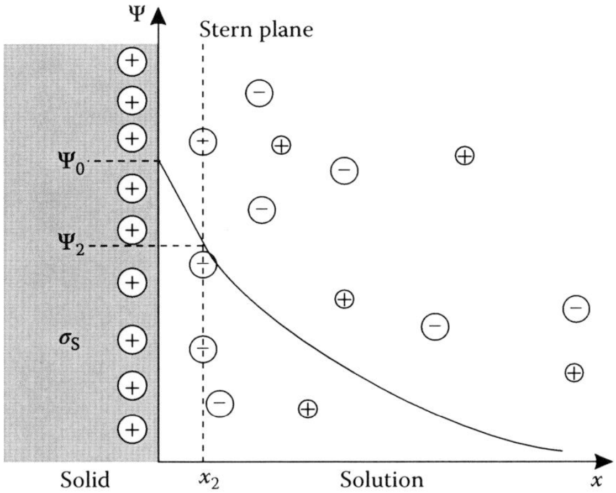

FIGURE 3.11 Schematic representation of the Gouy–Chapman–Stern theory. Compare with Figure 3.3: an additional plane, the Stern plane, is defined as the distance of closest ionic approach to the surface. The region between the Stern plane and the surface behaves as a dielectric.

$$\sigma_d = -\left(8c^{\infty}\varepsilon \boldsymbol{RT}\right)^{1/2} \sinh\left(\frac{z\boldsymbol{F}\psi_2}{2\boldsymbol{RT}}\right) \tag{3.43}$$

$$\psi = \psi_2 e^{-\kappa(x-x_2)} \tag{3.44}$$

where x_2 is the position of the Stern plane. In Equation 3.43, σ_d is the charge of the diffuse layer, which has sign opposite to the potential ψ_2. Thus, the diffuse layer capacitance becomes

$$C_d = \varepsilon \boldsymbol{F} \frac{\sum_i z_i c_i^{\infty} \exp\left(-\frac{z_i \boldsymbol{F}\psi_2}{\boldsymbol{RT}}\right)}{\left[2\boldsymbol{RT}\varepsilon \sum_i c_i^{\infty}\left[\exp\left(-\frac{z_i \boldsymbol{F}\psi_2}{\boldsymbol{RT}}\right) - 1\right]\right]^{1/2}} \tag{3.45}$$

C_i is a simple parallel-plate capacitor, so that its capacitance is related to the potential drop by

$$C_i = \frac{\sigma_S}{\psi_0 - \psi_2} \tag{3.46}$$

The above expressions are given in the framework of the GC theory—that is, planar geometry and z–z electrolytes—but the same changes can be made in other cases (as Equations 3.32 and 3.33), replacing ψ_0 by ψ_2 and x by $x - x_2$.

The Stern capacitance C_i is not expected to be dependent on the solution composition. Grahame (1947, 1954) assumed C_i to be independent on I and deduced its value from the highest NaF concentration available, 0.916 M (Figure 3.12, solid line), by calculating C_d from the GC theory (Equation 3.45) and computing C_i

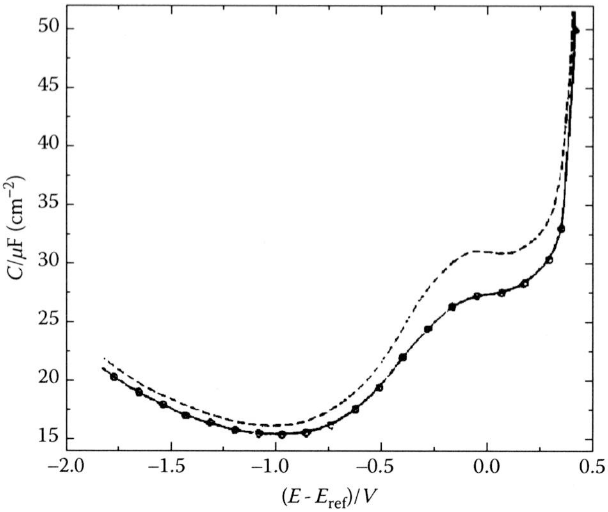

FIGURE 3.12 Differential capacitance of the interface Hg/0.916 M NaF. The potentials are measured against a normal calomel electrode. (—O—), Experimental results; (····) inner layer capacitance calculated by subtracting the diffuse layer capacitance following the GC theory. (Adapted from Grahame, D. C., *J. Am. Chem. Soc.*, 76, 4819–4823, 1954. With permission.)

from Equation 3.41, resulting in the curve showed as a dashed line in Figure 3.12. Now this C_i curve, taken constant as a function of electrode charge, was combined with the C_d calculated from GC theory at other NaF concentrations, down to 0.001 M. For intermediate values (0.1 and 0.01 M), the agreement was excellent. For 0.001 M, the dotted curve of Figure 3.10, labeled Method I, was obtained, also with good agreement. The curve labeled Method II, with slightly better agreement, was obtained by a thermodynamic argument, which will not be reproduced here (Grahame 1954).

Returning to Figure 3.10, it is seen a pronounced minimum in the curve; this minimum is identified as the point where the surface charge becomes null; here, the diffuse layer capacitance is determining the total surface capacitance (because in a series connection, the lower capacitance dominates the result), and as C_d shows a minimum when both ψ_2 and σ_d vanish, the potential (measured against a reference) of the electrode at this point should correspond to zero surface charge. The *potential of zero charge*, E_z, or more generally, the *point of zero charge*, is an important surface property, which will be addressed in Chapter 5. For the time being, we will note that the point of zero charge is determined by the potential on conducting electrodes under external control, but on nonconducting and/or isolated surfaces, it is determined by the medium conditions, usually the pH.

It was found that the Stern modification of the GC theory predicts well the interfacial capacitance of the interface between mercury electrodes and some electrolytes such as NaF or $NaClO_4$, but fails for others such as KCl or KI (Grahame 1947;

Grahame and Soderberg 1954; Parsons 1954; Grahame 1958; Grahame and Parsons 1961; Delahay 1965). This was taken into account by a further modification of the GC theory.

3.3.3 Grahame Modification of the Gouy–Chapman–Stern Theory

The discrepancy for some electrolytes is due, according to Grahame (1947), to the fact that some ions, especially polarizable anions such as Br^- and I^-, can reach the surface closer than the Stern plane depicted in Figure 3.11; the ions shown there retain their hydration layer, but others may lose this layer and get in direct contact with the surface. The Grahame model is depicted in Figure 3.13. Two planes of closest approach are defined: the *inner Helmholtz plane* (IHP), where the center of ions that are in direct contact with the surface lie, and the *outer Helmholtz plane* (OHP), where ions are separated from the surface by a layer of water molecules. The first case corresponds to *inner sphere complexes*, as it is termed in surface complexation studies, or *specific adsorption*, as it is commonly stated in electrochemistry and surface science. We will return to these concepts in detail in Chapter 4. The IHP is located at distance x_1 from the surface with potential ψ_1, whereas the OHP is at x_2 with potential ψ_2. However, the description of the capacitance and potential distribution is more complex and will not be treated here; the reader can refer to appropriate references (Grahame 1954; Parsons 1954; Delahay 1965; Lyklema 1991; Bockris, Reddy, and Gamboa-Aldeco 2001). Figure 3.14 shows schematically results computed by Grahame (1947) for NaCl 0.3 M. We will not go into details of the calculations made but will just examine the results. There are two straight segments

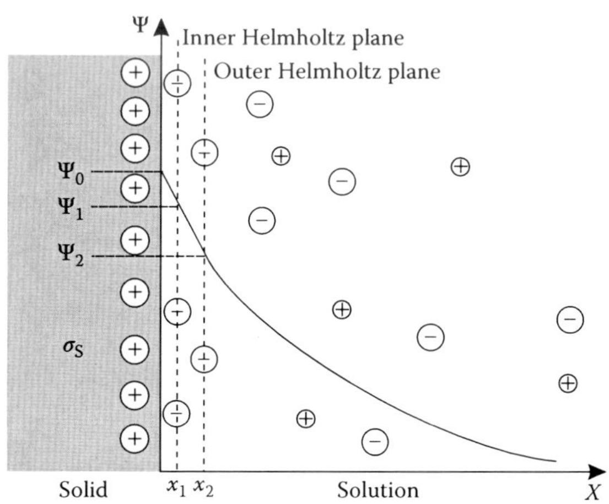

FIGURE 3.13 Schematic representation of the Gouy–Chapman–Stern–Grahame theory. Compare with Figure 3.11: Two planes of closest approach are defined, the outer Helmholtz plane, where ions that retain their hydration layer are adsorbed (nonspecific adsorption) and the inner Helmholtz plane, is defined as the distance where ions loosing the hydration layer (specifically adsorbed) to the surface.

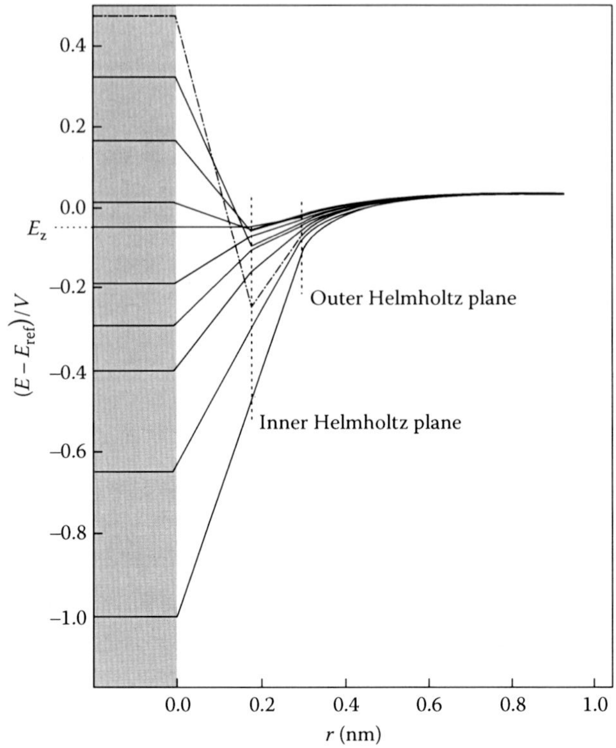

FIGURE 3.14 Qualitative description of the potential profile for a mercury electrode (grayed area) in contact with a 0.3 M NaCl solution. Potentials in the left axis are against an arbitrary reference; E_z is the zero charge potential. The profile at the highest positive potential of the electrode has been plotted as a dash-dot line for clarity. (Adapted from Grahame, D. C., *Chem. Rev.*, 41, 441–501, 1947. With permission.)

in the profile: between surface and IHP (0–1 segment) and between IHP and OHP (1–2 segment). At the most negative E values, both segments have the same slope, because there are no ions in the inner layer (Na^+ does not show specific adsorption); as the potential becomes more positive, the two segments have different slopes, indicative of adsorption at the IHP. It should be noted that at a high positive potential of the electrode (dash-dot line), the potential at the IHP becomes *negative*. That happens because the amount of Cl^- ions adsorbed at the IHP *exceeds* the surface charge, resulting in a net negative charge and, consequently, a negative potential; the net charge at the OHP and the diffuse layer will be consequently positive.

The preceding descriptions are based on a metallic electrode in laboratory conditions under potential control, a situation that is not, in principle, representative of what can be found in a complex natural system as a soil. However, we presented these results because liquid mercury is a surface as devoid of structure as possible, thus concentrating ourselves in the contributions to the charge from the solution side. The main elements presented here, namely the diffuse layer, outer sphere (OHP), and inner sphere (IHP) adsorption, all are present in the interaction between colloidal particles and solution and contribute to the charge distribution in the interfacial region. In the following chapters, we will develop the topic of the colloid/solution interface.

3.4 BEYOND THE POISSON–BOLTZMANN EQUATION*

Despite its widespread application, the PB equation has a number of limitations, as discussed in Section 3.1.2.2 (Kirkwood 1934; Blum 1977; Levine and Outhwaite 1978). Thus, a number of modifications and/or alternative approaches have been proposed. As the PB equation results are usually adequate for environmental purposes, here we will introduce some definitions and terminology. A model where the ions are assumed to be charges entities, either point-like or hard spheres, in a continuum medium of fixed dielectric constant ε, is termed the *primitive model* (PM); if all ions are of the same size, then we have the *restricted primitive model* (RPM). Models that consider the solvent structure, and consequently ion solvation effects, are *non-PMs*. Thus, the GC model is a RPM with point charges. The Gouy–Chapman–Stern–Grahame (GCSG) model treats the ions as hard spheres of finite size, but only in their interaction with the surface, not between themselves. According to Kirkwood (1934), the approximations of the GC theory, in the framework of classic statistical mechanics, consisted in neglecting an exclusion-volume term (i.e., neglecting ion–ion repulsion at short distances) and a fluctuation term. The approximations lead to inconsistencies, as pointed out by several authors (Blum 1977; Levine and Outhwaite 1978; Henderson 1983). There have been two main lines in the research to improve over the GCSG model (Levine, Outhwaite, and Bhuiyan 1981; Blum, Hernando, and Lebowitz 1983; Henderson 1983; Henderson and Boda 2009): several authors, mainly Outhwaite and coworkers, tried to include in the original PB equation the neglected effects, leading to several treatments usually known as *modified Poisson–Boltzmann* (MPB). Another line of thought was followed by Blum, Henderson, and coworkers, applying methods from the statistical mechanic theory of electrolytes (Blum 1977; Blum, Hernando, and Lebowitz 1983; Henderson 1983). More recently, the development of the density functional theory (DFT) opened new roads for the theoretical study of interfaces. The DFT approach developed by Rosenfeld (1993) has been used by Mier-y-Teran et al. (2001) and Boda et al. (2002), who have compared the DFT structural results for a planar DL for 1:1 and 2:1/1:2 RPM electrolytes with a new set of Monte Carlo (MC) simulations. Finally, computer simulations have been conducted by a number of authors, mainly using methods of either molecular dynamics, where the molecular motion equations are solved (Philpott, Glosli, and Zhu 1995; Mattke and Kecke 1998; Dimitrov and Raev 2000; Marquez and Balbuena 2001; Předota et al. 2004; Sugino et al. 2007; Zarzycki, Kerisit, and Rosso 2010), or MC, where phase space averages over random runs are made (Torrie and Valleau 1980; Torrie 1982; Valleau 1982; Valleau and Torrie 1984; Boda, Chan, and Henderson 1998; Henderson and Boda 2009). Computer simulations yield essentially exact results for a given model, independently of the physical correctness of such model. They serve as both a test of the model and a reference to assess analytical approximations.

* Section has more advanced or specific materials.

Outhwaite and coworkers (Outhwaite 1978; Bhuiyan, Outhwaite, and Levine 1979; Outhwaite, Bhuiyan, and Levine 1980; Outhwaite and Bhuiyan 1983) introduced a MPB approach, in which a modified PB equation was proposed in the RPM, always in the infinite plane surface geometry:

$$\frac{d^2\psi}{dx^2} = -\frac{F}{\varepsilon}\sum_i z_i c_i^\infty g_{0i}(x) \tag{3.47}$$

where $g_{0i}(x)$ is the ion-wall distribution function. This function gives the probability of finding an ion of species i at a distance x from the surface, relative to the solution bulk, and here, it is equal to $c_i(x)/c^\infty$. The concept of distribution functions has been first developed in the statistical mechanics of liquids and applied to interfacial problems afterward. It has been shown (Kirkwood 1934) that

$$g_{0i}(x) = \zeta_i(x)e^{\eta_i-(z_i F\psi(x)/RT)} \tag{3.48}$$

where $\zeta_i(x)$ is the exclusion volume due to ion i, and η_i is a fluctuation term related to ion–ion interactions. In the PB equation, $\zeta_i = 1$ and $\eta_i = 0$, which means neglecting both contributions. In the MPB approach, the problem is to find expressions for these two quantities to close Equation 3.47. We will not go into derivation details here, which can be found in the original papers (Levine and Outhwaite 1978; Bhuiyan, Outhwaite, and Levine 1979; Outhwaite, Bhuiyan, and Levine 1980); we will only recall that, after solving η_i in terms of ψ, four cases were considered for the exclusion volume ζ_i:

- MPB1:

$$\zeta_i = 1$$

- MPB2:

$$\zeta_i(x) = \begin{cases} 1 + \dfrac{9\pi N_A a^3}{8}\sum_j c_j^\infty\left(1-\dfrac{2x}{3a}\right)^2\left(1+\dfrac{2x}{3a}\right), & \dfrac{a}{2} \le x \le \dfrac{3a}{2} \\ 1, & x \ge \dfrac{3a}{2} \end{cases} \tag{3.49}$$

- MPB3 and MPB4:

$$\zeta_i(x) = 1 + B(x) + \frac{4\pi a^3 N_A}{3}\sum_j c_j^\infty \tag{3.50}$$

with different definitions of the function $B(x)$ for MPB3 and MPB4, the latter being more exact. Here, a is the ionic diameter; note that for $x < a/2$, no ions will be present (inner layer), and so, the potential profile is linear, as in the Gouy–Chapman–Stern (GCS) model.

Figures 3.15 and 3.16 show some numerical results found by Outhwaite, Bhuiyan, and Levine (1980) compared with the GCS theory, considering 1:1 electrolytes. For

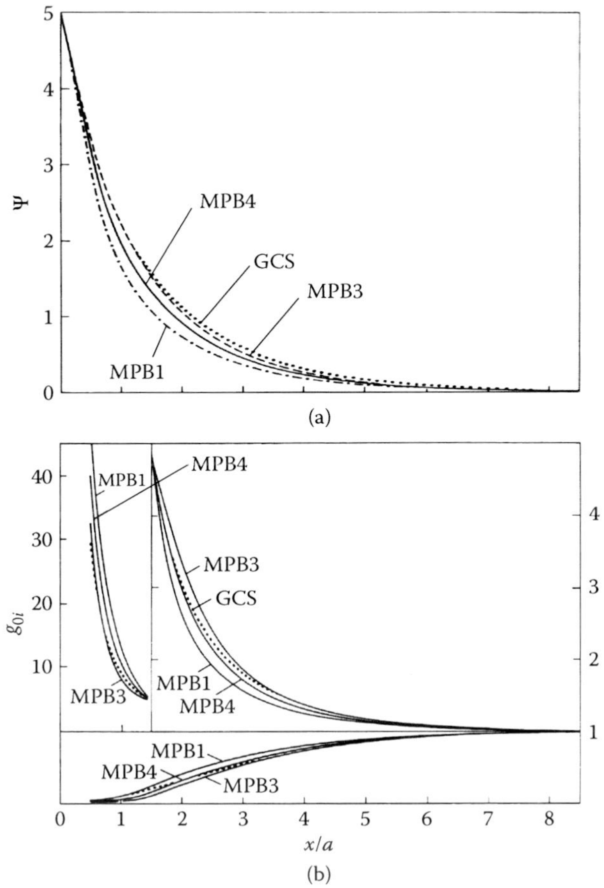

FIGURE 3.15 Dimensionless mean electrostatic potential (a) and surface-ion distribution function (b) as predicted by the Gouy–Chapman–Stern (GCS) and modified Poisson–Boltzmann (MPB) theories for a 1:1 electrolyte with a = 0.425 nm and c^∞ = 0.197 M. (Outhwaite, Bhuiyan, and Levine, 1980, Theory of the electric double layer using a modified Poisson–Boltzmann equation, *Journal of the Chemical Society, Faraday Transactions 2: Molecular and Chemical Physics*, 76, 1388–1408. Reproduced by permission of The Royal Society of Chemistry.)

ionic strengths not very high (Figure 3.15), the results are qualitatively similar to GCS; however, for high concentrations (about 2 M; Figure 3.16), the MPB theory predicts oscillations both in ψ and the concentrations, which depart significantly from the GCS theory.

The MPB theory was further improved, yielding the MPB5 formulation (Outhwaite and Bhuiyan 1983) and an alternative formulation that we will call MPB6 (Outhwaite and Lamperski 2001); both introduce new expressions for the excluded volume ζ_i. These theories have been compared (Bhuiyan and Outhwaite 2004) with DFT results (Boda et al. 2004) and MC results from the study by Boda et al. (2002). The reader is referred to these papers for details; here, we will

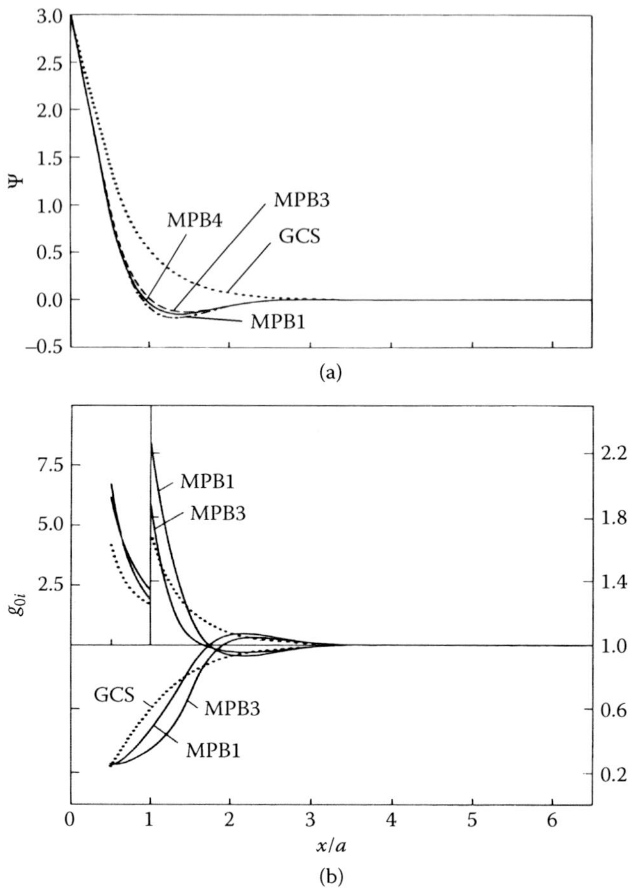

FIGURE 3.16 Dimensionless mean electrostatic potential (a) and surface-ion distribution function (b) as predicted by the Gouy–Chapman–Stern (GCS) and modified Poisson–Boltzmann (MPB) theories for a 1:1 electrolyte with a = 0.425 nm and c^{∞} = 1.97 M. (Outhwaite, Bhuiyan, and Levine, 1980, Theory of the electric double layer using a modified Poisson–Boltzmann equation, *Journal of the Chemical Society, Faraday Transactions 2: Molecular and Chemical Physics*, 76, 1388–1408. Reproduced by permission of The Royal Society of Chemistry.)

reproduce some comparisons. Figures 3.17 and 3.18 show the surface ion distribution function and the mean electrostatic potential, respectively, as a function of distance to the surface for a 2:1 electrolyte. The MPB profiles overall are close to the MC data. Damped oscillations are seen in the MPB, and in this respect, the MPB and DFT results are strikingly similar, with perhaps the MPB5 having a slight edge in reproducing the MC coion profile. Overall, the MPB5 approach seems to give good results, as confirmed for some more recent results (Bhuiyan, Outhwaite, and Henderson 2007).

The other line of thought is based in integral equation schemes to obtain the ion-wall distribution function rather than the electrostatic potential. They are based on

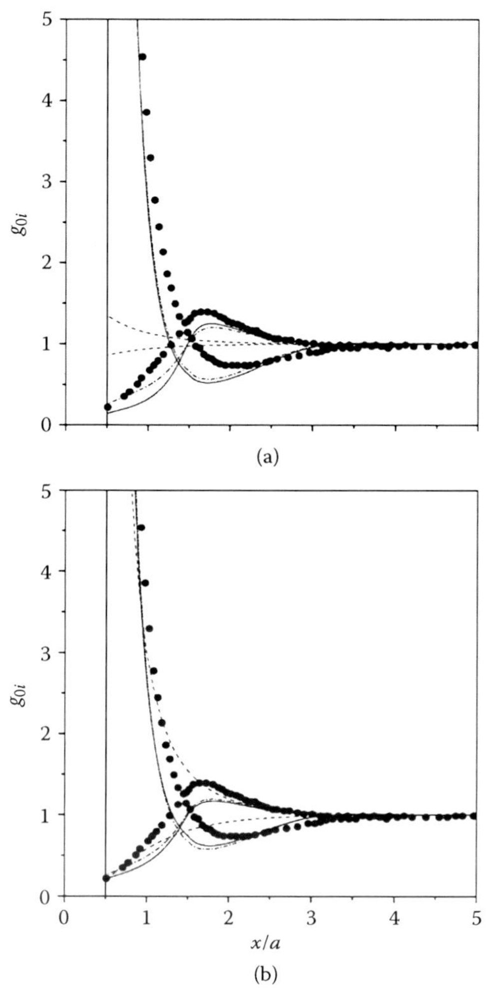

FIGURE 3.17 Surface-ion distribution function for a 2:1 electrolyte with $a = 0.30$ nm and $c^\infty = 1.0$ M for (a) a fixed dimensionless potential at the Helmholtz plane $\Psi(a/2) = -0.15$; (b) a fixed dimensionless surface charge $\sigma_S{}^* = \sigma_S a^2/q_e = 0.1685$. Symbols: MC simulations; solid line, MPB5; dash-dotted line, MPB6; dotted line, DFT; dashed line, GCS (Bhuiyan and Outhwaite, 2004, Comparison of the modified Poisson–Boltzmann theory with recent density functional theory and simulation results in the planar electric double layer, *Physical Chemistry Chemical Physics*, 6, no. 13, 3467. Reproduced by permission of The Royal Society of Chemistry.)

the Ornstein–Zernicke equation for an electrolyte in contact with a charged surface (Henderson, Abraham, and Barker 1976; Levine and Outhwaite 1978):

$$g_{0i}(x_i) = 1 + f_{0i}^C(x_i) + \sum_{j=1}^{n} c_j^\infty \int_V f_{ji}^C(\mathbf{r}_i, \mathbf{r}_j)\left[g_{0j}(x_j) - 1\right] d\mathbf{r}_j \tag{3.51}$$

where $f_{0i}{}^C(x_i)$ is the direct correlation function between the surface and ion i at a distance x_i for the surface, $f_{ji}{}^C(\mathbf{r}_i, \mathbf{r}_j)$ is the direct correlation function between species i

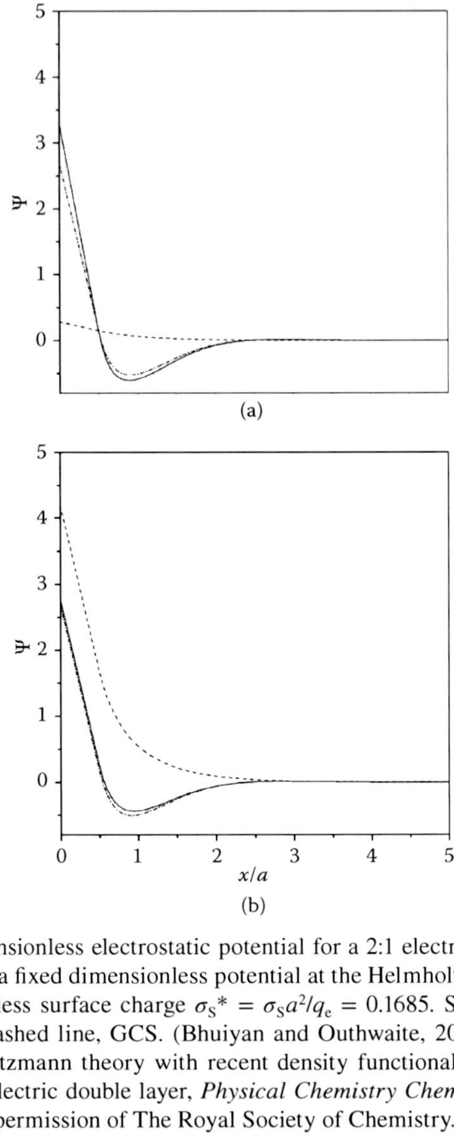

FIGURE 3.18 Dimensionless electrostatic potential for a 2:1 electrolyte with $a = 0.30$ nm and $c^{\infty} = 1.0$ M for (a) a fixed dimensionless potential at the Helmholtz plane $\Psi(a/2) = -0.15$; (b) a fixed dimensionless surface charge $\sigma_S{}^* = \sigma_S a^2/q_e = 0.1685$. Solid line, MPB5; dash-dotted line, MPB6; dashed line, GCS. (Bhuiyan and Outhwaite, 2004, Comparison of the modified Poisson–Boltzmann theory with recent density functional theory and simulation results in the planar electric double layer, *Physical Chemistry Chemical Physics*, 6, no. 13, 3467. Reproduced by permission of The Royal Society of Chemistry.)

and j at positions $\mathbf{r}_i$ and $\mathbf{r}_j$, respectively; the sum is over all species in the solution, and the integral is over the solution volume. One of the first theories of this type was the application of the mean spherical approximation (MSA) of fluids to the electrified interfaces by Blum (1977). The MSA can be regarded as an extension of the linearized GCS theory, but where the ionic radii are treated self-consistently, whereas in GCS, they are considered only for the surface–ion interaction. We will mention only the MSA result for the differential capacitance:

$$\frac{1}{C} = \frac{2\pi}{\varepsilon \Gamma_C} \tag{3.52}$$

where

$$\Gamma_C = \frac{\sqrt{1+2\kappa a}-1}{2a} \tag{3.53}$$

Expansion in power series of κ and retaining the first terms only gives

$$\frac{1}{C} \approx \frac{4\pi}{\varepsilon\kappa} + \frac{2\pi a}{\varepsilon} - \frac{\pi\kappa a^2}{\varepsilon} \tag{3.54}$$

which is consistent with the GCS theory (Equation 3.41) only in the linearized version and at small κ. This puts in question the validity of Equation 3.41. The MSA theory is in general an improvement over the GCS but restricted to small surface charge/potential due to linearization.

There have been other attempts to apply integral equations methods derived from Equation 3.51 or other similar expressions. One of them is the hypernetted chain (Henderson 1983), which is a generalization of the MSA theory, applicable to higher charge/potential values. It gives results comparable to the MPB ones but requires more extensive numerical evaluations. Another proposal is so-called dressed-ion theory of Kjellander and Mitchell (1994, 1997).

All the above effort was made within the PM; that is, considering water as a continuum. There has been comparatively little work on non-PMs with explicit solvent molecules. The simplest model of a solvent is a fluid of hard spheres with a background dielectric continuum. This is a crude model, but it does reflect the fact that water or another solvent consists of molecules, and it has been studied by simulations (Zhang, Davis, and White 1993; Boda, Chan, and Henderson 1998; Lamperski and Zydor 2007). It is easy to apply this model to the MSA (Carnie and Chan 1980; Blum and Henderson 1981). One interesting feature of this work is that the interfacial region of the solvent dipoles is not limited to some inner layer. The interfacial region of the solvent molecules turns out to be as broad as that of the ions. There have been integral equation calculations for more general solvent models (Torrie and Patey 1993; Vossen and Forstmann 1995; Lamperski and Outhwaite 1999). It should be noted that it is quite difficult to perform molecular simulations except at the high concentrations; simulations at 0.01 M ionic concentrations would require about 5500 water molecules for each ion, so that including a reasonable number of ions would need a huge number of water molecules.

3.5 APPROXIMATE SOLUTIONS OF THE PB EQUATION

A number of approximate analytical solutions to the PB equation have been proposed as summarized by Ohshima, Healy, and White (1982). These authors introduced quite accurate expressions for the potential profile and the surface charge–potential relationship around a spherical particle of radius a. For the surface charge density–surface potential relationship in a 1-1 electrolyte, they found

$$\sigma_S = \frac{2\boldsymbol{F}}{\varepsilon\kappa \boldsymbol{RT}}\sinh\left(\frac{\boldsymbol{F}\psi_S}{\boldsymbol{RT}}\right)\left\{1+\frac{2}{\kappa a\cosh^2\left(\boldsymbol{F}\psi_S/4\boldsymbol{RT}\right)}+\frac{8\ln\left[\cosh\left(\boldsymbol{F}\psi_S/4\boldsymbol{RT}\right)\right]}{\left[\kappa a\sinh\left(\boldsymbol{F}\psi_S/2\boldsymbol{RT}\right)\right]}\right\}^{1/2} \tag{3.55}$$

whereas for a 2-1 electrolyte, the approximate expression is

$$\sigma_S = \frac{\boldsymbol{F}}{\varepsilon\kappa \boldsymbol{RT}}\left\{p^2q^2+\frac{4}{\kappa a}\left[(3-p)q-3\right]+\frac{4}{(\kappa a)^2}\left[6\ln\left(\frac{q+1}{2}\right)+\ln(1-p)\right]\right\}^{1/2} \tag{3.56}$$

where

$$p = 1-e^{-\boldsymbol{F}\psi_S/\boldsymbol{RT}} \tag{3.57}$$

$$q = \left(\frac{2}{3}e^{-\boldsymbol{F}\psi_S/\boldsymbol{RT}}+\frac{1}{3}\right)^{1/2} \tag{3.58}$$

Other expressions are mentioned in the study by Ohshima, Healy, and White (1982).

3.6 SUMMARY

In this chapter, the general features of the electrostatic potential and ion concentration profiles near a charged surface have been introduced. The classic theory of Gouy–Chapman–Stern with Grahame modification (GCSG), based on the PB equation, has been presented. Even with its approximate nature, due to a number of simplifying assumptions, it is useful in describing the electrostatic structure of the interface. There are several modern approaches, such as the modified PB theory of Outhwaite and coworkers, which describe better the physics of the problem, but at the cost of complex expressions that in many cases have to be solved numerically. Thus, the GCSG theory continues to be used because it gives analytical expressions. In the case of soil systems, it is adequate in many circumstances.

The origin of the surface charge has not been deal with as yet. The charge is due to adsorption of ionic species and/or surface reactions; these matters will be analyzed in the next chapters.

REFERENCES

Atkins, P., and J. de Paula. 2009. *Physical Chemistry*. 9th ed. New York: W. H. Freeman.

Baldelli, S. 2008. Surface structure at the ionic liquid–electrified metal interface. *Accounts of Chemical Research* 41, no. 3: 421–431. doi:10.1021/ar700185h.

Bard, A. J., and L. R. Faulkner. 2000. *Electrochemical Methods: Fundamentals and Applications*. 2nd ed. New York: John Wiley & Sons.

Benham, C. J. 1983. The cylindrical Poisson–Boltzmann equation. I. Transformations and general solutions. *The Journal of Chemical Physics* 79: 1969–1973.

Berry, R. S., S. A. Rice, and J. Ross. 2000. *Physical Chemistry*. 2nd ed. New York: Oxford University Press.

Bhuiyan, L. B., and C. W. Outhwaite. 2004. Comparison of the modified Poisson-Boltzmann theory with recent density functional theory and simulation results in the planar electric double layer. *Physical Chemistry Chemical Physics* 6, no. 13: 3467. doi:10.1039/b316098j.

Bhuiyan, L. B., C. W. Outhwaite, and D. Henderson. 2007. Some simulation and modified Poisson-Boltzmann theory results for the contact values of an electrolyte near a charged electrode. *Journal of Electroanalytical Chemistry* 607, no. 1–2: 54–60. doi:10.1016/j.jelechem.2006.10.010.

Bhuiyan, L. B., C. W. Outhwaite, and S. Levine. 1979. Numerical solution of a modified Poisson-Boltzmann equation in electric double layer theory. *Chemical Physics Letters* 66, no. 2: 321–324. doi:10.1016/0009-2614(79)85026-5.

Blum, L. 1977. Theory of electrified interfaces. *The Journal of Physical Chemistry* 81, no. 2: 136–147.

Blum, L., and D. Henderson. 1981. Mixtures of hard ions and dipoles against a charged wall: The Ornstein–Zernike equation, some exact results, and the mean spherical approximation. *The Journal of Chemical Physics* 74, no. 3: 1902. doi:10.1063/1.441282.

Blum, L., J. Hernando, and J. L. Lebowitz. 1983. Numerical method and general discussion of integral equations for the primitive model of the electric interface. *The Journal of Physical Chemistry* 87, no. 15: 2825–2832. doi:10.1021/j100238a026.

Bockris, J. O'M., A. K. N. Reddy, and M. E. Gamboa-Aldeco. 2001. *Modern Electrochemistry 2A: Fundamentals of Electrodics*. 2nd ed. New York: Kluwer Academic/Plenum Publishers.

Boda, D., D. Henderson, P. Plaschko, and W. R. Fawcett. 2004. Monte Carlo and density functional theory study of the electrical double layer: The dependence of the charge/voltage relation on the diameter of the ions. *Molecular Simulation* 30, no. 2–3: 137–141.

Boda, D., K.-Y. Chan, and D. Henderson. 1998. Monte Carlo simulation of an ion-dipole mixture as a model of an electrical double layer. *The Journal of Chemical Physics* 109, no. 17: 7362. doi:10.1063/1.477342.

Boda, D., W. R. Fawcett, D. Henderson, and S. Sokołowski. 2002. Monte Carlo, density functional theory, and Poisson–Boltzmann theory study of the structure of an electrolyte near an electrode. *The Journal of Chemical Physics* 116, no. 16: 7170. doi:10.1063/1.1464826.

Carnie, S. L., and D. Y. C. Chan. 1980. The structure of electrolytes at charged surfaces: Ion–dipole mixtures. *The Journal of Chemical Physics* 73, no. 6: 2949. doi:10.1063/1.440468.

Chen, Z., and R. K Singh. 2002. General solution for Poisson-Boltzmann equation in semiinfinite planar symmetry. *Journal of Colloid and Interface Science* 245, no. 2: 301–306.

D'yachkov, L. G. 2005a. Screening of macroions in colloidal plasmas: Accurate analytical solution of the Poisson-Boltzmann equation. *Physics Letters A* 340, no. 5–6: 440–448.

D'yachkov, L. G. 2005b. Analytical solution of the Poisson-Boltzmann equation in cases of spherical and axial symmetry. *Technical Physics Letters* 31, no. 3: 204–207.

Das, T., D. Bratko, L. B. Bhuiyan, and C. W. Outhwaite. 1997. Polyelectrolyte solutions containing mixed valency ions in the cell model: A simulation and modified Poisson–Boltzmann study. *The Journal of Chemical Physics* 107, no. 21: 9197. doi:10.1063/1.475211.

Davies, C. W. 1962. *Ion Association*. London: Butterworths.

Delahay, P. 1965. *Double Layer and Electrode Kinetics*. New York: Interscience Publishers.

Dimitrov, D. I., and N. D. Raev. 2000. Molecular dynamics simulations of the electrical double layer at the 1 M KCl solution | Hg electrode interface. *Journal of Electroanalytical Chemistry* 486, no. 1: 1–8. doi:10.1016/S0022-0728(00)00105-4.

Dyshlovenko, P. E. 2002. Adaptive numerical method for Poisson-Boltzmann equation and its application. *Computer Physics Communications* 147, no. 1–2: 335–338. doi:10.1016/S0010-4655(02)00298-9.

Fawcett, W. R. 2009. Monte Carlo studies of ion size effects in the diffuse double layer. *Electrochimica Acta* 54, no. 22: 4997–5005. doi:10.1016/j.electacta.2009.02.025.

Feynman, R. P., R. B. Leighton, and M. L. Sands. 1970. *Mainly Electromagnetism and Matter. The Feynman Lectures on Physics*. Vol. 2 (3 Volume Set). Reading, MA: Addison Wesley Longman.

Grahame, D. C. 1947. The electrical double layer and the theory of electrocapillarity. *Chemical Reviews* 41, no. 3: 441–501. doi:10.1021/cr60130a002.

Grahame, D. C. 1954. Differential capacity of mercury in aqueous sodium fluoride solutions. I. Effect of concentration at 25°. *Journal of the American Chemical Society* 76, no. 19: 4819–4823. doi:10.1021/ja01648a014.

Grahame, D. C. 1958. Components of charge and potential in the non-diffuse region of the electrical double layer: Potassium iodide solutions in contact with mercury at 25°1. *Journal of the American Chemical Society* 80, no. 16: 4201–4210. doi:10.1021/ja01549a022.

Grahame, D. C., and B. A. Soderberg. 1954. Ionic components of charge in the electrical double layer. *The Journal of Chemical Physics* 22, no. 3: 449–460. doi:10.1063/1.1740089.

Grahame, D. C., and R. Parsons. 1961. Components of charge and potential in the inner region of the electrical double layer: Aqueous potassium chloride solutions in contact with mercury at 25°. *Journal of the American Chemical Society* 83, no. 6: 1291–1296. doi:10.1021/ja01467a009.

Guerrero-García, G. I., E. González-Tovar, and M. Olvera de la Cruz. 2010. Effects of the ionic size-asymmetry around a charged nanoparticle: Unequal charge neutralization and electrostatic screening. *Soft Matter* 6, no. 9: 2056. doi:10.1039/b924438g.

Henderson, D. 1983. Recent progress in the theory of the electric double layer. *Progress in Surface Science* 13, no. 3: 197–224. doi:10.1016/0079-6816(83)90004-7.

Henderson, D., F. F. Abraham, and J. A. Barker. 1976. The Ornstein-Zernike equation for a fluid in contact with a surface. *Molecular Physics: An International Journal at the Interface between Chemistry and Physics* 31, no. 4: 1291. doi:10.1080/00268977600101021.

Henderson, D., and D. Boda. 2009. Insights from theory and simulation on the electrical double layer. *Physical Chemistry Chemical Physics* 11, no. 20: 3822. doi:10.1039/b815946g.

Hiemenz, P. C., and R. Rajagopalan. 1997. *Principles of Colloid and Surface Chemistry*. 3rd ed. New York: Marcel Dekker.

James, A. E., and D. J. A. Williams. 1985. Numerical solution of the Poisson–Boltzmann equation. *Journal of Colloid and Interface Science* 107, no. 1: 44–59. doi:10.1016/0021-9797(85)90147-X.

Kirkwood, J. G. 1934. On the theory of strong electrolyte solutions. *The Journal of Chemical Physics* 2, no. 11: 767–781. doi:10.1063/1.1749393.

Kjellander, R., and D. J. Mitchell. 1994. Dressed-ion theory for electrolyte solutions: A Debye–Hückel-like reformulation of the exact theory for the primitive model. *The Journal of Chemical Physics* 101, no. 1: 603. doi:10.1063/1.468116.

Kjellander, R., and D. J. Mitchell. 1997. Dressed ion theory for electric double layer structure and interactions; an exact analysis. *Molecular Physics: An International Journal at the Interface between Chemistry and Physics* 91, no. 2: 173. doi:10.1080/00268979709482709.

Klein, B. K., C. F. Anderson, and M. T. Record. 1981. Comparison of Poisson-Boltzmann and condensation model expression for the colligative properties of cylindrical polyions. *Biopolymers* 20, no. 10: 2263–2280. doi:10.1002/bip.1981.360201018.

Lamm, G. 2003. The Poisson-Boltzmann equation. In *Reviews in Computational Chemistry*, ed. K. B. Lipkowitz, R. Larter, and T. R. Cundari, 19: 147–365. Hoboken, NJ: John Wiley & Sons.

Lamperski, S., and A. Zydor. 2007. Monte Carlo study of the electrode solvent primitive model electrolyte interface. *Electrochimica Acta* 52, no. 7 (February 1): 2429–2436. doi:10.1016/j.electacta.2006.08.045.

Lamperski, S., and C. W. Outhwaite. 1999. A non-primitive model for the electrode | electrolyte interface based on the Percus-Yevick theory. Analysis of the different molecular sizes, ion valences and electrolyte concentrations. *Journal of Electroanalytical Chemistry* 460, no. 1–2: 135–143.

Levine, I. 2008. *Physical Chemistry*. 6th ed. New York: McGraw-Hill Science/Engineering/Math.

Levine, S., and C. W. Outhwaite. 1978. Comparison of theories of the aqueous electric double layer at a charged plane interface. *Journal of the Chemical Society, Faraday Transactions 2: Molecular and Chemical Physics* 74: 1670–1689.

Levine, S., C. W. Outhwaite, and L. B. Bhuiyan. 1981. Statistical mechanical theories of the electric double layer. *Journal of Electroanalytical Chemistry* 123, no. 1: 105–119.

Lima, E. R. A., F. W. Tavares, and E. C. Biscaia Jr. 2007. Finite volume solution of the modified Poisson-Boltzmann equation for two colloidal particles. *Physical Chemistry Chemical Physics* 9, no. 24: 3174–3180.

Liu, L., and I. Neretnieks. 2008. Homo-interaction between parallel plates at constant charge. *Colloids and Surfaces A: Physicochemical and Engineering Aspects* 317, no. 1–3: 636–642. doi:10.1016/j.colsurfa.2007.11.055.

Lu, B. Z., Y. C. Zhou, M. J. Holst, and J. A. McCammon. 2008. Recent progress in numerical methods for the Poisson-Boltzmann equation in biophysical applications. *Communications in Computational Physics* 3, no. 5: 973–1009.

Lyklema, J. 1991. *Fundamentals of Interface and Colloid Science*. Amsterdam: Academic Press.

Marquez, A., and P. B. Balbuena. 2001. Molecular dynamics study of graphite/electrolyte interfaces. *Journal of The Electrochemical Society* 148, no. 6: A624–A635. doi:10.1149/1.1372216.

Mattke, T., and H.-J. Kecke. 1998. Molecular dynamic simulations of single, interacting, and sheared double layers: 1. Configuration of a double layer. *Journal of Colloid and Interface Science* 208, no. 2: 555–561. doi:10.1006/jcis.1998.5825.

McQuarrie, D. A., and J. D. Simon. 1997. *Physical Chemistry: A Molecular Approach*. Herndon, VA: University Science Books.

Mier-y-Teran, L., D. Boda, D. Henderson, and S. E. Quiñones-Cisneros. 2001. On the low temperature anomalies in the properties of the electrochemical interface. A non-local free-energy density functional approach. *Molecular Physics: An International Journal at the Interface between Chemistry and Physics* 99, no. 15: 1323. doi:10.1080/00268970110048383.

Morrison Jr., F. A., and J. F. Osterle. 1965. Electrokinetic energy conversion in ultrafine capillaries. *The Journal of Chemical Physics* 43: 2111.

Newman, J., and K. E. Thomas-Alyea. 2004. *Electrochemical Systems*. 3rd ed. Hoboken, NJ: Wiley-Interscience.

Nussbaum, A. 1965. *Electromagnetic Theory for Engineers and Scientists*. Englewood Cliffs, NJ: Prentice-Hall.

Ohshima, H. 2006. Chapter 1. Electrical double layer around a charged colloidal particle in an electrolyte solution. In H. Ohshima *Theory of Colloid and Interfacial Electric Phenomena*, 12: 1–38. Interface Science and Technology. San Diego: Academic Press.

Ohshima, H., T. W. Healy, and L. R. White. 1982. Accurate analytic expressions for the surface charge density/surface potential relationship and double-layer potential distribution for a spherical colloidal particle. *Journal of Colloid and Interface Science* 90, no. 1: 17–26. doi:10.1016/0021-9797(82)90393-9.

Olivares, W., and D. A. McQuarrie. 1985. Comments on the calculation of the potential inside a charged microcapillary. *The Journal of Physical Chemistry* 89, no. 13: 2966–2967.

Outhwaite, C. W. 1978. Modified Poisson-Boltzmann equation in electric double layer theory based on the Bogoliubov-Born-Green-Yvon integral equations. *Journal of the Chemical Society, Faraday Transactions 2: Molecular and Chemical Physics* 74: 1214–1221.

Outhwaite, C. W., and L. B. Bhuiyan. 1983. An improved modified Poisson–Boltzmann equation in electric-double-layer theory. *Journal of the Chemical Society, Faraday Transactions 2* 79, no. 5: 707. doi:10.1039/f29837900707.

Outhwaite, C. W., L. B. Bhuiyan, and S. Levine. 1980. Theory of the electric double layer using a modified Poisson-Boltzmann equation. *Journal of the Chemical Society, Faraday Transactions 2: Molecular and Chemical Physics* 76: 1388–1408.

Outhwaite, C. W., and S. Lamperski. 2001. A treatment of the exclusion volume term in the inhomogeneous Poisson-Boltzmann theory for an ion-dipole mixture. *Condensed Matter Physics* 4, no. 4: 739–748.

Parsons, R. 1954. Equilibrium properties of electrified interphases. In *Modern Aspects of Electrochemistry*. Vol. 1, ed. J. O'M. Bockris, 1: 103–179. London: Butterworths.

Philip, J. R., and R. A. Wooding. 1970. Solution of the Poisson–Boltzmann equation about a cylindrical particle. *The Journal of Chemical Physics* 52: 953.

Phillips, R. J. 1995. Calculation of multisphere linearized Poisson-Boltzmann interactions near cylindrical fibers and planar surfaces. *Journal of Colloid and Interface Science* 175, no. 2: 386–399. doi:10.1006/jcis.1995.1469.

Philpott, M. R., J. N. Glosli, and S.-B. Zhu. 1995. Molecular dynamics simulation of adsorption in electric double layers. *Surface Science* 335 (July 20): 422–431. doi:10.1016/0039-6028(95)00456-4.

Předota, M., A. V. Bandura, P. T. Cummings, J. D. Kubicki, D. J. Wesolowski, A. A. Chialvo, and M. L. Machesky. 2004. Electric double layer at the rutile (110) surface. 1. Structure of surfaces and interfacial water from molecular dynamics by use of ab initio potentials. *The Journal of Physical Chemistry B* 108, no. 32: 12049–12060. doi:10.1021/jp037197c.

Purcell, E. M. 1984. *Electricity and Magnetism. Berkeley Physics Course*. Vol. 2. 2nd ed. New York: McGraw-Hill Science/Engineering/Math.

Rice, C. L., and R. Whitehead. 1965. Electrokinetic flow in a narrow cylindrical capillary. *The Journal of Physical Chemistry* 69, no. 11: 4017–4024.

Rice, R. E. 1985. Exact solution to the linearized Poisson–Boltzmann equation in cylindrical coordinates. *The Journal of Chemical Physics* 82, no. 9: 4337–4340. doi:10.1063/1.448826.

Rice, R. E., and F. H. Horne. 1981. Analytical solution of the linearized Poisson–Boltzmann equation in cylindrical coordinates. *The Journal of Chemical Physics* 75: 5582.

Rice, R. E., and F. H. Horne. 1985. Analytical solutions to the linearized Poisson–Boltzmann equation in cylindrical coordinates for different ionic–strength distributions. *Journal of Colloid and Interface Science* 105, no. 1: 172–182.

Rosenfeld, Y. 1993. Free energy model for inhomogeneous fluid mixtures: Yukawa-charged hard spheres, general interactions, and plasmas. *The Journal of Chemical Physics* 98, no. 10: 8126. doi:10.1063/1.464569.

Sayyed-Ahmad, A., K. Tuncay, and P. J. Ortoleva. 2004. Efficient solution technique for solving the Poisson-Boltzmann equation. *Journal of Computational Chemistry* 25, no. 8: 1068–1074.

Shi, X., and P. Koehl. 2008. The geometry behind numerical solvers of the Poisson-Boltzmann equation. *Communications in Computational Physics* 3, no. 5: 1032–1050.

Shkel, I. A., O. V. Tsodikov, and M. T. Record. 2002. Asymptotic solution of the cylindrical nonlinear Poisson–Boltzmann equation at low salt concentration: Analytic expressions for surface potential and preferential interaction coefficient. *Proceedings of the National Academy of Sciences of the United States of America* 99, no. 5: 2597–2602. doi:10.1073/pnas.032480699.

Sigal, V. L., and Y. Y. Ginsburg. 1981. Solution of the nonlinear Poisson-Boltzmann equation for the inner problems of the electrical double-layer theory. *The Journal of Physical Chemistry* 85, no. 25: 3735–3736.

Singh, U., and G. Uehara. 1998. Electrochemistry of the double layer. Principles and applications to soils. In *Soil Physical Chemistry, Second Edition*, ed. D. L. Sparks. Boca Raton, FL: CRC.

Smagala, T. G., and W. R. Fawcett. 2007. Series approach to modeling ion size effects for symmetric electrolytes in the diffuse double layer. *The Journal of Physical Chemistry B* 111, no. 6: 1443–1448. doi:10.1021/jp067039w.

Spitzer, J. J. 2003. Colloidal interactions: Contact limiting laws, double-layer dissociation, and "non-DLVO" (Derjaguin–Landau–Verwey–Overbeek) forces. *Colloid and Polymer Science* 281, no. 6: 589–592. doi:10.1007/s00396-002-0836-3.

Stern, O. 1924. Zur theorie der elektrolytischen doppelschicht. *Zeitschrift fur Elektrochemie* 30: 508–516.

Sugino, O., I. Hamada, M. Otani, Y. Morikawa, T. Ikeshoji, and Y. Okamoto. 2007. First-principles molecular dynamics simulation of biased electrode/solution interface. *Surface Science* 601, no. 22: 5237–5240. doi:10.1016/j.susc.2007.04.208.

Téllez, G., and E. Trizac. 2006. Exact asymptotic expansions for the cylindrical Poisson–Boltzmann equation. *Journal of Statistical Mechanics: Theory and Experiment* 2006: P06018.

Torres, A., R. van Roij, and G. Téllez. 2006. Finite thickness and charge relaxation in double-layer interactions. *Journal of Colloid and Interface Science* 301, no. 1: 176–183. doi:10.1016/j.jcis.2006.04.058.

Torrie, G. M. 1982. Electrical double layers. II. Monte Carlo and HNC studies of image effects. *The Journal of Chemical Physics* 76, no. 9: 4615. doi:10.1063/1.443541.

Torrie, G. M., and G. N. Patey. 1993. Molecular solvent model for an electrical double layer: Asymmetric solvent effects. *The Journal of Physical Chemistry* 97, no. 49: 12909–12918. doi:10.1021/j100151a045.

Torrie, G. M., and J. P. Valleau. 1980. Electrical double layers. I. Monte Carlo study of a uniformly charged surface. *The Journal of Chemical Physics* 73, no. 11: 5807. doi:10.1063/1.440065.

Tournassat, C., Y. Chapron, P. Leroy, M. Bizi, and F. Boulahya. 2009. Comparison of molecular dynamics simulations with triple layer and modified Gouy-Chapman models in a 0.1 M NaCl-montmorillonite system. *Journal of Colloid and Interface Science* 339, no. 2: 533–541. doi:10.1016/j.jcis.2009.06.051.

Tuinier, R. 2003. Approximate solutions to the Poisson-Boltzmann equation in spherical and cylindrical geometry. *Journal of Colloid and Interface Science* 258, no. 1: 45–49. doi:10.1016/S0021-9797(02)00142-X.

Valleau, J. P. 1982. The electrical double layer. III. Modified Gouy–Chapman theory with unequal ion sizes. *The Journal of Chemical Physics* 76, no. 9: 4623. doi:10.1063/1.443542.

Valleau, J. P., and G. M. Torrie. 1984. Electrical double layers. V. Asymmetric ion–wall interactions. *The Journal of Chemical Physics* 81, no. 12: 6291. doi:10.1063/1.447535.

Van, H., and J. A. M. Smit. 1995. Approximative analytical solutions of the Poisson-Boltzmann equation for charged rods in the presence of salt: An analysis of the cylindrical cell model. *Journal of Colloid and Interface Science* 170, no. 1: 134–145. doi:10.1006/jcis.1995.1081.

Van Soestbergen, M., P. M. Biesheuvel, and M. Z. Bazant. 2010. Diffuse-charge effects on the transient response of electrochemical cells. *Physical Review E* 81, no. 2: 021503. doi:10.1103/PhysRevE.81.021503.

Varela, L. M., M. García, and V. Mosquera. 2003. Exact mean-field theory of ionic solutions: Non-Debye screening. *Physics Reports* 382, no. 1–2 (July): 1–111. doi:10.1016/S0370-1573(03)00210-2.

Verwey, E., and J. Th. G. Overbeek. 1948. *Theory of the Stability of Lyophobic Colloids*. 1st ed. Amsterdam: Elsevier.

Vossen, M., and F. Forstmann. 1995. Integral equation theory for the electrode-electrolyte interface with the central force water model. Results for an aqueous solution of sodium chloride. *Molecular Physics: An International Journal at the Interface between Chemistry and Physics* 86, no. 6: 1493. doi:10.1080/00268979500102881.

Wang, J. U. N., and R. A. Y. Luo. 2010. Assessment of linear finite-difference Poisson-Boltzmann solvers. *Journal of Computational Chemistry* 31, no. 8: 1689–1698.

Wang, Z. W., M. Y. Gu, G. Z. Li, and X. Z. Yi. 2005. Solution of the Poisson-Boltzmann equation about a cylindrical particle: Functional theoretical approach. *Journal of Dispersion Science and Technology* 26, no. 4: 517–519.

Wright, M. R. 2007. *An Introduction to Aqueous Electrolyte Solutions*. Hoboken, NJ: John Wiley & Sons.

Zachara, J. M., and J. C. Westall. 1998. Chemical modeling of ion adsorption in soils. In *Soil Physical Chemistry, Second Edition*, ed. D. L. Sparks, 47–96. 2nd ed. Boca Raton, FL: CRC Press.

Zarzycki, P., S. Kerisit, and K. M. Rosso. 2010. Molecular dynamics study of the electrical double layer at silver chloride–electrolyte interfaces. *The Journal of Physical Chemistry C* 114, no. 19: 8905–8916. doi:10.1021/jp9118666.

Zhang, L., H. T. Davis, and H. S. White. 1993. Simulations of solvent effects on confined electrolytes. *The Journal of Chemical Physics* 98, no. 7: 5793. doi:10.1063/1.464872.

Zuckerman, D. M., M. E. Fisher, and S. Bekiranov. 2001. Asymmetric primitive-model electrolytes: Debye-Hückel theory, criticality, and energy bounds. *Physical Review E* 64, no. 1: 011206. doi:10.1103/PhysRevE.64.011206.

4 Sorption onto Colloidal Particles

4.1 GENERAL CONCEPTS

Single molecules and ions interact with each other by various forces, ranging from strong chemical bonding and Coulomb interactions (for ions) to relatively weak dipolar and dispersion forces; these interactions come from the very nature of matter (Hiemenz and Rajagopalan 1997; McQuarrie and Simon 1997; Berry, Rice, and Ross 2000; Levine 2008; Atkins and Paula 2009; Israelachvili 2010), and such interactions result in consequences ranging from the formation of new molecules or complexes to the condensation of gases into liquids and solids. For the same reasons and with the same forces, small molecules and ions interact with surfaces and also with macromolecules and solid bodies such as colloidal particles. As a consequence of that, ions and molecules are bound to the surface of solid particles such as oxides and clay minerals, a phenomenon called *adsorption* (Delahay 1965; Lyklema 1991; Shaw 1992; Adamson and Gast 1997; Hiemenz and Rajagopalan 1997; Israelachvili 2010). Natural organic matter colloids such as humic substances behave more like permeable particles (Ohshima and Kondo 1993; Kuo and Hsu 1999; Duval et al. 2005), thus allowing small ions and molecules to penetrate and bind to internal sites, which constitutes, in some sense, *absorption*. *Sorption* is the general term embodying both the concepts. We refer mainly to sorption in a general sense but use more specific terms when appropriate; most of the treatments presented here apply equally to either case. In this chapter, we treat sorption phenomena in a rather general way, laying the foundations for more specific considerations regarding soil particles in upcoming chapters.

4.2 SORPTION ISOTHERMS

In a sorption process, a species from a fluid phase binds to the surface of (in the case of adsorption) or in the interior of (for absorption) another condensed phase. Thus, one can find sorption of gas molecules to liquids, adsorption of gas molecules to the surface of liquids or solids, or sorption of solute species (ions or molecules) from solution to solid particles. An *adsorption isotherm* or, more generally, a *sorption isotherm,* is an equilibrium relationship between the activity (or fugacity) of the species to be bound, the *sorbate*, in the bulk phase and the activity of the bound sorbate. The process can be regarded formally as a chemical reaction, for example, in the case of adsorption from aqueous solution

$$\mathrm{A(aq)} + \mathrm{S} \rightleftharpoons \mathrm{A(ads)} \tag{4.1}$$

where A is the adsorbate and S is a surface site. The amount of adsorbate bound is usually expressed in terms of surface excess or *surface concentration* Γ_A Section 2.4; as *sorbed quantity or amount*, which is moles of sorbate per unit mass of sorbent:

$$Q_A = \frac{n_A}{m_S} = \Gamma_A A_S \tag{4.2}$$

where n_A is the number of moles of adsorbate and m_S is the sorbent (solid phase) mass; or, in a relative way, as coverage degree θ

$$\theta_A = \frac{\Gamma_A}{\Gamma_S} = \frac{Q_A}{Q_S} \tag{4.3}$$

where Γ_S and Q_S represent the maximum amount bound in a single layer covering the surface (*saturation coverage*, number of sites per unit area, or *site density*). Thus, the adsorption isotherm for the adsorption process, Equation 4.1, is formally a function

$$\Gamma_A = f(a_A) \quad \text{or} \quad Q_A = f'(a_A) \quad \text{or} \quad \theta_A = f''(a_A) \tag{4.4}$$

where the primes mean formally different functions, not derivatives. We now discuss, in an empirical way, the different types of isotherms observed in the adsorption of gases onto solids and afterward the isotherms observed for adsorption of solutes in liquid phase onto solids. The case of gaseous adsorbates may not be the most relevant in soil chemistry, but adsorption from the atmosphere of compounds such as volatile organics (Ruiz, Bilbao, and Murillo 1998) into a dry portion of soil can occur, and it is a good way to introduce the topic. Although we mention occasionally some isotherm equations, such as the Langmuir expression, the theory leading to isotherm equations is discussed in the following sections.

4.2.1 Gas Phase Isotherm Types

When molecules in gaseous state adsorb onto the surface of a condensed phase, usually a solid, different behaviors can be observed depending on the adsorbate, the substrate, and the experimental conditions such as temperature. Brunauer et al. (1940) gave a classification with five main types which remains in use today, albeit extended with a sixth type (Sing 1985). These types cover the majority of situations, except in some cases supercritical fluids show different behavior (Donohue and Aranovich 1998). Here we discuss the six "classical" types.

Figure 4.1 shows typical plots of a type I isotherm and a schematic picture of the process. The relevant independent variable is the gas pressure (or adsorbate partial pressure, considering ideal behavior of the gas); in Figure 4.1a, the isotherm is plotted as a function of pressure, whereas in Figure 4.1b, as a function of the logarithm of pressure; this form is more informative, as, at least for ideal gas behavior, it is directly proportional to the chemical potential variation

$$\mu_{\text{ideal gas}}(T,P) = \mu^0_{\text{ideal gas}} + \boldsymbol{R}T \ln \frac{P}{P^0} \tag{4.5}$$

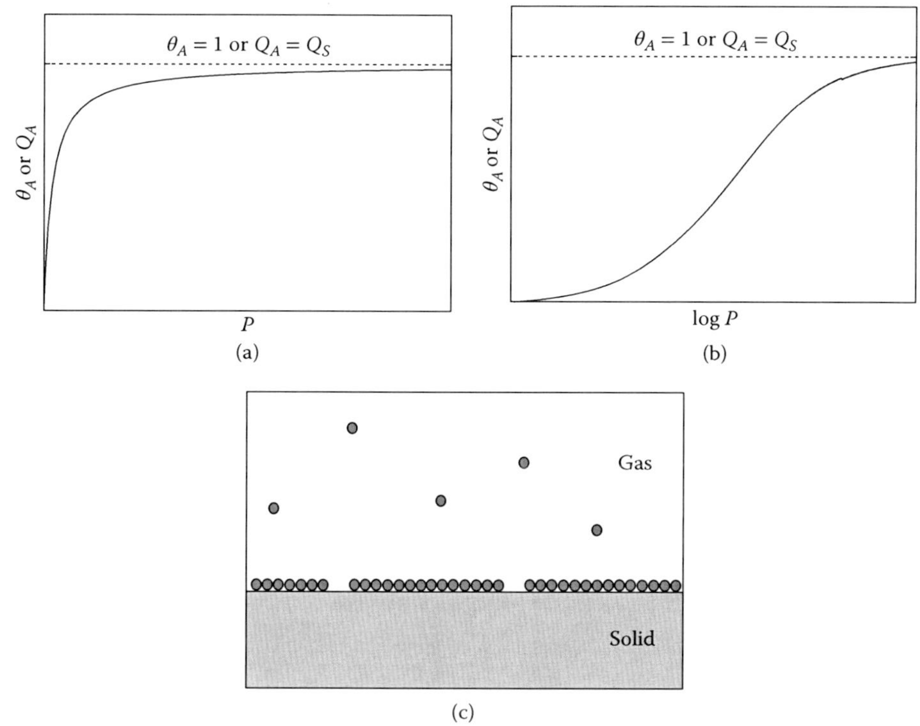

FIGURE 4.1 Schematic plot of a type I isotherm from IUPAC classification (Sing 1985) as a function of pressure (a) and the logarithm of pressure (b). The dashed line shows the point of monolayer saturation; that is, when the surface is covered by a layer just one molecule thick. In (c), a schematic drawing of the (ideal) situation is presented, showing an almost complete monolayer.

where μ^0 is the standard chemical potential and P^0 the standard pressure. The ordinate axis can be plotted as θ, Γ, or Q; often, in soil sorption studies, the last form is used, because it is not required to know neither the true surface area nor the saturation monolayer amount; these quantities in many cases are not easily accessible. Type I isotherms correspond to monolayer adsorption: the limiting value for high pressures is $\theta_A = 1$ (or $\Gamma_A = \Gamma_S$), which corresponds to the surface completely covered by adsorbate molecules forming a layer whose thickness is determined by the adsorbate molecular dimensions (Figure 4.1c). This isotherm type is generally described by the Langmuir equation, which is discussed later in Section 4.4.2. A classical example of this behavior is the adsorption of N_2 or O_2 on nonporous solids at temperatures over the critical temperature (Brunauer et al. 1940).

Figure 4.2 shows typical plots of a type II isotherm as a function of both the pressure and its logarithm. It is quite different from type I, in that the amount adsorbed does not reach saturation but diverges as the pressure approaches a saturation value, usually the adsorbate vapor pressure P_V. At this point, condensation is likely taking place. Figure 4.2c is a pictorial representation of the physical situation, showing the onset of multilayer adsorption; the interaction energy of an

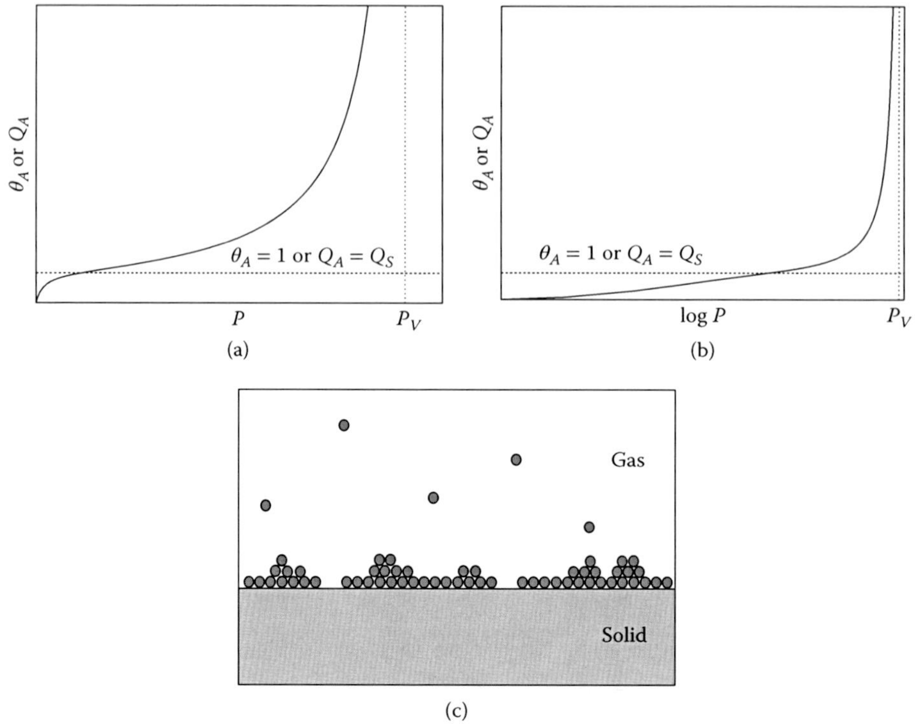

FIGURE 4.2 Schematic plot of a type II isotherm from IUPAC classification (Sing 1985) as a function of pressure (a) and the logarithm of pressure (b). The dashed line shows the point of monolayer saturation, and the dotted line marks the point of the adsorbate vapor pressure. In (c), a schematic drawing of the (ideal) situation is shown; multilayer adsorption takes place, but the adsorbate–surface interaction is stronger than adsorbate–adsorbate interaction, thus the first layer forms preferentially to multilayer growth.

adsorbate molecule with the surface is higher than the interaction between adsorbate molecules in successive layers, so that the first layer is almost complete when adsorption on further layers starts. This isotherm type is generally described by the Brunauer, Emmet, and Teller (BET) equation, as we shall see. A very common example is the adsorption of N_2 on solids such as iron, oxides, and so on at or around 77 K.

In Figure 4.3, in turn, typical plots of a type III isotherm as a function of both the pressure and its logarithm are presented. It differs from type II in that there is no slope decrease as the quantity adsorbed reaches a monolayer amount. Figure 4.3c is a pictorial representation of the physical situation, showing multilayer adsorption; the interaction energy of an adsorbate molecule with the surface is comparable to or lower than the interaction between adsorbate molecules in successive layers, so that the first layer is not complete (sometimes far from complete) when multilayer adsorption starts. In extreme cases, such as water on glass (McHaffie and Lenher 1925), the adsorption starts only when the pressure is near P_V. This isotherm type is

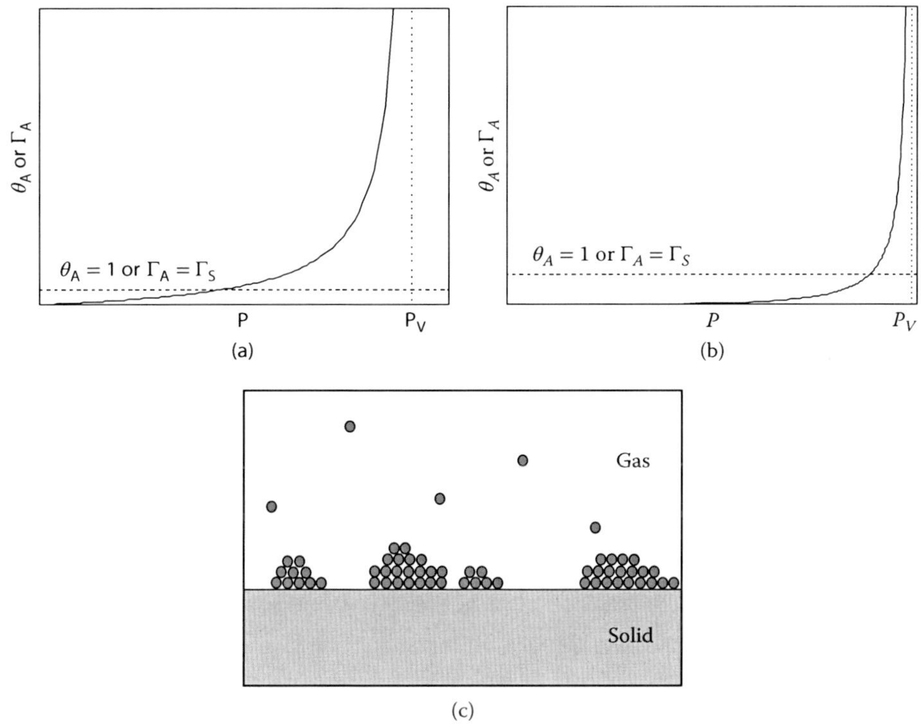

FIGURE 4.3 Schematic plot of a type III isotherm from IUPAC classification (Sing 1985) as a function of pressure (a) and the logarithm of pressure (b). The dashed line shows the point of monolayer saturation, and the dotted line marks the point of the adsorbate vapor pressure. Note the similitude with type II (Figure 4.2). In (c), a schematic drawing of the (ideal) situation is shown; multilayer adsorption takes place, and here the adsorbate–surface interaction is similar or weaker than the adsorbate–adsorbate interaction, thus multilayer growth takes place as soon as the first layer forms.

also described by the BET equation, as we shall see. An example is the adsorption of Br_2 and I_2 on silica gel (Reyerson and Cameron 1935).

Figure 4.4 depicts typical plots of a type IV isotherm as a function of both the pressure and its logarithm. In the lower pressure region, the graph is quite similar to that of type II. This is explained by the formation of monolayer followed by multilayer growth. However, a saturation level is reached at a pressure below (sometimes much below) the saturation vapor pressure. This is explained on the basis of capillary condensation of the adsorbate gas in the pores of the adsorbent at a pressure below the saturation (vapor) pressure of the gas, P_V. Figure 4.4c is a pictorial representation of the physical situation, where the pores are filled up with the adsorbate. This may be accompanied by adsorption on the outer surface, but at most forming a monolayer. It is interesting to note that if the substrate pores are being filled by the gas molecules, it may be a case of (or near to) *ab*sorption; nevertheless, it is usually treated as adsorption. A classical example is benzene on ferric oxide at 50°C (Lambert and Clark 1929).

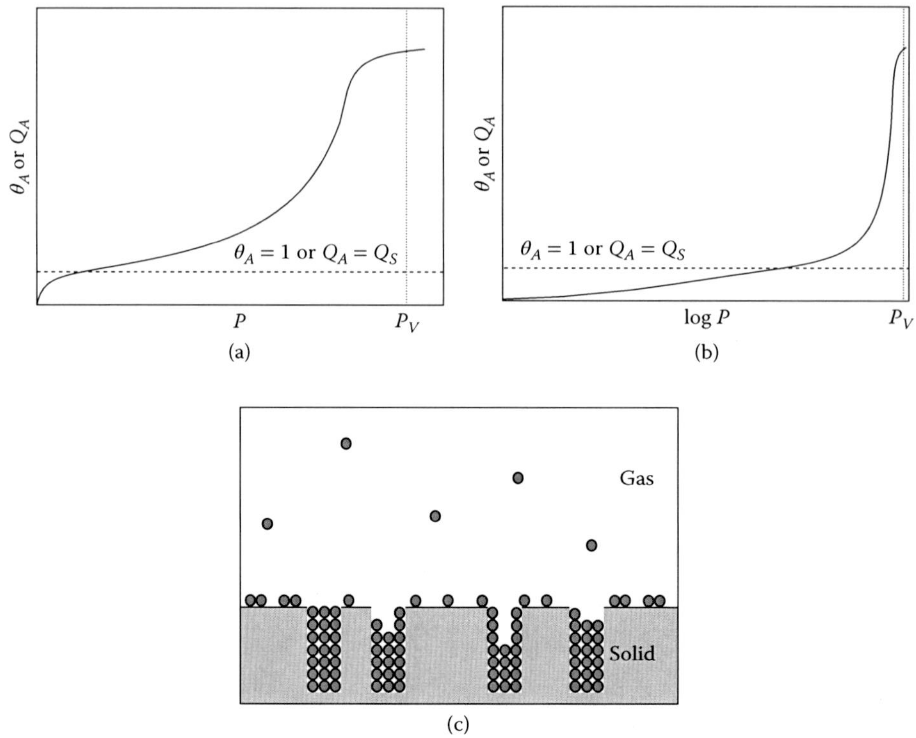

FIGURE 4.4 Schematic plot of a type IV isotherm from IUPAC classification (Sing 1985) as a function of pressure (a) and the logarithm of pressure (b). The dashed line shows the point of monolayer saturation, and the dotted line marks the point of the adsorbate vapor pressure. Note that this is similar to type II (Figure 4.2) but with a saturation level, which is reached at a pressure below P_V. In (c), a schematic drawing of the (ideal) situation is shown; pores are present, where capillary condensation occurs. In the outer surface at most a monolayer can be formed.

Figure 4.5 shows typical plots of a type V isotherm as a function of both the pressure and its logarithm. In the lower pressure region, the graph is quite similar to that of type III. This is explained by the formation of a monolayer followed by multilayer growth. However, a saturation level is reached below the vapor pressure. This is explained on the basis of capillary condensation of the adsorbate gas in the pores of the adsorbent at a pressure below the saturation (vapor) pressure of the gas, P_V, in a way similar as the case of type III isotherms. The situation is essentially the same as that shown in Figure 4.4c. A classical example is the adsorption of water on charcoal at 100°C (Coolidge 1927).

Finally, Figure 4.6 shows the type VI isotherm, which was not defined originally by Brunauer et al. but introduced much later by Sing (1985). It corresponds to stepwise multilayer adsorption. The step height represents the monolayer capacity for each adsorbed layer and, in the simplest case, remains nearly constant for two or three adsorbed layers. After the steps, it may show a divergent increase as types II and III. Classical examples are the adsorption of argon and krypton on graphitized carbon black (Hill 1955).

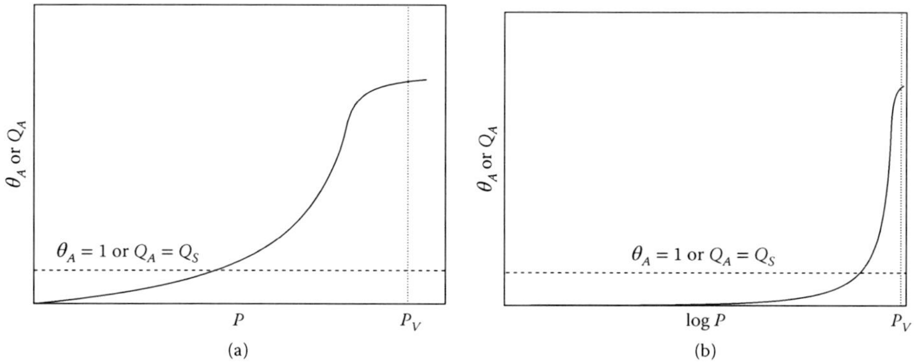

FIGURE 4.5 Schematic plot of a type V isotherm from IUPAC classification (Sing 1985) as a function of pressure (a) and the logarithm of pressure (b). The dashed line shows the point of monolayer saturation, and the dotted line marks the point of the adsorbate vapor pressure. Note that this is similar to type III (Figure 4.3) but with a saturation level, which is reached at a pressure below P_V.

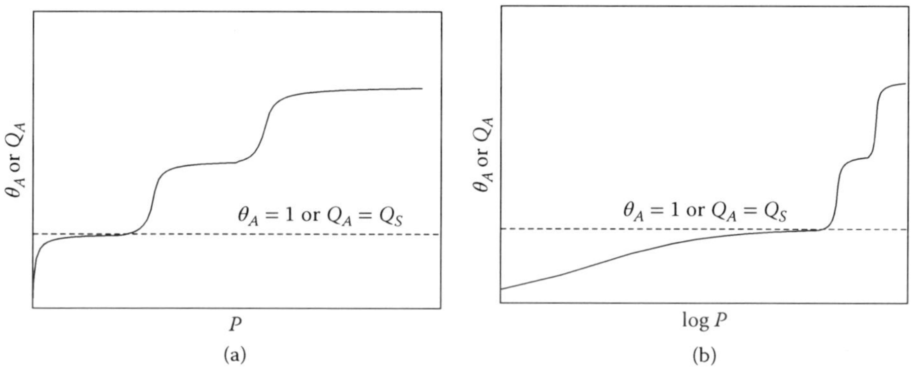

FIGURE 4.6 Schematic plot of a type VI isotherm from IUPAC classification (Sing 1985) as a function of pressure (a) and the logarithm of pressure (b).

4.2.2 Isotherm Classification for Sorption from Solution

When sorption processes occur in condensed phases, the usual case being sorption of a solute from a liquid phase (solution) onto a solid, one finds isotherm plots that are rather similar to those of gas phase (see Figures 4.7 through 4.10). However, it must be realized that this is a quite different situation: in gas phase, adsorption molecules from the gas hit the surface on empty sites, where they may eventually get bound. In a solution, solvent molecules are present almost everywhere, so that the adsorption of a solute molecule is actually an exchange process between solvent and solute. Figure 4.7 illustrates this: the molecule signaled by an arrow needs the removal of a solvent molecule from the surface in order to adsorb on it.

Isotherms for sorption from solution have been classified by Giles, Smith, and Huitson (1974) and reviewed by other authors, for example, Hinz (2001). They are

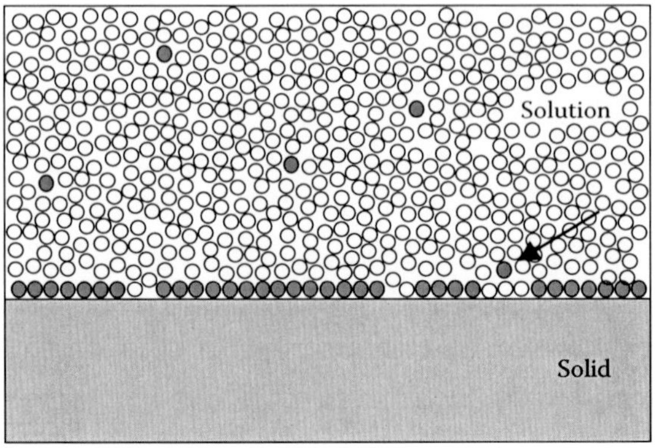

FIGURE 4.7 Schematic drawing of the situation for adsorption from solution; open circles represent solvent molecules, and gray filled circles are solute molecules. The binding of a solute molecule to the surface, such as that indicated by the arrow, requires removal of one or several solvent molecules from the surface.

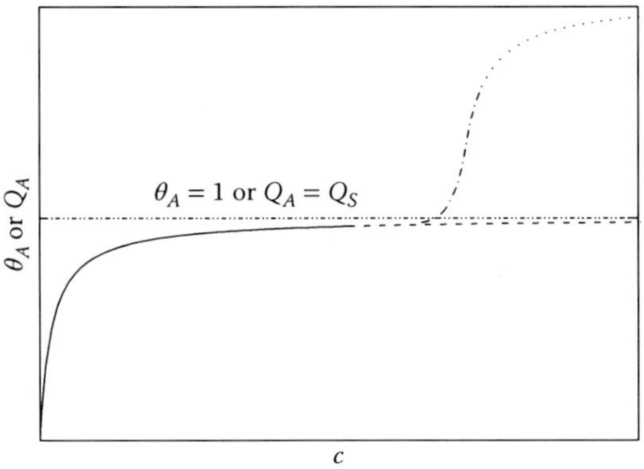

FIGURE 4.8 Example plot of a type L isotherm and its subtypes. Continuous line: subtype 1; dashed line: subtype 2; dash-dot line: subtype 3; dotted line: subtype 4.

primarily classified by the initial shape and slope and further divided into subtypes or subgroups. In the following the different types are shown schematically, along with a few examples. Here, they are plotted as a function of adsorbate equilibrium concentration as it is commonly done; strictly, the correct variable is the activity, but concentration is the experimentally controlled variable. Also, in some cases (e.g., with natural organic matter, whose exact composition is unknown), the activity is not accessible. Furthermore, it is better to put in the abscissas the logarithm of activity or concentration, as it is more directly connected with the chemical potential (Equation 3.2); this will be more clearly seen in the discussion of type H isotherm.

For each group, there are subgroups mainly defined by the extent of adsorption reached. It should be noted that often, due to the nature of the systems, the experimentally accessible concentration range is limited, thus the maximum amount sorbed may be far from saturation coverage.

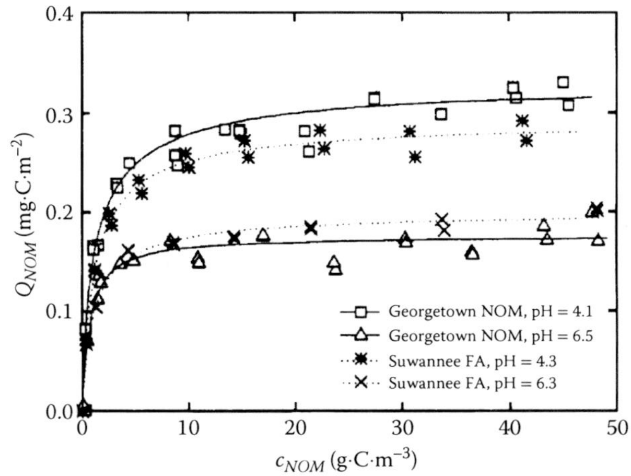

FIGURE 4.9 Adsorption isotherms of NOM onto hematite in NaCl 0.01 M. Lines are fittings to a modified Langmuir isotherm. (Reprinted with permission from Gu, B. et al., 1994, *Environmental Science & Technology*, 28, no. 1, 38–46. Copyright 1994 American Chemical Society.)

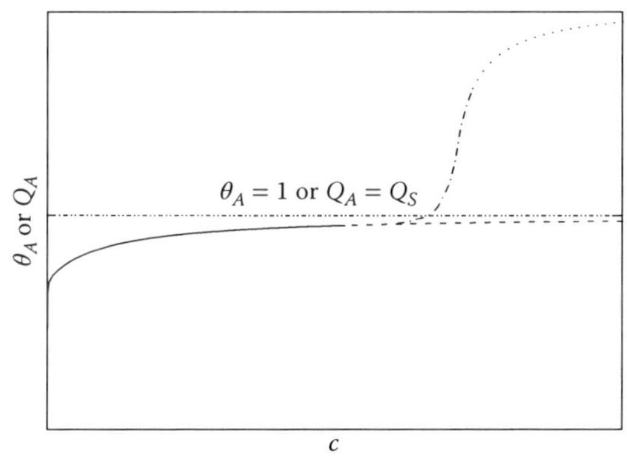

FIGURE 4.10 Example plot of a type H isotherm and its subtypes. Continuous line: subtype 1; dashed line: subtype 2; dash-dot line: subtype 3; dotted line: subtype 4.

Figure 4.8 shows the type (group) "L" isotherms (the L means low affinity type, as compared with the next group); as it is evident, it is same as type I in Brunauer's classification. The different line types refer to the subgroups: subgroup 1 (solid line) corresponds to cases where monolayer saturation is not reached within the experimentally accessible concentration range; subgroup 2 (dashed line) shows monolayer saturation, as in gas phase type I; subgroup 3 (dash-dotted line) shows increased sorption beyond monolayer; and subgroup 4 presents saturation for more than a monolayer. It should be noted that sorption corresponding to subgroups 3 and 4 gives plots equivalent to Brunauer types II and III, respectively. Examples for all groups and subgroups have been given by Giles, D'Silva, and Easton (1974).

An example of this type of behavior in soil systems is shown in Figure 4.9, for the adsorption of soil natural organic matter (Georgetown NOM) and river fulvic acid

(Suwannee FA) on iron oxides, plotted as the amount of bound NOM, Q_{NOM} versus its concentration c_{NOM}. The lines represent fittings to a modified Langmuir model (Fowler–Guggenheim [FG]), which is discussed in Section 4.4.6 (Gu et al. 1994). As can be seen, the isotherms are pH dependent, which illustrates that the medium always has an influence on adsorption. L-type isotherms are commonly observed in sorption processes in soils, including cation binding, both inorganic (Bradl 2004; Flogeac, Guillon, and Aplincourt 2004; Pagnanelli et al. 2006) and organic (Rytwo 2004), and also anion binding (Manning and Goldberg 1997).

Figure 4.10 shows schematically the group H of isotherms according to Giles. An example is the adsorption of As(III) and (V) species on iron oxide (Dixit and Hering 2003). It has been frequently stated that this type is similar to the group L but with a higher initial slope. However, it should be pointed out that, whereas type L usually corresponds to a single process (one type of site), H-type isotherms can be found for two different processes (two types of sites), one of relatively higher affinity and the other with a relatively lower affinity, such as that shown in Figure 4.11 in the adsorption of Cd on the surface of soil pseudomonas (Pabst et al. 2010); notice the high initial slope that changes after the third point, and also notice that a two-site equation is needed to fit the data.

This is also observed in the adsorption of organic cations to soils, where the two-site behavior is revealed in the logarithmic plot as seen in Figure 4.12 (Brownawell et al. 1990). Figure 4.12a shows the amount of dodecylpiridinium sorbed on a soil with relatively high C contents, Q, as a function of its concentration, c; a typical H-shaped isotherm is observed, which qualitatively looks similar to the L group. However, in a log Q versus log c plot (Figure 4.12b), two "humps" can be observed, one at log c approximately −8 and the other at log c approximately −4, corresponding to two

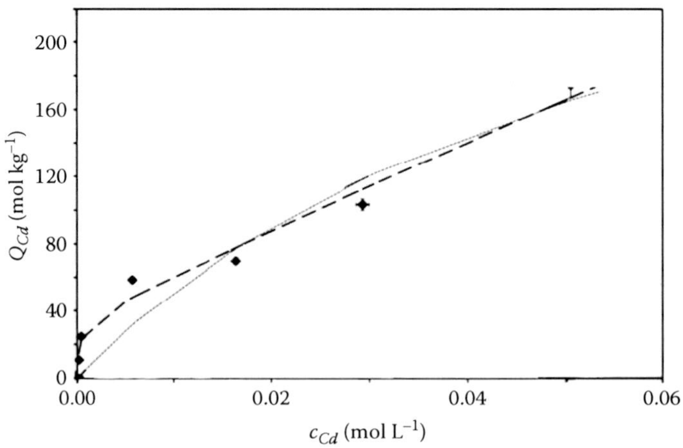

FIGURE 4.11 Sorption isotherms of Cd associated with exchangeable sites *Pseudomonas putida* KT2440. The points are averages of the experimental data ($n = 3$) with 95% confidence interval error bars. The dotted line is the one-site Langmuir model prediction. The dashed line is the two-site Langmuir model, which was used to describe satisfactorily the surface exchange behavior of Cd. (Reprinted from *Chemosphere*, 81, Pabst, M. W. et al., Defining the surface adsorption and internalization of copper and cadmium in a soil bacterium, *Pseudomonas putida*, 904–910. Copyright 2010, with permission from Elsevier.)

different processes (compare with Figure 4.1b). Thus, the logarithmic plot reveals the presence of two types of sites; in this case, one type corresponds to negatively charged sites, whereas the second type corresponds to uncharged sites.

Figure 4.13 shows typical isotherms belonging to group S, which is characterized by a low initial slope; also, S isotherms have a concave shape at low concentrations. These isotherms may have two causes. First, solute–solute attractive forces at the surface may cause cooperative adsorption, which leads to an increase of affinity and consequently to the S-shape. Second, the sorption of a solute may be inhibited by a competing reaction within the solution, such as a complexation reaction with a ligand. In many cases, the total amount of a solute, and not the activity of the different species, is measured. For example, a heavy metal may form a stable complex with a chelating agent. Measuring total metal concentration will lead to an isotherm describing the amount sorbed to the solid phase as a function of the total amount of metal in solution, instead of free metal

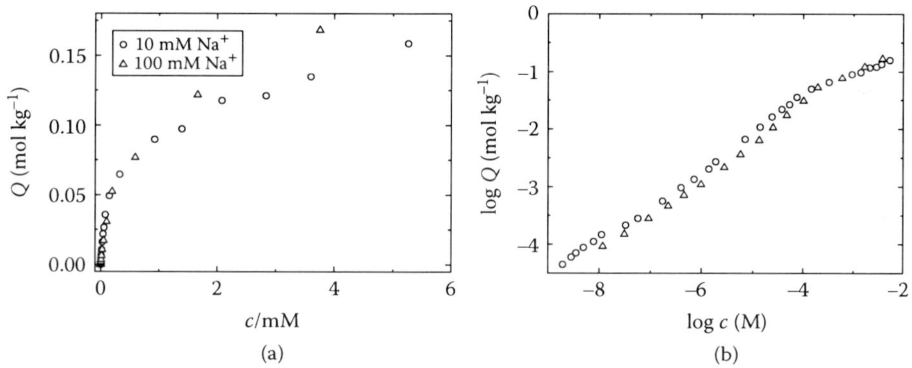

FIGURE 4.12 Sorption isotherms of dodecylpiridinium cation in a soil in the presence of 0.01 M NaN_3 (10 mM Na^+) and 0.01 M NaN_3 + 0.09 M NaCl (100 mM Na^+). Linear plot (a) and double logarithmic plot (b). (Reprinted with permission from Brownawell, B.J. et al., 1990, *Environmental Science & Technology*, 24, no. 8, 1234–1241. Copyright 1990 American Chemical Society.)

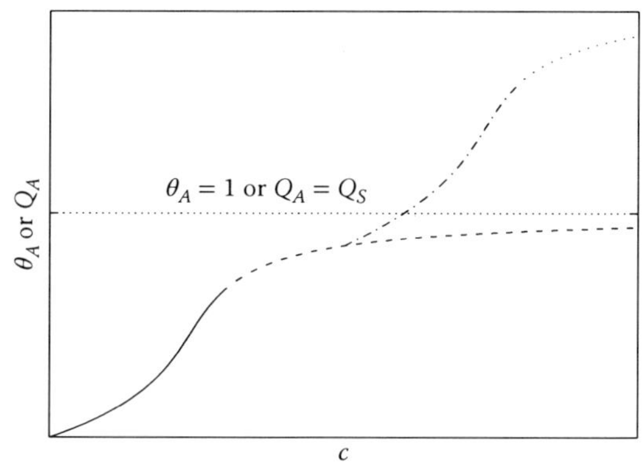

FIGURE 4.13 Example plot of a type S isotherm and its subtypes. Continuous line: subtype 1; dashed line: subtype 2; dash-dot line: subtype 3; dotted line: subtype 4.

activity (or concentration). Organic substances often show S-shaped isotherms, such as benzimidazole fungicides on clay surfaces (Aharonson and Kafkafi 1975).

Figure 4.14 shows typical isotherms belonging to group C, which is characterized by a constant slope; this has been attributed by Giles, Smith, and Huitson (1974) to the number of free sites remaining constant during the process, that is, to the surface being in expansion as the sorption proceeds. Giles et al. explained this, among other possible causes, as due to microporous solids where the pores open as the adsorbate molecules go into them. As most sorption processes behave linearly at low coverages, trace solutes sorption usually show linear isotherms. Many organic compounds show C isotherms, such as atrazine adsorption on Laurentian soil, as observed in Figure 4.15 (Wang, Gamble, and Langford 1992). It should be noted, however, that

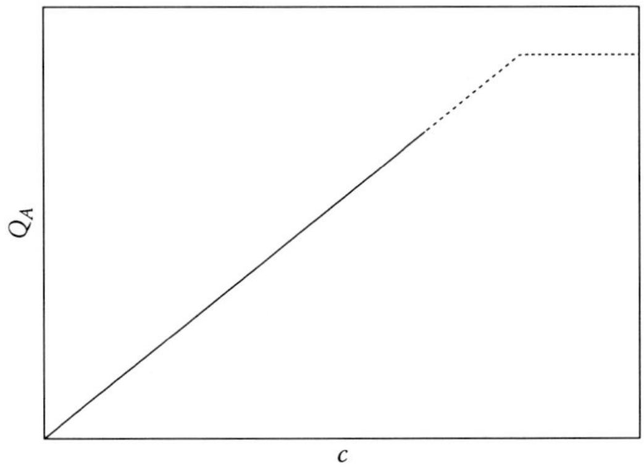

FIGURE 4.14 Example plot of a type C isotherm and its subtypes. Continuous line: subtype 1; dashed line: subtype 2.

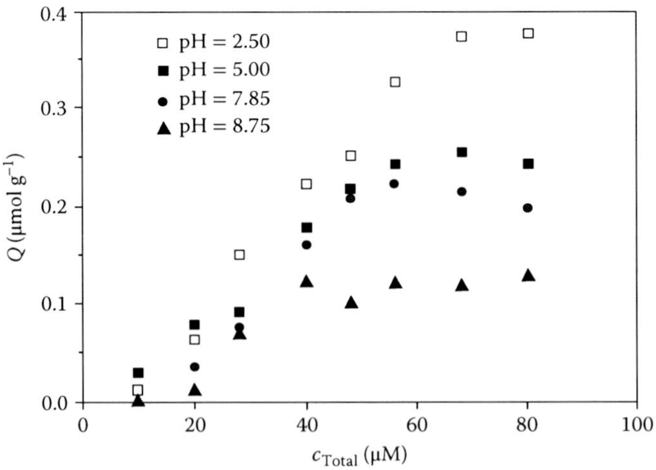

FIGURE 4.15 Sorption isotherms of atrazine bound to Laurentian soil as a function of total atrazine concentration at different pH values. (Reprinted with permission from Wang, Z. et al., 1992, *Environmental Science & Technology*, 26, no. 3, 560–565. Copyright 1992 American Chemical Society.)

the amount bound, Q, is plotted as a function of the *total* atrazine concentration and not the equilibrium one, as it should be. Notwithstanding that, due to the quantities employed (0.5 g soil in 25 mL solution), the maximum amount bound is a relatively small fraction of the total solute.

4.3 PHYSICAL NATURE OF ADSORPTION

4.3.1 Adsorption from Gas Phase

We now consider the process of adsorption from a microscopical point of view and, as above, start discussing adsorption from gas phase (Adamson and Gast 1997; Hiemenz and Rajagopalan 1997; Levine 2008). Figure 4.16 shows a simplified view of the process of interaction of a gas molecule such as Ar (black sphere) with a solid surface such as graphite (gray spheres).

In Figure 4.16a, the molecule is approaching the surface with a kinetic energy falling in the Boltzmann distribution (McQuarrie and Simon 1997; Berry, Rice, and Ross 2000; Levine 2008; Atkins and Paula 2009). In Figure 4.16b, the molecule has reached the surface, where it can bounce and go back to the gas phase or remain in the surface for some finite amount of time. The outcome depends on the energies involved, basically the temperature, the kinetic energy of the arriving molecule, and the molecule–surface interaction energy. This last factor depends on the nature of the adsorbate and the surface. If it is comparable to the thermal energy in magnitude, the molecule will stay for some time on the surface until it will desorb, returning to the gas phase. This is the case for the example of argon on graphite, which is given earlier in text: van der Waals (dispersion) forces will be the only interaction type present, and the adsorption is observed only at low temperature, near or below the argon critical temperature. On the contrary, in the case, for example, of nitrogen on iron, adsorption is observed at much higher temperatures (Emmett and Brunauer 1934), the interaction of nitrogen–iron surface being strong to the point that N_2 molecules become dissociated. In fact, this

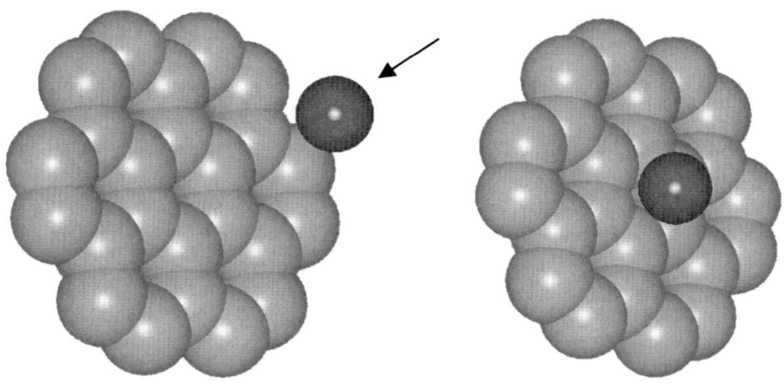

(a) (b)

FIGURE 4.16 Schematic view of the adsorption of an Ar molecule on a graphite surface.

is the way in which the Haber process for ammonia synthesis is catalyzed by iron or iron oxides. In this case, strong chemical bonding interactions are present. The two examples given so far apply to the two main types of adsorption: *physical adsorption* or *physisorption* and *chemical adsorption* or *chemisorption*. These are usually distinguishable by the magnitude of the *heat of adsorption*, ΔH_{ad}, which is strictly the enthalpy change upon *desorption*; clearly adsorption is exothermic whereas desorption is endothermic, thus the heat of adsorption is always given as positive. In physisorption processes, it is relatively low, similar to vaporization enthalpies; for example, for argon on graphite, ΔH_{ad} is approximately 11 kJ·mol^{-1} (Avgul and Kiselev 1958), which is comparable to the argon heat of vaporization of 6.45 kJ·mol^{-1}. Chemisorption processes have higher heats, typically one order of magnitude higher; for N_2 adsorption on iron at high temperatures (approximately 400°C), ΔH_{ad} is approximately 209 kJ·mol^{-1} (Brito and Ralek 1978), whereas its heat of vaporization is 2.8 kJ·mol^{-1}. It should be noted that at low temperatures (approximately 75 K), N_2 shows physical adsorption on iron with $\Delta H_{ad} \sim 4$ kJ·mol^{-1} (Kavtaradze and Zelyaeva 1970); at first, it would appear contradictory, because at lower temperatures, the more energetically favorable process, chemisorption, should prevail. However, this last process is a chemical reaction with an energy of activation (physical adsorption has no activation energy, in principle); at low temperatures, the adsorbing molecules are unable to overcome the activation barrier, and thus, chemisorption is kinetically prevented.

4.3.2 Adsorption from Solution

As said before, here the process is more complex due to the presence of the solvent (Figure 4.7). Now the adsorption is an exchange process, so that the enthalpy of adsorption is the balance between adsorbate–surface binding and solvent–surface unbinding. In the adsorption process, the adsorbate approaches the surface by diffusion in the condensed phase, which involves a random movement due to collisions with solvent molecules, with a free path much shorter than in the gas phase; in other words, one can think that the solute should "find its way" toward the surface in a crowded multitude of solvent molecules. The interaction between the adsorbate and the surface may have several contributions, basically dispersion forces, chemical bonding, and electrostatic forces. There is also an entropy change that depends on other factors, such as adsorbate solvation changes upon adsorption. In a similar way to gas phase, there are cases of relatively weak adsorption and others of strong chemical bonding; these (extreme) situations are usually termed *nonspecific adsorption* and *specific adsorption*, respectively. These terms, in fact, are quite equivalent to physisorption and chemisorption but came from the world of adsorption in condensed phase and electrochemistry. On the other hand, depending on the location of the interacting sorbate, especially in the case of ions, there is another terminology, as illustrated in Figure 4.17.

In Figure 4.17, the two horizontal groups of atoms illustrated correspond to two typical phyllosilicate layers (as described in Chapter 8). The sorbate may reach

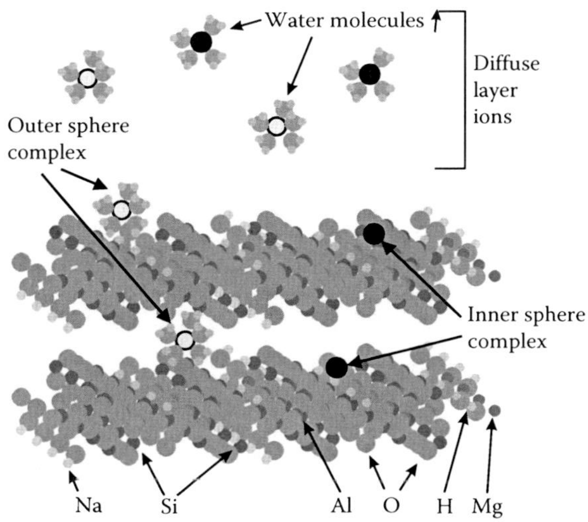

FIGURE 4.17 Schematic view of the different modes of adsorption, classified per the location of the adsorbate. The two horizontal groups of atoms illustrated correspond to two typical phyllosilicate layers (Chapter 8). The sorbate may reach close contact with the surface giving an inner-sphere complex, may retain its hydration layer yielding an outer-sphere complex, or stay farther in the diffuse layer region. (Adapted from Wang, J., and Gutierrez, M. S., *J. Nanomater.*, 1–13, 2010.)

close contact with the surface, loosing either totally or partially its hydration layer, giving an *inner-sphere complex*. Alternatively, it may retain its hydration layer, yielding an *outer-sphere complex*, where the adsorbate is separated from the surface for one water layer. Note that both the types are distinguished by the distance of closest approximation to the surface; usually this is treated as that the inner- and outer-sphere adsorbates are located in two distinct planes, both parallel to the surface. This is detailed in Chapter 12 of Part III. For ions only, they can also stay farther away in the diffuse layer region, which is a situation quite similar to the outer-sphere complex regarding the interaction with the surface (typically, a negatively charged surface that retains cations by some or all these mechanisms). As it is seen in Chapter 5, ions retained in the outer sphere and diffuse layer are the main subjects of ionic exchange, a process of fundamental importance in soils, which is discussed there.

It should be pointed out that the distinction between inner- and outer-sphere complexes is not the same as that of nonspecific versus specific sorption. In an outer-sphere complex, the adsorbate is separated from the surface for (at least) one layer of water molecules, preventing any type of bonding, thus this is always nonspecific adsorption. On the other hand, an adsorbate in an inner-sphere complex may form some sort of chemical bond, but not necessarily; for example, in micas K^+ ions are often found as inner-sphere complexes, but the interaction is mainly electrostatic; no chemical bond is formed. Thus, specific adsorption implies inner-sphere complex, but the opposite is not true.

4.3.3 Interaction between a Molecule or Ion and a Surface

Before trying to understand the molecular basis of interactions involving particles and surfaces, we will review the interactions between single ions and molecules; details can be found in the literature (Adamson and Gast 1997; Hiemenz and Rajagopalan 1997; McQuarrie and Simon 1997; Berry, Rice, and Ross 2000; Levine 2008; Atkins and Paula 2009; Israelachvili 2010). We will now enumerate the different contributions to such interactions, which are assumed to be independent.

4.3.3.1 Intermolecular Interactions

4.3.3.1.1 Ion–Ion Interactions

Ion–ion interactions are given by Coulomb's law; for the interaction energy between ions 1 and 2

$$u_{I-I}(r) = \frac{q_1 q_2}{4\pi\varepsilon r} \tag{4.6}$$

where u_{I-I} is the energy of interaction between two ions, q_1 and q_2 are the respective charges, $\boldsymbol{\varepsilon}$ is the medium permittivity, and r the distance between them. These are considered long-range interactions, due to the exponent −1 for the distance, and strong; the energy involved in separating two monovalent ions of opposite charge, bringing them from close contact to an infinite distance, is of the order of 300–500 $kJ{\cdot}mol^{-1}$.

4.3.3.1.2 Ion–Dipole Interactions

Ion–dipole interactions result from the interaction of the ionic charges with each charges of the dipole; the result will be, in general, dependent on the dipole rotation freedom. We will consider only the case where that freedom exists, so that, because ion–dipole interactions are considerably greater than $\boldsymbol{kT}$, the dipole will align itself such that the extreme facing the ion will be that of opposite charge; for an ion of charge q_1 and a dipole of dipolar moment p_2 we have thus:

$$u_{I-D}(r) = -\frac{|q_1|\, p_2}{4\pi\varepsilon r^2} \tag{4.7}$$

This interaction energy is attractive irrespective of the sign of the ion charge because of the favorable orientation adopted by the dipole. The exponent of r is now −2, resulting in shorter range compared with ion–ion interaction; this can be understood considering that, as the ion–dipole distance increases, the later is "seen" by the ion as two opposite charges increasingly merged and thus cancelling out. These interactions are not as strong as ion–ion, usually on the order of 30–100 $kJ{\cdot}mol^{-1}$ but higher than $\boldsymbol{kT} \cong 2.5$ $kJ{\cdot}mol^{-1}$ at room temperature.

4.3.3.1.3 Dipolar (van der Waals) Interactions

There are three different contributions to this type of interaction. We will first consider permanent dipole interactions. Again, we will restrict the analysis to freely rotating dipoles (as in a liquid medium); these dipoles will rotate due to thermal agitation but, due to mutual interaction, will have, on average, attractive orientations (opposite charges facing each other). This is the *Keesom interaction energy*, given by the equation

$$u_K(r) = -\frac{2}{3kT}\left(\frac{p_1 p_2}{4\pi\varepsilon}\right)^2 \frac{1}{r^6} \tag{4.8}$$

The presence of the temperature arises from averaging all possible orientations. It should be noted, as pointed out by Israelachvili (Israelachvili 2010), that, unlike previous contributions, Equation 4.8 gives the total internal energy, which is related of the Helmholtz free energy F through the well-known thermodynamic relationship:

$$U = F - T\left.\frac{\partial F}{\partial T}\right)_V \tag{4.9}$$

The Keesom free energy of interaction for a pair of molecules, f_K, is half of the internal energy

$$f_K(r) = -\frac{1}{3kT}\left(\frac{p_1 p_2}{4\pi\varepsilon}\right)^2 \frac{1}{r^6} \tag{4.10}$$

because there is an entropic component due to the dipole rotation. For other contributions that are independent of temperature, of course the internal and free energy are the same.

When a polar molecule and a nonpolar one are in proximity, the first will induce a transient dipole in the second, giving rise to a permanent dipole-induced dipole interaction. This is known as *Debye interaction*, given by

$$u_D(r) = -\frac{p_1^2\alpha_2}{(4\pi\varepsilon)^2 r^6} \tag{4.11}$$

where α_2 is the electronic polarizability of the nonpolar molecule. Of course, if two permanent dipoles interact, each one will induce a transient dipole on the other, so that in this case the Debye interaction will be given by

$$u_D(r) = -\frac{p_1^2\alpha_2 + p_2^2\alpha_1}{(4\pi\varepsilon)^2 r^6} \tag{4.12}$$

Note that Equations 4.8 through 4.12 result in short-range interactions, due to the −6 exponent on the dipole–dipole distance. We will now consider the *London or dispersion* interactions. These interactions, also known as induced dipole–induced dipole, arise from quantum mechanical results for a system composed of two "vibrating dipoles" due to symmetrical vibration of electrons. The London interaction energy is given approximately by (Mahanty and Ninham 1977; Hiemenz and Rajagopalan 1997; Israelachvili 2010) the following:

$$u_L(r) = -\frac{3h\nu_1\nu_2\alpha_1\alpha_2}{2(\nu_1+\nu_2)(4\pi\varepsilon)^2 r^6} \tag{4.13}$$

where ν_1 and ν_2 are the frequency of oscillation of the electrons for the free (noninteracting) molecules, and h is Plank's constant. These are sometimes approximated through the ionization energies I_1 and I_2 of the molecules:

$$\frac{h\nu_1\nu_2}{(\nu_1+\nu_2)} \approx \frac{I_1 I_2}{I_1+I_2} \tag{4.14}$$

Other equivalent expressions have been given (Hiemenz and Rajagopalan 1997). As with the previously discussed dipole interactions, Equation 4.13 shows a short-range interaction, with also an exponent −6 in the distance. Because the London interactions do not require the presence of charges and depends solely on polarizabilities (i.e., the mobility of electrons) all ions, atoms, molecules, or particles are affected by them.

These three contributions (Keesom, Debye, and London) are collectively known as *van der Waals interactions*, and the general expression for the interaction between particles 1 and 2 is

$$u_{12}(r) = -\left[\frac{2p_1^2 p_2^2}{3kT} + p_1^2\alpha_2 + p_2^2\alpha_1 + \frac{3h\nu_1\nu_2\alpha_1\alpha_2}{2(\nu_1+\nu_2)}\right]\frac{1}{(4\pi\varepsilon)^2 r^6} = -\frac{u'_{12}}{r^6} \tag{4.15}$$

where u'_{12} collects all quantities independent of r.

The overall magnitude is similar to that of ion–dipole interactions; the weight of each contribution depends on the particular molecule. Table 4.1 presents some examples for pure substances (i.e., interactions between two '1' molecules). In most cases,

TABLE 4.1
Relative Contributions of the Debye, Keesom, and London Interactions to the van der Waals Attraction for Some Compounds

Compound	u'_{11} 10^{-77} (J m^6)	Keesom (%)	Debye (%)	London (%)
CCl4	4.41	0.0	0.0	100.0
Ethanol	3.40	42.6	9.7	47.6
Thiophene	3.90	0.3	1.3	98.5
t-Butanol	5.46	23.1	9.7	67.2
Benzene	4.29	0.0	0.0	100.0
Fluorobenzene	5.09	10.6	7.5	81.9
Phenol	6.48	14.5	8.6	76.9
Aniline	7.06	13.6	8.5	77.9
Diphenylamine	14.25	1.5	3.7	94.7
Water	2.10	84.8	4.5	10.5

Source: Hiemenz, P. C., and Rajagopalan, R., *Principles of Colloid and Surface Chemistry*, 3rd ed., Marcel Dekker, New York, 1997.

the London term is the largest contribution to the overall interaction. In the case of highly polar molecules, such as ethanol and water, the Keesom part is important and predominant for water; in these cases, however, there is additionally an important contribution from hydrogen bonding, not considered here. Equation 4.13 (and consequently 4.15) is approximate by several reasons, such as considering only one frequency (ionization potential) and not being able to handle interactions in a third medium (solvent). A more general theory has been given by McLachlan in 1963–1965, including in a single expression all contributions to van der Waals forces. The interaction free energy for two molecules 1 and 2 immersed in a medium 3 is given by

$$f_{123}(r) = -\frac{6kT}{(4\pi\varepsilon_0)^2 r^6}\left[\frac{\alpha_{e1}(0)\alpha_{e2}(0)}{2\varepsilon_{r3}^2(0)} + \sum_{n=1}^{\infty}\frac{\alpha_{e1}(i\nu_n)\alpha_{e2}(i\nu_n)}{\varepsilon_{r3}^2(i\nu_n)}\right] \tag{4.16}$$

where α_{e1} is the excess polarizability of molecule i and $\boldsymbol{\varepsilon}_{r3}$ the relative permittivity of medium 3. All are function of the *imaginary* frequencies $i\nu_n$ with ν_n given by

$$\nu_n = \frac{2\pi nkT}{h} \tag{4.17}$$

The polarizabilities and the medium permittivity are, however, real quantities. Its exact form depends on the molecular details of the three substances. For the case of isolated (gas phase) molecules, $\varepsilon_{r3}(i\nu) = 1$, and if there is only one electronic absorption frequency, the ionization frequency ν_I, the excess polarizabilities are equal to the total ones, which are shown to be

$$\alpha_{ti}(\nu_n) = \frac{p_i^2}{3kT\left(1+\dfrac{\nu_n}{\nu_{rot,i}}\right)} + \frac{\alpha}{1+\kappa\dfrac{\nu_n}{\nu_{I,i}} + \left(\dfrac{\nu_n}{\nu_{I,i}}\right)^2} \tag{4.18}$$

where α is the normal electronic polarizability and $\nu_{rot,i}$ is some average rotational frequency for the molecule (typically $\nu_{rot} \approx 10^{11}\ \mathrm{s}^{-1}$), and κ is a small damping factor (usually <<1, so that it can be ignored). When the molecule is inserted in a medium 3, the excess polarizability is given by

$$\alpha_{e,i}(\nu) = 4\pi\varepsilon_0\varepsilon_{r3}(\nu)a_i^3\left(\frac{\varepsilon_{r1}(\nu)-\varepsilon_{r3}(\nu)}{\varepsilon_{r1}(\nu)+2\varepsilon_{r3}(\nu)}\right) \tag{4.19}$$

where a_i is the radius of molecule i. If the dielectric behavior of the substances can be described, in the condensed state, by a single electronic absorption peak at frequency ν_e, the permittivity can be expressed by

$$\varepsilon_{r,i}(i\nu) = 1 + \frac{n_i^2 - 1}{1+\left(\dfrac{\nu}{\nu_{e,i}}\right)} \tag{4.20}$$

where n_i is the refractive index of substance i. Because the frequencies in the sum of Equation 4.16 are close together, it can be replaced by an integral, resulting, combining with Equation 4.19, in

$$f_{123}(r) = -\frac{3a_1^3a_2^3}{r^6}\left[kT\left(\frac{\varepsilon_{r1}(0)-\varepsilon_{r3}(0)}{\varepsilon_{r1}(0)+2\varepsilon_{r3}(0)}\right)\left(\frac{\varepsilon_{r2}(0)-\varepsilon_{r3}(0)}{\varepsilon_{r2}(0)+2\varepsilon_{r3}(0)}\right)\right.$$
$$\left.+\frac{h}{\pi}\int_0^\infty\left(\frac{\varepsilon_{r1}(iv)-\varepsilon_{r3}(iv)}{\varepsilon_{r1}(iv)+2\varepsilon_{r3}(iv)}\right)\left(\frac{\varepsilon_{r2}(iv)-\varepsilon_{r3}(iv)}{\varepsilon_{r2}(iv)+2\varepsilon_{r3}(iv)}\right)\right]\mathrm{d}v \quad (4.21)$$

Equation 4.21 includes de Debye and Keesom contributions (first term in square brackets) and the London interactions (integral term). We will not include further details of McLachlan theory here, the reader is referred to the literature (McLachlan 1963a, b, 1965; Mahanty and Ninham 1977; Israelachvili 2010). We will only mention some features that can be deduced by analysis of Equation 4.21 (Israelachvili 2010):

1. Usually the dispersion (London) contribution is higher than the permanent dipole (Debye–Keesom) one. Some exceptions are fluorocarbons and low-molecular-weight alkanes in water.
2. The van der Waals forces suffer an important reduction in a medium compared with gas phase, typically 1–2 orders of magnitude.
3. The dispersion force between identical molecules in a medium is always attractive, but different molecules can attract or repel each other, depending on the relative magnitudes of the respective diffraction indexes.

It is important to remark here that dispersion interactions are *not* additive; that is, the London force interaction of molecule 1 with molecules 2 and 3 is not, strictly, the sum of London energies 1–2 and 2–3, because the third molecule polarizes molecules 1 and 2, thus $u_{L,12}(r)$ is not the same in presence or absence at short distances of other molecules. Therefore, the use of sums of pairwise interactions—that is, writing the interactions in a system of many molecules as sum of individual u_{ij} interactions, as is very frequently done—is only an approximation.

4.3.3.1.4 Retardation of London Forces

The electric field involved in London interactions propagates with the speed of light from one particle (molecule, ion, or colloid) to another; if the particles are far apart, there is a time retardation (equivalent to a phase lag) between the vibrations of both the bodies. The importance of the retardation increases as the distance approaches the wavelength of the electric field, λ.

The effect was considered by Casimir and Polder (1948) by quantum electrodynamics. The result is that, for $r >> \lambda$, the dispersion interaction becomes

$$u_R(r) = \frac{23hc\alpha_1\alpha_2}{8\pi^2(4\pi\varepsilon)^2 r^7} \quad (4.22)$$

where c is the speed of light. Now we find a dependence with r^{-7}, even more rapidly decaying than the van der Waals forces. Measurements (Israelachvili 2010) show that the crossover from r^{-6} to r^{-7} occur in the range 10–100 nm. At such distances, interactions involving molecules are negligible; this effect is only important when considering the interactions between macroscopic particles, as we shall discuss in Chapter 6.

4.3.3.2 Interactions with a Macroscopic Particle

Except at close contact, there are two main contributions to the interaction between a small particle (ion or neutral molecule) and a colloidal particle: electrostatic and van der Waals. Strictly speaking, electrostatic interactions involve localized charges on the particle surface, thus the ion–particle interaction would be given by the sum of pairwise Coulomb interactions (Equation 4.6). The problem requires detailed knowledge of the location of individual charges, which makes the solution extremely difficult. To solve it, the surface charge is assumed to be smeared out or averaged, leading to the theory presented in Chapter 3. We will return to this question in more detail in Chapter 5.

4.3.3.2.1 Molecule–Surface van der Waals Interactions

As discussed in Section 4.3.3.1, attractive van der Waals interaction follows the expression for the intermolecular pair potential energy:

$$u_{12}(r) = -\frac{u'_{12}}{r^6} \tag{4.23}$$

At short distances, a repulsive interaction arises due to the overlapping of the electronic clouds; this is nearly exponential but is often approximated by a term of the form u''/r^n, with $n > 9$; in the Lennard–Jones potential, it is taken as 12 for mathematical convenience (Hill 1986; Hiemenz and Rajagopalan 1997; Levine 2008; Israelachvili 2010). Now, we will concentrate in the attractive interactions. If a molecule A is near the surface of a solid formed by molecules of B (Figure 4.18), the interaction of A with the solid will be, assuming that additivity is valid, the sum of the interaction with all B molecules

$$u_{A-S}(d) = \int_{z=d}^{z=\infty}\int_{x=-\infty}^{x=\infty}\int_{y=-\infty}^{y=\infty} u_{A-B}(r)\,\mathrm{d}x\,\mathrm{d}y\,\mathrm{d}z \tag{4.24}$$

which can be easily shown to result in Israelachvili (2010)

$$u_{A-S}(d) = -\frac{\pi u'_{A-B}\rho_B}{6d^3} \tag{4.25}$$

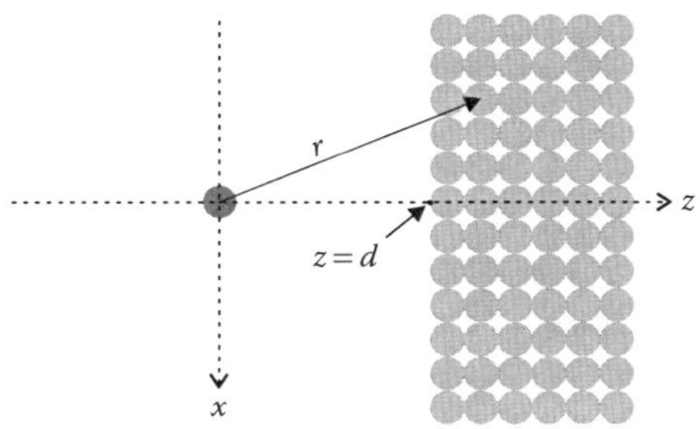

FIGURE 4.18 The interaction of a molecule A (dark circle) with a solid formed by molecules B (gray circles).

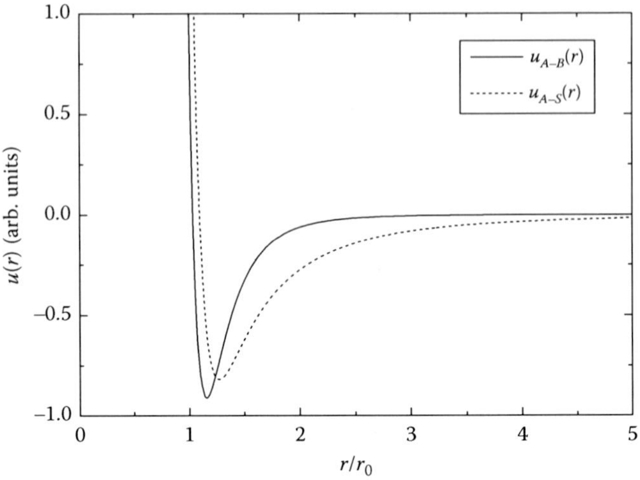

FIGURE 4.19 Qualitative comparison of the intermolecular potential u_{A-B} (Equation 4.26) and the molecule-surface potential u_{A-S} (Equation 4.27).

where ρ_B is the density of solid B, and u_{A-S} is the interaction potential energy between molecule A and the surface. It should be noted that from a short-range interaction in Equation 4.23 (r^{-6}), we get a longer range one (r^{-3}). Figure 4.19 compares qualitatively the shapes of the intermolecular potential u_{A-B} and the molecule-surface potential u_{A-S}, both combined with a repulsive term (r^{-12}) (Hill 1986)

$$u_{A-B}(r) = -\frac{u'}{r^6} + \frac{u''}{r^{12}} = 4u_0\left[-\left(\frac{r_0}{r}\right)^6 + \left(\frac{r_0}{r}\right)^{12}\right] \tag{4.26}$$

$$u_{A-S}(r) = -\frac{\pi u'\rho}{6d^3} + \frac{u''}{d^{12}} = u_0\left[-\frac{2\pi}{3}\left(\frac{r_0}{d}\right)^3 + 4\left(\frac{r_0}{d}\right)^{12}\right] \tag{4.27}$$

where r_0 is the distance where the Lennard-Jones potential (Equation 4.26) is equal to zero. In Equation 4.27, the density has been taken as $\rho \approx 2/r_0^3$, corresponding to a closed packed solid (Israelachvili 2010). The longer range (slower decrease) of Equation 4.27 is noticeable; this potential predicts a minimum (equilibrium position) at a distance closer to that of an intermolecular interaction; that is, a distance of molecular dimensions. The minimum depth is of the order of 10^{-21} J (Hill 1986), leading, as expected, to physical, nonspecific interactions. For specific, chemical interactions, a particular treatment for each problem should, in principle, be done.

4.3.3.2.2 *General Considerations for Molecule–Surface Interactions*

In the above considerations, it can be seen that contributions to the van der Waals energy of interaction (Section 4.3.3.1.3) at a fixed distance can be described roughly by

$$u_{AA} = -k' P_A^2 \tag{4.28}$$

or

$$u_{A-B} = -k'' P_A P_B \tag{4.29}$$

where P_i is some property that can be either p_i^2 or α_i or I_i (see Equations 4.8 through 4.15), and k' and k'' are proportionality constants. Now if we have, as in Figure 4.20a, two molecules A and B in a medium C that become bound, detaching two solvent molecules each one, the energy change in the process will be

$$\Delta u \propto -P_A P_B - 2P_C^2 + 2P_A P_C + 2P_B P_C = -(P_A - P_C)(P_B - 2P_C) \tag{4.30}$$

Essentially the same can be deduced for the case of a molecule A adsorbing onto a surface B in a solvent C (Figure 4.20b). Equation 4.30 predicts that the change of van der Waals energy on adsorption can be positive or negative, depending on the relative values of P_A, P_B, and P_C; if the later is higher than P_A and P_B (solvent–solvent interactions are stronger than solute–solute interaction) or, conversely, P_C is lower than the other two, adsorption of A onto B will be favored but disfavored otherwise. It should be noted that this conclusion does not apply to other type of interactions (electrostatic, hydrogen bonding, and chemical bonding) nor to entropic effects. However, as the van der Waals interactions are always present, whereas the other types may or may not be present, it is important to take into account this conclusion.

Regarding ion binding, electrostatic interactions, with the exception of some special cases, will be important. As discussed in Section 4.3.3.1.1, Coulombic interactions are strong and long-range, so that one would expect at first that they will dominate over other contributions. However, in a medium of high dielectric constant like water ($\varepsilon_r \cong 78$), and in the presence of dissolved ions that screen the interactions, its relative importance is strongly reduced, becoming comparable to dipolar interactions. The interaction energy of an ion of charge q near a charged surface at distance r is given by $q\psi(r)$, where $\psi(r)$ is the electrostatic potential at a distance r from the surface. In Chapter 3, the general properties of charged interfaces have been studied, and expressions predicting the behavior of the potential have been developed. In the

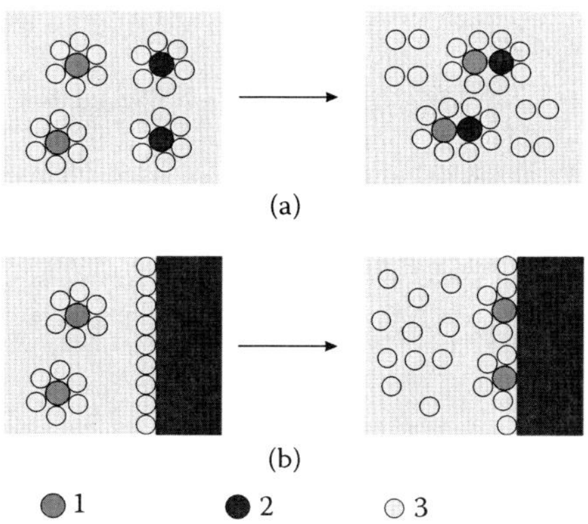

FIGURE 4.20 (a) Schematic depiction of the bonding of two molecules 1 and 2 in a solvent 3; (b) the same for the adsorption of molecules of 1 on a solid 2 in a solvent 3.

next chapter, we shall advance in the analysis of the structure of charged interfaces. Here, we will only note that dipolar (van der Waals) and electrostatic interactions are important contributions to the free energy of ion sorption, thus they will be included within the sorption equilibrium constants, which are found in Section 4.4.

4.4 THEORY OF ADSORPTION ISOTHERMS

4.4.1 LIMITING (HENRY) ISOTHERM

We will consider briefly the limiting behavior that one should expect when the adsorbate activity is low enough. According to Equation 4.4, the isotherm expression should be a function of adsorbate activity. This function can be expanded in Taylor series around $a_A = 0$, yielding

$$\Gamma_A = f(0) + \left.\frac{\partial f}{\partial a_A}\right)_{a_A=0} a_A + \frac{1}{2}\left.\frac{\partial^2 f}{\partial a_A^2}\right)_{a_A=0} a_A^2 + \ldots \tag{4.31}$$

For sufficiently low values of a_A, all terms for powers greater than 1 vanish; also, obviously it should be $f(0) = 0$, so that for $a_A \to 0$ one gets

$$\Gamma_A = \left.\frac{\partial f}{\partial a_A}\right)_{a_A=0} a_A = K_h a_A \tag{4.32}$$

This is a linear relationship between activity and amount sorbed, having the same form as Henry's law in solution chemistry, so that it is often termed *Henry isotherm*, and K_h is the Henry constant. In principle, it is expected that isotherm equations should approximate to the Henry isotherm for low sorbate activities.

4.4.2 IDEAL MONOLAYER: LANGMUIR ISOTHERM

The Langmuir isotherm was proposed by Irving Langmuir nearly a century ago (Langmuir 1916, 1918), using a kinetic reasoning for the adsorption of gases on solids. It has been since revisited on statistical mechanics grounds (Hill 1986; Hiemenz and Rajagopalan 1997). Here, we will follow the kinetic approach. Consider the situation depicted in Figure 4.21, where a fragment of a surface (considered as essentially infinite, in comparison with molecular dimensions) is drawn (gray spheres) with some adsorbate molecules (black spheres) either in gaseous state (1–3) or adsorbed (4 and 5).

The Langmuir model is based on some fundamental assumptions:

1. The surface is composed of localized *sites*, where the adsorbate molecules will eventually adsorb. In Figure 4.21, some empty sites are marked with a white circle. Two sites are shown occupied by adsorbate molecules.
2. These sites are all equivalent and independent, meaning that the adsorbate–surface interaction energy is the same for all sites, and it is not affected by the state (empty or occupied) of neighboring sites.

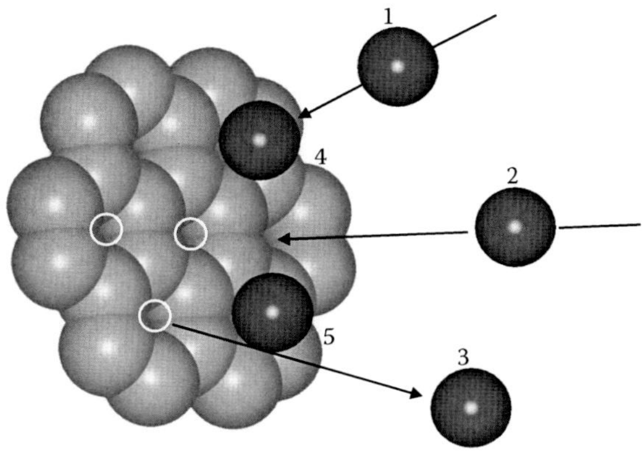

FIGURE 4.21 Schematic picture of the adsorption of a gas onto a solid. White circles indicate (some) empty sites. Molecule 1 will bounce against an already adsorbed molecule (4) so that it cannot adsorb. Molecule 2 could adsorb. Molecule 3 has just desorbed. Molecule 5 is still adsorbed.

3. The adsorbate–surface potential energy is of the type depicted in Figure 4.19; this means, in kinetic terms, that the adsorption has no activation barrier, and conversely the activation energy for desorption is equal in magnitude to the adsorption energy.
4. Adsorption of an incoming molecule can proceed only on empty sites.

In Figure 4.21, these considerations are illustrated. Molecule 1 is about to hit the surface on a site already occupied (by molecule 4); thus, it cannot adsorb and will bounce back to the gas phase. Molecule 2 is going to hit on an empty site, so that it can be adsorbed with a certain probability, and molecule 3 has just been desorbed; this process should occur with a temperature-dependent probability.

From the kinetic theory, it can be found that the number of gas molecules hitting the surface per unit of time and area is (Hill 1986; Adamson and Gast 1997; Hiemenz and Rajagopalan 1997; McQuarrie and Simon 1997; Levine 2008)

$$n_g = \frac{N_A P_A}{\sqrt{2\pi M_r \mathbf{R} T}} \tag{4.33}$$

where N_0 is Avogadro's number, P_A is the gas pressure of A, and M_r is the molar mass of the adsorbate. If each of these molecules can become adsorbed with a probability α, the rate of adsorption (moles adsorbed per unit of time and area) will be

$$\nu_{ads} = \frac{\alpha}{\sqrt{2\pi M_r \mathbf{R} T}}(1-\theta_A)P_A = k_{ads}(1-\theta_A)P_A \tag{4.34}$$

where $1-\theta_A$ is the fraction of surface uncovered (unoccupied), and k_{ads} is the adsorption rate constant. Conversely, the rate of desorption will be given by

$$\nu_{des} = \beta e^{\frac{E_{ads}}{RT}} \Gamma_A = \beta e^{\frac{E_{ads}}{RT}} \Gamma_S \frac{\Gamma_A}{\Gamma_S} = k_{des}\theta_A \tag{4.35}$$

where β is a desorption probability (related to the frequency of vibration of the adsorbed molecule), and E_{ads} the adsorption energy ($\cong \Delta H_{ads}$ and <0, because it is an exothermic process). In equilibrium, the rates given by Equations 4.34 and 4.35 should be equal, so that

$$k_{ads}(1-\theta_A)P_A = k_{des}\theta_A \tag{4.36}$$

Thus,

$$K_A P_A = \frac{k_{ads}}{k_{des}} P_A = \frac{\theta_A}{1-\theta_A} = \frac{\Gamma_A}{\Gamma_S - \Gamma_A} = \frac{Q_A}{Q_S - Q_A} \tag{4.37}$$

which can also be written as

$$\theta_A = \frac{K_A P_A}{1 + K_A P_A} \tag{4.38}$$

and

$$Q_A = Q_S \frac{K_A P_A}{1 + K_A P_A} \tag{4.39}$$

and similarly for Γ_A. Equations 4.37 through 4.39 (in any form) represent the Langmuir adsorption isotherm; and

$$K_A = \frac{\alpha e^{-\frac{\Delta H_{ads}}{RT}}}{\beta \Gamma_S \sqrt{2\pi M_r \boldsymbol{R}T}} \tag{4.40}$$

is the adsorption constant, which can be regarded as an equilibrium constant, as it applies to the heterogeneous equilibrium between the gas phase and the adsorbed layer. It can be easily seen that Equation 4.37 predicts the behavior observed in Figure 4.1, with surface saturation (θ_A approaching unity) as the pressure increases. On the contrary, for sufficiently low pressures, Equation 4.38 becomes

$$\theta_A = K_A P_A \tag{4.41}$$

predicting the Henry isotherm.

We will not enter here on the derivation, but on statistical thermodynamics grounds, it can be found quite straightforwardly from the assumptions above that the chemical potential of the adsorbed layer is given by the following (Hill 1986; Hiemenz and Rajagopalan 1997)

$$\mu_{ads} = \mu_{ads}^0 + \boldsymbol{R}T \ln \frac{\theta_A}{1-\theta_A} \tag{4.42}$$

where μ_{ads}^0 is the standard chemical potential of the adsorbed layer (note that the standard state corresponds to $\theta_A = 0.5$). If it is in equilibrium with an ideal gas, the equality of the chemical potentials requires that

$$\mu_g^0 + RT\ln\frac{P_A}{P^0} = \mu_{ads} = \mu_{ads}^0 + RT\ln\frac{\theta_A}{1-\theta_A} \tag{4.43}$$

and from Equation 4.43, Equation 4.37 follows easily. Now if we consider a solid in contact with a solution where the solute is adsorbed, the solution is diluted *and* the assumptions supporting the Langmuir isotherm are still valid; in equilibrium Equation 4.42 can be equated to the chemical potential of the solute (Equation 2.5)

$$\mu_A^0 + RT\ln a_A = \mu_{ads}^0 + RT\ln\frac{\theta_A}{1-\theta_A} \tag{4.44}$$

leading to

$$K_A a_A = e^{-\frac{\mu_{ads}^0 - \mu_A^0}{RT}} a_A = \frac{\theta_A}{1-\theta_A} = \frac{Q_A/Q_S}{1-Q_A/Q_S} \tag{4.45}$$

again the Langmuir equation. It should be remarked that the basic assumptions given above should be still fulfilled, at least in principle. In solution phase, as discussed in Section 4.2.2, the situation depicted in Figure 4.21 no longer holds because of the presence of the solvent. However, it suffices if ΔH_{ads} is the same or similar for all sites and if it is greater than other interaction energies, such as adsorbate–adsorbate interactions. Similarly, the kinetic approach can still be applied under these conditions, but the interpretation of k_{ads} is different. Type L isotherms (Figure 4.8) obviously also follow the Langmuir expression, and sometimes the H group also does. Figure 4.22 shows an example of application to the complexation of Cu(II) to particles of a French calcareous soil from the Champagne–Ardenne region (Flogeac, Guillon, and Aplincourt 2004). Figure 4.22a shows the isotherm plot, indicating a type L isotherm. To verify the fulfillment of the Langmuir isotherm, Equation 4.45 (using concentration instead of activity) can be rearranged in a linear form, for example, as

$$\frac{c_{\text{Cu}}}{Q_{\text{Cu}}} = \frac{c_{\text{Cu}}}{Q_S} + \frac{1}{K_{\text{Cu}}Q_S} \tag{4.46}$$

Figure 4.22b shows a plot of the data in Figure 4.22a in the form of Equation 4.46, showing a good agreement, indicating that the data can be described satisfactorily by the Langmuir isotherm. It should be noted that the fitting to one or another isotherm equation in general tells little about the molecular nature of the particular adsorption process.

The Langmuir isotherm (and the Langmuir chemical potential, Equation 4.42) is the ideal adsorption case, similar in some aspects to the ideal gas law. Despite this, it is found to describe satisfactorily the experimental data in many cases, either in gas phase or in solution, including many cases in soil science. Presumably, in a number of cases, a compensation of effects takes place.

4.4.3 Competitive and Multisite Langmuir Isotherms

When two or more adsorbates are present (A, B,...) in contact with a solid, both will try to adsorb on the empty sites, whereas those occupied can hold either adsorbate. This is a situation of *competitive adsorption*, and the Langmuir isotherm

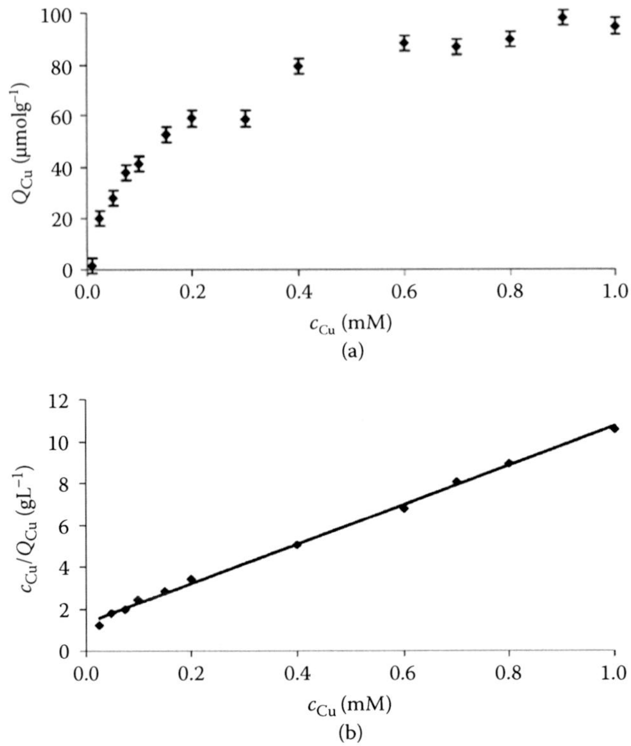

FIGURE 4.22 Sorption of Cu(II) in soil particles (Champagne–Ardenne region, France) at 293 K and pH = 6.2 in KNO_3 0.1 M. (a) Isotherm plot showing the presence of a limiting coverage. (b) Langmuir isotherm fitting plot (Equation 4.46) showing good agreement; linear fitting yielded $R^2 = 0.9974$. (Reprinted with permission from Flogeac, K. et al., 2004, *Environmental Science & Technology*, 38, no. 11, 3098–3103. Copyright 2004 American Chemical Society.)

can easily be extended to this case (Langmuir 1918; Hiemenz and Rajagopalan 1997; Atkins and Paula 2009). For two adsorbates A and B, the rate of adsorption of A is

$$v_{ads,A} = k_{ads,A}\left(1-\theta_A-\theta_B\right)P_A \tag{4.47}$$

and similarly for B. The rate of desorption is still given by Equation 4.35, so that in equilibrium

$$K_A P_A = \frac{\theta_A}{1-\theta_A-\theta_B} \tag{4.48}$$

and also

$$K_B P_B = \frac{\theta_B}{1-\theta_A-\theta_B} \tag{4.49}$$

Inverting Equations 4.48 and 4.49 together, it is found

$$\theta_A = \frac{K_A P_A}{1 + K_A P_A + K_B P_B} \tag{4.50}$$

$$\theta_B = \frac{K_B P_B}{1 + K_A P_A + K_B P_B} \tag{4.51}$$

Equations 4.48 through 4.51 are different forms of the Langmuir competitive adsorption isotherm; it can easily be extended to more than two adsorbates. They have been written for gas phase, but equally they can apply to adsorption from solution, replacing the pressure by activity or (approximately) concentration of each solute.

On the other hand, if the surface sites are not all equal, but of different kinds, which essentially means having substantially different adsorption energies, the Langmuir isotherm can be extended easily if *all sites of either type are independent.* Let the surface be composed of Q_1 sites of type 1 and Q_2 sites of type 2, so that $Q_S = Q_1 + Q_2$, and let $f_i = Q_i/Q_S$; the total amount adsorbed on the surface, because of the site independence, will be given by the following (Langmuir 1918; Adamson and Gast 1997; Hiemenz and Rajagopalan 1997):

$$Q_A = Q_S \left(f_1 \frac{K_{1,A} a_A}{1 - K_{1,A} a_A} + f_2 \frac{K_{2,A} a_A}{1 - K_{2,A} a_A} \right) \tag{4.52}$$

$$\theta_A = f_1 \frac{K_{1,A} a_A}{1 - K_{1,A} a_A} + f_2 \frac{K_{2,A} a_A}{1 - K_{2,A} a_A} \tag{4.53}$$

Expressions like Equations 4.52 and 4.53 explain some type H isotherms, such as those shown in Figures 4.11 (Pabst et al. 2010) and 4.12 (Brownawell et al. 1990).

4.4.4 Ideal Multilayer (BET) Isotherm

We will now consider the simplest model for multilayer adsorption, due to Brunauer, Emmet, and Teller (1938), hence the acronym BET. The basic assumptions involved are illustrated in Figure 4.23. The molecules adsorbing in multiple layers are assumed to be stacked, each molecule standing on top of another belonging to the previous layer, except, of course, those in the first layer (compare with Figure 4.2c, where a more closely packed arrangement is depicted). In other words, each site can have 0, 1, 2, and so on molecules adsorbed on it. The sites are still considered independent, so that each molecule interacts only with those immediately below and above it in the same site (stack); molecules in the first layer interact with the surface and with the next molecule on top of it, if present. Interactions of longer range are neglected.

In the following, we will write θ_0 for the fraction of sites that are empty, and θ_1, θ_2, and so on for the fraction of sites occupied by 1, 2, and so on molecules, respectively. Recalling Equations 4.35 and 4.36, we can write under the same considerations the adsorption and desorption rates of layer i

$$v_{ads,i} = k_{ads,i} \theta_{i-1} P_A \quad i > 0 \tag{4.54}$$

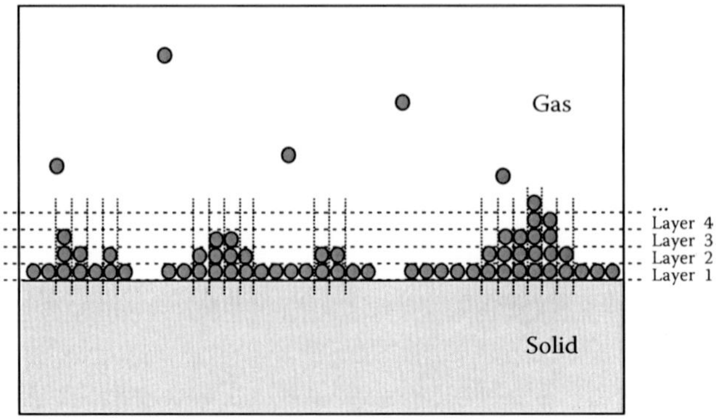

FIGURE 4.23 Schematic drawing of the ideal multilayer adsorption, as assumed in the BET model; multilayer adsorption takes place, but in this example, the adsorbate–surface interaction is stronger than adsorbate–adsorbate interaction, thus the first layer forms preferentially to multilayer growth. The adsorbate molecules are assumed to form "stacks" independent of each other, so that every molecule has only "up and down" interactions, not lateral ones.

$$v_{des,i} = k_{des,i}\theta_i \quad i > 0 \tag{4.55}$$

so that in equilibrium

$$K_{A,i}P_A = \frac{\theta_i}{\theta_{i-1}} \tag{4.56}$$

The total number of sites is fixed, thus

$$1 = \theta_0 + \theta_1 + \theta_2 + \ldots = \sum_{i=0}^{\infty} \theta_i \tag{4.57}$$

Note that the number of layers has no upper bound; an essentially infinite number of layers means adsorbate condensation. Now, there is an assumption introduced by Brunauer et al.: the adsorbing molecules interact only with the nearest ones above and below, so that beyond layer 1, all interactions, and hence adsorption constants, are equal and also the same as in the adsorbate liquid: that is, $K_{A,2} = K_{A,3} = \ldots$. With this condition, and after some algebra that will not be reproduced here (Brunauer, Emmett, and Teller 1938; Adamson and Gast 1997; Hiemenz and Rajagopalan 1997), it is found that

$$Q_A = Q_S \frac{c\dfrac{P_A}{P_V}}{\left(1 - \dfrac{P_A}{P_V}\right)\left[1 + (c-1)\dfrac{P_A}{P_V}\right]} \tag{4.58}$$

where $c = K_{A,1}/K_{A,2}$. Equation 4.58 is the BET isotherm equation; it can be written in terms of Γ or θ immediately. In the original paper by Brunauer, Emmett, and Teller (1938), the quantity adsorbed Q_A as well as the monolayer coverage Q_S were written as v and v_m respectively, which are the volumes of gas adsorbed at P_A and forming a monolayer respectively, at specified temperature and pressure; this is because the volume change is the usual way to measure the amount of gas adsorbed. The BET equation is often still written in this way.

Figure 4.24 shows plots of the BET Equation 4.58 for different c values, indicated in the figure. Note that as $P_A \rightarrow P_V$, Q_A diverges; that is, condensation occurs. For $c > 1$ the plots correspond to type II in Brunauer's classification, whereas for $c \leq 1$ the type III is predicted. In the first case, the interaction of the adsorbate with the surface is stronger than with itself, thus the first layer is populated with some preference to subsequent layers, and so the curves show a "hump" around $\theta = 1$. For $c < 1$ the opposite preference occurs, the layers grow as soon as the first molecules adsorb, and thus no hump is seen in the plot.

As usual, it has been assumed above that an infinite number of layers can be adsorbed. If this number is finite, N, the isotherm obtained reads (Brunauer, Emmett, and Teller 1938)

$$Q_A = Q_S \frac{cx}{1-x}\left[\frac{1-(N-1)x^N + Nx^{N+1}}{1+(c-1)x - cx^{N+1}}\right] \tag{4.59}$$

where $x = P_A/P_V$; in general, except for low N, Equation 4.59 results turn to be almost indistinguishable from Equation 4.58. Also, a more general equation was developed for the case of capillary condensation (types IV and V). That will not be treated here; the reader is referred to the original paper by Brunauer et al. (1940). Statistical derivation of the BET and related isotherms can be found elsewhere (Fowler and Guggenheim 1939; Ruthven 1984; Hill 1986).

The BET isotherm has little, if any, application in problems of adsorption in the soil but is important, because it constitutes the basis of a commonly applied method to determine surface areas, as we shall see in Chapter 7. There are a number of cases where adsorption beyond a monolayer is observed, especially for transition metal

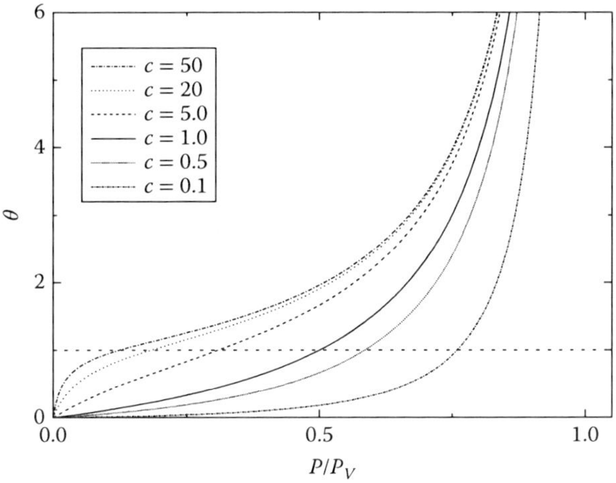

FIGURE 4.24 The BET isotherm plot for different values of the parameter c.

cations on mineral particles at relatively high pH values; however, this is generally due to *surface precipitation*, which is a different process (Robarge 1998; McBride 1999).

4.4.5 Beyond Ideal Isotherms

There are several causes for the "ideal" (Langmuir) equation to break down; the most important are as follows:

1. Few solid surfaces are homogeneous at the molecular level.
   The assumption of surface homogeneity is essential in the Langmuir equation; otherwise a different value of K would apply to Equations 4.37 through 4.45 at various places, and eventually at every site, on the surface. For a few different values of K, the treatment of Section 4.4.3 can be applied. Attempts to deal with surface heterogeneity have been undertaken, and some of these will be detailed in Section 4.4.7.
2. Few monolayers are ideal.
   In many cases lateral interactions—that is, interactions between neighboring adsorbates—are present, causing those adsorbates that adsorb to a nonempty surface to find different conditions compared with those arriving to a "clean" surface (in the sense of absence of adsorbates other than the solvent). The interactions can be attractive, sometimes giving rise to S-type isotherms as discussed in Section 4.2.2, or repulsive, which can result in H-type isotherms, as the incoming adsorbates find it more "difficult" to adsorb as the surface becomes more covered.
3. The solutions may not be diluted.
   The adsorption process in condensed media is a complex one involving several different kinds of interactions: solvent–solute, solvent–adsorbent, and solute–adsorbent. In adsorption form solution, as said before, what happens actually is an exchange process between the solute and the solvent; a requirement for the Langmuir isotherm to remain valid is that the solvent–surface interaction is weaker than the solute–surface (Hiemenz and Rajagopalan 1997). Also, if the solution is not diluted, the solvent concentration may be comparable to that of the "solute" and not constant, thus its chemical potential should be included in Equation 4.44, and its activity should show up in the isotherm expression (Bartell and Sloan 1929).

It is important to realize that from the adsorption equilibrium data (i.e., the experimental isotherm), generally it cannot be distinguished between the effects of surface heterogeneity and lateral interactions. Consider the adsorption equilibrium constant K; it can be related to the Gibbs standard free energy of adsorption:

$$\Delta G^0_{ads} = -\boldsymbol{R}T\ln K \tag{4.60}$$

If a surface is heterogeneous, different sites will have different adsorption energies and consequently different ΔG^0_{ads} and K values. But in an homogeneous surface, if there is an interaction energy w between two adsorbates in a pair of neighboring sites, then the first adsorbate will adsorb with a free-energy change equal to ΔG^0_{ads}, but another adsorbate molecule binding at a neighbor site will adsorb with a free-energy change (approximately) equal to $\Delta G^0_{ads} + w$ and result in a different effective value of K. The isotherm alone does not allow to distinguish between these two circumstances; other experimental information (spectroscopic) will be needed.

The above discussion remarks that isotherm analysis alone yields usually little information about the molecular nature of adsorption, simply because it is not sensitive enough to molecular details. It should not be surprising, thus, that in many complex systems (such as the Cu(II) in soil example in Figure 4.22), simple isotherms, such as the Langmuir one, describe the experimental results well.

4.4.6 Adsorption with Interactions: The Fowler–Guggenheim Isotherm

We now discuss a simple way to account for lateral interactions that was first proposed by Fowler and Guggenheim (1939). Here it will be developed in an approximate way for adsorption from solution; complete treatments can be found in Fowler and Guggenheim's book as well as in other sources (Hill 1986). Consider Figure 4.25; the gray spheres form a surface fragment with its sites occupied by solvent molecules (light gray) and solute molecules adsorbed (dark spheres); between parts (a) and (b), the solvent molecule at the center is replaced by one solute molecule, which thus becomes adsorbed. This center site is surrounded by six other sites in the immediate vicinity: these are called *nearest neighbors*, and its number, six in this example, is the *coordination number*, C_n, of the surface. Of course beyond these there are second nearest neighbors, and so on (not shown). It is assumed in the FG isotherm that each molecule at the surface interacts only with the nearest neighbors, and it is assumed also that such interaction can be written as a sum of pair potentials

$$u_i = \sum_{j=1}^{C_n} u_{ij} \tag{4.61}$$

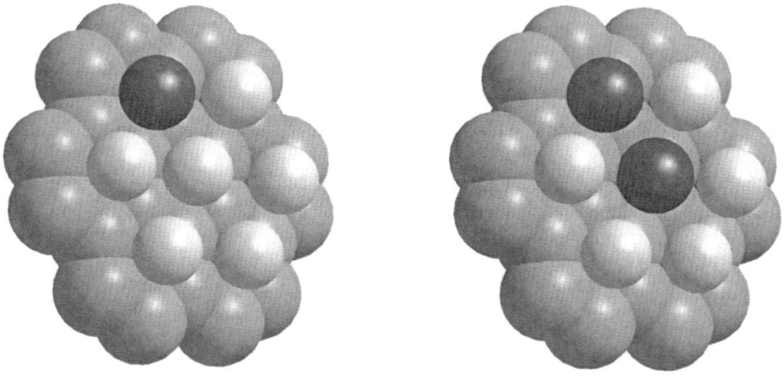

(a) (b)

FIGURE 4.25 Schematic of the process of adsorption from solution: the gray spheres represent atoms of a surface; light gray spheres represent solvent molecules; and black spheres, solute (adsorbate) molecules. From (a) to (b), one solute molecule has adsorbed, replacing a solvent molecule.

where u_{ij} are the potential energies of interaction of molecule i with each neighbor j. These potential energies will be fluctuating in time, as each molecule vibrates around the equilibrium position; a further assumption will be to consider only the average values (i.e., those corresponding to all molecules resting in their equilibrium position). Let us call w_{11} and w_{22} the interaction energies of solvent and solute molecules, respectively, with themselves, and w_{12} the interaction between solute and solvent molecules. The total interaction energy of the center molecule in Figure 4.25a is

$$u_a = 5w_{11} + w_{12} = C_n\left[\left(1 - f_2\right)w_{11} + f_2 w_{12}\right] \tag{4.62}$$

where f_2 is the fraction of the nearest neighbor sites, occupied by solute molecules, in this case 1/6. In Figure 4.25b the total energy results

$$u_b = C_n\left[\left(1 - f_2\right)w_{12} + f_2 w_{22}\right] \tag{4.63}$$

The change of energy upon adsorption is found to be from Equations 4.62 and 4.63:

$$\Delta u = u_b - u_a = w_{12} - w_{11} + f_2\left(w_{11} + w_{22} - 2w_{12}\right) = w_0 + f_2 w \tag{4.64}$$

This interaction is of course dependent on the distribution of solute and solvent molecules. The important assumption introduced here is that the distribution of adsorbate molecules on the surface sites is at random, just as in the ideal case. This means that the interactions do not affect the distribution, which in turn implies relatively weaker interactions; strong attractive interactions, for example, would cause the adsorbate molecules to group preferentially around themselves, thus losing randomness. Because of this assumption, the *average* fraction of solute molecules around a site will be equal to θ_A, the fraction of sites occupied by the solute (adsorbate). Another consequence is that the entropy change upon adsorption will be the same as in the ideal case; thus, we can write for the standard Gibbs free energy change, in molar units

$$\Delta G^0_{ads} = \Delta G^0_{ads,id} + W_0 + \theta_A W \tag{4.65}$$

where $\Delta G^0_{ads,id}$ is the ideal free energy change, whereas W_0 and W are the molar equivalents of w_0 and w. The invariance of the entropy leads to an equation with the same form of Langmuir's with the adsorption constant, through Equations 4.60 and 4.65, replaced by $K\mathrm{e}^{-\frac{\theta_A W}{RT}} = K\mathrm{e}^{-B_{FG}\theta_A}$ resulting in (Fowler and Guggenheim 1939; Hill 1986)

$$K_A \mathrm{e}^{-B_{FG}\theta_A} a_A = \frac{\theta_A}{1 - \theta_A} \tag{4.66}$$

or more usually written as

$$K_A a_A = \frac{\theta_A}{1 - \theta_A}\mathrm{e}^{B_{FG}\theta_A} \tag{4.67}$$

resulting in a modified Langmuir isotherm (with $B_{FG} = 0$, that equation is recovered). This is usually known as an FG isotherm, first published in 1939 (Fowler and Guggenheim 1939), but was also termed Bragg-Williams by Hill (1986) and also known by electrochemists as a Frumkin isotherm (Parsons 1964; Delahay 1965).

Figure 4.26 shows the effect of the interaction parameter B_{FG} on the isotherm shape. Positive values of B_{FG}, corresponding to repulsive interactions (see Equation 4.65), cause the coverage, for fixed activity, to decrease as B_{FG} increases. On the contrary, negative (attractive) B_{FG} cause faster adsorption as the activity increases; notice that for $B_{FG} \leq -4$, a sharp rise in θ_A is observed, which can be interpreted as a bidimensional phase transition (Hill 1986); even when for localized adsorption on a solid surface, this concept is questionable, it means here that the adsorbate molecules are binding preferentially next to others already adsorbed. Of course, one can question for these cases the validity of the assumption of random adsorbate distribution, but nevertheless the prediction is important, as this is the simplest isotherm equation predicting such phenomena.

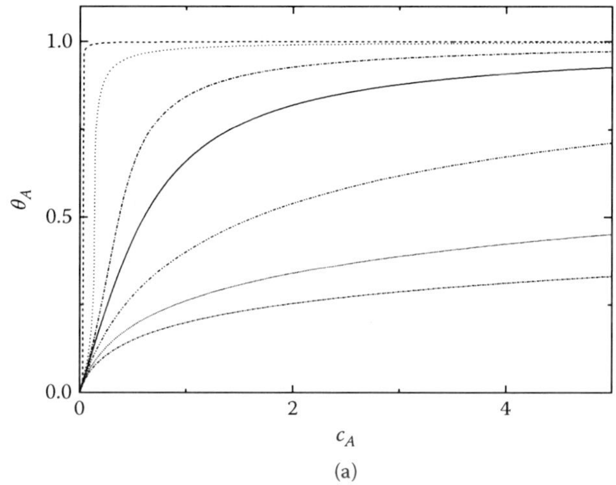

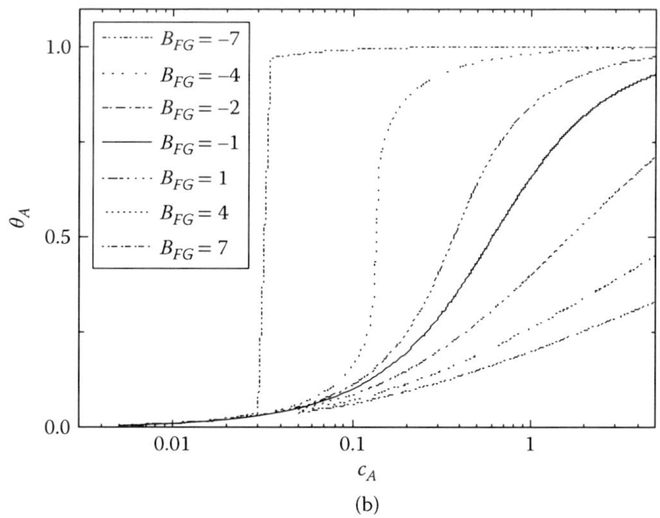

FIGURE 4.26 Plots of the Fowler–Guggenheim isotherm for different values of the interaction parameter B_{FG} in (a) linear and (b) logarithmic concentration scale.

An example of application of the FG isotherm was presented in Figure 4.9 for the adsorption of NOM on iron oxides (Gu et al. 1994).

4.4.7 Surface Heterogeneity

A sorbent, especially in natural environments like soils, is usually not composed of equal and equivalent sites but of a number of different types, each with its own characteristics. For the case of a relatively simple patchwise surface (a few site types), expressions like Equation 4.52 will be applicable. However, for highly heterogeneous surfaces, a high number of different adsorption energies will be present; in these cases, it is better to treat the surface as composed of sites belonging to a continuous distribution of adsorption energies. The general expression for the amount adsorbed becomes

$$\begin{aligned} Q_A &= Q_S \int_{-\infty}^{\infty} \theta(\Delta G^0_{ads,A}, a_A) f(\Delta G^0_{ads,A}) \mathrm{d}\Delta G^0_{ads,A} \\ &= Q_S \int_{-\infty}^{\infty} \theta(\log K_A, a_A) f(\log K_A) \mathrm{d}\log K_A \end{aligned} \tag{4.68}$$

Because ΔG^0 is proportional to ln *K* (Equation 4.60) or log *K*, the last side of Equation 4.68 is equivalent and is the form usually employed. Here, $\theta(\Delta G^0_{ads,A}, a_A)$ or $\theta(\log K_A, a_A)$ is the *local isotherm*—that is, the expression giving the degree of coverage at a given adsorbate activity for sites with a particular value of K_A (or of the standard Gibbs free energy of adsorption)—and *f* is the fraction of sites with free energy between $\Delta G^0_{ads,A}$ and $\Delta G^0_{ads,A} + \mathrm{d}\Delta G^0_{ads,A}$. Equation 4.68 is a quite general expression but is, at best, of difficult solution. We shall return to this topic more extensively in Chapter 11, but here we address an important case.

4.4.7.1 Sips Distribution Function

In most cases, the local isotherm in Equation 4.68 is Langmuir's, leading to

$$Q_A = Q_S \int_{-\infty}^{\infty} \frac{K_A a_A}{1 + K_A a_A} f(\log K_A) \mathrm{d}\log K_A \tag{4.69}$$

For a random distribution of log K_A values, the Gaussian distribution is a logical choice for *f*. However, Equation 4.69 with a Gaussian function has no analytical solution. Sips (1948) has shown that the quasi-Gaussian distribution function

$$f(\log K_A) = \frac{\ln(10)}{2\pi} \times \frac{\sin(\pi n)}{\cosh\left[n \ln(10)\left(\log K_m - \log K_A\right)\right] + \cos(\pi n)} \tag{4.70}$$

leads to an analytical solution for Equation 4.69. Here, K_m is the value of K_A corresponding to the maximum (center) of the distribution, and *n* is a parameter defining the width of the distribution; we will return to it a little later.

Figure 4.27a compares the Sips distribution, taking $n = 0.5$, with the Gaussian distribution with $w = 2.0$ in

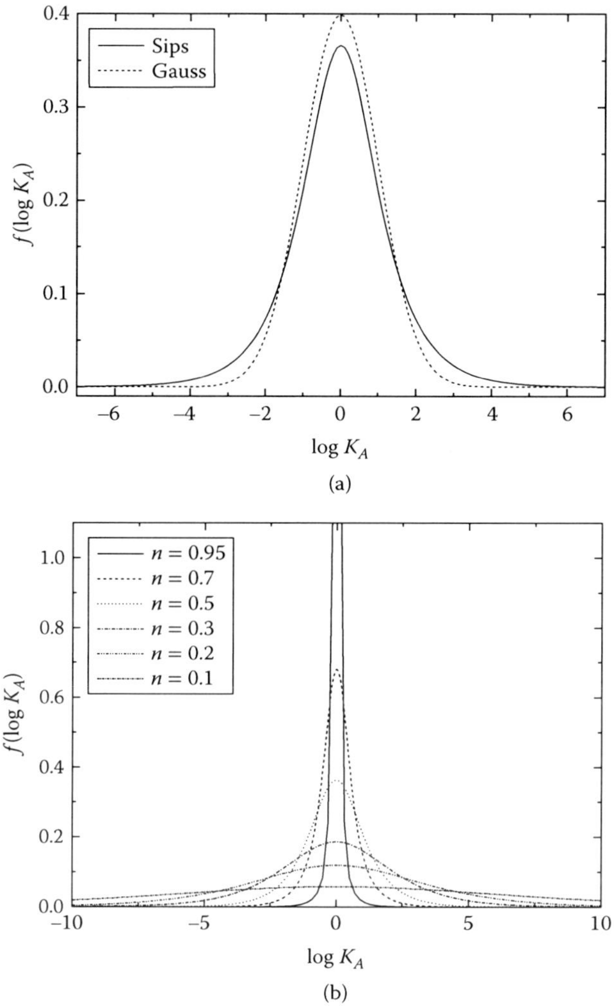

FIGURE 4.27 (a) Comparison of the (normalized) Gauss and Sips distributions. (b) Influence of the parameter n on the shape of the Sips distribution.

$$f(\log K_A) = \frac{1}{w\sqrt{\pi/2}} e^{-\frac{2(\log K_m - \log K_A)^2}{w^2}} \tag{4.71}$$

As it can be seen, the shape of Equation 4.70 closely matches that of Equation 4.71, thus it is considered quasi-Gaussian. The parameter n defines the width (Figure 4.27b); $n = 1$ gives a very narrow peak at $\log K_A = \log K_m$, approaching a Dirac's delta, and lower values give broader peaks.

4.4.7.2 Langmuir–Freundlich and Freundlich Isotherms

As it was said above, it was shown (Sips 1948) that with the function given in Equation 4.70, the integral in Equation 4.69 has an analytical result; it is

$$\theta_A = \frac{Q_A}{Q_S} = \frac{(K_m a_A)^n}{1+(K_m a_A)^n} \quad (4.72)$$

This is an isotherm differing from Langmuir's in the exponent n affecting the activity. For $a_A \to 0$, Equation 4.72 simplifies to

$$Q_A = Q_S (K_m a_A)^n = Q_S K_m^n a_A^n = B_F a_A^n \quad (4.73)$$

where B_F is another constant. The rightmost side of Equation 4.73 has been known for long time as the *Freundlich isotherm* (Freundlich 1926), originally proposed on purely empirical grounds. Thus, Equation 4.72 became known (Sips 1948) as the *Langmuir–Freundlich isotherm* (LF) and is one of the most commonly employed expressions in adsorption studies. In soil science in particular, it is underlying some modern models of interaction between ions and molecules with soil particles.

In Figure 4.28 the shape of the isotherm for different values of n are presented. Of course, $n = 1$ is the pure Langmuir isotherm. As n decreases, the initial slope increases, but the approach to the asymptotic value is slower, as it is more apparent in Figure 4.28b. This behavior is due to the underlying distribution: for lower n values, there is a greater fraction of sites with high and low affinities; high-affinity sites are rapidly populated as the concentration increases, but low-affinity sites require greater concentrations (activities) to be occupied. One limitation of the LF isotherm is that it does not reduce to the Henry equation as $\theta_A \to 0$, because Equation 4.32 cannot be applied for $n < 1$. Töth (see Jaroniec and Töth 1976 and references therein) proposed an equation with a different form under similar assumptions

$$\theta_A = \frac{K_A a_A}{\left[1+(K_A a_A)^n\right]^{1/n}} \quad (4.74)$$

which at low activities reduces to the Henry equation but is no longer compatible with the Freundlich isotherm.

4.5 OTHER ISOTHERM EQUATIONS

From nearly a century ago, from the earlier works of Langmuir and Freundlich, many different models and isotherm equations have been proposed and compared (Parsons 1964; Delahay 1965; Kinniburgh 1986; Tóth 1995; Aranovich and Donohue 1996; Rocha et al. 1997; Donohue and Aranovich 1998; Marczewski 2002), especially in food science (Van den Berg and Bruin 1981; Iglesias and Chirife 1982).

Table 4.2 summarizes the isotherm equations discussed so far plus some others, added for comparison. The first four have been discussed earlier in the text. The Guggenheim–Anderson–de Boer isotherm (GAB) is a multilayer sorption equation, a modified form of the BET equation that has found wide application in food science and technology (Timmermann 2003 and references therein). Basically, it does not consider the layers beyond the first as the liquid adsorbate but as an intermediate

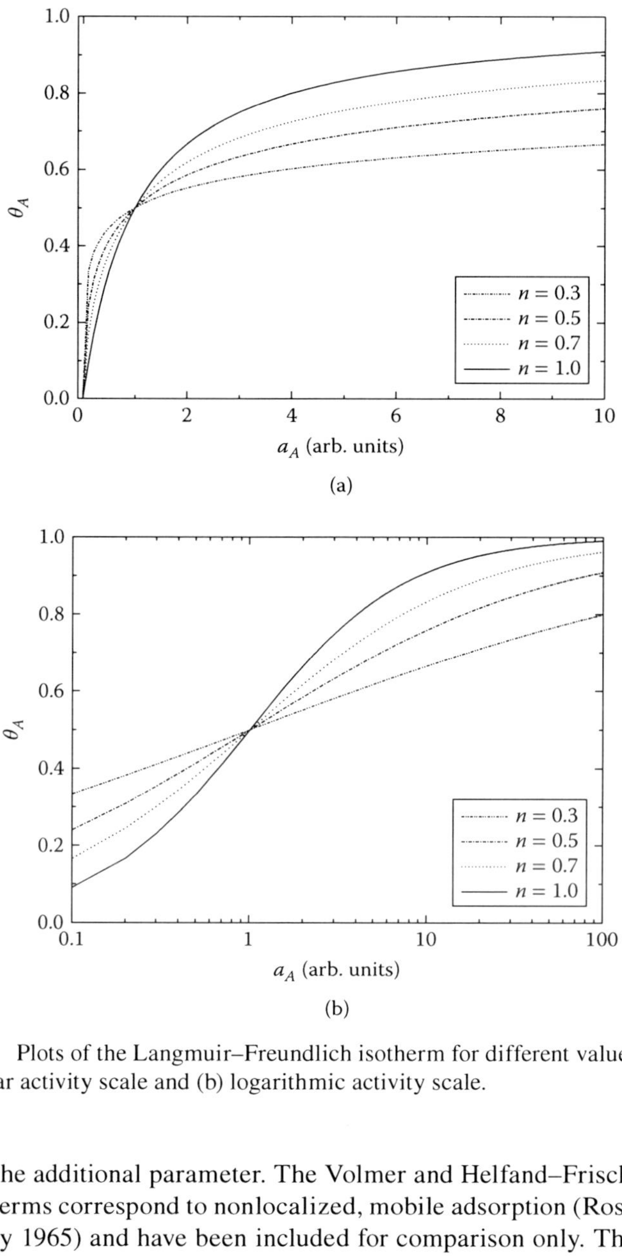

FIGURE 4.28 Plots of the Langmuir–Freundlich isotherm for different values of the exponent n: (a) linear activity scale and (b) logarithmic activity scale.

form, hence the additional parameter. The Volmer and Helfand–Frisch–Lebowitz–Parsons isotherms correspond to nonlocalized, mobile adsorption (Ross and Olivier 1964; Delahay 1965) and have been included for comparison only. The Freundlich and LF have already been considered. The Tóth isotherm is a modified form of the LF equation, which reduces to Henry's law at low coverage; it was introduced from conditions of thermodynamic consistency (Jaroniec and Töth 1976). The generalized Langmuir (GL or Marczewski–Jaroniec) isotherm is yet another generalization of the LF isotherm (Marczewski and Jaroniec 1983; Jaroniec and Marczewski 1984), allowing for asymmetric free energy distributions: the parameter m characterizes

TABLE 4.2
Some Common Isotherm Equations

Name	Isotherm	Basic Assumptions
Henry	$\theta_A = \frac{Q_A}{Q_S} = K_A a_A$	$\theta_A \to 0$
Langmuir	$\theta_A = \frac{K_A a_A}{1 + K_A a_A}$	Localized adsorption, identical sites, no lateral interactions
Fowler–Guggenheim–Frumkin	$K_A a_A = \frac{\theta_A}{1-\theta_A} e^{B_{FG}\theta_A}$	Localized adsorption, identical sites, lateral interactions with random distribution
Brunauer–Emmet–Teller (BET)	$\theta_A = \frac{c a_A}{(1-a_A)[1+(c-1)a_A]}$ $a_A = P_A/P_V$	Localized adsorption, identical sites, no lateral interactions, multilayer, liquid-like from second layer up
Guggenheim–Anderson–de Boer (GAB)	$\theta_A = \frac{CK_A a_A}{(1-K_A a_A)(1-K_A a_A + CK_A a_A)}$ $a_A = P_A/P_V$	Localized adsorption, identical sites, no lateral interactions, multilayer not liquid-like from second layer up
Volmer	$K_A a_A = \frac{\Gamma_A}{1-b\Gamma_A} e^{b\Gamma_A/(1-b\Gamma_A)}$	Mobile adsorption, no lateral interactions, except for excluded volume
Helfand–Frisch–Lebowitz–Parsons	$K_A a_A = \frac{1}{\Gamma_S}\frac{\theta_A}{1-\theta_A} \times \exp\left[B_{HL}\theta_A + \frac{\theta_A}{1-\theta_A} + \frac{1}{(1-\theta_A)^2} - 2\right]$	Mobile adsorption, with lateral interactions and excluded volume
Langmuir–Freundlich	$\theta_A = \frac{(K_A a_A)^n}{1+(K_A a_A)^n}$ $0 < n \le 1$	Langmuir isotherm with quasi-Gaussian surface heterogeneity (energy distribution)
Freundlich	$\theta_A = K a_A^n$ $0 < n \le 1$	Langmuir–Freundlich for $a_A \to 0$; formerly empirical
Tóth	$\theta_A = \frac{K_A a_A}{\left[1+(K_A a_A)^n\right]^{1/n}}$ $0 < n \le 1$	Modified Langmuir–Freundlich with Henry limiting form at low coverage
Generalized Langmuir	$\theta_A = \left[\frac{(K_A a_A)^n}{1+(K_A a_A)^n}\right]^{\frac{m}{n}}$	Heterogeneous sorbent with asymmetric distribution. Extension of LF

the high-energy part of the distribution, whereas the parameter n does the same for the low-energy part. Also, depending on these parameters, the GL equation reduces to other forms: for $m = n = 1$, the Langmuir isotherm is obtained; for $m = n \le 1$, one gets the LF equation; for $n = 1$ and $m \le 1$, a "generalized Freundlich" (Jaroniec and Marczewski 1984) form is found; and for $m = 1$ and $n \le 1$, the Tóth isotherm is recovered. However, the GL isotherm is based, partially, on empirical grounds.

4.6 ISOTHERM COMPARISON AND ASSIGNMENT

Given the high number of available isotherm equations, one can wonder which one will represent more appropriately a particular adsorption phenomenon, thus the problem of isotherm assignation to a given set of data arises. Usually the first attempt is to fit the experimental data to different isotherm equations. To do this, in some cases the equations can be linearized, as for the Langmuir isotherm, as shown in Equation 4.46 and Figure 4.22. The Freundlich isotherm (Equation 4.73) can also readily brought to linear form taking logarithms

$$\log Q_A = \log B_F + n \log a_A \tag{4.75}$$

so that a linear plot of log Q_A versus log θ_A would verify the consistency of data with the Freundlich isotherm. Other equations, such as LF or FG, cannot be treated in the same way, unless the monolayer coverage, Q_S is independently known, because they have three parameters. In those cases, one can resort to nonlinear least squares fitting, which, despite the wide availability of software tools, is a nontrivial matter. We will not treat this topic here; the reader is referred to the literature (Draper and Smith 1998; Seber and Wild 2003; Bates and Watts 2007).

In Figure 4.29 an example of curve fitting is shown. Here, adsorption data were simulated using the Tóth equation (Equation 4.74) with the addition of random deviations around ±8%, which is a reasonable estimation for this type of measurements. The parameter values were (in arbitrary units) $Q_S = 3.0$, $K_A = 0.50$, and $n = 0.80$. The results of fitting these data to the Langmuir, LF, FG, and Töth isotherms are shown in Figure 4.29 and Table 4.3. As it is clearly seen, all four equations fit equally well; moreover, whereas for the last three equations, the fitting parameters have a reasonable agreement with the original ones, the Langmuir equation gives clearly different results. In Figure 4.30, a similar procedure is done with actual data, for

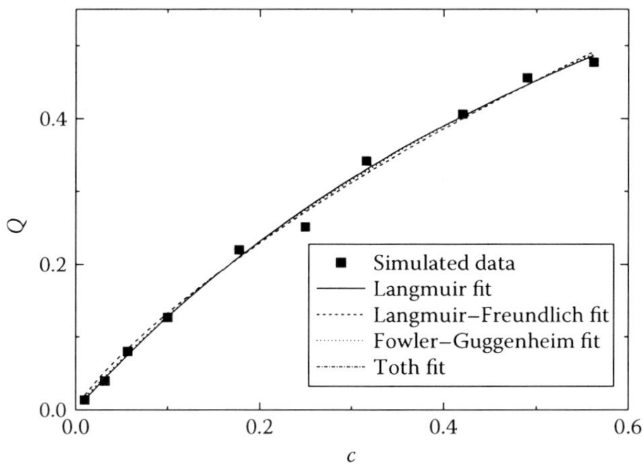

FIGURE 4.29 Fitting of different isotherms to simulated data, generated with the Tóth isotherm with random deviations added.

TABLE 4.3
Parameter Results for Fitting of Simulated Data

Isotherm	Q_S	K_A	Other	R^2
Langmuir	1.24	1.15	N/A	0.996
Langmuir–Freundlich	3.1	0.24	$n = 0.83$	0.994
Fowler–Guggenheim	3.0	0.49	$B_{FG} = 2.1$	0.995
Tóth	3.2	0.53	$n = 0.59$	0.995

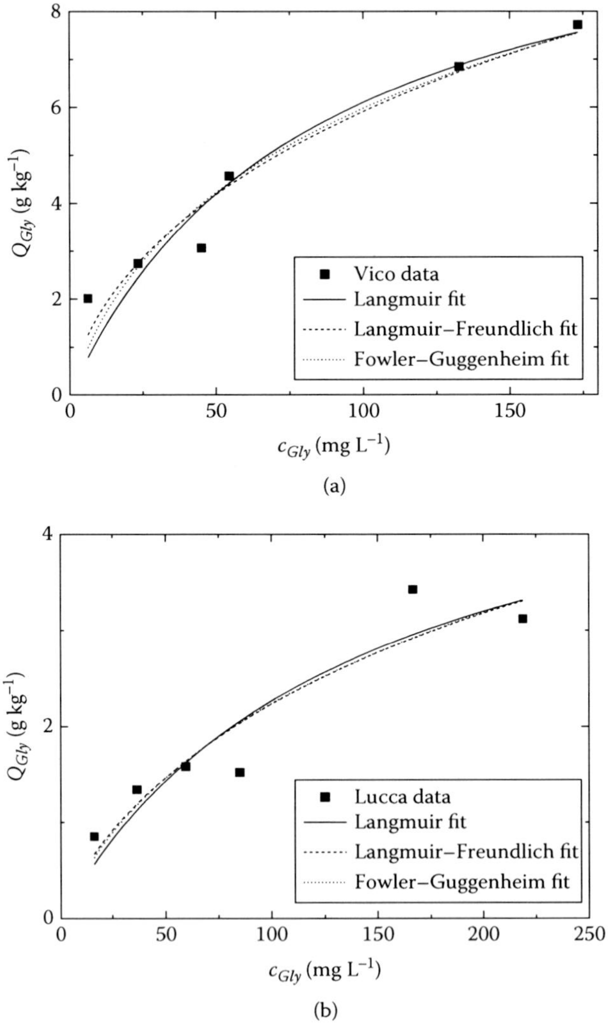

FIGURE 4.30 Fitting of different isotherms to data for glyphosate sorption on humic acids extracted from Italian soils: (a) from a volcanic soil near lake Vico (near Rome) and (b) a peat soil near Lucca. In 0.01 M $CaCl_2$, pH = 3. (Adapted with permission from Piccolo, A. et al., 1996, *Journal of Agricultural and Food Chemistry*, 44, no. 8, 2442–2446. Copyright 1996 American Chemical Society.)

TABLE 4.4
Parameter Results for Fitting of Glyphosate Sorption to Humic Acids

Isotherm	Q_S/g (kg^{-1})	K_A/L (mg^{-1})	Other	R^2
		Vico humic acid		
Langmuir	11	0.012	N/A	0.89
Langmuir–Freundlich	21	2.3×10^{-3}	$n = 0.66$	0.93
Fowler–Guggenheim	23	8.0×10^{-3}	$B_{FG} = 3.4$	0.88
		Lucca humic acid		
Langmuir	5.4	7.2×10^{-3}	N/A	0.84
Langmuir–Freundlich	7.2	3.7×10^{-3}	$n = 0.81$	0.86
Fowler–Guggenheim	12	4.0×10^{-3}	$B_{FG} = 2.9$	0.80

glyphosate sorption on humic acids from Italian soils (Piccolo, Celano, and Conte 1996); Table 4.4 summarizes the results. Here, the data have a relatively high dispersion, and the fitting quality is obviously poorer but is rather similar for the three isotherms tested.

Thus, returning to the initial question about which isotherm should be assigned, there is no simple answer. On the basis of fitting results, a wise election is to choose the simplest equation (the one with lesser number of adjustable parameters) giving a good fit, which indicates the Langmuir isotherm in both examples. However, in the case of Figure 4.29, one would expect to recover the Tóth isotherm, as the data were generated with it; moreover, the adsorption parameters yielded by the fitting to the Langmuir equation are quite different from those originally used to generate the data. That discrepancy is because the fitting parameters are valid *in the range of the measured data*, that is, in the concentration range studied. It is always advisable to cover a concentration range as wide as possible, trying to cover several orders of magnitude, if it can be done, in a logarithmic scale. This is connected with the fact that logarithmic concentration (or activity) plots are more informative than linear plots. Turning to the case in Figure 4.30, dispersion is difficult to decide due to the data, although the LF isotherm appears to be slightly better. An important consideration should always be the physical nature of the system under study. In this case, at the experiment pH (3.0), glyphosate will be partially ionized to anion form, thus repulsive interactions might be present; this would give some support to choosing the FG isotherm, which explicitly includes interactions, in this case repulsive (due to $B_{FG} > 0$) as expected. However, to give sound conclusions, additional independent experimental information (mainly spectroscopic) should be obtained. It should be noted that in the cases examined, and many others in the literature, even for sorption in complex systems such as soils, the Langmuir isotherm fits the data well. That may lead to consideration of a relatively simple behavior; however, the range of sorbate concentration examined is often limited (even less than one order of magnitude), and so only part of the problem is examined. For example, in the case shown in Figure 4.12, approximately six orders of magnitude are covered, observing two Langmuir-type sites; if the experiments had covered, say, for 10^{-2}–10^{-4} M, only one site would respond, and data could have been fitted by a single-site isotherm.

4.7 SUMMARY

In this chapter, the phenomenon of sorption, mainly in a phenomenological way, has been examined. The concept of adsorption isotherm and its different types have been described. Some important isotherm equations, such as Langmuir, LF, and FG, have been considered and studied. A basic discussion of the problem of isotherm assignment has been presented. In all this, we have not made any distinction between charged or uncharged adsorbates. In the specific case of ion adsorption, consideration to the electrostatic problem should be done to attain a better understanding of the process. In the next chapter, we will join the study of the electrified interface (Chapter 3) with the present chapter, and will introduce the different contributions to the surface charge. That will allow us to develop the concepts regarding the important issues of ion sorption and exchange. Afterward, in Part III the more advanced models of ion binding to soil colloids will be developed.

REFERENCES

Adamson, A. W., and A. P. Gast. 1997. *Physical Chemistry of Surfaces*. 6th ed. Wiley-Interscience.

Aharonson, N., and U. Kafkafi. 1975. Adsorption of benzimidazole fungicides on montmorillonite and kaolinite clay surfaces. *Journal of Agricultural and Food Chemistry* 23, no. 3: 434–437. doi:10.1021/jf60199a069.

Aranovich, G. L., and M. D. Donohue. 1996. Universal singularities of multilayer adsorption isotherms and determination of surface area of adsorbents. *Journal of Colloid and Interface Science* 181, no. 1: 313–318. doi:10.1006/jcis.1996.0383.

Atkins, P., and J. de Paula. 2009. *Physical Chemistry*. 9th ed. New York: W. H. Freeman.

Avgul, N. N., and A. V. Kiselev. 1958. Energy of adsorption forces and heat of adsorption of simple molecules on graphite. *Bulletin of the Academy of Sciences of the USSR Division of Chemical Science* 6, no. 2: 241–243. doi:10.1007/BF01170560.

Bartell, F. E., and C. K. Sloan. 1929. Adsorption by pure carbon from non-aqueous binary systems over the entire concentration range. *Journal of the American Chemical Society* 51, no. 6: 1643–1656. doi:10.1021/ja01381a004.

Bates, D. M., and D. G. Watts. 2007. *Nonlinear Regression Analysis and Its Applications*. Hoboken, NJ: Wiley-Interscience.

Berry, R. S., S. A. Rice, and J. Ross. 2000. *Physical Chemistry*. 2nd ed. New York: Oxford University Press.

Bradl, H. B. 2004. Adsorption of heavy metal ions on soils and soils constituents. *Journal of Colloid and Interface Science* 277, no. 1: 1–18. doi:10.1016/j.jcis.2004.04.005.

Brito, V., and M. Ralek. 1978. Calorimetric determination of the heat of adsorption of nitrogen on a technical iron catalyst for ammonia synthesis at 400°C. *Reaction Kinetics and Catalysis Letters* 9, no. 1: 15–21. doi:10.1007/BF02070362.

Brownawell, B. J., H. Chen, J. M. Collier, and J. C. Westall. 1990. Adsorption of organic cations to natural materials. *Environmental Science & Technology* 24, no. 8: 1234–1241. doi:10.1021/es00078a011.

Brunauer, S., L. S. Deming, W. E. Deming, and E. Teller. 1940. On a theory of the van der Waals adsorption of gases. *Journal of the American Chemical Society* 62, no. 7: 1723–1732. doi:10.1021/ja01864a025.

Brunauer, S., P. H. Emmett, and E. Teller. 1938. Adsorption of gases in multimolecular layers. *Journal of the American Chemical Society* 60, no. 2: 309–319. doi:10.1021/ja01269a023.

Casimir, H. B. G., and D. Polder. 1948. The influence of retardation on the London-van der Waals forces. *Physical Review* 73, no. 4: 360. doi:10.1103/PhysRev.73.360.

Coolidge, A. S. 1927. The adsorption of water vapor by charcoal. *Journal of the American Chemical Society* 49, no. 3: 708–721. doi:10.1021/ja01402a014.

Delahay, P. 1965. *Double Layer and Electrode Kinetics*. New York: Interscience Publishers.

Dixit, S., and J. G. Hering. 2003. Comparison of arsenic(V) and arsenic(III) sorption onto iron oxide minerals: Implications for arsenic mobility. *Environmental Science & Technology* 37, no. 18: 4182–4189. doi:10.1021/es030309t.

Donohue, M. D., and G. L. Aranovich. 1998. Classification of Gibbs adsorption isotherms. *Advances in Colloid and Interface Science* 76–77: 137–152. doi:10.1016/S0001-8686(98)00044-X.

Draper, N. R., and H. Smith. 1998. *Applied Regression Analysis*. 3rd ed. Hoboken, NJ: Wiley-Interscience.

Duval, J. F. L., K. J. Wilkinson, H. P. Van Leeuwen, and J. Buffle. 2005. Humic substances are soft and permeable: Evidence from their electrophoretic mobilities. *Environmental Science and Technology* 39, no. 17: 6435–6445.

Emmett, P. H., and S. Brunauer. 1934. The adsorption of nitrogen by iron synthetic ammonia catalysts. *Journal of the American Chemical Society* 56, no. 1: 35–41. doi:10.1021/ja01316a011.

Flogeac, K., E. Guillon, and M. Aplincourt. 2004. Surface complexation of copper(II) on soil particles: EPR and XAFS studies. *Environmental Science & Technology* 38, no. 11: 3098–3103. doi:10.1021/es049973f.

Fowler, R. H., and E. A. Guggenheim. 1939. *Statistical Thermodynamics*. Cambridge, UK: Cambridge University Press.

Freundlich, H. 1926. *Colloid and Capillary Chemistry*. 3rd ed. New York: E. P. Dutton and Company.

Giles, C. H., A. P. D'Silva, and I. A. Easton. 1974. A general treatment and classification of the solute adsorption isotherm. Part II: Experimental interpretation. *Journal of Colloid and Interface Science* 47, no. 3: 766–778. doi:10.1016/0021-9797(74)90253-7.

Giles, C. H., D. Smith, and A. Huitson. 1974. A general treatment and classification of the solute adsorption isotherm. I. Theoretical. *Journal of Colloid and Interface Science* 47, no. 3: 755–765. doi:10.1016/0021-9797(74)90252-5.

Gu, B., J. Schmitt, Z. Chen, L. Liang, and J. F. McCarthy. 1994. Adsorption and desorption of natural organic matter on iron oxide: Mechanisms and models. *Environmental Science & Technology* 28, no. 1: 38–46. doi:10.1021/es00050a007.

Hiemenz, P. C., and R. Rajagopalan. 1997. *Principles of Colloid and Surface Chemistry*. 3rd ed. New York: Marcel Dekker.

Hill, T. L. 1955. Corresponding states in multilayer step adsorption. *The Journal of Physical Chemistry* 59, no. 10: 1065–1067. doi:10.1021/j150532a017.

Hill, T. L. 1986. *An Introduction to Statistical Thermodynamics*. Dover: Courier Dover Publications.

Hinz, C. 2001. Description of sorption data with isotherm equations. *Geoderma* 99, no. 3–4: 225–243. doi:10.1016/S0016-7061(00)00071-9.

Iglesias, H. A., and J. Chirife. 1982. *Handbook of Food Isotherms: Water Sorption Parameters for Food and Food Components*. New York: Academic Press.

Israelachvili, J. N. 2010. *Intermolecular and Surface Forces*. 3rd ed. Amsterdam: Academic Press.

Jaroniec, M., and A. W. Marczewski. 1984. Physical adsorption of gases on energetically heterogeneous solids I. Generalized Langmuir equation and its energy distribution. *Monatshefte für Chemie–Chemical Monthly* 115, no. 8–9: 997–1012. doi:10.1007/BF00798768.

Jaroniec, M., and J. Töth. 1976. Adsorption of gas mixtures on heterogeneous solid surfaces: I. Extension of Tóth isotherm on adsorption from gas mixtures. *Colloid and Polymer Science [Kolloid Zeitschrift & Zeitschrift für Polymere]* 254, no. 7: 643–649. doi:10.1007/BF01753693.

Kavtaradze, N. N., and E. A. Zelyaeva. 1970. Adsorption of nitrogen on iron and manganese. *Bulletin of the Academy of Sciences of the USSR Division of Chemical Science* 19, no. 11: 2465–2467. doi:10.1007/BF00859103.

Kinniburgh, D. G. 1986. General purpose adsorption isotherms. *Environmental Science & Technology* 20, no. 9: 895–904. doi:10.1021/es00151a008.

Kuo, Y.-C., and J.-P. Hsu. 1999. Double-layer properties of an ion-penetrable charged membrane: Effect of sizes of charged species. *Journal of Physical Chemistry B* 103, no. 44: 9743–9748.

Lambert, B., and A. M. Clark. 1929. Studies of gas-solid equilibria. Part II. Pressure-concentration equilibria between benzene and (a) ferric oxide gel, (b) silica gel, directly determined under isothermal conditions. *Proceedings of the Royal Society A* 122: 497–512. doi:10.1098/rspa.1929.0036.

Langmuir, I. 1916. The constitution and fundamental properties of solids and liquids. Part I. solids. *Journal of the American Chemical Society* 38, no. 11: 2221–2295. doi:10.1021/ja02268a002.

Langmuir, I. 1918. The adsorption of gases on plane surfaces of glass, mica and platinum. *Journal of the American Chemical Society* 40, no. 9: 1361–1403. doi:10.1021/ja02242a004.

Levine, I. 2008. *Physical Chemistry*. 6th ed. New York: McGraw-Hill Science/Engineering/Math.

Lyklema, J. 1991. *Fundamentals of Interface and Colloid Science*. Amsterdam: Academic Press.

Mahanty, J., and B. W. Ninham. 1977. *Dispersion Forces*. San Diego, CA: Academic Press.

Manning, B. A., and S. Goldberg. 1997. Arsenic(III) and arsenic(V) adsorption on three California soils. *Soil Science* 162, no. 12: 886–895.

Marczewski, A. W. 2002. A practical guide to isotherms of adsorption on heterogeneous surfaces. http://adsorption.org/awm/index_m.htm.

Marczewski, A. W., and M. Jaroniec. 1983. A new isotherm equation for single-solute adsorption from dilute solutions on energetically heterogeneous solids. *Monatshefte für Chemie—Chemical Monthly* 114, no. 6–7: 711–715. doi:10.1007/BF01134184.

McBride, M. B. 1999. Chemisorption and precipitation reactions. In *Handbook of Soil Science*, ed. M. E. Sumner, B265–B302. 1st ed. Boca raton, FL: CRC Press.

McHaffie, I. R., and S. Lenher. 1925. CCVIII. The adsorption of water from the gas phase on plane surfaces of glass and platinum. *Journal of the Chemical Society, Transactions* 127: 1559. doi:10.1039/ct9252701559.

McLachlan, A. D. 1963a. Retarded dispersion forces between molecules. *Proceedings of the Royal Society of London. Series A. Mathematical and Physical Sciences* 271, no. 1346: 387–401. doi:10.1098/rspa.1963.0025.

McLachlan, A. D. 1963b. Retarded dispersion forces in dielectrics at finite temperatures. *Proceedings of the Royal Society of London. Series A. Mathematical and Physical Sciences* 274, no. 1356: 80–90. doi:10.1098/rspa.1963.0115.

McLachlan, A. D. 1965. Effect of the medium on dispersion forces in liquids. *Discussions of the Faraday Society* 40: 239–245.

McQuarrie, D. A., and J. D. Simon. 1997. *Physical Chemistry: A Molecular Approach*. Herndon, VA: University Science Books.

Ohshima, H., and T. Kondo. 1993. Electrophoretic mobility of a "soft particle" with a non-uniformly charged surface layer as a model for cells. *Biophysical Chemistry* 46, no. 2: 145–152. doi:10.1016/0301-4622(93)85021-9.

Pabst, M. W., C. D. Miller, C. O. Dimkpa, A. J. Anderson, and J. E. McLean. 2010. Defining the surface adsorption and internalization of copper and cadmium in a soil bacterium, Pseudomonas putida. *Chemosphere* 81, no. 7: 904–910. doi:10.1016/j.chemosphere.2010.07.069.

Pagnanelli, F., L. Bornoroni, E. Moscardini, and L. Toro. 2006. Non-electrostatic surface complexation models for protons and lead(II) sorption onto single minerals and their mixture. *Chemosphere* 63, no. 7: 1063–1073. doi:10.1016/j.chemosphere.2005.09.017.

Parsons, R. 1964. The description of adsorption at electrodes. *Journal of Electroanalytical Chemistry (1959)* 7, no. 2: 136–152. doi:10.1016/0022-0728(64)85007-5.

Piccolo, A., G. Celano, and P. Conte. 1996. Adsorption of glyphosate by humic substances. *Journal of Agricultural and Food Chemistry* 44, no. 8: 2442–2446. doi:10.1021/jf950620x.

Reyerson, L. H., and A. E. Cameron. 1935. Studies on the sorption of the halogens by silica gel and charcoal. *Journal of Physical Chemistry* 39, no. 1: 181–190.

Robarge, W. P. 1998. Precipitation/dissolution reactions in soils. In *Soil Physical Chemistry*, ed. D. L. Sparks, 193–238. 2nd ed. Boca Raton, FL: CRC Press.

Rocha, M. S., K. Iha, A. C. Faleiros, E. J. Corat, and M. E. V. Suárez-Iha. 1997. Freundlich's isotherm extended by statistical mechanics. *Journal of Colloid and Interface Science* 185, no. 2: 493–496. doi:10.1006/jcis.1996.4588.

Ross, S., and J. P. Olivier. 1964. *On Physical Adsorption*. New York: Interscience Publishers.

Ruiz, J., R. Bilbao, and M. B. Murillo. 1998. Adsorption of different VOC onto soil minerals from gas phase: Influence of mineral, type of VOC, and air humidity. *Environmental Science & Technology* 32, no. 8: 1079–1084. doi:10.1021/es9704996.

Ruthven, D. M. 1984. *Principles of Adsorption and Adsorption Processes*. Hoboken, NJ: Wiley-Interscience.

Rytwo, G. 2004. Applying a Gouy-Chapman-Stern model for adsorption of organic cations to soils. *Applied Clay Science* 24, no. 3–4: 137–147. doi:10.1016/j.clay.2003.01.001.

Seber, G. A. F., and C. J. Wild. 2003. *Nonlinear Regression*. Hoboken, NJ: Wiley-Interscience.

Shaw, D. 1992. *Introduction to Colloid and Surface Chemistry*. 4th ed. Oxford: Butterworth-Heinemann.

Sing, K. S. W. 1985. Reporting physisorption data for gas/solid systems with special reference to the determination of surface area and porosity (Recommendations 1984). *Pure and Applied Chemistry* 57, no. 4: 603–619. doi:10.1351/pac198557040603.

Sips, R. 1948. On the structure of a catalyst surface. *The Journal of Chemical Physics* 16, no. 5: 490–495. doi:10.1063/1.1746922.

Timmermann, E. O. 2003. Multilayer sorption parameters: BET or GAB values? *Colloids and Surfaces A: Physicochemical and Engineering Aspects* 220, no. 1–3: 235–260.

Tóth, J. 1995. Uniform interpretation of gas/solid adsorption. *Advances in Colloid and Interface Science* 55: 1–239. doi:10.1016/0001-8686(94)00226-3.

Van den Berg, C., and S. Bruin. 1981. Water activity and its estimation in food systems: Theoretical aspects. In *Water Activity and Its Estimation in Food Systems: Theoretical Aspects*, ed. L. B. Rockland and G. F. Stewards, 1–61. New York: Academic Press.

Wang, Z., D. S. Gamble, and C. H. Langford. 1992. Interaction of atrazine with Laurentian soil. *Environmental Science & Technology* 26, no. 3: 560–565. doi:10.1021/es00027a019.

Wang, J., and M. S. Gutierrez. 2010. Molecular structural transformation of 2:1 clay minerals by a constant-pressure molecular dynamics simulation method. *Journal of Nanomaterials* 2010: 1–13. doi:10.1155/2010/795174.

5 Surface Charge of Colloidal Particles

5.1 CONTRIBUTIONS TO SURFACE CHARGE

As it was already stated in Chapter 3, particles suspended in a polar liquid medium such as water show an electrical charge. This charge can be verified and determined by methods such as electrophoretic mobility measurement, discussed in Section 5.4.1. In the following, we discuss how the charge arises from a number of sources, some due to the particle itself and some due to the interaction between the particle and the medium (Hiemenz and Rajagopalan 1997; Sparks 2002; Sposito 2008).

5.1.1 PERMANENT CHARGE

Consider a (hypothetical) simple neutral molecule of silicon dioxide, SiO_2; the Si atom is in its normal +4 oxidation state, whereas the two oxygen atoms are in the –2 state. Now if we replace the Si atom by an Al atom, the species can no longer be neutral, because aluminum has an oxidation state of +3, and we would end up with the aluminate anion AlO_2^{-1}. Now if we consider a (crystalline) solid with the stoichiometry SiO_2 and a *few* Si atoms are replaced with Al, we will get a particle with a fixed negative charge, which must be compensated by cations on the outer side. Likewise, if in a solid with the stoichiometry $Al(OH)_3$ (such as gibbsite or bayerite), some Al^{3+} species are substituted with a +2 cation (such as Mg^{2+} or Fe^{2+}); the solid will have a fixed negative charge. If the number of substituting species is small, the crystalline structure of the unsubstituted solid will remain unchanged, the foreign atoms being forced to adopt the host structure. This situation is termed "isomorphic substitution," and the resulting negative charge of the particle is a fixed, permanent one. It has been termed the "fixed" (Zachara and Westall 1998) or "structural" (Sposito 1999) or "permanent" (Sparks 2002) particle charge q_f. Because there are, in general, particles of many different sizes, the absolute particle charge is not significant (nor, usually, accessible), so that the preferred quantity, on fundamental grounds, is the *fixed or permanent surface charge density* σ_f (C m^{-2}). More frequently, because also the surface area is difficult to determine, the practical quantity is the *fixed or permanent specific particle charge* ϑ_f, (C kg^{-1}), usually also expressed as mole of charge (equivalent) per unit of mass (mol kg^{-1}; some authors employ the notation mol$_c$ kg^{-1}, the subscript *c* indicating charge, which here is not used, as it should be clear from the context); the ratio between the two units is obviously the Faraday constant, $\boldsymbol{F}$. It should be noted that the examples given here are simple, hypothetical cases for illustrative purposes; in practice, this effect is normally found for some layered phyllosilicate minerals. Isomorphic substitution is discussed in detail in Chapter 8.

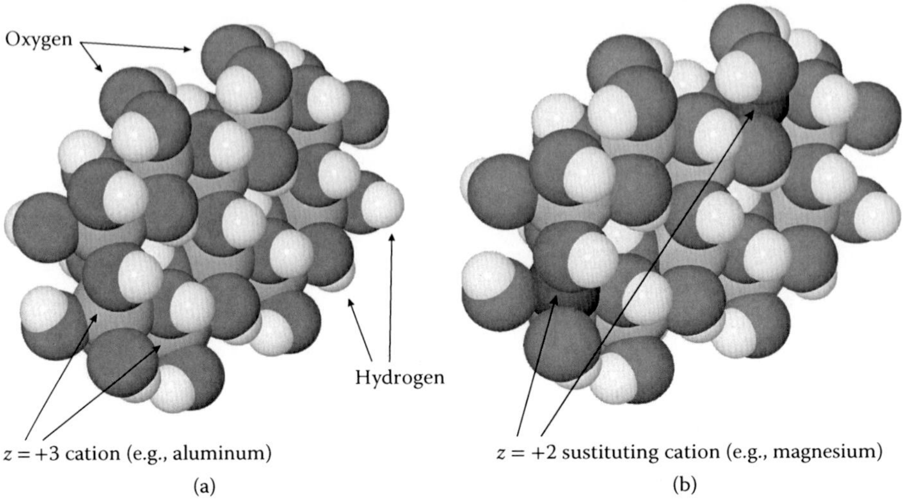

FIGURE 5.1 (**See color insert.**) (a) Molecular model representation (not to scale) of a fragment of a gibbsite-like layer. The light violet spheres represent aluminum atoms, the red ones represent oxygen, and the white ones hydrogen. (b) The blue spheres represent divalent atoms (such as magnesium or iron) substituting trivalent aluminum atoms, resulting in a permanent charge $q_f = -2$.

Figure 5.1 shows a simple example of this; in part, a fragment of a layer with a gibbsite structure is shown (gibbsite crystals are made up of many stacked layers); there are three OH^- (red/white) for each Al^{3+} (light violet), so that the structure is neutral. In Figure 5.1b, two Al^{3+} have been replaced by cations of charge $z = +2$ (such as Mg^{2+}); thus, in this example, we have now a permanent charge (in units of elementary charge) $q_f = -2$.

5.1.2 Protonic and Intrinsic Charge

Most soil colloids (either mineral or organic) have –OH surface groups, capable of getting ionized in contact with water by reactions such as

$$\text{M-OH} + \text{H}^+ \rightleftharpoons \text{M-OH}_2^+ \tag{5.1}$$

$$\text{M-OH} \rightleftharpoons \text{M-O}^- + \text{H}^+ \tag{5.2}$$

Thus, these groups can have either positive or negative charge, depending on medium pH. At a given pH, a number of surface groups of a soil particle can be positively charged, others neutral, and some others negatively charged (because the surface ionizable groups do not necessarily have the same value of equilibrium constant for reactions 5.1 and 5.2). The resulting charge is termed the "net proton surface charge density" σ_H, given by

$$\sigma_H = \boldsymbol{F}\left(\Gamma_H - \Gamma_O\right) \tag{5.3}$$

where Γ_H and Γ_O are the surface concentrations of positively and negatively charged sites (protonated and ionized groups), respectively. Likewise, the *net proton particle charge* q_H is

$$q_H = A_p \boldsymbol{F}(\Gamma_H - \Gamma_O) \tag{5.4}$$

where A_p is the particle area. The net proton-specific charge is

$$\vartheta_H = z_{H^+} Q_H + z_{OH^-} Q_O \tag{5.5}$$

where Q_H and Q_O are the specific amounts of positively and negatively charged groups, the corresponding z being their charges, +1 and −1 respectively. Figure 5.2 shows the structure of Figure 5.1b, where two protons are ionized giving negatively charged groups, so that, in the same units as given in the preceding discussion, in this example, we have $q_H = -2$. Some authors (such as Zachara and Westall 1998; Sposito 1999) group the fixed and protonic charges into an *intrinsic* surface charge density

$$\sigma_{in} = \sigma_f + \sigma_H \tag{5.6}$$

and similarly for q_{in} and ϑ_{in}.

5.1.3 Inner and Outer Sphere Complex Charges

As discussed in Chapter 4, ions bind to colloidal particles in several ways; depending on the plane where the bound species is located, there are inner and outer sphere bound ions. Both types contribute, of course, to the surface charge. Figure 5.3 shows the surface fragment of Figure 5.2 with two cations (assumed to be of $z = +1$, such as K^+) sorbed as inner sphere complexes. The charge thus added to the particle is the *inner sphere sorption particle charge* q_I, whereas the corresponding charge density and specific charge are denoted by σ_I and ϑ_I, respectively. Figure 5.4 shows the same as Figure 5.3 plus one cation (assumed univalent, such as Na^+) sorbed as an outer sphere complex, giving the *outer sphere sorption particle charge* q_O (or σ_O, ϑ_O). In the examples given in the figures, we have $q_I = +2$ and $q_O = +1$.

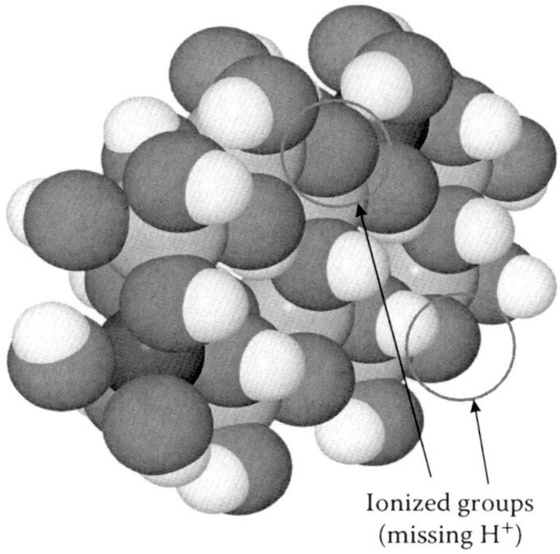

FIGURE 5.2 **(See color insert.)** The gibbsite layer fragment from Figure 5.1b, which has lost two hydrogen atoms (due to acid–base interactions with the surrounding solution) at locations indicated by the green circles, resulting in a net proton charge of $q_H = -2$.

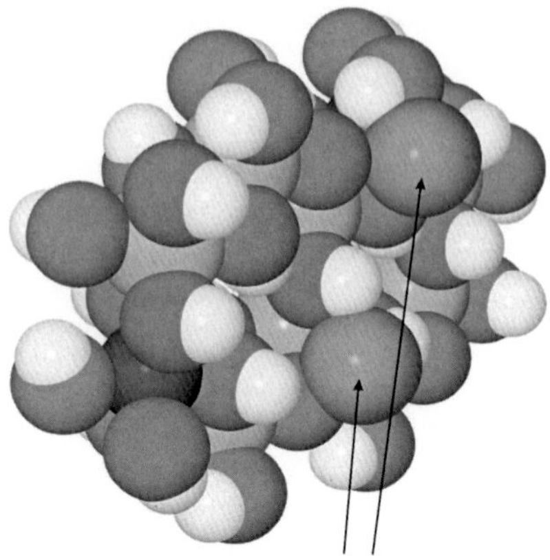

FIGURE 5.3 **(See color insert.)** The gibbsite layer fragment from Figure 5.2 with two inner sphere complexed cations (light violet spheres), assumed of charge +1 each, giving an inner sphere sorption charge $q_I = +2$.

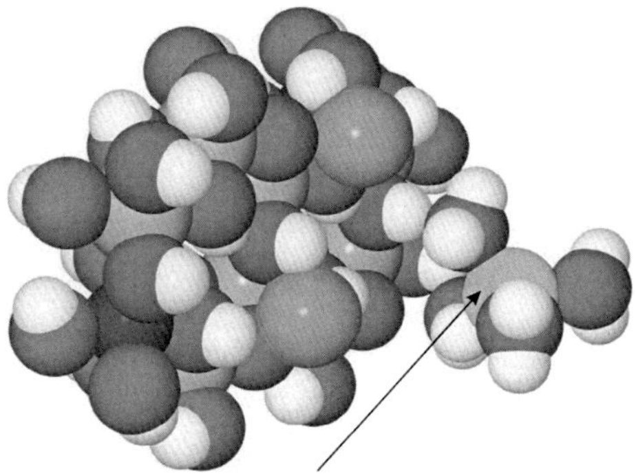

FIGURE 5.4 **(See color insert.)** The gibbsite layer fragment with inner sphere sorption shown in Figure 5.3 with an outer sphere bound cation, represented by a green sphere with some water molecules. Free water molecules are not represented. This cation is assumed to be of charge $z = 1$, giving the outer sphere sorption charge $q_O = 1$.

5.1.4 Particle Charge Balance

The contributions detailed in the preceding discussion combine to give the resulting particle surface charge density σ_p:

$$\sigma_p = \sigma_f + \sigma_H + \sigma_I + \sigma_O \tag{5.7}$$

(Of course, the same for q_p and ϑ_p; we will omit these in the following). In the simple example developed in Figures 5.1 through 5.4, the total charge is

$$q_p = q_f + q_H + q_I + q_O = -2 - 2 + 2 + 1 = -1 \tag{5.8}$$

The particle charge is counterbalanced by the diffuse layer charge, as described in Chapter 3, so that the charge balance is

$$\sigma_p + \sigma_d = \sigma_f + \sigma_H + \sigma_I + \sigma_O + \sigma_d = 0 \tag{5.9}$$

The same is true of for q_p and ϑ_p. Figure 5.5 completes the example developed in the previous figures, showing in the diffuse layer part (beyond the outer sphere complex plane) two cations and one anion (of course, there are also water molecules filling up the space, which are not shown, for clarity), all univalent, thus resulting in a diffuse layer charge $q_d = 1$, compensating the negative net particle charge given by Equation 5.8.

By exchange experiments (see Section 5.3), it is possible to measure the total amount of anions and cations sorbed; thus we can introduce (Chorover and Sposito 1995) the *total adsorbed cation surface charge* σ_+ and the corresponding *total adsorbed anion surface charge* σ_-. The balance between both is the *net adsorbed ion surface charge*, which is also related to other contributions:

$$\sigma_{ad} = \sigma_+ - \sigma_- = \sigma_I + \sigma_O + \sigma_d \tag{5.10}$$

Note that, from Equation 5.10, the charge balance Equation 5.9 can be written as

$$\sigma_{in} + \sigma_{ad} = 0 \tag{5.11}$$

5.1.5 Soil Charge

In practice, it is not very usual to find in a single colloidal substance (either mineral or organic) all the contributions detailed in the preceding discussion (Equation 5.7). Only a particular type of mineral (some phyllosilicates, such as micas or smectites) show significant amounts of fixed charge. In other cases, inner sphere complexes are not important. On the contrary, almost all soil colloids (and most other colloids

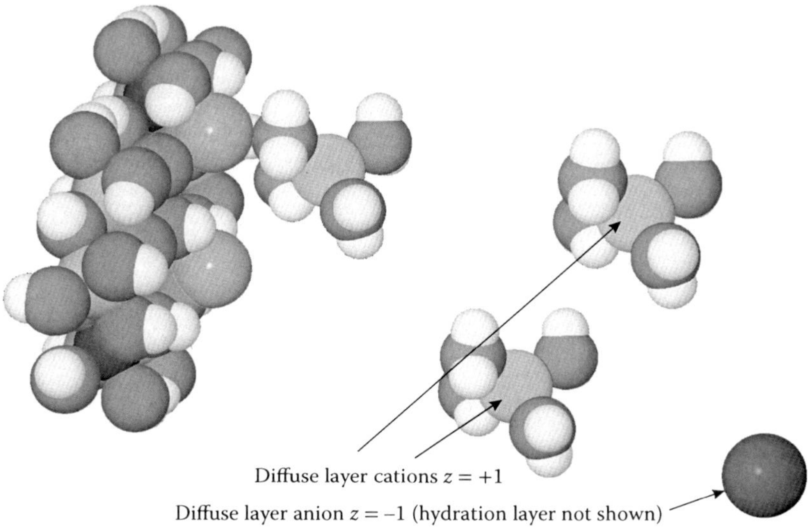

FIGURE 5.5 (**See color insert.**) The example given in Figure 5.4 is completed with the diffuse layer charge $q_d = +1$, resulting in this example from two cations and one anion (dark purple, extreme right). Solvent water molecules are not shown.

as well) develop protonic charge—that is, show a pH-dependent intrinsic charge—but this contribution is more important for oxides and organic matter than for clay minerals. Also (except at the particular pH where σ_p happens to become zero), all have a diffuse layer in the solution side, compensating the particle charge. The different cases will be discussed in Part II, but we will note here that in soils there is a net charge resulting from the contribution of the different components present. The behavior of the whole soil in general depends, of course, on the superposition and mutual interaction of the individual components, behavior, so that one can expect a wide range of soil charge behaviors. It happens, however, that most soils can be classified into two broad types: *constant charge soils* and *variable charge soils* (Yu 1997; Singh and Uehara 1998). Constant charge soils are found mainly in temperate regions and are dominated by clay minerals having an important amount of permanent charge; thus, they show little response to pH changes. It should be noted that a *strictly* constant charge soil does not exist, but many show an approximately constant charge in certain pH range (usually at low pHs). Variable charge soils are those dominated by oxides and organic matter or clay minerals with low fixed charge such as kaolinite and are found mainly in tropical or subtropical regions. Evidently, both types contain components with both fixed and variable charges, so there is no clear distinction between them.

5.2 ION SORPTION AND SURFACE CHARGE DEVELOPMENT

Recalling the contributions to the particle charge in Equation 5.7, the mechanisms leading to the surface charge development of soil colloidal particles can be understood. The permanent surface charge σ_f is due to isomorphic substitution, which is produced at the time of particle formation and growth and cannot be changed by the external medium in the soil. Particles having only this contribution to the intrinsic charge (Equation 5.6) will have permanent constant charge compensated by inner sphere and/or outer sphere sorbed counterions plus the diffuse layer charge. However, almost all types of soil colloids have, to some extent, pH-dependent proton charge, arising from acid–base reactions (Equations 5.1 and 5.2); once the equilibrium is established, the protonic charge is defined; the modeling of these processes will be treated in Part III. Once the proton charge is established, from it and the permanent charge (if any), the intrinsic charge is determined. On top, further ion sorption may take place, as shown in Figures 5.3 and 5.4 as inner and outer sphere complexes. Inner sphere sorption may be due, but not necessarily, to specific adsorption, as discussed in Chapter 4. For example, phosphate ions sorb on the surface of oxides such as goethite by a ligand exchange mechanism, clearly resulting in specific adsorption. On the contrary, potassium ions sorb in the interlayer of mica clays (Chapter 8) as inner sphere complexes, but no chemical bond is formed, thus constituting nonspecific adsorption. The importance of the difference is that specific adsorption, being a chemical reaction, can cause a reversal of the surface charge sign, whereas nonspecific adsorption, being essentially driven by electrostatic forces, cannot overturn the intrinsic charge. When no specific adsorption is present, the final state of the interface is determined by the intrinsic charge, which in turn is fixed by the medium pH (except in rare cases). If the surface charge is determined, so is the surface potential, and it will decay more or less pronouncedly toward the bulk solution. Thus, it is said that the proton is the *potential determining ion* (PDI), because the surface potential value arises from the acid–base reactions (Equations 5.1 and 5.2). In cases where strong specific adsorption exists, the adsorbing ion may become the PDI, but in most cases, H^+ is the PDI. Ions sorbing as inner and outer sphere complexes are also under electrostatic effect, but different from proton binding/dissociation.

Figure 5.6 illustrates schematically the situation, similar to that exemplified in Figure 5.4. The surface plane (assumed to be located at x_0) is where the protonic charge, due to either binding or dissociation of H^+ ions, is assumed to lie. The inner plane (at x_1) is where the inner sphere complexes reside; these can be due to specific

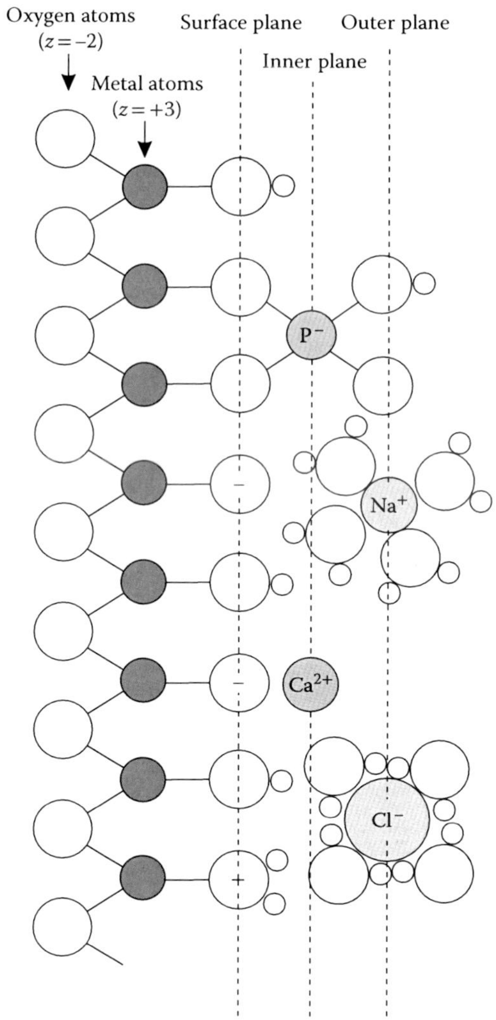

FIGURE 5.6 Schematic representation of the solid–solution interface, indicating the different planes where the contributions to the surface charge are assumed to be located.

adsorption, such as the phosphate ion shown near the top, or strong electrostatic interaction, like the Ca^{2+} ion near the center. Note that the charge shown in the figure for phosphate (–1) is a net charge, due to dissociation of an external OH group. Beyond the outer plane, the diffuse layer region begins. Note that the inner and outer sphere planes are equivalent to the inner and outer Helmholtz planes (IHPs and OHPs) in the Gouy–Chapman–Stern–Grahame model (Chapter 3).

It should be remarked now that the arrangement of charges on "planes" is a simplification. The planes are assumed as if charges were averaged and "smeared out" over the plane. However, the true nature of the interface is a *discrete* charge distribution; not only ions are actually more like localized charges, but also there is no guarantee that they will all be located at the same distance from the surface. Solving the electrostatic problem of a discrete charge distribution, combined with chemical reactions of proton and specific ion adsorption, would be a very difficult task; moreover, as discussed in Chapter 3, most probably it should be solved by lengthy numeric procedures. Thus, all current models employ, in one form or another, the concept of planes where the charge contributions are located. Depending on different cases, several models have been proposed, which will be discussed in Part III.

5.3 ION EXCHANGE

Ion exchange is a process involving essentially nonspecifically sorbed ions. Consider the ions at and beyond the outer plane. It is clear that they are bound only for charge compensation, and any ions of the same sign will do the same job. Thus, if in a suspension of colloidal particles the solution phase of, for example, NaCl is replaced by another one of KNO_3, then in the interfacial region of the particles, sodium ions will be replaced by potassium ones and chloride by nitrate (Helfferich 1995; Sposito 1999; Sparks 2002; Evangelou and Phillips 2005). These processes can be written as chemical reactions such as

$$S^-Na^+ + K^+(aq) \rightleftharpoons S^-K^+ + Na^+(aq) \tag{5.12}$$

$$S^+Cl^- + NO_3^-(aq) \rightleftharpoons S^+NO_3^- + Cl^-(aq) \tag{5.13}$$

where S^- and S^+ represent positively and negatively charged sites at the solid particle (termed the *exchanger phase*). Thus, ion exchange is the replacement of sorbed ions by others with charge of the same sign. Despite what may be implied by the equations, it is worth to remark that the ions need not be fixed where the opposite charges are localized (recall the discussion about the diffuse layer in Chapter 3); merely there should be equal number of charges of each sign, but on average, they will be close to the opposite fixed charges, of course. Also, in many cases, one does not have positively and negatively charged sites at a solid (but there are cases where this happens, see Chapter 8); usually, soil particles and soils as a whole are negatively charged (often there are sites of both signs, but those negatively charged exceed largely the positive ones) and act as cation exchangers, so in the following, we will refer mainly to this case. Of course, for anion exchange, the same concepts are applicable; often there are, at different places, also positively charged sites, so there is anion exchange behavior coexisting with the cation exchange.

The retention of ions by exchange can be studied as any other sorption process, and a saturation value (in the general case dependent on pH) is always observed; usually, the Langmuir isotherm describes the process well. The maximum amount of monovalent ions (i.e., equivalents of charge) of a given sign that can be retained is the *exchange capacity*; thus, there is a *cation exchange capacity*, ϑ_{CE} (or CEC, as it is frequently symbolized) and an *anion exchange capacity*, ϑ_{AE} (or AEC); as almost always, they are expressed as moles of charge (equivalents) per unit of exchanger mass (mol kg^{-1}); we will employ consistently the symbols given. Most soils have an important ϑ_{CE}, whereas ϑ_{AE}, if present, is generally lower than the cationic one. Both ion exchange capacities are in general pH-dependent, because many soil colloids have acid–base reactive groups that cause such dependence. As discussed in Section 5.1.5, soils from temperate regions formed by minerals with an important amount of fixed negative charge will be constant charge soils and thus have constant ϑ_{CE}; soils with high contents of organic matter and/or oxides from tropical or subtropical areas will have a charge strongly dependent on pH, and the same is true for ϑ_{CE}. ϑ_{AE}, if present, always is pH-dependent. It should be noted that constant charge soils have truly constant charge only in a limited pH, range; above a given pH value (and sometimes below another threshold value), the charge becomes pH-dependent. It is interesting to note that the charge variability of soils was not truly realized until the 1960s, even when it was discovered as early as 1931 (Mattson 1931). Sparks (2006) has given an interesting account of the progress on this matter.

The ion exchange capacity (either cationic or anionic) can be measured by various methods (Sumner and Miller 1996; Zelazny, He, and Vanwormhoudt 1996); one of the most commonly employed is the method of the *index ion* that consists in saturating the soil colloid (or a whole soil sample) with a salt of a given ion (the index ion, typically Ba^{2+}); analysis of the displaced solution can be performed to find the composition of the sorbed cations. Next, the sample is treated with $MgSO_4$, which precipitates $BaSO_4$ and forces the exchange of Ba^{2+} by Mg^{2+}; the amount of Mg^{2+} consumed in the process gives ϑ_{CE}. A similar procedure can be employed for ϑ_{AE}, or it can be estimated in the cation experiment (Sumner and Miller 1996).

5.3.1 Thermodynamics of Ion Exchange: Vanselow Convention

Consider the following general equation for the exchange of cations A^{u+} and B^{v+}

$$vS_uA + uB^{v+} \rightleftharpoons uS_vB + vA^{u+} \tag{5.14}$$

where S represents sites with a single negative charge, and u and v are the stoichiometric coefficients resulting from electroneutrality. A general expression for the thermodynamic equilibrium is

$$K_{ex} = \frac{\left(a_A^W\right)^v \left(a_B^S\right)^u}{\left(a_A^S\right)^v \left(a_B^W\right)^u} \tag{5.15}$$

where K_{ex} is the thermodynamic exchange equilibrium constant; the superscripts W and S refer to the solution (water) and solid (exchanger) phases, respectively. Now the meaning of the activities has to be considered; in an aqueous phase, they are well known, being expressed in terms of molalities (or molarities) and the respective activity coefficients (see Equations 2.6 and also 3.2). In the solid (exchanger) phase, there are several choices; the Vanselow convention is the same choice as for heterogeneous bulk equilibria, being the use of mole fractions and Raoult's activity coefficients (McQuarrie and Simon 1997; Berry, Rice, and Ross 2000; Levine 2008; Atkins and Paula 2009), as first proposed by Vanselow (1932):

$$a_i^S = x_i \gamma_i^R \tag{5.16}$$

Thus, Equation 5.15 becomes

$$K_{ex} = \frac{\left(m_A \gamma_A^m\right)^v \left(x_B \gamma_B^R\right)^u}{\left(x_A \gamma_A^R\right)^v \left(m_B \gamma_B^m\right)^u} \approx \frac{\left(c_A \gamma_A^m\right)^v \left(x_B \gamma_B^R\right)^u}{\left(x_A \gamma_A^R\right)^v \left(c_B \gamma_B^m\right)^u} \tag{5.17}$$

where γ^m means molality scale activity coefficient (Chapter 2), where we have omitted the superscripts S and W for clarity. A difficulty arises here: whereas the activity coefficients in aqueous solution can be computed with good approximation using several expressions (Sparks 2002; Evangelou and Phillips 2005; Levine 2008; Atkins and Paula 2009) derived from Debye–Hückel theory (see also Chapter 3), the activity coefficients for the exchanger phase are more difficult to deal with. At present, there is no predictive theory allowing its calculation in advance. Thus, several alternative expressions using conditional constants, known as *selectivity coefficients*, have been proposed. Vanselow (1932) used an expression derived from Equation 5.17 but ignoring the soil phase activity coefficients:

$$K_V = \frac{\left(m_A \gamma_A^m\right)^v x_B^u}{\left(m_B \gamma_B^m\right)^u x_A^v} \approx \frac{\left(c_A \gamma_A^m\right)^v x_B^u}{\left(c_B \gamma_B^m\right)^u x_A^v} \tag{5.18}$$

where K_V stands for Vanselow selectivity coefficient. Clearly, K_V is related to the thermodynamic constant by

$$K_{ex} = K_V \frac{\left(\gamma_B^R\right)^u}{\left(\gamma_A^R\right)^v} \tag{5.19}$$

Based on Equation 5.19, a method to obtain exchanger phase activity coefficients was proposed as follows (Argersinger, Davidson, and Bonner 1950): we can rewrite it as

$$\ln K_{ex} - \ln K_V = u \ln \gamma_B^R - v \ln \gamma_A^R \tag{5.20}$$

and differentiating

$$-d \ln K_V = u d \ln \gamma_B^R - v d \ln \gamma_A^R \tag{5.21}$$

Now, derived from the second law of thermodynamics, the Gibbs–Duhem equation relates the activities of the components in a phase (McQuarrie and Simon 1997; Berry, Rice, and Ross 2000; Levine 2008; Atkins and Paula 2009). At constant temperature and pressure, for a two-component system, it reads

$$x_A d \ln \gamma_A^R + x_B d \ln \gamma_B^R = 0 \tag{5.22}$$

From Equations 5.21 and 5.22, we get

$$d\ln\gamma_A^R = \frac{1}{v}\left(\frac{vx_B}{ux_A + vx_B}\right)d\ln K_V \tag{5.23}$$

$$d\ln\gamma_B^R = -\frac{1}{u}\left(\frac{ux_A}{ux_A + vx_B}\right)d\ln K_V \tag{5.24}$$

Now, we will introduce the *equivalent fraction* of an ionic species i in a multicomponent ionic system as

$$y_i = \frac{z_i n_i}{\sum_j z_j n_j} = \frac{z_i c_i}{\sum_j z_j c_j} = \frac{z_i x_i}{\sum_j z_j x_j} \tag{5.25}$$

where the sum is carried by all ions of the same sign as ion i. Note that this is the fraction of the total charge (either positive or negative) carried by ions of species i. Note also that for a solution composed of all univalent ions, $y_i = x_i$. With this definition, Equations 5.23 and 5.24 result in

$$d\ln\gamma_A^R = \frac{y_B}{v}d\ln K_V \tag{5.26}$$

$$d\ln\gamma_B^R = -\frac{1-y_B}{u}d\ln K_V \tag{5.27}$$

Integrating Equations 5.26 and 5.27 and replacing in 5.20, it is found that

$$\ln K_{ex} = \int_{y_A=0}^{y_A=1} \ln K_V \, dy_A \tag{5.28}$$

Thus, measuring K_V for different y_A values and numerically integrating Equation 5.28, one can obtain the thermodynamic exchange equilibrium constant K_{ex}. Also, numerically integrating Equations 5.26 and 5.27, the exchanger phase activity coefficients for a given composition (y_A, y_B) can be obtained (Bond 1995).

$$\ln\gamma_A^R(y_B) = \frac{y_B}{v}\ln K_V(y_B) - \frac{1}{v}\int_0^{y_B} \ln K_V(y'_B)\mathrm{d}y'_B \tag{5.29}$$

$$\ln\gamma_B^R(y_A) = -\frac{y_A}{u}\ln K_V(y_A) + \frac{1}{u}\int_{y_B}^{1} \ln K_V(y'_B)\mathrm{d}y'_B \tag{5.30}$$

Determining the thermodynamic equilibrium constant at different temperatures enables obtaining other thermodynamic quantities through well-known relations such as

$$\Delta G_{ex}^0 = -\boldsymbol{R}T\ln K_{ex} \tag{5.31}$$

$$\left(\frac{\partial \ln K_{ex}}{\partial T}\right)_{P,n_i} = \frac{\Delta H_{ex}^0}{\boldsymbol{R}T^2} \tag{5.32}$$

$$\Delta G_{ex}^0 = \Delta H_{ex}^0 - T\Delta S_{ex}^0 \tag{5.33}$$

where ΔH_{ex}^0 and ΔS_{ex}^0 are, respectively, the standard enthalpy and entropy changes upon ion exchange.

5.3.2 Other Conventions and Selectivity Coefficient Expressions

In the Vanselow approach, the selectivity coefficient (conditional constant) given by Equation 5.18 is used, which effectively assumes ideality in the exchanger phase, in an attempt to overcome the lack of ability to obtain the corresponding activity coefficients. However, Vanselow selectivity coefficients show a dependence on the composition, especially for heterovalent exchange (i.e., exchange between ions of different charge, such as Na^+–Ca^{2+}). Other forms of selectivity coefficients have been proposed, as detailed in the following discussion. Table 5.1 summarizes these coefficients.

5.3.2.1 Kerr Convention

The Kerr convention was actually the first chronologically (Sparks 2002; Evangelou and Phillips 2005), and followed what is customary for homogeneous diluted aqueous phase reactions, using concentrations for activities in solution and sorbed amounts in the solid phase; the Kerr conditional constant is thus

$$K_K = \frac{c_A^v Q_B^u}{c_B^u Q_A^v} \tag{5.34}$$

In this convention, the relation to the thermodynamic activity is

$$a_i^S = Q_i \gamma_i^Q \tag{5.35}$$

where γ_i^Q is the exchanger phase activity coefficient in the sorbed amount scale. In Table 5.1, the relation between K_K and K_{ex} is shown. In most cases, Kerr selectivity coefficients show dependence on composition (Sparks 2002).

5.3.2.2 Gaines and Thomas Convention

An alternative proposal to that of Vanselow was the use of equivalent fractions (Equation 5.25) instead of mole fractions to express exchanger phase composition (Gaines and Thomas 1953); from a thermodynamic point of view, the use of equivalent instead of moles is questionable, because the physical entities are ions and not charges. However, it has been reported that, at least for some cases, thermodynamics parameters deduced using the Gaines and Thomas (GT) convention are coincident with those computed using the Vanselow convention (Ogwada and Sparks 1986). In the GT convention, the activity in the exchanger phase is

$$a_i^S = y_i^S \gamma_i^y \tag{5.36}$$

where γ_i^y is the activity coefficient in the equivalent fraction scale^y. The GT selectivity coefficient is

$$K_{GT} = \frac{y_B^v \left(c_A \gamma_A^m\right)^u}{\left(c_B \gamma_B^m\right)^v y_A^u} \tag{5.37}$$

TABLE 5.1
Ion Exchange Activity Conventions and Binary Exchange Conditional Constants[a]

Name	Activity Convention[b] Aqueous phase $a_i^W =$	Activity Convention[b] Exchanger phase $a_i^S =$	Relation to Equilibrium Constant[c]	Conditional Constants or Selectivity Coefficients: 1:1 exchange	Conditional Constants or Selectivity Coefficients: 1:2 exchange[d]	References
Vanselow	$c_i\gamma_i^m$	x_i	$K_{ex} = K_V \dfrac{(\gamma_B^R)^u}{(\gamma_A^R)^v}$	$K_V = \dfrac{x_B c_A \gamma_A^m}{c_B \gamma_B^m x_A}$	$K_V = \dfrac{x_B^2 c_A \gamma_A^m}{(c_B \gamma_B^m)^2 x_A}$	Vanselow (1932) and Argersinger, Davidson, and Bonner (1950)
Kerr	c_i	Q_i	$K_{ex} = K_K \dfrac{(\gamma_B^Q)^u (\gamma_A^m)^v}{(\gamma_A^Q)^v (\gamma_B^m)^u}$	$K_K = \dfrac{Q_B c_A}{c_B Q_A}$	$K_K = \dfrac{Q_B^2 c_A}{c_B^2 Q_A}$	Sparks (2002)
Gaines–Thomas	$c_i\gamma_i^m$	y_i	$K_{ex} = K_{GT} \dfrac{(\gamma_B^y)^u}{(\gamma_A^y)^v}$	$K_{GT} = \dfrac{y_B c_A \gamma_A^m}{c_B \gamma_B^m y_A}$	$K_{GT} = \dfrac{y_B^2 c_A \gamma_A^m}{(c_B \gamma_B^m)^2 y_A}$	Gaines and Thomas (1953)
Krishnamoorty–Overstreet	$c_i\gamma_i^m$	w_i	$K_{ex} = K_{KO} \dfrac{(\gamma_B^w)^u}{(\gamma_A^w)^v}$	$K_{KO} = \dfrac{w_B c_A \gamma_A^m}{c_B \gamma_B^m w_A}$	$K_{KO} = \dfrac{w_B^2 c_A \gamma_A^m}{(c_B \gamma_B^m)^2 w_A}$	Krishnamoorthy and Overstreet (1949)
Gapon	$c_i\gamma_i^m$	Q_i'	—	$K_K = \dfrac{Q_B' c_A}{c_B Q_A'}$	$K_K = \dfrac{Q_B' c_A^{1/2}}{c_B Q_A'}$	Gapon (1933)
Cernik	$c_i\gamma_i^m$	y_i	$K_{ex} = K_C \dfrac{\gamma_B^y}{\gamma_A^y}$	$K_C = \dfrac{y_B c_A \gamma_A^m}{c_B \gamma_B^m y_A}$	$K_C = \dfrac{y_B c_A \gamma_A^m}{(c_B \gamma_B^m)^2 y_A}$	Cernik, Borkovec, and Westall (1996)

[a] For a binary exchange reaction as Equation 5.14, except for Gapon and Cernik conventions.
[b] Here, c_i is concentration in aqueous phase; molality m is approximate by molar concentration c; and x_i, y_i, and w_i are molar, equivalent, and weighted fractions, respectively, in the exchanger phase. Q_i and Q_i' are, respectively, moles per exchanger unit mass of adsorbate and sites occupied. Superscripts W and S are omitted for simplicity.
[c] Note that K_{ex} does not have the same numerical value in the different conventions, due to the different standard states.
[d] Assumed A^{2+} and B^+.

Following a procedure similar to that outlined in Section 5.3.2.1, the following expressions can be found for thermodynamic quantities (Bond 1995):

$$\ln K_{ex} = u - v + \int_0^1 \ln K_{GT}\, dy_B \tag{5.38}$$

$$\ln \gamma_A^y(y_B) = \frac{y_B}{v}\left[\ln K_{GT}(y_B) - (u - v)\right] - \frac{1}{v}\int_0^{y_B} \ln K_{GT}(y'_B)\,dy'_B \tag{5.39}$$

$$\ln \gamma_B^y(y_A) = \frac{y_A}{u}\left[u - v - \ln K_{GT}(y_A)\right] + \frac{1}{u}\int_{y_B}^{1} \ln K_{GT}(y'_B)\,dy'_B \tag{5.40}$$

This convention has been widely employed in soil chemistry (Bond 1995; Vulava et al. 2000; Escudey et al. 2001, 2002; Appel et al. 2003; Ma, Dong, and Zhou 2005; Gacitúa, Antilén, and Briceño 2008).

5.3.2.3 Krishnamoorthy and Overstreet Convention

Krishnamoorthy and Overstreet (1949) proposed a different way to express activities in the solid phase. They considered that ions were adsorbed onto localized charged sites; under this assumption, it should be taken into account that multivalent ions (such as Ca^{2+} or Al^{3+}) would occupy several adjacent sites; both the Vanselow and Gaines–Thomas conventions, using fractions, implicitly assume that an ion such as Al^{3+} would neutralize an equivalent number of charges irrespective of their localization. Krishnamoorthy and Overstreet assumed that the charged sites were arranged in a lattice, where each "site" has Z adjacent (nearest-neighbors) sites; Z is termed the lattice coordination index. They concluded that the activity of a sorbed ion i of charge z_i would be proportional to a "weighted" mole fraction defined by

$$w_i = \frac{n_i}{\sum_j r_j n_j} \tag{5.41}$$

where the sum is carried over all ions j of the same sign and r_j is a lattice factor defined from

$$r_j = z_j - \frac{2z_j}{Z} + \frac{2}{Z} \tag{5.42}$$

In their paper (Krishnamoorthy and Overstreet 1949), they considered reasonable for clay minerals a coordination index $Z = 4$, thus r_j will have a value of 1 for univalent ions, 1.5 for divalent ions, and 2 for trivalent ions. Consequently, for a binary exchange K^+–Mg^{2+}, the weighted fractions will be

$$w_{K^+} = \frac{n_{K^+}}{n_{K^+} + 1.5n_{Ca^{2+}}} \tag{5.43}$$

$$w_{Ca^{2+}} = \frac{n_{Ca^{2+}}}{n_{K^+} + 1.5n_{Ca^{2+}}} \tag{5.44}$$

Apart from this definition, the form of the Krishnamoorthy–Overstreet conditional constant would be similar to those of Vanselow and Gaines–Thomas (see Table 5.1); it should be noted, by inspection of Equations 5.43 and 5.44, that the

w_i is not like molar or equivalent fractions, because its sum is lower than unity. The Krishnamoorthy–Overstreet approach, although it is mentioned in the literature (Sparks 2002; Evangelou and Phillips 2005), has not been frequently employed.

5.3.2.4 Gapon Selectivity Coefficient

Gapon defined the exchange reaction at variance with Equation 5.14, as follows (Gapon 1933; Sparks 2002):

$$S^{-}A_{1/u} + \frac{1}{v}B^{v+} \rightleftharpoons S^{-}B_{1/v} + \frac{1}{u}A^{u+} \tag{5.45}$$

Note that it is written as the exchange of a single positive charge, irrespective of the valence of the ions; the Gapon selectivity coefficient is written as

$$K_G = \frac{\vartheta_B c_A^{1/u}}{c_B^{1/v}\vartheta_A} = \frac{y_B c_A^{1/u}}{c_B^{1/v} y_A} \tag{5.46}$$

where ϑ_i is the specific charge (or equivalent amount) sorbed of ion i (also, exchanger charges neutralized by ion i):

$$\vartheta_i = z_i Q_i \tag{5.47}$$

As it has been pointed out before (Evangelou and Phillips 1988), the Gapon selectivity coefficient is not thermodynamically consistent, thus an equation relating it to K_{ex} has not been given in Table 5.1. Despite this, the Gapon formulation has been widely employed, along the Vanselow and Gaines–Thomas formulations (Reuss 1983; Feigenbaum et al. 1991; Vulava et al. 2000; Endo et al. 2002; Agbenin and Yakubu 2006; Ghosh and Debnath 2010). A corrected Gapon coefficient was introduced (Sposito 1977), which is consistent with the Vanselow formulation; it was shown, however, that in Na^+–Ca^{2+} exchange, the traditional expression and the correct one will not be very different numerically if the exchangeable sodium percentage is below 20%. This was confirmed by Evangelou and Phillips (2005). Under some circumstances, it has been shown to be more constant than other coefficients (Evangelou and Coale 1987).

5.3.2.5 Cernik et al. Convention

More recently, Cernik, Borkovec, and Westall (1996), introduced a description based on a different scheme; they wrote the binary exchange reaction (albeit for the 1–2 exchange case) as follows:

$$SA_v + uB^{v+} \rightleftharpoons SB_u + vA^{u+} \tag{5.48}$$

Note that no "individual" sites are considered, just the surface taken as an exchanger. The activity convention is to use equivalent fractions, as in the Gaines–Thomas case; thus, the corresponding selectivity coefficient is given by

$$K_C = \frac{y_B\left(c_A\gamma_A^m\right)^u}{\left(c_B\gamma_B^m\right)^v y_A} \tag{5.49}$$

Other relations are given in Table 5.1. Until today, the Cernik convention seems to have found little application in the literature (Vulava et al. 2000).

5.3.2.6 Other Aspects

Table 5.1 collects the different conventions discussed in the preceding paragraphs, including the equations linking the selectivity coefficients with the true equilibrium constant, whenever possible. It should be remarked now that the numerical value of K_{ex} and, of course, of the selectivity coefficients depends on the convention chosen. This should be clear considering Equation 5.33 and the general equation for the chemical potential (Equation 2.5):

$$\mu_i = \mu_i^0 + \boldsymbol{RT} \ln a_i \tag{5.50}$$

Because the chemical potential is a thermodynamic property of species i, it must have the same value irrespective of the particular convention chosen for a_i. It follows that the value of the standard chemical potential μ_i^0, and consequently the values of he standard Gibbs free energy ΔG_{ex}^0 and equilibrium constant K_{ex}, should depend on that choice. As μ_i^0 is the value of μ_i for unit activity, the state having such activity in the scale chosen is the standard state. There are IUPAC recommendations for most common situations (Cox et al. 1982). For the solutes in the solution phase activities a_i^W in all cases collected in Table 5.1, the recommended and commonly employed Henry law convention is used

$$\gamma_i^m \rightarrow 1 \quad \text{as} \quad c_i \rightarrow 0 \quad (\text{or strictly} \;\; m_i \rightarrow 0) \tag{5.51}$$

which means that the standard state is a hypothetical solution of unit concentration behaving as at infinite dilution. For solid phases, the use of mole fractions and the Raoult state is recommended, which is defined by

$$\gamma_i^{\mathrm{R}} \rightarrow 1 \quad \text{as} \quad x_i \rightarrow 1 \tag{5.52}$$

that is, the standard state is the pure substance. On this basis, the Vanselow convention fits perfectly in a natural way, considering the homoionic exchanger (the exchanger saturated with a single ionic species), as a "pure" phase. The use of equivalent fractions, as in the Gaines–Thomas and Cernik conventions, albeit not strictly thermodynamics, leads also to this standard state, as the homoionic exchanger will also have an equivalent fraction equal to unity. As it has been pointed out (Goulding 1983), the Gaines–Thomas equilibrium constants differ from the Vanselow ones only in a constant factor (i.e., an additive constant in ΔG^0) independent of cations and exchanger, thus not affecting comparisons made on the basis of these parameters. When amounts sorbed (moles per unit of mass) are employed, a Henry type standard state would be the natural choice.

Thus, it is in principle important to take into account the standard state when computing thermodynamic parameters from ion exchange data, which not always is done. Also, when comparing different standard states, the possible differences in water activity should be taken into account (Evangelou and Phillips 2005). The correction values, however, are usually small and can be neglected in many practical cases (Sposito 1981; Bond 1995).

5.3.3 Ion Exchange Analysis and Parameter Computation

5.3.3.1 Argensinger Method

We will now consider some actual examples of ion exchange to illustrate the concepts discussed so far and introduce some more aspects. Most of the examples are of binary exchange, as presented in Equation 5.14; despite that in nature soil colloids are always in the presence of a soil solution containing a number of different anions and cations (see Chapter 7), most studies have been confined to binary cases. In general, the sample is first *homoionized*, that is, saturated with a single ion, either by repeatedly placing the sample in contact with fresh solutions of the ion of interest or by column leaching with an appropriate solution of such ion.

We begin with homovalent exchange; Figure 5.7a shows the Na^+–K^+ exchange isotherm (Bond 1995) for Brucedale soil, a soil with a high clay contents (Smiles and Gardiner 1982). The soil was previously homoionized with Na^+. In the figure, as it is customary plotted, the fraction of a given cation in the exchanger phase (here both $x^S_{Na^+}$ and $x^S_{K^+}$) is plotted as a function of the corresponding fraction of the same or another cation in the solution phase, here $x^W_{Na^+}$. Note that this is an apparent mole fraction, because water is not included, that is,

$$x^W_{Na^+} = \frac{n^W_{Na^+}}{n^W_{Na^+} + n^W_{K^+}} \tag{5.53}$$

Because here we have univalent cations, equivalent and molar fractions are the same, but for heterovalent exchange, both fraction types can be found (see below). Of course both plots are mirror images around the diagonal, because $x_{Na^+} + x_{K^+} = 1$. The dashed line represents the *nonpreference isotherm*, which is that obtained assuming that the equilibrium constant has unit value *and* assuming that the activity coefficients of both ions, in each phase separately, are equal, that is,

$$K_{ex} = 1 = \frac{x^S_{K^+}\gamma^R_{K^+}c^W_{Na^+}\gamma^m_{Na^+}}{x^S_{Na^+}\gamma^R_{Na^+}c^W_{K^+}\gamma^m_{K^+}} = \frac{x^S_{K^+}x^W_{Na^+}}{x^S_{Na^+}x^W_{K^+}} = \frac{\left(1 - x^S_{Na^+}\right)x^W_{Na^+}}{x^S_{Na^+}\left(1 - x^W_{Na^+}\right)} \tag{5.54}$$

From Equation 5.54, it is readily found that $x^S_{Na^+} = x^W_{Na^+}$, which is plotted as a dotted line. Note that the Na^+ isotherm is below that line, whereas the K^+ one is over the nonpreference line, which indicates that there is preference of K^+ over Na^+, because the amount of bonded K^+ is higher than the nonpreference prediction, whereas the amount of Na^+ is correspondingly lower; the possible causes will be discussed in Section 5.3.4.2.

We will now discuss the computation of K_{ex} from these results. As we have seen in Section 5.3.1, following Argersinger, Davidson, and (Bonner 1950), the following integral should be computed:

$$\ln K_{ex} = \int_{y^S_{K^+}=0}^{y^S_{K^+}=1} \ln K_V \, dy^S_{K^+} \tag{5.55}$$

Figure 5.7b shows a plot of $\ln K_V$ as a function of the fraction of K^+ in the exchanger (open squares). Now there is a difficulty: the integral should be computed in the entire range of mole fractions, but K_V is obviously not directly available for mole fractions of 0 or 1 (the homoionic exchanger with either ion). Furthermore, obtaining

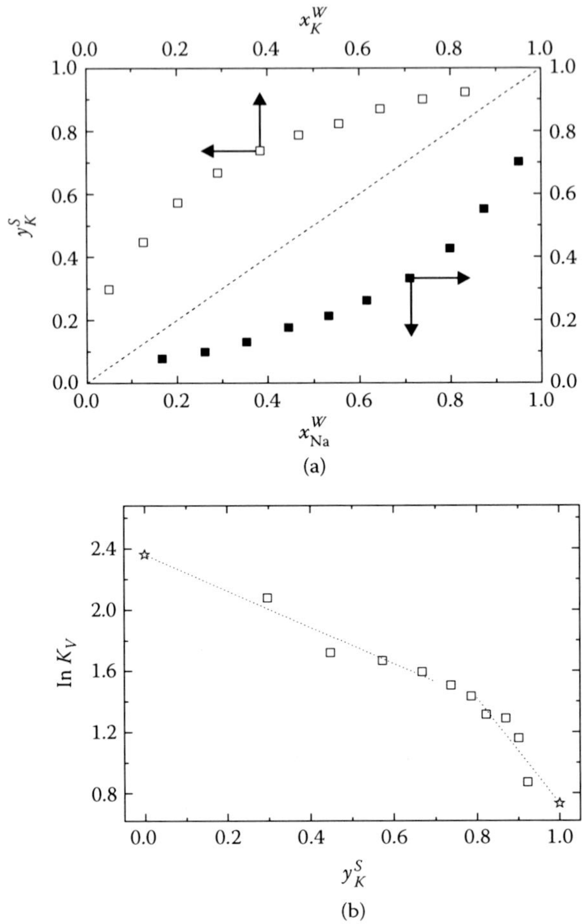

FIGURE 5.7 Homovalent K^+–Na^+ exchange on a Brucedale subsoil (Bond 1995) at $I = 0.05$ M. (a) Measured exchange isotherms; open symbols—K^+ mole fraction, closed symbols—Na^+ mole fraction. Dashed line: nonpreference isotherm. (b) ln K_V plot to determine the equilibrium constant K_{ex}. The dotted lines are extrapolations and the stars represent extrapolated points.

experimentally values for mole fractions close to 0 or 1 is difficult and inaccurate (Vulava et al. 2000). Thus, extrapolation is required; in Figure 5.7b, the dotted lines show the linear extrapolations for both extremes, and the stars are the extrapolated points. Integrating results in $K_{ex} = 5.62$, which compares well with the results obtained in the original paper (Bond 1995) of 5.57 by the same method and 5.82 by a different one, introduced later on. Clearly, the process is not very accurate, due to the extrapolations.

Before advancing further, we will show an example of heterovalent exchange, between Ca^{2+} and Na^+ in a Brucedale soil sample homoionized with Ca^{2+}:

$$S_2^-Ca^{2+} + 2Na^+ \rightleftharpoons 2S^-Na^+ + Ca^{2+} \tag{5.56}$$

The results (Bond 1995) are shown in Figure 5.8a. Here the Na^+ isotherm only is plotted, as an equivalent fraction in the solid and as a function of both mole fraction and equivalent fraction in solution phase; the second way is often found in the literature. Obtaining the nonpreference isotherm is a bit more complex here, starting from the equilibrium expression for reaction 5.56 *in terms of the Gaines–Thomas selectivity coefficient*, because the exchanger phase activity coefficients are not known.

$$K_{GT} = 1 = \frac{\left(y^S_{Na^+}\right)^2 c^W_{Ca^{2+}} \gamma^m_{Ca^{2+}}}{y^S_{Ca^{2+}} \left(c^W_{Na^+} \gamma^m_{Na^+}\right)^2} \tag{5.57}$$

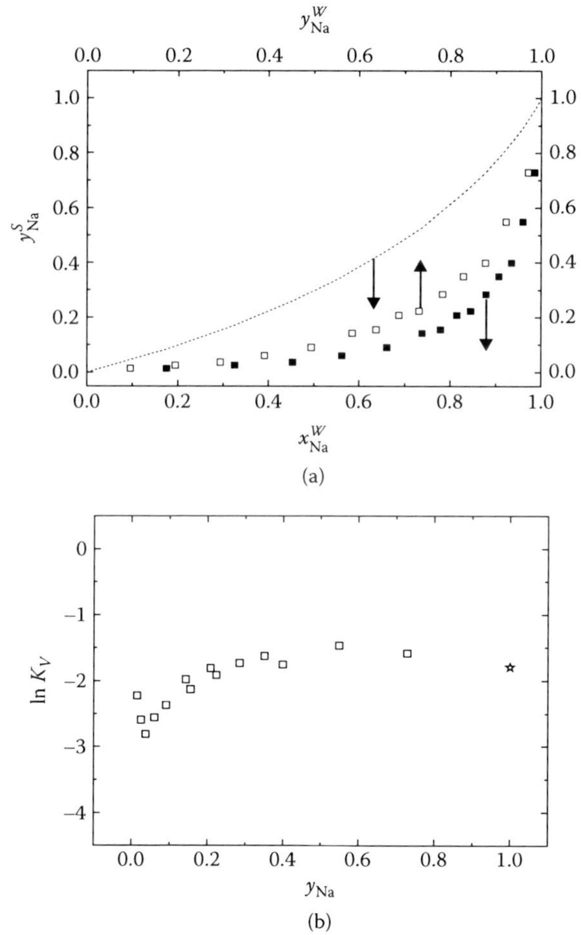

FIGURE 5.8 Heterovalent Ca^{2+}–Na^+ exchange on a Brucedale subsoil (Bond 1995) at a total equivalent concentration $c_{eq} = \sum z_i c_i = 0.2$ M. (a) Measured exchange isotherms as Na^+ equivalent fraction in the exchanger phase; closed symbols—as a function of Na^+ mole fraction in aqueous phase (bottom scale); open symbols—as a function of Na^+ equivalent fraction in aqueous phase (top scale). Dashed line: nonpreference isotherm, as a function of Na^+ mole fraction. (b) $\ln K_V$ plot to determine the equilibrium constant K_{ex}. The star represents an extrapolated point.

Considering Equation 5.53, we can write

$$1=\frac{\left(y_{\mathrm{Na}^+}^S\right)^2 c_{\mathrm{Ca}^{2+}}^W \gamma_{\mathrm{Ca}^{2+}}^m}{y_{\mathrm{Ca}^{2+}}^S\left(c_{\mathrm{Na}^+}^W \gamma_{\mathrm{Na}^+}^m\right)^2}=\frac{\left(y_{\mathrm{Na}^+}^S\right)^2 x_{\mathrm{Ca}^{2+}}^W \gamma_{\mathrm{Ca}^{2+}}^m}{\left(1-y_{\mathrm{Na}^+}^S\right)\left(x_{\mathrm{Na}^+}^W\right)^2\left(c_{\mathrm{Ca}^{2+}}^W+c_{\mathrm{Na}^+}^W\right)\left(\gamma_{\mathrm{Na}^+}^m\right)^2} \tag{5.58}$$

which can be solved for $y_{\mathrm{Na}^+}^S$ giving

$$y_{\mathrm{Na}^+}^S=\frac{\sqrt{A^2+4A}-A}{2} \tag{5.59}$$

with

$$A=\frac{\left(c_{\mathrm{Ca}^{2+}}^W+c_{\mathrm{Na}^+}^W\right)\left(\gamma_{\mathrm{Na}^+}^m\right)^2\left(x_{\mathrm{Na}^+}^W\right)^2}{\gamma_{\mathrm{Ca}^{2+}}^m\left(1-x_{\mathrm{Na}^+}^W\right)} \tag{5.60}$$

Equation 5.59 is plotted in Figure 5.8a as a dashed line. The Vanselow coefficients can be calculated and are plotted in Figure 5.8b; here only one extrapolated point is required for $y_{\mathrm{Na}^+}^S = 1$. The integration yields $K_{ex} = 0.175$, comparing reasonably well with values of 0.182 and 0.197 reported by Bond (1995).

5.3.3.2 Rothmund–Kornfeld Method

Rothmund and Kornfeld observed (Bond 1995) that ion exchange data followed approximately logarithmic relationships. Thus, they postulated fulfillment of an equation, which, for a general binary exchange reaction such as Equation 5.14, reads,

$$k_{RK}=\frac{\left(vQ_B\right)^u}{\left(uQ_A\right)^v}\left[\frac{\left(uc_A\right)^v}{\left(vc_B\right)^u}\right]^{n_{RK}} \tag{5.61}$$

where k_{RK} and n_{RK} are empirical constants. If Equation 5.61 is followed, then a plot of the equation

$$\log\left[\frac{\left(vQ_B\right)^u}{\left(uQ_A\right)^v}\right]=\log k_{RK}+n_{RK}\log\left[\frac{\left(vc_B\right)^u}{\left(uc_A\right)^v}\right] \tag{5.62}$$

should give a straight line with slope n_{RK} and intercept log k_{RK}. It was afterward verified (Bond 1995) that same relations hold for all the conventions to describe the exchange experiments. Thus, for the Vanselow convention, the Rothmund–Kornfeld equation can be written as

$$k_V=\frac{\left(x_B^S\right)^u}{\left(x_A^S\right)^v}\left[\frac{a_A^v}{a_B^u}\right]^{n_V} \tag{5.63}$$

and for the Gaines–Thomas case,

$$k_{GT}=\frac{\left(y_B^S\right)^u}{\left(y_A^S\right)^v}\left[\frac{a_A^v}{a_B^u}\right]^{n_{GT}} \tag{5.64}$$

It has been proved (Högfeldt 1955) that the thermodynamic equilibrium constant can be expressed in terms of the Rothmund–Kornfeld approximation parameters. For the Vanselow convention, it is expressed as

$$K_{ex} = k_V^{1/n_V} \tag{5.65}$$

whereas the activity coefficients can be obtained by

$$\gamma_i^R = \left(x_i^S\right)^{1/n_V - 1} \tag{5.66}$$

For the Gaines–Thomas convention, the corresponding expressions are

$$K_{ex} = k_{GT}^{1/n_{GT}} \boldsymbol{e}^{\frac{u-v}{n_{GT}}} \tag{5.67}$$

$$\gamma_i^y = \left(y_i^S\right)^{1/n_{GT} - 1} \boldsymbol{e}^{\frac{\left(y_j^S\right)\left(1 - z_i/z_j\right)}{n_{GT}}} \tag{5.68}$$

where j indicates the other exchanged ion.

For the example shown in Figure 5.7, Na^+–K^+ exchange, Equation 5.63 can be written as

$$\log\left(\frac{y_{K^+}}{y_{Na^+}}\right) = \log k_V + n_V \log\left(\frac{a_{K^+}}{a_{Na^+}}\right) \tag{5.69}$$

whereas for the Ca^{2+}–Na^+ exchange we get for Equation 5.64,

$$\log\left(\frac{y_{Na^+}^2}{y_{Ca^{2+}}}\right) = \log k_{GT} + n_{GT} \log\left(\frac{a_{Na^+}^2}{a_{Ca^{2+}}}\right) \tag{5.70}$$

Figure 5.9a and b show the plots of Equations 5.69 and 5.70, respectively, where a good linearity is found if points near the ends are discarded. The resulting Rothmund–Kornfeld parameters are given in Table 5.2, where the results obtained for K_{ex} are presented and compared with those from the original paper (Bond 1995). Even with the same data, noticeable differences are observed in the results obtained with the Argensinger method, due to the uncertainties in the extrapolations. The Rothmund–Kornfeld method has been employed and compared by other researchers (Bond and Verburg 1997; Escudey et al. 2001; Alves and Lavorenti 2003; Araujo et al. 2003; Gacitúa, Antilén, and Briceño 2008) with satisfactory results.

5.3.4 Ion Exchange in Soils

Ion exchange in a soil is very important since it determines the capacity of a soil to retain ions in a form that they are available for plant uptake and not susceptible to leaching in the soil profile. Nutrient uptake by plants is achieved through cation exchange, where the root hairs pump H^+ ions into the soil through proton pumps (Marschner 1995; Epstein and Bloom 2005; Barker and Pilbeam 2007). These pumped out hydrogen ions create an electrochemical potential gradient and

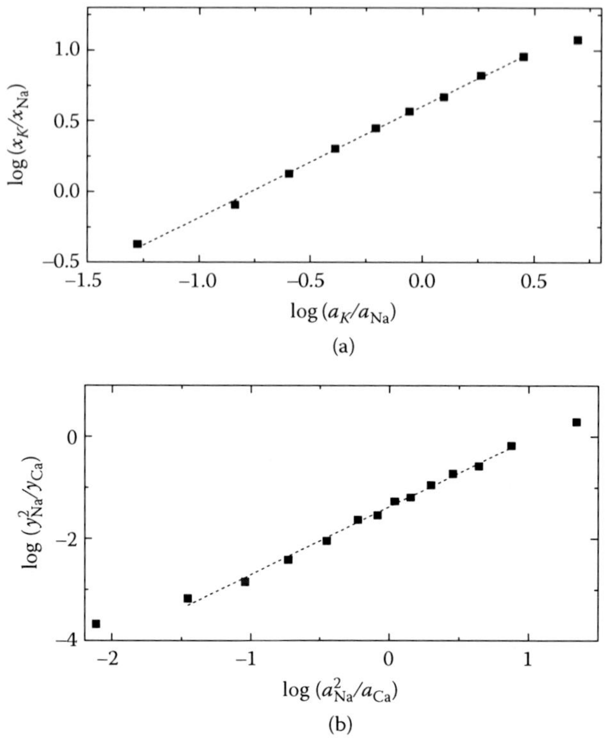

FIGURE 5.9 Rothmund–Kornfeld plots for Na^+–K^+ exchange of Figure 5.7 (a) and for Ca^{2+}–Na^+ exchange of Figure 5.8 (b). Squares: experimental points; dashed lines: fitting to Equations 5.69 and 5.70.

TABLE 5.2
Equilibrium Exchange Constants Evaluated by Different Methods

		Parameters from Rothmund–Kornfeld Method			K_{ex} Literature (Bond 1995)	
Exchange	K_{ex} Calculated by Argensinger Method	k	n	K_{ex}	Argensinger method	Rothmund–Kornfeld method
Na^+–K^+	5.62	4.013	0.785	5.87	5.57	5.82
$Ca2^+$–Na^+	0.175	0.0421	1.330	0.196	0.182	0.197

Source: Data from Bond, W. J., *Soil Sci. Soc. Am. J.*, 59,436–443, 1995.

also displace *exchangeable* cations (mainly NH_4^+, K^+, Ca^{2+}, and Mg^{2+}) retained by negative charges in the soil, allowing them to be taken by the roots. Anions are incorporated by an anion–proton cotransport mechanism (Marschner 1995). Of course, cations and anions not bound to negatively or positively (respectively) charged soil particles can be readily taken, but can be also easily *leached* into the soil. Leaching is the process through which water, from either rain or irrigation, pouring through into deeper soil levels, carries unbound solutes, making them unavailable to plants.

Thus, ion retention and exchange by charged soil colloids is important for plant nutrient retention and has many environmental implications; as an example, NO_3^- is important for plant growth, but if it leaches, it will move below the plant root zone and leach into groundwater, where it is deleterious to human health. If a soil has a significant anion exchange capacity, nitrate can be held so it is available to plants.

5.3.4.1 Examples of Ion Exchange in Soils

We shall not dig further into the formal and thermodynamics aspects of ion exchange; the interested reader is referred to the abundant accounts in the literature (see, e.g., McBride 1994; Sposito 1999; Sparks 2002; Evangelou and Phillips 2005). Instead, we will present now a few examples of ion exchange studies in soils to discuss its main features. Most of them address the K^+–Ca^{2+} exchange, one of the most important cases, which has been extensively studied because it affects the plant's available K^+ (Beckett 1964; Beckett 1972; Evangelou, Wang, and Phillips 1994).

Figure 5.10 shows Ca–K exchange isotherms for two horizons in an acidic soil from Guinea, Nigeria (Agbenin and Yakubu 2006). This is mainly a kaolinitic soil, with the oxide fraction being dominated by iron oxides. The dashed lines show the nonpreference isotherms; it can be seen that the surface horizon shows a small preference for potassium, whereas the subsurface shows a nearly nonpreferential behavior.

Figure 5.11 shows results for a highly weathered soil from the north of São Paulo state, Brazil (Alves and Lavorenti 2003). This is a clayey soil, having predominance of oxides (mainly hematite and gibbsite) over kaolinite. Here, a strong preference of K^+ over Ca^{2+} is clearly seen.

Figure 5.12 shows the same exchange for seven soils from north Chile (Gacitúa, Antilén, and Briceño 2008). These are all sandy aridisols, most of them dominated by quartz; the exceptions are Canchones and La Tirana, with a predominance of halite, and Coposa, mainly composed of sodium anorthite. Here, a range of behaviors is observed, from mostly K^+ preference (Coposa) to strong Ca^{2+} preference (Alto Hospicio); it should be noted, however, that most soil samples cross the nonpreference line at some point in the x_{K^+} range (notably Pica soil, near the middle of the range). In the last example, Figure 5.13 shows a case of "ideal" behavior, for the Ca(II)–Fe(II) exchange on an MX80 natural clay sample material (commercial Wyoming montmorillonite, a layered swelling clay) in chloride medium (Charlet and Tournassat 2005). Note that the results are plotted in a different way than the previous figures, as amount sorbed as a function of equivalent Fe(II) fraction; the symbols are experimental points, whereas the lines are the nonpreference isotherms. It is interesting to mention here that the nonpreferential behavior is observed even when, on this type of minerals, divalent cations are partially sorbed as ion pairs with chloride, such as $FeCl^+$. Other divalent ion pairs, notably Ca^{2+}–Mg^{2+}, show nonpreferential behavior; monovalent cations, on the contrary, usually do not (see, e.g., Figure 5.7).

With just the few examples given here one should realize that a simple view of purely electrostatic interaction between point charges is not adequate to understand the phenomena of ion exchange. The first question here is what types of sorbed ions

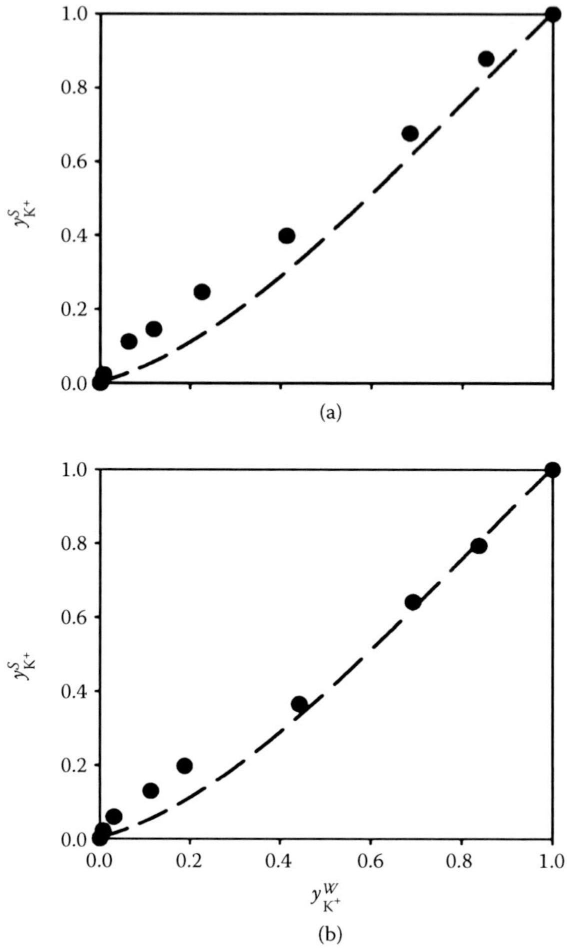

FIGURE 5.10 Ca–K exchange isotherms in an acid savanna Nigerian soil (pH in water, 4.5–4.7) at constant charge concentration of 0.2 N. (a) Surface horizon (0–15 cm); (b) subsurface horizon (15–30 cm). Circles: experimental points; lines: nonpreferential isotherms. (Reprinted from *Geoderma*, 136, Agbenin and Yakubu, Potassium-calcium and potassium-magnesium exchange equilibria in an acid savanna soil from northern Nigeria, 542–554. Copyright 2006, with permission from Elsevier.)

(from those discussed in Section 5.1) are able to participate in ion exchange; the usual answer includes diffuse layer and outer sphere bound ions (Sposito 1999; Sparks 2002). Inner sphere bound ions are assumed to not exchange; however, this is not guaranteed because, for one side, there is nothing in an ion exchange experiment preventing, from a fundamental point of view, a full exchange of all bound ions, if these ions are in true (i.e., dynamical) equilibrium with the solution phase. For an equilibrium reaction such as Equation 5.14, assuming monovalent exchange for simplicity,

$$SA + B^+ \rightleftharpoons SB + A^+ \tag{5.71}$$

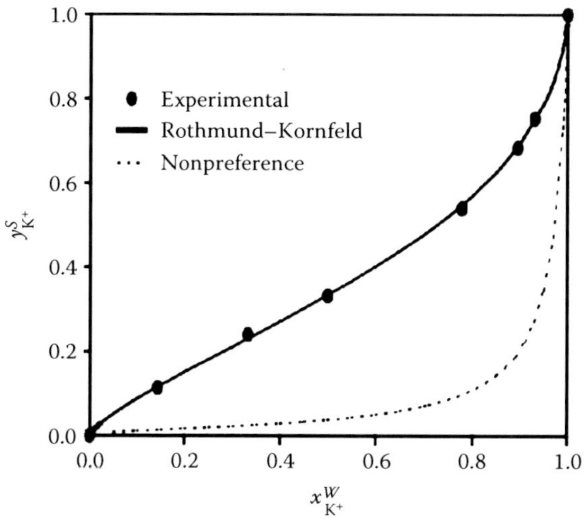

FIGURE 5.11 Ca–K exchange isotherms in horizon B of a Rhodic Acrudox (Soil Survey Staff 1998) soil from the northern part of São Paulo state, Brazil. (From Alves, M. E., A. Lavorenti, *Aust. J. Soil Res.* 41, 1423–1438, 2003. With kind permission of CSIRO Publishing, Australia.)

If A^+ is strongly bound to the exchanger phase, the activity (concentration) of A^+ in solution will be small but still can be "washed" away by B^+ if a sufficiently large volume/concentration is employed; if A^+ is removed, some bound A will be released to reestablish equilibrium and so forth. Eventually, all bound A will be removed; this may also be limited by the kinetics of Reaction 5.71. Thus, with limited exchange (i.e., a limited amount of B^+), A^+ will be seen as not exchangeable, but after long soil leaching in a natural environment, it could finally be removed. For the following discussion, we will restrict ourselves to consider only diffuse and outer sphere ions.

5.3.4.2 Ion Preference

One of the most important aspects of ion exchange is the definite preference for particular ions of a given sample, either of some soil components or of whole natural soils. There has been many studies on ion exchange on soils (Sparks 2001), but this issue is not yet settled (Teppen and David 2006; Rotenberg et al. 2009). It is sometimes, but not always, observed that ion preference is described by the lyotropic series, and an example, from highest to lowest preference, is (Helfferich 1995)

$$Ba^{2+} > Pb^{2+} > Sr^{2+} > Ca^{2+} > Ni^{2+} > Cd^{2+} > Cu^{2+} > Co^{2+} > Mg^{2+} > Ag^{+} > Cs^{+} > Rb^{+} > K^{+} > NH^{4+} > Na^{+} > Li^{+}$$

Actually, the series often depends on the nature of the adsorbing surface, as in the examples given in Section 5.3.4.1. Several models have been proposed in the literature (Eisenman 1962; Sullivan 1977; Xu and Harsh 1992; Auboiroux et al. 1998), but with limited success. A number of factors have been proposed to influence ion

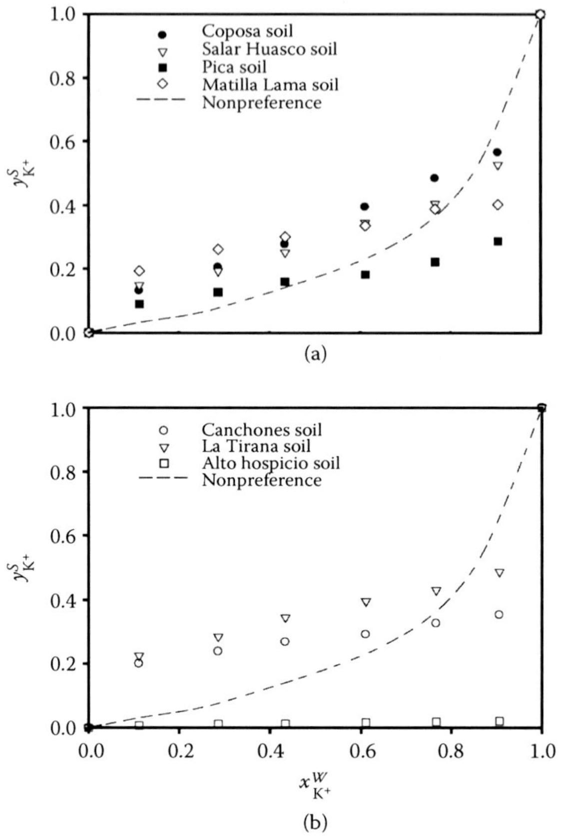

FIGURE 5.12 K–Ca exchange isotherms and no preference isotherm at seven Aridisol soils from Iquique province, northern Chile, mostly composed of quartz. (From Gacitúa, M. et al., *Aust. J. Soil Res.* 46, 745–750, 2008. With kind permission of CSIRO Publishing, Australia.)

exchange, especially for cations; the most-often cited are mentioned here (Evangelou and Phillips 2005); they include *ion-related factors*:

1. Hydration: For two ions with the same valence, the cation with the smaller hydrated radius, or least negative heat of hydration, is preferred (e.g., K^+ over Na^+).
2. Polarization: Given two cations with different valence (e.g., Na^+ and Ca^{2+}), the cation with the higher valence (Ca^{2+}) is preferred because of its greater polarization capability. For the case of anions, the opposite is true, larger anions show greater polarizability than smaller anions. Thus, larger anions would be preferred by a positively charged surface.
3. Lewis acidity: Comparing two cations with the same valence but one being a stronger Lewis acid than the other (e.g., Cu^{2+} vs. Ca^{2+}), the cation with the stronger Lewis acidity would be preferred if the surface behaves as a relatively strong Lewis base. A similar statement can be formulated for anions (Lewis bases) and a positive (Lewis acid) surface.

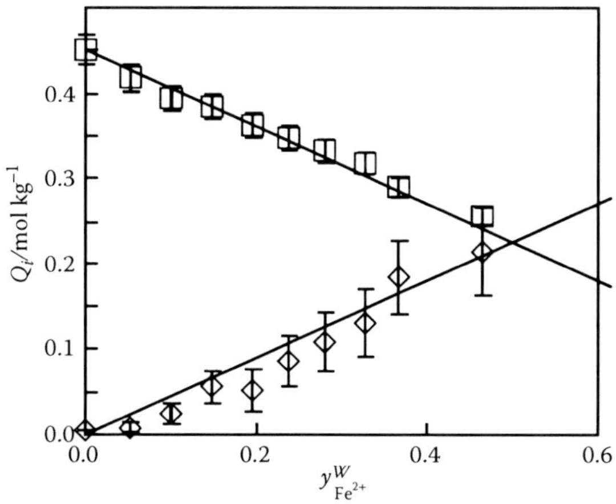

FIGURE 5.13 Ca–Fe(II) exchange isotherms (plotted as amount sorbed) on montmorillonite, a layered clay mineral (see Chapter 8). (□) Ca^{2+} sorbed; (◊) Fe(II) sorbed; lines: non-preference isotherms. (With kind permission from Springer Science+Business Media: *Aquat. Geochem.*, Charlet, L., C. Tournassat, Fe(II)-Na(I)-Ca(II) cation exchange on montmorillonite in chloride medium: Evidence for preferential clay adsorption of chloride – metal ion pairs in seawater, 11, 2005, 115–137.)

4. Ion pair formation: For a given cation in the presence of two different anions, the anion with the highest potential to form neutral ion pairs with the cation controls the cation adsorption potential, assuming that the anions do not react with the surface, and vice versa. For example, Ca^{2+} in the presence of Cl^- exhibits greater adsorption potential than Ca^{2+} in the presence of SO_4^{2-} because of the latter's greater preference to form neutral $CaSO_4$ pairs.
5. Counterion adsorption: For a metal cation in the presence of two different anions, the anion with the highest potential to form surface complexes controls the metal's adsorption potential. For example, Ni^{2+} in the presence of $Ca(NO_3)^2$ exhibits greater adsorption potential than Ni^{2+} in the presence of $CaSO_4$ because of the sulfate's potential to react with the surface and produce sites with higher specificity for Ca^{2+}.

And *exchanger (surface)-related factors*:

6. Different adsorption modes: A surface capable to form inner sphere complexes with certain cations (e.g., K^+ or NH^{4+}) shows stronger preference for these over any other cation (e.g., some layered clay minerals such as vermiculite show strong preference for K^+ over Na^+ or Ca^{2+}).
7. Surface electric potential: A surface that does not have the potential to form inner sphere complexes prefers cations with higher valence. This preference depends on the magnitude of the surface electrical potential—the higher the surface electrical potential, the higher the preference for divalent cation (such as Ca^{2+}) over monovalent cations (e.g., Na^+).

8. pH influence: Surfaces that exhibit pH-dependent electrical potential show variable selectivity for a given cation. For example, kaolinite at high pH (where the surface has a high negative charge and so a high negative surface electrical potential) shows increasing preference for divalent cations over monovalent cations, while at low pH (low negative electrical potential), kaolinite or kaolinitic soils show increasing preference for monovalent cations rather than divalent cations.
9. Conformational flexibility: Exchangers having the potential to undergo conformational changes (e.g., humic acids) prefer higher valence cations (such as Ca^{2+}) to lower valence cations (such as K^+).
10. Surfaces acidity: Exchangers exhibiting weaker acid behavior (high pK_a) show stronger preference for heavy metals than for hard metals in comparison to surfaces with stronger acid behavior (low pK_a).

However, some of these commonly mentioned factors have been challenged recently, because the ion hydration energy has been found, for some soil components (such as montmorillonite), to be an important factor determining ion preference (Teppen and David 2006; Rotenberg et al. 2009). It should be remarked, also, that the preceding points refer generally to *individual* soil components (clay minerals, humic substances, etc.), but in the whole soil, these components interact among themselves and with other species; as a result, the response of the soil is not the simple sum of individual behaviors. Furthermore, the great majority of studies deal with binary exchange, whereas in the soil many ions are present together; few studies deal with ternary exchange, as yet (Feigenbaum et al. 1991; Bond and Verburg 1997; Escudey et al. 2002). Other aspects not usually considered are (1) the stoichiometry (always assumed to be 1 to 1 in terms of charge), deviations from this rule are usually considered as the result of complex formation or simultaneous adsorption of anions and cations; (2) complete reversibility (fast exchange) is always assumed, but for strongly adsorbed cations that assumption may not hold (Evangelou and Phillips 2005).

As the soil behavior toward ion exchange depends on the behavior of the individual components, we will not progress further at this point; in Part III, the ion exchange response of such components is discussed, and finally some aspects of soil ion exchange are considered.

5.4 DETERMINATION OF SURFACE CHARGE

There are a number of experimental methods to measure particle charge, but most, if not all of them, fall into one of two categories: those based on electrophoresis, which measure the net particle charge, and acid–base titrations, which determine proton charge.

5.4.1 Electrophoretic Methods

Electrophoresis is the movement of charged particles under an electric field; these particles can be small ions, proteins, DNA and other biomolecules, and colloidal particles such as oxides or clay minerals. The basic idea is that if an electric field E is applied to a particle with charge q, an electric force arises, given by

$$\tilde{F}_{el} = qE \tag{5.72}$$

In fact, both the field and the force are vectors; here the absolute values are given, the force having the same direction as the field, and pointing in either the same or the opposite direction, depending on the signs of q and E (Nussbaum 1965; Feynman, Leighton, and Sands 1970; Purcell 1984). Figure 5.14a shows a schematic representation, where a positively charged particle is under a constant electric field E, thus experimenting a driving force given by Equation 5.72. As soon as the particle starts moving, friction with the surrounding liquid media gives rise to an opposing viscous force $\tilde{F}_v$, which, for a spherical particle large enough that the solution behaves effectively as a continuum, is given by Stokes's law,

$$\tilde{F}_v = 6\pi\eta r_P v \tag{5.73}$$

where η is the viscosity coefficient of the solution (usually taken as that of water), r_P the particle radius, and v its velocity. Under the electric field, the particle accelerates, so the viscous force increases until it levels with the electric one, and the particle

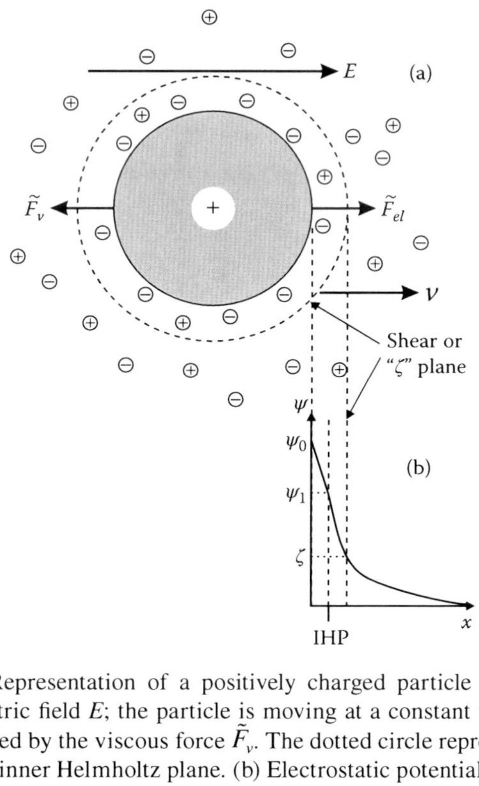

FIGURE 5.14 (a) Representation of a positively charged particle under electrophoretic motion under an electric field E; the particle is moving at a constant velocity v, the electric force $\tilde{F}_{el}$ being balanced by the viscous force $\tilde{F}_v$. The dotted circle represents the shear or "ζ" plane, and IHP is the inner Helmholtz plane. (b) Electrostatic potential profile.

from then on moves at constant speed. This process usually happens in a short time, and for all practical purposes, the particles move always at constant speed; in such a case, we have $\tilde{F}_{el} = \tilde{F}_v$, thus

$$v = \frac{qE}{6\pi\eta r_P} \tag{5.74}$$

or

$$v = \mu E \tag{5.75}$$

Here, μ is the particle *electrophoretic mobility*. It should be noted that, whereas Equation 5.74 is valid for a spherical particle satisfying Stokes's law, the definition of μ given by Equation 5.75 is more general; for large nonspherical particles, other numerical coefficients would appear in Equation 5.73, and for smaller particles and ions, this law losses validity, but always a mobility can be measured.

The mobility of particles is determined by electrophoretic experiments by several techniques, which are covered in numerous texts (Kruyt 1949; Lyklema 1995; Birdi 1997; Hiemenz and Rajagopalan 1997; Kosmulski 2009) and will not be discussed here.

Once the mobility is measured, one wants to know the particle charge. Now, complications arise, as simple expressions such as Equation 5.74 cannot be employed. Hydrodynamics dictates that a liquid film must move along the particle, thus there will be (e.g., at top and bottom of particle in Figure 5.14a) a thin film that is carried by the particle at its speed, and, at increasing distances from it, the speed of the fluid will be decreasing until vanishing in the bulk. This situation is idealized, assuming the presence of a *shear* or *slipping plane* that divides the solution between a layer carried by the particle and the static bulk solution. Thus, the effectively "moving particle" is composed by the true particle plus the accompanying liquid layer, and the charge "seen" by the electric field, which determines the electric force, is the net charge of the whole moving entity, that is, the net charge at the shear plane. The consequence is that to find the net *surface* charge, some calculations should be made, and to this end, some theoretical analysis is needed. It is important to remember that the true experimental parameter is the mobility; all other quantities that are discussed in the following are calculated applying some theoretical assumptions.

It turns out to be that an important parameter is the electric potential at the shear plane; it is known as *zeta potential*, ζ, and thus the shear plane also is termed sometimes the zeta plane. Its importance resides in the fact that, once it is known, the surface charge can be computed by applying the double-layer theories. The problem of relating the mobility to zeta potential is not simple, because the particle creates a distortion in the applied electric field, as shown in Figure 5.15. There are several theoretical approaches to this problem. A very general one is that due to Henry (1931). We will not go into the mathematical details here, but only present the results. Henry considered a dilute suspension (i.e., neglected particle–particle interactions), assumed validity of the Gouy–Chapman (GC) theory, and used the linearized form of the Poisson–Boltzmann (PB) equation. These last assumptions are reasonable in most cases: as shown in Figure 5.14a, the ζ potential is expected to lie within the diffuse

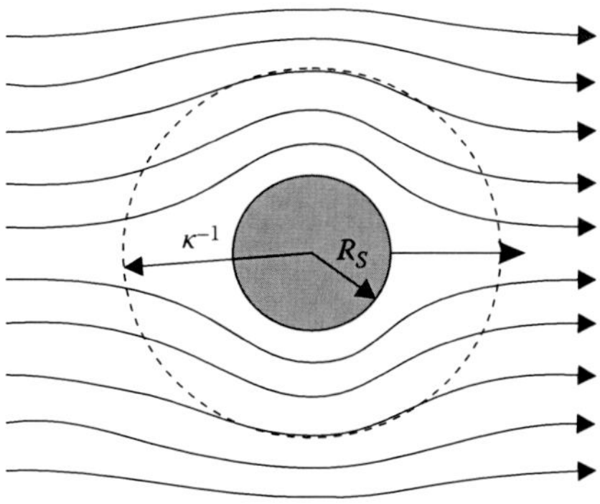

FIGURE 5.15 Electric field lines around a spherical particle of radius R_S; the interaction with the diffuse layer (its "thickness" κ^{-1} is marked) causes a distortion of the field lines.

layer (its exact position, of course, depends on electrodynamic conditions, particle charge, and solution ionic strength), thus the GC theory applies, and its absolute value is certainly lower than the surface potential. The Henry theory also neglects the retardation and relaxations effects, which distort the double layer and consequently retard the particle movement. The Henry equation is (Hiemenz and Rajagopalan 1997)

$$\mu = \frac{2\varepsilon\zeta}{3\eta} f(\alpha) \tag{5.76}$$

where $\alpha = \kappa r_P$ and

$$f(\alpha) = \begin{cases} 1 + \dfrac{\alpha^2}{16} - \dfrac{5\alpha^3}{48} - \dfrac{1}{96}(\alpha^4 + \alpha^5) - \left[\dfrac{\alpha^4}{8} - \dfrac{\alpha^6}{96}\right] e^{\alpha} \displaystyle\int_{\infty}^{\alpha} \dfrac{e^{-t}}{t} dt & \alpha < 50 \\ \dfrac{3}{2} - \dfrac{\alpha^{-1}}{2} + \dfrac{75}{2}\alpha^{-2} - 330\alpha^{-3} & \alpha > 40 \end{cases} \tag{5.77}$$

The former expression in principle is valid for all α values, but is impractical for high α; the second one is an approximation for high α. The Henry equation has the following limits: for $\alpha >> 1$ (large particles and/or high ionic strengths; i.e., double layer, and small compared with particle radius), we get the *Smoluchowski equation*

$$\mu = \frac{\varepsilon\zeta}{\eta} \tag{5.78}$$

whereas in the opposite case (double layer and larger than the particle), the *Hückel equation* is found.

$$\mu = \frac{2\varepsilon\zeta}{3\eta} \tag{5.79}$$

Both are limiting cases of Henry's equation. Equation 5.78 has been widely employed; however, it is valid only for large particles. For sufficiently small particles, Equation 5.79 holds, and for intermediate cases (which may be the greatest part), the full Henry Equation 5.77 should be applied. Even in that case, it is only valid for low potentials. Figure 5.16 shows the behavior of the μ–ζ relationship as α varies for different surface potentials; the Henry equation is the low potential limit. In the same plot, numerical results for the problem for different values of ζ, including relaxation effect (Wiersema, Loeb, and Overbeek 1966), have been included. Generally, to obtain ζ with reasonable accuracy, it is preferable to work in one of the two limiting cases given in the preceding discussion; otherwise an iterative procedure, with more sources of possible error, has to be performed. Currently, most of the time, these determinations are carried out using automated instruments that compute the zeta potential from mobility measurements (Kosmulski 2009), but the preceding considerations should be borne in mind.

Once ζ is known, one may want to obtain the particle charge from it, which now presents another problem, because the slipping plane position is difficult to know. One can readily find the net charge density at the zeta plane, σ_ζ, assuming that this plane lies within the diffuse layer, which is generally correct, from GC theory (Equation 3.28); then, for a symmetrical electrolyte,

$$\sigma_\zeta = \left(8c^\infty \varepsilon \boldsymbol{R}T\right)^{1/2} \sinh\left(\frac{z\boldsymbol{F}\zeta}{2\boldsymbol{R}T}\right) \quad (5.80)$$

For a spherical particle, under the assumption of low potentials (Debye–Hückel approximation, Equation 3.17), the charge enclosed by the ζ plane is given by

$$q_\zeta = 4\pi\varepsilon r_\zeta (1 + \kappa r_\zeta)\zeta \quad (5.81)$$

where r_ζ is the position of the shear plane. If the particle radius, r_P, is large compared with the double layer thickness (large values of radius and/or ionic strength), r_ζ can be approximated by r_P. Often, it is assumed that the shear plane is coincident with the OHP, thus one has at once that $q_\zeta = q_P$, the net particle charge (or $\sigma_\zeta = \sigma_P$); this assumption is reasonable under relatively high ionic strengths. When the preceding assumptions do not hold, it is necessary to solve the PB problem for the particular case, most of the time recurring to numerical procedures, or alternatively using approximate

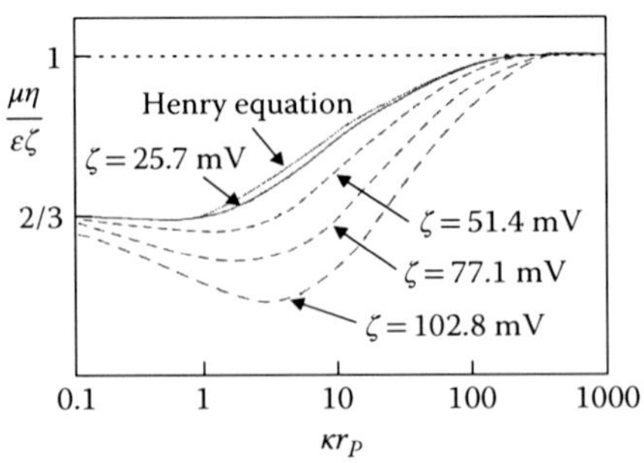

FIGURE 5.16 Transition from the Smoluchowski limit (low κr_p) to the Hückel limit (high κr_p) according to the Henry equation (dotted line) and numerical calculations (Wiersema, Loeb, and Overbeek 1966) including relaxation effect at various potentials (continuous and dashed lines).

expressions. For the case of spherical particles in a symmetric electrolyte, Ohshima and coworkers (Ohshima, Healy, and White 1982; Ohshima 2001; Makino and Ohshima 2010) developed fairly accurate approximate expressions. For the particle mobility as a function of ζ, it was found the following equation (with $\alpha = \kappa r_P$)

$$\mu = \frac{2\varepsilon\zeta}{3\eta}\left\{\begin{array}{l} 1+\dfrac{1}{2\left[1+\dfrac{2.5}{\alpha\left(1+2e^{-\alpha}\right)}\right]^3} \\ -\left(\dfrac{zF\zeta}{RT}\right)^2\dfrac{\alpha\left(\alpha+1.3e^{-0.18\alpha}+2.5\right)}{2\left(\alpha+1.2e^{-7.4\alpha}+4.8\right)^3}+\left(\dfrac{f_+ + f_-}{2}\right)\dfrac{9\alpha\left(\kappa r_P+5.2e^{-3.9\alpha}+5.6\right)}{8\left(\alpha-1.55e^{-0.32\alpha}+6.02\right)^3} \end{array}\right\} \tag{5.82}$$

where f_+ and f_- are the dimensionless ionic drag coefficients, defined by

$$f_i = \frac{2\varepsilon RT}{3\eta z \lambda_i^0} \tag{5.83}$$

where λ_i^0 is the limiting molar conductivity of ion i. Equation 5.82 is a good approximation for $zF\zeta/RT \leq 4$. For the particle charge, the following expression was proposed (Ohshima, Healy, and White 1982):

$$\sigma_P = \frac{2\varepsilon\kappa RT}{zF}\sinh\left(\frac{zF\zeta}{2RT}\right)\left\{1+\frac{2}{\alpha\cosh^2\left(\dfrac{zF\zeta}{4RT}\right)}+\frac{8\ln\left[\cosh\left(\dfrac{zF\zeta}{4RT}\right)\right]}{\left(\kappa r_P\right)^2\sinh^2\left(\dfrac{zF\zeta}{2RT}\right)}\right\}^{1/2} \tag{5.84}$$

This expression is fairly accurate for values of κr_P down to 0.1. These expressions can be used, provided the validity conditions are met, to find the particle zeta potential and net charge from mobility measurements. Other approximate expressions have been proposed, but we will not go further here (see Makino and Ohshima 2010 and references therein).

Figure 5.17 shows zeta potential results for goethite particles (Boily et al. 2001). As it can be seen, these measurements have been done at three different ionic strengths, and all three intersect, within experimental error, at $\zeta = 0$, at pH = 9.4; this is the *isoelectric point* (IEP) of the particles, that is, the pH where they do not move under an electric field (Table 5.3). The use of at least three different electrolyte concentrations ensures that the chosen salt does not affect the IEP. Unless the shear plane falls within the "Stern layer," which is a rare case, this point can be regarded as *point of zero charge (pzc)*, pH_0. Usually the shear plane is coincident with, or near to, the OHP—that is, inside the diffuse layer. Due to the nature of the potential profile in the diffuse layer (see Chapter 3), if the potential is zero at some point, then it is zero everywhere in that layer, thus σ_d should be zero and the same for σ_P. To obtain the surface charge from these results, some expression valid for the present conditions should be used,

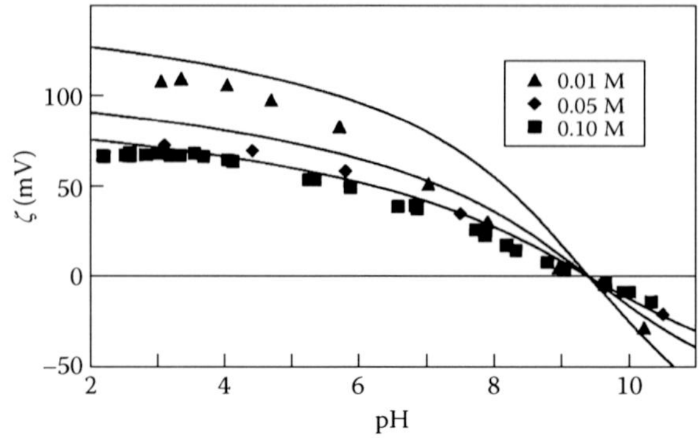

FIGURE 5.17 Zeta potential as a function of pH for goethite particles (needles of about 100–200 nm wide by 1000–200 nm long, A_S = 23 m² g⁻¹) in $NaNO_3$ suspension (concentrations indicated in the figure) at 298.15 K; the lines are model simulations under the assumption that $\zeta = \psi_2$. (Reprinted from *Colloids Surf. A Physicochem. Eng. Asp.*, 179, Boily, J.-F. et al., Modeling proton binding at the goethite ([alpha]-FeOOH)-water interface, 11–27. Copyright 2001, with permission from Elsevier.)

TABLE 5.3
Different Points of Zero Charge

Name	Abbreviation	Symbol[a]	Condition
Point of zero charge	pzc	pH_0	$\sigma_p = 0$
Point of zero net proton charge	pznpc	$pH_{H,0}$	$\sigma_H = 0$
Point of zero net intrinsic charge	pznc	$pH_{in,0}$	$\sigma_{in} = 0$
Point of zero salt effect	pzse	$pH_{se,0}$	$\frac{\partial \sigma_H}{\partial I} = 0$
Isoelectric point	iep	pH_{iep}	$\mu = 0$

[a] Assuming that H^+ is the PDI.

or other measurements should be performed; here it was assumed (as it is often done) that the zeta potential is coincident with the OHP potential ($\zeta = \psi_2$), thus σ_P can be obtained from Equation 5.80; modeling the interface under such assumption leads to the ζ potential predictions plotted as lines in Figure 5.17, showing fair agreement.

5.4.2 Titration Methods

For colloids that undergo acid–base exchanges in aqueous media (as the vast majority of soil colloids do), there is a contribution to the surface charge given by Equations 5.3 and 5.4, which can be determined rather easily by means of acid–base titrations. Figure 5.18 shows an example of the procedure: a titration of the colloidal suspension in an appropriate supporting electrolyte (one which does not show specific adsorption onto the colloidal particles) is compared with the same electrolyte alone. The

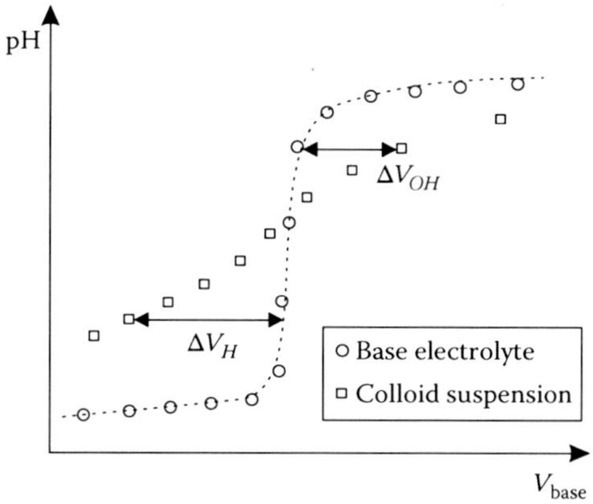

FIGURE 5.18 The determination of proton surface charge from potentiometric data by comparing a titration of the colloidal suspension (squares) with one of the support electrolyte alone (circles).

differences in titrant volume between both curves at a given pH correspond to equivalent amounts of excess H^+ of OH^- required by the colloid to reach the same pH as the electrolyte. In the example of Figure 5.18, a known amount of acid was added to both samples, and then they were titrated with base. The crossover point of both curves is, ideally, the natural pH of the colloidal suspension in the absence of any other acid–base active species, and, also ideally, the pH where the particle charge vanishes, or *point of zero net proton charge* (pznpc, see Table 5.3) $pH_{H,0}$ (Sposito 1999). In practice, often the solid particles carry some minor amounts of acidic or basic impurities, which give, when suspended, a different pH from the true natural pH (Kosmulski 2009).

In the ideal case, the volume difference ΔV_{OH}, the amount of additional base solution employed to bring the colloidal particles to a given pH, is directly related to the OH^- sorbed by the colloid, thus

$$\Gamma_{OH} = \frac{\Delta V_{OH} c_B}{A} \tag{5.85}$$

and, conversely, the difference ΔV_H (in absolute value) gives the amount of H^+ sorbed,

$$\Gamma_H = \frac{\Delta V_H c_B}{A} \tag{5.86}$$

where A is the total area of the colloid and c_B the concentration of the basic titrant. However, as discussed in the preceding discussion in general, these are not the true surface excesses but apparent ones due to the possible presence of remaining acid or base. Now, calling ΔV to the difference in volume spent in the presence and in the absence of particles, and noting that ΔV_H is actually a negative difference, we get from Equation 5.3 that the apparent surface net proton charge is

$$\sigma_{H,ap} = -\frac{\Delta V c_B \boldsymbol{F}}{A} \tag{5.87}$$

The sign would be reversed in a titration with an acid reagent; there are also several variants in the experimental procedure (Kosmulski 2009). An alternative to the use of blank titrations, if the solution composition is well known, is to deduce σ_H from a charge balance, such as (Boily et al. 2001),

$$\sigma_H = \boldsymbol{F}\frac{c_a - [\mathrm{H}^+] + K_W/[\mathrm{H}^+]}{A_S c_P} \tag{5.88}$$

where c_a is the total concentration of strong acid added, A_S is the specific surface area, and c_P is the mass concentration of particles.

Taking into account the presence of acid or base impurities, the true surface charge is

$$\sigma_H = \sigma_{H,ap} - \sigma_{ex} \tag{5.89}$$

where σ_{ex} is the extra charge observed due to acid or base impurities. To determine the $\mathrm{pH}_{H,0}$ and the true charges, one can consider the fact that when $\sigma_P = \sigma_d = 0$, there is no effect of the ionic strength on the electric surface properties (there is no potential gradient, thus the surface potential is equal to that of the solution). In the absence of permanent charge and specific adsorption, when $\sigma_H = 0$, then $\sigma_I = \sigma_O = 0$, and thus $\sigma_d = 0$ also. Thus, under such conditions, curves of charge versus pH should intersect at the pznpc, $\mathrm{pH}_{H,0}$. Figure 5.19 shows an example of such experiment, where charge curves of γ-Al_2O_3 for several ionic strength are shown that intersect, within experimental error, at pH = 8.1, thus determining that this is $\mathrm{pH}_{H,0}$. Here, it is observed that the curves intersect at $\sigma_H = 0$, thus the extra charge is also zero. It should be noted that not always do curves for different I values intersect at a common point

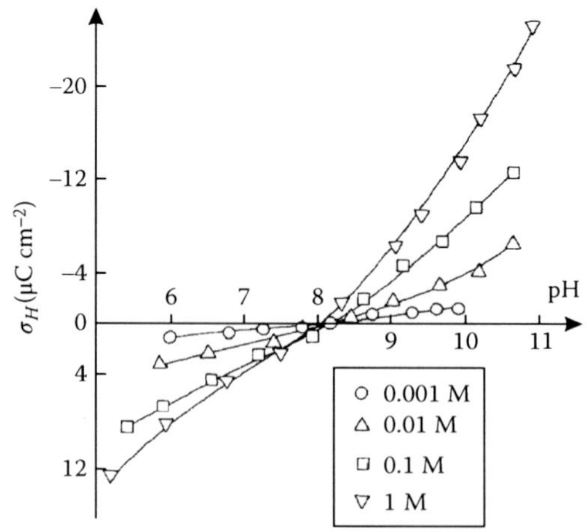

FIGURE 5.19 Variation of surface charge as a function of solution pH for γ-Al_2O_3 at different concentrations of NaCl. (Reprinted from *J. Colloid Interface Sci.*, 127, Sprycha, R., Electrical double layer at alumina/electrolyte interface: I. Surface charge and zeta potential, 1–11. Copyright 1989, with permission from Elsevier.)

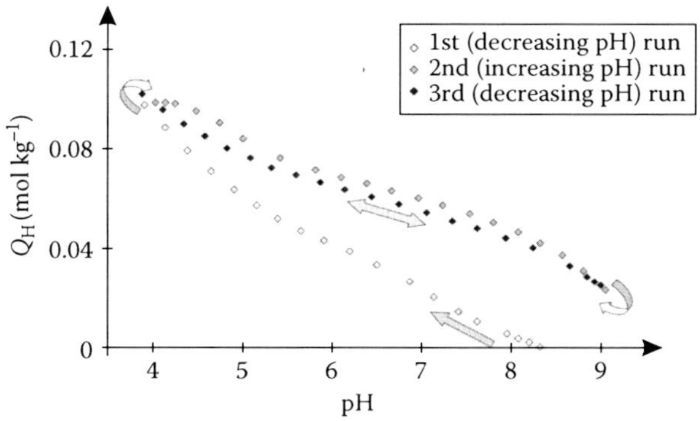

FIGURE 5.20 Experimental net proton consumption curves for Wyoming Na–montmorillonite dialyzed against, and diluted with, 0.01 M NaCl solutions at room temperature. The points were calculated from the data of an equilibrium titration cycle to test the reversibility of acid–base processes: first a backward (open symbols) with 0.1 M HCl solution, then a forward (grey symbols) with 0.1 M NaOH solution, and finally a backward titration (black symbols) again. (Reprinted from *Appl. Clay Sci.*, 27, Tombácz, E., M. Szekeres, Colloidal behavior of aqueous montmorillonite suspensions: The specific role of pH in the presence of indifferent electrolytes, 75–94. Copyright 2004, with permission from Elsevier.)

(see, e.g., Cao, Bryant, and Williams 1996); in such a case, the pznpc is not defined, probably because the electrolyte is not indifferent, having some specific adsorption (Kosmulski 2009).

It is worth mentioning here that, besides other experimental aspects, which are reported elsewhere (Kosmulski 2009), the reproducibility of titration curves should be checked by conducting "back and forth" titrations; Figure 5.20 shows an example, where a montmorillonite sample needs two titrations before reaching a rather reproducible state. This verification is important also for humic substances.

An alternative to standard acid–base titration is the *mass titration*, originally proposed by Noh and Schwarz (1989), consisting in the addition of the solid colloid to an electrolyte solution. As the solid is added, the solution pH changes, until it reaches a limiting value pH_∞; if the sample is pure, then this value corresponds to $pH_{H,0}$. However, if it contains acid or base impurities, these get dissolved and affect pH_∞, which no longer is equal to pH_0. Later on, Kallay and coworkers (Zâlac and Kallay 1992; Preočanin and Kallay 2006) extended the method to account for the presence of impurities, combining the mass titration with a following acid–base titration; in principle, both the $pH_{H,0}$ and the amount of impurities can be determined.

5.5 CHARGE CONTRIBUTIONS AND POINTS OF ZERO CHARGE

As should be evident from all the topics previously discussed, the charge of colloidal particles has a complex dependence on the surrounding solution composition (and this is especially true for soil colloids, as it will be seen in Part III). Also, determining accurately the net surface charge and its contributions is in general difficult. An important parameter, which is relatively easier to find, is the *point of zero charge*, that is, the conditions leading to vanishing particle charge. In fact, there are several possible definitions of point of zero charge; some have been introduced in Section 5.4, and Table 5.3 collects the relevant definitions, which are discussed in the following sections. These are defined as pH values, because most usually the H^+ is the PDI; however, under some circumstances (such as presence of specific adsorption), other ions determine or contribute to determine the surface charge/potential.

5.5.1 Indifferent Electrolytes

In the presence of indifferent electrolytes (usually $NaNO_3$, KNO_3, $NaClO_4$, but often also NaCl and KCl are employed), the zero charge points are solely determined by the proton charge and the permanent one, if present. The pzc (also, point of zero net charge) is the pH value that makes σ_P (and consequently σ_d) vanish. This is observed as the pH at which the mobility (or ζ potential) also vanishes; Figure 5.21a shows such measurements for hematite particles (Schudel et al. 1997). As it can be seen, these measurements have been done at two different ionic strengths, which intersect, giving an IEP at pH = 9.4; that is, the pH where they do not move under an electric field (Table 5.3). As said in the preceding discussion, this point can be regarded as *point of zero charge* (pzc), pH_0. Here, also acid–base titrations were conducted, and their results are shown in Figure 5.21b. These measurements were conducted at four different ionic strength values, and all curves intersect at a common intersection point; this ensures that the electrolyte does not affect significantly the behavior (Kosmulski 2009). The intersection point of proton charging curves is always the *point of zero salt effect* (Table 5.3). In this case, it lies at the axis $\sigma_H = 0$, so it can be concluded that it is also the *pznpc* (Table 5.3), which is at pH = 9.2. Because the IEP is coincident with the pznpc, it can be concluded that the pzc is also at $pH_0 = 9.2$; no permanent charge is present and no significant ion adsorption occurs at the pzc. All these conclusions are the usual case with common iron and aluminum oxides. The lines shown in the figures are modeling results assuming electrostatic ion adsorption in the Stern layer (outer sphere complexes). This modeling is discussed in Chapter 12. It is worth noting than iron oxides present usually "nice" proton charge curves, as those shown in Figure 5.21b. Figure 5.22 shows typical curves for silica, where a common intersection point is not found, but the curves tend to merge together asymptotically; this behavior is due to the absence of appreciable protonation of the silica surface, thus there is no positive charge in all the pH range.

Hydrotalcite ($Mg_6Al_2(CO_3)(OH)_{16}\cdot 4(H_2O)$) is a secondary mineral, resembling talc, found in serpentinite rocks. Hydrotalcite-like compounds, which present a positive permanent charge due to isotopic substitution, have found interest in various applications (Hou et al. 2001). Figure 5.23 shows acid–base titration results for such a colloid. It is seen that the curves have a common intersection point at pH = 10.75, which is the pzse (Table 5.3) but with a negative surface proton charge (note that it is a positive value of $\Gamma_{OH} - \Gamma_H$). That corresponds to a positive permanent surface charge. This is analyzed as follows (Hou et al. 2001), with a reasoning essentially equal for the case of variable charge soils (Uehara and Gillman 1980). In the absence of inner and outer sphere adsorption, the net particle charge is given by the sum of permanent and proton contributions:

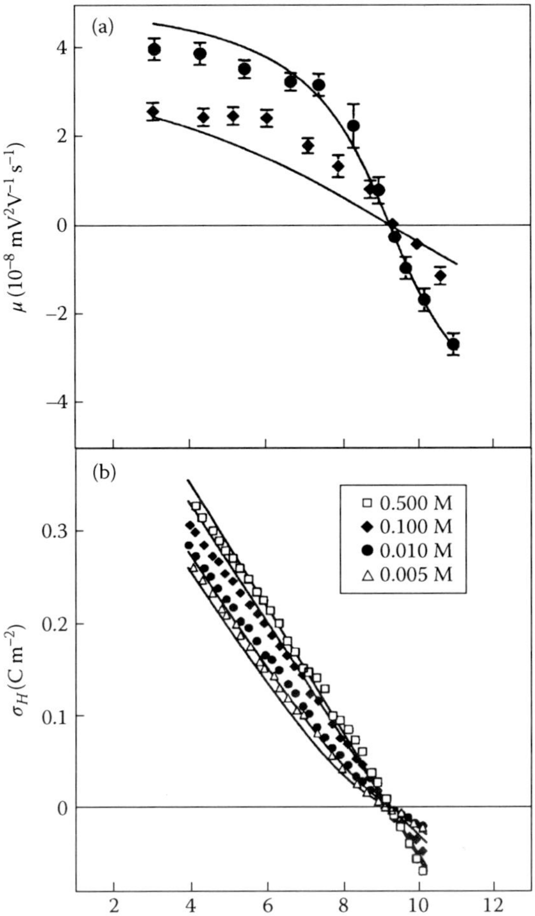

FIGURE 5.21 Electrophoretic mobilities and net proton charge as a function of pH for hematite particles (approximately spherical particles of about 50 nm diameter, $A_S = 15$ m^2 g^{-1}, Brunauer–Emmet–Teller theory [BET] measurement) suspended (30 mg L^{-1}) in $NaNO_3$ solutions (concentrations indicated in the figure) at 298.15 K. (a) Electrophoretic mobility; points are averages of 3–10 experimental data points, and the error bars correspond to standard deviations. The lines are model simulations under the assumption that $\zeta = \psi_2$. (b) Net proton charge density. (Reprinted from *J. Colloid Interface Sci.*, 196, Schudel, M. et al., Absolute aggregation rate constants of hematite particles in aqueous suspensions: A comparison of two different surface morphologies, 241–253. Copyright 1997, with permission from Elsevier.)

$$\sigma_P = \sigma_f + \sigma_H \tag{5.90}$$

The proton charge is given by Equation 5.3, whereas the particle charge is of the opposite sign to the diffuse charge, for a symmetric electrolyte and a plane surface (as hydrotalcite has a layered structure) given by the GC theory (Equation 3.28):

$$\sigma_P = \left(8c^{\infty}\varepsilon \boldsymbol{R}T\right)^{1/2} \sinh\left(\frac{z\boldsymbol{F}\psi_0}{2\boldsymbol{R}T}\right) \tag{5.91}$$

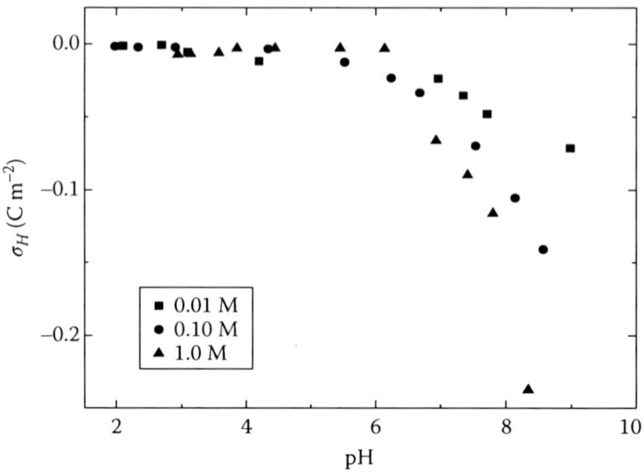

FIGURE 5.22 Surface charges of pyrogenic SiO_2 particles in KCl solutions of different concentrations. (Data from Sahai, N., D. A. Sverjensky, *Geochim. Cosmochim. Acta*, 61, 2801–2826, 1997; Abendroth, R. P., J. Colloid Interface Sci., 34, 591–596, 1970.)

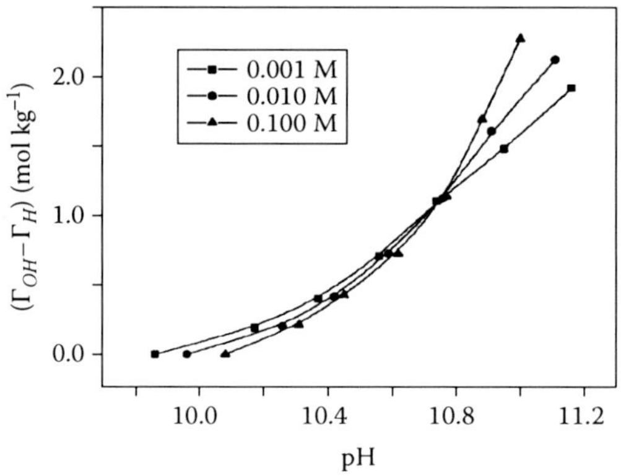

FIGURE 5.23 Surface proton balance for a zinc hydrotalcite at different NaCl concentrations. (Reprinted with permission from Hou, W. et al., 2001, *Langmuir*, 17, no. 6, 1885–1888. Copyright 2001 American Chemical Society.)

Note that it is assumed that $\psi_0 = \psi_2$, that is, the Stern layer is disregarded. If the acid–base reactions (Equations 5.1 and 5.2) are at equilibrium, the surface potential is given by the Nernst equation,

$$\psi_0 = \frac{\boldsymbol{RT}}{z\boldsymbol{F}} \ln\left(\frac{a_{H^+}}{a^0_{H^+}}\right) \tag{5.92}$$

where $a^0_{H^+}$ is the proton activity in the bulk solution when the surface potential is zero. This can be rewritten as

$$\psi_0 = \frac{2.303\boldsymbol{RT}}{z\boldsymbol{F}}(pH_0 - pH) \tag{5.93}$$

where we have written pH_0 in advance. Combining Equations 5.91 and 5.93, we get

$$\sigma_P = \left(8c^{\infty}\varepsilon \boldsymbol{R}T\right)^{1/2} \sinh\left[1.152\left(pH_0 - pH\right)\right] \tag{5.94}$$

which shows that effectively pH_0 is the pzc and is the point where σ_P is independent of the ionic strength. Now, the net proton charge at this point is not zero, as shown in Figure 5.23, so its value must compensate the fixed charge:

$$\sigma_f = -\sigma_H(pH_0) \tag{5.95}$$

Note that when permanent charge is present, the pznpc, $pH_{H,0}$, is now ionic strength dependent, being between $pH_{H,0} \approx 9.85$ and 10.1 in Figure 5.23. Also, from Equations 5.94 and 5.90, it is found that

$$\sigma_f = \left(8c^{\infty}\varepsilon \boldsymbol{R}T\right)^{1/2} \sinh\left[1.152\left(pH_0 - pH_{H,0}\right)\right] \tag{5.96}$$

Finally, the *point of zero net intrinsic charge*, the pH where $\sigma_{in} = \sigma_f + \sigma_H = 0$, is in the preceding examples coincident with pH_0, because no adsorption is present at pH_0.

The methods shown in the preceding discussion work well for isolated, pure soil colloids, in the absence of acid or basic impurities. In experiments with whole soil samples, often a common intersection point is not found. Chorover and Sposito (Chorover and Sposito 1995; Sposito 2008) have employed a method based in the so-called *Chorover plot*. Returning to Equation 5.11, we can write

$$0 = \sigma_{in} + \sigma_{ad} = \sigma_f + \sigma_H + \sigma_{ad} \tag{5.97}$$

so that

$$\sigma_{ad} = -\sigma_H - \sigma_f \tag{5.98}$$

Thus, if the net ionic charge σ_{ad} is measured by exchange experiments, and σ_H is measured by titration, a plot of σ_{ad} against σ_{H} should give a straight line of slope −1 and both *x*- and *y*-intercepts equal to $-\sigma_H$. Figure 5.24 (Chorover and Sposito 1995) shows an example for a tropical soil where ϑ_{ad} is plotted against ϑ_H; both intercepts lie at +12.5 ± 0.8 millimoles of charge per kilogram, in excellent agreement with ϑ_f = −12.50 ± 0.04 mmol kg^{-1} found by the independent CsCl method. This method works in the absence of strong specific adsorption (see Section 5.5.2) and is important, because it is the only known which allows determination of the amount of acid–base impurities in the solid.

5.5.2 Charge in the Presence of Specific Adsorption

When species that interact strongly with the particle surface are present, specific adsorption takes place, and this causes marked changes in the colloid charging properties. Figure 5.25 shows zeta potential measurements for goethite in the absence and presence of phosphate. As it is clearly seen, this ion causes a strong shift of the IEP toward lower pH. That is a consequence of the specific adsorption of PO_4^{3-} on the oxide (as it is schematically shown in Figure 5.6; more details are given in Chapter

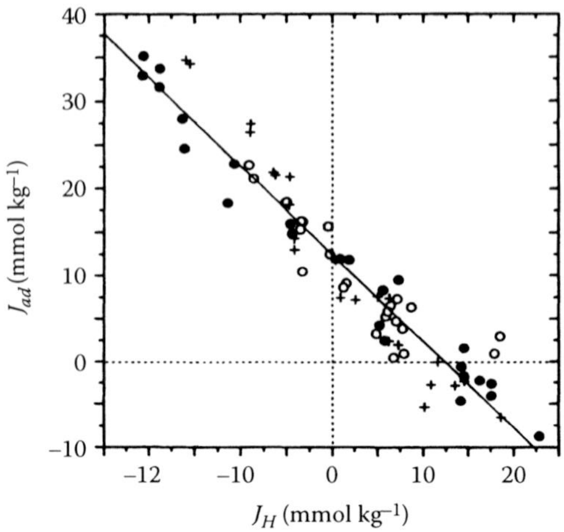

FIGURE 5.24 Chorover–Sposito plot for a cultivated kaolinitic Oxisol (Manaus, Brazil) in LiCl solutions at several ionic strengths: (○) 1 mM, (+) 5 mM, (●) 10 mM, and pH 2–6. The intersections with horizontal and vertical axes are equal, thus confirming the charge balance. (Reprinted from *Geochim. Cosmochim. Acta*, 59, Chorover, J., G. Sposito, Surface charge characteristics of kaolinitic tropical soils, 875–884. Copyright 1995, with permission from Elsevier.)

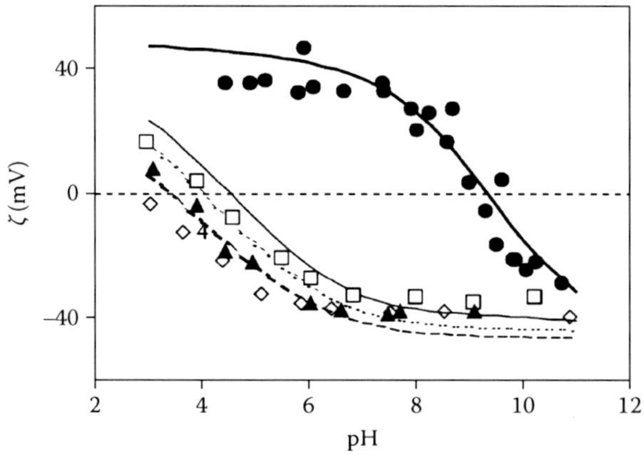

FIGURE 5.25 Zeta potential as a function of pH for goethite particles in absence and presence of phosphate ions, in 0.01 M KNO_3. (●, thick solid line) absence of phosphate; (□, thin solid line) 10^{-5} M PO_4^{3-}; (▲, dotted line) 10^{-4} M PO_4^{3-}; (◊, dashed line) 10^{-3} M PO_4^{3-}. (Reprinted from J. *Colloid Interface Sci.*, 285, Antelo, J. et al., Effects of pH and ionic strength on the adsorption of phosphate and arsenate at the goethite-water interface, 476–486. Copyright 2005, with permission from Elsevier.)

9) that incorporates negative charges, thus shifting the pzc to more acidic pH values. Note that this is an inner-sphere complex but not easily exchangeable, due to its stability. The titration curves are also clearly affected, as observed in Figure 5.26. The titration in the presence of phosphate requires an increased amount of base, due to both the acid–base equilibria of solution PO_4^{3-} and the desorption of phosphate from

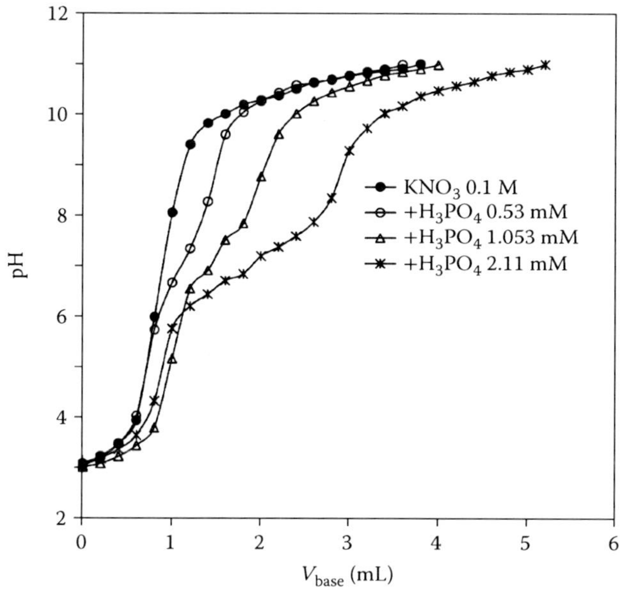

FIGURE 5.26 Potentiometric titration curves of MnO_2 (3.33 g L^{-1}, 214 m^2 g^{-1}) in the presence of different concentrations of phosphate. The pH was initially adjusted to 3.0 with HNO_3. (Reprinted from *J. Colloid Interface Sci.*, 301, Mustafa, S. et al., pH effect on phosphate sorption by crystalline MnO2, 370–375. Copyright 2006, with permission from Elsevier.)

the surface; this desorption is promoted by the increased negative charges of surface and adsorbate with increased pH.

Generally speaking, to obtain the surface charge composition in the presence of specific adsorption, the preceding methods alone usually do not suffice. Modeling of the interface with the adsorption reactions is required, and independent measurements, for example, of adsorption isotherms, are very useful if not unavoidable. These topics will be considered in Part III.

5.6 SUMMARY

We have analyzed the processes leading to the development of an electrically charged interface between colloidal particles (in particular, soil colloidal particles) and the surrounding solution. The visualization as "layers" of charge (permanent, protonic, inner and outer sphere, diffuse), albeit being a simplification, helps in the understanding of the interfacial behavior. Next, the important topic of ion exchange was examined. Being in principle a consequence of unspecific, unlocalized ion adsorption, it is a process of fundamental importance in soil chemistry; it was treated with some extension, including the thermodynamics underlying this phenomenon. Finally, we studied the issue of points of zero charge (an important parameter, among other causes because from its knowledge the particle charge sign in a particular situation can be quickly inferred) and the determination of charge curves in the presence of indifferent electrolytes.

This chapter is built on the foundations laid by its predecessors, especially Chapters 3 and 4, and all together serve as basis for the next chapter, where we leave the interaction between a particle and the solution to enter the study of the interaction between particles themselves. Besides, the concepts developed here and in Chapter 3 are necessary to understand further topics related to ion binding in Part III.

REFERENCES

Abendroth, R. P. 1970. Behavior of a pyrogenic silica in simple electrolytes. *Journal of Colloid and Interface Science* 34, no. 4 (December): 591–596. doi:10.1016/0021-9797(70)90223-7.

Agbenin, J. O., and S. Yakubu. 2006. Potassium-calcium and potassium-magnesium exchange equilibria in an acid savanna soil from northern Nigeria. *Geoderma* 136, no. 3–4: 542–554. doi:10.1016/j.geoderma.2006.04.008.

Alves, M. E., and A. Lavorenti. 2003. Potassium–calcium exchange in electropositive oxisols: Description of exchange sites. *Australian Journal of Soil Research* 41, no. 8: 1423–1438.

Antelo, J., M. Avena, S. Fiol, R. López, and F. Arce. 2005. Effects of pH and ionic strength on the adsorption of phosphate and arsenate at the goethite-water interface. *Journal of Colloid and Interface Science* 285, no. 2: 476–486. doi:10.1016/j.jcis.2004.12.032.

Appel, C., L. Q. Ma, R. D. Rhue, and W. Reve. 2003. Selectivities of potassium-calcium and potassium-lead exchange in two tropical soils. *Soil Science Society of America Journal* 67, no. 6: 1707. doi:10.2136/sssaj2003.1707.

Araujo, R. J., S. Likitvanichkul, Y. Thibault, and D. C. Allan. 2003. Ion exchange equilibria between glass and molten salts. *Journal of Non-Crystalline Solids* 318, no. 3: 262–267. doi:10.1016/S0022-3093(02)01888-4.

Argersinger, W. J., A. W. Davidson, and O. D. Bonner. 1950. Thermodynamics and ion exchange phenomena. *Transactions of the Kansas Academy of Science (1903)* 53, no. 3: 404–410. doi:10.2307/3626157.

Atkins, P., and J. de Paula. 2009. *Physical Chemistry*. 9th ed. New York: W. H. Freeman.

Auboiroux, M., F. Melou, F. Bergaya, and J. C. Touray. 1998. Hard and soft acid-base model applied to bivalent cation selectivity on a 2:1 clay mineral. *Clays and Clay Minerals* 46, no. 5: 546–555.

Barker, A. V., and D. J. Pilbeam. 2007. *Handbook of Plant Nutrition*. Boca Raton, FL: CRC/Taylor & Francis.

Beckett, P. H. T. 1964. Studies on soil potassium. I. Confirmation of the ratio law: Measurement of potassium potential. *Journal of Soil Science* 15, no. 1: 1–8. doi:10.1111/j.1365-2389.1964.tb00239.x.

Beckett, P. 1972. Critical cation activity ratios. In *Advances in Agronomy*, ed. N. C. Brady, 24: 379–412. New York: Academic Press. http://www.sciencedirect.com/science/article/B7CSX-4S987WT-F/2/6f2f8ba84204e848b02d62d993fc3c39.

Berry, R. S., S. A. Rice, and J. Ross. 2000. *Physical Chemistry*. 2nd ed. New York: Oxford University Press.

Birdi, K. S. 1997. *Handbook of Surface and Colloid Chemistry*. 1st ed. Boca Raton, FL: CRC Press.

Boily, J.-F., J. Lützenkirchen, O. Balmès, J. Beattie, and S. Sjöberg. 2001. Modeling proton binding at the goethite ([alpha]-FeOOH)-water interface. *Colloids and Surfaces A: Physicochemical and Engineering Aspects* 179, no. 1: 11–27. doi:10.1016/S0927-7757(00)00712-3.

Bond, W. J. 1995. On the Rothmund-Kornfeld description of cation exchange. *Soil Science Society of America Journal* 59, no. 2: 436–443.

Bond, W. J., and K. Verburg. 1997. Comparison of methods for predicting ternary exchange from binary isotherms. *Soil Science Society of America Journal* 61, no. 2: 444–454.

Cao, E., R. Bryant, and D. J. A. Williams. 1996. Electrochemical properties of Na-attapulgite. *Journal of Colloid and Interface Science* 179, no. 1: 143–150. doi:10.1006/jcis.1996.0196.

Cernik, M., M. Borkovec, and J. C. Westall. 1996. Affinity distribution description of competitive ion binding to heterogeneous materials. *Langmuir* 12, no. 25: 6127–6137. doi:10.1021/la960008f.

Charlet, L., and C. Tournassat. 2005. Fe(II)-Na(I)-Ca(II) cation exchange on montmorillonite in chloride medium: Evidence for preferential clay adsorption of chloride–metal ion pairs in seawater. *Aquatic Geochemistry* 11, no. 2: 115–137.

Chorover, J., and G. Sposito. 1995. Surface charge characteristics of kaolinitic tropical soils. *Geochimica et Cosmochimica Acta* 59, no. 5: 875–884. doi:10.1016/0016-7037(94)00357-2.

Cox, J. D., S. Angus, G. T. Armstrong, R. D. Freeman, M. Laffitte, G. M. Schneider, G. Somsen, C. B. Alcock, and P. W. Gilles. 1982. Notation for states and processes, significance of the word standard in chemical thermodynamics, and remarks on commonly tabulated forms of thermodynamic functions. *Pure and Applied Chemistry* 54, no. 6: 1239–1250.

Eisenman, G. 1962. Cation selective glass electrodes and their mode of operation. *Biophysical Journal* 2, no. 2, Part 2: 259–323. doi:10.1016/S0006-3495(62)86959-8.

Endo, T., S. Yamamoto, T. Honna, and A. E. Eneji. 2002. Sodium-calcium exchange selectivity as influenced by clay minerals and composition. *Soil Science* 167, no. 2: 117–125.

Epstein, E., and A. J. Bloom. 2005. *Mineral Nutrition of Plants: Principles and Perspectives.* Sunderland, MA: Sinauer Associates, Inc.

Escudey, M., P. Díaz, J. E. Förster, C. Pizarro, and G. Galindo. 2001. Gaines–Thomas and Rothmund–Kornfeld descriptions of potassium–calcium exchange on variable surface charge soils. *Communications in Soil Science and Plant Analysis* 32, no. 19: 3087–3097.

Escudey, M., P. Diaz, J. E. Förster, C. Pizarro, G. Galindo, and L. Beltrán. 2002. Prediction of K-Ca-Mg ternary exchange from binary isotherms in volcanic soils using the Rothmund-Kornfeld approach. *Australian Journal of Soil Research* 40, no. 5: 781–790.

Evangelou, V. P., and F. J. Coale. 1987. Dependence of the Gapon coefficient on exchangeable sodium for mineralogically different soils. *Soil Science Society of America Journal* 51, no. 1: 68–72.

Evangelou, V. P., and R. E. Phillips. 1988. Comparison between the Gapon and Vanselow exchange selectivity coefficients. *Soil Science Society of America Journal* 52, no. 2: 379–382.

Evangelou, V. P., and R. E. Phillips. 2005. Cation exchange in soils. In *Chemical Processes in Soils*, eds. M. A. Tabatabai and D. L. Sparks, 343–410. Madison, WI: Soil Science Society of America.

Evangelou, V. P., J. Wang, and R. E. Phillips. 1994. New developments and perspectives on soil potassium quantity/intensity relationships. In *Advances in Agronomy*, ed. D. L. Sparks, 52:173–227. New York: Academic Press. http://www.sciencedirect.com/science/article/B7CSX-4S987WV-9/2/2d114c45add9280c27d84f69cf710871.

Feigenbaum, S., A. Bar-Tal, R. Portnoy, and D. L. Sparks. 1991. Binary and ternary exchange of potassium on calcareous montmorillonitic soils. *Soil Science Society of America Journal* 55, no. 1: 49–56.

Feynman, R. P., R. B. Leighton, and M. L. Sands. 1970. *The Feynman Lectures on Physics: Mainly Electromagnetism and Matter*. Vols. 2. and 3. Reading, MA: Addison Wesley Longman.

Gacitúa, M., M. Antilén, and M. Briceño. 2008. K–Ca–Mg binary cation exchange in saline soils from the north of Chile. *Australian Journal of Soil Research* 46, no. 8: 745–750.

Gaines, G. L., and H. C. Thomas. 1953. Adsorption studies on clay minerals. II. A formulation of the thermodynamics of exchange adsorption. *Journal of Chemical Physics* 21, no. 4: 714–718. doi:10.1063/1.1698996.

Gapon, Y. N. 1933. On the theory of exchange adsorption on solids. *Journal of General Chemistry of the USSR (English translation)* 3: 144–160.

Ghosh, D., and A. Debnath. 2010. Study on the threshold values of soil potassium parameters for release and fixation: A prognostic approach to improve the use efficiency of soil and fertilizer potassium. *Communications in Soil Science and Plant Analysis* 41, no. 22: 2661. doi:10.1080/00103624.2010.518047.

Goulding, K. W. T. 1983. Adsorbed ion activities and other thermodynamic parameters of ion exchange defined by mole or equivalent fractions. *Journal of Soil Science* 34, no. 1: 69–74. doi:10.1111/j.1365-2389.1983.tb00813.x.

Högfeldt, E. 1955. On ion exchange equilibria III. An investigation of some empirical equations. *Acta Chemica Scandinavica* 9: 151–165. doi:10.3891/acta.chem .scand.09-0151.

Helfferich, F. 1995. *Ion Exchange*. Mineola, NY: Dover Publications.

Henry, D. C. 1931. The cataphoresis of suspended particles. Part I. The equation of cataphoresis. *Proceedings of the Royal Society of London. Series A* 133, no. 821: 106–129. doi:10.1098/rspa.1931.0133.

Hiemenz, P. C., and R. Rajagopalan. 1997. *Principles of Colloid and Surface Chemistry*. 3rd ed. New York: Marcel Dekker.

Hou, W.-G., Y.-L. Su, De-J. Sun, and C.-G. Zhang. 2001. Studies on zero point of charge and permanent charge density of Mg–Fe hydrotalcite-like compounds. *Langmuir* 17, no. 6: 1885–1888. doi:10.1021/la0008838.

Kosmulski, M. 2009. *Surface Charging and Points of Zero Charge*. 1st ed. Surfactant Science Series 145. Boca Raton, FL: CRC Press, May 14.

Krishnamoorthy, C., and R. Overstreet. 1949. Theory of ion-exchange relationships. *Soil Science* 68, no. 4: 307.

Kruyt, H. R., ed. 1949. *Colloid Science Vol. I, Irreversible Systems*. Amsterdam: Elsevier.

Levine, I. 2008. *Physical Chemistry*. 6th ed. New York: McGraw-Hill Science/Engineering/Math.

Lyklema, J. 1995. *Fundamentals of Interface and Colloid Science: Solid-Liquid Interfaces*. Amsterdam: Academic Press, December 18.

Ma, L. Q., Y. Dong, and Q. Zhou. 2005. Relation of relative colloid stability ratio and colloid release in two lead-contaminated soils. *Water, Air, and Soil Pollution* 160, no. 1–4: 343–355. doi:10.1007/s11270-005-3386-8.

Makino, K., and H. Ohshima. 2010. Electrophoretic mobility of a colloidal particle with constant surface charge density. *Langmuir* 26, no. 23: 18016–18019. doi:10.1021/la1035745.

Marschner, H. 1995. *Mineral Nutrition of Higher Plants*. San Diego, CA: Academic Press.

Mattson, S. 1931. The laws of soil colloidal behavior: VI. Amphoteric behavior. *Soil Science* 32, no. 5. http://journals.lww.com/soilsci/Fulltext/1931/11000/The_Laws_of_Soil_Colloidal_Behavior__Vi_.2.aspx.

McBride, M. B. 1994. *Environmental Chemistry of Soils*. New York: Oxford University Press.

McBride, M. B. 1999. Chemisorption and precipitation reactions. In *Handbook of Soil Science*, ed. M. E. Sumner, B265–B302. 1st ed. Boca Raton, FL: CRC Press.

McQuarrie, D. A., and J. D. Simon. 1997. *Physical Chemistry: A Molecular Approach*. Herndon, VA: University Science Books.

Mustafa, S., M. I. Zaman, and S. Khan. 2006. pH effect on phosphate sorption by crystalline MnO_2. *Journal of Colloid and Interface Science* 301, no. 2: 370–375. doi:10.1016/j.jcis.2006.05.020.

Noh, J. S., and J. A. Schwarz. 1989. Estimation of the point of zero charge of simple oxides by mass titration. *Journal of Colloid and Interface Science* 130, no. 1: 157–164. doi:10.1016/0021-9797(89)90086-6.

Nussbaum, A. 1965. *Electromagnetic Theory for Engineers and Scientists*. Englewood Cliffs, NJ: Prentice-Hall.

Ogwada, R. A., and D. L. Sparks. 1986. Use of mole or equivalent fractions in determining thermodynamic parameters for potassium exchange in soils. *Soil Science* 141, no. 4: 268–273.

Ohshima, H. 2001. Approximate analytic expression for the electrophoretic mobility of a spherical colloidal particle. *Journal of Colloid and Interface Science* 239, no. 2: 587–590. doi:10.1006/jcis.2001.7608.

Ohshima, H., T. W. Healy, and L. R. White. 1982. Accurate analytic expressions for the surface charge density/surface potential relationship and double-layer potential distribution for a spherical colloidal particle. *Journal of Colloid and Interface Science* 90, no. 1: 17–26. doi:10.1016/0021-9797(82)90393-9.

Preočanin, T., and N. Kallay. 2006. Point of zero charge and surface charge density of TiO_2 in aqueous electrolyte solution as obtained by potentiometric mass titration. *Croatica Chemica Acta* 79, no. 1: 95–106.

Purcell, E. M. 1984. *Electricity and Magnetism*. 2nd ed. Vol. 2. Berkeley Physics Course. New York: McGraw-Hill Science/Engineering/Math.

Reuss, J. O. 1983. Implications of the calcium-aluminum exchange system for the effect of acid precipitation on soils. *Journal of Environmental Quality* 12, no. 4: 591–595.

Rotenberg, B., J.-P. Morel, V. Marry, P. Turq, and N. Morel-Desrosiers. 2009. On the driving force of cation exchange in clays: Insights from combined microcalorimetry experiments and molecular simulation. *Geochimica et Cosmochimica Acta* 73, no. 14: 4034–4044. doi:10.1016/j.gca.2009.04.012.

Sahai, N., and D. A. Sverjensky. 1997. Evaluation of internally consistent parameters for the triple-layer model by the systematic analysis of oxide surface titration data. *Geochimica et Cosmochimica Acta* 61, no. 14: 2801–2826.

Schudel, M., S. H. Behrens, H. Holthoff, R. Kretzschmar, and M. Borkovec. 1997. Absolute aggregation rate constants of hematite particles in aqueous suspensions: A comparison of two different surface morphologies. *Journal of Colloid and Interface Science* 196, no. 2: 241–253. doi:10.1006/jcis.1997.5207.

Singh, U., and G. Uehara. 1998. Electrochemistry of the double layer. Principles and applications to soils. In *Soil Physical Chemistry, Second Edition*, ed. D. L. Sparks. Boca Raton, FL: CRC Press.

Smiles, D. E., and B. N. Gardiner. 1982. Hydrodynamic dispersion during unsteady, unsaturated water flow in a clay soil. *Soil Science Society of America Journal* 46, no. 1: 9–14.

Sparks, D. L. 2001. Elucidating the fundamental chemistry of soils: Past and recent achievements and future frontiers. *Geoderma* 100, no. 3–4: 303–319. doi:10.1016/S0016-7061(01)00026-X.

Sparks, D. L. 2002. *Environmental Soil Chemistry, Second Edition*. San Diego, CA: Academic Press.

Sparks, D. L. 2006. Milestones in soil chemistry. *Soil Science* 171, no. 1: S47–S50. doi:10.1097/01.ss.0000228050.62345.96.

Sposito, G. 1977. The Gapon and the Vanselow selectivity coefficients. *Soil Science Society of America Journal* 41, no. 6: 1205–1206.

Sposito, G. 1981. *The Thermodynamics of Soil Solution*. New York: Oxford University Press.

Sposito, G. 1999. Ion exchange phenomena. In *Handbook of Soil Science*, ed. M. E. Sumner, B241–B264. 1st ed. CRC Press, August 31.

Sposito, G. 2008. *The Chemistry of Soils*. 2nd ed. New York: Oxford University Press, USA.

Sprycha, R. 1989. Electrical double layer at alumina/electrolyte interface: I. Surface charge and zeta potential. *Journal of Colloid and Interface Science* 127, no. 1: 1–11. doi:10.1016/0021-9797(89)90002-7.

Sullivan, P. J. 1977. The principle of hard and soft acids and bases as applied to exchangeable cation selectivity in soils. *Soil Science* 124, no. 2: 117–121.

Sumner, M. E., and W. P. Miller. 1996. Cation exchange capacity and exchange coefficients. In *Methods of Soil Analysis. Part 3. Chemical Methods*, ed. D. L. Sparks, 1201–1230. Soil Science Society of America Book Series 5. Madison, WI: American Society of Agronomy-Soil Science Society of America.

Teppen, B. J., and D. M. Miller. 2006. Hydration energy determines isovalent cation exchange selectivity by clay minerals. *Soil Science Society of America Journal* 70, no. 1: 31–40.

Tombácz, E., and M. Szekeres. 2004. Colloidal behavior of aqueous montmorillonite suspensions: The specific role of pH in the presence of indifferent electrolytes. *Applied Clay Science* 27, no. 1–2: 75–94. doi:10.1016/j.clay.2004.01.001.

Uehara, G., and G. P. Gillman. 1980. Charge characteristics of soils with variable and permanent charge minerals: I. Theory. *Soil Science Society of America Journal* 44, no. 2: 250–252.

Vanselow, A. P. 1932. Equilibria of the base-exchange reactions of bentonites, permutites, soil colloids, and zeolites. *Soil Science* 33, no. 2: 95–114.

Vulava, V. M., R. Kretzschmar, U. Rusch, D. Grolimund, J. C. Westall, and M. Borkovec. 2000. Cation competition in a natural subsurface material: Modeling of sorption equilibria. *Environmental Science & Technology* 34, no. 11: 2149–2155. doi:10.1021/es990214k.

Wiersema, P. H., A. L. Loeb, and J. Th. G. Overbeek. 1966. Calculation of the electrophoretic mobility of a spherical colloid particle. *Journal of Colloid and Interface Science* 22, no. 1: 78–99. doi:10.1016/0021-9797(66)90069-5.

Xu, S., and J. B Harsh. 1992. Alkali cation selectivity and surface charge of 2:1 clay minerals. *Clays and Clay Minerals* 40: 567–567.

Yu, T. R. 1997. *Chemistry of Variable Charge Soils*. New York: Oxford University Press.

Zachara, J. M., and J. C. Westall. 1998. Chemical modeling of ion adsorption in soils. In *Soil Physical Chemistry, Second Edition*, ed. D. L. Sparks, 47–96. 2nd ed. Boca Raton, FL: CRC Press.

Zâlac, S., and N. Kallay. 1992. Application of mass titration to the point of zero charge determination. *Journal of Colloid and Interface Science* 149, no. 1: 233–240. doi:10.1016/0021-9797(92)90408-E.

Zelazny, L. W., L. He, and A. M. Vanwormhoudt. 1996. Charge analyses of soils and anion exchange. In *Methods of Soil Analysis. Part 3. Chemical Methods*, ed. D. L. Sparks, 1231–1254. Soil Science Society of America Book Series 5. Madison, WI: American Society of Agronomy-Soil Science Society of America.

6 Interparticle Interactions and Colloid Stability

One of the most important aspects in colloid chemistry in general and in soil science in particular is that of *colloid stability*. That refers to the processes schematically depicted in Figure 6.1. Colloid stability is the ability of a colloidal suspension to remain as such, even when the spontaneous tendency should be (as we shall discuss in this chapter) to aggregate; that is, the kinetics of the forward process of Figure 6.1 is slow enough. On the contrary, dispersion, which is the reverse process affecting aggregate stability, can be caused by a change in medium conditions, mainly ionic strength, leading to mechanical instability, which can be the root cause for some landslides (Rankka et al. 2004; Eilertsen et al. 2008). In the following sections, we develop the basis of particle–particle interactions and colloid stability.

6.1 THERMODYNAMICS OF PARTICLE–PARTICLE INTERACTION

Interactions between macroscopic particles can be classified into several groups. The first and most important group includes long-range interactions, namely van der Waals (VW), electrostatic, and magnetic forces; electrostatic interactions include both image forces and electrical double layer forces. The second group includes very short range interactions that can act only after the establishment of a contact area. These include chemical bonds of all types and hydrogen bonds. Also, some form of Born repulsion at very short distances is always present, precluding the two bodies to penetrate each other; recent research suggests the presence of other types of short-range repulsive forces such as oscillatory. There are also other interactions, but they are important in adhesion problems and will not be considered here (Fowlkes and Robinson 1989; Israelachvili 2010).

We begin the theoretical analysis of particle coagulation by considering thermodynamics aspects. Even when thermodynamic equilibrium is not always attained, in the sense of a dynamic equilibrium, the system behavior should reflect the spontaneous tendency. The Gibbs free energy change of contact, when two (in the general case, dissimilar) surfaces 1 and 2 are brought to contact from infinity in vacuum is (Israelachvili 2010)

$$\Delta G_{con} = \gamma_{12} - \gamma_1 - \gamma_2 = -w_{12} \tag{6.1}$$

where γ_{12} is the 1–2 interfacial energy, and γ_1 and γ_2 are the original surface energies, respectively; w_{12} is the *reversible work of adhesion* between surfaces 1 and 2. Figure 6.2a shows the process in an idealized way; in practice, especially with natural particles that are expected to be rough and irregular, only part of a given particle surface will be in close contact with another one, but the qualitative features are

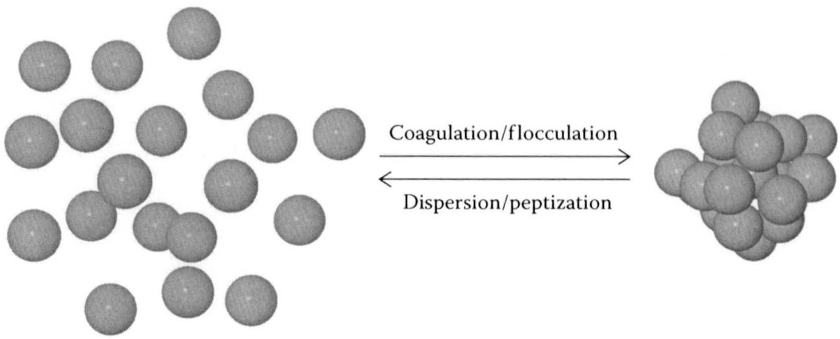

FIGURE 6.1 Colloids in suspension are in the dispersed or *peptized* state shown at left. Eventually the particles would aggregate into larger entities, which will sediment. The reverse process is possible, which can lead to landslides.

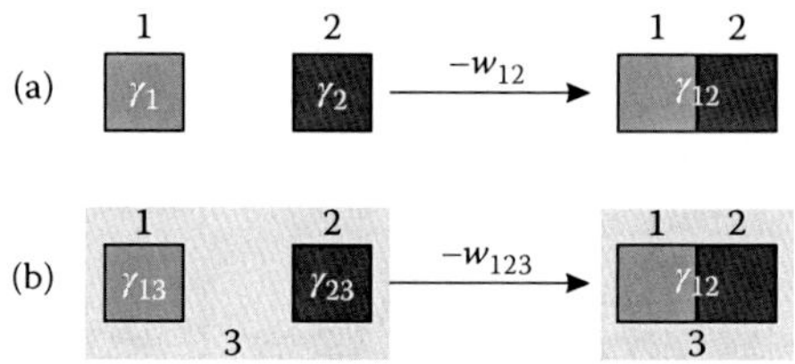

FIGURE 6.2 (a) Idealized illustration of the contact or adhesion between two surfaces 1 and 2 in vacuum; (b) the same for the two surfaces in a third medium.

unchanged. ΔG_{con} is thus the difference between the free energy spent in forming a unit area of the 1–2 interface and the free energy released by the removal of the 1-vacuum and 2-vacuum unit area surfaces. If the two surfaces are of the same material, then Equation 6.1 becomes

$$\Delta G_{con} = 0 - \gamma_1 - \gamma_1 = -w_{11} \tag{6.2}$$

because now in the right-hand side of Figure 6.2a there is no interface; or

$$\gamma_1 = \frac{w_{11}}{2} \tag{6.3}$$

where w_{11} is the reversible *work of cohesion* of phase 1. Note that both works, cohesion and adhesion, are defined as the reversible work needed to break apart the two phases (adhesion) or a single phase into two (cohesion).

The Gibbs free energy change per unit area when two surfaces, immersed in a third medium (3), attach each other is given by the difference between the energy spent forming the unit area 1–2 interface and that released for the removal of corresponding 1–3 and 2–3 interfaces as schematized in Figure 6.2b (Israelachvili 2010):

$$\Delta G_{con} = \gamma_{12} - \gamma_{13} - \gamma_{23} = -w_{123} \tag{6.4}$$

where γ_{12}, γ_{13}, and γ_{23} are the respective interfacial free energies. The surface energy of each interface is, in turn, the sum of several contributions, as discussed at the beginning of this section, namely VW forces (including dispersion and polarization), electrostatic double layer (el) and image (im) forces, and short range (sr) and very short range chemical (chem) interactions, so that

$$\gamma_i = \gamma_{i,VW} + \gamma_{i,el} + \gamma_{i,im} + \gamma_{i,sr} + \gamma_{i,chem} \tag{6.5}$$

Then, ΔG_{con} should, in general, be affected by all these contributions, which are usually assumed to be independent (i.e., can be calculated independently of each other). However, except at short distances, only the two first terms in the right-hand side of Equation 6.5 are important; image forces are present only for metal surfaces and the remainder act only at very short distances.

The theory of VW interactions between macroscopic bodies has been extensively developed, on the basis of the work by Hamaker and Lifshitz (Israelachvili 2010). Electrostatic interactions between charged surfaces have also been extensively studied in the context of colloidal particles homocoagulation and heterocoagulation (Verwey and Overbeek 1948; Derjaguin 1954; Hogg, Healy, and Fuerstenau 1966; Usui 1972). Image forces arise when a charged body approaches a conducting surface, developing by field reflection an image charge of opposite sign and hence an attractive interaction (Fowlkes and Robinson 1989). In general, the quantitative treatment of the two last types in Equation 6.5 is very difficult because they are primarily specific to each case, being dependent on the particle and surface materials. The following sections thus cover mainly the treatments of the long-range interactions.

As it will be seen in the next sections, there are basically two approaches: analysis of interaction forces and analysis of interaction energies. In the second view, the interaction energy $U(d)$ of two particles at a distance d apart can be found from the free energy as

$$U(d) = \Delta G(d) - \Delta G(\infty) \tag{6.6}$$

where $\Delta G(d)$ is the free energy change in bringing the particles at a distance d, whereas $\Delta G(\infty)$ is the same when the particles are at infinite distance.

6.2 VAN DER WAALS PARTICLE–PARTICLE INTERACTIONS

The VW interactions between two plane surfaces can be readily found by extension of the molecule–surface interaction deduced in Section 4.3.3.2.1; Figure 6.3a shows the calculation: we consider a thin slab of molecules 1 at a distance z from the surface formed by molecules 2 and thickness dz; for each molecule in the slab, the interaction energy with the other surface is given by Equation 4.25, so that the total energy of interaction per unit area between planes 1 and 2 is

$$U^S_{12,pl-pl}(d) = \int_{z=d}^{z=\infty} \int_{x=0}^{x=\infty} -\frac{\pi u'_{12}\,\rho_2}{6z^3}\,\rho_1\,dx\,dz = -\frac{\pi u'_{12}\,\rho_1\rho_2}{12d^2} = -\frac{A_{12}}{12\pi d^2} \tag{6.7}$$

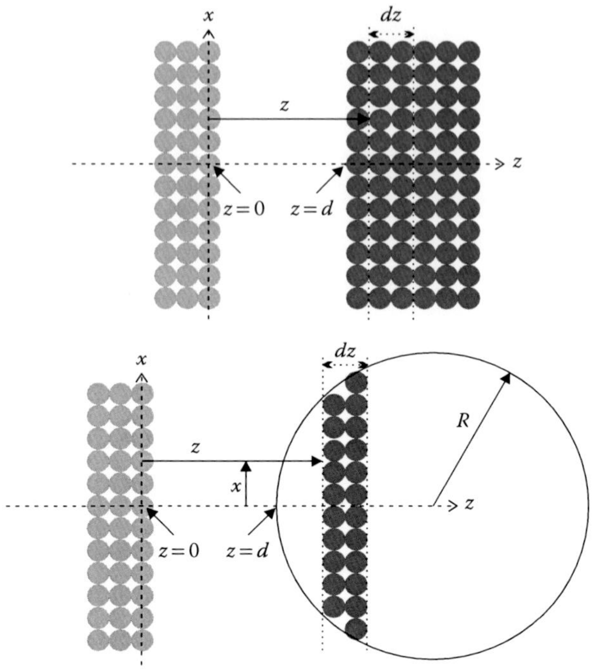

FIGURE 6.3 Method of integration of the van der Waals forces over macroscopic bodies. (a) The interaction of two planar surfaces formed by molecules A (light spheres) and B (dark spheres) at a distance d apart; (b) the same for a sphere and a surface.

where A_{12}, known as the Hamaker constant for the interaction between 1 and 2, is a function of the static and dynamic dielectric constants of the surfaces and the intervening medium, if any. ρ_1 and ρ_2 are the respective densities, and the superscript S indicates that it is energy per unit area. For a sphere 1 of radius R_1 and a surface 2, Figure 6.3b shows the geometric construction: now the integration is carried over x and z, taking into account the area of the slab as a function of z. It is found that for this case that if $d << R_1$, the total energy of interaction is

$$U_{12,sp-pl}(d) = -\frac{\pi^2 u'_{12}\,\rho_1\rho_2 R_1}{6d} = -\frac{A_{12}R_1}{6d} \tag{6.8}$$

and if $d >> R$,

$$U_{12,sp-pl}(d) = -\frac{2A_{12}R_1}{8d^3} \tag{6.9}$$

For the general case of two spheres of radii R_1 and R_2 at a distance d apart, Hamaker (1937) derived the following general equation:

$$U_{12,sp-sp}(d) = -\frac{A_{12}}{6}\left[\frac{2R_1R_2}{(2R_1+2R_2+d)d} + \frac{2R_1R_2}{(2R_1+d)(2R_2+d)} + \ln\frac{(2R_1+2R_2+d)d}{(2R_1+d)(2R_2+d)}\right] \tag{6.10}$$

It may be verified (Israelachvili 2010) that Equation 6.10 reduces to Equations 6.8 and 6.9 under appropriate conditions. For small distances, Equation 6.10 can be approximated by

$$U_{12,sp-sp}(d) = -\frac{A_{12}}{6d}\left(\frac{R_1 R_2}{R_1 + R_2}\right) \tag{6.11}$$

The important parameter characterizing the VW interactions is the Hamaker constant. In Equation 6.8, a simplified form is given:

$$A_{12} = \pi^2 u'_{12}\, \rho_1 \rho_2 \tag{6.12}$$

However, it is based on the approximations already discussed in Chapter 4, especially the additivity of pair interactions, which does not hold for London interactions. The alternative Lifshitz theory (Dzyaloshinskii, Lifshitz, and Pitaevskii 1961), based on quantum field theory, treats the large bodies as continuous media, avoiding totally the consideration of atomic structure but using bulk properties such as dielectric permittivity and refractive indices instead. It has been further developed more recently and alternative, simpler treatments have been given (Mahanty and Ninham 1977; Hough and White 1980; Parsegian 2006; Israelachvili 2010). It can be shown that the bulk or volume polarizability of a planar dielectric medium 1 in a medium 3 is given by

$$\alpha_1 = \frac{2\varepsilon_0 \varepsilon_{r3}\left(\varepsilon_{r1} - \varepsilon_{r3}\right)}{\rho_1\left(\varepsilon_{r1} + \varepsilon_{r3}\right)} \tag{6.13}$$

where ε_0 is the vacuum permittivity and the ε_r are the relative static dielectric constants. Thus, the Hamaker constant (nonretarded contribution) can be now expressed in terms of McLachlan equation (Equation 4.16), replacing the sum by an integral. It is found that A is approximately given, for two surfaces 1 and 2 immersed in a medium 3, considering only the nonretarded contribution, by

$$A_{123} = \frac{3}{4}\left\{kT\left(\frac{\varepsilon_{r1} - \varepsilon_{r3}}{\varepsilon_{r1} + \varepsilon_{r3}}\right)\left(\frac{\varepsilon_{r2} - \varepsilon_{r3}}{\varepsilon_{r2} + \varepsilon_{r3}}\right) + \frac{h}{\pi}\int_{\nu_1}^{\infty}\left[\frac{\varepsilon_{r1}(i\nu) - \varepsilon_{r3}(i\nu)}{\varepsilon_{r1}(i\nu) + \varepsilon_{r3}(i\nu)}\right]\left[\frac{\varepsilon_{r2}(i\nu) - \varepsilon_{r3}(i\nu)}{\varepsilon_{r2}(i\nu) + \varepsilon_{r3}(i\nu)}\right] d\nu\right\} \tag{6.14}$$

Here, $\varepsilon_{r1}(i\nu)$, $\varepsilon_{r2}(i\nu)$, and $\varepsilon_{r3}(i\nu)$ are the corresponding frequency dependent parts of the relative dielectric constants. This expression applies to any of the Equations 6.7 through 6.10. In general, a knowledge of the dielectric properties of a given material is required to calculate A. However, usually they can be approximately represented, for a dielectric, by an expression similar to that given in Equation 4.18 for the polarizability (Mahanty and Ninham 1977; Parsegian 2006):

$$\varepsilon_{ri}(i\nu) = 1 + \frac{\varepsilon_{ri}(0) - n_i}{1 + \dfrac{\nu}{\nu_{rot,i}}} + \frac{n_i^2 - 1}{1 + \dfrac{\nu^2}{\nu_{e,i}^2}} \tag{6.15}$$

where n_i is the refractive index, $\nu_{\mathrm{rot},i}$ is a characteristic rotational relaxation frequency, and $\nu_{\mathrm{e},i}$ is the main electronic absorption frequency for medium i. Usually, because $\nu_{\mathrm{rot}} \approx 10^{11}$ Hz and the minimum frequency (ν_1) in the McLachlan expression (Equation 4.16) is in the UV range ($\approx 4 \times 10^{13}$ Hz), the second term in the right-hand side of Equation 6.15 can be omitted. For a metal, on the contrary, $\varepsilon_{ri}(i\nu)$ is given by

$$\varepsilon_{ri}(i\nu) = 1 + \frac{\nu_{e,i}^2}{\nu^2} \tag{6.16}$$

where $\nu_{e,i}$ is the plasma frequency of the free electron gas, about 4×10^{15} Hz. In Equation 6.14, the integration is carried out in the optical frequency range, so the lower limit is usually taken as ν_1. Equation 6.14 includes, to a first approximation, the zero frequency interaction, with the Debye and Keesom contributions (first term) as well as the London dispersion energy (second term). If the medium is vacuum, or a diluted gas, then $\varepsilon_{r3}(0) = \varepsilon_{r3}(i\nu) = 1$. If the absorption frequencies of the three media are assumed equal, it is found (Israelachvili 2010) that

$$A_{123} \approx \frac{3}{4}\left[kT\left(\frac{\varepsilon_{r1}-\varepsilon_{r3}}{\varepsilon_{r1}+\varepsilon_{r3}}\right)\left(\frac{\varepsilon_{r2}-\varepsilon_{r3}}{\varepsilon_{r2}+\varepsilon_{r3}}\right) + \frac{h\nu_e}{2\sqrt{2}} \times \frac{(n_1^2-n_3^2)(n_2^2-n_3^2)}{\sqrt{n_1^2+n_3^2}\sqrt{n_2^2+n_3^2}\left(\sqrt{n_1^2+n_3^2}+\sqrt{n_2^2+n_3^2}\right)}\right] \tag{6.17}$$

where the n_i are the respective diffraction indexes. There are some interesting aspects that stem from Equations 6.14 and 6.17:

1. The London–VW interactions between similar particles of surfaces (identical material) is always attractive (A positive), whereas for dissimilar materials it can be positive or negative, depending on the values of the ε_r and n; that is, the interactions can be *repulsive.*
2. If the surrounding medium (3) is vacuum or air ($\varepsilon_{r3} = n_3 = 1$), the interaction is always attractive.
3. The Hamaker constant for two similar bodies interacting across a medium is the same if both materials are interchanged.
4. The dispersion contribution (second term in Equations 6.14 and 6.17) can be very high if one of the media has a high refractive index, such as a metal or other conducting media. On the contrary, the zero frequency contribution cannot exceed $(3/4)kT$ ($\approx 3 \times 10^{-21}$ J at 300 K).

Some calculated and experimental values of Hamaker constants for several surfaces are given in Table 6.1. It is also seen in Table 6.1 that interactions involving metallic surfaces are stronger.

There are a number of other approaches to calculate VW interactions that are not covered here; more exact expressions are given, especially, for the case of two spheres (Langbein 1970; Clayfield, Lumb, and Mackey 1971; Langbein 1974). Recently, more accurate calculations have been reported, using the coupled dipole method for nanocolloids (Kim and Kent 2009) and multipole expansions using classic electrostatics and statistical mechanics methods (Stenhammar et al. 2010); the results are compared with the classical Hamaker expression.

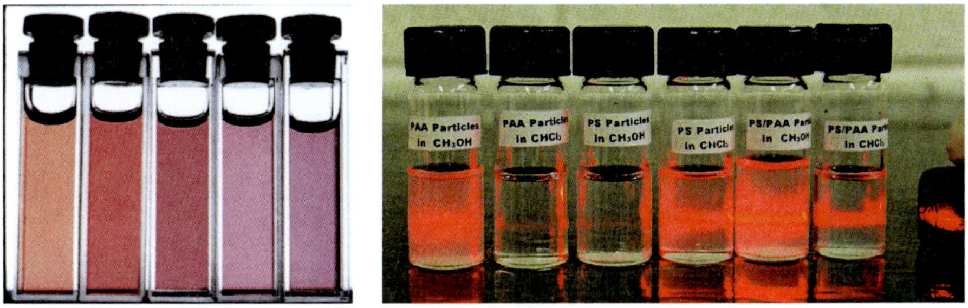

(a) (b)

FIGURE 2.16 (a) Photograph of five sols of colloidal gold prepared in water and in mixtures of butyl acetate and carbon disulfide. Refractive indices of the solutions increase from left to right. (Reprinted with permission from Underwood, S., and P., Mulvaney, 1994, *Langmuir*, 10, no. 10, 3427–3430. Copyright 1994 American Chemical Society.) (b) Tyndall light-scattering experiment shows the difference between stable and unstable suspensions of polymer-coated silica particles in different solvents; those which disperse the light contain particles in suspension. (Reprinted with permission from Li, D. et al., 2005, *Journal of the American Chemical Society*, 127, no. 17, 6248–6256. Copyright 2005 American Chemical Society.)

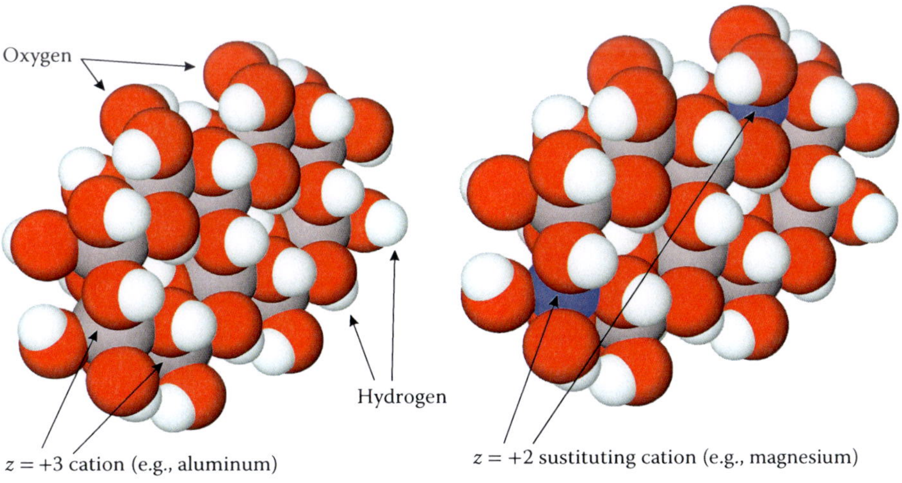

(a) (b)

FIGURE 5.1 (a) Molecular model representation (not to scale) of a fragment of a gibbsite-like layer. The light violet spheres represent aluminum atoms, the red ones represent oxygen, and the white ones hydrogen. (b) The blue spheres represent divalent atoms (such as magnesium or iron) substituting trivalent aluminum atoms, resulting in a permanent charge $q_f = -2$.

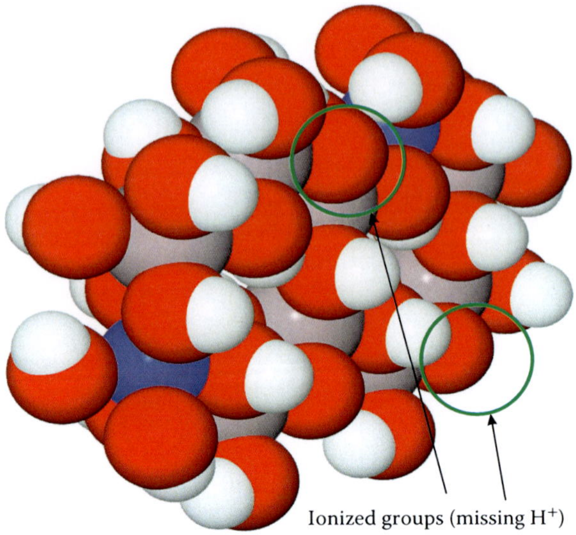

FIGURE 5.2 The gibbsite layer fragment from Figure 5.1b, which has lost two hydrogen atoms (due to acid–base interactions with the surrounding solution) at locations indicated by the green circles, resulting in a net proton charge of $q_H = -2$.

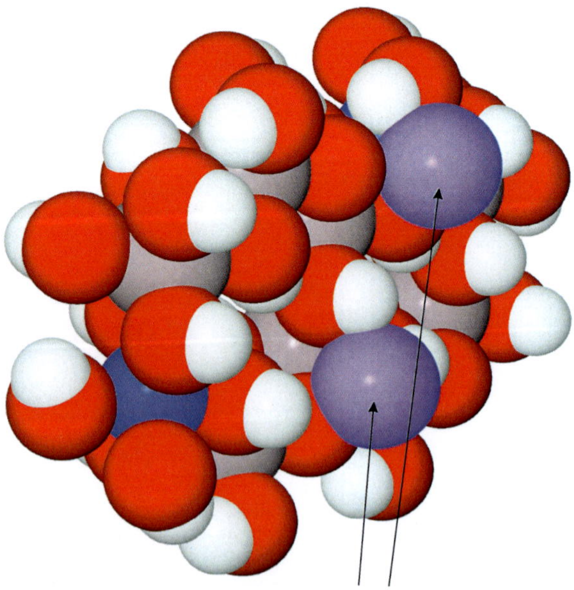

FIGURE 5.3 The gibbsite layer fragment from Figure 5.2 with two inner sphere complexed cations (light violet spheres), assumed of charge +1 each, giving an inner sphere sorption charge $q_1 = +2$.

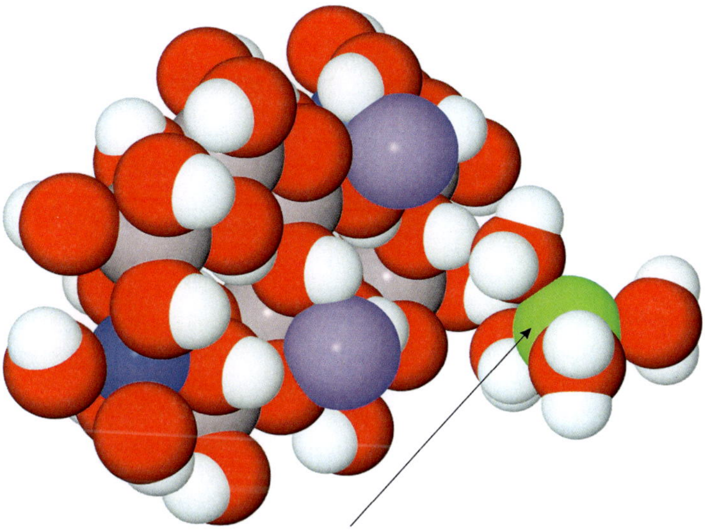

FIGURE 5.4 The gibbsite layer fragment with inner sphere sorption shown in Figure 5.3 with an outer sphere bound cation, represented by a green sphere with some water molecules. Free water molecules are not represented. This cation is assumed to be of charge $z = 1$, giving the outer sphere sorption charge $q_O = 1$.

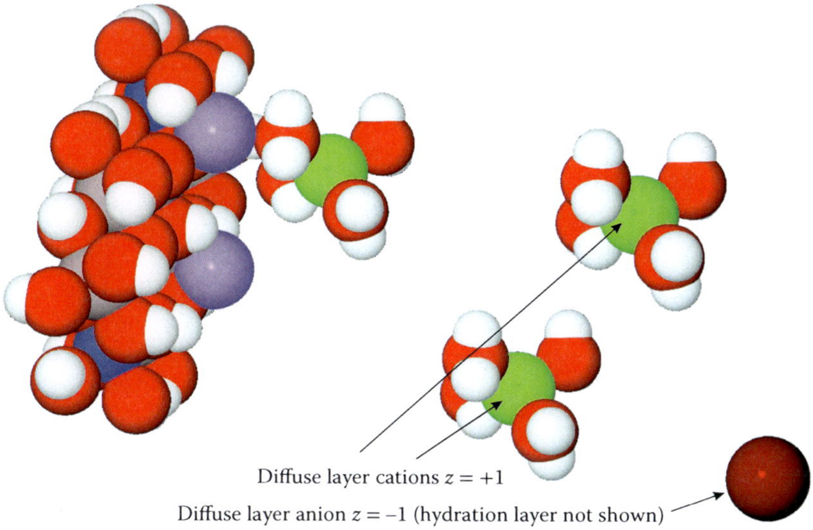

FIGURE 5.5 The example given in Figure 5.4 is completed with the diffuse layer charge $q_d = +1$, resulting in this example from two cations and one anion (dark purple, extreme right). Solvent water molecules are not shown.

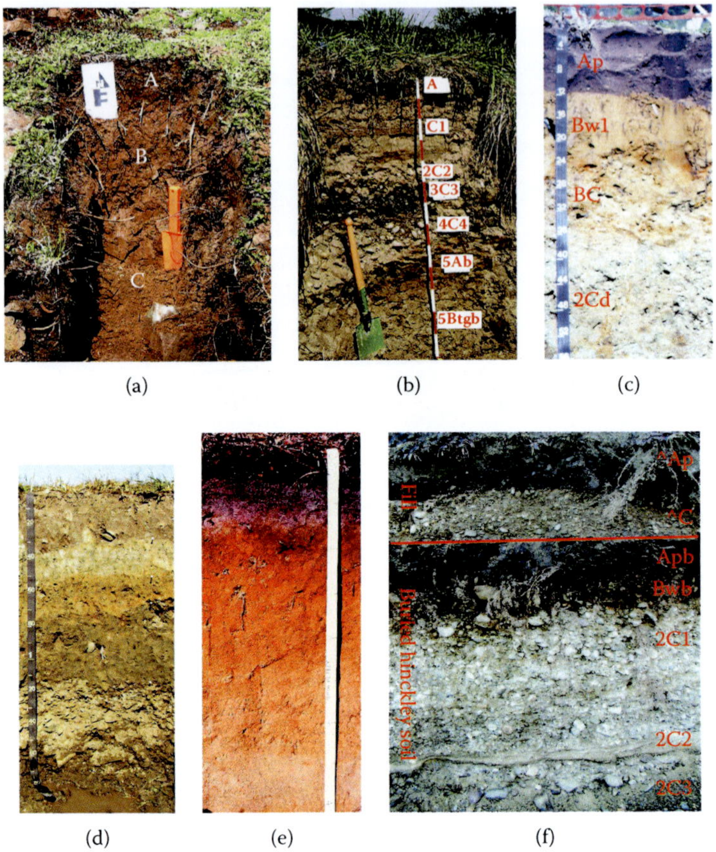

FIGURE 7.1 Examples of soil profiles: (a) soil section in a turf mound of a typical stone polygon on Trinchera Peak, a subglacial region in the Rocky Mountains, Colorado. Well-developed A, B, and C soil horizons are observed. (Copyright 2003 J.S. Aber, reprinted with permission from the author.) (b) Soil profile from the Paint Branch creek in Maryland. The different horizons are labeled in detail. (Courtesy of Ray R. Weil, University of Maryland, College Park, Maryland.) (c) Birchwood soil profile. Birchwood soils are moderately well drained soils formed in sandy eolian and/or fluvial material underlain by loamy dense till. This profile has the following morphology: 0–12 in., Ap horizon—loamy sand; 12–20 in., Bw1 horizon—loamy sand; 20–34 in., BC horizon—gravelly loamy sand with many redoximorphic features (seasonal high water table within 24 in.); 34–65 in., 2Cd—gravelly fine sandy loam, very firm basal till, and many redoximorphic features. (Courtesy of Rob Tunstead, reprinted from www.nesoil.com with permission from James Turenne.) (d) Raypole (like) soil profile, poorly drained loess over fluvial material, from Newport, Rhode Island. (Reprinted from www.nesoil.com with permission from James Turenne.) (e) Soil profile of a Carver soil from Massachusetts. Carver soils are very deep, excessively drained soils formed in layers of coarse and very coarse sand that contain less than 20% rock fragments, most of which are fine gravel. Carver soils are level to steep soils on pitted and dissected outwash plains and moraines. (Reprinted from www.nesoil.com with permission from James Turenne.) (f) Profile of an anthropogenic soil. This profile shows a series of fill layers that were placed over a Hinckley soil series. The "^" symbol is the new horizon prefix for human transported material (HTM). (Reprinted from www.nesoil.com with permission from James Turenne.)

TABLE 6.1
Some Calculated and Experimental Values of the Hamaker Constant

Interacting Media			$A/10^{-20}$ J			
1	3	2	Equation 6.14	Exact Calculation	Experimental	References
Water	Air	Water	3.7	3.70		Israelachvili (2010)
Polystyrene	Water	Polystyrene	1.4	1.3		Israelachvili (2010)
Mica	Water	Mica	2.0	2.0	2.2	Israelachvili (2010)
Fused quartz	Tetradecane	Air	−0.4		−0.5	Israelachvili (2010)
Fused quartz	Water	Air	−0.87	−1.0		Israelachvili (2010)
Fused quartz	Water	Fused quartz	0.63	0.5–1.0		Israelachvili (2010)
α-Alumina	Water	α-Alumina	4.2	2.7–5.2	6.7	Israelachvili (2010)
Metal (Au, Ag, Cu)	Air	Metal (Au, Ag, Cu)	25–40			Israelachvili (2010)
Au, Ag, Cu	Water	Au, Ag, Cu	20–40	30–40	40 (Au)	Israelachvili (2010)
Pt	KCl solution	Pt			3.5–4.5	Derjaguin and Voropayeva (1964)
Au	KCl solution	Au			14	Derjaguin and Voropayeva (1964)
Hg	KF solution	Hg			12	Usui, Yamasaki, and Shimoizaka (1967)
Hg	Water	Hematite	12			Andrade, Molina, and Posadas (1999)
Hg	Air	Hematite	17			Andrade, Molina, and Posadas (1999)

6.3 ENERGIES AND FORCES OF INTERACTION

We are treating interactions in terms of energies, but they can also be treated as forces, because all contributions to intermolecular interactions studied in Section 4.3.3 cause forces to act on the molecules; the relation between force and potential energy is well known:

$$\tilde{F}(d) = -\frac{\partial U(d)}{\partial d} \tag{6.18}$$

where $\tilde{F}(d)$ is the force between two bodies separated by a distance d. From a thermodynamic point of view, one is interested in energies, but nowadays with the widespread diffusion of force measuring methods, such as atomic force microscopy (Balnois, Papastavrou, and Wilkinson 2007), the force approach is also interesting. Besides, some theoretical treatments have been developed in terms of interaction forces (Derjaguin 1954; Lima et al. 2007; Quesada-Pérez, Hidalgo-Álvarez, and Martín-Molina 2009). We begin noting that for the interaction of a sphere with a plane at short distances, Equation 6.8, we get

$$\tilde{F}_{12,sp-pl}(d) = -\frac{\partial U_{12}(d)}{\partial d} = -\frac{A_{12}R_1}{6d^2} \tag{6.19}$$

which, comparing with Equation 6.7, leads to

$$\tilde{F}_{12,sp-pl}(d) = 2\pi R_1 U^S_{12,pl-pl}(d) \tag{6.20}$$

which is a useful equation for sphere–plane interactions. Now, we can derive an approximate expression for sphere–sphere interaction forces as follows. Consider Figure 6.4, where we have two spheres of radii R_1 and R_2 at a distance d apart, where $d << R_i$; if we take a small circular section of infinitesimal thickness dx on each sphere facing each other at a distance $z = d + z_1 + z_2$, the force between them will be

$$\tilde{F}_{12,sp-sp}(d) = \int_{z=d}^{z=\infty} 2\pi x \tilde{F}^S_{12,pl-pl}(z)\,dx \tag{6.21}$$

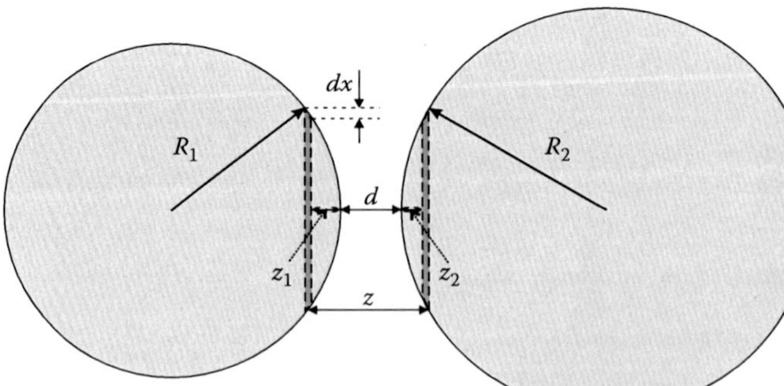

FIGURE 6.4 The Derjaguin approximation geometry. The force between the two spheres is calculated from the interaction between two circular regions (dark segments), assumed flat, using the expression for interaction between plane surfaces.

where $\tilde{F}^{S}_{12,pl-pl}(z)$ is the force per unit area between parallel planes at a distance z. After some algebra, it is found that

$$\tilde{F}_{12,sp-sp}(d) = 2\pi\left(\frac{R_1R_2}{R_1+R_2}\right)U^{S}_{12,pl-pl}(d) \tag{6.22}$$

where $U^{S}_{12,pl-pl}(d)$ is given by Equation 6.7. This result is known as the *Derjaguin approximation* (Derjaguin 1934; Israelachvili 2010) and in principle is applicable to any type of force law, as long as d is small compared with the sphere radii. Some conclusions can be drawn from Equation 6.22:

1. If one sphere is much larger than the other one ($R_2 >> R_1$), it is found again in the expression for a sphere near a plane surface, Equation 6.20; if both spheres have the same radius, half that value is obtained: $\tilde{F}_{12}(d) = \pi R_1 U^{S}_{12}(d)$.
2. For two spheres in contact ($d = d_0$, the collision diameter), the interaction energy may be equated to -2γ, where γ is the surface energy (Chapter 2),

$$\tilde{F}_{12,sp-sp}(d_0) = -\frac{4\pi\gamma R_1R_2}{R_1+R_2} \tag{6.23}$$

   which is also the *adhesion* force between both spheres (Israelachvili 2010).
3. One interesting point is that Equation 6.22 predicts a different distance dependence for the force between two curved surfaces compared with two planes. This is schematized in Figure 6.5, comparing force and energy

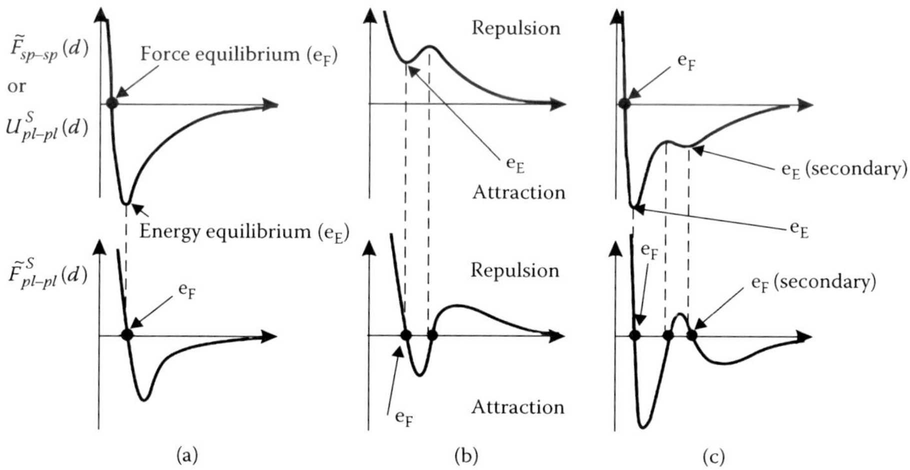

FIGURE 6.5 Comparison between force laws for curved surfaces (top row) and plane surfaces (bottom row). The points marked with "e$_F$" represent equilibrium points (null force and negative slope) for the force curves, and those marked with "e$_E$" represent the equivalent points (minima) in the energy curves. (a) A curve with a single minimum: the force minimum for curved surfaces is the zero force point for flat surfaces; (b) a purely repulsive force for curves gives an attractive force range for planes; (c) a purely attractive force for curved surfaces gives a repulsive force range for planes. (Adapted from *Intermolecular and Surface Forces*, 3rd ed., Israelachvili, J.N. Copyright 2010, with permission from Elsevier.)

curves; it can be seen, for example, that a repulsive interaction between curved surfaces can be transformed (for certain distances) into attractive forces for planar surfaces (Figure 6.5b). In Figure 6.5c, an example of a curve with a *secondary minimum* in the energy curve, which results in a point of quasi-equilibrium (true equilibrium is at the absolute, or primary, minimum), is shown. We shall return to the different curve types later on.

6.4 ELECTROSTATIC INTERACTIONS

The electrostatic interactions between surfaces and/or particles immersed in a medium of high dielectric constant are characterized by the overlapping and interaction of the respective double layers. The surfaces develop charge, as discussed in Chapter 5, and the surface charge, which in the general case is determined through reactions with the solution, in turn determines the potential profile, as developed in Chapter 3. As it was seen there, under appropriate assumptions, the relationship between ionic distribution and potential in these systems is given by the Poisson–Boltzmann (PB) equation:

$$\nabla^2\psi(\mathbf{r}) = -\frac{\boldsymbol{F}}{\varepsilon}\sum_i z_i c_i^{\infty} \boldsymbol{e}^{-\frac{z_i \boldsymbol{F}\psi(\mathbf{r})}{RT}} \tag{6.24}$$

Now, this equation is applied to the interaction of two particles immersed in an electrolyte solution. For plane surfaces, the situation is depicted in Figure 6.6: if the surfaces are far away (Figure 6.6a), the potential decays to nearly zero in the middle region, and interactions can be neglected. When the surfaces are at shorter distances (of the order of κ^{-1}), the diffuse layers overlap and interact with each other. To solve the electrostatic problem, we have to set boundary conditions at the surfaces. Here, several possibilities arise:

1. Constant potential: The potential of both surfaces are assumed to remain constant as d decreases. In practice, this case arises when the surface is in equilibrium with the solution through some ion-binding reaction, such ion being the potential determining ion, usually H^+. This case is the most often considered (Overbeek 1949a; Derjaguin 1954; Hogg, Healy, and Fuerstenau 1966; Attard 2001), as most soil colloids are equilibrated with the soil solution; for example, if charging is attained through a reaction like

$$\mathrm{SOH} \rightleftharpoons \mathrm{SO^-} + \mathrm{H^+} \tag{6.25}$$

equilibrium requires that

$$\tilde{\mu}_{\mathrm{SOH}} = \tilde{\mu}_{\mathrm{SO^-}} + \tilde{\mu}_{\mathrm{H^+}} \tag{6.26}$$

where S is a surface hydroxylated site, or

$$\mu_{\mathrm{SOH}}^S = \mu_{\mathrm{SO^-}}^S - \boldsymbol{F}\psi^S + \mu_{\mathrm{H^+}}^W + \boldsymbol{F}\psi^W \tag{6.27}$$

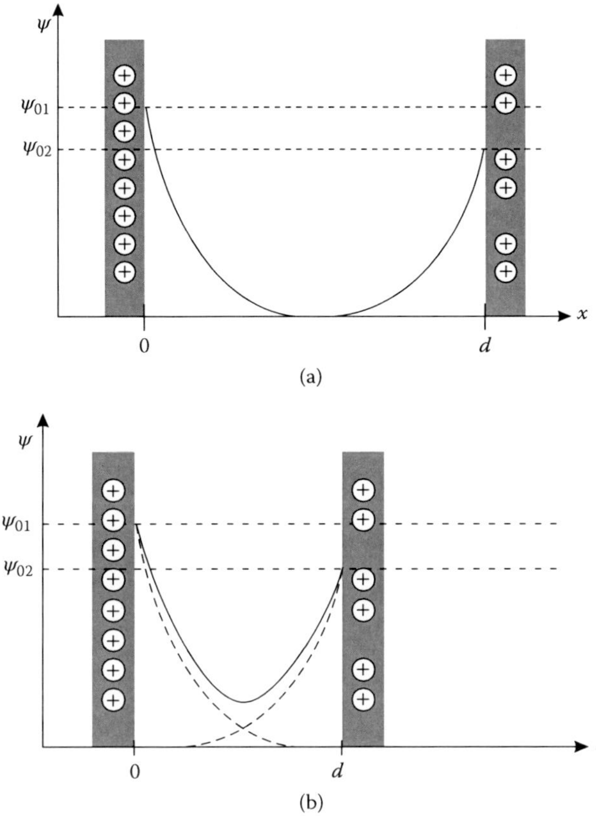

FIGURE 6.6 The interaction between parallel plane double layers in an aqueous electrolyte solution. (a) Two plane identical slabs bearing positive charge, not necessarily equal, when far away, have independent diffuse layers, with the electrostatic potential decaying with an approximately exponential profile; (b), as the distance d diminishes, the diffuse layers overlap, and the potential in the middle region no longer decays to zero but to an intermediate value.

where ψ^W is the potential in the bulk solution, which is conventionally taken as zero. We find from Equation 6.27 for the surface potential

$$\psi^S = \frac{\mu^S_{SO^-} + \mu^W_{H^+} - \mu^S_{SOH}}{\boldsymbol{F}} \tag{6.28}$$

which will remain constant as long as the H^+ activity (or pH) remains constant.

2. Constant charge: Under some circumstances, the equilibrium at the particle surface cannot be attained because of slow kinetics or of a dominating permanent charge. In such cases, the constant surface charge density is the boundary condition (Overbeek 1949a; Gregory 1975; McCormack, Carnie, and Chan 1995).
3. Charge regulation: In this case, neither the potential nor the charge is assumed constant but adjust along with the surface reaction equilibria as the particles approach each other. This condition is proposed to provide a more accurate

representation of the behavior of amphoteric surfaces than the constant charge or the constant potential models. Typical examples of such surfaces are metal oxides (Ettelaie and Buscall 1995; Dan 2002; Chan et al. 2006).

Also, depending on the attainment of equilibrium, the Gibbs free energy per unit of area can be found from (Overbeek 1949a; Kihira and Matijevic 1992)

$$\Delta G_S = -\int_0^{\psi_s} \sigma_S \mathrm{d}\psi' \tag{6.29}$$

where the integration is carried over the surface potential, and ψ_s is the actual surface potential, usually taken at the outer plane (ψ_2) as the relevant interaction is of the diffuse layers. Note that this applies, in principle, to constant potential interactions, as discussed in Section 6.4. On the contrary, if equilibrium is not maintained at all, under constant charge, the free energy is given by (Overbeek 1949a)

$$\Delta G_S = \int_0^{\sigma_s} \psi_S \mathrm{d}\sigma' \tag{6.30}$$

6.4.1 Constant Potential Conditions

Under the assumption of constant potential, the boundary conditions for the problem of Figure 6.6 are

$$\psi(0) = \psi_{0,1} \tag{6.31}$$

$$\psi(d) = \psi_{0,2} \tag{6.32}$$

However, the mathematical difficulties here are even higher than in the case of a single diffuse layer (Chapter 3), so, as it was done there, diverse simplifying assumptions were introduced, resulting in a number of different theories.

6.4.1.1 Linear Approximation: The Hogg–Healy–Fuerstenau Theory

The simplest approximation is of the Debye–Hückel type, first proposed by Hogg, Healy, and Fuerstenau (1966), assuming low potentials ($z_i F\psi \ll RT$) and linearizing Equation 6.24, leading to (see Section 3.2.1)

$$\nabla^2\psi(x) = \frac{F}{\varepsilon}\sum_i \frac{z_i^2 c_i^\infty F\psi(x)}{RT} = \kappa^2\psi(x) \tag{6.33}$$

which has been written in one dimension according to Figure 6.6. Equations 6.31 through 6.33 are straightforwardly solved, resulting in (Hogg, Healy, and Fuerstenau 1966):

$$\psi(x) = \psi_{0,1}\cosh(\kappa x) + \left[\frac{\psi_{0,2} - \psi_{0,1}\cosh(\kappa d)}{\sinh(\kappa d)}\right]\sinh(\kappa x) \tag{6.34}$$

and, from this expression and Equations 6.6, 6.29, and 3.22, it is found that the potential energy (per unit area) of two plane parallel double layers, which are at a distance d apart, is

$$U_{el}(d) = \frac{\varepsilon\kappa}{2}\left[\left(\psi_{0,1}^2 + \psi_{0,2}^2\right)(1 - \coth(\kappa d)) + 2\psi_{0,1}\psi_{0,2}\mathrm{cosech}(\kappa d)\right] \qquad (6.35)$$

For two spheres of radii R_1 and R_2 at a distance d, the following expression is obtained using the Derjaguin approximation:

$$U_{el}(d) = \frac{\pi\varepsilon R_1 R_2\left(\psi_{0,1}^2 + \psi_{0,2}^2\right)}{R_1 + R_2}\left\{\frac{2\psi_{0,1}\psi_{0,2}}{\psi_{0,1}^2 + \psi_{0,2}^2}\ln\left[\frac{1 + \exp(-\kappa d)}{1 - \exp(-\kappa d)}\right] + \ln\left[1 - \exp(-2\kappa d)\right]\right\} \qquad (6.36)$$

In fact, from Equation 6.36, other cases such as sphere–plane interaction can be easily found.

6.4.1.2 Identical Particles: Verwey–Overbeek Results

The classic treatment for identical surfaces involves, as in most treatments of particle–particle interaction, a symmetric ($z_+ = z_- = z$) electrolyte, where the PB equation reduces to

$$\frac{d^2\psi}{dx^2} = \frac{\boldsymbol{R}T\kappa^2}{z\boldsymbol{F}}\sinh\left(\frac{z\boldsymbol{F}\psi}{\boldsymbol{R}T}\right) \qquad (6.37)$$

which, for a single surface, leads to the Gouy–Chapman model (Chapter 3). Furthermore, Verwey and Overbeek assumed univalent electrolytes, so that here $z = 1$. They considered the situation shown in Figure 6.7a and assumed low potential values but not for all the range, only for the potential in the middle plane, ψ_m; thus, they treated the whole potential curve as the superposition of two simple Gouy–Chapman-like potential profiles (which is known as the *weak potential approximation*). The mathematical details can be found in the original work (Verwey and Overbeek 1948). They found for the electrostatic interaction energy (per unit area) between two plane parallel double layers

$$U_{el}(d) = \frac{64c^{\infty}\boldsymbol{R}T}{\kappa}\tanh^2\left(\frac{z\boldsymbol{F}\psi_0}{4\boldsymbol{R}T}\right)\boldsymbol{e}^{-\kappa d} \qquad (6.38)$$

and for two spheres of radius R_S

$$U_{el}(d) = \frac{64R_S c^{\infty}\boldsymbol{R}T}{\kappa^2}\tanh^2\left(\frac{z\boldsymbol{F}\psi_0}{4\boldsymbol{R}T}\right)\boldsymbol{e}^{-\kappa d} \qquad (6.39)$$

It should be noted in these equations, and in those of Section 6.4.1.1, that always the exponential decay of the form $\boldsymbol{e}^{-\kappa d}$ is predicted.

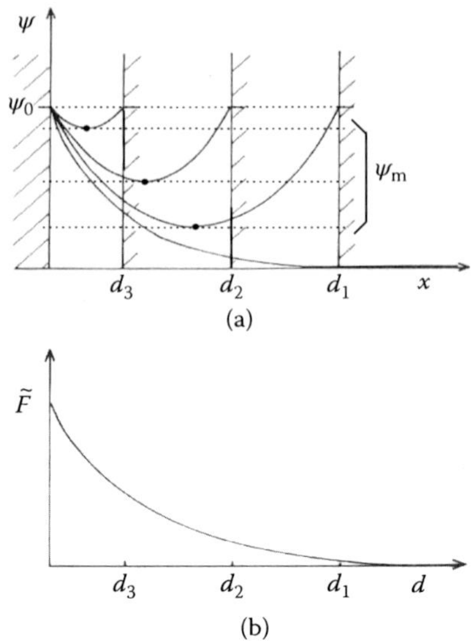

FIGURE 6.7 Schematic view of the potential profile evolution (a) and resulting interaction force (b) for two identical plane double layers. (Adapted from *Adv. Colloid Interface Sci.*, 42, Kihira, H., E. Matijevic, An assessment of heterocoagulation theories, 1–31. Copyright 1992, with permission from Elsevier.)

6.4.1.3 Derjaguin Theory

The linearization of the PB equation works reasonably well for low potentials and separations not too small, but leads to an overestimation of the electrostatic interactions when the surface potentials are higher than 50–60 mV (Urrutia et al. 1983). Also, experimental data on heterocoagulation kinetics showed great discrepancies with predictions based on this approximation (Marshall and Kitchener 1966; Hull and Kitchener 1969; Hansen and Matijevic 1980). Several treatments have been proposed to overcome these limitations (Kihira and Matijevic 1992; Ettelaie and Buscall 1995; Stankovich and Carnie 1996; Attard 2001; Chan et al. 2006). However, a previous work by Derjaguin (1954) with the help of V. G. Levich, proposed a more general theory, based on the force approach. He analyzed the potential profile and the interaction force as two plane parallel surfaces approached each other for dissimilar surfaces (different potentials). The interaction pressure (force per unit of surface) considering symmetric ($z_+ = z_- = z$) electrolytes is given (Verwey and Overbeek 1948; Kihira and Matijevic 1992; Hiemenz and Rajagopalan 1997; Israelachvili 2010) by a balance of two contributions, one osmotic

$$P_{os} = 2zc^{\infty}\boldsymbol{RT}\left[\cosh\left(\frac{\boldsymbol{F}\psi_0}{\boldsymbol{RT}}\right) - 1\right] \tag{6.40}$$

and the other electrostatic, coming from the interaction between the surfaces and the diffuse layers, known as *Maxwell tension* or *Maxwell stress*:

$$P_{el} = c^{\infty} \boldsymbol{RT} \left(\frac{z\boldsymbol{F}}{\boldsymbol{RT}} \right)^2 \frac{(\psi_{0,1} - \psi_{0,2})^2 \exp(-\kappa d)}{1 + \left(\dfrac{z\boldsymbol{F}}{\boldsymbol{RT}} \right)^2 \dfrac{(\psi_{0,1} + \psi_{0,2})^2}{4} \operatorname{csch}^2 \left(\dfrac{\kappa d}{2} \right)} - 2 \tag{6.41}$$

The solution found by Derjaguin was not in the form of an expression for the interaction force between the surfaces as a function of d but the opposite, determining d as a function of the force for various parameter values, such as surface potentials and ionic charges; furthermore, there are several expressions valid under different conditions (attractive and repulsive forces). The mathematics is rather involved, but the results can be represented as follows (Kihira and Matijevic 1992). First, we will consider the case of $\psi_{0,1} = \psi_{0,2}$; in Figure 6.7a, the evolution of the potential profile as the surfaces approach each other from d_1 to d_3 is shown, compared with the curve for an isolated surface (bottom curve); the profile has always the same shape, and only the potential at the middle point increases. Note that the charges at both surfaces, given by the slope of the potential profile

$$\sigma_S = -\varepsilon \left. \frac{d\psi}{dx} \right)_{x=0} \tag{6.42}$$

are equal and constant as the surfaces approach. The repulsion force between them is shown in Figure 6.7b, which increases steadily as the distance decreases. Similarly, the interaction energy, related to the force by Equation 6.18, increases also, showing increased repulsion. Of course, these results are not valid at short distances, as it will be seen later on, because the primitive model (continuum solvent and point charge ions) breaks down at distances of the order of a few molecular diameters.

For the case of dissimilar surfaces (here, different surface potentials), the behavior is more complex, as shown in Figure 6.8. Now, surface 1 has higher potential than surface 2; as the two surfaces approach, the profile changes its shape. At distance d_1, both surfaces have positive values of potential and charge, and the potential profile resembles that observed for equal potentials in Figure 6.7. At distance d_2, the slope of the potential profile at surface 2 vanishes, so that the surface charge is null. The interaction force, which is so far repulsive as in Figure 6.7b, reaches a maximum. At shorter distances, there is no change in the potential slope between the two surfaces; thus, now the second one has *negative* charge so that the electrostatic part of the interaction is attractive. At a particular value of the distance between the two surfaces, d_3 in Figure 6.8, the potential profile matches the case of an isolated surface, so that the force vanishes. Finally, for distances shorter than d_3, the overall force is *attractive*. These results, the charge and force inversion as distance decreases, may be at first surprising, but the charge reversal can be rationalized, considering that the surface more strongly charged (higher potential in absolute value) polarizes, at short distances, the

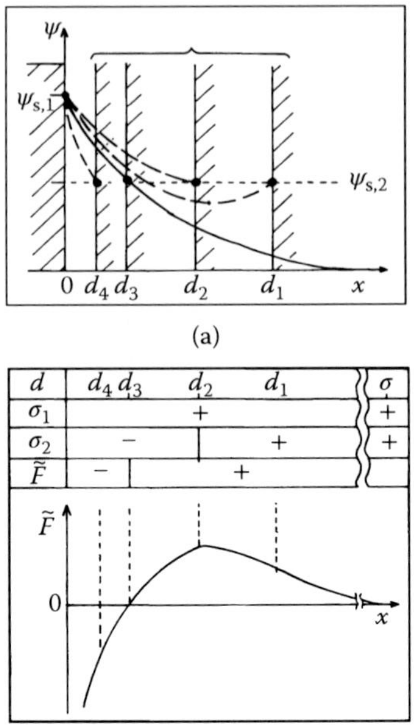

(b)

FIGURE 6.8 Schematic view of the potential profile evolution (a, dashed lines) and resulting interaction force (b) for two dissimilar plane double layers. The upper pane in (b) shows the signs of surface charges and interaction force for different values of the separation distance *d*. In (a), the closed circles indicate the potential value of the second surface; and the continuous line, the profile for an isolated surface. (Adapted from *Adv. Colloid Interface Sci.*, 42, Kihira, H., E. Matijevic, An assessment of heterocoagulation theories, 1–31. Copyright 1992, with permission from Elsevier.)

second one, forcing it to change its sign. Charge reversal has been experimentally observed (Kékicheff et al. 1993; Toikka, Hayes, and Ralston 1997; Hartley and Scales 1998); it is interesting to note that the Derjaguin treatment predicted charge reversal from basic arguments, with no mention of the possible mechanisms that would actually take place. In fact, a number of different explanations have been proposed, and the matter is not settled; Lyklema (2006) has reviewed the situation quite recently.

Extensive numerical results were presented for the case of plane surfaces by Devereux and De Bruyn (1963); however, some results have been challenged (Lyklema 2006).

6.4.2 Constant Charge Conditions and Comparison with Constant Potential

Here we start from the symmetrical electrolyte PB expression, Equation 6.37. It can be integrated once, leading to

$$\left(\frac{d\psi}{dx}\right)^2 = \left(\frac{\boldsymbol{R}T\kappa}{z\boldsymbol{F}}\right)^2 \left[2\cosh\left(\frac{z\boldsymbol{F}\psi}{\boldsymbol{R}T}\right) + C\right] \tag{6.43}$$

where C is an integration constant, dependent on the boundary conditions (see, e.g., McCormack, Carnie, and Chan 1995). Clearly, the final solution is strongly dependent on these conditions. For constant charge conditions, they are

$$\sigma_1 = -\varepsilon \left.\frac{d\psi}{dx}\right)_{x=0} \tag{6.44}$$

$$\sigma_2 = -\varepsilon \left.\frac{d\psi}{dx}\right)_{x=d} \tag{6.45}$$

whereas for constant potential they are given in Equations 6.31 and 6.32. The linearization of the PB equation, as done in the Hogg–Healy–Fuerstenau treatment for the constant potential case, gave inaccurate results. This is so because at constant charge the surface potentials grow as the surfaces approach each other; thus, the low potential condition (essential in the linearization) no longer holds. A rather complete numerical treatment of this case, along with the constant potential condition, was given by Bell and Peterson (1972). They derived simple formulas involving quadratures for the interaction free energy between two planes at both constant potential and constant charge, and the difference formula of Frens (1968) was generalized. Transformations to standard elliptic integrals were given for the various constant charge situations. The case of sphere–sphere interactions was solved using the Derjaguin approximation. It was shown that the behavior of the electrical forces between two dissimilar surfaces at small separations can be complex, and many unusual effects may arise, depending on which boundary condition is applicable. They also studied the small potential approximation, also finding it to be less effective at constant charge density than at constant potential. Approximate analytical expressions for surfaces at constant charge were proposed by Gregory (1975). He considered the interaction force per unit area (Equations 6.40 and 6.41). The interaction pressure under constant charge was found to be

$$P_{CC} = c^{\infty}\boldsymbol{RT}\left\{2\left[1+\left(\frac{z\boldsymbol{F}}{\boldsymbol{RT}}\right)^2\frac{\left(\psi_{0,1}+\psi_{0,2}\right)^2}{4}\operatorname{csch}^2\left(\frac{\kappa d}{2}\right)\right]^{1/2}\right.$$
$$\left. -\left(\frac{z\boldsymbol{F}}{\boldsymbol{RT}}\right)^2\frac{\left(\psi_{0,1}-\psi_{0,2}\right)^2\exp(-\kappa d)}{1+\left(\frac{z\boldsymbol{F}}{\boldsymbol{RT}}\right)^2\frac{\left(\psi_{0,1}+\psi_{0,2}\right)^2}{4}\operatorname{csch}^2\left(\frac{\kappa d}{2}\right)}-2\right\} \tag{6.46}$$

where $\psi_{0,1}$ and $\psi_{0,2}$ are the potentials of both *isolated* surfaces, that is, when they are at infinite distance. For comparison, the corresponding expressions for the interaction pressure for other cases were also derived (Gregory 1975); for the linear PB approximation at constant charge, it is found

$$P_{CC,l} = \frac{c^{\infty}(zF)^2}{RT}\left[\frac{\psi_{0,1}^2 + \psi_{0,2}^2 + 2\psi_{0,1}\psi_{0,2}\cosh(\kappa d)}{\sinh^2(\kappa d)}\right] \tag{6.47}$$

and at constant potential it reads

$$P_{CP,l} = \frac{c^{\infty}(zF)^2}{RT}\left[\frac{2\psi_{0,1}\psi_{0,2}\cosh(\kappa d) - \psi_{0,1}^2 - \psi_{0,2}^2}{\sinh^2(\kappa d)}\right] \tag{6.48}$$

Finally, the *linear superposition approximation* (LSA) is a simple way to try to take into account that real interfaces most often behave neither as constant potential surfaces nor as constant charge ones; LSA is a simple average between those two cases; for the linear PB, it results in

$$P_{LSA} = 64c^{\infty}RT\tanh\left(\frac{zF\psi_{0,1}}{RT}\right)\tanh\left(\frac{zF\psi_{0,2}}{RT}\right)e^{-\kappa d} \tag{6.49}$$

In Figure 6.9, the case for equal surface conditions is presented; different results are compared, including constant potential and constant charge cases, both approximate and numerical computations (Devereux and De Bruyn 1963; Bell and Peterson 1972) and the LSA results. In Figure 6.10, the case of different surface charges (potentials) of the same sign and different magnitude is presented, whereas Figure 6.11 is for different sign of the surface charges. A general observation is that the constant potential and charge cases have very different behavior, especially for unequally charged surfaces. It is seen also that the linear PB approximation gives poor results at constant charge but behaves reasonably well at constant potentials. The approximate expression given by Gregory for constant charge (Equation 6.46) behaves quite well under most conditions, except when the two surfaces have charges of opposite sign and equal magnitude. Charge reversal is predicted for equal sign charges at constant potential and for unequal sign under constant charge conditions. From the force expressions found, the energy of interaction can be deduced (Gregory 1975).

More recently, the interaction forces and energies have been computed for two-sphere (and sphere–plate) cases from the nonlinear PB equation (Carnie, Chan, and Stankovich 1994; Stankovich and Carnie 1996) and results compared with the linear approximation. Figure 6.12 shows some results for equal size spheres, in terms of the dimensionless potential, $\Psi = F\psi/RT$ and dimensionless interaction free energy, $G^* = G/\varepsilon\kappa^{-1}\left(RT/F\right)^2$ further scaled by Ψ^2. It is observed that the linear approximation gives reasonable results at low potentials as expected, but also for constant charge conditions, at variance with the parallel plane surfaces examined earlier. Charge inversion is found in the same cases as given in the preceding discussion, but only at small distances.

6.4.3 Charge Regulation

As discussed in the previous section, often neither constant potential nor constant charge conditions describe well the surface behavior, especially at small separation

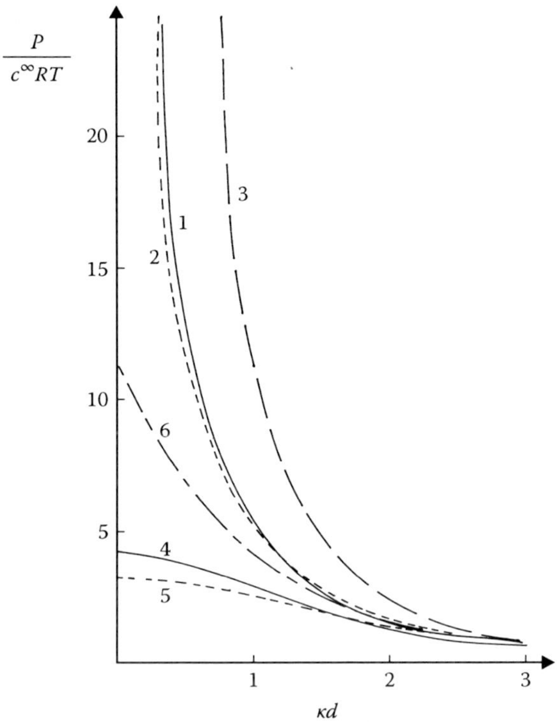

FIGURE 6.9 Force between plane parallel surfaces as a function of distance between them. Initial potentials at infinite separation: $\psi_{0,1} = \psi_{0,2} = 1.8$ ***RT/F***. (1) "Exact" results for constant charge (Bell and Peterson 1972); (2) approximate results, P_{CC}, Equation 6.46; (3) results from linear PB approximation, constant charge, $P_{CC,l}$, Equation 6.47; (4) "exact" constant potential results (Devereux and De Bruyn 1963); (5) linear PB approximation, constant potential, $P_{CP,l}$, Equation 6.48; (6) linear superposition results, P_{LSA}, Equation 6.49. (Adapted from *J. Colloid Interface Sci.*, 51, Gregory, J., Interaction of unequal double layers at constant charge, 44–51. Copyright 1975, with permission from Elsevier.)

distances. The process taking place is an adjustment of the surface equilibria as the particles approach each other, which causes change of both surface potential and charge. This is termed *charge regulation.* In the general case, the interaction with regulation involves solving a set of transcendental equations for self-consistent values of surface charge and potential at all separations for any given set of bulk parameters. Clearly, each solid–solution interface would, in general, have different surface reactions and hence different behavior under charge regulation. The subject has been addressed in a number of studies (Ninham and Parsegian 1971; Chan, Healy, and White 1976; Prieve and Ruckenstein 1976; Pashley 1981; Carnie and Chan 1993; Reiner and Radke 1993; Ettelaie and Buscall 1995; Hsu and Kuo 1997; Dan 2002; Chan et al. 2006). Because its treatment requires concepts to be introduced in Chapters 11 and 12, in Chapter 12 we show the results of a rather recent study (Chan et al. 2006).

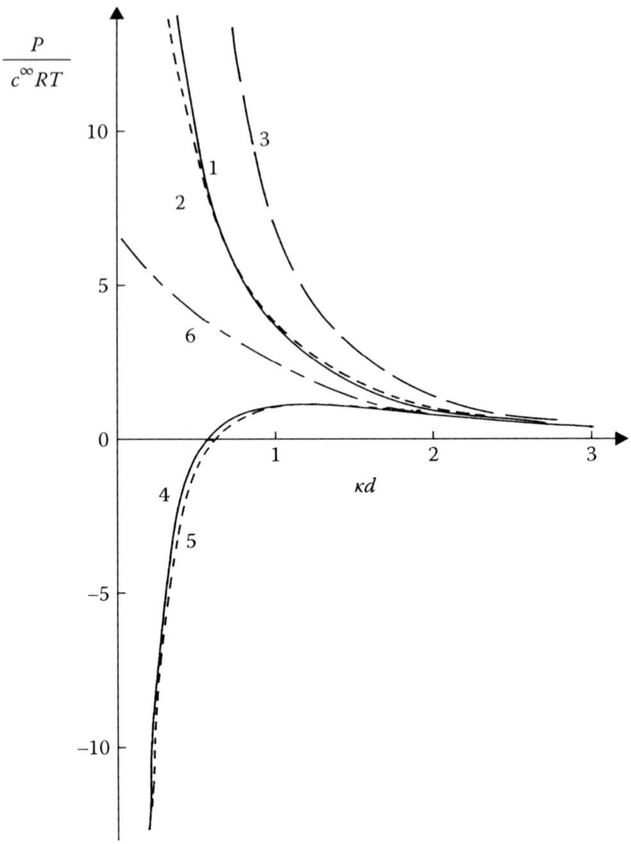

FIGURE 6.10 Similar to Figure 6.9, but with $\psi_{0,1} = 1.8\ \boldsymbol{RT/F}$, $\psi_{0,2} = 1.0\ \boldsymbol{RT/F}$. (Adapted from *J. Colloid Interface Sci.*, 51, Gregory, J., Interaction of unequal double layers at constant charge, 44–51. Copyright 1975, with permission from Elsevier.)

6.5 TOTAL LONG-RANGE INTERACTIONS

6.5.1 Similar Surfaces: Classic DLVO Theory

Considering only VW and electrostatic double layer interactions for identical surfaces, we have the famous Derjaguin–Landau–Verwey–Overbeek (DLVO) theory of colloid stability (Derjaguin and Landau 1941; Verwey and Overbeek 1948; Lyklema 1995; Hiemenz and Rajagopalan 1997; Israelachvili 2010). The stability of colloidal suspensions (i.e., the capability of remaining in the disperse state, Figure 6.1, left) is explained by this theory as a result of the balance between these two contributions: the dispersion interactions, being always attractive between similar particles, favors flocculation, whereas the repulsive electrostatic interactions prevents or, more precisely, delays this process, sometimes to the extent that some suspensions remain unchanged for years. This process is affected by the ionic strength (which can be regarded as salinity) of the medium. This is quite extensively discussed in the classic book by Verwey and Overbeek (1948), so we show here a few examples only.

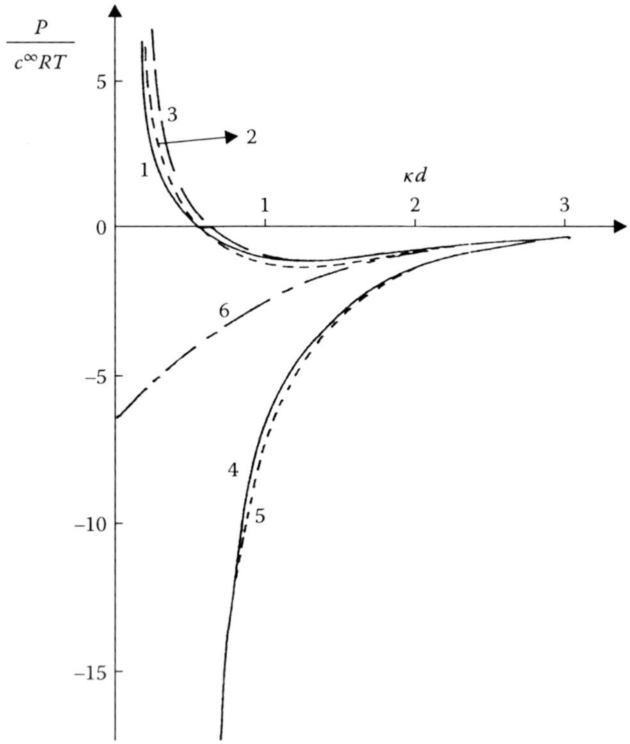

FIGURE 6.11 Similar to Figure 6.9, but with $\psi_{0,1}$ = 1.8 ***RT/F***, $\psi_{0,2}$ = −1.0 ***RT/F***. (Adapted from *J. Colloid Interface Sci.*, 51, Gregory, J., Interaction of unequal double layers at constant charge, 44–51. Copyright 1975, with permission from Elsevier.)

Figure 6.13a shows typical curves of dispersion and electrostatic interaction energies (in units of ***kT***) for two spherical particles obtained, for the sake of illustration, using Equations 6.10 and 6.36, respectively; the last one (based on a linear approximation) is not quantitatively valid for potentials of 35 mV as used here, but it does not depart significantly. The VW interaction (solid curve) is always attractive, whereas the electrostatic interaction (broken curves) is always repulsive, but its strength and effective length are dependent on the ionic strength, as it can be seen in the figure. Summing the two, different possibilities arise, as shown in Figure 6.13b; we will now discuss these curves from the bottom up. At high ionic strength, the electrostatic interaction is quite weak, the dispersion forces dominate, and the total interaction is attractive in almost all the range. It is not shown here, but at short distances other interactions and phenomena arise (as discussed in Section 6.6), most of them repulsive, and, in the limit of very short interparticle distances, Born repulsion should appear due to the overlapping of electronic clouds; thus, a minimum in the interaction curve should exist at short distances, hence this region is labeled "primary minimum." In fact, the DLVO theory is no longer valid at short distances, of the order of a few water molecular diameters, but for colloid stability the long-range interactions are the most important, so we ignore other interactions for the time being. Returning to Figure 6.13b, as the ionic strength decreases, the repulsion increases (see also Figure 6.14); the intermediate curve shows also the primary minimum at short distances, but then a local repulsive maximum is observed, and beyond it a shallow secondary minimum is found; this situation is quite likely to occur in normal conditions. Particles flocculated at the primary minimum are often irreversibly bound; when there is no barrier (high ionic strength), all the particles will rapidly flocculate. Particles bound at the secondary minimum are reversibly attached, and its residence time is greatly affected by the hydrodynamic conditions. Depending on the height of the barrier relative to the temperature, particles at or near the secondary minimum will slowly cross the barrier and become irreversibly bound. Sometimes particles adsorbed at the secondary minimum attach irreversibly, even when they do not cross the barrier, by the action of other

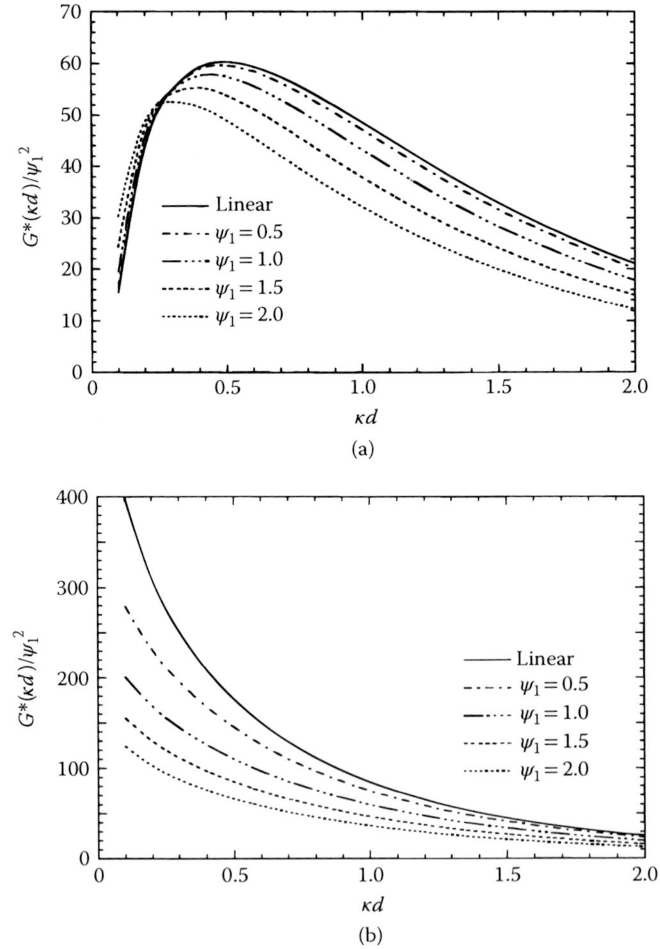

FIGURE 6.12 The scaled interaction free energy between two equal-sized spheres, with $\kappa a = 10$ (a being the sphere radius) for several values of the dimensionless surface potential Ψ_1 and at various ratios of the surface potentials. Results are shown for (a) constant potential interaction with $\Psi_2/\Psi_1 = 3$, (b) constant charge interaction with $\Psi_2/\Psi_1 = 3$.

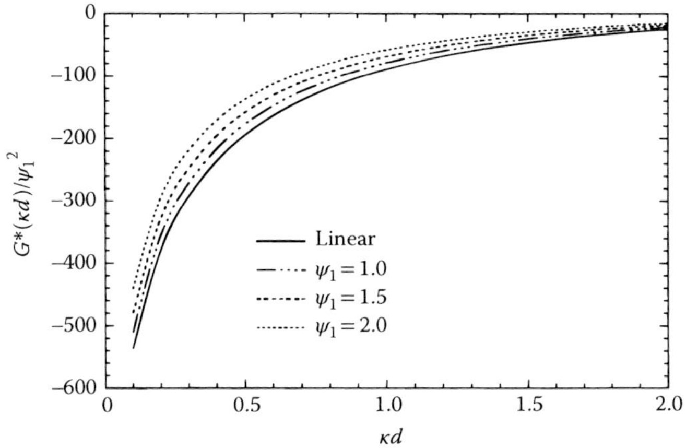

(c)

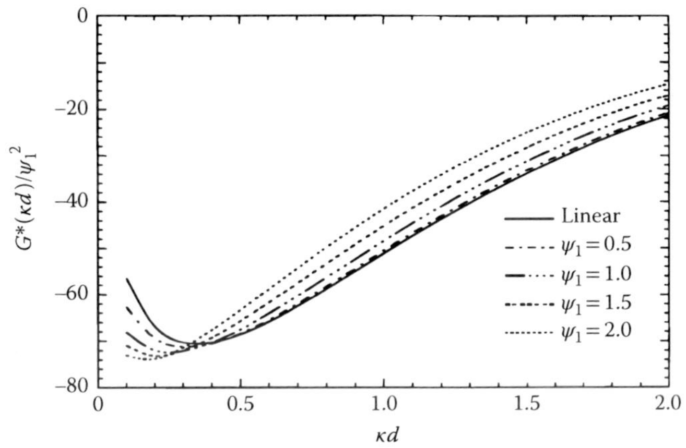

(d)

FIGURE 6.12 (*Continued*) The scaled interaction free energy between two equal-sized spheres, with $\kappa a = 10$ (a being the sphere radius) for several values of the dimensionless surface potential Ψ_1 and at various ratios of the surface potentials. Results are shown for (c) constant potential interaction with $\Psi_2/\Psi_1 = -3$, and (d) constant charge interaction with $\Psi_2/\Psi_1 = -3$. (Adapted with permission from Stankovich, J., S.L. Carnie, 1996, *Langmuir*, 12, no. 6, 1453–1461. Copyright 1996 American Chemical Society.)

forces, or because the secondary minimum is deep enough. This is often the case of bacteria, which can develop structures attaching themselves to other particles or surfaces. In this example, the height of the repulsion maximum is of the order of $\boldsymbol{kT}$, so that the particles can pass, with some kinetic delay, through it to fall in the primary minimum. Of course, a change of the thermodynamic conditions (surface potentials, ionic strength, etc.) will affect the interaction curves and can provoke repeptization of particles, even when they had fallen into the primary minimum. Finally, for the topmost curve in Figure 6.13b, even when there is a primary minimum, so that the stable state is the coagulated colloid, the process will likely be slow; moreover, for

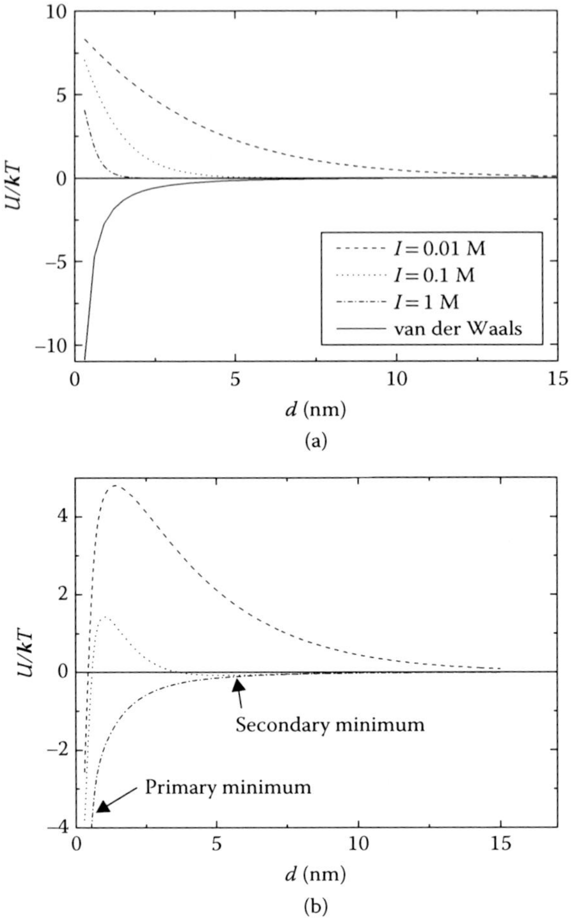

FIGURE 6.13 Schematic representation of the relative interaction strength of the different contributions (a), and the resulting total DLVO interaction (b), at different ionic strengths for two identical spheres at constant potential conditions with $a_1 = a_2 = 10$ nm, $\psi_{0,1} = \psi_{0,2} \approx$ 35 mV, and $A_{11} \approx 10^{-20}$ J.

sufficiently high local maxima, the flocculation will be so slow that the suspension may remain stable for years (some gold suspensions made by Faraday remain stable [Faraday 1857; Das and Marsili 2010] still today). Note that this is stability in the kinetic sense, as opposed to thermodynamic stability, which is only attained at the equilibrium (here, it should be at the primary minimum, if any equilibrium exists).

6.5.2 Dissimilar Particles

When considering dissimilar particles, even taking into account only VW and electric interactions, there are various possible situations that may arise for the energy/distance curve, depending on the nature of interactions, dielectric constant, ionic strength, potential of both surfaces, and so on. Working similarly to the example shown in

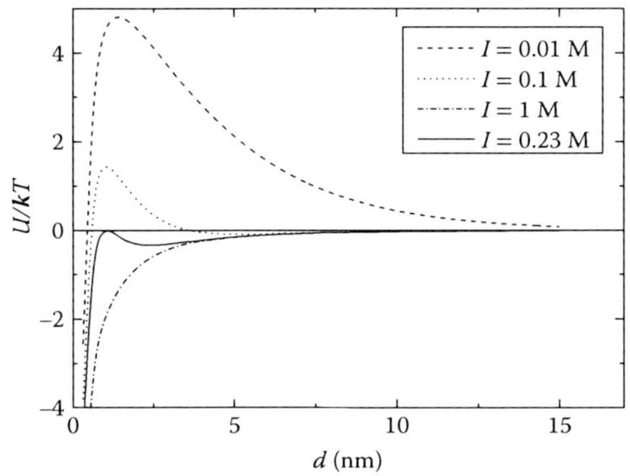

FIGURE 6.14 The same as Figure 6.13 but including the curve that has its local maximum at $U_{el} = 0$ for a particular value of ionic strength.

Figure 6.13, but assuming different surface potentials and different particle sizes, one can find under constant potential conditions a rather similar behavior, as shown in Figure 6.15 (see also Adamczyk and Van De Ven 1984; Kihira and Matijevic 1992; Israelachvili 2010). The electrostatic part shows attractive interaction at short distances, which overlaps with, and is smaller than, the dispersion contribution, so that the total curve has the same characteristics as in Figure 6.13. However, different situations can arise. For example, the electrostatic force curve may be nonmonotonic, which in turn means that the corresponding energy curve will have an attractive part in a range of distances. Possible resulting curves at several ionic strengths under charge regulation conditions are shown schematically in Figure 6.16a, similarly to the results of Chan et al. (2006); it shows a minimum separated by a barrier from the bulk solution and getting repulsive again at shorter distances. When this type of curve is added to the dispersion contribution, curves such as those in Figure 6.16b result. Here, a rather deep secondary minimum is found. At high ionic strength, this minimum is separated from the bulk by a moderate height barrier; thus, particles can slowly pass it to get united at the secondary minimum, and afterward they can go into the primary minimum more rapidly, because the barrier here is lower. As the ionic strength decreases, the barriers grow, so that the ability to flocculate diminishes; also, as the primary barrier grows faster than the secondary, it is possible that the particles get permanently bound in the secondary minimum. As discussed in Section 6.5.1, a number of other situations can arise. The curves are dependent on the nature of the particles, surface potential, and temperature, among other factors. Also, they can be affected by the application of magnetic fields (Haque et al. 1990a, 1990b). The subject of heteroaggregation of colloidal particles was reviewed some time ago (Islam, Chowdhry, and Snowden 1995).

6.5.3 Colloidal Suspension Stability

From what has been discussed in Section 6.5.1, the DLVO theory predicts that colloid flocculation-dispersion is governed, among other factors, by the ionic strength of

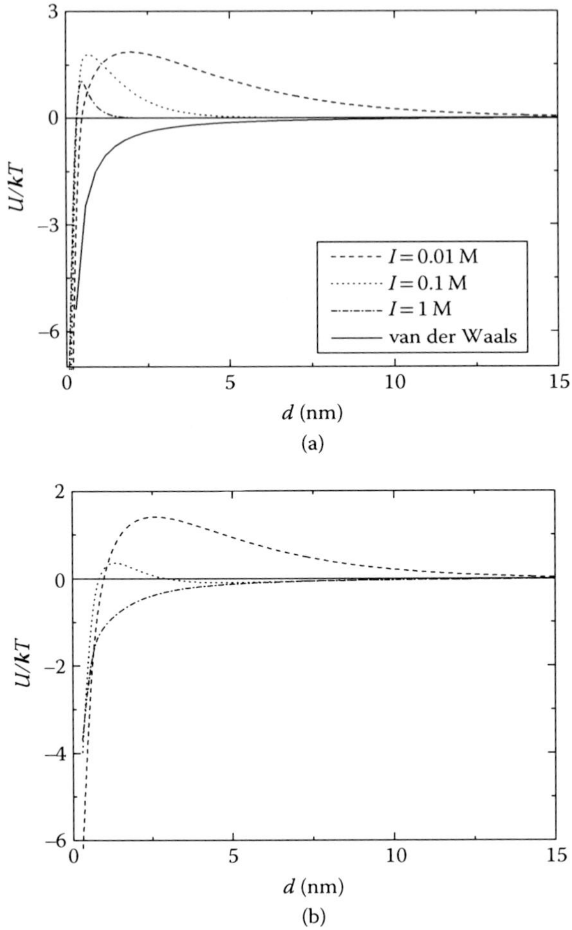

FIGURE 6.15 Schematic representation of the relative interaction strength of the different contributions (a), and the resulting total DLVO interaction (b), at different ionic strengths for two dissimilar spheres at constant potential with a_1 = 10 nm, a_2 = 100 nm, $\psi_{0,1} \approx$ 35 mV, $\psi_{0,2} \approx$ 10 mV, and $A_{11} \approx 10^{-20}$ J. The presence of both primary and secondary minimum for the curve at intermediate I is indicated.

the medium, a fact well known, for example, in chemical analysis when filtering colloidal precipitates (Kolthoff et al. 1969). When working in 1-1 electrolytes, electrolyte concentration is coincident with ionic strength; from the discussion in Sections 6.5.1 and 6.5.2, it is clear that for some value of concentration, the suspension will undergo a transition from "stable" to "unstable." This particular concentration value is known as the *critical coagulation concentration* (CCC). Figure 6.14 shows that for the hypothetical example presented in Figure 6.13; here, for a 1-1 electrolyte, the CCC is 0.23 M, but for other electrolytes, it is much lower. It is known since the end of the 19th century that the CCC decreases with the counterion charge approximately as z^6, which is known as the Schulze–Hardy rule (Schulze 1883; Hardy 1899;

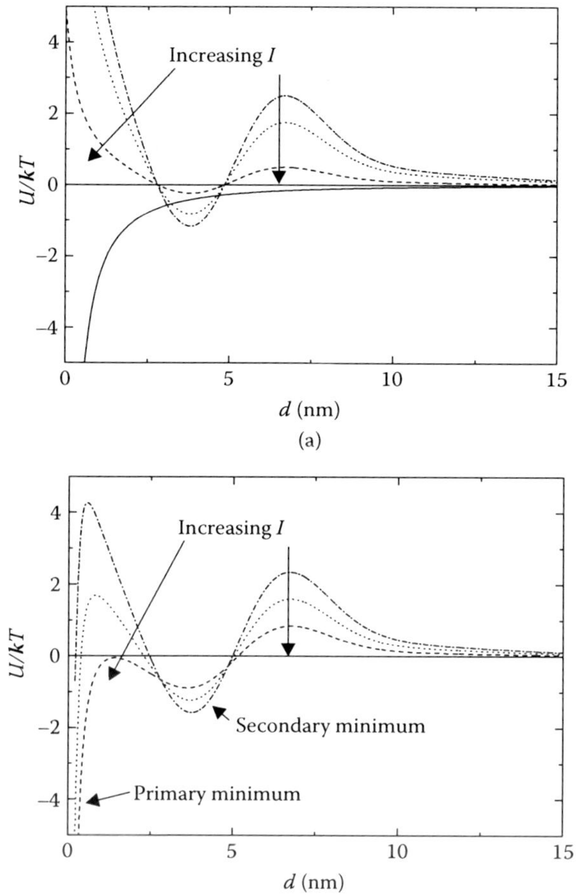

FIGURE 6.16 Schematic qualitative representation of the relative interaction strength of the different contributions (a), and the resulting total DLVO interaction (b), at different ionic strengths for two dissimilar spheres under charge regulation conditions.

Overbeek 1949b). This can be interpreted in terms of the DLVO theory as follows. Considering two identical spherical particles at not far distances, the total long-range interaction energy can be given by the sum of dispersion and electrostatic contributions (Equations 6.11 and 6.39, respectively)

$$U(d) = U_{el}(d) + U_{11,sp-sp}(d) = \frac{64 R_s c^{\infty} \boldsymbol{RT}}{\kappa^2} \tanh^2\left(\frac{z\boldsymbol{F}\psi_s}{4\boldsymbol{RT}}\right) e^{-\kappa d} - \frac{A_{11} R_s}{12d} \quad (6.50)$$

where A_{11} is the Hamaker constant. The condition for the transition curve (the solid line in Figure 6.14) touching the axis implies that

$$U_{el} = U_{sp-sp} \quad (6.51)$$

and

$$\frac{\mathrm{d}U_{el}}{\mathrm{d}d} = -\frac{\mathrm{d}U_{sp-sp}}{\mathrm{d}d} \tag{6.52}$$

Applying this condition to Equation 6.50, it is found that $\kappa d = 1$ and

$$\frac{64R_s c^\infty \boldsymbol{RT}}{\kappa^2} y_0^2 e = \frac{A_{11}R_s\kappa}{12} \tag{6.53}$$

where $y_0 = \tanh^2\left(z\boldsymbol{F}\psi_s/4\boldsymbol{RT}\right)$. Taking the bulk concentration c^∞ as the CCC and rearranging, we get

$$c_{CC} = \frac{A_{11}\kappa^3}{768e\boldsymbol{RT}y_0^2} \tag{6.54}$$

or

$$c_{CC}^2 = \left(\frac{A_{11}}{768e\boldsymbol{RT}y_0^2}\right)^2 \kappa^6 \tag{6.55}$$

and recalling Equation 3.15, with $I = z^2c^\infty$ (z–z electrolyte)

$$\kappa = \sqrt{\frac{2\boldsymbol{F}^2c^\infty z^2}{\varepsilon \boldsymbol{RT}}} \tag{6.56}$$

we get finally

$$c_{CC} = \frac{ky_0^4}{z^6} \tag{6.57}$$

where k collects constants and other parameters such as dielectric constant and temperature. Equation 6.57 predicts the Shulze–Hardy rule if y_0 is constant. This poses a difficulty, because ψ_s is not really the potential at the surface but at the diffuse layer limit, ψ_2, which is also dependent on the ionic strength. For high potentials, $\tanh\left(z\boldsymbol{F}\psi_s/4\boldsymbol{RT}\right) \to 1$, so that it will be approximately constant. For low potentials, $\tanh\left(z\boldsymbol{F}\psi_s/4\boldsymbol{RT}\right) \to z\mathrm{F}\psi_s/4\boldsymbol{RT}$, so that the z^6 dependence is lost; on the contrary, ψ_2 is a function of the ionic strength, so the Shulze–Hardy rule may regain validity, or show a less marked dependence on z (Hiemenz and Rajagopalan 1997; Israelachvili 2010). Table 6.2 shows some examples; as it is seen, the rule is followed quite well; however, there is some dependence on the electrolyte, either on the specific counterion (see, e.g., As_2S_3, univalent cations) or, even, on the coion (e.g., Al_2O_3, univalent anions); thus the rule is not exact, as it should be expected, because other ion properties, besides valence, are expected to have influence; we will return to this topic in Section 6.6. Nevertheless, the reasonable prediction of the Schulze–Hardy rule was at the time of its formulation an important success of the DLVO theory.

The dependence of the CCC on the ionic strength may be related to diverse soil colloidal phenomena, such as landslides. A particular example is that of quick clays. The term "quick clay" refers to a clay whose structure collapses completely at remolding and whose shear strength is thereby reduced almost to zero (Rankka

TABLE 6.2
Some Values of Critical Coagulation Concentration (mM)

	Colloid Composition and Charge Sign							
	As_2S_3, Negative			Al_2O_3, Positive				
Counterion Valency	**Electrolyte**	**CCC**	**Ratio to $z = 1$**	**Electrolyte**	**CCC**	**Ratio to $z = 1$**	**Expected Ratio**
1	LiCl	58		KCl	46		
	NaCl	51		NaCl	43.5		
	KNO_3	50		KNO_3	60		
	Mean value[a]	55	1	Mean value[a]	52	1	1
2	$MgCl_2$	0.72		K_2SO_4	0.30		
	$MgSO_4$	0.81		K_2CrO_4	0.95		
	$ZnCl_2$	0.69		$K_2Cr_2O_7$	0.63		
	Mean value[a]	0.69	0.013	Mean value[a]	0.63	0.012	0.016
3	$AlCl_3$	0.093					
	$Al_2(SO_4)_3$	0.096[b]					
	$Ce(NO_3)_3$	0.080					
	Mean value[a]	0.091	1.7×10^{-3}	$K_3Fe(CN)_6$	0.080	1.5×10^{-3}	1.4×10^{-3}

Source: Data from Overbeek, J.G. *Colloid Science, Vol. I, Irreversible Systems*, Elsevier, Amsterdam, 1949b.

[a] Including other electrolytes.

[b] Concentration of Al^{3+}.

et al. 2004). Remolding is the change that occurs in clay that has been disturbed and lost shearing strength but gained compressibility; usually remolding leads to compaction of a clay sediment, but in quick clays, it may result in collapse. They are deposits of clay minerals of marine origin, thus with high saline contents that will shield the repulsive electrostatic interactions. These deposits have voids containing water and/or air, the water in turn having high ionic concentrations. The attractive forces between the particles will then dominate and may cause the particles to flocculate, whereby large aggregates of particles will sediment without any preferred orientation and thereby create a clay with a high void ratio. When flooded, or wetted by rain or other sources, salt is slowly leached out. The ionic concentration will thereby be lower and each clay particle has to have a large diffuse double layer to obtain a neutral charge. A reflocculation after remolding of the soil is thereby not possible, but a material may be created with clay particles suspended in the pore water. If the void ratio in the undisturbed clay is high, the water content may then be significantly higher than the liquid limit, which means that the remolded clay terrain behaves more like a liquid, and a massive landslide can occur. Massive landslides of this origin occur with some regularity in northern countries, such as Sweden (Rankka et al. 2004) and Norway (Eilertsen et al. 2008). Presumably, even for other types of clay deposits, albeit less saline, the effect of soil salinity (ionic strength) on interactions may also have a role in landslides and similar phenomena.

6.6 BEYOND DLVO*

6.6.1 DLVO Theory Limitations

The DLVO theory was established more than half a century ago, and has been, and is still, being widely applied to many problems in surface and colloid chemistry with reasonable results at not-too-short separation distances (Anderson et al. 2010; Schäfer et al. 2010). However, there are many cases where the experimental behavior appears to contradict DLVO, and it has also been criticized, sometimes strongly (Ninham 1999; McBride and Baveye 2002; Spitzer 2003; Ninham 2006; Boinovich 2010). In fact, there are a number of experimental results that contradict the DLVO theory; we mention some of them here. The first one is in fact known for a long time. Under some circumstances, colloidal particles form layered suspensions, known as Schiller layers (Overbeek 1949c), that are well differentiated from ordinary isotropic suspensions in that they are denser and show light interference patterns; the common explanation is that the particles are attracted by gravity and repelled by electrostatics interactions. However, β-FeOOH only form Schiller layers at very low pH, where the particles have a definite positive charge; as the pH is raised, the particles are disordered forming a normal suspension, which is in disagreement with DLVO theory: if the electrostatic force was repulsive, raising pH toward neutral charge would force the particles to sediment quickly. The evidence suggests strongly that the electrostatic force here is *attractive*; in general, these systems tend to show the coexistence of a Schiller ordered phase with a disordered one, which is usually interpreted as due to the presence of long-range non-VW attractive forces. Several studies on clay minerals support this view (McBride and Baveye 2002). More recent studies also are at variance with DLVO concepts, such as the interactions across thin liquid layers (Boinovich 2007), critical coagulation of sulfidic colloidal sediments (Macdonald et al. 2009), macroionic interactions (Ise 2010), and numerous experiments involving bacterial interaction and adhesion (see, e.g., Poortinga et al. 2002; Ninham 2006).

It should be not surprising to find such discrepancies. In the first place, this theory was developed to address the problem of the stability of hydrophobic colloids; thus, it only considered from the start long-range interactions. And even within this restriction, there are a number of approximations involved such as the following (the list is not exhaustive):

1. The electrostatic and dispersion contributions are not exactly separable in independent terms (Ninham 1999).
2. The electrostatic part is based on the PB equation that implies a number of assumptions (see Section 3.1.2), including treating the solvent as a continuum, and the fact that it does not comply with Maxwell's electromagnetic theory; the linearized PB does, but it has limited validity.
3. In the dispersion interaction, the Lifshitz theory also treats the electrolyte solution like a continuum, characterized by its frequency-dependent dielectric function, but devoid of any structure.

* Section has more advanced or specific materials.

Considering the liquid as a continuum (primitive model) amounts to spatially averaging the solution properties; this could be acceptable in the bulk, but fails near the surfaces because they cause changes in the solvent structure, which extend several molecular diameters beyond the surface. Figure 6.17 shows schematically how the liquid structure changes near a boundary, as it has been found by simulations and, more recently, experiments (Israelachvili 2010 and references therein). Here, r_0 is the molecular diameter, ρ_z is the number density of the liquid in the direction normal to the interface, and ρ_∞ is the bulk liquid density (note that $\rho_z/\rho_\infty = g_{0s}(z)$, the wall–molecule distribution function; see Chapter 3, Section 3.4). In Figure 6.17a, the density profile at a vapor–liquid interface is represented: a relatively smooth density curve is found, which reflects that some molecules are allowed to "stick out" beyond the limit (in a statistical sense) of the liquid phase. When a hard wall is present, the situation is quite different; if it is assumed to be ideally smooth, as in Figure 6.17b, it (along with the liquid molecular interactions) forces the liquid molecules to order into "quasi-discrete" ordered layers, but this order lasts only for a distance of a few molecular diameters, after which the disordered nature of the liquid prevails. The density at the wall position is zero, and at a distance $r_0/2$, a maximum in ρ is observed, corresponding to the first liquid layer; this density, ρ_s, is known as the *contact value* of the density. The midplane density ρ_m may or may not approach the bulk density, depending on the wall–wall distance d. Experiments have shown that for water in the presence of mica surfaces, there are about four quasi-ordered water layers, covering a distance of about 1 nm from

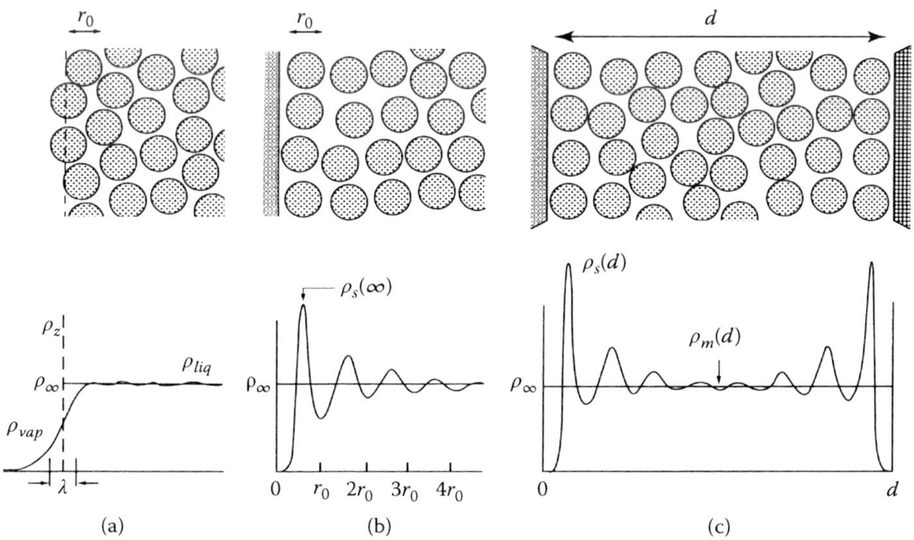

FIGURE 6.17 Schematic representation of liquid number density profiles: (a) at a vapor–liquid interface; λ is a measure of the molecular-scale surface roughness; (b) in the vicinity of a wall–liquid interface; (c) between two hard walls at a distance d. (Reprinted from *Intermolecular and Surface Forces*, 3rd ed., Israelachvili, J.N. Copyright 2010, with permission from Elsevier.)

the surface. Of course, surfaces are not mathematically smooth but have their own order at the atomic level; this order is thus reflected in the ordering of the adjacent liquid layer. When there are two surfaces close enough, as in Figure 6.17c, the effect is similar but considerably more marked. In the middle of the space between the walls, as they approximate each other, part of the liquid molecules are expulsed and the remaining forced to reorder; the resulting wall–liquid interactions are termed by some authors as *solvation forces*. Because the DLVO theory treats the electrolyte like a continuum and uses the bulk water properties, clearly it fails where the density departs significantly from the bulk value; that is, when the solvent shows a distinct structure.

Another important point is that of ion effects. In the DLVO theory, the only ionic property considered is its charge. However, it has been known for a long time (see Table 6.2, with data already reviewed by Overbeek in 1949) that the Schulze–Hardy rule is predicted only semiquantitatively: the CCC depends not only on the counterion (for negatively charged particles, Na^+ is not the same as K^+) but also on the coion ($MgCl_2$ does not give the same CCC as $MgSO_4$ for a negative colloid). In more concentrated solutions, the differences tend to be more marked (Boström, Deniz et al. 2006). Finally, is quite well known that the DLVO theory is not accurate for multivalent ions (Ninham 2006; Dahirel and Jardat 2010).

In general, it is observed that DLVO works reasonably well for ionic strengths no higher than about 0.01 M and interparticle distances not less than a few nanometers. Outside these limits, the theory shows important discrepancies with experiments. Fortunately, in soil-related problems, these conditions hold; thus, DLVO can give, most of the time, correct predictions; however, in some cases, such as highly saline soils, it can fail severely. In view of the discrepancies, there have been diverse proposals to improve/replace the DLVO theory, as mentioned in the following sections.

6.6.2 DLVO Extensions

One important group of researchers (Israelachvili [2010] and references therein) has proposed additional "forces" acting between colloidal particles in addition to those of the original DLVO. That amounts to considering in Equation 6.5 additional additive terms of short-range interaction. These forces include mainly three categories, two of them specific for aqueous media:

1. Solvation oscillatory forces: As described in previous section, they arise from the alterations on the solvent structure as the particles approach. In fact, several experimental measurements of interactions at very short distances have shown an oscillatory behavior (Horn and Israelachvili 1980; Horn and Israelachvili 1981; Christenson and Horn 1983), which were interpreted in terms of electrostatic and oscillatory forces. The oscillations in these measurements clearly are due to the solvent structure discussed in Section 6.6.1, but its treatment as a force, as in other DLVO interactions, has been questioned (Ninham 1999).
2. Repulsive hydration forces: These include steric-hydration forces between amphiphilic surfaces and repulsive hydration forces between hydrophilic

surfaces. Experimental measurements have shown the presence of strongly repulsive forces in water between surfactants and lipid layers and between hydrophilic clay surfaces such as mica, montmorillonite, and other hydrophilic minerals. These are mostly monotonic, but some oscillatory behavior has been observed. In the case of micas and similar negatively charged clay mineral colloids, it has been observed that these forces are dependent on the cation present, following a Hofmeister series (Kunz, Henle, and Ninham 2004; Zhang and Cremer 2006): $Mg^{2+} > Ca^{2+} > Li^{+} \approx Na^{+} > K^{+} > Cs^{+}$ (Israelachvili 2010). These forces have been interpreted as due to cation adsorption onto the surfaces, which in turn would hydrate the surfaces by retaining some of their hydration layers. However, there is no confirmation of such effect, and there is no clear explanation for these results.

3. Attractive "hydrophobic" forces: Some nonpolar surfaces show rather strong attractive interactions; this applies, for example, to surfaces covered by hydrocarbon or fluorocarbon layers, methylated surfaces, and siloxane (or similar) surfaces. For silanated silica surfaces, the presence of strong long-range attractions, much higher than those predicted by dispersion theory (Yoon, Flinn, and Rabinovich 1997), has been reported. It appears to be a force of surface origin (as opposed to body origin, like dispersion ones). As in the previous case, its origin is not known as yet (Israelachvili 2010).

Other more specific studies include the "acid–base" interactions, actually Lewis acid–base or electron acceptor–donor interactions, having relatively short range (~2–3 nm) and considered mainly in the case of bacteria and biomolecules (Hermansson 1999; Farahat et al. 2009; Farahat, Hirajima, and Sasaki 2010).

Finally, "steric" forces have been mentioned by some authors, due to the presence of filament structures on the particle surface, for example, to account for the colloidal behavior of growing silica particles (Schoeman 1998; Liang et al. 2007).

6.6.3 Alternative Models

Several models that depart partially or totally from DLVO have been proposed. One such model is the Spitzer "dissociative electrical double layer" theory (Spitzer 1984, 2003 and references therein). It essentially uses the linearized PB equation (which is consistent with Maxwellian electromagnetism) along with a "coion exclusion" boundary, which prevents ions of the same charge as the surface becomes too close; in practice, this avoids negative concentrations, which will be predicted by the linear PB theory. It also includes a "double layer association" parameter α, which gives the fractions of counterions that are associated to the surface forming the Stern layer (Spitzer 1992). This theory, however, has not been further developed.

Another line of thought has been developed by Ninham and coworkers (Ninham and Yaminsky 1997; Ninham 1999; Boström, Williams, and Ninham 2001; Boström, Deniz et al. 2006; Boström, Tavares et al. 2006; Ninham 2006; Parsons and Ninham 2010), consisting essentially of the introduction of ion-specific interactions through ion dispersion energies. It considers ionic dispersion and image potentials, giving an ion-specific potential energy $U_i(x)$, which is deduced from the Lifschitz theory (Parsons and Ninham 2010). The ion concentration is given here by

$$c_i(x) = c_i^{\infty} \exp\left(-\frac{z_i \boldsymbol{F} \psi(x)}{\boldsymbol{RT}} + U_i(x)\right) \tag{6.58}$$

where ψ is the potential satisfying the nonlinear PB equation. The system is solved in a self-consistent way. This approach has been shown to explain, at least qualitatively, many experiments not accounted by the DLVO theory (Dahirel and Jardat 2010).

Finally, we mention a different approach developed by Sogami, Ise, Smalley, and coworkers (Sogami 1983; Sogami and Ise 1984; Smalley 1990; Sogami, Shinohara, and Smalley 1991; Sogami, Shinohara, and Smalley 1992; Smalley 1994), termed the Coulombic attraction theory (CAT). This theory does not consider the van der Waals attraction; instead a long-range electrostatic *attractive* interaction between the particles and the counterion diffuse layer, balanced by an osmotic interaction, is proposed. Experimental evidence supporting long-range electrostatic attraction has been accumulated for several years (see Crocker and Grier 1996; Nagornyak, Yoo, and Pollack 2009 and references therein). The CAT ideas are not new; in fact, one of the founding fathers of surface science, Irving Langmuir, already proposed that (Langmuir 1938). These ideas were shortly afterward discarded and later recovered (see McBride and Baveye 2002 for a short review), but a considerable debate arose between DLVO and CAT supporters (see Smalley 1994, 1995; Hall 1996 and references therein). Essentially the CAT approach includes three terms in the adiabatic (Helmholtz) potential energy for two parallel plates at a distance d:

$$U(d) = U_{el}(d) + U_i^0(d) + U_o^0(d) \tag{6.59}$$

The term $U_i^0(d)$ is purely repulsive, given by the osmotic pressure of the small counterions between (*inside*) the two surfaces. The third term, $U_o^0(d)$, is purely attractive, given by the osmotic pressure generated by ions *outside* the plates; these two terms together give a potential energy similar to the DLVO theory. The term $U_{el}(d)$, which has no equivalent in DLVO, comes from the accumulation of counterions between the plates, accounts for the interactions between the charged plates and the ions, and leads to long-range electrical attraction between the plates.

From this short summary, it should be evident that the whole subject of what may be termed "post DLVO" theories remains controversial, and publications supporting these two different approaches continue to be published (Lian and Ma 2007; Liu, Moreno, and Neretnieks 2009; Ise 2010; Lee and Ng 2010). As discussed in Section 6.6.1 for many problems in soil science, however, the DLVO theory provides adequate predictions.

6.7 SUMMARY

In this chapter, the interaction in colloidal particle systems, based on the principles developed in Chapters 3 through 5, has been studied. The DLVO theory, the mainstream formalism yet employed to rationalize colloidal behavior, has been developed. The shortcomings of DLVO as well as proposed improvements or different theories have been discussed. The matter of how to address theoretically problems that depart from DLVO validity conditions, such as cases with high ionic strength and or very short distances, is still controversial. For moderate ionic strength situations (no greater than ~0.01 M), the DLVO theory can be applied. The importance of colloidal behavior in a number of soil problems has been discussed.

REFERENCES

Adamczyk, Z., and T. G. M. Van De Ven. 1984. Kinetics of particle accumulation at collector surfaces. I. Approximate analytical solutions. *Journal of Colloid and Interface Science* 97, no. 1: 68–90. doi:10.1016/0021-9797(84)90276-5.

Anderson, T. H., S. H. Donaldson, H. Zeng, and J. N. Israelachvili. 2010. Direct measurement of double-layer, van der Waals, and polymer depletion attraction forces between supported cationic bilayers. *Langmuir* 26, no. 18: 14458–14465. doi:10.1021/la1020687.

Andrade, E. M., F. V. Molina, and D. Posadas. 1999. Adhesion of colloidal hematite onto mercury in water-ethanol media. *Journal of Colloid and Interface Science* 215, no. 2: 370–380.

Attard, P. 2001. Recent advances in the electric double layer in colloid science. *Current Opinion in Colloid & Interface Science* 6, no. 4 (August): 366–371. doi:10.1016/S1359-0294(01)00102-9.

Balnois, E., G. Papastavrou, and K. J. Wilkinson. 2007. Force microscopy and force measurements of environmental colloids. In *Environmental Colloids and Particles: Behaviour, Separation and Characterisation*, eds. J. R. Lead and K. J. Wilkinson, 405–467. 1st ed. Chichester, UK: Wiley.

Bell, G. M., and G. C. Peterson. 1972. Calculation of the electric double-layer force between unlike spheres. *Journal of Colloid and Interface Science* 41, no. 3: 542–566. doi:10.1016/0021-9797(72)90377-3.

Boinovich, L. 2010. DLVO forces in thin liquid films beyond the conventional DLVO theory. *Current Opinion in Colloid & Interface Science* 15, no. 5: 297–302. doi:10.1016/j.cocis.2010.05.003.

Boinovich, L. B. 2007. Long-range surface forces and their role in the progress. *Russian Chemical Reviews* 76, no. 5: 471–488. doi:10.1070/RC2007v076n05ABEH003692.

Boström, M., V. Deniz, G. V. Franks, and B. W. Ninham. 2006. Extended DLVO theory: Electrostatic and non-electrostatic forces in oxide suspensions. *Advances in Colloid and Interface Science* 123–126: 5–15. doi:10.1016/j.cis.2006.05.001.

Boström, M., F. W. Tavares, D. Bratko, and B. W. Ninham. 2006. Ion specific interactions between pairs of nanometer sized particles in aqueous solutions. *Progress in Colloid and Polymer Science* 133: 74–77.

Boström, M., D. R. M. Williams, and B. W. Ninham. 2001. Specific ion effects: Why DLVO theory fails for biology and colloid systems. *Physical Review Letters* 87, no. 16: 168103. doi:10.1103/PhysRevLett.87.168103.

Carnie, S. L., and D. Y. C. Chan. 1993. Interaction free energy between plates with charge regulation: A linearized model. *Journal of Colloid and Interface Science* 161, no. 1: 260–264. doi:10.1006/jcis.1993.1464.

Carnie, S. L., D. Y. C. Chan, and J. Stankovich. 1994. Computation of forces between spherical colloidal particles: Nonlinear Poisson-Boltzmann theory. *Journal of Colloid and Interface Science* 165, no. 1: 116–128. doi:10.1006/jcis.1994.1212.

Chan, D. Y. C., T. W. Healy, T. Supasiti, and S. Usui. 2006. Electrical double layer interactions between dissimilar oxide surfaces with charge regulation and Stern-Grahame layers. *Journal of Colloid and Interface Science* 296, no. 1: 150–158. doi:10.1016/j.jcis.2005.09.003.

Chan, D., T. W. Healy, and L. R. White. 1976. Electrical double layer interactions under regulation by surface ionization equilibria-dissimilar amphoteric surfaces. *Journal of the Chemical Society, Faraday Transactions 1* 72: 2844–2865. doi:10.1039/f19767202844.

Christenson, H. K., and R. G. Horn. 1983. Direct measurement of the force between solid surfaces in a polar liquid. *Chemical Physics Letters* 98, no. 1: 45–48. doi:10.1016/0009-2614(83)80199-7.

Clayfield, E. J., E. C. Lumb, and P. H. Mackey. 1971. Retarded dispersion forces in colloidal particles—Exact integration of the Casimir and Polder equation. *Journal of Colloid and Interface Science* 37, no. 2: 382–389. doi:10.1016/0021-9797(71)90306-7.

Crocker, J. C., and D. G. Grier. 1996. When like charges attract: The effects of geometrical confinement on long-range colloidal interactions. *Physical Review Letters* 77, no. 9: 1897. doi:10.1103/PhysRevLett.77.1897.

Dahirel, V., and M. Jardat. 2010. Effective interactions between charged nanoparticles in water: What is left from the DLVO theory? *Current Opinion in Colloid & Interface Science* 15, no. 1–2: 2–7. doi:10.1016/j.cocis.2009.05.006.

Dan, N. 2002. Interactions between charge-regulating surface layers. *Langmuir* 18, no. 9: 3524–3527. doi:10.1021/la011147m.

Das, S. K., and E. Marsili. 2010. A green chemical approach for the synthesis of gold nanoparticles: Characterization and mechanistic aspect. *Reviews in Environmental Science and Bio/Technology* 9, no. 3: 199–204. doi:10.1007/s11157-010-9188-5.

Derjaguin, B. 1934. Untersuchungen über die reibung und adhäsion, IV. *Kolloid-Zeitschrift* 69, no. 2: 155–164. doi:10.1007/BF01433225. Studies on the friction and adhesion, IV, Colloid journal.

Derjaguin, B. V. 1954. A theory of the heterocoagulation, interaction and adhesion of dissimilar particles in solutions of electrolytes. *Discussions of the Faraday Society* 18: 85. doi:10.1039/df9541800085.

Derjaguin, B. V., and L. D. Landau. 1941. Theory of the stability of strongly charged lyophobic sols and of the adhesion of strongly charged particles in solution of electrolytes. *Acta Physicochimica URSS* 14: 633–662.

Derjaguin, B. V., and T. N. Voropayeva. 1964. Surface forces and the stability of colloids and disperse systems. *Journal of Colloid Science* 19, no. 2: 113–135. doi:10.1016/0095-8522(64)90023-6.

Devereux, O. F., and P. L. De Bruyn. 1963. *Interaction of Plane-Parallel Double Layers*. Cambridge, MA: The MIT Press.

Dzyaloshinskii, I. E., E. M. Lifshitz, and L. P. Pitaevskii. 1961. The general theory of van der Waals forces. *Advances in Physics* 10, no. 38: 165–209. doi:10.1080/00018736100101281.

Eilertsen, R. S., L. Hansen, T. H. Bargel, and I.-L. Solberg. 2008. Clay slides in the Målselv valley, northern Norway: Characteristics, occurrence, and triggering mechanisms. *Geomorphology* 93, no. 3–4: 548–562. doi:10.1016/j.geomorph.2007.03.013.

Ettelaie, R., and R. Buscall. 1995. Electrical double layer interactions for spherical charge regulating colloidal particles. *Advances in Colloid and Interface Science* 61: 131–160. doi:10.1016/0001-8686(95)00263-P.

Faraday, M. 1857. The Bakerian lecture: Experimental relations of gold (and other metals) to light. *Philosophical Transactions of the Royal Society of London* 147: 145–181. doi:10.1098/rstl.1857.0011.

Farahat, M., T. Hirajima, and K. Sasaki. 2010. Adhesion of ferroplasma acidiphilum onto pyrite calculated from the extended DLVO theory using the van Oss-Good-Chaudhury approach. *Journal of Colloid and Interface Science* 349, no. 2: 594–601. doi:10.1016/j.jcis.2010.05.091.

Farahat, M., T. Hirajima, K. Sasaki, and K. Doi. 2009. Adhesion of *Escherichia coli* onto quartz, hematite and corundum: Extended DLVO theory and flotation behavior. *Colloids and Surfaces B: Biointerfaces* 74, no. 1: 140–149. doi:10.1016/j.colsurfb.2009.07.009.

Fowlkes, W. Y., and K. S. Robinson. 1989. The electrostatic force on a dielectric sphere resting on a conducting substrate. In *Particles on Surfaces 1: Detection, Adhesion, and Removal*, ed. K. L. Mittal, 143–155. 1st ed. New York: Plenum Press.

Frens, G. 1968. *The Reversibility of Irreversible Colloids*. Thesis, Utrecht.

Gregory, J. 1975. Interaction of unequal double layers at constant charge. *Journal of Colloid and Interface Science* 51, no. 1: 44–51. doi:10.1016/0021-9797(75)90081-8.

Hall, D. G. 1996. Interaction of highly charged plates in an electrolyte. An attempt to clarify issues and correct mistakes. *Langmuir* 12, no. 17: 4308–4310. doi:10.1021/la960244q.

Hamaker, H. C. 1937. The London-van der Waals attraction between spherical particles. *Physica* 4, no. 10: 1058–1072.

Hansen, F. K., and E. Matijevic. 1980. Heterocoagulation. Part 5. Adsorption of a carboxylated polymer latex on monodispersed hydrated metal oxides. *Journal of the Chemical Society, Faraday Transactions 1* 76: 1240–1262. doi:10.1039/f19807601240.

Haque, M. F., N. Kallay, V. Privman, and E. Matijevic. 1990a. Magnetic effects in particle adhesion: I. Kinetics of hematite particle deposition on stainless steel. *Journal of Colloid and Interface Science* 137, no. 1: 36–47. doi:10.1016/0021-9797(90)90041-L.

Haque, M. F., N. Kallay, V. Privman, and E. Matijevic. 1990b. Magnetic effects in particle adhesion: II. The influence of particle composition and size on deposition in a magnetic field. *Journal of Adhesion Science and Technology* 4, no. 1: 205–220. doi:10.1163/156856190X00225.

Hardy, W. B. 1899. A preliminary investigation of the conditions which determine the stability of irreversible hydrosols. *Proceedings of the Royal Society of London* 66, no. 424–433: 110–125. doi:10.1098/rspl.1899.0081.

Hartley, P. G., and P. J. Scales. 1998. Electrostatic properties of polyelectrolyte modified surfaces studied by direct force measurement. *Langmuir* 14, no. 24: 6948–6955. doi:10.1021/la9804841.

Hermansson, M. 1999. The DLVO theory in microbial adhesion. *Colloids and Surfaces B: Biointerfaces* 14, no. 1–4: 105–119. doi:10.1016/S0927-7765(99)00029-6.

Hiemenz, P. C., and R. Rajagopalan. 1997. *Principles of Colloid and Surface Chemistry*. 3rd ed. New York: Marcel Dekker.

Hogg, R., T. W. Healy, and D. W. Fuerstenau. 1966. Mutual coagulation of colloidal dispersions. *Transactions of the Faraday Society* 62: 1638. doi:10.1039/tf9666201638.

Horn, R. G., and J. N. Israelachvili. 1980. Direct measurement of forces due to solvent structure. *Chemical Physics Letters* 71, no. 2: 192–194. doi:10.1016/0009-2614(80)80144-8.

Horn, R. G., and J. N. Israelachvili. 1981. Direct measurement of structural forces between two surfaces in a nonpolar liquid. *The Journal of Chemical Physics* 75, no. 3: 1400. doi:10.1063/1.442146.

Hough, D. B., and L. R. White. 1980. The calculation of Hamaker constants from liftshitz theory with applications to wetting phenomena. *Advances in Colloid and Interface Science* 14, no. 1: 3–41. doi:10.1016/0001-8686(80)80006-6.

Hsu, J.-P., and Y.-C. Kuo. 1997. Adsorption of a charge-regulated particle to a charged surface. *Langmuir* 13, no. 16: 4372–4376. doi:10.1021/la961018j.

Hull, M., and J. A. Kitchener. 1969. Interaction of spherical colloidal particles with planar surfaces. *Transactions of the Faraday Society* 65: 3093–3104.

Ise, N. 2010. Like likes like: Counterion-mediated attraction in macroionic and colloidal interaction. *Physical Chemistry Chemical Physics* 12, no. 35: 10279. doi:10.1039/c000729c.

Islam, A. M., B. Z. Chowdhry, and M. J. Snowden. 1995. Heteroaggregation in colloidal dispersions. *Advances in Colloid and Interface Science* 62, no. 2–3: 109–136. doi:10.1016/0001-8686(95)00276-V.

Israelachvili, J. N. 2010. *Intermolecular and Surface Forces*. 3rd ed. Amsterdam: Academic Press.

Kékicheff, P., S. Marčelja, T. J. Senden, and V. E. Shubin. 1993. Charge reversal seen in electrical double layer interaction of surfaces immersed in 2:1 calcium electrolyte. *The Journal of Chemical Physics* 99, no. 8: 6098. doi:10.1063/1.465906.

Kihira, H., and E. Matijevic. 1992. An assessment of heterocoagulation theories. *Advances in Colloid and Interface Science* 42: 1–31. doi:10.1016/0001-8686(92)80017-R.

Kim, H.-Y., and P. R. C. Kent. 2009. Van der Waals forces: Accurate calculation and assessment of approximate methods in dielectric nanocolloids up to 16 nm. *The Journal of Chemical Physics* 131, no. 14: 144705. doi:10.1063/1.3244645.

Kolthoff, I. M., E. B. Sandell, E. J. Meehan, and S. Bruckenstein. 1969. *Quantitative Chemical Analysis*. 4th ed. New York: Macmillan Pub Co.

Kunz, W., J. Henle, and B. W. Ninham. 2004. 'Zur Lehre von der Wirkung der Salze' (about the science of the effect of salts): Franz Hofmeister's historical papers. *Current Opinion in Colloid & Interface Science* 9, no. 1–2: 19–37. doi:10.1016/j.cocis.2004.05.005.

Langbein, D. 1970. Retarded dispersion energy between macroscopic bodies. *Physical Review B* 2, no. 8: 3371. doi:10.1103/PhysRevB.2.3371.

Langbein, D. 1974. *Theory of van der Waals Attraction*. Berlin: Springer-Verlag.

Langmuir, I. 1938. The role of attractive and repulsive forces in the formation of tactoids, thixotropic gels, protein crystals and coacervates. *The Journal of Chemical Physics* 6, no. 12: 873. doi:10.1063/1.1750183.

Lee, C.-L., and S.-K. Ng. 2010. Studies on electrostatic interactions of colloidal particles in two dimensions: A modeling approach. *The Journal of Chemical Physics* 133, no. 8: 084504. doi:10.1063/1.3474805.

Lian, Z. J., and H. R. Ma. 2007. Effective interaction of nonuniformly charged colloid spheres in a bulk electrolyte. *The Journal of Chemical Physics* 127, no. 10: 104507. doi:10.1063/1.2772266.

Liang, Y., N. Hilal, P. Langston, and V. Starov. 2007. Interaction forces between colloidal particles in liquid: Theory and experiment. *Advances in Colloid and Interface Science* 134–135 (October 31): 151–166. doi:10.1016/j.cis.2007.04.003.

Lima, E. R. A., E. C. Biscaia, M. Boström, and F. W. Tavares. 2007. Ion-specific forces between a colloidal nanoprobe and a charged surface. *Langmuir* 23, no. 14: 7456–7458.

Liu, L., L. Moreno, and I. Neretnieks. 2009. A dynamic force balance model for colloidal expansion and its DLVO-based application. *Langmuir* 25, no. 2: 679–687. doi:10.1021/la8026573.

Lyklema, J. 1995. *Fundamentals of Interface and Colloid Science: Solid-Liquid Interfaces*. London: Academic Press.

Lyklema, J. 2006. Overcharging, charge reversal: Chemistry or physics? *Colloids and Surfaces A: Physicochemical and Engineering Aspects* 291, no. 1–3: 3–12. doi:10.1016/j.colsurfa.2006.06.043.

Macdonald, B. C. T., J. K. Reynolds, A. S. Kinsela, R. J. Reilly, P. van Oploo, T. D. Waite, and I. White. 2009. Critical coagulation in sulfidic sediments from an east-coast Australian acid sulfate landscape. *Applied Clay Science* 46, no. 2: 166–175. doi:10.1016/j.clay.2009.07.020.

Mahanty, J., and B. W. Ninham. 1977. *Dispersion Forces*. San Diego, CA: Academic Press.

Marshall, J. K., and J. A. Kitchener. 1966. The deposition of colloidal particles on smooth solids. *Journal of Colloid and Interface Science* 22, no. 4: 342–351.

McBride, M. B., and P. Baveye. 2002. Diffuse double-layer models, long-range forces, and ordering in clay colloids. *Soil Science Society of America Journal* 66, no. 4: 1207. doi:10.2136/sssaj2002.1207.

McCormack, D., S. L. Carnie, and D. Y. C. Chan. 1995. Calculations of electric double-layer force and interaction free energy between dissimilar surfaces. *Journal of Colloid and Interface Science* 169, no. 1 (January): 177–196. doi:10.1006/jcis.1995.1019.

Nagornyak, E., H. Yoo, and G. H. Pollack. 2009. Mechanism of attraction between like-charged particles in aqueous solution. *Soft Matter* 5, no. 20: 3850. doi:10.1039/b905080a.

Ninham, B. W. 1999. On progress in forces since the DLVO theory. *Advances in Colloid and Interface Science* 83, no. 1–3: 1–17. doi:10.1016/S0001-8686(99)00008-1.

Ninham, B. W. 2006. The present state of molecular forces. *Progress in Colloid and Polymer Science* 133: 65–73.

Ninham, B. W., and V. A. Parsegian. 1971. Electrostatic potential between surfaces bearing ionizable groups in ionic equilibrium with physiologic saline solution. *Journal of Theoretical Biology* 31, no. 3: 405–428. doi:10.1016/0022-5193(71)90019-1.

Ninham, B. W., and V. Yaminsky. 1997. Ion binding and ion specificity: The Hofmeister effect and Onsager and Lifshitz theories. *Langmuir* 13, no. 7: 2097–2108. doi:10.1021/la960974y.

Overbeek, J. Th. G. 1949a. VI: The interaction between colloidal particles. In *Colloid Science Vol. I, Irreversible Systems*, ed. H. R. Kruyt, 245–277. Amsterdam: Elsevier.

Overbeek, J. Th. G. 1949b. II: Phenomenology of lyophobic systems. In *Colloid Science Vol. I, Irreversible Systems*, ed. H. R. Kruyt. Amsterdam: Elsevier.

Overbeek, J. Th. G. 1949c. VIII: Stability of hydrophobic colloids and emulsions. In *Colloid Science VOL. I, Irreversible Systems*, ed. H. R. Kruyt, 245–277. Amsterdam: Elsevier.

Parsegian, V. A. 2006. *Van der Waals Forces: A Handbook for Biologists, Chemists, Engineers, and Physicists*. New York: Cambridge University Press.

Parsons, D. F., and B. W. Ninham. 2010. Charge reversal of surfaces in divalent electrolytes: The role of ionic dispersion interactions. *Langmuir* 26, no. 9: 6430–6436. doi:10.1021/la9041265.

Pashley, R. M. 1981. DLVO and hydration forces between mica surfaces in Li+, Na+, K+, and Cs+ electrolyte solutions: A correlation of double-layer and hydration forces with surface cation exchange properties. *Journal of Colloid and Interface Science* 83, no. 2: 531–546. doi:10.1016/0021-9797(81)90348-9.

Poortinga, A. T., R. Bos, W. Norde, and H. J. Busscher. 2002. Electric double layer interactions in bacterial adhesion to surfaces. *Surface Science Reports* 47, no. 1: 1–32. doi:10.1016/S0167-5729(02)00032-8.

Prieve, D. C., and E. Ruckenstein. 1976. The surface potential of and double-layer interaction force between surfaces characterized by multiple ionizable groups. *Journal of Theoretical Biology* 56, no. 1: 205–228. doi:10.1016/S0022-5193(76)80053-7.

Quesada-Pérez, M., R. Hidalgo-Álvarez, and A. Martín-Molina. 2009. Effect of ionic van der Waals forces on the diffuse potential of model colloids. *Colloid and Polymer Science* 288, no. 2: 151–158. doi:10.1007/s00396-009-2139-4.

Rankka, K., Y. Andersson-Sköld, C. Hultén, R. Larsson, V. Leroux, and T. Dahlin. 2004. *Quick Clay in Sweden*. Linköping, Sweden: Swedish geotechnical institute. http://www.swedgeo.se/upload/publikationer/rapporter/pdf/sgi-r65.pdf.

Reiner, E. S., and C. J. Radke. 1993. Double layer interactions between charge-regulated colloidal surfaces: Pair potentials for spherical particles bearing ionogenic surface groups. *Advances in Colloid and Interface Science* 47: 59–147. doi:10.1016/0001-8686(93)80014-3.

Schäfer, B., M. Hecht, J. Harting, and H. Nirschl. 2010. Agglomeration and filtration of colloidal suspensions with DVLO interactions in simulation and experiment. *Journal of Colloid and Interface Science* 349, no. 1: 186–195. doi:10.1016/j.jcis.2010.05.025.

Schoeman, B. J. 1998. Analysis of the nucleation and growth of TPA-silicalite-1 at elevated temperatures with the emphasis on colloidal stability. *Microporous and Mesoporous Materials* 22, no. 1–3: 9–22.

Schulze, H. 1883. Antimony trisulfide in aqueous solution. [In German.] *Practical Journal for Chemistry* 27, no. 1: 320–332. doi:10.1002/prac.18830270116.

Smalley, M. V. 1990. Electrostatic interaction in macro-ionic solutions and gels. *Molecular Physics: An International Journal at the Interface Between Chemistry and Physics* 71, no. 6: 1251. doi:10.1080/00268979000102471.

Smalley, M. V. 1994. Electrical theory of clay swelling. *Langmuir* 10, no. 9: 2884–2891. doi:10.1021/la00021a009.
Smalley, M. V. 1995. On the Helmholtz free energy of highly charged plates in an electrolyte; reply to the criticisms. *Langmuir* 11, no. 5: 1813–1816. doi:10.1021/la00005a061.
Sogami, I. 1983. Effective potential between charged spherical particles in dilute suspension. *Physics Letters A* 96, no. 4: 199–203.
Sogami, I., and N. Ise. 1984. On the electrostatic interaction in macroionic solutions. *The Journal of Chemical Physics* 81, no. 12: 6320–6332. doi:10.1063/1.447541.
Sogami, I. S., T. Shinohara, and M. V. Smalley. 1991. Effective interaction of highly charged plates in an electrolyte. *Molecular Physics: An International Journal at the Interface between Chemistry and Physics* 74, no. 3: 599. doi:10.1080/00268979100102451.
Sogami, I. S., T. Shinohara, and M. V. Smalley. 1992. Adiabatic pair potential of highly charged plates in an electrolyte. *Molecular Physics: An International Journal at the Interface between Chemistry and Physics* 76, no. 1: 1. doi:10.1080/00268979200101121.
Spitzer, J. J. 1984. A re-interpretation of hydration forces near charged surfaces. *Nature* 310, no. 5976: 396–397.
Spitzer, J. J. 1992. Theory of dissociative electrical double layers: The limit of close separations and 'hydration' forces. *Langmuir* 8, no. 6: 1659–1662. doi:10.1021/la00042a027.
Spitzer, J. J. 2003. Colloidal interactions: Contact limiting laws, double-layer dissociation, and 'non-DLVO' (Derjaguin–Landau–Verwey–Overbeek) forces. *Colloid and Polymer Science* 281, no. 6: 589–592. doi:10.1007/s00396-002-0836-3.
Stankovich, J., and S. L. Carnie. 1996. Electrical double layer interaction between dissimilar spherical colloidal particles and between a sphere and a plate: Nonlinear Poisson–Boltzmann theory. *Langmuir* 12, no. 6: 1453–1461. doi:10.1021/la950384k.
Stenhammar, J., P. Linse, H. Wennerström, and G. Karlström. 2010. An exact calculation of the van der Waals interaction between two spheres of classical Dipolar Fluid. *The Journal of Physical Chemistry B* 114, no. 42: 13372–13380. doi:10.1021/jp105754t.
Toikka, G., R. A. Hayes, and J. Ralston. 1997. Evidence of charge reversal from direct force measurements involving dissimilar metal sulfides in aqueous electrolyte. *Journal of the Chemical Society, Faraday Transactions* 93, no. 19: 3523–3528. doi:10.1039/a703260i.
Urrutia, G. A., S. I. Passaggio, A. J. G. Maroto, and M. A. Blesa. 1983. Model of the deposition of colloidal crud particles on the fuel elements of nuclear power plants. *Nuclear Science and Engineering* 84, no. 2: 120–130.
Usui, S. 1972. Heterocoagulation. In *Progress in Surface and Membrane Science*, ed. J. F. Danielli, 5:223. San Diego, CA: Academic Press.
Usui, S., T. Yamasaki, and J. Shimoizaka. 1967. Coalescence of mercury droplets in aqueous solutions, with special reference to the examination of double-layer interaction. *The Journal of Physical Chemistry* 71, no. 10: 3195–3202. doi:10.1021/j100869a010.
Verwey, E., and J. Th. G. Overbeek. 1948. *Theory of the Stability of Lyophobic Colloids*. 1st ed. Amsterdam: Elsevier.
Yoon, R.-H., D. H. Flinn, and Y. I. Rabinovich. 1997. Hydrophobic interactions between dissimilar surfaces. *Journal of Colloid and Interface Science* 185, no. 2: 363–370. doi:10.1006/jcis.1996.4583.
Zhang, Y., and P. S. Cremer. 2006. Interactions between macromolecules and ions: The Hofmeister series. *Current Opinion in Chemical Biology* 10, no. 6: 658–663. doi:10.1016/j.cbpa.2006.09.020.

Part II

Soil Components

7 Soil Composition and Characterization

7.1 GENERAL CHARACTERISTICS

7.1.1 SOIL STRUCTURE

Soils are the result of complex and perennial geological and environmental processes, and as such they have a complex structure and composition. Soils are typically formed by the aggregation of solid particles of widely varying composition, embodying liquid (aqueous solution) and gas (air) phases inside. One of the main outcomes of the soil formation processes (termed "pedogenic" processes) is its layered structure. "Pedogenesis" is the continuous process causing soil formation and evolution; this process includes geological and climatological transformations, biological action, and so on. Any digging in soils reveals several layers, often clearly distinguishable by their color and morphology, called "soil horizons." Figure 7.1 shows several examples of soil profiles, revealing the different horizons. Figure 7.1a shows an example in which the basic horizon types (A, B, and C) are identified (see the following discussion for a definition of soil horizon types). Figure 7.1b and c shows other, more complex examples with a number of different horizons showing up. Figure 7.1d and e shows examples of a poorly drained soil and an excessively drained one, respectively. The first one is composed of small particles that are closely packed, thus preventing water leaching through its structure, whereas in the second case there are mostly coarse particles that allow fast water leaching. Finally, Figure 7.1f shows an example of an anthropogenic soil, or human transported material (HTM) type. HTM is a type of soil that has been formed or modified by human activity. Note the two topmost horizons in Figure 7.1f, identified by a "^" sign: these are the HTM layers. Below them lies the native soil.

The basic soil structure is that shown in Figure 7.1a: the A horizon is the top layer containing a high proportion of organic matter; the B horizon is the mid layer composed mainly of partially "weathered" (modified by climatic action) material; and the C horizon contains relatively undisturbed material. There are seven other horizon types; a basic classification was given by McDonald et al. (1990) as follows (from top to bottom):

1. O horizons are dominated by organic matter that has accumulated on the surface of the soil. O horizons are subdivided according to the degree of organic material decomposition. These horizons are not common and are mostly restricted to moist or cool environments (e.g., alpine areas, swamps, and rainforests).
2. A horizons consist of one or more surface mineral horizons with some accumulation of organic materials (less than that of O horizons). A horizons are usually darker than those underlying, but they may also be horizons that

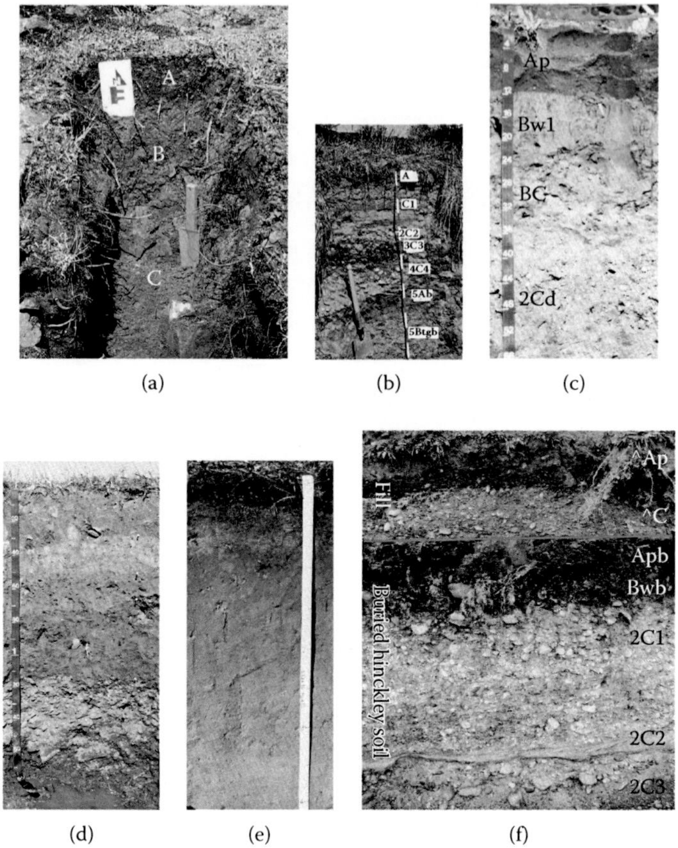

FIGURE 7.1 **(See color insert.)** Examples of soil profiles: (a) soil section in a turf mound of a typical stone polygon on Trinchera Peak, a subglacial region in the Rocky Mountains, Colorado. Well-developed A, B, and C soil horizons are observed. (Copyright 2003 J.S. Aber, reprinted with permission from the author.) (b) Soil profile from the Paint Branch creek in Maryland. The different horizons are labeled in detail. (Courtesy of Ray R. Weil, University of Maryland, College Park, Maryland.) (c) Birchwood soil profile. Birchwood soils are moderately well drained soils formed in sandy eolian and/or fluvial material underlain by loamy dense till. This profile has the following morphology: 0–12 in., Ap horizon—loamy sand; 12–20 in., Bw1 horizon—loamy sand; 20–34 in., BC horizon—gravelly loamy sand with many redoximorphic features (seasonal high water table within 24 in.); 34–65 in., 2Cd—gravelly fine sandy loam, very firm basal till, and many redoximorphic features. (Courtesy of Rob Tunstead, reprinted from www.nesoil.com with permission from James Turenne.) (d) Raypole (like) soil profile, poorly drained loess over fluvial material, from Newport, Rhode Island. (Reprinted from www.nesoil.com with permission from James Turenne.) (e) Soil profile of a Carver soil from Massachusetts. Carver soils are very deep, excessively drained soils formed in layers of coarse and very coarse sand that contain less than 20% rock fragments, most of which are fine gravel. Carver soils are level to steep soils on pitted and dissected outwash plains and moraines. (Reprinted from www.nesoil.com with permission from James Turenne.) (f) Profile of an anthropogenic soil. This profile shows a series of fill layers that were placed over a Hinckley soil series. The "^" symbol is the new horizon prefix for human transported material (HTM). (Reprinted from www.nesoil.com with permission from James Turenne.)

are lighter colored or have a lower content of "clay" when compared with underlying horizons. There are three types of A horizons:

a. The A1 horizon, which is a mineral horizon at or near the soil surface that has some accumulation of organic matter. It is usually darker than the underlying horizons and is the zone of maximum biological activity.
b. The A2 horizon, which is a mineral horizon that has less organic matter, sesquioxides, or clay than the horizons above or below it. It is a pale horizon and is common throughout Australia. Various degrees of bleaching are recognized with white or nearly white layers being referred to as sporadically or conspicuously bleached depending on its extent.
c. The A3 horizon, which is a transitional horizon between A and B horizons.

3. P horizons are dominated by organic materials in various stages of decomposition that have accumulated either under water or in very wet areas. They are often referred to as peat layers.
4. B horizons have one or more of the following:
   a. Concentration of clay, Fe, or Al oxides or organic material.
   b. A structure or consistency unlike the A horizon above and different from the horizons below.
   c. Stronger colors than the horizons above or below.
   d. The B1 horizon is a transitional layer between the A and B horizons, but it is more like the B horizon. Similarly, a B3 horizon is a transitional layer to the underlying material.
5. C horizons are layers below the A and B horizons, which are composed of consolidated or unconsolidated materials. These materials are usually partially weathered, and geological features are often evident. C horizons can be dug by hand when they are moist.
6. D horizons are soil layers below the A and B horizons, which differ in general character and are not C horizons. They cannot be reliably described as buried soils but have a contrasting pedological organization to the overlying horizons.
7. R horizons are continuous masses of rock that are usually too strong to dig with hand tools and constitute the basement for the soil structure.

On the other hand, soils are often discussed in terms of "topsoil" and "subsoil." These terms are not rigorously defined but denote the following:

- Topsoil: the surface zone, including the zone of accumulation of organic material (usually the A horizons). A topsoil can be modified by management practices such as ploughing and addition of fertilizers.
- Subsoil: layers underlying the topsoil (usually B and C horizons), which cannot usually be modified except by drainage.

7.1.2 Gas, Liquid, and Solid Soil Phases

The soil is a complex heterogeneous system, including solid, liquid, and gas phases. Figure 7.2a shows a simple scheme depicting the main features of soil: it can be viewed as an assortment of solid particles that often are of colloidal sizes

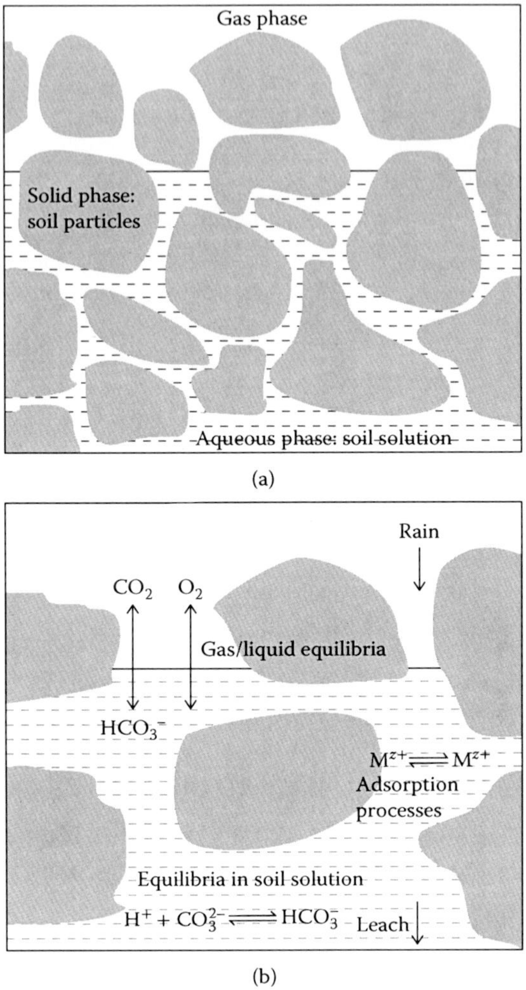

FIGURE 7.2 Schematic view of the interplay between gas, liquid, and solid phases in soil: (a) soil can be viewed as an assortment of solid particles (often of colloidal sizes), with the pores between them partially filled by a liquid phase (the soil solution) and a gas phase (essentially air). The proportion of liquid and gas phases is dependent on rain and draught periods; rain fills up the pores, which are afterward emptied by leaching toward lower horizons. (b) In a close-up, many processes (mainly chemical reactions) take place in the soil solution, including gas dissolution–separation, homogeneous chemical equilibria, and heterogeneous processes, including adsorption and surface chemical reactions.

(below ~10 μm), but much coarser particles can also be found. The unoccupied spaces between particles define pores, which are partially filled by a liquid phase (the soil solution) and a gas phase (essentially air). The proportion of liquid and gas phases is subject to dynamic conditions under climatic influence: rain fills up the pores, eventually totally flooding the soil under heavy rain; the soil solution then leaches toward lower horizons, usually carrying along solutes (either ionic or molecular species). These processes have a considerable influence on the availability of both nutrients and pollutants. In a zoomed view of the system (imaginary, not physically possible) (Figure 7.2b), many processes (mainly chemical reactions) take place in the soil solution, including gas dissolution–separation, homogeneous chemical equilibria, and heterogeneous processes, such as adsorption and surface chemical reactions, dissolution–precipitation reactions, and so on (Sumner 1999; Sparks 2002; White 2005; Brady and Weil 2007; Sposito 2008).

7.1.3 Soil Particles

The range of "soil particles" is operationally defined as those with a diameter (for spherical or similarly shaped particles) not higher than 2 mm. This is usually accomplished by sieving; thus, in the case of irregular particles, the largest dimension should not be larger than 2 mm. The particles (and the soils containing them) are classified as clay, comprising particles smaller (in its largest dimension) than 2 μm; "silt" for particles between 2 and 50 μm; and "sand" for particles larger than 50 μm up to 2 mm. The last two classes are further subdivided, for example, into fine, medium, coarse, or even more fractions, but the fractions and boundaries are not universally agreed upon (Skopp 1999). According to the particle size distribution in a soil, it is classified as one of several categories or *textural classes* as shown, for example, in Figure 7.3. There are other classifications as well (see the study by Skopp [1999] and references therein).

7.1.4 Soil Solution

The soil solution is the center place where chemical reactions occur (see Figure 7.2). A number of processes are present in and around the soil solution (Suarez 1998; Schwab 1999; Sparks 2002; Sposito 2008):

- The soil solution interacts with the air contained in the not-flooded part of soil pores, exchanging mainly O_2 and CO_2 and, depending on the conditions, other gases including pollutants.
- The soil solution also interacts with the solid phase, that is, soil particles, through sorption–desorption processes and surface chemical reactions (dissolution–precipitation, oxidation–reduction, etc.).
- Plants get their nutrients from the soil solution, mostly as inorganic species.
- Microorganisms also interact with the soil solution, taking their food from it and releasing organic substances.
- Living organisms, when they die, release inorganic and organic compounds to the soil solution.
- The soil solution is the vehicle for transport of ions and neutral species, either beneficial or pollutants.
- Finally, in the soil solution, a number of chemical reactions, mainly ionic equilibria, take place, determining the chemical speciation of solutes.

The soil solution is essential for soil chemical and biological activity. In the absence of water, soil organisms become dormant (if they do not die) and weathering, mineral

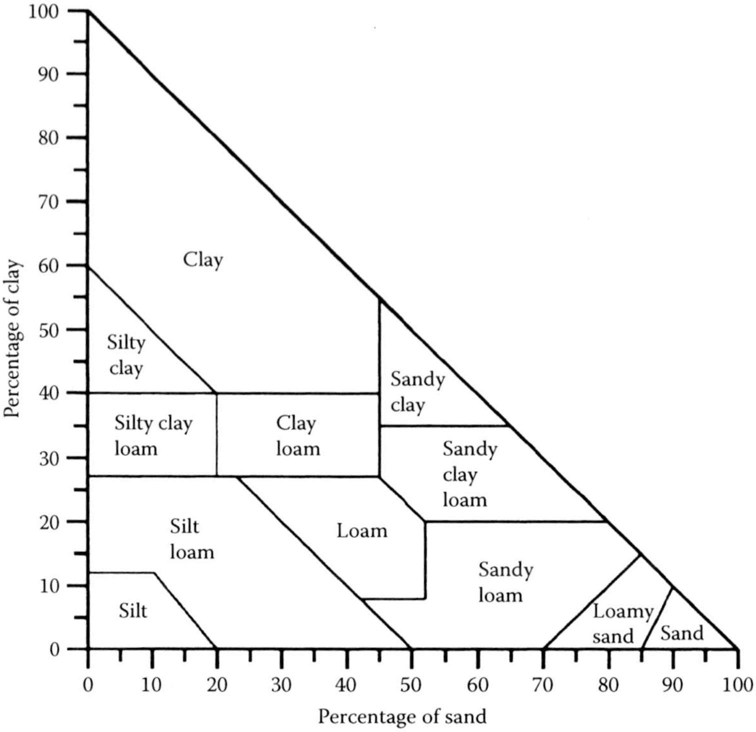

FIGURE 7.3 A simplified textural triangle using sand and clay soil contents to find texture class name. (Reprinted from Elgharmy, W., and M. Elashkar, *Soil Sci. Soc. Am. J.*, 26, 612–613, 1962. With permission from the Soil Science Society of America.)

transformations, and chemical processes in general become extremely slow, if they occur at all. The homogeneous chemical reactions include acid–base reactions, such as the carbonate equilibria:

$$CO_2(ac) + H_2O \rightleftharpoons HCO_3^- + H^+ \tag{7.1}$$

$$HCO_3^- \rightleftharpoons CO_3^{2-} + H^+ \tag{7.2}$$

or metal cation hydrolysis:

$$Al^{3+} + H_2O \rightleftharpoons AlOH^{2+} + H^+ \tag{7.3}$$

Complex formation such as the binding of transition metals with organic matter also occurs:

$$Pb^{2+} + R\text{–}COOH \rightleftharpoons R\text{–}COOPb^+ + H^+ \tag{7.4}$$

where R–COOH represents a carboxylic acid such as acetic acid or a carboxylic group of a larger molecule such as fulvic acid (FA) or humic acid (HA) (see Chapter 10). Complex formation with inorganic ligands also occurs, for example,

$$Cu^{2+} + Cl^- \rightleftharpoons CuCl^+ \quad (7.5)$$

Redox reactions, although they occur more commonly on the surface of Fe and Mn oxides, also can be present in the homogeneous phase, such as the oxidation of trivalent As to pentavalent by chromate:

$$3H_3AsO_3 + 2CrO_4^{2-} + 3H^+ \rightleftharpoons 2Cr(OH)_2^+ + 3H_2AsO_4^- + H_2O \quad (7.6)$$

The soil solution is involved as well in heterogeneous reactions such as precipitation–dissolution of calcium carbonate:

$$Ca^{2+} + CO_3^{2-} \rightleftharpoons CaCO_3(s) \quad (7.7)$$

or solid–solution redox reactions, such as the oxidation of As by Mn oxides:

$$H_3AsO_3 + MnO_2(s) + H^+ \rightleftharpoons H_2AsO_4^- + Mn^{2+} + H_2O \quad (7.8)$$

(Note that this example also involves reductive dissolution of MnO_2.)

7.2 SOIL COMPOSITION

As part of the earth's crust, soil chemical composition is similar but not identical to that of the whole crust, and this composition determines, to a large extent, the types of minerals present (Huang 1999; Sparks 2002; Sposito 2008). Table 7.1 shows the main chemical elements found in soil and its contents (in parts per million or milligram per kilogram) compared with those in the crust; some minor elements, known to be pollutants, are included, such as As, Cd, Cu, and Pb. Figure 7.4 shows the concentrations of the most abundant elements, both in mass and molar units. Note that it is a logarithmic scale, so the differences are important; for example, for those shown in Figure 7.4b, there are about three orders of magnitude between oxygen and fluorine. In most cases the soil abundance follows that of the crust, with the notorious exceptions of C and N; these differences are due to the accumulation of these elements in the organic-matter fraction of the soil.

7.2.1 Soil Minerals

From the molar elemental contents in Figure 7.4b, it is noted that O_2 and Si are the most abundant; they are followed by Al and C, and these, in turn, are followed closely by Fe, Ca, K, Mg, and Na. It should be no surprise then that Si and Al minerals are the most abundantly found, followed by Fe minerals. Si, Al, and Fe oxides are frequently found along with "aluminosilicates," minerals composed of O_2, Si, Al, and

TABLE 7.1
Common Elements in Soils

Element	Mean Soil Contents (ppm)	Earth's Crust Contents (ppm)	Mean Soil Contents ($mmol \cdot kg^{-1}$)	Earth's Crust Contents ($mmol \cdot kg^{-1}$)
Al	7.1×10^4	8.2×10^4	2600	3000
As	6	1.5	0.08	0.02
B	20	10	1.8	0.91
Ba	500	500	3.6	3.6
C	2.0×10^4	480	1670	40
Cd	0.35	0.11	0.00875	0.00275
Ca	1.5×10^4	4.1×10^4	370	1000
Cl	100	130	2.8	3.7
Cr	70	100	1.35	1.92
Cu	30	50	0.47	0.79
F	200	950	10.5	50
Fe	4.0×10^4	4.1×10^4	720	735
K	1.4×10^4	2.1×10^4	360	540
Mg	5000	2.3×10^4	210	960
Mn	1000	950	18	17
N	2000	25	140	1.8
Na	5000	2.3×10^4	220	1000
O	4.9×10^5	4.7×10^5	3.06×10^4	2.94×10^4
P	800	1000	26	32
Pb	35	14	0.17	0.07
S	700	260	22	8.1
Si	3.3×10^5	2.8×10^5	1.2×10^4	1.0×10^4
Sr	250	370	2.9	4.2
Ti	5000	5600	100	120
U	2	2.4	0.008	0.010
Zn	90	75	1.4	1.15
Zr	400	190	4.4	2.1

Source: Sparks, D. L., *Environmental Soil Chemistry*, Academic Press, San Diego, California, 2002.

countercations, which comprise the main part of soil clays, and have an important role in soil behavior, including ion binding (Sparks 2002; Sposito 2008; Tan 2009).

7.2.1.1 Primary Minerals

Soil minerals are classified as "primary" and "secondary." Primary minerals are formed under igneous conditions and found in volcanic and similar rocks. Many of them strictly are not thermodynamically stable (or the most stable) under surface conditions; thus they evolve slowly in the presence of water toward more stable structures, which are the secondary minerals. However, some primary minerals (such as quartz and micas) do remain stable and are found intermixed with secondary ones.

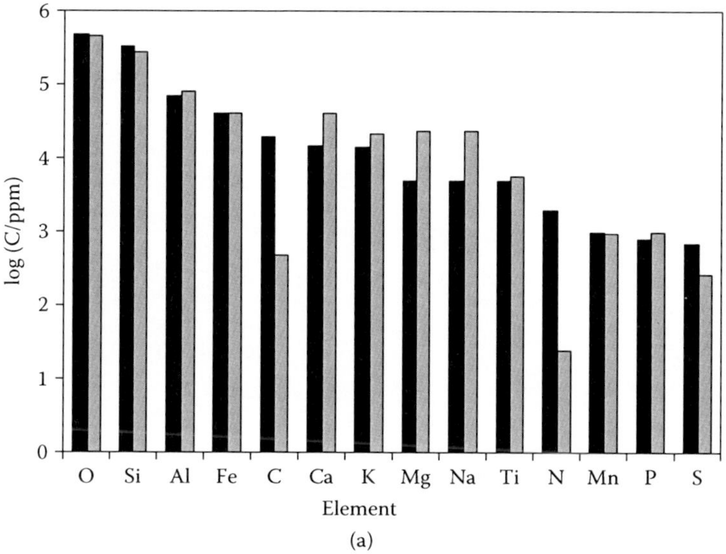

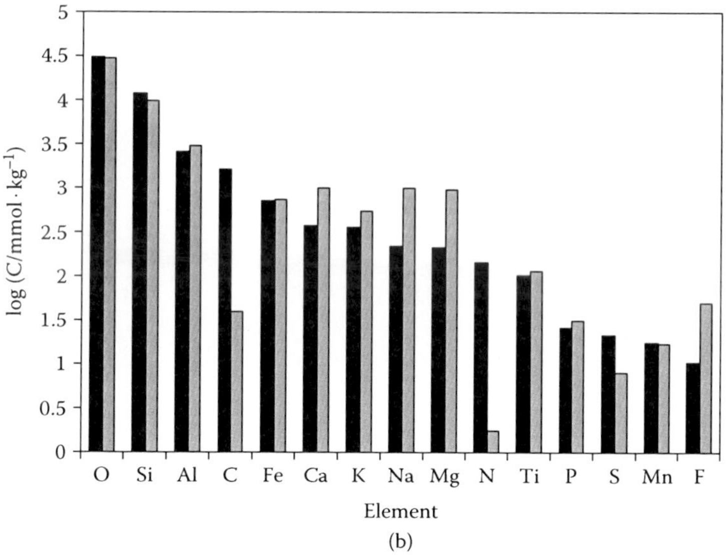

FIGURE 7.4 Abundance of the most important elements in soils (black bars) and in the earth's crust (grey bars) in (a) mass units (logarithm of concentration, in parts per million) and (b) molar units (logarithm of concentration in millimoles per kilogram).

Table 7.2 shows some examples of primary minerals. Feldspars are among the most important and prevalent primary minerals in soils; they are common in the sand and silt fractions of soils and can also be found in the clay fraction. Metamorphic rocks also contain feldspars. K feldspars are important sources of K in soils (Sparks 2002). Pyroxenes and olivines are some of the "accessory minerals" found in the high-density fraction. Olivines are green nesosilicates in which Mg^{2+} and Fe^{2+} are

TABLE 7.2
Some Important Primary Minerals

Type	Name	Chemical Formula
Oxides	Quartz	SiO_2
	Rutile	TiO_2
Silicates		
Pyroxenes	Enstatite	$MgSiO_3$
	Rhodonite	MnSiO3
	Diopside	$CaMg(SiO_3)_2$
Nesosilicates	Zircon	$ZrSO_4$
	Forsterite	$MgSiO_4$
	Fayalite	$FeSiO_4$
	Olivines	$(Mg,Fe)(SiO_4)_2$
Silicoaluminates		
Feldspars	Orthoclase	$KAliSiO_8$
	Albite	$NaAlSiO_8$
Micas	Muscovite	$KA_{12}(AlSi_3O_{10})(OH)_2$
	Biotite	$K(Mg,Fe)_3(AlSi_3O_{10})(OH)_2$

octahedrally coordinated by O atoms; they are one of the several "mineral series": the pure Mg^{2+} and Fe^{2+} nesosilicates, forsterite and fayalite, respectively, are the end members, and in general olivines have both cations in a variable ratio but with the same crystal structure, forming solid solutions of both end members.

7.2.1.2 Secondary Minerals

Primary minerals are formed under volcanic conditions, under high temperatures and pressures. When, due to geological processes, they come in contact with the atmosphere and surface waters, they undergo slow dissolution and reprecipitation reactions. This process is termed "chemical weathering," and its outcome is the formation of secondary minerals. Different types of secondary minerals, with some important examples, are presented in Table 7.3. Phyllosilicates (also called clay minerals) are arguably the most important class of soil minerals and play a key role in soil chemistry, including ion binding and nutrient retention (Olson et al. 1999). These minerals have a complex layered structure, which is described in Chapter 8. For the moment, it will suffice to note that Si and Al atoms are located in separate layers and that they frequently present isomorphic substitution, thus carrying a permanent charge as described in Chapter 5. It is worth noting that micas are structurally phyllosilicates and that there are primary and secondary micas. Allophane and imogolite are less crystalline materials formed from volcanic ashes; they are also important components described in Chapter 8.

Besides clay minerals, Al, Fe, and Mn oxides (including both anhydrous oxides and oxohydroxides) are very important in soil chemistry, even if they are not very abundant as phyllosilicates. They usually have high surface areas and are capable of adsorbing ions, particularly anions, unlike clay minerals, which more often adsorb

TABLE 7.3
Some Important Secondary Minerals

Family	Name	Chemical Formula
Clay minerals (phyllosilicates)		
	Kaolinite	$Si_4Al_4O_{10}(OH)_8$
	Montmorillonite	M_x, $(Al,Fe(II),Mg)_4\ Si_8O_{20}(OH)_4$ (M = interlayer cation)
	Vermiculite	$(Al,Fe(III),Mg)_4\ (Si,Al)_8O_{20}(OH)_4$
Al oxides		
	Gibbsite	$Al(OH)_3$
	Bohemite	γ-AlOOH
	Corundum	α-Al_2O_3
Fe oxides		
	Goethite	FeOOH
	Hematite	α-Fe_2O_3
	Maghemite	γ-Fe_2O_3
Mn oxides		
	Pyrolusite	β-MnO_2
	Birnessite	δ-MnO_2
Limes		
	Calcite	$CaCO_3$
	Dolomite	$CaMg(CO_3)_2$
Evaporites (soluble minerals)		
	Halite	NaCl
	Gypsum	$CaSO_4{\cdot}2H_2O$
Poorly crystalline minerals		
	Allophane	$Si_3Al_4O_{12}{\cdot}nH_2O$
	Imogolite	$Si_2Al_4O_{10}{\cdot}5H_2O$

cations than anions. Iron and manganese oxides, in particular, have key roles in soil redox reactions. Oxides in general are described in Chapter 9. Carbonates, sulfates, and other soluble minerals such as halites are found only in arid and semiarid regions and have less importance with respect to ion binding.

7.2.1.3 Structural Elements of Soil Minerals

The structure of chemical compounds in general, and of minerals in particular, is dictated to a large extent by geometrical constraints imposed on the arrangement of individual atoms comprising a given substance by the size of these atoms. Chemical compounds are formed by the interaction of different elements, which have different properties, such as electronegativity and size. Being the most abundant element in soil (and in earth's crust), O is present in the vast majority of compounds and, being highly electronegative, acts with an effective negative charge. In ionic compounds, it is found as the oxide ion, O^{2-}; setting aside fluorine, all other elements are less electronegative and react with oxygen with a positive charge. In many cases, however, including most soil minerals, the chemical bonds are not ionic but mainly covalent, albeit with an important degree of ionic character. Nevertheless, the following reasoning applies equally to purely ionic compounds and to partially ionic ones. When a cation (or electropositive atom) interacts with an anion (or electronegative atom), it will try to be surrounded by as many of them as possible; the limit is determined by the sizes of both species. The behavior is summarized in the "Pauling rules" (Pauling 1929), which were originally formulated for ionic compounds. An important concept here is the "coordination number," Z, which is the maximum number of neighbors that can be accommodated around a given central atom, a cation in this case; another parameter is the "radius ratio," R_R, which is the quotient between the cation and the anion radii, r_+ and r_-, respectively:

$$R_R = \frac{r_+}{r_-} \tag{7.9}$$

The Pauling rules state the following:

1. A coordinated polyhedron of anions is formed around each cation, the distance being determined by the sum of the ionic radii of both and the coordination number by the radius ratio.
2. In a stable coordination structure, the total strength of the valence bonds of an anion with all neighboring cations is equal to the anion charge.
3. The existence of edges, and particularly of faces, common to the anion polyhedra in a coordinated structure decreases its stability. This effect is larger for cations with high valency and small coordination number.
4. In a crystal containing different cations, those of high valency and small coordination number tend not to share polyhedron elements with each other.
5. The number of essentially different kinds of constituents in a crystal tends to be small.

As a result of these rules (Sparks 2002), the coordination number for an atom or a cation surrounded by O is essentially due to the radius of the central cation, which determines how many O species can be accommodated around it. Table 7.4 and Figure 7.5 show this for several important elements found in soil. Si acts with a +4 oxidation state (strictly, it is not an ion); thus it always has a tetrahedral environment (due to symmetry reasons, if the surrounding atoms are equal, the polyhedron must be regular). Al has a borderline size between $Z = 4$ and $Z = 6$; it usually has an octahedral environment, but it can replace Si in a tetrahedron, thus giving isomorphic substitution. Likewise, there are several divalent cations (e.g., Mg^{2+} and Fe^{2+}) that also fit in an octahedral configuration, thus being able to replace Al producing also isomorphic substitution with charge imbalance, as described in Chapter 5. The mineral structures and the consequences of substitution are discussed in Chapters 8 and 9.

7.2.2 Soil Organic Matter

Soil organic matter (SOM) can be defined (Baldock and Nelson 1999) as the sum of all natural and thermally altered biologically derived organic material found in the soil, or on the soil surface, irrespective of its source or stage of decomposition and

TABLE 7.4
Ionic Radius, Radius Ratio to Oxide, and Coordination Number of Common Cations

Cation	Radius (r_i/nm)	Radius Ratio ($R_R = r_i/r_{O^{2-}}$)	Coordination Number (Z)	Geometrical Arrangement
Si^{4+}	0.039	0.279	4	4: Tetrahedron:
Al^{3+}	0.051	0.364	4, 6	Figure 7.5a
Fe^{3+}	0.064	0.457	6	6: Octahedron:
Mg^{2+}	0.066	0.471	6	Figure 7.5b
Ti^{4+}	0.068	0.486	6	
Fe^{2+}	0.074	0.529	6	
Mn^{2+}	0.080	0.571	6	
Na^{+}	0.097	0.693	8	8: Cube:
Ca^{2+}	0.099	0.707	8	Figure 7.5c
K^{+}	0.133	0.950	8, 12	12: Cuboctahedron:
Ba^{2+}	0.134	0.957	8, 12	Figure 7.5d

Source: Sparks, D. L., *Environmental Soil Chemistry*, Academic Press, San Diego, CA, 2002.

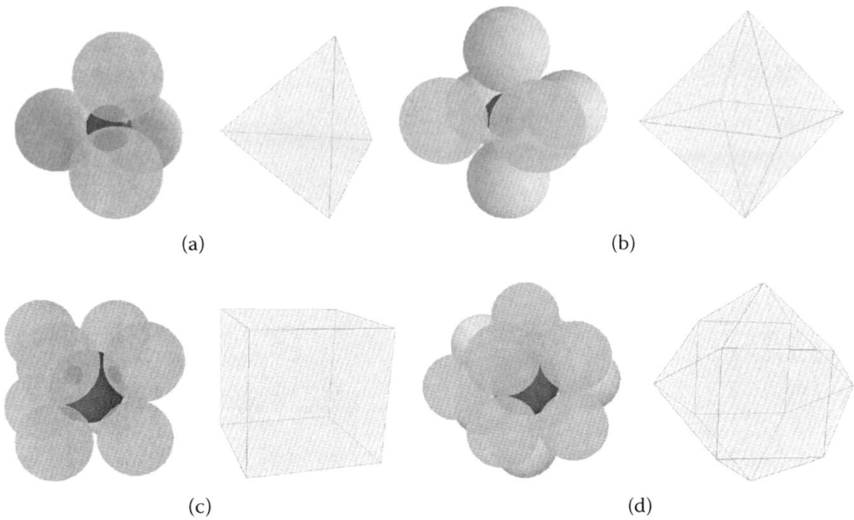

FIGURE 7.5 The basic chemical structures found around common elements surrounded by oxygen atoms in soil minerals, for different sizes of the central atom. Left: the arrangement of oxygen atoms or oxide ions (light spheres) around a central cation (dark sphere); the front spheres have been drawn partially transparent for clarity. Right: the geometrical body defined by the oxygen atoms. (a) Tetrahedron, found with smaller central species such as Si^{4+} or Al^{3+}; (b) octahedron, a commonly found geometry in many cases, such as Al^{3+}, Mg^{2+}, $Fe^{2+/3+}$, Mn^{2+}, and so on; and (c) cube and (d) cuboctahedron, forms found with larger cations such as Na^{+}, K^{+}, or Ba^{2+}.

whether it is living or dead, but excluding the aboveground portion of living plants. It should be noted that this definition includes, among other biological components, plant roots; however, several soil studies start by sieving the soil through a 2 mm sieve, so many studies are conducted only with the below-2-mm particulate portion. In turn, this implies that the macroscopic organic structures (roots and plant debris) are in practice not included in SOM. In fact, there are other definitions of SOM in the literature (MacCarthy, Clapp, and Malcolm 1990).

SOM has a number of particular properties (Senesi and Miano 1994; Senesi and Loffredo 1998; Baldock and Nelson 1999; Sparks 2002):

Chemical properties: SOM provides pH buffering capacity for slightly acidic (pH ~4–5) or alkaline (pH ~6–8) soils, thus helping to keep the soil pH within the range acceptable to plants; SOM has an important metal chelating capacity, forming stable complexes with many metals, thus contributing to its fixation, which is particularly important in the case of pollutants. SOM also provides cation-exchange capacity, due to its acidic character, which presents negative charge on dissociation and contributes to the retention and availability of cationic nutrients; also, SOM interacts with xenobiotics such as pesticides, altering its behavior in soil.

Biological properties: Organic matter is a source of nutrients to plants, due to its cation-exchange capacity; it provides metabolic energy for soil biological processes (e.g., bacterial) and contributes to ecosystem resilience, because it enhances the ecosystem's ability to recover from natural or anthropogenic perturbations.

Physical properties: SOM significantly contributes to the stabilization of soil structure, because it binds to soil mineral particles; thus, it is capable of binding these particles together into water-resistant aggregates. It has an important contribution to soil water retention because of its ability to absorb up to 20 times its mass in water; the low solubility of SOM adds to its stability in upper horizons, without leaching, and the dark color of SOM contributes to soil thermal properties.

SOM is classified into two main categories: "humic" and "nonhumic" substances (MacCarthy, Clapp, and Malcolm 1990; Senesi and Loffredo 1998; Baldock and Nelson 1999; Sparks 2002; Brady and Weil 2007).

7.2.2.1 Humic Substances

The term humic substances (HSs) refers to a collection of refractory, relatively high-molecular-weight, dark-colored substances formed from the debris of plants and other biological organisms by degradation and secondary synthesis reactions (Senesi and Loffredo 1998). HSs constitute the major fraction of SOM, up to 80%. These materials are distinctive to the soil and different from undecayed plant and animal biological materials. HSs are usually associated with mineral particles, other organic compounds, and/or charge-neutralizing or chelated cations. HSs are not well-defined chemical compounds described by a common chemical structure but a chemically heterogeneous collection of compounds. In practical terms, HSs are defined, obtained, and classified through extraction and separation procedures; thus, one finds the "operational definition" of HS and its fractions through these procedures. They are usually divided in the literature into three groups:

1. FAs: the fraction that is soluble in water at any pH value
2. HAs: the fraction that is soluble at pH > 1
3. Humin: the fraction insoluble in water at any pH

Other groups include, for example, hymatomelanic acid, the alcohol-soluble fraction (Senesi and Loffredo 1998). Clearly, these fractions are loosely defined and can include many different compounds. Usually, HSs are obtained from the soil through alkaline extraction, which separates FA and HA but not humin, albeit other procedures have also been proposed (Swift 1996). The structure and properties of HSs are considered in Chapter 10.

7.2.2.2 Nonhumic Substances

Nonhumic substances are those recognizable as known biochemical compounds, such as amino acids, carbohydrates, fats, organic acids, resins, and waxes; these are usually under degradation conditions and are thus relatively short lived (Senesi and Loffredo 1998; Baldock and Nelson 1999). The main categories of nonhumic materials are the following biopolymers:

Carbohydrates: About 5%–25% of SOM are estimated to be carbohydrates, forming usually the largest fraction of nonhumic SOM. This includes simple sugars to polysaccharides of high molecular weights. They influence soil physical conditions, interact with cations through exchange and complexation, and take part in C metabolism and other biological activities. Figure 7.6 shows a very common polysaccharide, cellulose, which is part of the structure of many plants and is thus found in soil as a result of degradation of plant debris.

Polypeptides and other N-containing compounds: N is a major nutrient and is unique in that organic N accounts for more than 90% of the total N in topsoil. A variety of nitrogen-containing compounds can be found in soils, including amino acids and amino acid–derived substances (constituting about

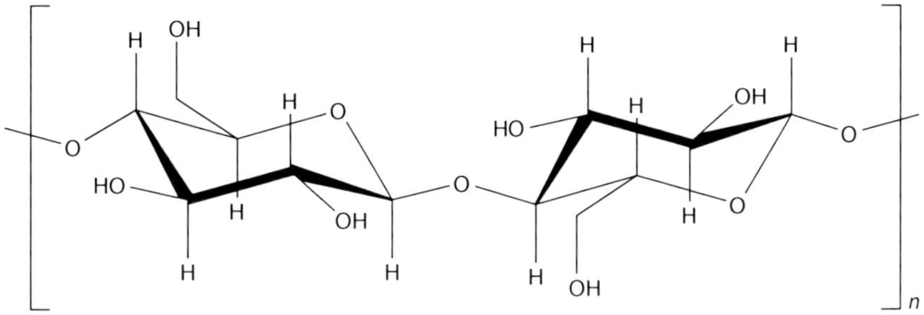

FIGURE 7.6 The structure of cellulose, a typical polysaccharide, which is a structural constituent of most plants: it is composed of D-glucopyranose units.

30%–40% of soil N), amino sugars and teichoic acids, chitin, vitamins, and other compounds. Figure 7.7 shows some examples of an amino acid chain and chitin, a polymer of *N*-acetylglucosamine, an amino sugar. In isolation, most of these substances are quite reactive, but several of them are able to bind with other organic constituents (HAs, lignins, polyphenols, etc.) to form biologically resistant products. Also, they can form stable complexes with cations such as Fe(III) and Al(III), also producing biologically resistant products.

Alkyl compounds: It has been found (Baldock et al. 1997) that a significant part of soil C is in aliphatic form, accounting for between 15% and 40% of the total organic C, depending on soil type and humification degree. It is found mainly in free and bound lipids (fatty acids and waxes), insoluble polyesters, and other macromolecules produced by soil microorganisms. The solvent-extractable lipids include waxes, organic acids, steroids and terpenoids, carotenoids, porphyrins, glycerides, and phospholipids. These compounds are usually readily decomposed by soil microorganisms (Senesi and Loffredo 1998; Baldock and Nelson 1999). Figure 7.8 shows the general

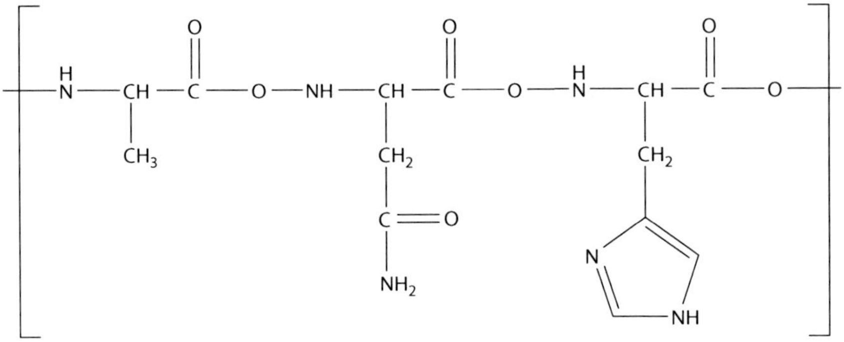

(a)

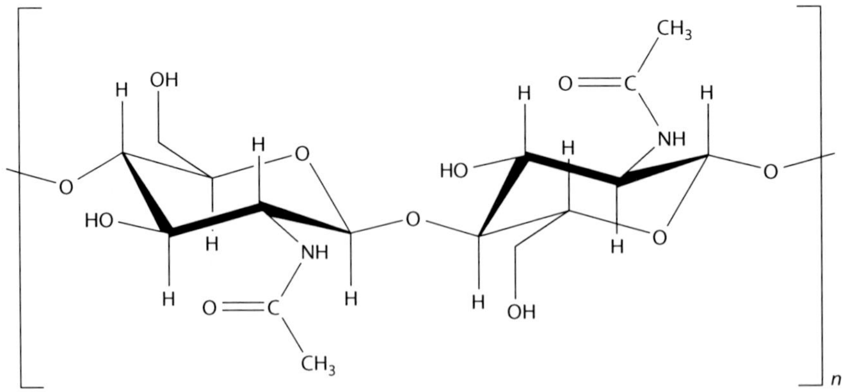

(b)

FIGURE 7.7 Some examples of nitrogen-containing compounds found in soils: (a) a segment of a polypeptide chain. (b) A structural unit of chitin, an amino sugar polymer found in plant structures and in insect exoskeleton.

structure of polyester chains and some monomers typically found in cutin and suberin, the main components of plant cuticle.

Lignins: Lignins are polyphenols mainly constituted by phenylpropane units of different nature, depending on their origin (some examples are shown in Figure 7.9, involving C–O linkages; C–C linkages are also found). These

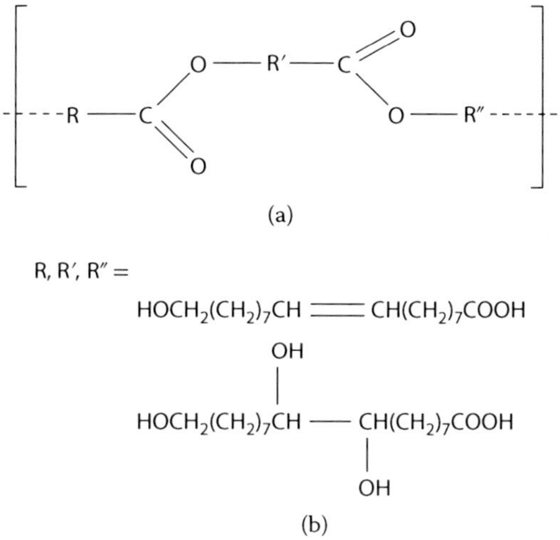

FIGURE 7.8 The general structure of a polyester chain (a) and monomers commonly found in soil polyesters (b).

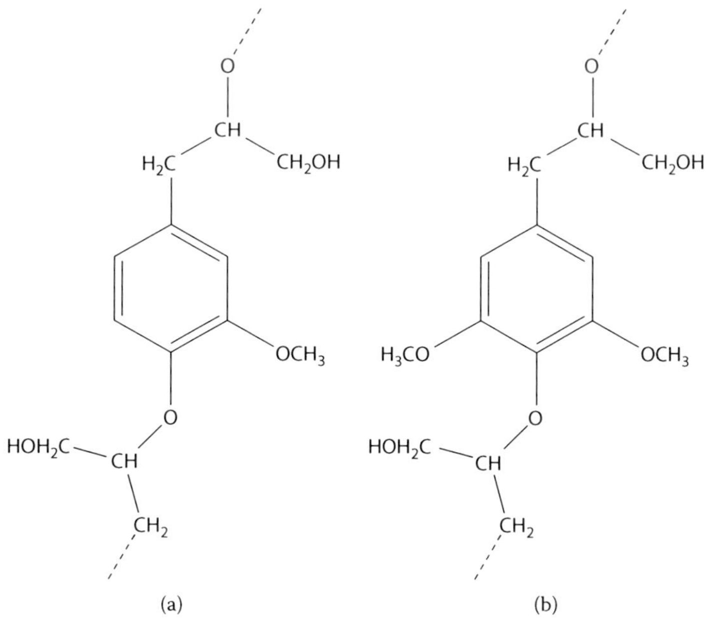

FIGURE 7.9 Typical structures of lignin polymers based on polyphenols: (a) guaiacyl unit (derived from coniferyl alcohol) and (b) syringyl unit, from sinapyl alcohol.

polymers are constituents of plant structures along with cellulose and represent the second most abundant organic material in plant residues after cellulose.

7.3 X-RAY-BASED CHARACTERIZATION METHODS

The use of x radiation for analytical purposes has found wide application in environmental sciences, especially with the advent of high-intensity synchrotron sources, which allowed the development of many sophisticated methods that are not possible with conventional sources (Bertsch and Hunter 2001; de Groot 2001; Gaillard 2007; Matsuda 2008; Luo and Zhang 2010; Meirer et al. 2010). X radiation is composed of high-energy photons (see Figure 7.10), corresponding to the binding energy of core electrons and wavelengths on the order of lattice spacing; these facts give rise to most x-ray applications.

7.3.1 X-Ray Diffraction

Crystalline solids have a regular three-dimensional (3-D) geometric arrangement of atoms in a lattice, with spacing on the order of a few angstroms falling in the x-ray range (Figure 7.10). Thus, diffraction phenomena, consisting of radiation dispersion by lattice atoms and interference of dispersed rays, can occur. Figure 7.11 illustrates an example of constructive interference, resulting in Bragg's law:

$$2d\sin\theta = n\lambda \tag{7.10}$$

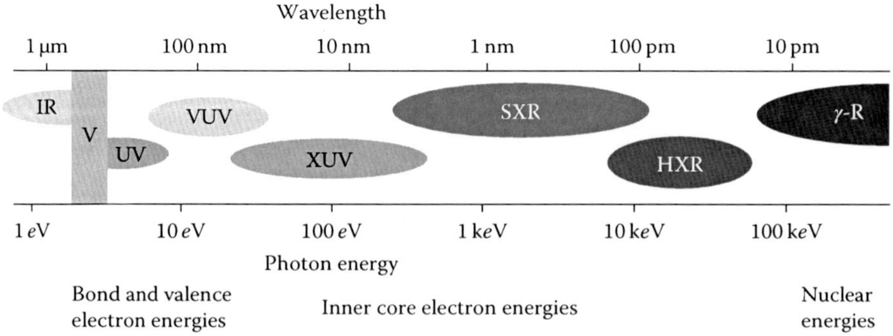

FIGURE 7.10 The electromagnetic spectrum, from infrared to gamma rays, in wavelength and photon energy scales. Legend: IR, infrared region (from 700 nm up); V, visible range (400–700 nm); UV, ultraviolet region (~200–400 nm); VUV, vacuum ultraviolet (~ 200–100 nm); XUV, extreme ultraviolet (~10–100 nm); SXR, soft x-ray (~200 pm – 10 nm); and HXR, hard x-ray (~10–200 pm). Below 10 pm is the γ-ray region (γ-R), which corresponds to nuclear energy levels, extending far into the megaelectron volt range; beyond the IR region is the radio wavelength region. The visible–UV region photons have energies comparable to those of bond and outer orbital levels, whereas core electrons of most elements have binding energies over 100 eV, lying in the extreme UV and x-ray regions.

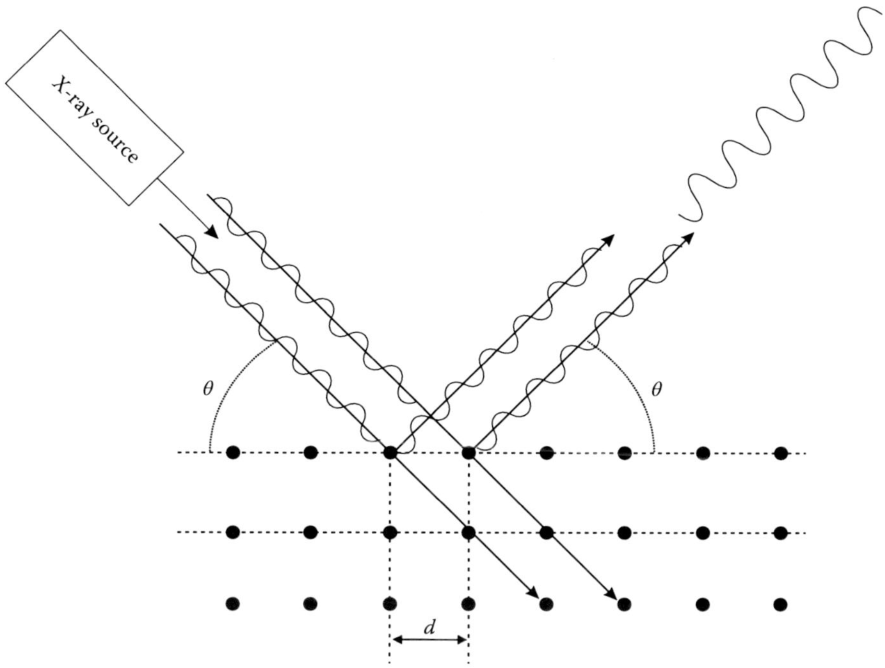

FIGURE 7.11 Schematic representation of x-ray diffraction in a lattice with constructive interference, because the difference in optical paths is equal to an integer number of wavelengths (just one in this example).

where λ is the radiation wavelength, d the lattice spacing, θ the incidence angle, and n an integer giving the order of interference (equal to 1 in Figure 7.11). X-ray diffraction (XRD) is the technique essentially based on Bragg's law (Woolfson 1997; Ladd and Palmer 2003). Thus, by measuring the θ values for which constructive interference occurs, the lattice spacing distances can be determined. Because each crystalline solid has its own structure, a unique pattern of intensities as a function of the angle, or an "x-ray diffractogram," will be produced; XRD is thus a qualitative and, usually, semiquantitative technique for studying crystalline solids. An x-ray diffractometer is in instrument that measures the intensity of the diffracted beam as a function of the angle θ (or the total angle 2θ), presenting the XRD diffractogram. Figure 7.12 shows examples of diffractograms from soil samples where some minerals have been identified; in the bottom difractogram, Fe oxides have been removed. More examples are presented in Chapter 9. XRD is one of the most widely used techniques in the study of minerals, and soil minerals in our case.

7.3.2 X-Ray Absorption–Based Methods

When electromagnetic energy interacts with matter, two main processes are possible: absorption and scattering. From these, several phenomena can take place, the main ones being schematically represented in Figure 7.13. The absorption of x

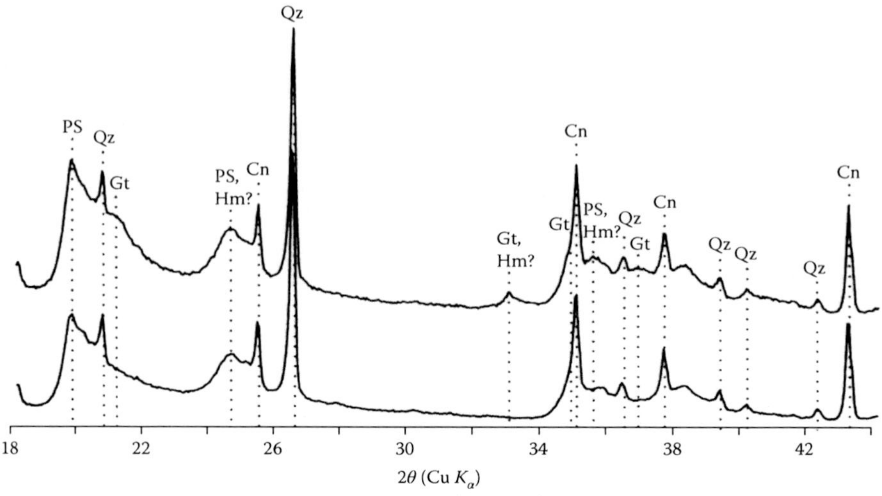

FIGURE 7.12 X-ray powder diffraction patterns of a soil clay fraction untreated (top) and treated (bottom) with dithionite-citrate-bicarbonate attack, which removes all secondary Fe oxides: Cn = α-Al_2O_3 (corundum); Gt = goethite; Hm = hematite; PS = phyllosilicates; and Qz = quartz. Hematite identification is uncertain because of strong overlap with phyllosilicate and goethite peaks, hence the question marks. Samples contained α-Al_2O_3 as an internal standard. (Reprinted from Schulze, D. G., *Clays Clay Miner.* 34, 6, 681–685, 1986. With permission from the Clay Minerals Society.)

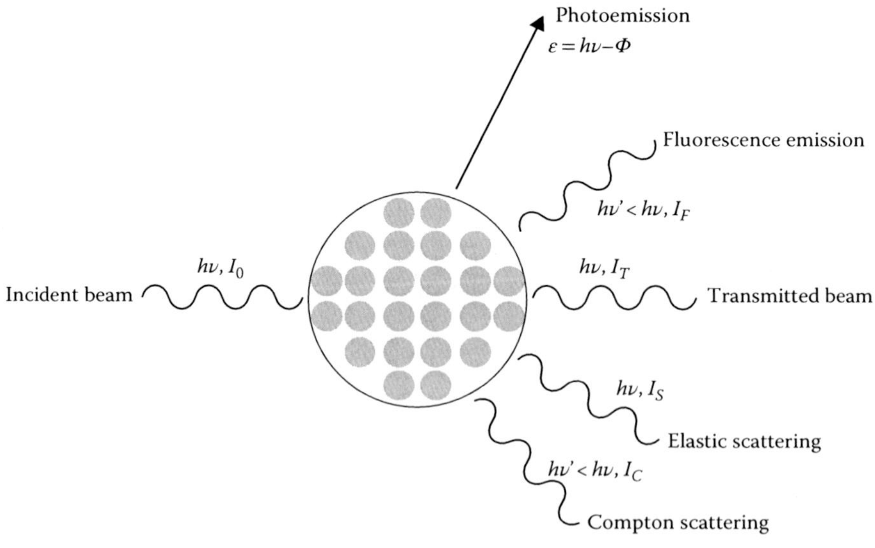

FIGURE 7.13 Schematic representation of the main processes involved in the interaction of x-ray radiation with matter: there are also other processes (i.e., Auger electrons and electron–positron pair formation) that are usually minor and not represented here.

radiation results in the excitation of a core electron belonging to inner energy levels, which can be promoted to an unoccupied state of higher energy or emitted into the continuum. This photoemitted electron can be ejected from the sample with a kinetic energy equal to the difference between the photon and the binding energies or be backscattered by other atoms. The photoelectron energy can be analyzed to find the binding energy, a method known as "x-ray photoelectron spectroscopy" (XPS), which is discussed in Section 7.3.4. The electron emission leaves an empty state in the original level, which eventually is occupied by another electron falling from upper levels; this transition produces the emission of energy, also in the x-ray range but of lower energy than the absorbed photon. This results in x-ray emission or, as it is more frequently called, x-ray fluorescence (XRF).

The backscattering of the photoelectron can result in the production of secondary electrons or fluorescence emission. The measurement of transmitted intensity constitutes x-ray absorption spectroscopy (XAS). When the transmitted energy of an x-ray is scanned near the energy necessary to excite an electron, responses such as those shown in Figure 7.14 are obtained. The sharp increase in absorption at the electron energy is termed the absorption edge. From that on, several fluctuations and oscillations are observed; due to the scattering of photoelectrons by surrounding atoms, interference between outgoing and backscattered photoelectron waves is observed. The energy region next to the absorption edge is usually (and somewhat arbitrarily) divided in two parts (Figure 7.14): x-ray absorption near-edge structure (XANES), also known as near-edge x-ray absorption fine structure, and extended x-ray absorption fine structure (EXAFS). We do not discuss here the rather involved physical principles governing these phenomena but only mention some key points. The reader is referred to the literature for further insight (Sayers, Stern, and Lytle 1971; Koningsberger and Prins 1988; Fendorf and Sparks 1996; Rehr and Albers 2000;

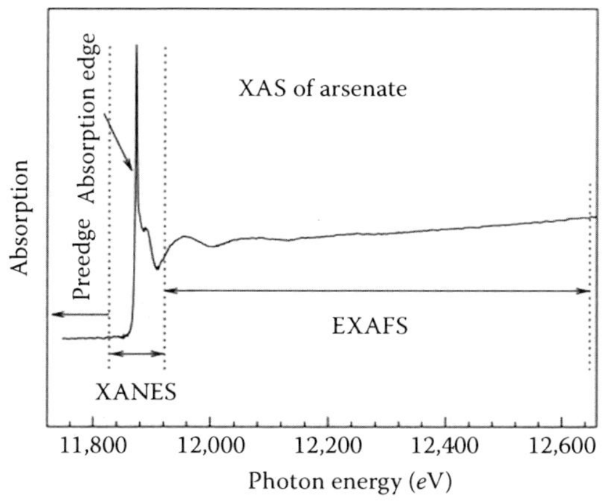

FIGURE 7.14 X-ray absorption spectroscopy (XAS) spectrum of arsenate with the x-ray absorption near-edge structure (XANES) and extended x-ray absorption fine structure (EXAFS) regions indicated. (Reprinted from Luo, L., and S. Zhang, *Sci. China: Chem.* 53, 12, 2529–2538, 2010. With permission from Science China Press.)

Rehr and Ankudinov 2001). It is worth mentioning at this point that these techniques often require high-intensity x-ray beams, which can only be obtained from a synchrotron facility; thus, its application is limited by the availability of such equipment.

In the case of EXAFS, in this region sinusoidal oscillations are produced by the interferences, as represented in Figure 7.15. Qualitatively, the amplitude is proportional to the coordination number, and so is the frequency to the inverse of bond length, whereas the phase and the amplitude are related to the scatterer. In Figure 7.15, $x(k)$ is the fractional modulation of the absorption coefficient (Sayers, Stern, and Lytle 1971; Penner-Hahn 1999):

$$x(k) = \frac{\mu(E) - \mu_0(E)}{\Delta\mu_0(E)} \tag{7.11}$$

where $\mu(E)$ is the measured absorption coefficient, $\mu_0(E)$ a smooth background function (single atom absorption), $\Delta\mu_0(E)$ the jump in absorption at the edge, and k the photoelectron wavevector

$$k = \sqrt{\frac{2m(E - E_0)}{\hbar^2}} \tag{7.12}$$

where m is the electron mass, E its energy, and E_0 the binding energy; $x(k)$ can be described as the superposition of various oscillations caused by the surrounding atoms:

$$x(k) = -S_0^2 \sum_i \frac{N_i}{kR_i^2} \left| f_i(k) \right| e^{-2\sigma_i^2 k^2} e^{-\frac{2R_i}{\lambda(k)}} \sin\left[2kR_i + \phi_{ij}(k) \right] \tag{7.13}$$

Here, S_0^2 is the passive electron reduction factor; N_i, the number of neighbors of type i at a distance R_i; $f_i(k)$, the amplitude of the scattering wave from the neighbor of type i; σ_i a factor describing static and dynamic disorders; $\lambda(k)$ the mean free path of the photoelectron, a phenomenological term accounting for inelastic losses; and $f_{ij}(k)$

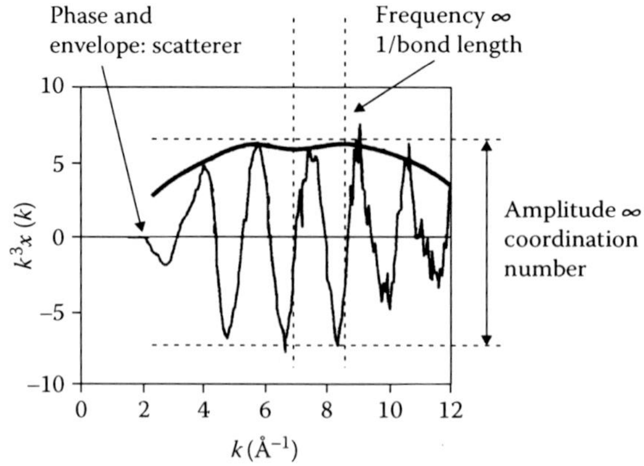

FIGURE 7.15 Schematic representation of an x-ray absorption spectrum in the EXAFS region, indicating the diverse pieces of information obtained.

is the phase shift between the central atom j and the neighbors i. The sum is carried over, in principle, all neighbors. Equation 7.13 can be used, through appropriate fitting, to determine the coordination number, type of neighbors, and bond distances of the coordination environment of a metal atom. The energy dependence of $f_i(k)$ and the phase shift $\varphi_{ij}(k)$ are the most important parameters containing information about the scattering atom. However, there are several limitations; in the first place, there is a weak dependence on the nature of the atom so that it is not possible, for example, to distinguish between N and O. Also, in practice only first and second neighbors can usually be determined. Nevertheless, EXAFS has been applied in a number of problems of ion binding to soils and soil components (Xia, Bleam, and Helmke 1997; Karlsson, Persson, and Skyllberg 2006; Ghabbour, Scheinost, and Davies 2007; Gustafsson et al. 2007).

Another interesting and convenient way of qualitatively analyzing EXAFS signals is by means of the Fourier transform, which provides a pseudoradial distribution function, allowing to visualize the "position" of different scatterers (Gaillard 2007).

In the XANES region, the same principles apply, but there is greater sensitivity to a number of factors such as distance to the scatterer. Thus, this part of the spectrum will be sensitive to multiple scattering events and, thus, to the geometry of the scatterer, but the calculations are more complicated. The XANES region is sensitive to the geometry of the element; thus it is able to provide information on the coordinating species (Gaillard 2007). Also, the position of the peak is slightly dependent on the oxidation state of the element (see Figure 7.16), thus enabling speciation studies under favorable conditions (Strawn and Baker 2008).

Additionally, there may appear preedge features, which are also related to the oxidation state of the element, arising from bound-state transitions and indicating the crystalline geometry (Westre et al. 1997; Wilke et al. 2001).

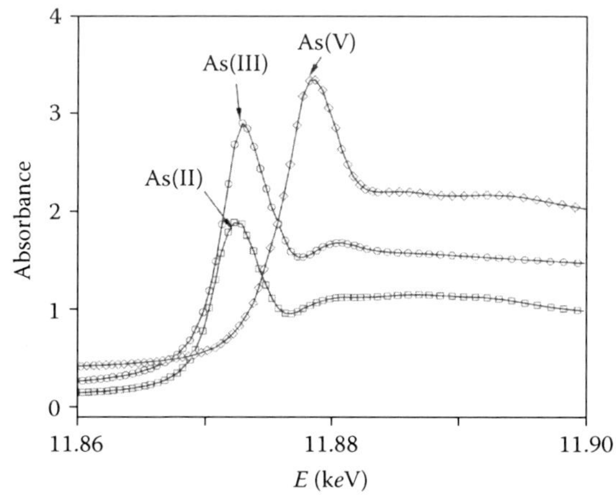

FIGURE 7.16 Changes in XANES spectra for various oxidation states and site geometry: note the difference between the different resonance features. (From Gaillard, J. –F/Wilkinson, K. J., and J. R. Lead: *Environmental Colloids and Particles: Behaviour, Separation and Characterisation*. 2007. 613–666. Copyright Wiley-VCH Verlag GmbH & Co. KGaA. Reproduced with permission.)

XAS methods have become a valuable tool to study the coordination environment of metal ions in soils, sediments, and other environmental problems. The key points are as follows: first, of course, getting a good quality spectrum can be difficult if the metal concentration is low, even with synchrotron light (Gaillard 2007). In the second place, the analysis is relatively involved, because fitting to Equation 7.13 or numerical analysis from first principles requires much computational effort; nowadays, there are several software packages that simplify the matters considerably (Gaillard 2007).

7.3.3 X-Ray Fluorescence

As shown in Figure 7.13 and discussed in Section 7.3.2, one of the possible outcomes of x-ray absorption is the emission of fluorescence (XRF). The emitted photon has an energy corresponding to the difference between the lower level (typically 1s) and the upper level (e.g., 3p in the case of Mn(II) and similar species) of the electronic transition, thus being characteristic of the element (de Groot 2001). XRF is thus a technique useful for identifying elements in complex samples and has found ample application in environmental studies, including soil analysis.

There are two types of XRF spectrometers, wavelength dispersive and energy dispersive, which are both illustrated in Figure 7.17. Wavelength-dispersive instruments (Figure 7.17a) are somewhat similar to a conventional spectrophotometer: the emitted x-ray beam is collimated and directed to a dispersing element, where an appropriate crystal selects one wavelength from the incident beam through Bragg diffraction (Section 7.3.1). The crystal is rotated and the detector is moved according to Bragg's law in a circle around it (Figure 7.17a). In an energy-dispersive instrument (Figure 7.17b), a solid-state detector, traditionally lithium-doped Si cooled with liquid N_2, generates for each arriving photon a pulse whose amplitude is proportional to the photon energy. The output of this detector is fed to a multichannel analyzer, which discriminates the different pulses and counts pulses with the same amplitude. These detectors are mechanically simple because they do not need the crystal-detector rotating assembly but require liquid N_2 to work properly. These detectors are fast and convenient, especially when doing multielement analysis. On the other hand, wavelength-dispersive instruments have better overall resolution, and also better signal-to-noise ratios, and do not require continuous N_2 cooling. Modern devices employ a Si drift detector (SDD) cooled through the Peltier effect (Strüder, Lechner, and Leutenegger 1998). With SDD it is possible to build portable XRF instruments, allowing field determinations through energy-dispersive measurements (Goldstein, Slemmons, and Canavan 1996; Carr et al. 2007; Terzano et al. 2008; Jang 2009; Morgenstern et al. 2009).

7.3.4 X-Ray Photoelectron Spectroscopy

The absorption of x-ray photons frequently results in photoelectron emission (Figure 7.13); these electrons when not scattered by other atoms leave the material, according to the photoelectric effect, with a kinetic energy equal to the difference between the photon energy and that needed to remove it from the target atom, which is the binding energy or work function:

$$\varepsilon = h\nu - F \tag{7.14}$$

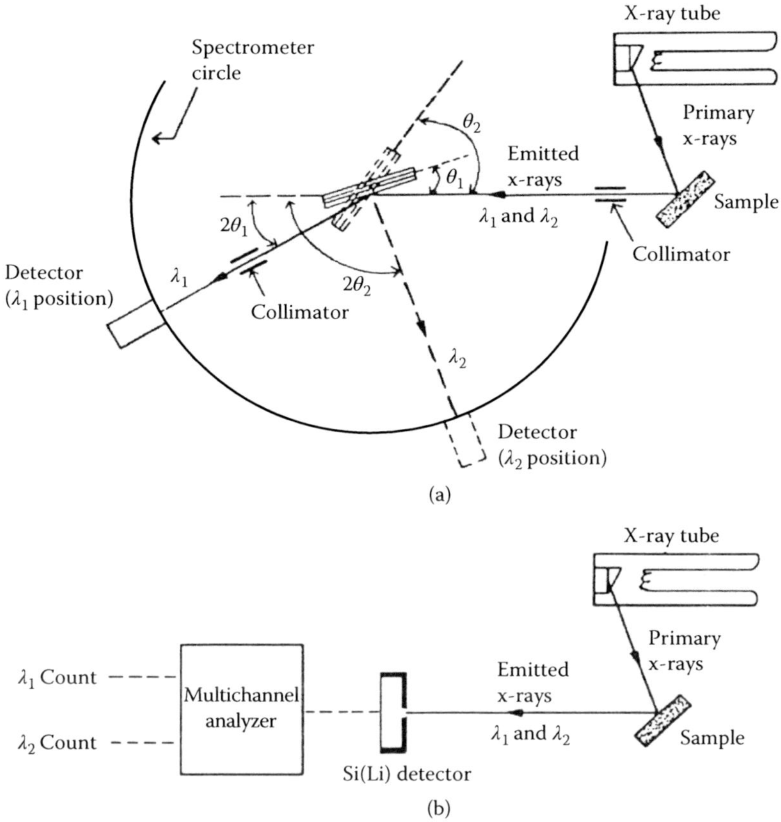

FIGURE 7.17 Schematic arrangement of (a) wavelength-dispersive and (b) energy-dispersive x-ray fluorescence spectrometers. (Reprinted from Karathanasis, A. D., and B. F. Hajek, *Methods of Soil Analysis. Part 3. Chemical Methods*, American Society of Agronomy–Soil Science Society of America, Madison, Wisconsin, 1996. With permission from the Soil Science Society of America.)

where ε is the electron kinetic energy, ν the frequency of the incident radiation, and Φ the work function. As the incident radiation is known (usually K_α emission from Al or Mg), by analyzing ε, it is possible to determine the binding energy Φ and, consequently, identify the element that the photoelectron comes from. XPS is widely used in the study of surfaces and has found application in environmental and geochemical sciences, where it is important to determine the surface composition of solid samples because it can differ from the bulk composition due to processes such as adsorption, dissolution, and so on (Vempati, Hess, and Cocke 1996; Seyama and Soma 2003). XPS can also be applied to the solid–solution interface, by using the fast-freezing technique applied to slurries or pastes obtained by centrifugation (Shchukarev 2006). This technique is applicable to suspensions of inorganic and organic colloids and thus can be employed to study the solid–solution interface of soils and sediments.

Figure 7.18 shows a typical wide XPS spectrum from a soil sample, revealing the peaks corresponding to the main elements present. The binding energy of the inner electrons is slightly dependent on the atom binding state, so different states of the same element can be detected. For example, Figure 7.19 shows the decomposition of a C 1s band into two contributions due to differently bound C atoms (Yuan et al. 1998);

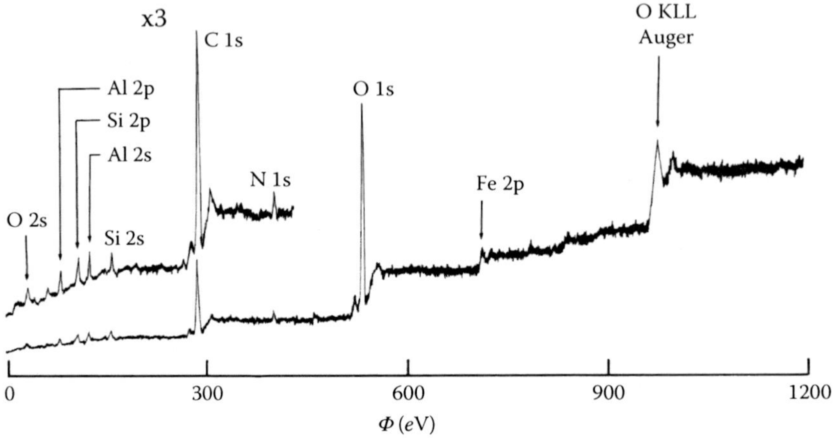

FIGURE 7.18 Wide-scan x-ray photoelectron spectrum of a podzolic soil from British Columbia, Canada, excited by Al K_α radiation: the upper band is the enlargement of the left part of the spectrum. (Reprinted from *Geoderma*, 86, 3–4, Yuan, G. et al., Assessing the surface composition of soil particles from some podzolic soils by x-ray photoelectron spectroscopy, 169–181. Copyright 1998, with permission from Elsevier.)

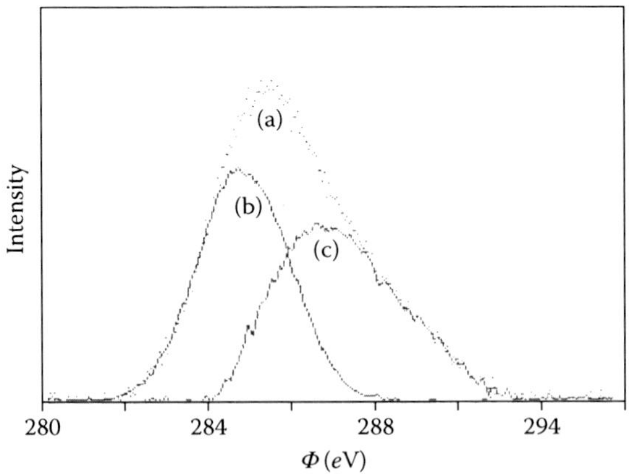

FIGURE 7.19 Observed C 1s spectrum of (a) the soil shown in Figure 7.18 with two computer-fit components: (b) band indicative of C 1s spectra due to hydrocarbon C and (c) the same for nonhydrocarbon C (e.g., carbonyl group). (Reprinted from *Geoderma*, 86, 3–4, Yuan, G. et al., Assessing the surface composition of soil particles from some podzolic soils by x-ray photoelectron spectroscopy, 169–181. Copyright 1998, with permission from Elsevier.)

in another example, different N types can be discerned in organic matter (Abe and Watanabe 2004). There are also several examples of XPS application to mineral components (Gerin et al. 2003; Wang and Mulligan 2008; Kawai 2010).

7.4 OTHER SPECTROSCOPIC METHODS

7.4.1 UV–Visible Molecular Spectrometries

UV–visible light usually interacts with valence electrons; thus the outcome of this interaction is simpler than that in the case of the higher energy x-ray excitation discussed in Section 7.3.2. Figure 7.20 shows the main events associated with the interaction of UV and visible radiation. On absorbing a photon of appropriate energy, a valency electron is promoted from the fundamental to an excited level. The nonabsorbed photons form the transmitted beam; as it is well known (Skoog et al. 2004; Harris 2010), the incident and transmitted intensities are related by the Lambert–Beer law (see Figure 7.20):

$$A = \log\frac{I_0}{I_T} = \varepsilon bc \tag{7.15}$$

where A is the absorbance at a given wavelength, ε the molar absorptivity at that wavelength, b the optical path in the sample, and c the absorbent concentration. Equation 7.15 is the basis of applying UV–visible spectrophotometry to the quantitative determination of many chemical species. Typical examples for metal ions are intensely colored species such as MnO_4^-, which shows strong absorption around 520 nm, and appropriate complexes of metal cations with ligands such as 4-(2-pyridylazo) resorcinol. This technique is frequently employed on a routine basis for the determination of metallic species after appropriate solubilization (Sparks 1996). Unsaturated organic compounds absorbing in the near-UV region are often determined with this technique as well. On the other hand, this spectroscopy is not suitable for qualitative identification, because the spectra usually show wide and featureless bands.

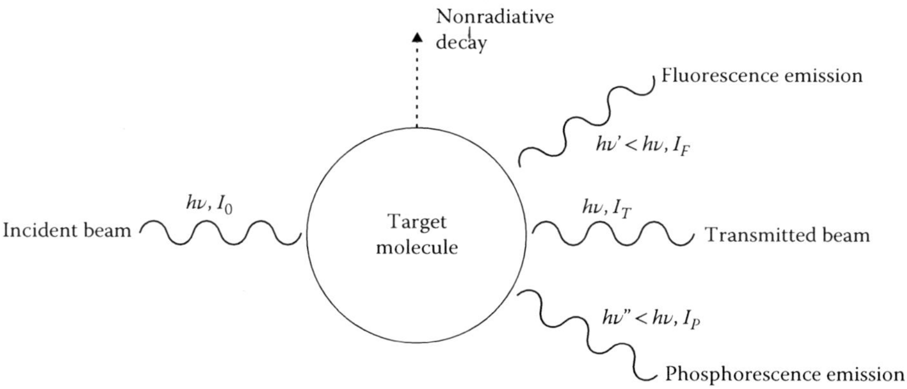

FIGURE 7.20 Schematic representation of the main processes involved in the interaction of UV–visible radiation with matter.

Turning back to the processes shown in Figure 7.20, the excited state will return to the fundamental one in one of several ways. The most common is nonradiative thermal deactivation. In other cases, the initial excited state can nonradiatively decay to a somewhat lower one and from this to the fundamental state by photon emission. This emission, of a lower energy than the incident, is the fluorescence that occurs a short time after excitation. Another possible outcome is "intersystem crossing," when the first excited state, which has the same multiplicity as the fundamental, usually a singlet state, evolves to a state of different multiplicity (usually a triplet). This state will eventually decay to the fundamental one, but this occurs at a much lower rate due to the different spin multiplicities; this is the "phosphorescence" emission. Fluorescence can be applied for quantitative purposes, especially for organic compounds, for low concentrations. Phosphorescence, on the other hand, has low intensity and is generally not useful for such purposes.

7.4.2 UV–Visible Atomic Spectrometry

In this family of methods, samples are atomized (by means of a flame or some other energy source) so that the elements of interest interact with radiation in an atomic state. The main advantages are increased sensitivity and high selectivity; the main disadvantage is that the speciation is lost in the atomization process. There are in principle three different ways of interaction that can be employed for analytical purposes (Wright and Stuczynski 1996): absorption, fluorescence, and atomic emission. In the first case, atoms in the vapor phase absorb photons from the incident light (resulting in electronic excitation), and the absorbance is related to concentration through the Lambert–Beer law (Equation 7.15) just as with molecular spectrometry. The light source is usually made from the excitation of the same element, to get very narrow lines of the precise wavelength needed (Skoog et al. 2004; Harris 2010). This technique, known as atomic absorption spectrometry (AAS) is widely employed in quantitative determination of elements (Wright and Stuczynski 1996; Skoog et al. 2004; Harris 2010). The sample is often vaporized in a flame, although a graphite furnace is also frequently employed; in this case, the sample (a few microliters) is placed in a hollow, electrically heated graphite bar where it is vaporized. This technique, known as graphite-furnace AAS, has higher sensitivity and requires a lower amount of sample. There are other special techniques, such as "hydride generation" AAS, that are useful for elements capable of forming volatile hydrides such as As, antimony, bismuth, germanium, Pb, Sa, tellurium, and tin, which result in improved sensitivity. The cold vapor (CV) method has been specifically developed for Hg, which gives very low detection limits for Hg.

Atomic emission occurs when an atom gets excited by nonradiative means (usually in a flame) and emits a photon, returning to the ground state. This technique has been, and is still, used for the determination of alkaline and alkaline earth elements, which have efficient emission at normal flame temperatures (Wright and Stuczynski 1996). For most other elements, higher temperatures are required, which can be obtained in inductively coupled plasma (ICP), methods known as inductively coupled plasma–atomic emission spectrometry (ICP-AES) and inductively coupled plasma–mass spectrometry (ICP-MS). An ICP is obtained by passing ionized Ar through a quartz torch inside a Cu coil; the coil is fed with a high-power radio-frequency signal (27.1 MHz). This signal generates oscillating magnetic fields. Electrons and ions pass through the oscillating electromagnetic field, and as a result of their trajectories, electrons are accelerated, colliding with other atoms and ions, resulting in ohmic heating with gas temperatures (spectroscopically measured) in the range of 6,000–10,000 K. In ICP-AES, the emission of the analyte at a characteristic wavelength is measured; in ICP-MS, the gas produced in the flame is fed to a mass spectrometer for analysis. The torch configuration is different in each case (Figure 7.21). ICP has unique physical properties, being an excellent source for vaporization, atomization, ionization, and excitation of many elements that are difficult to analyze by other spectrometric methods. ICP has become a very useful technique for environmental samples, among other uses (Soltanpour et al. 1996).

7.4.3 Infrared Spectroscopy

In the infrared (IR) portion of the spectrum (Figure 7.10) photon energies are incapable of exciting electrons, but they do match vibrational energies; thus, IR spectroscopy provides information about molecular vibrations, which, in turn, are related to the molecular structure (Bellamy 1975; McQuarrie and Simon 1997; Berry, Rice, and Ross 2000; Levine 2008; Atkins and Paula 2009). IR spectroscopy has been extensively used to identify soil constituents, elucidate structural features

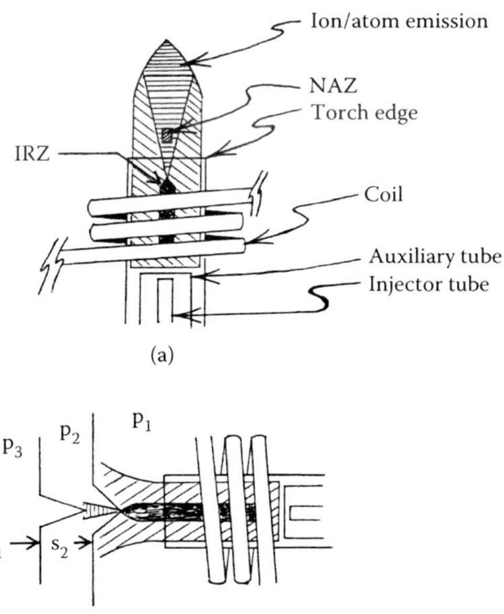

FIGURE 7.21 Schematic representation of inductively coupled plasma (ICP) torches: (a) for ICP-AES, NAZ is the normal analytical zone and IRZ is the initial radiation zone; (b) for ICP-MS, p_1–p_3 are zones of decreasing pressure, $p_1 = 760$ Torr, $p_2 \approx 1$ Torr, and $p_3 \approx 10^{-4}$–10^{-7} Torr; s_1: sampling cone; and s_2: skimmer cone.

of inorganic and organic components, and study the adsorption of ions and small molecules onto soil colloids through techniques such as attenuated total reflection IR spectroscopy (Johnston and Aochi 1996). A closely related technique is Raman scattering spectroscopy; here, visible light dispersed from a sample is analyzed around the incident light wavenumber (or wavelength). Most of the scattered photons have their energy unchanged; but a few undergo inelastic scattering, thus exchanging energy with the sample molecules. The most common type of Raman spectroscopy corresponds to Stokes dispersion, where the scattered photons have a lower energy than the incident ones, corresponding to an absorption by the sample. Thus, in the Raman spectrum there are small bands near the incident energy, and the energy difference respective to the main band corresponds to the vibrational energies. A Raman spectrum, plotted as amplitude against energy difference (usually in wavenumbers), is comparable to an IR absorption spectrum in absorbance units. Raman spectra provide information complementary to IR because, depending on the molecular geometry, there are vibrational bands that are inactive in IR and others inactive in Raman spectra.

IR spectrometers initially were of the dispersive type (the same principle as UV spectrometers), but the advent of Fourier transform IR (FTIR) spectroscopy, in which an interference pattern from a Michelson-type interferometer is processed using Fourier transform analysis, has brought better response and simpler instruments (Harris 2010). IR spectroscopy has been applied to soil study in several ways (Senesi and Calderoni 1988; Cabaniss 1991; Patterson et al. 1992; Sun and Doner 1996; Celi, Schnitzer, and Nègre 1997; Elkins and Nelson 2001; Orsetti, Quiroga, and Andrade 2006; Hay and Myneni 2007; Du and Zhou 2008). In recent years, near-IR spectroscopy has been investigated for a fast, nondestructive method to analyze soil C, minerals, and other soil constituents. Near-IR diffuse reflectance spectroscopy (NIRDRS) has been shown to be extremely versatile for the rapid analysis of many agricultural materials. More recently, mid-IR diffuse reflectance spectroscopy (DRIFTS) and NIRDRS have come under comparative study for this purpose. It has been demonstrated that for the determination of soil C, DRIFTS is often more accurate and produces more robust calibrations than NIRDRS when analyzing ground, dry soils under laboratory conditions. However, DRIFTS is more affected by moisture and sample preparation than NIRDRS, making it impractical for fast analysis, especially in field applications. Thus, there is ongoing research on the application of NIRDRS to soil analysis (Reeves III 2010; Bellon-Maurel and McBratney 2011).

Figure 7.22 shows examples of the interaction between HA and metal cations as observed with FTIR (Alvarez-Puebla, Valenzuela-Calahorro, and Garrido 2004). All spectra show characteristic variations in their bands due to metal ion binding: an intensity increase of the O–H stretching band (around 3400 cm^{-1}) due to the absorption of water of the retained aquo complexes; a decrease in intensity of the C=O stretching band of –COOH (near 1720 cm^{-1}), and C–O stretching and O–H rocking bands (near 1240 cm^{-1}); an intensity increase of the bands assigned to the asymmetric and symmetric $-COO^-$ stretching bands (around 1600 and 1380 cm^{-1}, respectively); and a decrease in intensity of the shoulder at about 2600 cm^{-1}, assigned to

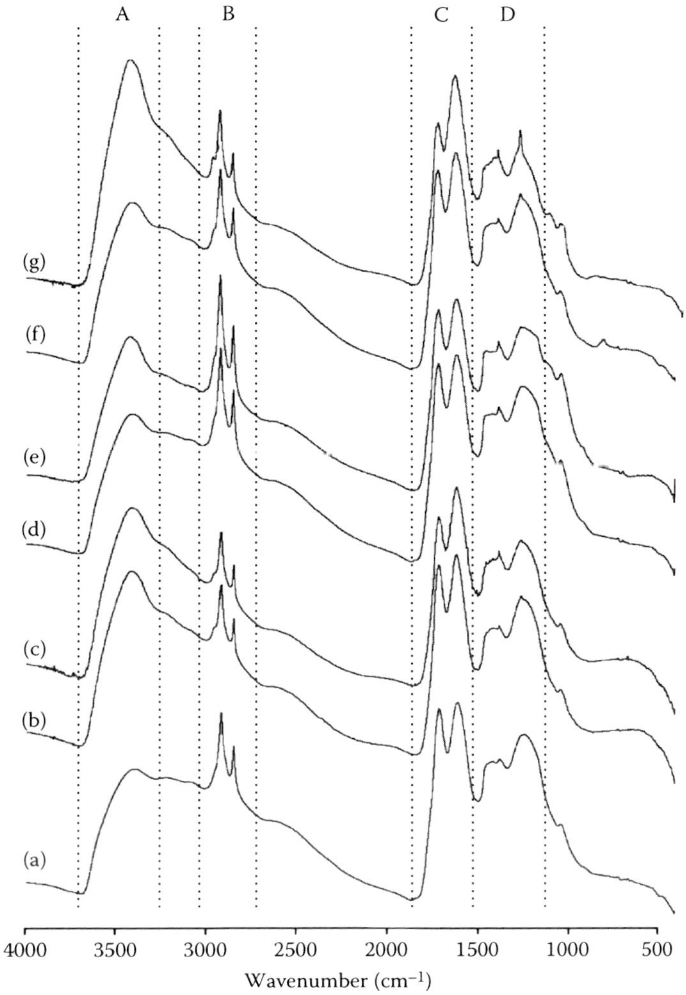

FIGURE 7.22 Fourier transform infrared (FTIR) spectra of brown humic acid (BHA) obtained and purified from a commercial humic substance (HS) (Acros Organics): (a) BHA alone, (b) loaded with 80 mmol·kg^{-1} Co^{2+}, (c) loaded with 1280 mmol·kg^{-1} Co^{2+}, (d) and (e) the same with Ni^{2+}, and (f) and (g) the same with Cu^{2+}. (Reprinted with permission from Alvarez-Puebla, R.A. et al., 2004, *Langmuir*, 20, no. 9, 3657–3664. Copyright 2004 American Chemical Society.)

H-bonded –COOH. The higher variations are consistently found in samples doped with a greater amount of metal; there is an increase in the bands at about 3420, 1620, and 1380 cm^{-1} and a decrease in intensity in those near 1710 and 1240 cm^{-1}, which indicates that the metal is bound by surface complexation (specific adsorption). The samples doped with 80 mmol·kg^{-1} show lower variations, suggesting that electrostatic retention plays a role in ion binding.

7.4.4 Nuclear Magnetic Resonance

Nuclear magnetic resonance (NMR) is a powerful technique for investigating the structure of chemical compounds, specially organic matter (Fyfe 1983; Steelink et al. 1990; Wilson 1990; Sanders and Hunter 1993).

When nuclei with a nonzero spin are placed in a magnetic field, they align themselves such that some have their spin magnetic moment vector parallel to the field vector and are at slightly lower energies than the remaining nuclei that are antiparallel in the field. The difference in energy between the two spin states is dependent on the external magnetic field strength and is always very small. If an oscillating magnetic field is superimposed on a steady magnetic field with a perpendicular magnetic vector, then for a particular strength of the steady magnetic field absorption of radiation will occur at certain frequencies of oscillation, allowing transitions between different energy spin states. This resonance condition can be met by varying the magnitude of either the steady-state field or the oscillating field. In Fourier transform NMR spectroscopy, the sample is subjected to a pulse of radio-frequency radiation comprising all frequencies. Strong magnetic fields are necessary for NMR spectroscopy. The earth's magnetic field is not constant but is ~10^{-4} T (T stands for tesla, the international unit for magnetic flux) at ground level. Modern NMR spectrometers use powerful magnets having fields on the order of 1–20 T. Even with such high fields, the energy difference between the two spin states is less than 0.4 $J{\cdot}mol^{-1}$.

In a molecular environment, the atoms surrounding a nucleus partially shield it from the magnetic field, so that the frequency of the oscillating field required to achieve resonance is changed. As this shielding is a function of the chemical environment associated with nuclei, the resonance frequencies of atoms are able to give information about the chemical structure of the molecule. Common nuclei used for the studies with NMR are usually those of spin ½ such as ^{1}H, ^{13}C, ^{19}F, and ^{31}P; these have dipolar magnetic moments, whereas others may exhibit higher polar orders, such as ^{35}Cl and ^{37}Cl, which are quadrupoles.

In NMR, the frequency of a nucleus is measured relative to that of a standard compound (typically tetramethylsilane). The resonant frequency is then given as the difference, or "chemical shift" (δ), between these two frequencies, expressed as parts per million. The study of a single element is at present referred to as 1-D NMR (see Figure 7.23 and related text); the most commonly studied elements are ^{1}H and ^{13}C. Tables 7.5 and 7.6 provide δ values for some common types of H and C atoms in organic compounds. Also, ^{15}N has been increasingly investigated in recent times (Thorn and Cox 2009).

The most commonly used method involves dissolving the sample compound in an NMR-inactive liquid (deuterated solvents are usually employed). There are a number of problems associated with liquid-state NMR, such as the amount of sample required (100–200 mg), lack of solubility, interference from water, and the long analysis times required (2–12 hours are common for ^{1}H-NMR and up to 1 week for ^{13}C NMR). In some situations, the dissolution process interferes with the analysis. The development of solid-state NMR has avoided these problems as well as achieving a higher signal-to-noise ratio, generally giving greater sensitivity in the spectra. However, the technical problems in getting good spectra are much greater.

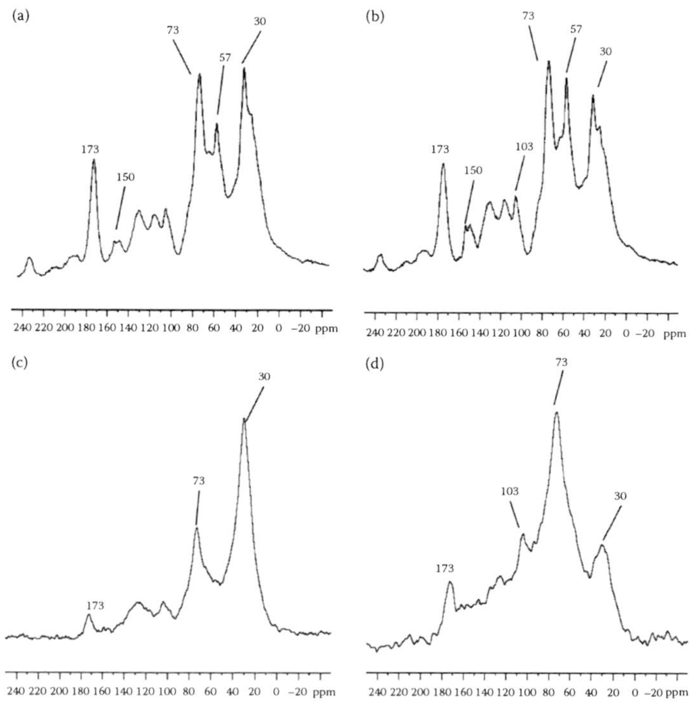

FIGURE 7.23 Solid-state CP/MAS 13C nuclear magnetic resonance (NMR) spectra of four humin concentrates from French soils: (a) humin–Marais Poitevin 30–50 cm (MP35), (b) humin–Marais Poitevin 70–90 cm (MP79), (c) humin–Marais de Gironde (CAL), and (d) humin–Marais de Rochefort (MAT). (Reprinted from *Organic Geochemistry* 37, 6, Deport, C., L. Lemée, and A. Amblès, Comparison between humic substances from soil and peats using TMAH and TEAAc thermochemolysis, 649–664, Copyright 2006, with permission from Elsevier.)

TABLE 7.5
Major Proton NMR Resonances of SOM

Proton Type	δ (ppm)
Carboxylic acid protons	13.0
Hydroxyl protons	10.0
Aromatic protons	6.0–7.5
Lactone protons	4.0–5.5
Methoxyl protons	3.7
Aliphatic protons adjacent to a benzene ring	2.6–0.9
Amine protons	1.0–5.0
Carbonyl compounds	2.0–2.7

Source: Swift, R. S., *Methods of Soil Analysis. Part 3—Chemical Methods*, American Society of Agronomy-Soil Science Society of America, Madison, Wisconsin, 1996.

Note: Spectra measured in deuterodimethylsulfoxide [$(CD_3)_2SO$].

TABLE 7.6
Major Chemical Shift Assignments in the ^{13}C NMR Spectra of SOM

Possible Assignments	δ (ppm)
Unsubstituted saturated aliphatic C atoms	0–50
Terminal methyl groups	10–20
Methylene groups in alkyl chains	15–50
Methine groups in alkyl chains	25–50
Methylene C α, β, γ, ε from terminal methyl group	29–33
Methylene C atoms of branched alkyl chains	35–50
α-C in aliphatic acids	41–42
R_2 NCH_3	45–46
Aliphatic C singly bonded to one O or N atom	50–95
Aliphatic esters and ether, methoxy, ethoxy	51–61
Carbon in CH_2OH groups; C_6 in polysaccharides	57–65
Carbon in CH(OH) groups; ring C atoms of polysaccharides; ether-bonded aliphatic C	65–85
Carbon singly bonded to two O atoms; C, anomeric in polysaccharides, acetal or ketal	90–110
Aromatic and unsaturated C	110–160
Protonated aromatic C, aryl H	110–120
Aromatic C ortho to O-substituted aromatic C	118–122
Unsubstituted and alkyl-substituted aromatic C	120–140
Aromatic C substituted by O and N; aromatic ether, phenol, aromatic amines	140–160
Carbonyl, carboxyl, amide, ester C atoms	160–230
Largely carboxyl C atoms	160–190
Carbonyl C atoms	190–230

Source: Swift, R. S., *Methods of Soil Analysis. Part 3—Chemical Methods*, American Society of Agronomy–Soil Science Society of America, Madison, Wisconsin, 1996.

NMR is mostly applied to the study of organic compounds, such as HSs (Cardoza et al. 2004). There are diverse techniques that have been commonly employed (Swift 1996; Baldock and Nelson 1999). Figure 7.23 shows, as an example, the solid-state ^{13}C NMR spectra of humin extracted from French soils (Deport, Lemée, and Amblès 2006). The presence of a signal due to alkyl Cs near 30 ppm is observed in all cases. Methoxyl Cs are detected at 57 ppm in two of the humin samples, and only as a shoulder in the other cases. The peak at 150 ppm, which is clearly observed only in the samples from Marais Poitevin, France, indicates the presence of the C atom in phenolic groups, whereas the carboxyl signal at 173 ppm suggests the presence of acid and ester groups. The peaks at 73 and 103 ppm are attributable to O-alkyl Cs from carbohydrate units of cellulose and hemicellulose. In summary, these spectra indicate a high aliphatic contribution to humin in the samples and a significant occurrence of cellulose and hemicellulose moieties.

Two-dimensional NMR (2-D NMR) allows revealing the relationship between different atoms; in a 2-D NMR experiment, a simple series of one-dimensional (1-D) measurements is collected with different timings. In general, 2-D NMRs can be divided into two basic types, homonuclear and heteronuclear, and each type can in turn provide either through-bond (COSY-type) or through-space (NOESY-type) coupling information. A 2-D frequency correlation map is produced after a Fourier transform in both dimensions. The result is intensity as a function of two frequencies and is usually plotted as contour maps. There are a number of 2-D and 3-D NMR techniques that can be used to advantage in environmental samples (Simpson, Kingery, and Hatcher 2003; Cardoza et al. 2004 and references therein). In a study of plant-derived structures in humic materials, Simpson, Kingery, and Hatcher (2003) extensively employed these techniques. Figure 7.24 shows a 2-D heteronuclear multiple quantum coherence (HMQC) (a 2-D inverse H–C correlation technique that allows for the determination of C to H connectivity) spectrum of a FA extracted from a pine forest soil; in this case, H–C connectivity is difficult to interpret due to the high overlap, so structures cannot be identified. In a 3-D NMR experiment, as shown in Figure 7.25a, HMQC is combined with total correlation spectroscopy (TOCSY), which creates correlations between all the protons within a given spin system, not just between geminal or vicinal protons. Correlations are seen between distant protons as long as there are couplings between every intervening proton. This is extremely useful for identifying protons in complex structures such as in sugar rings or amino acids. TOCSY information is contained in the *y–z* plane (F2–F3 in Figure 7.25a) and HMQC data in the *x–y* (F1–F3) plane. The *x–z* (F1–F2) plane contains both TOCSY and HMQC information. By taking a slice through the cube, 2-D spectra can be created that contain detailed coupling and chemical shift information for individual resonances. Taking slices through the F1–F2 plane is most productive. To create a slice, a point on the F3 dimension (proton) is selected. In Figure 7.25b, an F1–F2

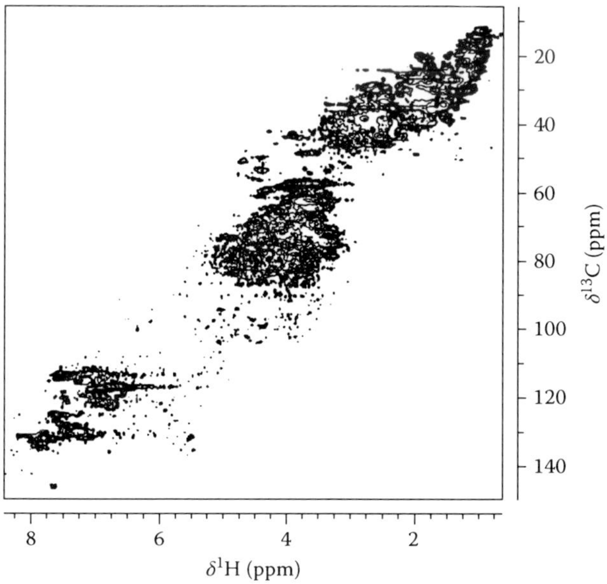

FIGURE 7.24 A 2-D ^{1}H-^{13}C heteronuclear multiple quantum coherence (HMQC) spectrum of a fulvic acid (FA) isolated from the surface soil horizon of a pine forest. (Reprinted with permission from Simpson, A.J. et al., 2003, *Environmental Science & Technology* 37, no. 2, 337–342. Copyright 2003 American Chemical Society.)

slice though the largest peak in the proton spectrum is shown, which results from main-chain CH_2 units and occurs at 1.3 ppm. The resulting F1–F2 slice produces a spectrum in which the cross peaks describe the C and proton chemical shifts of the CH_2 units themselves and all other units with which they couple. Standard interpretations become much easier, resulting in the assignments shown in Figure 7.25c.

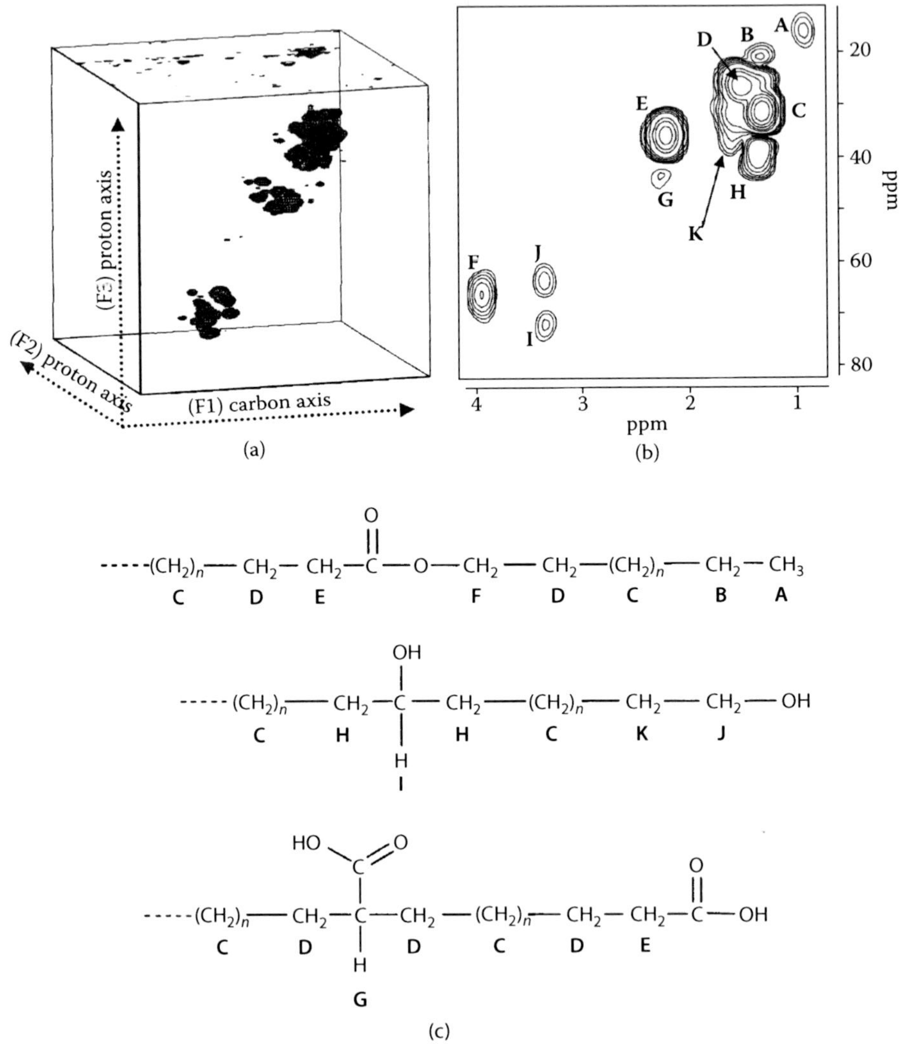

FIGURE 7.25 (a) A 3-D HMQC-TOCSY spectrum of the pine forest FA: the F2–F3 plane contains TOCSY information. (b) A 2-D spectrum produced from an F1–F2 slice through the 3-D HMQC-TOCSY spectrum at 1.3 ppm on the F3 (proton) axis. Labels on cross peaks correspond to the C–H structures in aliphatic chains shown in (c). (Reprinted with permission from Simpson, A.J. et al., 2003, *Environmental Science & Technology*, 37, no. 2, 337–342. Copyright 2003 American Chemical Society.)

7.5 MICROSCOPIES

Microscopic observation of soil samples is useful not only for observing the morphology of soil particles but also for providing information about their structure and composition. In Sections 7.5.1 through 7.5.3, the most important methods—namely, optical, electronic, and scanning probe microscopy—are described.

7.5.1 Optical Microscopy

The simplest method is optical microscopy, in which visible light (photons) is used to observe a sample. It has a resolution limit around 0.25–0.5 μm, which is on the order of $\lambda/2$, where λ is the wavelength of incident light. From a strict colloid science point of view, it lies near the upper limit of colloid particle sizes and appears to be of limited utility. However, it is of great help in the identification of minerals, because it allows observation of crystal habits (the shape and size of crystals, which are determined by their internal symmetry). With experience, many minerals can be identified in a soil sample under a microscope, even from simple inspection. A unique feature of optical microscopy is the availability of polarized light, which is handy in distinguishing minerals or even different crystal types of the same compound (Bullock et al. 1985; Cady, Wilding, and Drees 2010).

7.5.2 Electron Microscopy

As stated by de Broglie's duality principle, all particles, especially electrons, can behave as waves under appropriate circumstances. This principle states in its simplest form that a particle of mass m moving at a speed v has a wavelength given by

$$\lambda = \frac{h}{mv} = \frac{h}{p} \tag{7.16}$$

where p is the linear momentum. Whereas for macroscopic bodies the wavelengths given by Equation 7.16 have little, if any, physical meaning, for electrons the equation results in wavelengths down to a fraction of a nanometer, depending on the kinetic energy. Thus, electrons can be used to observe much smaller dimensions (up to five orders of magnitude) than visible light. Figure 7.26 shows schemes of the two main types of electronic microscope compared with an optical one (Mavrocordatos, Perret, and Leppard 2007). In transmission electron microscope (TEM), electrons are accelerated by electrostatic fields and focused by magnetic fields; an electron beam passes through a sample specimen where electrons are absorbed according to the thickness and electronic absorption–dispersion behavior of the sample materials: metals absorb electrons more strongly, whereas organic matter absorbs them more weakly. Thus, organic-rich thick samples can show the same density to TEM as a thin metal-rich material. The transmitted electrons are detected by a fluorescent screen or a camera. In scanning electron microscope (SEM), the specimen is usually opaque; the sample-dispersed electrons are collected by a detector. To form an image, the incident electron beam is forced to scan the sample by magnetic fields. The specimen is usually metalized to make it conductive; in isolating materials, the incident electrons absorbed cause a buildup of negative charge that prevents further impacts of incident electrons. Chapter 2 presents several examples of SEM and TEM observations of particles and soil materials.

Coupled with electron microscopy observation, other techniques provide additional information. Two commonly employed techniques are x-ray energy-dispersive spectroscopy (X-EDS) and electron energy-loss spectrometry (EELS). Both are based on the fact that, with enough energy, incident electrons can remove electrons from the inner levels of the sample on collision; electrons from the upper levels fall to the empty levels, emitting photons of x-ray energy. In the first technique (Goldstein et al. 2003), the x-rays are analyzed by an energy-dispersive analyzer (see Section 7.3.3.) placed over the sample at a side, as shown in Figure 7.26; in the figure, it is shown in combination with TEM observation, but it is also often coupled to SEM instruments. In EELS (Egerton 1996), the sample is illuminated with a monochromatic (in practice, with a narrow kinetic energy dispersion) electron beam, and the energy loss of transmitted electrons is measured. This loss can be due to several interactions: low-energy losses (below 100 *e*V) are related to plasmon oscillations and high-energy losses (over 100 *e*V) are due to atom ionization as described earlier for EDS, the two techniques being complementary. An important difference is the higher resolution (down to 0.2 *e*V) and useful energy range of 0–1000 *e*V of EELS, compared with the minimum of 150 *e*V for EDS.

7.5.3 Atomic Force Microscopy

Probe scanning techniques, atomic force microscopy (AFM) in particular, are being currently used for a great number of applications since the introduction of scanning tunneling microscopy (STM) in 1982. STM requires a (at least partially) conductive surface and so it is not suitable for environmental applications. AFM, on the other hand, is widely employed in environmental research (Balnois, Papastavrou, and Wilkinson 2007). The basic principle of AFM is illustrated in Figure 7.27; a nanometer-sized tip is attached to a cantilever, which scans the surface, moved by piezoelectric actuators (not shown); and a laser beam reflected by the cantilever is received by a position-sensitive detector, allowing the detection of deflection under surface–tip forces. AFM tips and cantilevers are made from Si or Si_3N_4. Forces are not measured directly but calculated by measuring the deflection of the lever and knowing the stiffness of the cantilever, using Hook's law. This flexible setup is employed in a variety of methods, termed "modes":

Contact mode: Being the most basic mode of operation, contact mode is widely used. The tip is raster-scanned across the surface, where it is deflected following surface corrugation. In constant-force mode, the tip is constantly adjusted to maintain a fixed deflection, thus keeping a constant height above the surface; this adjustment along the z axis is displayed as data. However, the ability to track the surface in this way is limited by the response of the

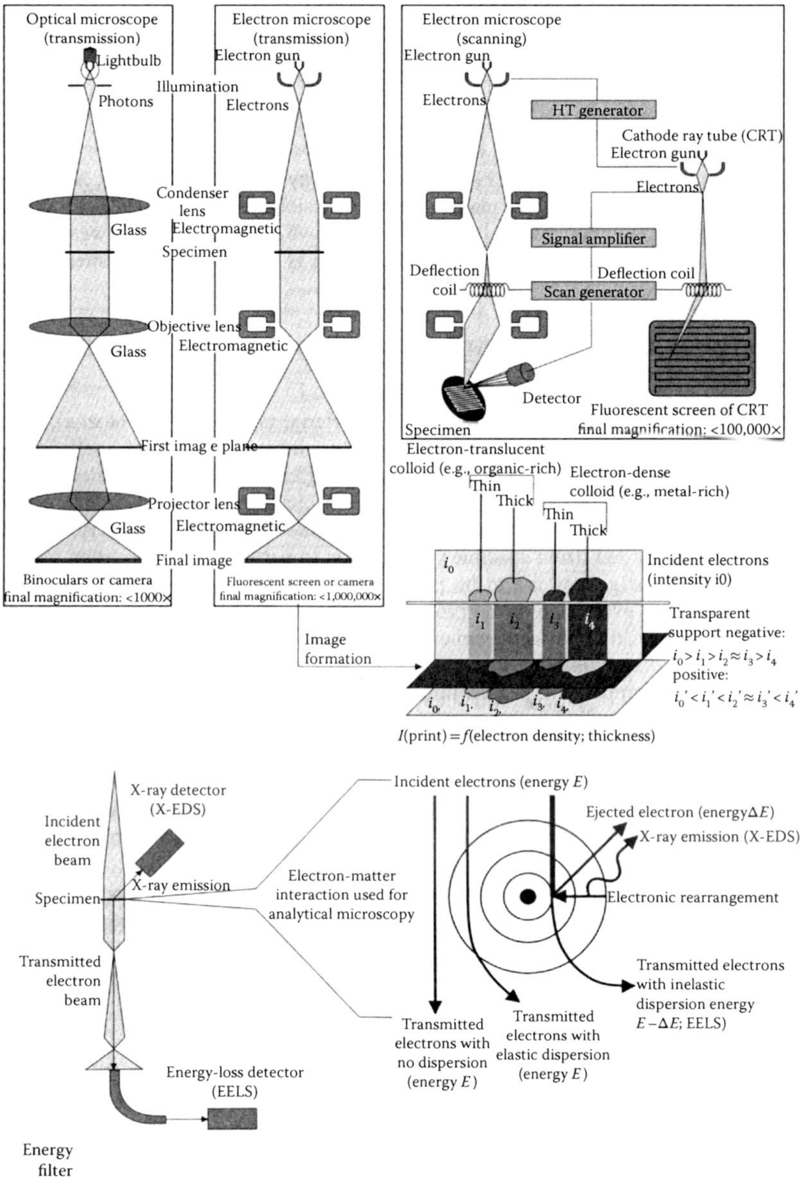

FIGURE 7.26 (Top) comparison of an optical microscope (left) with a transmission electron microscope (TEM) (center) and a scanning electron microscope (SEM) (right): the scheme below the TEM picture shows that the final image from diverse particles is the combined result of their composition and thickness. (Bottom) analytical TEM can be performed by measuring the x-rays emitted due to electronic rearrangement in the sample atoms due to collisions between microscope and sample electrons. (From Mavrocordatos, D., D. Perret, and G. G. Leppard/ K. J. Wilkinson and J. R. Lead: *Environmental Colloids and Particles: Behaviour, Separation and Characterisation*. 345–404. 2007. Copyright Wiley-VCH Verlag GmbH & Co. KGaA. Reproduced with permission.)

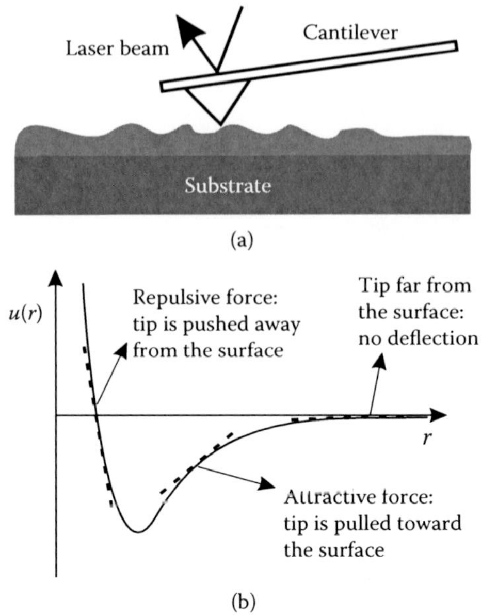

FIGURE 7.27 (a) Schematic view of the atomic force microscope (AFM): a nanometric tip attached to a cantilever scans the surface, moved by piezoelectric actuators (not shown); a laser beam reflected by the cantilever is received by a motion detector, allowing the detection of deflection under surface–tip forces. (b) Schematic plot of the tip–surface interaction potential energy as a function of distance; depending on the tip position, the cantilever may be deflected toward the surface (attractive force) or away from it (repulsive force).

feedback circuitry. Sometimes the tip is allowed to scan without this adjustment, and one measures only the deflection. This obviously is feasible for very small roughness and is useful for small, high-speed atomic resolution scans, and it is known as the "variable-deflection mode."

Lateral force microscopy (LFM): This mode measures the frictional forces on a surface. By measuring the "twist" of the cantilever, rather than merely its deflection, one can qualitatively determine areas of higher and lower friction.

Noncontact mode: noncontact mode belongs to a family of alternating current modes, where an oscillating cantilever is employed. A stiff cantilever is oscillated in the attractive regime, which means that the tip is quite close to the sample but not touching it (hence, noncontact). The forces between the tip and the sample are quite low, on the order of piconewtons (10^{-12} N). The detection scheme is based on measuring changes to the resonant frequency or amplitude of the cantilever.

Dynamic force or tapping mode: commonly referred to as "tapping mode," it is also referred to as intermittent contact or, more generally, dynamic force mode (DFM). A stiff cantilever is oscillating at a short distance from the surface, closer than in noncontact mode. The oscillation range includes the repulsive region of the curve, so the tip intermittently touches or taps the surface; very stiff cantilevers are typically used. The advantage of tapping the surface is improved lateral resolution on soft samples. Lateral forces such as drag, common in contact mode, are virtually eliminated. For poorly adsorbed specimens on a substrate surface, the advantage is clearly seen.

Force modulation: In force modulation, material properties are probed through sample–tip interactions. The tip (or the sample) is oscillated at a high frequency and pushed into the repulsive regime. The slope of the force–distance curve is measured and is correlated with the sample elasticity. Data can be acquired along with topography, allowing the comparison of height and material properties.

Phase imaging: In phase mode imaging, the phase shift of the oscillating cantilever relative to the driving signal is measured. This phase shift can be correlated with specific material properties that affect sample–tip interaction. The phase shift can be used to differentiate areas on a sample with such differing properties as friction, adhesion, and viscoelasticity. Because the technique can be used simultaneously with DFM, topography can also be measured.

7.6 MISCELLANEOUS METHODS

We discuss in this section some methods that do not fall into a common category but are nonetheless important in soil studies. More methods can be found in the ample literature (Sparks 1996, 1998; Sumner 1999; Wilkinson and Lead 2007; Klute 2010).

7.6.1 Neutron Activation Analysis

Neutron activation analysis (NAA) is a method of elemental analysis based on the principles of nuclear structure and behavior, nuclear reactions, and the interaction of radiations with matter (Helmke 1996). NAA provides a highly sensitive and accurate method for the analysis of more than 70 elements. About 30 elements can be readily determined in soil samples. In general, this technique has the important advantage of being relatively free from interferences and the effects of contamination. It is also a very efficient method of analysis for elements, especially those with isotopes having short half-lives, such as Na, K, Cl, and Mn, and 50 or more samples can be analyzed in several hours. The results are not affected by the chemical state of an element because NAA is a nuclear technique. Instrumental NAA (in the absence of postirradiation) also has the advantage of being nondestructive and free from the problems associated with sample dissolution and chemical treatment, an important consideration when determining total element concentrations in chemically complex materials such as soils. The sensitivity of NAA can be increased more than thousandfold by postirradiation radiochemical separations, but this approach is labor intensive and used only when justified. The main disadvantages are that chemical speciation is obviously not determined, access to a nuclear reactor is required, and some isotopes require several days between irradiation and analysis (Helmke 1996).

NAA is based on the production of radioisotopes of the element of interest (this is the activation step), which is done by neutron capture, hence the name neutron activation analysis. The nuclear reactions usually occur with energy release in the form of γ-rays, which are analyzed to determine the elements present. Most of the neutrons in nuclear reactors are in thermal equilibrium with the surrounding atoms, and capture of these neutrons by common elements produces suitable radioisotopes. An example of a neutron capture reaction is as follows:

$$^{41}_{19}\mathrm{K} + ^{1}_{0}n \longrightarrow ^{42}_{19}\mathrm{K} + \gamma \tag{7.17}$$

where n indicates a neutron and γ a gamma photon. The emission that usually arises as a result of the capture process, from excess energy of the new radioisotope, is normally not studied in NAA. Instead, the emission resulting from the decay of the new radioisotope is studied. In the aforementioned example, isotope $^{42}_{19}\mathrm{K}$ decays as follows:

$$^{42}_{19}\mathrm{K} \longrightarrow ^{42}_{20}\mathrm{Ca} + \beta^- + \bar{\nu} \tag{7.18}$$

In Equation 7.18, β^- is an electron and $\bar{\nu}$ a neutrino. The product isotope $^{42}_{20}\mathrm{Ca}$ can be obtained in one of several energy levels, with 82% yield in the ground state and nearly 18% yield in an excited state 1.525 MeV above the ground one (Helmke 1996); nuclei in this state will eventually decay to the ground state with the emission of 1.525 M*e*V photons, which are detected and measured to quantify K in the sample.

NAA is a powerful technique for elemental analysis in complex samples such as soils, despite the aforementioned limitations, and has been applied in a number of studies (Steinnes 2000; Grosheva, Zaichick, and Zaichick 2007; Srivastava et al. 2011). In the past years, NAA applications have diminished somewhat in favor of other techniques such as ICP-MS (Tsukada et al. 2005).

7.6.2 Electrophoretic Separations

Electrophoresis is the movement of charged particles under the action of an electric field, known as migration. In a mixture of different particles, they migrate according to their respective mobilities as per the following equation (see Section 5.4.1):

$$v_i = \mu_i E \tag{7.19}$$

where E is the electric field, v_i is the velocity of particles of type i, and μ_i their mobility; this is essentially a function of particle charge and size, with smaller and/or more charged particles migrating faster. Electrophoresis may play an important role as an analytical or separation technique for characterization of natural colloids such as SOM; however, to date it has been mostly employed for extracting natural organic matter (Schmitt-Koplin and Junkers 2007). There are several electrophoretic techniques that in most cases differ in the support used for separation: free solution, paper, gel, and isoelectric focusing were employed until the 1980s, and more recently capillary electrophoresis techniques (which are simpler to implement and require minimum sample quantities) are commonly employed (Schmitt-Koplin and Junkers 2007); soft colloids such as humic particles require a special analysis of the electrophoretic results (Duval 2007).

7.6.3 Field-Flow Fractionation

Field-flow fractionation (FFF) is a family of flow-driven separation techniques that provide a continuous separation of particles whose sizes range from about 1 nm to 50 μm. An external field applies over the particles a force perpendicular to the separation channel, forcing them against a wall. A nonuniform distribution results, due to the different particle sizes and properties, and thus they are separated by the laminar flow in the channel (Hassellöv, Kammer, and Beckett 2007). This family of techniques is classified according to the field applied: sedimentation FFF employs a centrifugal force to separate particles according to their buoyant mass, whereas flow FFF uses a cross flow as force and achieves separation according to the particles' diffusion coefficient. A number of different detection systems, as in chromatographic methods, have been employed (Hassellöv, Kammer, and Beckett 2007).

7.6.4 Surface Area Measurement

The determination of specific surface area of soils and soil colloids is of great importance in characterizing the reactivity of a sample, among other factors. However, it is actually an operational concept, because the A_S value depends on the experimental method employed, as it will be shown in Section 7.6.4.1. The underlying fact is that the effective area available for a particular reaction or process is dependent on the reactants and/or external factors involved. The experimental methods can be broadly classified into three categories (Sposito 1984): physical methods, positive adsorption methods, and negative adsorption methods.

7.6.4.1 Physical Methods

Physical methods classically included XRD (Section 7.3.1) and electron microscopy (Section 7.5.1) to determine the shapes and dimensions of typical particles; compute their area, A_p; and calculate the "crystallographic specific surface area," A_S^{cryst}, knowing the density of the particles, ρ:

$$A_S^{\text{cryst}} = \frac{A_p}{\rho V_p} \tag{7.20}$$

where V_p is the particle volume. This approach depends on how representative the chosen particles are of the sample material; clearly, it may be reasonable for single components (e.g., kaolinite) but difficult at best for whole soil (Sposito 1984). More recently, AFM measurements have been employed to determine the specific surface area of clay particles (Macht et al. 2011). In these methods, the geometric particle area is in principle employed; however, the particles have roughness, which is not taken into account in Equation 7.20. The use of fractal geometry is a way to consider particle microroughness and also particle size distribution (Borkovec et al. 1993; Ersahin et al. 2006). More recently, synchrotron-radiation-based x-ray microtomography was applied to the quantitative analysis of 3-D pore-space geometry analyses of small-scale soil aggregates (Peth et al. 2008). A related measurement is soil roughness, which can be accomplished by a variety of methods such as roller chain, laser scan, or photogrammetry (Jester and Klik 2005; Koiter and Lobb 2008).

Among other methods, one that lies between physical and adsorption classes is the Hg porosimeter, in which the sample is placed in contact with Hg, pressure is applied to fill the pores with the liquid metal, and the amount of Hg required is measured (Arnepalli et al. 2007).

7.6.4.2 Positive Adsorption Methods

In positive adsorption methods, the amount adsorbed of a given chemical species (either molecular or ionic) on an initially free surface is determined and, in principle, from the knowledge of the area covered by that species, the total solid area can be computed. These methods require three conditions (Sposito 1984):

1. It should be a reaction between the surface and the adsorbate resulting in the accumulation of the adsorbate ("positive adsorption"). The adsorbate may be in gas phase or in solution.
2. The amount of adsorbate corresponding to one monolayer must be determined.
3. The packing area of the adsorbate (equivalent area occupied per molecule) must be known.

Under such conditions, the adsorption specific surface area, $A_{S,ad}$, can be calculated as follows:

$$A_{S,ad} = Q_{ad} N_A a_{ad} \tag{7.21}$$

where Q_{ad} is the specific amount adsorbed and a_{ad} the adsorbate packing area. The monolayer amount of adsorbate is almost always found through adsorption measurements by fitting to some isotherm equation.

Most commonly, the adsorbate is in gaseous state, at a temperature below the critical point, and adsorbs in multilayer, and the Brunauer–Emmet–Teller (BET) isotherm (Chapter 4, Section 4.4.4) is applied; because one parameter is the monolayer coverage, fitting the BET equation brings the desired result. The isotherm is (Chapter 4, Equation 4.58) as follows:

$$Q_{ad} = Q_S \frac{c\dfrac{P}{P_V}}{\left(1-\dfrac{P}{P_V}\right)\left[1+(c-1)\dfrac{P}{P_V}\right]} \tag{7.22}$$

where P_V is the vapor pressure of the adsorbate at the experiment temperature, P its pressure, Q_S the specific monolayer coverage, and c an isotherm constant. Equation 7.22 can be rearranged in linear form as follows:

$$\frac{1}{Q_{ad}\left(\frac{P_V}{P}-1\right)}=\frac{c-1}{cQ_S}\frac{P}{P_V}+\frac{1}{cQ_S} \tag{7.23}$$

Thus, a plot of $1/\{Q_{ad}[(P_V/P) - 1]\}$ versus P/P_V allows the determination of Q_S. Figure 7.28a shows plots of the amount of N_2 adsorbed onto three soils with high SOM content, presenting the shape of a BET isotherm (Chapter 4, Section 4.4.4). Figure 7.28b shows the fittings of some curves to Equation 7.23 with good linearity; from the slope $[(c - 1)/cQ_S]$ and y intercept $(1/cQ_S)$, Q_S can be obtained (Chiou, Lee, and Boyd 1990). Other gases apart from N_2 can be used as well, such as Ar (at 77 K) or water and ethylene glycol monoethyl ether (EGME) (at temperatures around 273–293 K); see Table 7.7. The selection of the adsorbate (probe molecule) is very important in the study of soils and particularly of clay minerals, as discussed in Chapter 8, Section 8.6.3. More recently, the use of CO_2 to characterize nanoporosity has been introduced. The CO_2 adsorption isotherm on micro- and nanopores of solids has been successfully modeled using Monte Carlo simulations and density functional theory (Vishnyakov, Ravikovitch, and Neimark 1999), and CO_2 was proposed as a probe gas to study nanoporosity in carbonaceous solids and soils (among other materials) combined with N_2 and/or Ar (Ravikovitch et al. 2000). This methodology has been applied to soil surface area studies (Pignatello, Kwon, and Lu 2006; Mukherjee, Zimmerman, and Harris 2011). A point worth noting here is that there are discrepancies on the packing area of most adsorbates; this should not be surprising because, besides experimental uncertainties, a given substance may interact differently with different substrates, leading to different Q_S values and, consequently, different packing areas. In the fourth column of Table 7.7, the "ideal" areas are collected and calculated from molecular dimensions, whereas in the fifth column the range of used vales is placed.

The other common way is using an adsorbate dissolved in water or another solvent. In these cases, usually the probe molecule adsorbs as a single monolayer; thus, the Langmuir isotherm (Chapter 4, Section 4.4.2) is employed, which in the linearized form reads as follows (Equation 4.46):

$$\frac{c_{ad}}{Q_{ad}}=\frac{c_{ad}}{Q_S}+\frac{1}{KQ_S} \tag{7.24}$$

where c_{ad} is the adsorbate concentration in solution. In Table 7.7, some substances employed as adsorbates are shown; there are many compounds that have been used for this purpose, but only a few are presented here. Figure 7.29 shows the plot, according to Equation 7.24, of the adsorption of three organic compounds on an Argentine topsoil (Ferreiro, De Bussetti, and Helmy 2002). From the slopes and molecular areas on Table 7.7, the specific surface areas can be calculated.

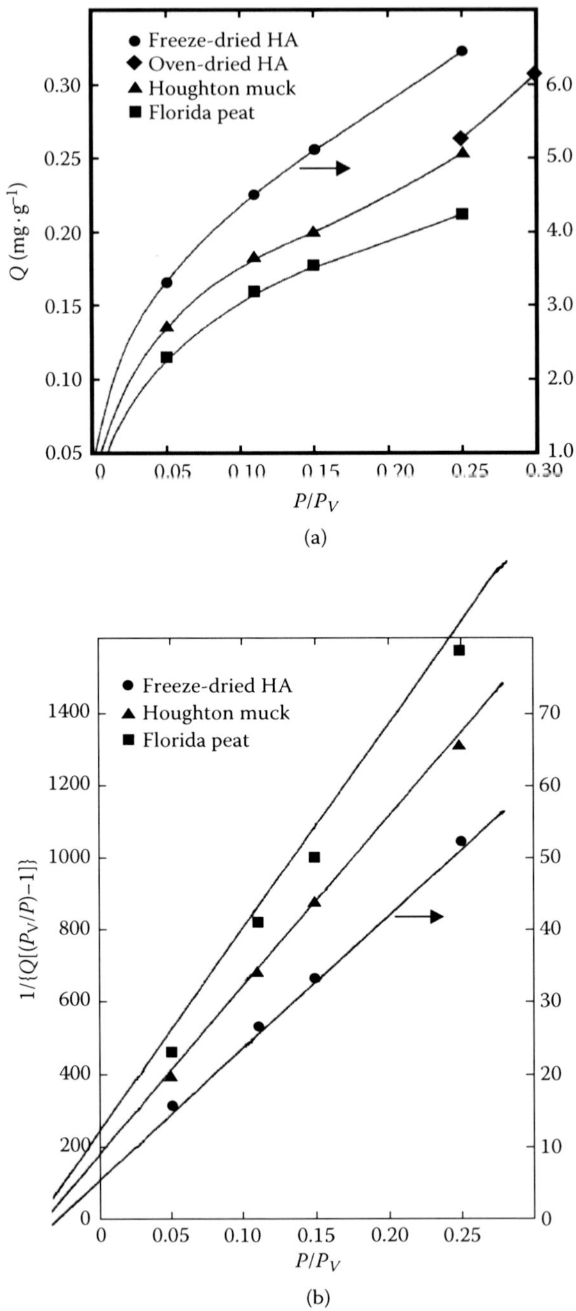

FIGURE 7.28 Adsorption of N_2 on soils with high organic matter content at 77 K: (a) plot of the specific amount adsorbed as a function of N_2 pressure, showing a Brunauer–Emmet–Teller (BET) isotherm shape. (b) Plot of the data as Equation 7.23. (Reprinted with permission from Chiou, C.T. et al., 1990, *Environmental Science & Technology*, 24, no. 8, 1164–1166. Copyright 2003 American Chemical Society.)

TABLE 7.7
Adsorbates Commonly Employed to Determine Surface Areas

Adsorbate	Method	T (K)	a_{ad} (Ideal/nm^2)	a_{ad} (Observed/ used/ nm^2)	Comments
Nitrogen[a]	BET	77	0.162	0.162 (0.13–0.20)	
Argon[a]	BET	77	0.138	0.167 (0.13–0.18)	
Carbon dioxide[b]	MC[c]	273	n/a	n/a	Based on sites occupied on the surface
Water[a]	BET	293[d]	0.105	0.106 (0.08–0.2)	
Ethylene glycol[a]	BET	293[d]	0.224	0.332 (0.22–0.33)	
EGME[a]	BET	293[d]	0.323	0.523 (0.4–0.6)	
MB[e]	Langmuir	298[d]	1.30	1.30 (1.2–2.0)	
Orthophenathroline[f]	Langmuir	298[d]	0.60	0.60	
2,2′-Bipyridyl[g]	Langmuir	298[d]	0.529	0.715	
Quinoline[h]	Langmuir	298[d]	0.63	0.58–0.63	

Note: BET = gas-phase adsorption with BET isotherm fitting; Langmuir = liquid-phase adsorption with Langmuir isotherm fitting.

[a] Sposito (1984).
[b] Ravikovitch et al. (2000).
[c] Isotherm analysis by grand canonical Monte Carlo simulation and density functional theory calculations.
[d] May vary; near room temperature.
[e] Graham (1955); Hang and Brindley (1970); Avena et al. (2001).
[f] De Bussetti (1980); Ferreiro, de Bussetti, and Helmy (2002).
[g] Ferreiro, De Bussetti, and Helmy (2002).
[h] Helmy (1983); Ferreiro, de Bussetti, and Helmy (2002).

The results are 52 $m^2 \cdot g^{-1}$ for bipyridyl, 83 $m^2 \cdot g^{-1}$ for orthophenanthroline, and 51 $m^2 \cdot g^{-1}$ for quinoline. The different values reflect the different interactions of these compounds with the soil surface.

The most widely employed method is N_2 adsorption; however, it is relatively involved, requiring special equipment. An alternative is to employ gases with higher boiling points, water, ethylene glycol, and EGME; the last compound is often employed in laboratory measurements, because it provides a simpler procedure (Carter, Mortland, and Kemper 2010). The other alternative, as said earlier, is to work with adsorption from solution. Methylene blue (MB) is one of the most common substances employed for this purpose; although its use has been criticized (Sposito 1984), more recent research supports its application (Avena et al. 2001; Zhang, Luo, and Deng 2002; Yukselen and Kaya 2008). To simplify the process, a procedure has been devised to avoid detailed adsorption study; this procedure consists of titration of the sample with MB until no more is adsorbed (or exchanged). There are two variants, one termed "MB spot test" and the other "MB titration" (Santamarina et al. 2002); it should be noted that both methods are

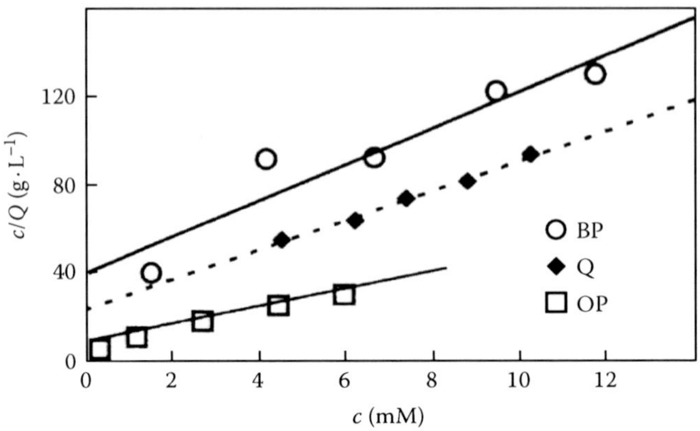

FIGURE 7.29 Linear Langmuir plot for the adsorption of 2,2′-bipyridyl (BP), orthophenanthroline (OP), and quinoline (Q) on the A horizon of a calcic chernozem soil from the southwestern Buenos Aires province, Argentina. (Reprinted from Ferreiro, E. A., S. G. de Bussetti, and A. K. Helmy: Determination of surface area of soil by adsorption methods. *European Journal of Soil Science* 53. 2002. 3. 475–480. Copyright Wiley-VCH Verlag GmbH & Co. KGaA. Reproduced with permission.)

titrations, and they only differ in the determination of the end point: in the spot test it is done by a simple drop test on paper, whereas in the other it is determined by spectrophotometry.

7.6.4.3 Negative Adsorption Methods

In negative adsorption methods, a charged solid surface is placed in suspension in an electrolyte solution, where the ions of like charge are repelled (desorbed) from the surface, producing a region deleted of such ions, as developed and discussed in Chapter 3. This effect is measured by placing the solution in two chambers separated by a dialysis membrane permeable to water and ions but not to the colloid. Let X be the ion of interest of charge z and initial concentration c. After a mass m of solid is suspended in one chamber and equilibrium across the membrane is established, the concentration of X in the chamber not containing solid will rise to a new value c', satisfying the following relationship (Sposito 1984):

$$c'V_{ex} = \frac{(c'-c)V}{m} \tag{7.25}$$

where V is the solution volume in the chamber, and V_{ex} is the "specific exclusion volume," the volume per unit mass of solid depleted of the ion X, given as follows:

$$V_{ex} = A_{ex}d_{ex}(c) \tag{7.26}$$

In Equation 7.26, A_{ex} is the solid area per unit of mass where the ion is excluded and d_{ex} is the "exclusion distance," the (effective) distance from the surface where the solution is depleted of X. It is a function of concentration and can be evaluated applying the diffuse layer theory (Chapter 3). The following can be found for 1:1 electrolytes and under the assumption of low potentials:

$$d_{ex}(c) = \frac{2}{\sqrt{\beta c}}\left(1 - e^{F\psi(\delta)/2RT}\right) \approx \frac{2}{\sqrt{\beta c}} - \delta \quad (7.27)$$

where $\beta = 2F^2/\varepsilon_0\varepsilon RT$, δ is the distance from the surface where the net particle charge is equal to that of the ions repelled from the surface, and $\psi(\delta)$ is the potential at that point. Combining Equations 7.26 and 7.27, the following is found:

$$V_{ex} = \frac{2A_{ex}}{\sqrt{\beta c}} - \delta A_{ex} \quad (7.28)$$

Thus, a plot of V_{ex} versus $c^{-1/2}$ will be a straight line, and A_{ex} can be evaluated from the slope. The negative adsorption method has been applied to some phyllosilicates (Bolt and Warkentin 1958; Edwards, Posner, and Quirk 1965a, b) but has received less attention than its positive adsorption counterpart.

7.6.5 Thermal Analysis Methods

Under the general term "thermal analysis," several techniques based on sample changes on heating are grouped (Tan, Hajek, and Barshad 2010). In general, sample heating leads to thermal reactions, causing changes that are useful for sample qualitative and quantitative characterization. The reactions taking place on heating of mineral phases mainly involve loss of water. There are two types of reactions: the first one refers to H_2O molecules that are either part of the crystal structure (water of hydration) or "wetting" the solid, known as "water of adsorption," their removal being termed dehydration and desorption, respectively; the second type is the removal of –OH groups (or OH^- ions) in which two such groups react to give a –O- atom (or an O^{2-} ion) liberating an H_2O molecule, termed "dehydroxylation." The heating can be performed under a controlled atmosphere; in the presence of oxygen, some materials undergo oxidation reactions incorporating O atoms, or burning in the case of organic matter. Other changes may be crystal lattice reorganization, either due to water elimination or not. Organic matter, on the other hand, in the presence of oxygen is completely removed. All of these changes involve heat exchange between the sample and the equipment, which can be either endo- or exothermic and in most cases, also a mass change (usually loss). These changes can be measured, and from the results one can obtain information about the sample. The most used methods in soil studies are differential thermal analysis (DTA), differential scanning calorimetry (DSC), and thermogravimetric analysis or thermogravimetry. In Sections 7.6.5.1 through 7.6.5.3, we discuss examples applied to minerals; organic matter is discussed in Chapter 10, Section 10.3.4.3.

7.6.5.1 Differential Thermal Analysis

In DTA, a controlled rate heating is applied simultaneously to a sample and to a reference material, and the differences in temperature are recorded. If at a certain temperature the sample undergoes an endothermic reaction, its temperature will fall below the reference, whereas for an exothermic process a temperature rise above the reference will be observed, as schematically shown in Figure 7.30. Dehydration, desorption, dehydroxylation, decomposition, and phase change are examples of endothermic processes, whereas others such as oxidation (especially organic matter) or crystal formation are exothermic. Albeit rare, it is possible that an exothermic reaction and an endothermic reaction occur simultaneously, thus cancelling each other; this is believed to occur for some clay minerals (Martin 1954). Figure 7.31 shows

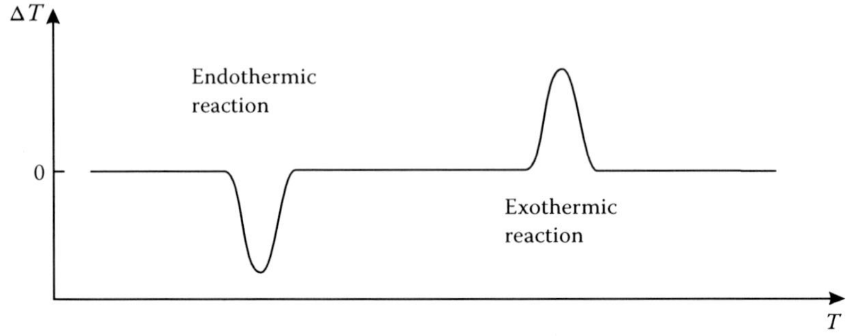

FIGURE 7.30 Ideal differential thermal analysis (DTA) response: the temperature difference between the sample and the reference is plotted as a function of the reference temperature, showing endothermic and exothermic reaction peaks.

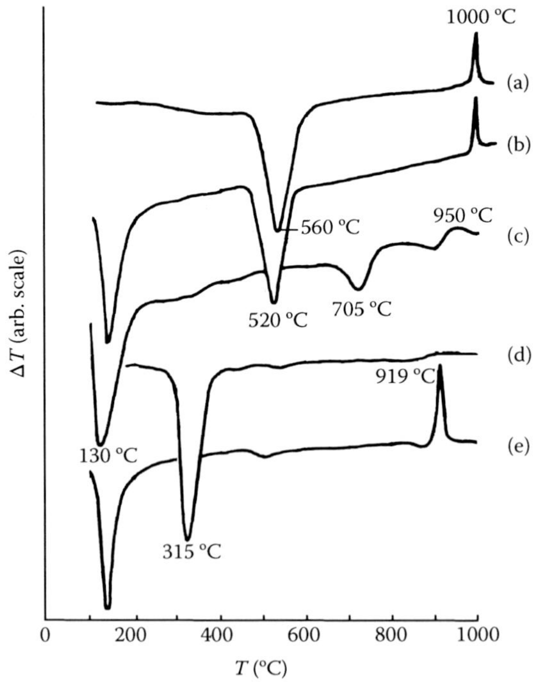

FIGURE 7.31 DTA curves for some important soil minerals: (a) kaolinite, (b) halloysite, (c) montmorillonite, (d) gibbsite, and (e) allophane. (Reprinted from Tan et al., *Methods of Soil Analysis. Part 1. Physical and Mineralogical Methods*, American Society of Agronomy–Soil Science Society of America, Madison, Wisconsin, 2010. With permission from the Soil Science Society of America.)

the DTA curves for some representative soil minerals. Kaolinite shows a strong endothermic peak at about 500°C due to dehydroxylation and a strong exothermic peak near 1000°C, which is attributed to the formation of γ-Al_2O_3. Halloysite presents, in addition, an endothermic peak at around 130°C due to the loss of adsorbed (interlayer) water. Montmorillonite shows the same water desorption low-temperature peak, another endothermic peak near 700°C due to dehydroxylation and an exothermic peak at 950°C due to recrystallization. Gibbsite (as well as other hydroxides and oxohydroxides such as goethite) presents a single endothermic peak due to dehydroxylation, at 315°C. Allophane presents a low-temperature peak due to water desorption and a high-temperature one due to γ-alumina formation. Figure 7.32 shows the DTA curves for several soil samples. Figure 7.32a corresponds to a whole soil sample with a noticeable amount of organic matter, which gives rise to strong exothermic peaks around 300°C; these features obscure the rest of the curve, making analysis difficult at best. In Figure 7.32b, a soil with a lower organic content

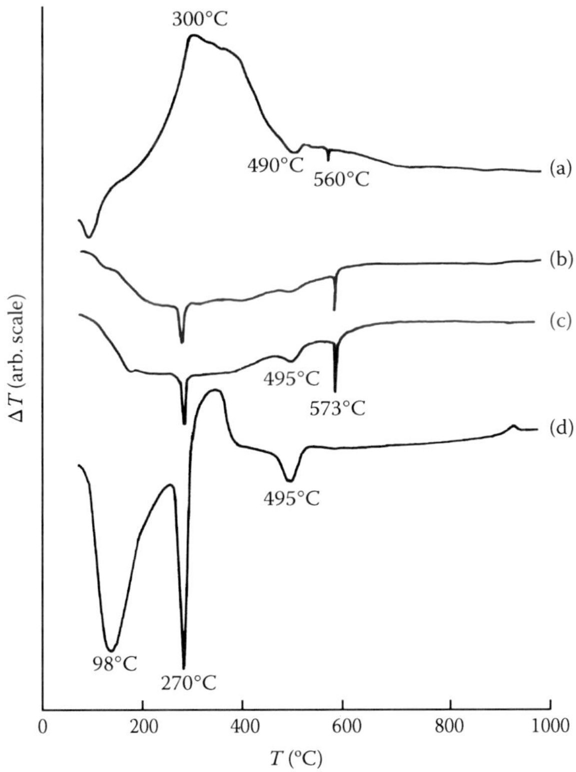

FIGURE 7.32 DTA curves for whole soils and clay fraction: (a) nontreated Madison surface soil, 2% organic matter; (b) nontreated Red Bay surface soil, 1% organic matter; (c) Red Bay soil pretreated with 30% H_2O_2; and (d) clay fraction of Red Bay surface soil. (Reprinted from Tan et al., *Methods of Soil Analysis. Part 1. Physical and Mineralogical Methods*, American Society of Agronomy–Soil Science Society of America, Madison, Wisconsin, 2010. With permission from the Soil Science Society of America.)

lacks the large exothermic peaks and shows more features, which nevertheless are more distinguishable when organic matter is previously removed with H_2O_2 treatment (Figure 7.32c). In Figure 7.32c and d, the 573°C peak corresponds to quartz (α–β transition); the 270°C peak presumably is due to a hydroxide or oxohydroxide, and the 495°C one probably comes from kaolinite, or a similar mineral. Examining different soil fractions is also useful: in Figure 7.32d, the clay fraction of the Red Bay soil of curves b and c shows more marked 270°C and 495°C peaks (indicating the presence of the corresponding minerals in this fraction) and the appearance of a new peak at 98°C (water desorption) that was obscured in the soil curve and the absence of the quartz peak, which is expected, because this mineral is usually present in higher size fractions. The DTA peaks, when they are well defined, can be used for quantitative analysis (Dixon 1966).

7.6.5.2 Differential Scanning Calorimetry

DSC is similar to DTA, but here the measured quantity is the amount of energy (heat, q) required to keep the sample and the reference at the same temperature and the difference between them, δq, is determined and plotted as a function of temperature. Thus, endo- and exothermic processes give oppositely oriented peaks, similar to DTA. Figure 7.33 shows some examples of DSC curves; here, along the δq versus T curves, the derivatives dq/dT are also plotted. Curves in Figure 7.33a and c correspond to two untreated soils; the low-temperature peaks (about 200°C) allow the identification of montmorillonite, whereas the peak at about 550°C corresponds to kaolinite. The "double peak" distinguishable in the derivative curves corresponds to the quartz peak.

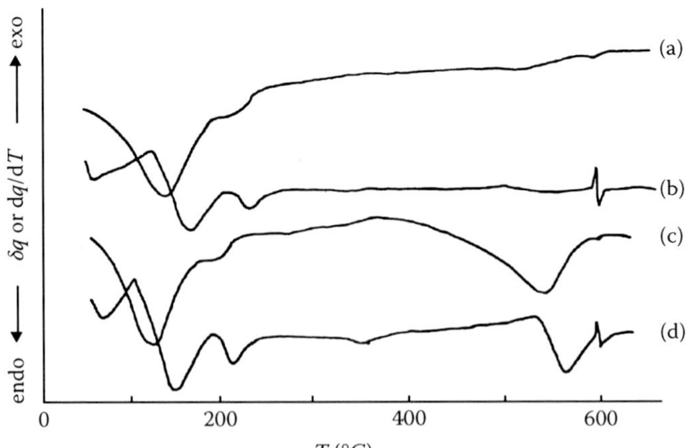

FIGURE 7.33 Differential scanning calorimetry curve (a, c) and their derivatives (b, d) for soil samples of (a, b) AC horizon of Houston city and (c, d) Houston black clay: the reference was an empty pan, the heating rate was 2.0°C/min, and the atmosphere was pure N_2. (Reprinted from Tan et al., *Methods of Soil Analysis. Part 1. Physical and Mineralogical Methods*, American Society of Agronomy–Soil Science Society of America, Madison, Wisconsin, 2010. With permission from the Soil Science Society of America.)

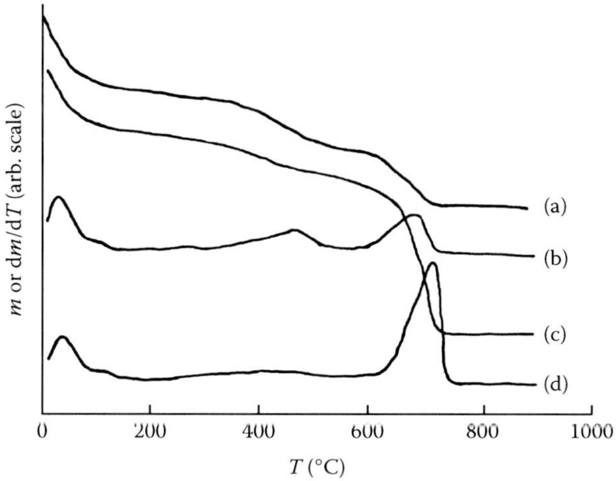

FIGURE 7.34 Thermogravimetric (a, c) and derivative (b, d) curves for the AC horizon of a Houston clay soil (a, b) and the A horizon of a Houston black clay soil (c, d): the heating was conducted in N_2 atmosphere by heating at 10°C/min. (Reprinted from Tan et al., *Methods of Soil Analysis. Part 1. Physical and Mineralogical Methods*, American Society of Agronomy–Soil Science Society of America, Madison, Wisconsin, 2010. With permission from the Soil Science Society of America.)

7.6.5.3 Thermogravimetry

Here, the sample mass is continuously weighted as it is heated at a constant rate, and the mass, or its time (or temperature, which is equivalent) derivative, is recorded as a function of temperature. As the temperature rises, reactions leading to gas evolution (e.g., water elimination) result in mass losses characteristic of the component involved and its contents in the sample. Figure 7.34 shows some examples for whole soil in N_2 atmosphere where oxidation of organic matter is precluded. There is an initial mass loss at ~100°C due to pore and/or adsorbed water elimination. The peak observed in the derivative for first soil at ~500°C corresponds to kaolinite; the peak at about 700°C is indicative of montmorillonite.

7.7 CHEMICAL CHARACTERIZATION

The chemical characterization of soils and soil components include a number of methods, ranging from simple measurements such as pH or electrical conductivity to elemental analysis after total dissolution by digestion with HNO_3/HCl/HF (Sparks 1996). Table 7.8 summarizes the most important analysis methods for elements of interest in soils. Many of these methods have been covered in previous sections. Some of them (e.g., chromatography, colorimetry) are common, well-known analytical techniques (Skoog et al. 2004; Harris 2010).

7.8 SOIL ANALYSIS

Most of the methods described so far are immediately applicable to separated soil components; however, in the analysis of a whole soil sample, matters are obviously more complicated because of the high number of components. There are several

TABLE 7.8
Commonly Employed Methods of Chemical Analysis for Elements in Soil

Elements	ICP	AE	AAS	NAA	XRF	IC	Color	Other	Comments	Reference
Li, Na, K, Rb, Cs	X	X		X	X	X				(Helmke and Sparks 1996)
Be, Mg, Ca, Sr, Ba	X	X	X		X	X		Titration		(Suarez 1996)
B	X						X	Potentiometry/ mannitol		(Keren 1996)
Al	X	X	X		X	X	X	Titration		(Bertsch and Bloom 1996)
Si	X		X				X			(Jones and Dreher 1996)
P						X (as H_2PO_4-)	X	^{31}P NMR	Speciation is important	(Kuo 1996)
As	X		X (Hydride)	X	X		X		Speciation is important	(Huang and Fujii 1996)
S						X (as SO_4^{2-})	X	Gravimetry, turbidimetry (as SO_4^{2-})	Organic/inorganic speciation is important	(Tabatabai 1996)
Se	X		X					Fluorimetry		(Huang and Fujii 1996)
F, Cl, Br						X	X (Br by microdiffusion)	Potentiometry, gravimetry, titration		(Frankengerger et al. 1996)
Fe	X		X				X			(Loeppert and Inskeep 1996)
Mn	X		X				X			(Gambrell 1996)
Cr	X		X			X	X	Chemiluminescence, voltammetry	Speciation is important	(Bartlett and James 1996)
Hg	X		X-CV							(Crock 1996)
Other transition metals	X		X			X	X			(Amacher 1996; Reed and Martens 1996; Sims 1996)

Note: AE = atomic emission; AAS = atomic absorption; IC = ionic chromatography; Color = colorimetry.

approaches that we describe very briefly here (see the study by Klute [2010] for a thorough coverage). In the first place, from the type of a soil sample and its origin (either geographic origin or depth, that is, horizon, studied) the main components are usually expected beforehand. For surface horizons, organic matter can be extracted and characterized separately (Swift 1996) (see also Chapter 10). The mineral part can in principle be characterized by XRD (Section 7.3.1), but usually only the components that are more abundant and/or of better crystallinity can be distinguished and identified; it is rather common to find quartz and little more in a XRD spectrum of the whole soil. Here, optical microscopy study can identify some minerals and give clues as to the others. Thus, the usual solution is separation; selective chemical dissolution can first destroy organic matter and afterward reduce oxides to isolate the phyllosilicates, which pose special difficulties for XRD identification (see Chapter 8). In very difficult cases, manual separation under the microscope can be done; it is a laborious and time-consuming task, but sometimes it may be the only way.

Spectroscopic analysis (such as IR) can give useful information, but again the most abundant components will dominate the spectra and make it difficult to observe the minor components. Electron microscopy is required to observe the smaller fraction (below ~2 μm) and can provide elemental composition information through EDS and/or EELS; however, the spatial resolution of these techniques is lower and may not differentiate between very small particles. XRF and NAA, as well as chemical analysis, provide only the overall elemental composition.

7.9 SUMMARY

In this chapter, soil structure and composition are introduced. A survey (not exhaustive by any means) of the most relevant characterization methods is given here. The main nonbiological components—namely, aluminosilicate and oxide minerals—and organic matter are analyzed and their characterization and ion binding behavior discussed in Chapters 8 through 10.

REFERENCES

Abe, T., and A. Watanabe. 2004. X-ray photoelectron spectroscopy of nitrogen functional groups in soil humic acids. *Soil Science* 169, no. 1: 35–43.

Alvarez-Puebla, R. A., C. Valenzuela-Calahorro, and J. J. Garrido. 2004. Retention of Co(II), Ni(II), and Cu(II) on a purified brown humic acid. Modeling and characterization of the sorption process. *Langmuir* 20, no. 9: 3657–3664.

Amacher, M. C. 1996. Nickel, cadmium and lead. In *Methods of Soil Analysis. Part 3. Chemical Methods*, ed. D. L. Sparks, 739–768. Soil Science Society of America Book Series 5. Madison, WI: American Society of Agronomy-Soil Science Society of America.

Arnepalli, D. N., S. Shanthakumar, B. Hanumantha Rao, and D. N. Singh. 2007. Comparison of methods for determining specific-surface area of fine-grained soils. *Geotechnical and Geological Engineering* 26, no. 2: 121–132. doi:10.1007/s10706-007-9152-5.

Atkins, P., and J. de Paula. 2009. *Physical Chemistry*. 9th ed. New York: W. H. Freeman.

Avena, M. J., L. E. Valenti, V. Pfaffen, and C. P. De Pauli. 2001. Methylene blue dimerization does not interfere in surface-area measurements of kaolinite and soils. *Clays and Clay Minerals* 49, no. 2: 168–173. doi:10.1346/CCMN.2001.0490206.

Baldock, J. A., and P. N. Nelson. 1999. Soil organic matter. In *Handbook of Soil Science*, ed. M. L. Sumner, B75–B84. 1st ed. Boca Raton, FL: CRC.

Baldock, J. A., J. M. Oades, P. N. Nelson, T. M. Skene, A. Golchin, and P. Clarke. 1997. Assessing the extent of decomposition of natural organic materials using solid-state 13C NMR spectroscopy. *Soil Research* 35, no. 5: 1061–1084.

Balnois, E., G. Papastavrou, and K. J. Wilkinson. 2007. Force microscopy and force measurements of environmental colloids. In *Environmental Colloids and Particles: Behaviour, Separation and Characterisation*, ed. J. R. Lead and K. J. Wilkinson, 405–467. 1st ed. Chichester, UK: Wiley.

Bartlett, R. J., and B. R. James. 1996. Chromium. In *Methods of Soil Analysis. Part 3. Chemical Methods*, ed. D. L. Sparks, 683–702. Soil Science Society of America Book Series 5. Madison, WI: American Society of Agronomy-Soil Science Society of America.

Bellamy, L. J. 1975. *The Infra-Red Spectra of Complex Molecules*. 3rd ed. New York: Wiley.

Bellon-Maurel, V., and A. McBratney. 2011. Near-infrared (NIR) and mid-infrared (MIR) spectroscopic techniques for assessing the amount of carbon stock in soils - Critical review and research perspectives. *Soil Biology and Biochemistry* 43, no. 7: 1398–1410. doi:16/j.soilbio.2011.02.019.

Berry, R. S., S. A. Rice, and J. Ross. 2000. *Physical Chemistry*. 2nd ed. New York: Oxford University Press.

Bertsch, P. M., and D. B. Hunter. 2001. Applications of synchrotron-based x-ray microprobes. *Chemical Reviews* 101, no. 6: 1809–1842. doi:10.1021/cr990070s.

Bertsch, P. M., and P. A. Bloom. 1996. Aluminum. In *Methods of Soil Analysis. Part 3. Chemical Methods*, ed. D. L. Sparks, 517–550. Soil Science Society of America Book Series 5. Madison, WI: American Society of Agronomy-Soil Science Society of America.

Bolt, G. H., and B. P. Warkentin. 1958. The negative adsorption of anions by clay suspensions. Colloid and Polymer Science (*Kolloid-Zeitschrift)* 156, no. 1: 41–46. doi:10.1007/BF01812361.

Borkovec, M., Q. Wu, G. Degovics, P. Laggner, and H. Sticher. 1993. Surface area and size distributions of soil particles. *Colloids and Surfaces A: Physicochemical and Engineering Aspects* 73: 65–76. doi:10.1016/0927-7757(93)80007-2.

Brady, N. C., and R. R. Weil. 2007. *The Nature and Properties of Soils*. 14th ed. Upper Saddle River, NJ: Prentice Hall.

Bullock, P., N. Fedoroff, A. Jongerius, G. Stoops, and T. Tursina. 1985. *Handbook for Soil Thin Section Description*. Wolverhampton, UK: Waine Research Publications.

Cabaniss, S. E. 1991. Carboxylic acid content of a fulvic acid determined by potentiometry and aqueous Fourier transform infrared spectrometry. *Analytica Chimica Acta* 255, no. 1: 23–30.

Cady, J. L., L. P. Wilding, and L. R. Drees. 2010. Petrographic microscope techniques. In *Methods of Soil Analysis. Part 1. Physical and Mineralogical Methods*, ed. A. Klute, 185–218. Soil Science Society of America Book Series 5. Madison, WI: American Society of Agronomy-Soil Science Society of America.

Cardoza, L. A., A. K. Korir, W. H. Otto, C. J. Wurrey, and C. K. Larive. 2004. Applications of NMR spectroscopy in environmental science. *Progress in Nuclear Magnetic Resonance Spectroscopy* 45, no. 3–4: 209–238. doi:10.1016/j.pnmrs.2004.06.002.

Carr, R., C. Zhang, N. Moles, and M. Harder. 2007. Identification and mapping of heavy metal pollution in soils of a sports ground in Galway City, Ireland, using a portable XRF analyser and GIS. *Environmental Geochemistry and Health* 30, no. 1: 45–52. doi:10.1007/s10653-007-9106-0.

Carter, D. L., M. M. Mortland, and W. D. Kemper. 2010. Specific surface. In *Methods of Soil Analysis. Part 1. Physical and Mineralogical Methods*, ed. A. Klute. Soil Science Society of America Book Series 5. Madison, WI: American Society of Agronomy-Soil Science Society of America.

Celi, L., M. Schnitzer, and M. Nègre. 1997. Analysis of carboxyl groups in soil humic acids by a wet chemical method, Fourier-transform infrared spectrophotometry, and solution-state carbon-13 nuclear magnetic resonance. A comparative study. *Soil Science* 162, no. 3: 189–197.

Chiou, C. T., J. F. Lee, and S. A. Boyd. 1990. The surface area of soil organic matter. *Environmental Science & Technology* 24, no. 8: 1164–1166. doi:10.1021/es00078a002.

Crock, J. G. 1996. Mercury. In *Methods of Soil Analysis. Part 3. Chemical Methods*, ed. D. L. Sparks, 769–792. Soil Science Society of America Book Series 5. Madison, WI: American Society of Agronomy-Soil Science Society of America.

De Bussetti, S. G. 1980. Adsorption of 1,10-phenanthroline by some clays and oxides. *Clays and Clay Minerals* 28, no. 2: 149–154. doi:10.1346/CCMN.1980.0280212.

de Groot, F. 2001. High-resolution X-ray emission and X-ray absorption spectroscopy. *Chemical Reviews* 101, no. 6: 1779–1808. doi:10.1021/cr9900681.

Deport, C., L. Lemée, and A. Amblès. 2006. Comparison between humic substances from soil and peats using TMAH and TEAAc thermochemolysis. *Organic Geochemistry* 37, no. 6: 649–664. doi:10.1016/j.orggeochem.2006.01.014.

Dixon, J. B. 1966. Quantitative analysis of kaolinite and gibbsite in soils by differential thermal and selective dissolution methods. *Clays and Clay Minerals* 14, no. 1: 83–89. doi:10.1346/CCMN.1966.0140107.

Du, C., and J. Zhou. 2008. Evaluation of soil fertility using infrared spectroscopy: A review. *Environmental Chemistry Letters* 7, no. 2: 97–113. doi:10.1007/s10311-008-0166-x.

Duval, J. F. L. 2007. Electrophoresis of soft colloids: Basic principles and applications. In *Environmental Colloids and Particles: Behaviour, Separation and Characterisation*, ed. K. J. Wilkinson and J. R. Lead, 315–344. 1st ed. Chichester, UK: Wiley.

Edwards, D. G., A. M. Posner, and J. P. Quirk. 1965a. Repulsion of chloride ions by negatively charged clay surfaces. Part 1—Monovalent cation Fithian illites. *Transactions of the Faraday Society* 61, no. 0: 2808–2815. doi:10.1039/TF9656102808.

Edwards, D. G., A. M. Posner, and J. P. Quirk. 1965b. Repulsion of chloride ions by negatively charged clay surfaces. Part 2—Monovalent cation montmorillonites. *Transactions of the Faraday Society* 61, no. 0: 2816–2819. doi:10.1039/TF9656102816.

Egerton, R. F. 1996. *Electron Energy-Loss Spectroscopy in the Electron Microscope*. New York: Springer.

Elgharmy, W., and M. Elashkar. 1962. Simplified textural classification triangles. *Soil Science Society of America Journal* 26: 612–613.

Elkins, K. M., and D. J. Nelson. 2001. Fluorescence and FT-IR spectroscopic studies of Suwannee river fulvic acid complexation with aluminum, terbium and calcium. *Journal of Inorganic Biochemistry* 87, no. 1–2: 81–96. doi:10.1016/S0162-0134(01)00318-X.

Ersahin, S., H. Gunal, T. Kutlu, B. Yetgin, and S. Coban. 2006. Estimating specific surface area and cation exchange capacity in soils using fractal dimension of particle-size distribution. *Geoderma* 136, no. 3–4: 588–597. doi:10.1016/j.geoderma.2006.04.014.

Fendorf, S. E., and D. L. Sparks. 1996. X-ray absorption fine structure spectroscopy. In *Methods of Soil Analysis. Part 3. Chemical Methods*, ed. D. L. Sparks, 377–416. Soil Science Society of America Book Series 5. Madison, WI: American Society of Agronomy-Soil Science Society of America.

Ferreiro, E. A., S. G. de Bussetti, and A. K. Helmy. 2002. Determination of surface area of soil by adsorption methods. *European Journal of Soil Science* 53, no. 3: 475–480. doi:10.1046/j.1365-2389.2002.00436.x.

Frankengerger, W. T., M. A. Tabatabai, D. C. Adriano, and H. E. Doner. 1996. Bromine. chlorine and fluorine. In *Methods of Soil Analysis. Part 3. Chemical Methods*, ed. D. L. Sparks, 833–868. Soil Science Society of America Book Series 5. Madison, WI: American Society of Agronomy-Soil Science Society of America.

Fyfe, C. 1983. *Solid State NMR for Chemists*. Guelph, ON: CFC Press.

Gaillard, J. -F. 2007. Probing environmental colloids and particles with X-rays. In *Environmental Colloids and Particles: Behaviour, Separation and Characterisation*, ed. K. J. Wilkinson and J. R. Lead, 613–666. 1st ed. Chichester, UK: Wiley.

Gambrell, R. P. 1996. Manganese. In *Methods of Soil Analysis. Part 3. Chemical Methods*, ed. D. L. Sparks, 665–682. Soil Science Society of America Book Series 5. Madison, WI: American Society of Agronomy-Soil Science Society of America.

Gerin, P. A., M. J. Genet, A. J. Herbillon, and B. Delvaux. 2003. Surface analysis of soil material by X-ray photoelectron spectroscopy. *European Journal of Soil Science* 54, no. 3: 589–604. doi:10.1046/j.1365-2389.2003.00537.x.

Ghabbour, E. A., A. C. Scheinost, and G. Davies. 2007. XAFS studies of cobalt (II) binding by solid peat and soil-derived humic acids and plant-derived humic acid-like substances. *Chemosphere* 67, no. 2: 285–291.

Goldstein, J., D. E. Newbury, D. C. Joy, C. E. Lyman, P. Echlin, E. Lifshin, L. Sawyer, and J. R. Michael. 2003. *Scanning Electron Microscopy and X-ray Microanalysis*. 3rd ed. New York: Springer.

Goldstein, S. J., A. K. Slemmons, and H. E. Canavan. 1996. Energy-dispersive X-ray fluorescence methods for environmental characterization of soils. *Environmental Science & Technology* 30, no. 7: 2318–2321. doi:10.1021/es950744q.

Graham, D. 1955. Characterization of physical adsorption systems. III. The separate effects of pore size and surface acidity upon the adsorbent capacities of activated carbons. *The Journal of Physical Chemistry* 59, no. 9: 896–900. doi:10.1021/j150531a022.

Grosheva, E., D. Zaichick, and V. Zaichick. 2007. Application of INAA in the assessment of chemical elements in soils of the Khamar-Daban mountain range. *Journal of Radioanalytical and Nuclear Chemistry* 271 (March): 565–572. doi:10.1007/s10967-007-0307-x.

Gustafsson, J. P., I. Persson, D. B. Kleja, and J. W. J. van Schaiks. 2007. Binding of Iron (III) to organic soils: EXAFS spectroscopy and chemical equilibrium modeling. *Environmental Science & Technology* 41, no. 4: 1232–1237.

Hang, P. T., and G. W. Brindley. 1970. Methylene blue absorption by clay minerals. Determination of surface areas and cation exchange capacities (Clay-organic studies XVIII). *Clays and Clay Minerals* 18, no. 4: 203–212. doi:10.1346/CCMN.1970.0180404.

Harris, D. C. 2010. *Quantitative Chemical Analysis*. 8th ed. New York: W. H. Freeman.

Hassellöv, M., F. V. D. Kammer, and R. Beckett. 2007. Characterization of aquatic colloids and macromolecules by filed-flow fractionation. In *Environmental Colloids and Particles: Behaviour, Separation and Characterisation*, ed. K. J. Wilkinson and J. R. Lead, 223–276. 1st ed. Chichester, UK: Wiley.

Hay, M. B., and S. C. B. Myneni. 2007. Structural environments of carboxyl groups in natural organic molecules from terrestrial systems. Part 1: Infrared spectroscopy. *Geochimica et Cosmochimica Acta* 71, no. 14: 3518–3532. doi:10.1016/j.gca.2007.03.038.

Helmke, P. A. 1996. Neutron activation analysis. In *Methods of Soil Analysis. Part 3. Chemical Methods*, ed. D. L. Sparks, 141–160. Soil Science Society of America Book Series 5. Madison, WI: American Society of Agronomy-Soil Science Society of America.

Helmke, P. A., and D. L. Sparks. 1996. Lithium, sodium, potassium, rubidium and cesium. In *Methods of Soil Analysis. Part 3. Chemical Methods*, ed. D. L. Sparks, 551–574. Soil Science Society of America Book Series 5. Madison, WI: American Society of Agronomy-Soil Science Society of America.

Helmy, A. K. 1983. Adsorption of quinoline from aqueous solutions by some clays and oxides. *Clays and Clay Minerals* 31, no. 1: 29–36. doi:10.1346/CCMN.1983.0310105.

Huang, P. M. 1999. The chemical composition of soils. In *Handbook of Soil Science*, ed. M. E. Sumner, B3–B24. 1st ed. Boca Raton, FL: CRC.

Huang, P. M., and R. Fujii. 1996. Selenium and arsenic. In *Methods of Soil Analysis. Part 3. Chemical Methods*, ed. D. L. Sparks, 793–832. Soil Science Society of America Book Series 5. Madison, WI: American Society of Agronomy-Soil Science Society of America.

Jang, M. 2009. Application of portable X-ray fluorescence (pXRF) for heavy metal analysis of soils in crop fields near abandoned mine sites. *Environmental Geochemistry and Health* 32, no. 3: 207–216. doi:10.1007/s10653-009-9276-z.

Jester, W., and A. Klik. 2005. Soil surface roughness measurement—methods, applicability, and surface representation. *CATENA* 64, no. 2–3: 174–192. doi:10.1016/j.catena. 2005.08.005.

Johnston, C. T., and Y. O. Aochi. 1996. Fourier transform infrared and Raman spectroscopy. In *Methods of Soil Analysis. Part 3. Chemical Methods*, ed. D. L. Sparks, 269–321. Soil Science Society of America Book Series 5. Madison, WI: American Society of Agronomy-Soil Science Society of America.

Jones, R. L., and G. B. Dreher. 1996. Silicon. In *Methods of Soil Analysis. Part 3. Chemical Methods*, ed. D. L. Sparks, 627–638. Soil Science Society of America Book Series 5. Madison, WI: American Society of Agronomy-Soil Science Society of America.

Karathanasis, A. D., and B. F. Hajek. 1996. Elemental analysis by X-ray fluorescence spectroscopy. In *Methods of Soil Analysis. Part 3. Chemical Methods*, ed. D. L. Sparks. Soil Science Society of America Book Series 5. Madison, WI: American Society of Agronomy-Soil Science Society of America.

Karlsson, T., P. Persson, and U. Skyllberg. 2006. Complexation of copper(II) in organic soils and in dissolved organic matter – EXAFS evidence for chelate ring structures. *Environmental Science & Technology* 40, no. 8 (April 1): 2623–2628. doi:10.1021/es052211f.

Kawai, J. 2010. Total reflection X-ray photoelectron spectroscopy: A review. *Journal of Electron Spectroscopy and Related Phenomena* 178–179, no. C: 268–272.

Keren, R. 1996. Boron. In *Methods of Soil Analysis. Part 3. Chemical Methods*, ed. D. L. Sparks, 603–626. Soil Science Society of America Book Series 5. Madison, WI: American Society of Agronomy-Soil Science Society of America.

Klute, A., ed. 2010. *Methods of Soil Analysis. Part 1. Physical and Mineralogical Methods.* Soil Science Society of America Book Series 5. Madison, WI: American Society of Agronomy-Soil Science Society of America.

Koiter, A. J., and D. A. Lobb. 2008. Characterization of soil surface properties following disturbance of a clay soil in southern Manitoba. *Soil and Tillage Research* 100, no. 1–2: 99–107. doi:10.1016/j.still.2008.05.002.

Koningsberger, D. C., and R. Prins, eds. 1988. *X-Ray Absorption: Principles, Applications, Techniques of EXAFS, SEXAFS and XANES.* 1st ed. New York: Wiley-Interscience.

Kuo, S. 1996. Phosphorus. In *Methods of Soil Analysis. Part 3. Chemical Methods*, ed. D. L. Sparks, 869–920. Soil Science Society of America Book Series 5. Madison, WI: American Society of Agronomy-Soil Science Society of America.

Ladd, M. F. C., and R. A. Palmer. 2003. *Structure Determination by X-Ray Crystallography.* 4th ed. New York: Kluwer Academic/Plenum Publishers.

Levine, I. 2008. *Physical Chemistry.* 6th ed. New York: McGraw-Hill Science/Engineering/Math.

Loeppert, R. L., and W. P. Inskeep. 1996. Iron. In *Methods of Soil Analysis. Part 3. Chemical Methods*, ed. D. L. Sparks, 639–664. Soil Science Society of America Book Series 5. Madison, WI: American Society of Agronomy-Soil Science Society of America.

Luo, L., and S. Zhang. 2010. Applications of synchrotron-based X-ray techniques in environmental science. *Science China Chemistry* 53, no. 12: 2529–2538. doi:10.1007/s11426-010-4085-x.

MacCarthy, P., C. E. Clapp, and R. L. Malcolm. 1990. *Humic Substances in Soil and Crop Sciences: Selected Readings: Proceedings of a Symposium Cosponsored by the International Humic Substances Society.* Madison WI: American Society of Agronomy, December.

Macht, F., K. Eusterhues, G. J. Pronk, and K. U. Totsche. 2011. Specific surface area of clay minerals: Comparison between atomic force microscopy measurements and bulk-gas (N2) and -liquid (EGME) adsorption methods. *Applied Clay Science* 53, no. 1: 20–26. doi:10.1016/j.clay.2011.04.006.

Martin, R. T. 1954. Reference chlorite characterization for chlorite identification in soil clays. *Clays and Clay Minerals* 3, no. 1: 117–145. doi:10.1346/CCMN.1954.0030113.

Matsuda, Y. H. 2008. Synchrotron X-ray experiments in pulsed high magnetic fields. *Analytical Sciences* 24, no. 1: 3–10.

Mavrocordatos, D., D. Perret, and G. G. Leppard. 2007. Strategies and advances in the characterization of environmental colloids by electron microscopy. In *Environmental Colloids and Particles: Behaviour, Separation and Characterisation*, ed. K. J. Wilkinson and J. R. Lead, 345–404. 1st ed. Chichester, UK: Wiley.

McDonald, R. C., R. F. Isbell, J. G. Speigth, J. Walker, and M. S. Hopkins. 1990. *Australian Soil and Land Survey - Field handbook*. 2nd ed. Melbourne: Inkata press.

McQuarrie, D. A., and J. D. Simon. 1997. *Physical Chemistry: A Molecular Approach*. Herndon, VA: University Science Books.

Meirer, F., A. Singh, P. Pianetta, G. Pepponi, F. Meirer, C. Streli, and T. Homma. 2010. Synchrotron radiation-induced total reflection X-ray fluorescence analysis. *TrAC Trends in Analytical Chemistry* 29, no. 6: 479–496. doi:10.1016/j.trac.2010.04.001.

Morgenstern, P., L. Brüggemann, R. Meissner, J. Seeger, and R. Wennrich. 2009. Capability of a XRF method for monitoring the content of the macronutrients Mg, P, S, K and Ca in agricultural crops. *Water, Air, & Soil Pollution* 209, no. 1–4: 315–322. doi:10.1007/s11270-009-0200-z.

Mukherjee, A., A. R. Zimmerman, and W. Harris. 2011. Surface chemistry variations among a series of laboratory-produced biochars. *Geoderma* 163, no. 3–4: 247–255. doi:10.1016/j.geoderma.2011.04.021.

Olson, G. C., M. L. Thompson, and M. A. Wilson. 1999. Phyllosilicates. In *Handbook of Soil Science*, ed. M. E. Sumner, F77–F123. 1st ed. Boca Raton, FL: CRC.

Orsetti, S., M. M. Quiroga, and E. M. Andrade. 2006. Binding of Pb(II) in the system humic acid/goethite at acidic pH. *Chemosphere* 65, no. 11: 2313–2321.

Patterson, H. H., C. S. Cronan, S. Lakshman, B. J. Plankey, and T. A. Taylor. 1992. Comparison of soil fulvic acids using synchronous scan fluorescence spectroscopy, FTIR, titration and metal complexation kinetics. *Science of The Total Environment* 113, no. 1–2: 179–196. doi:10.1016/0048-9697(92)90024-M.

Pauling, L. 1929. The principles determining the structure of complex ionic crystals. *Journal of the American Chemical Society* 51, no. 4: 1010–1026. doi:10.1021/ja01379a006.

Penner-Hahn, J. E. 1999. X-ray absorption spectroscopy in coordination chemistry. *Coordination Chemistry Reviews* 190–192 (September): 1101–1123. doi:16/S0010-8545(99)00160-5.

Peth, S., R. Horn, F. Beckmann, T. Donath, J. Fischer, and A. J. M. Smucker. 2008. Three-dimensional quantification of intra-aggregate pore-space features using synchrotron-radiation-based microtomography. *Soil Science Society of America Journal* 72, no. 4: 897. doi:10.2136/sssaj2007.0130.

Pignatello, J. J., S. Kwon, and Y. Lu. 2006. Effect of natural organic substances on the surface and adsorptive properties of environmental black carbon (char): Attenuation of surface activity by humic and fulvic acids. *Environmental Science & Technology* 40, no. 24: 7757–7763. doi:10.1021/es061307m.

Ravikovitch, P. I., A. Vishnyakov, R. Russo, and A. V. Neimark. 2000. Unified approach to pore size characterization of microporous carbonaceous materials from N2, Ar, and CO2 adsorption isotherms. *Langmuir* 16, no. 5: 2311–2320. doi:10.1021/la991011c.

Reed, S. T., and D. C. Martens. 1996. Copper and Zinc. In *Methods of Soil Analysis. Part 3. Chemical Methods*, ed. D. L. Sparks, 703–722. Soil Science Society of America Book Series 5. Madison, WI: American Society of Agronomy-Soil Science Society of America.

Reeves III, J. B. 2010. Near- versus mid-infrared diffuse reflectance spectroscopy for soil analysis emphasizing carbon and laboratory versus on-site analysis: Where are we and what needs to be done? *Geoderma* 158, no. 1–2: 3–14. doi:16/j.geoderma.2009.04.005.

Rehr, J. J., and R. C. Albers. 2000. Theoretical approaches to x-ray absorption fine structure. *Reviews of Modern Physics* 72, no. 3 (July 1): 621. doi:10.1103/RevModPhys.72.621.

Rehr, J. J., and A. L. Ankudinov. 2001. New developments in the theory of X-ray absorption and core photoemission. *Journal of Electron Spectroscopy and Related Phenomena* 114–116, (March): 1115–1121. doi:16/S0368-2048(00)00341-8.

Sanders, J. K. M., and B. K. Hunter. 1993. *Modern NMR Spectroscopy: A Guide for Chemists*. 2nd ed. New York: Oxford University Press, USA.

Santamarina, J. C., K. A. Klein, Y. H. Wang, and E. Prencke. 2002. Specific surface: Determination and relevance. *Canadian Geotechnical Journal* 39, no. 1: 233–241.

Sayers, D. E., E. A. Stern, and F. W. Lytle. 1971. New technique for investigating noncrystalline structures: Fourier analysis of the extended X-Ray—absorption fine structure. *Physical Review Letters* 27, no. 18: 1204. doi:10.1103/PhysRevLett.27.1204.

Schmitt-Koplin, P., and J. Junkers. 2007. Modern electrophoretic techniques for the characterization of natural organic matter. In *Environmental Colloids and Particles: Behaviour, Separation and Characterisation*, ed. K. J. Wilkinson and J. R. Lead, 277–313. 1st ed. Chichester, UK: Wiley.

Schulze, D. G. 1986. Correction of mismatches in 2θ scales during differential X-ray diffraction. *Clays and Clay Minerals* 34, no. 6: 681–685. doi:10.1346/CCMN.1986.0340610.

Schwab, A. P. 1999. The soil solution. In *Handbook of Soil Science*, ed. M. E. Sumner, B85–B120. 1st ed. Boca Raton, FL: CRC.

Senesi, N., and G. Calderoni. 1988. Structural and chemical characterization of copper, iron and manganese complexes formed by paleosol humic acids. *Organic Geochemistry* 13, no. 4–6: 1145–1152.

Senesi, N., and E. Loffredo. 1998. The chemistry of soil organic matter. In *Soil Physical Chemistry*, ed. D. L. Sparks, 239–370. 2nd ed. Boca Raton, FL: CRC Press.

Senesi, N., and T. M. Miano. 1994. *Humic Substances in the Global Environment and Implications on Human Health*. 1st ed. Amsterdam: Elsevier Science.

Seyama, H., and M. Soma. 2003. Surface-analytical studies on environmental and geochemical surface processes. *Analytical Sciences* 19, no. 4: 487–497.

Shchukarev, A. 2006. XPS at solid-aqueous solution interface. *Advances in Colloid and Interface Science* 122, no. 1–3: 149–157. doi:16/j.cis.2006.06.015.

Simpson, A. J., W. L. Kingery, and P. G. Hatcher. 2003. The identification of plant derived structures in humic materials using three-dimensional NMR spectroscopy. *Environmental Science & Technology* 37, no. 2: 337–342. doi:10.1021/es025956j.

Sims, J. L. 1996. Molybdenum and cobalt. In *Methods of Soil Analysis. Part 3—Chemical Methods*, ed. D. L. Sparks, 723–738. Soil Science Society of America Book Series 5. Madison, WI: American Society of Agronomy-Soil Science Society of America.

Skoog, D. A., D. M. West, F. James Holler, and S. R. Crouch. 2004. *Fundamentals of Analytical Chemistry*. 8th ed. New York: Brooks/Cole.

Skopp, J. M. 1999. Physical properties of primary particles. In *Handbook of Soil Science*, ed. M. E. Sumner. 1st ed. Boca Raton, FL: CRC Press.

Soltanpour, P. N., G. W. Johnson, S. M. Workman, J. B. Jones Jr., and R. O. Miller. 1996. Inductively coupled plasma emission spectrometry and inductively coupled plasma-mass spectrometry. In *Methods of Soil Analysis. Part 3. Chemical Methods*, ed. D. L. Sparks, 91–139. Soil Science Society of America Book Series 5. Madison, WI: American Society of Agronomy-Soil Science Society of America.

Sparks, D. L., ed. 1996. *Methods of Soil Analysis. Part 3. Chemical Methods*. Soil Science Society of America Book Series 5. Madison, WI: American Society of Agronomy-Soil Science Society of America, January 1.

Sparks, D. L. 1998. *Soil Physical Chemistry*. 2nd ed. Boca Raton, FL: CRC.

Sparks, D. L. 2002. *Environmental Soil Chemistry*. 2nd ed. San Diego, CA: Academic Press.

Sposito, G. 1984. *The Surface Chemistry of Soils*. New York: Oxford University Press.

Sposito, G. 2008. *The Chemistry of Soils*. 2nd ed. New York: Oxford University Press, USA.

Srivastava, A., G. S. Bains, R. Acharya, and A. V. R. Reddy. 2011. Study of seleniferous soils using instrumental neutron activation analysis. *Applied Radiation and Isotopes* 69, no. 5: 818–821. doi:10.1016/j.apradiso.2011.01.027.

Steelink, C., R. L. Wershaw, K. A. Thorn, and M. A. Wilson. 1990. Chapter 10: Application of liquid-state NMR spectroscopy to humic substances. In *Humic Substances II: In Search of Structure*, ed. M. H. B. Hayes, R. L. Malcolm, P. MacCarthy, and R. S. Swift. 1st ed. Hoboken, NJ: John Wiley & Sons.

Steinnes, E. 2000. Neutron activation techniques in environmental studies. *Journal of Radioanalytical and Nuclear Chemistry* 243, no. 1: 235–239. doi:10.1023/A: 1006764624405.

Strawn, D. G., and L. L. Baker. 2008. Speciation of Cu in a contaminated agricultural soil measured by XAFS, μ-XAFS, and μ-XRF. *Environmental Science & Technology* 42, no. 1: 37–42. doi:10.1021/es071605z.

Strüder, L., P. Lechner, and P. Leutenegger. 1998. Silicon drift detector—the key to new experiments. *Naturwissenschaften* 85, no. 11: 539–543. doi:10.1007/s001140050545.

Suarez, D. L. 1996. Beryllium, magnesium, calcium, strontium and barium. In *Methods of Soil Analysis. Part 3. Chemical Methods*, ed. D. L. Sparks, 575–602. Soil Science Society of America Book Series 5. Madison, WI: American Society of Agronomy-Soil Science Society of America.

Suarez, D. L. 1998. Thermodynamics of the soil solution. In *Soil Physical Chemistry, Second Edition*, ed. D. L. Sparks, 97–134. 2nd ed. Boca Raton, FL: CRC Press.

Sumner, M. E., ed. 1999. *Handbook of Soil Science*. 1st ed. Boca Raton, FL: CRC, August 31.

Sun, X., and H. E. Doner. 1996. An investigation of arsenate and arsenite bonding structures on goethite by FTIR. *Soil Science* 161, no. 12: 865–872.

Swift, R. S. 1996. Organic matter characterization. In *Methods of Soil Analysis. Part 3—Chemical Methods*, ed. D. L. Sparks. Soil Science Society of America Book Series 5. Madison, WI: American Society of Agronomy-Soil Science Society of America.

Tabatabai, M. A. 1996. Sulfur. In *Methods of Soil Analysis. Part 3. Chemical Methods*, ed. D. L. Sparks, 921–960. Soil Science Society of America Book Series 5. Madison, WI: American Society of Agronomy-Soil Science Society of America.

Tan, K. H. 2009. *Environmental Soil Science*. 3rd ed. Boca Raton, FL: CRC Press.

Tan, K. H., B. F. Hajek, and I. Barshad. 2010. Thermal analysis techniques. In *Methods of Soil Analysis. Part 1. Physical and Mineralogical Methods*, ed. A. Klute, 151–183. Soil Science Society of America Book Series 5. Madison, WI: American Society of Agronomy-Soil Science Society of America.

Terzano, R., Z. A. Chami, B. Vekemans, K. Janssens, T. Miano, and P. Ruggiero. 2008. Zinc distribution and speciation within rocket plants (*Eruca vesicaria* L. Cavalieri) grown on a polluted soil amended with compost as determined by XRF microtomography and Micro-XANES. *Journal of Agricultural and Food Chemistry* 56, no. 9: 3222–3231. doi:10.1021/jf073304e.

Thorn, K. A., and L. G. Cox. 2009. N-15 NMR spectra of naturally abundant nitrogen in soil and aquatic natural organic matter samples of the International Humic Substances Society. *Organic Geochemistry* 40, no. 4: 484–499. doi:10.1016/j.orggeochem. 2009.01.007.

Tsukada, H., A. Takeda, H. Hasegawa, S. Ueda, and T. Iyogi. 2005. Comparison of NAA and ICP-MS for the determination of major and trace elements in environmental sample. *Journal of Radioanalytical and Nuclear Chemistry* 263: 773–778. doi:10.1007/s10967-005-0656-2.

Vempati, R. K., T. R. Hess, and D. L. Cocke. 1996. X-ray photoelectron spectroscopy. In *Methods of Soil Analysis. Part 3. Chemical Methods*, ed. D. L. Sparks, 357–376. Soil Science Society of America Book Series 5. Madison, WI: American Society of Agronomy-Soil Science Society of America.

Vishnyakov, A., P. I. Ravikovitch, and A. V. Neimark. 1999. Molecular level models for CO2 sorption in nanopores. *Langmuir* 15, no. 25: 8736–8742. doi:10.1021/la990726c.

Wang, S., and C. N. Mulligan. 2008. Speciation and surface structure of inorganic arsenic in solid phases: A review. *Environment International* 34, no. 6: 867–879. doi:16/j.envint.2007.11.005.

Westre, T. E., P. Kennepohl, J. G. DeWitt, B. Hedman, K. O. Hodgson, and E. I. Solomon. 1997. A multiplet analysis of Fe K-Edge 1s → 3d pre-edge features of iron complexes. *Journal of the American Chemical Society* 119, no. 27: 6297–6314. doi:10.1021/ja964352a.

White, R. E. 2005. *Principles and Practice of Soil Science: The Soil as a Natural Resource*. 4th ed. Malden, MA: Wiley-Blackwell.

Wilke, M., F. Farges, P.-E. Petit, G. E. Brown Jr., and F. Martin. 2001. Oxidation state and coordination of Fe in minerals: An Fe K-XANES spectroscopic study. *American Mineralogist* 86, no. 5–6: 714–730.

Wilkinson, K. J., and J. R. Lead. 2007. *Environmental Colloids and Particles: Behaviour, Separation and Characterisation*. 1st ed. Chichester, UK: Wiley.

Wilson, M. A. 1990. Chapter 11: Solid state nuclear magnetic resonance spectroscopy of humic substances: Basic concepts and techniques. In *Humic Substances II: In Search of Structure*, ed. M. H. B. Hayes, P. MacCarthy, R. L. Malcolm, and R. S. Swift. 1st ed. Hoboken, NJ: John Wiley & Sons.

Woolfson, M. M. 1997. *An Introduction to X-ray Crystallography*. 2nd ed. Cambridge, UK: Cambridge University Press.

Wright, R. J., and T. Stuczynski. 1996. Atomic absorption and flame emission spectrometry. In *Methods of Soil Analysis. Part 3. Chemical Methods*, ed. D. L. Sparks, 65–90. Soil Science Society of America Book Series 5. Madison, WI: American Society of Agronomy-Soil Science Society of America.

Xia, K., W. Bleam, and P. A. Helmke. 1997. Studies of the nature of Cu2+ and Pb2+ binding sites in soil humic substances using X-ray absorption spectroscopy. *Geochimica et Cosmochimica Acta* 61, no. 11: 2211–2221. doi:10.1016/S0016-7037(97)00079-3.

Yuan, G., M. Soma, H. Seyama, B. K. G. Theng, L. M. Lavkulich, and T. Takamatsu. 1998. Assessing the surface composition of soil particles from some podzolic soils by X-ray photoelectron spectroscopy. *Geoderma* 86, no. 3–4: 169–181. doi:16/S0016-7061(98)00049-4.

Yukselen, Y., and A. Kaya. 2008. Suitability of the methylene blue test for surface area, cation exchange capacity and swell potential determination of clayey soils. *Engineering Geology* 102, no. 1–2: 38–45. doi:10.1016/j.enggeo.2008.07.002.

Zhang, G., Y. Luo, and X. Deng. 2002. Determination of surface area of red mud and beringite using methylene blue method. *Pedosphere* 12, no. 2: 189–192.

8 Silicate-Based Minerals

Silicon being the second most abundant element in Earth's crust, it is no surprise that most minerals are based on this element. There are several major classes of silicate-based minerals, including quartz (which is considered in Chapter 9), simple silicates, and more complex compounds. The largest and more important class is constituted by those which can be grouped, in a general way, as *aluminosilicates*, comprising many types of structures, including both primary and secondary minerals.

8.1 STRUCTURE AND CLASSIFICATION OF SILICATE-BASED MINERALS

With regard to composition, silicate minerals can be classified as shown in Table 8.1, whereas from a mineralogical point of view, they are generally classified as presented in Table 8.2 (Dixon and Schulze 2002). In Table 8.1, silicates are divided in increasing complexity order. All silicates are built from SiO_4 tetrahedra, this being the geometry required for Si–O structures, as discussed in Section 7.2.1.3. This basic building block is shown in Figure 8.1, as the isolated SiO_4 tetrahedron in different perspectives (a and b) and as the $Si(OH)_4$ (orthosilicic acid) molecule. In the simplest form, this tetrahedron is present as the SiO_4^{4-} ion in orthosilicate salts such as Na_4SiO_4. In minerals, this is present in the family known as *nesosilicates* (Table 8.2), which are chemically orthosilicates. The condensation of two $Si(OH)_4$ units, losing a water molecule, leads to the formation *pyrosilicic* acid and pyrosilicate ($Si_2O_7^{6-}$) compounds (Table 8.1), which constitute the *sorosilicate* mineral subclass (Table 8.2); Figure 8.2 schematically shows the $Si_2O_7^{6-}$ ion.

Further condensation of SiO_4 tetrahedra leads to extended structures sharing two O atoms, having the minimum formula SiO_3^{2-}, giving the *metasilicate* type of substances (Table 8.1). There are basically two types of structures: cycles formed by a few SiO_3^{2-} units (usually four to six; between three and eight have been observed) and linear chains of these units. The former constitutes the family of *cyclosilicates*, whereas the later is found in the *single-chain inosilicate* subclass of minerals (Table 8.2). An example of a six-member metasilicate ring, found in cyclosilicate minerals such as beryl or dioptase, is shown in Figure 8.3. Single-chain inosilicates, also known usually as *pyroxenes* (however, in some sources, the term pyroxenes refer to a particular group of these minerals), have the generic composition $Si_nO_{3n+2}^{(2n+2)-}$. Figure 8.4 shows schemes of the single-chain inosilicate structure, found in minerals like diopside or jadeite. Double-chain inosilicates, or *amphiboles*, have the minimum formula $Si_4O_{11}^{6-}$. Figure 8.5 shows an example of this structure; note that this is a chain of six-member rings like that shown in Figure 8.3.

An extension of the structure of Figure 8.5 in two dimensions results in a sheet formed by six-member rings (Figure 8.6), resulting in the sharing of three O atoms per Si tetrahedron; the shared oxygens are all located in the same plane (see Figure 8.6c); these O atoms are termed "basal oxygens" due to their location, whereas the remaining, non-shared O atoms are termed "apical oxygens," which play an important role in the mineral

TABLE 8.1
Composition-Classified Silicate-Based Minerals

Type	Minimum Chemical Unit	Generic Name	Examples
"Pure" silicates	SiO_4^{4-}	Orthosilicate (silicate)	Olivines: $(Fe, Mg)_2SiO_4$ Zircon: $ZrSiO_4$
	$Si_2O_7^{6-}$	Pyrosilicates	Barysilite: $Pb_8Mn(Si_2O_7)_3$ Hardystonite: $Ca_2ZnSi_2O_7$
	$Si_2O_3^{2-}$	Metasilicates	Enstatite: $MgSiO_3$ Beryl: $Be_3Al_2(SiO_3)_6$
	$Si_4O_{11}^{6-}$	Amphiboles	Grunerite: $Fe_7Si_8O_{22}(OH)_2$
	SiO_2	Silica	Quartz
Phyllosilicates	$Si_2O_5^{2-}$	Disilicates	Aluminosilicates: Kaolinite: $Al_2Si_4O_{10}(OH)_2$ Mg-Al Biotite: $K(Mg,Fe)_3(AlSi_3O_{10})(OH)_2$ Magnesium silicates: Talc: $Mg_3Si_4O_{10}(OH)_2$ Antigorite: $Mg_3Si_2O_5(OH)_4$
Other aluminosilicates	$Si_3O_8^{4-}$	Feldespars	Albite: $NaAlSi_3O_8$
	$Si_2O_6^{4-}$	Zeolites	Analcime: $NaAlSi_2O_6{\cdot}H_2O$
Poorly crystalline aluminosilicates	various		Imogolite: $Al_2SiO_5{\cdot}2H_2O$

TABLE 8.2
Mineralogical Classification of Silicate-Based Minerals

Subclass	Structure	Example Minerals (Typical Idealized Formulas)
Nesosilicates	Orthosilicates: isolated SiO_4 tetrahedra Figure 8.1	Olivines: $(Mg, Fe)SiO_4$ Sonolite: $Mn_9(SiO_4)_4(OH)_2$ Spurrite: $Ca_5(SiO_4)_2CO_3$ Andalusite, Kyanite: Al_2SiO_5
Sorosilicates	Pyrosilicates: one shared O between two tetrahedra Figure 8.2	Junitoite: $CaZn_2Si_2O_7{\cdot}H_2O$ Ferroaxinite: $Ca_2FeAl_2BSi_4O_{15}(OH)$ Epidote: $Ca_2(Al,Fe)_3(Si_2O_7)(SiO_4)(O,OH)_2$
Cyclosilicates	Cyclic metasilicates: two shared O per tetrahedron	Benitoite (three-Si rings): $BaTi(SiO_3)_3$ Baotite (four-Si rings): $Ba_4Ti_4(Ti,Nb,Fe)_4(SiO_3)_4O_{16}Cl$ Fluor-buergerite (six-Si rings): $NaFe_3Al_6(BO_3)_3(SiO_3)_6(O,F,OH)_4$
Single chain inosilicates	Linear metasilicates: two shared O per tetrahedron	*Pyroxene group:* Ferrosilite: $FeSiO_3$ Hedenbergite: $CaFeSi_2O_6$ *Pyroxenoid group:* Wollastonite: $CaSiO_3$

TABLE 8.2 *(Continued)*
Mineralogical Classification of Silicate-Based Minerals

Subclass	Structure	Example Minerals (Typical Idealized Formulas)
Phyllosilicates	Disilicates: sheets of six-member rings: three shared O per Si atom; bound to octahedral sheets	**1:1 layers**
		Kaolin-serpentine group:
		Dioctahedral:
		Kaolinite, Halloysite: $Al_2Si_2O_5(OH)_4$
		Trioctahedral (serpentines):
		Chrysotile: $Mg_3Si_2O_5(OH)_4$
		Lizardite: $[Mg_{2.8}Al_{0.2}](Si_{1.8}Al_{0.2})O_5(OH)_4$
		2:1 layers
		Pyrophilite-Talc group:
		Dioctahedral:
		Pyrophillite $Al_2Si_4O_{10}(OH)_2$
		Trioctahedral:
		Talc $Mg_3Si_4O_{10}(OH)_2$
		Mica group:
		Dioctahedral:
		Muscovite: $KAl_2(AlSi_3)O_{10}(OH)_2$
		Ferroceladonite: $K_2Fe(II)_2Fe(III)_2(Si_8O_{20})(OH)_4$
		Illite: $K_{0.75}Al_{1.75}[Mg,Fe(II)]_{0.5}(Al_{0.5}Si_{3.5})O_{10}(OH)_2$
		Trioctahedral:
		Eastonite: $KMg_2Al_3Si_2O_{10}(OH)_2$
		Norrishite: $KMn(III)_2LiSi_4O_{12}$
		Smectite group:
		Dioctahedral:
		Montmorillonite: $Na_{0.2}Ca_{0.1}Al_2Si_4O_{10}(OH)_2 \cdot n(H_2O)$
		Beidellite: $Na_{0.5}Al_{2.5}Si_{3.5}O_{10}(OH)_2 \cdot n(H_2O)$
		Trioctahedral:
		Hectorite: $Na_{0.4}Mg_{2.7}Li_{0.3}Si_4O_{10}(OH)_2$
		Saponite: $Ca_{0.25}(Mg,Fe)_3((Si,Al)_4O_{10})(OH)_2 \cdot n(H_2O)$
		Vermiculite group:
		Trioctahedral: $Mg_{0.6\text{-}0.9}\,nH_2O\ Mg_3(Al,Si)_4O_{10}(OH)_2$
		Dioctahedral: $Mg_{0.6\text{-}0.9}\,nH_2O\ Al_2(Al,Si)_4O_{10}(OH)_2$
		Chlorite group:
		Generalized chlorite: $(Mg,Fe)_3(Si,Al)_4O_{10}(OH)_2 \cdot (Mg,Fe)_3(OH)_6$
		Chamosite: $Fe(II)_3Mg_{1.5}AlFe(III)_{0.5}Si_3AlO_{12}(OH)_6$
		Nimite: $(Ni,Mg,Al)_6((Si,Al)_4O_{10})(OH)_8$
Tectosilicates	Tridimensional, four shared O per Si tetrahedron	*Silica group:*
		Quartz, Tridymite, Cristobalite: SiO_2
		Feldspars group:
		Microcline, orthoclase: $KAlSi_3O_8$
		Albite: $NaAlSi_3O_8$
		Andesine: $(Na,Ca)(Si,Al)_4O_8$ (Na:Ca 3:2)
		Zeolite group:
		Natrolite: $Na_2Al_2Si_3O_{10} \cdot 2H_2O$
		Heulandite: $CaAl_2Si_7O_{18} \cdot 6H_2O$

Source: Dixon, J.B. and Schulze, D.G., *Soil Mineralogy with Environmental Applications*. Soil Science Society of America, Madison, WI, 2002.

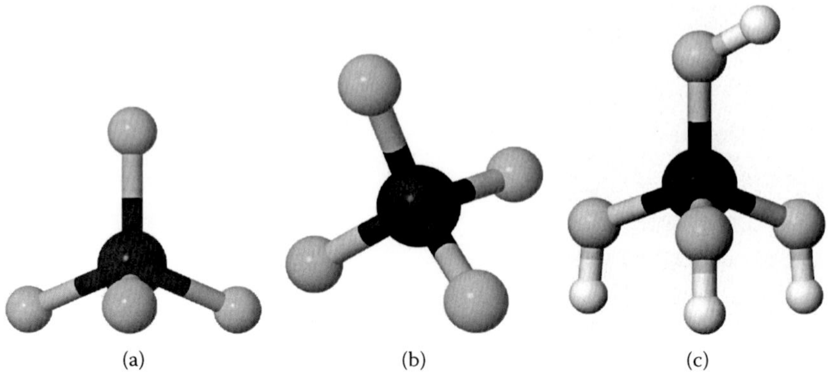

FIGURE 8.1 (a) and (b) Schematic representation (in ball and stick format) of a basic SiO_4 tetrahedron, corresponding to an isolated SiO_4^{4-} ion. (c) $Si(OH)_4$ (silicic acid) molecule. The black sphere represents a Si atom, the light gray spheres represent O atoms and the small white spheres represent H atoms.

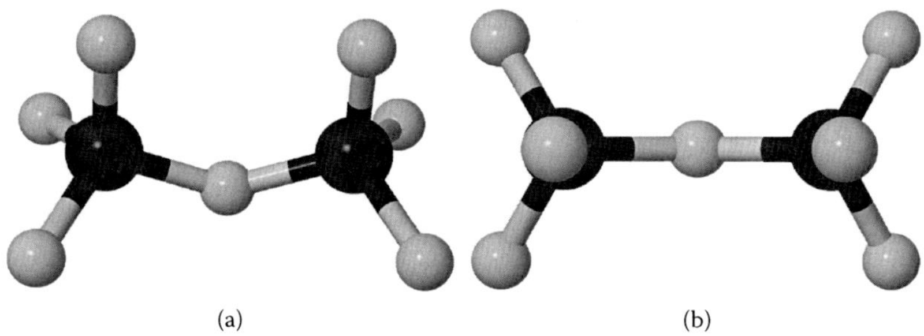

FIGURE 8.2 Schematic representation of the structure of $Si_2O_7^{6-}$ ions, found in pyrosilicate compounds and sorosilicate minerals; (a) side view and (b) top view. The black spheres represent Si atoms and the light gray spheres represent O atoms.

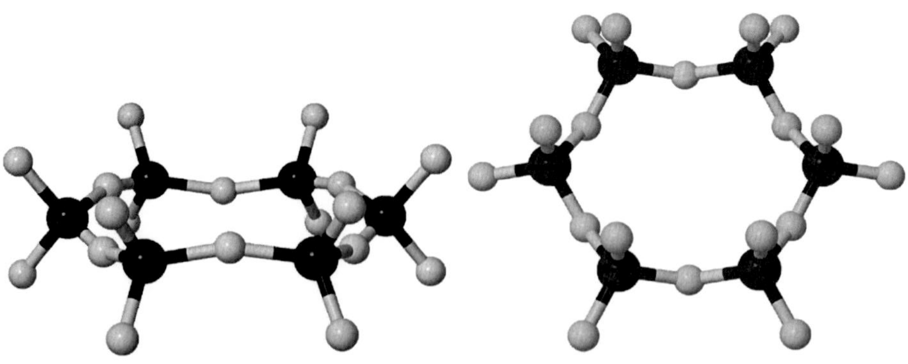

FIGURE 8.3 A six-member metasilicate ring ($Si_6O_{18}^{12-}$) found in cyclosilicate minerals, in side (a) and nearly top (b) views. The black spheres represent Si atoms and the light gray spheres represent O atoms.

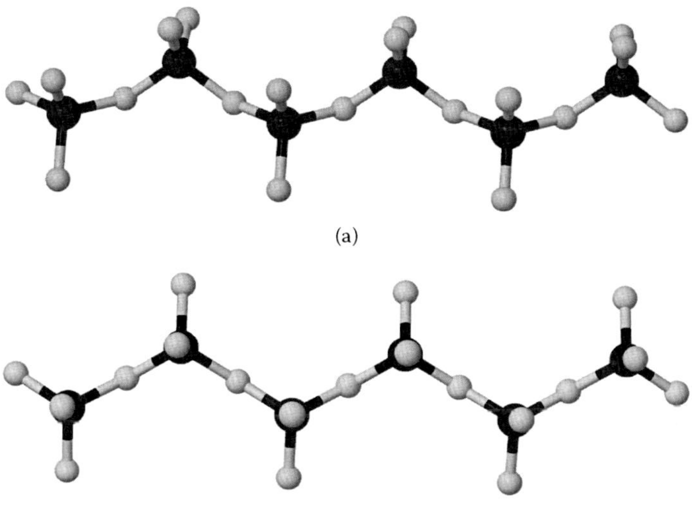

(a) (b)

FIGURE 8.4 Schematic example of a single-chain inosilicate unit, in side (a) and top (b) views. The black spheres represent Si atoms and the light gray spheres represent O atoms.

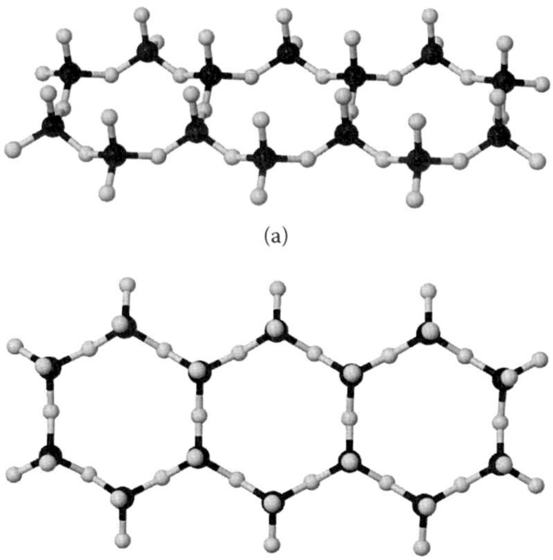

(a) (b)

FIGURE 8.5 Schematic example of a double-chain inosilicate unit, in side (a) and top (b) views. The black spheres represent Si atoms and the light gray spheres represent O atoms.

structure, as will be described shortly. This structure is known as a *tetrahedral sheet*. Finally, sharing all four O atoms between SiO_4^{4-} tetrahedra results in three-dimensional structures with the minimum formula SiO_2, including quartz and minerals such as feldspars and zeolites. Minerals having this structure (Figure 8.7) are termed *tectosilicates*.

The other important structure present in many minerals is that of an octahedron, which arises around central cations slightly larger than Si(IV) (Section 7.2.1.3), which

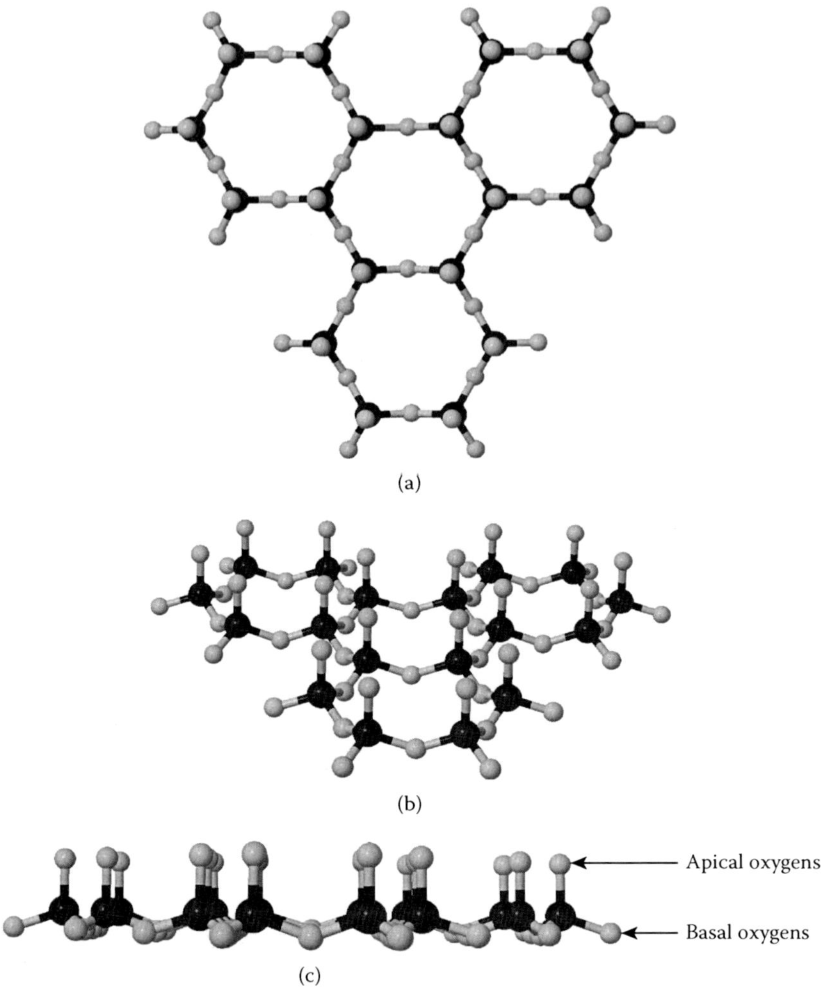

FIGURE 8.6 (a) Top view of a fragment of a 2D sheet formed by condensation of SiO_4 tetrahedra, known as "tetrahedral layer." (b) Perspective view of the same. (c) Side view, showing the non-shared O atoms at the top, termed "apical oxygens," and at the bottom, the O atoms shared between three tetrahedra, known as "basal oxygens." The black spheres represent Si atoms and the light gray spheres represent O atoms.

can accommodate six O^{2-} ions around them, usually Al(III), but also Mg(II), Fe(II), and so on (see Table 7.4). Figure 8.8 shows a central atom, such as Al(III) surrounded by six O atoms; in fact, around Al(III) one usually finds OH^- ions, as in an (hypothetical) ion $Al(OH)_6^{3-}$; here, H atoms are not shown for the sake of clarity. Actually, octahedra are always condensed sharing the OH^- units between two aluminum atoms; thus, if all OH^- units are shared, we find the stoichiometry $Al(OH)_3$ such as in gibbsite and related minerals (see Chapter 9). Similar to SiO_4^{4-} tetrahedra, the octahedra link together forming layers; Figure 8.9 shows the condensation of three $Al(OH)_6^{3-}$ octahedra; note that condensation leads to the sharing of OH^- units between two Al(III)

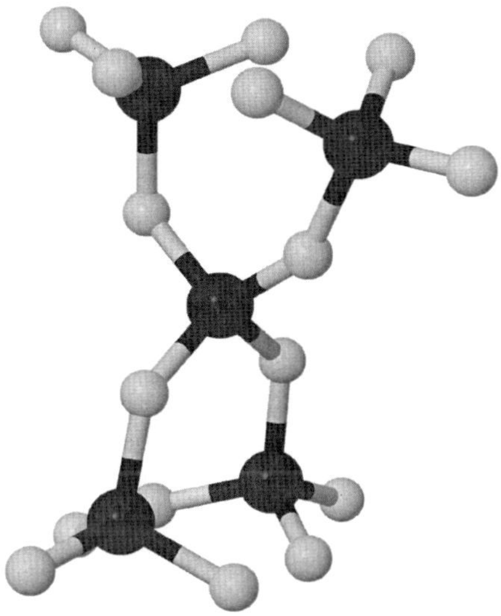

FIGURE 8.7 Schematic view of the structure of a tectosilicate, where all O atoms are shared between two Si atoms each. The black spheres represent Si atoms and the light gray spheres represent O atoms.

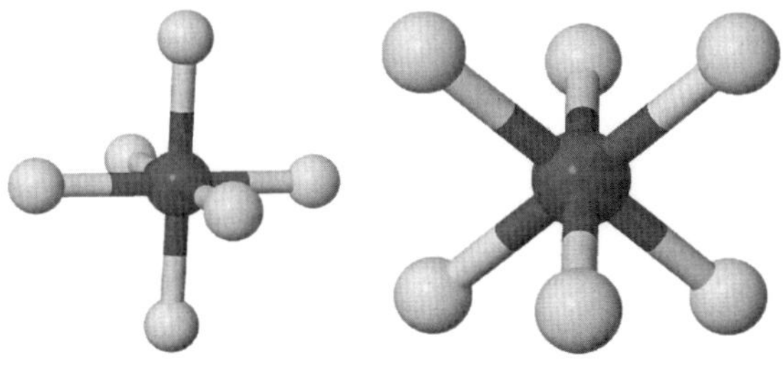

(a) (b)

FIGURE 8.8 Schematic view of an octahedral arrangement around an Al(III) center (or other cation such as Mg(II)) of six O atoms, usually part of an OH^- group (hydrogen atoms not shown). (a) Upright view; (b) side view, as it is found in octahedral layers. The dark gray sphere represents an Al atom and the light gray spheres represent O atoms.

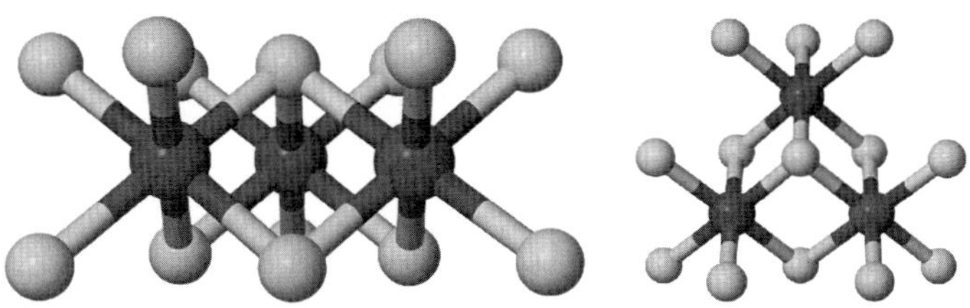

(a) (b)

FIGURE 8.9 Side (a) and top (b) views of the condensation of three octahedral units. The dark gray spheres represent Al atoms and the light gray spheres represent O atoms.

centers. Figure 8.10 shows a fragment of an *octahedral sheet* formed in such a way; here, all OH^- groups are shared between two Al centers, thus the minimum formula will be $Al(OH)_3$ as in gibbsite, which precisely have this structure; many oxides have this type of structure, as will be discussed in Chapter 9. Note that there are "holes" in the structure; these are required by the stoichiometry. For other cases, such as Mg(II), these holes are occupied by metal atoms, and all OH groups are shared between three metal centers, thus resulting in the stoichiometry $Mg(OH)_2$. The structure shown in Figure 8.10 is termed a *dioctahedral* sheet, whereas having all "sites" occupied (such as with Mg(II)) is a *trioctahedral* sheet, the difference being dictated by the stoichiometry. The three types of sheets, the tetrahedral and the two octahedral, are shown in Figure 8.11; note the difference between the two octahedral sheets.

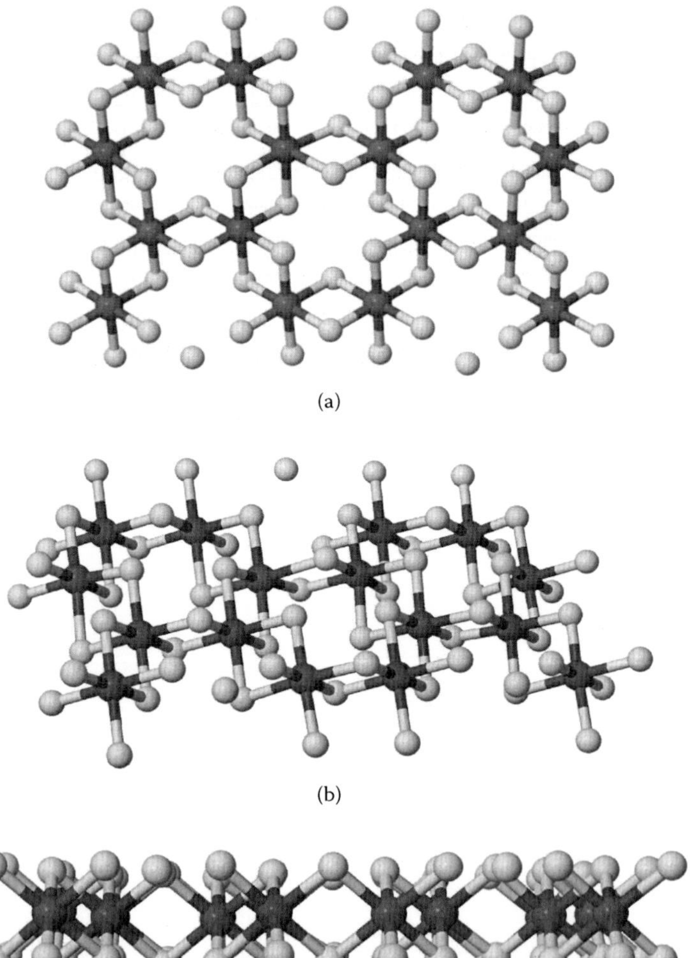

(a)

(b)

(c)

FIGURE 8.10 Fragment of a dioctahedral layer, shown in top (a), tilted (b), and side (c) views. The dark gray spheres represent Al atoms and the light gray spheres represent O atoms.

In the important subclass of *phyllosilicate* minerals, tetrahedral and octahedral sheets are bound by sharing the tetrahedral apical O atoms with the octahedral sheet; these atoms replace OH groups, forming layers characteristic of this type of minerals. This is shown in Figure 8.12; there are two types of structures: 1:1 layers, where one tetrahedral sheet is bound to an octahedral; and 2:1 layers, where one octahedral sheet is "sandwiched" between two tetrahedral sheets. Figure 8.13 shows a detail of the binding between sheets in kaolinite, a 1:1 phyllosilicate. Note that the octahedral sheet contains OH^-, except for the O atoms shared between the two sheets. In these types of layers, the octahedral sheet face has OH groups exposed, thus being termed *hydroxyl surface*, whereas the tetrahedral one has only O atoms exposed, being known as *siloxane surface*. On the other hand, 2:1 phyllosilicates have only siloxane surfaces exposed. It is important to remark that the bonding between tetrahedral and octahedral sheets forces distortions in both, because the distances between

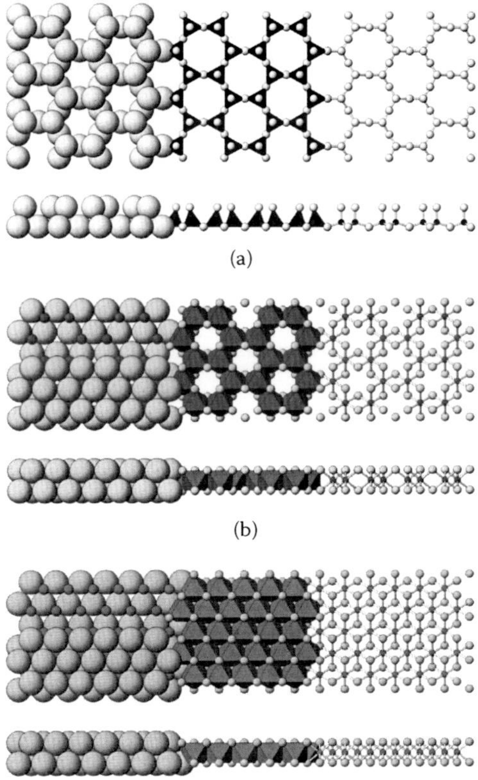

(a)

(b)

(c)

FIGURE 8.11 Schematic view of the layer types constituting phyllosilicate minerals, in different geometrical representations (from left to right): spheres, regular polihedra and ball-and-stick. (a) Tetrahedral layer; (b) dioctahedral layer; (c) trioctahedral layer. The Si atoms and tetrahedra are black; dark gray spheres and bodies represent Al and Mg atoms, respectively; and the light gray spheres represent O atoms. (Reprinted from Schulze, D. G., *Soil Mineralogy with Environmental Applications*, Soil Science Society of America, Madison, WI, 2002, with permission).

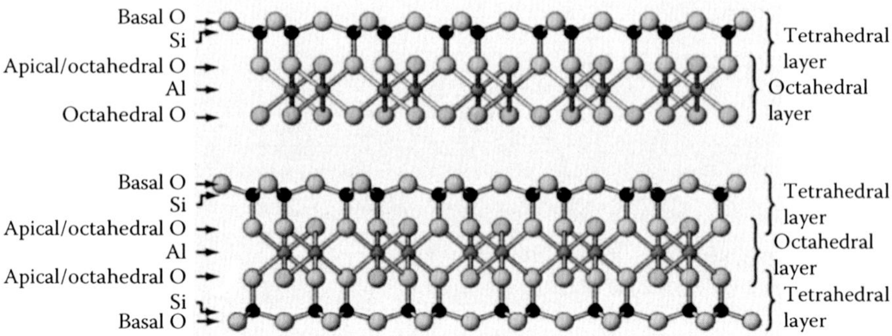

FIGURE 8.12 Schemes of a 1:1 phyllosilicate layer (top) and a 2:1 layer (bottom). The Si atoms are black; dark gray spheres represent Al and Mg atoms and the light gray spheres represent O atoms or OH groups (drawn slightly darker). (Adapted from Schulze, D. G., *Soil Mineralogy with Environmental Applications*, Soil Science Society of America, Madison, WI, 2002, with permission).

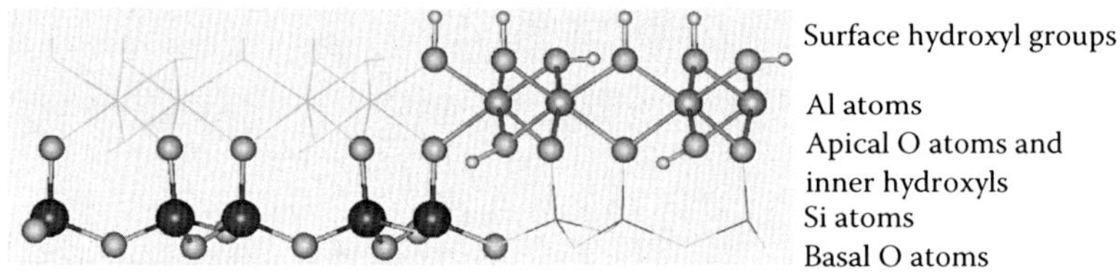

FIGURE 8.13 Detail of a kaolinite layer, a 1:1 phyllosilicate showing the sharing of the tetrahedral apical O atoms with the octahedral sheet. Note that, except for the shared O, all the remaining octahedral anions are OH^- ions. The Si atoms are black; dark gray spheres represent Al atoms; the light gray spheres represent O atoms; and the small light gray spheres represent H atoms. (Reprinted with permission from Tunega, D. et al., 2004, *J. Phys. Chem. B*, 108, 19, 5930–5936. Copyright 2004 American Chemical Society).

apical oxygen atoms in an undistorted tetrahedral sheet are different from those in an octahedral sheet, especially in the dioctahedral case.

Many soil minerals are phyllosilicates, several of them having important ion-binding properties. The crystals are formed by a number of layers, bound by different types of interaction including mainly van der Waals, hydrogen bonding and electrostatic forces.

There are conventions in writing the chemical formulas of phyllosilicates to distinguish the octahedral and tetrahedral sheets; in the most common, one first writes the non-structural, charge-compensating cations (if the mineral has permanent charge), then the cations present in the octahedral sheet, following between parenthesis the cations in the tetrahedral sheet(s); and finally the anionic lattice, which is almost always $O_5(OH)_4$ for 1:1 phyllosilicates and $O_{10}(OH)_2$ in the 2:1 case. Thus, kaolinite, shown in Figure 8.13, having equal amounts of Al(III) and Si(IV), has the formula unit (minimum formula) of $Al_2Si_2O_5(OH)_4$ (the parentheses around Si are omitted because there is no ambiguity); but if, just to say, one of four Si were replaced by Al(III), resulting in a net negative charge of −0.5 per formula unit, it would be written as $K_{0.5}Al_2(Si_{1.5}Al_{0.5})O_5(OH)_4$, assuming that the resulting permanent charge was compensated by K^+ ions. An alternate way is to place both octahedral and tetrahedral cation groups between parentheses, and distinguishing each one by the coordination number in roman numerals as superscript; the above example would be in this convention: $K_{0.5}(Si_{1.5}Al_{0.5})^{IV}(Al_2)^{VI}O_5(OH)_4$. In the following, we will employ, unless otherwise noted, the first convention.

The phyllosilicate layers are usually treated as infinite, but this is not actually so, as the layers have borders, and often also cracks, imperfections, and so on. Thus, the sheet structures are broken at some places, as schematically depicted in Figure 8.14; this breaking gives either "dangling" O atoms or "exposed" Si^{4+}/Al^{3+} centers. In the first case, the oxygen atoms will have negative charge and in the presence of water will take an H^+ ion to form an OH group. In the second case, the exposed center is a Lewis acid site with positive charge, which will react with a water molecule forming again an OH group and releasing an H^+ ion. Border OH groups bound to a Si center are known as "silanol" groups, whereas those bound to an Al center are termed

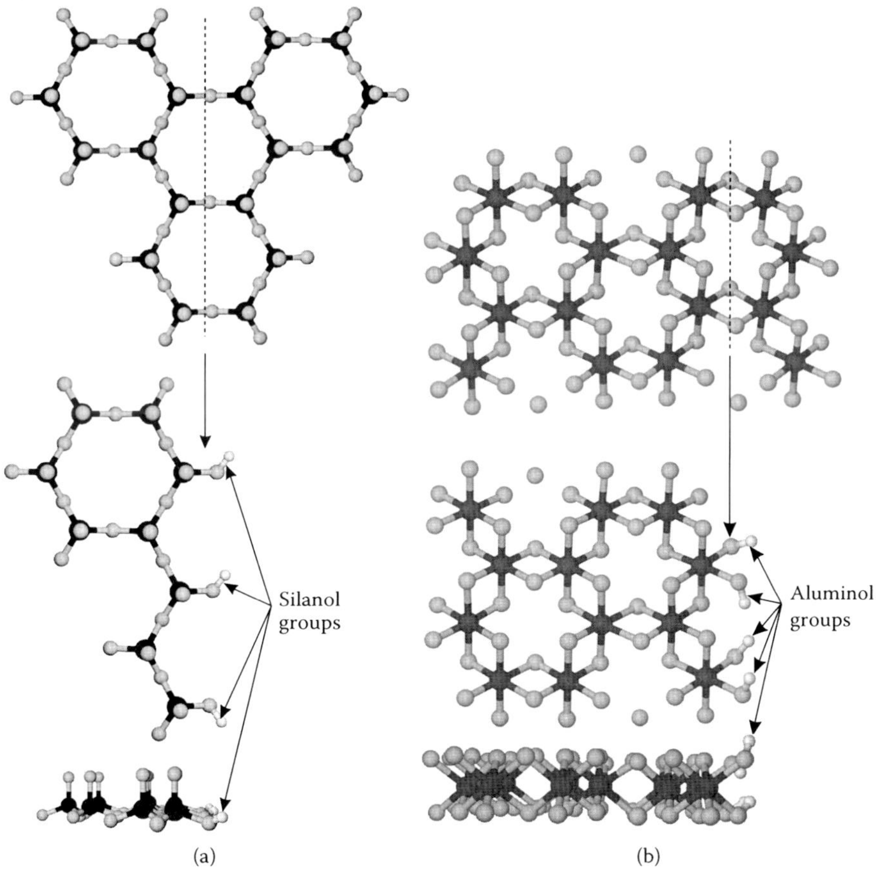

FIGURE 8.14 (a) Top: representation of a fragment of a tetrahedral sheet (as in Figure 8.6) with the dotted line showing the (arbitrary) surface plane. Bottom: the "dangling" basal O atoms form, in the presence of water, OH groups known as silanol, as shown in both top and lateral views. (b) The same for an octahedral sheet, showing the formation of "aluminol" groups. The Si atoms are black; dark gray spheres represent Al atoms, the light gray spheres represent O atoms, and the small light gray spheres represent H atoms.

aluminol groups. Both types of groups are reactive, having acid–base properties, thus acting as pH-dependent charge sites, and can specifically adsorb small chemical species, mainly metal ions. Aluminol groups can either bound or release an H^+ ion to acquire a positive or negative charge, respectively, obviously depending on medium pH. Silanol groups can lose an H^+ to give a negative charge, but are too acidic to become positively charged, at least in the normal pH range of a soil.

In the following sections, the most important, from the point of view of soil properties, silicate minerals groups are described. Table 8.3 collects some important

TABLE 8.3
Selected Structural and Chemical Properties of Silicate Minerals[a]

				Interlayer Distances Determined by XRD after Treatment at Different Temperatures, nm			
				K+ Homoionized		Mg^{2+} Homoionized	
Name	Specific Surface Area ($m^2\ g^{-1}$)	Anion Exchange Capacity ($mol\ kg^{-1}$)	Cation Exchange Capacity ($mol\ kg^{-1}$)	298 K	823 K	298 K	383 K
Kaolinite	7–30	0.01–0.2[b,c]	0.03–0.15	0.715	None[d]	0.715	0.715
Halloysite	10–45	0.08[f]	0.1–0.4	0.715 0.96–1.0[e]			
Pyrophillite	65–80		0.02	0.914	0.914	0.914	0.914
Talc	65–80		0.02	0.926	0.926	0.926	0.926
Muscovite	60–100		0.1–0.4	1.0	1.0	1.0	1.0
Biotite	40–100		0.1–0.4	1.0	1.0	1.0	1.0
Dioctahedral vermiculite	50–800		0.1–1.5	1.4	1.0	1.4	1.4
Trioctahedral vermiculite	600–800	0.04[b,g]	1.0–2.0	1.4	1.0	1.4	1.4
Montmorillonite	600–800	0.2–0.3[b,g]	0.8–1.5	1.4	1.0	1.82	1.42
Chlorite	25–150		0.1–0.4	1.4	1.4	1.4	1.4
Palygorskite	140–190		0.05–0.30	Different structure, main XRD line 1.05 nm			
Sepiolite	230–380		0.20–0.45	Different structure, main XRD line 1.21 nm			
Allophane	100–800		0.05–3.5	Not layered; broad peaks at 0.33 and 0.23 nm			

[a] Data from Olson, Thompson, and Wilson (1999); Sparks (2002); Dixon and Schulze (2002) unless otherwise noted.
[b] Hofmann (1955).
[c] Yukselen and Kaya (2008).
[d] Kaolinite decomposes above 573 K.
[e] Water-interlayered halloysite.
[f] Cope, Fuller, and Willetts (1983).
[g] Abd-Allah, El Hussaini, and Mahdy (2007).

properties that will be discussed in the following sections. It should be noted that some minerals show little changes in composition between different locations/soil types (such as kaolinite), whereas many present a range of compositions in soil samples; for example, the smectites. Furthermore, phyllosilicates are frequently interstratified; that is, a particle may consist of a mixture of layers corresponding to different types (Schulze 2002), as we will seen in Section 8.4.5. More detailed information can be found in the literature (Olson, Thompson, and Wilson 1999; Dixon and Schulze 2002; Sparks 2002 and references therein) and in the World Wide Web (Mindat.org 1993a; Mineral gallery 1995; Minerals.net 1997).

8.2 KAOLIN–SERPENTINE GROUP (1:1) CLAYS

The 1:1 phyllosilicate minerals are the simplest group in this class; this group has also been generically termed the serpentine group (Mindat.org 1993b). They are divided into two subgroups: kaolin and serpentine minerals. The first subgroup, whose main member is kaolinite, has an Al(III) dioctahedral sheet, whereas the second one, including mainly chrysotile and lizardite, has a trioctahedral sheet (White and Dixon 2002). Figure 8.15 shows a scheme of kaolinite layer, other minerals in the group having similar structures.

The kaolinite subgroup all share the formula $Al_2Si_2O_5(OH)_4$ with very little variation in actual soil materials: these minerals show almost no isomorphic substitution; only Fe substituting 0.25–1% of Al atoms has been usually reported; the differences are in arrangement of layers, leading to different crystalline habits. The main member is kaolinite and the other less-common minerals are dickite, nacrite, and halloysite. The 1:1 layers are held together by H bonding, except some halloysites that have water molecules in the interlayer. In general, their surfaces have little reactivity except at the edges (or crystal failures and cracks), where exposed aluminol and silanol groups can react with ions and other small species. The cation-exchange capacity (CEC) is rather low (Table 8.3) due to the almost inexistent substitution. Kaolinite is a mineral commonly found in soils, usually in the form of plate-like crystals with relatively low border-to-surface ratio (see Figure 2.12), thus having generally low reactivity. Halloysite is also found with some frequency, usually with a tubular shape.

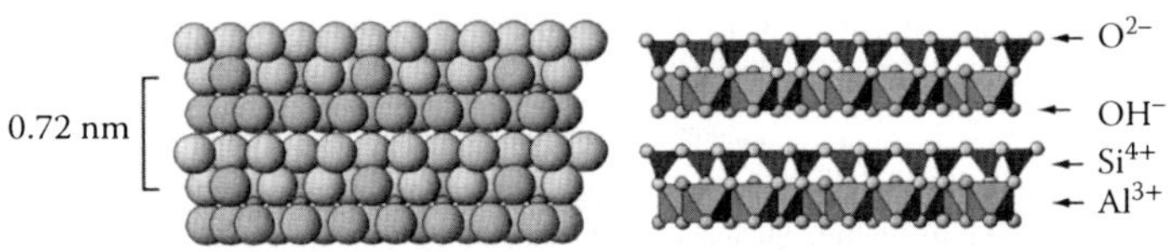

FIGURE 8.15 Lateral view of a scheme of two kaolinite layers, each comprised of a Si tetrahedral sheet and one Al dioctahedral sheet. Serpentines have Mg trioctahedral sheets with some substitution. Left: space-filling spheres representation; right: regular polyhedra representation. The Si atoms are indicated by black tetrahedra, Al atoms are indicated by gray octahedra, O atoms are light gray and OH groups are slightly darker than O atoms. (Adapted from Schulze, D. G., *Soil Mineralogy with Environmental Applications*, Soil Science Society of America, Madison, WI, 2002, with permission).

On the other hand, serpentine minerals have more variability in their composition, the trioctahedral sheet containing cations such as Mg(II), Fe(II), Fe(III), Mn(II), Zn(II), and Al(III), besides other minor components. Also, Al(III) substitution in the tetrahedral sheet is common. The most common member is chrysotile, which has the ideal formula $Mg_3Si_2O_5(OH)_4$ but commonly presents, in low proportion, substitution by Fe(II) and Fe(III) in the octahedral sheet and Al(III) in the tetrahedral one; chrysotile is further subdivided by crystallization in clinochrysotile, orthochrysotile, and parachrysotile, the first one being by far the most abundant. Serpentines usually show platelike crystals as in the kaolin group, but some of them, mainly chrysotile, are typically fibrous. On a side note, for that reason these serpentines are used commercially in asbestos, a material nowadays recognized as carcinogenic, so it has been banned in many countries. Other members of this subgroup include lizardite, having some Al substitution in both layers ($[Mg_{2.8}Al_{0.2}](Si_{1.8}Al_{0.2})O_5(OH)_4$) and the "modulated" minerals (such as antigorite and bementite), where the tetrahedral sheet periodically inverts their apical oxygen atoms sharing them alternately with octahedral sheets above and below the plane of Si atoms. Modulated serpentines are quite rare, or at least have been identified in rare cases. In general, serpentines share the same characteristics of the kaolin subgroup regarding reactivity, albeit they tend to have some isomorphic substitution and thus some permanent charge.

An important difference between the two subgroups is that kaolinites (and related minerals) are secondary minerals, known to form rather easily from weathering of feldspars, and are stable in ambient conditions (which explain its abundance). On the other hand, serpentines are not so stable, found in rocks as metamorphic alteration products, and usually have not been observed to form under low-temperature conditions, thus are more akin to primary minerals.

8.3 LOW PERMANENT CHARGE 2:1 PHYLLOSILICATES: PYROPHYLLITE-TALC GROUP

In the description of 2:1 phyllosilicates, they have been divided into two broad parts; in this section, we will consider those minerals that have a relatively low permanent charge, which does not suffer noticeable changes upon heating or other processes such as ionic exchange. These minerals form the pyrophyllite-talc group.

Pyrophyllites are the closest to "ideal" 2:1 phyllosilicates (Zelazny, Thomas, and Lawrence 2002). Figure 8.16 shows the general structure, comprising two tetrahedral sheets enclosing one octahedral sheet, with no isomorphic substitution, thus having neutral surfaces, with low reactivity; the theoretical formula is $Al_2Si_4O_{10}(OH)_2$. Talc is similar, but with an Mg(II) trioctahedral sheet: $Mg_3Si_4O_{10}(OH)_2$. The mineral layers are held together only by weak van der Waals interactions; because of this, these minerals are among the softest known. In actual field analyses, a small amount of substitution is found, usually about 1–2%, which is compensated by an equivalent amount of cations in the interlayer (see Section 8.4.1). Dioctahedral minerals in this group are pyrophyllite itself and ferripyrophyllite, where the octahedral sheet is built with Fe(III) cations instead of Al; also, some Fe(III) substitution in the tetrahedral layer is observed. Trioctahedral minerals include, besides talc, willemseite,

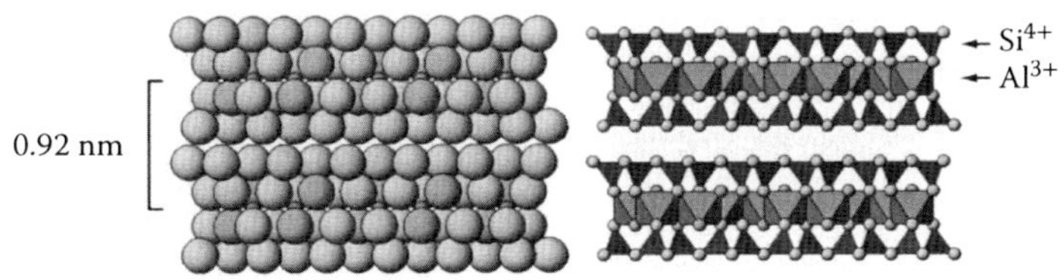

FIGURE 8.16 Lateral view of a scheme of two pyrophyllite layers, comprised of two Si tetrahedral sheets and one Al dioctahedral sheet. Left: space-filling spheres representation; right: regular polyhedra representation. The Si atoms are indicated by black tetrahedra; Al atoms are indicated by gray octahedra, O atoms are light gray and OH groups are slightly darker than O atoms. (Adapted from Schulze, D. G., *Soil Mineralogy with Environmental Applications*, Soil Science Society of America, Madison, WI, 2002, with permission).

where Ni(II) is the main cation of the octahedral sheet, and minnesotaite, with Fe(II) in the octahedral sites. Other more rare cases include a water interlayer, just as in some forms of halloysite.

All these minerals are primary minerals, thus are rather uncommon, and found more frequently in rocks than in soils. They have low reactivity, as observed by their low values of surface area and exchange capacity (Table 8.3), almost only due to edge OH groups, just as the minerals in the kaolin–serpentine group.

8.4 PERMANENTLY CHARGED 2:1 PHYLLOSILICATE MINERALS

A range of secondary 2:1 phyllosilicates have a considerable permanent charge, arising from important isomorphic substitution with heterovalent ions. Most of them have wider interlayer spacing as compared with those discussed in the previous section. The main groups falling in this category are micas, smectites, vermiculites, and chlorites. All of them are characterized by the presence in the interlayer of positively charged cations or octahedral sheets; the interlayer spacing can in many cases undergo changes rather easily due to diverse processes such as ion exchange and heating.

8.4.1 Micas

Micas are 2:1 phyllosilicate minerals, where one out of four Si atoms is replaced by Al; thus, the tetrahedral sheet has a permanent negative charge, which is balanced by dehydrated monovalent ions in the interlayer. This is the maximum substitution allowable according to Lowenstein's rule, which states that two Al(III) tetrahedra could not be adjacent due to repulsion of the negative charges. Originally, an empirical rule (Löwenstein 1954), it has been supported by recent theoretical calculations (Yang, Mora-Fonz, and Catlow 2011); thus, micas are the minerals with the highest permanent charge. The most common cation is K^+, which fits well in the voids of the siloxane surface. These ions constitute the main force that holds the layers together. Figure 8.17 shows a lateral view of a typical mica mineral, muscovite, showing two layers bound by K^+ ions. These minerals are not expandable due to the strong interactions; thus, these ions are difficult to exchange, as it is the ingress of water or other substances in the interlayer. The mica group comprises, as usually considered (Thompson and Ukrainczyk 2002), both the

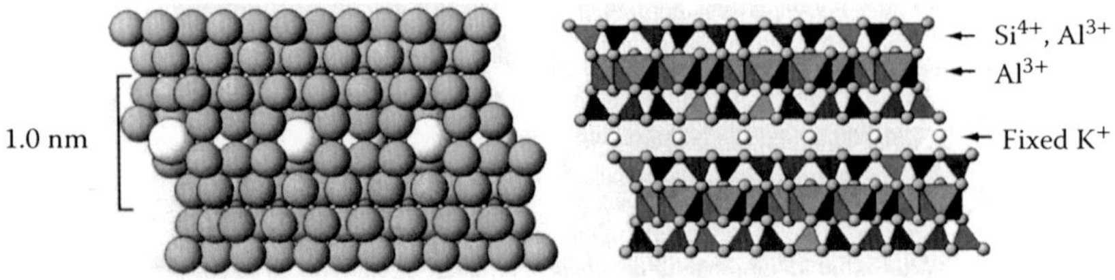

FIGURE 8.17 Lateral view of a scheme of two muscovite layers, comprised of two Si tetrahedral sheets, with isomorphic Al(III) substitution, and one Al dioctahedral sheet. The layers are bound together by K^+ ions, which compensate the negative charge of the tetrahedral sheets. Left: space-filling spheres representation; right: regular polyhedra representation. The Si atoms are indicated by black tetrahedra, Al atoms are indicated by gray octahedra, O atoms are light gray, OH groups are slightly darker than O atoms, and K ions are white. (Adapted from Schulze, D. G., *Soil Mineralogy with Environmental Applications*, Soil Science Society of America, Madison, WI, 2002, with permission).

primary micas such as muscovite or biotite and secondary ones such as illite, albeit some authors place illite as a separate group.

There are both dioctahedral and trioctahedral micas. Dioctahedral micas have the ideal formula $KAl_2(AlSi_3)O_{10}(OH)_2$, thus having a net permanent charge of −1 per formula unit; the most common members being muscovite (primary) and illite and glauconite (secondary). Actual muscovite specimens are close to the ideal, showing only small amounts of structural Fe. Muscovite is found in many soils, mainly in the sand and silt fractions; when muscovite particles are found in the clay fraction, they are typically part of larger size grains, broken in weathering. Illite and glauconite are secondary micas formed by weathering of primary minerals such as feldspars; its formation in soils where large amount of K have been added for fertilization has also been reported (Thompson and Ukrainczyk 2002 and references therein). Illite has the ideal formula $K_{0.75}Al_{1.75}[Mg,Fe(II)]_{0.25}(Al_{0.5}Si_{3.5})O_{10}(OH)_2$, thus having only −0.75 net permanent charge per formula unit, as a result of lesser Al substitution in the tetrahedral sheets but some divalent substitution in the octahedral one. Glauconite has still less Al and more Fe(II) plus some Fe(III), resulting in the average formula $K[Al,Fe(III)]_{1.33}[Mg,Fe(II)]_{0.67}(Si_{3.67}Al_{0.33})O_{10}(OH)_2$; note that the charge of −1 is recovered here due to substitution in the octahedral layer. These secondary minerals (especially glauconite) frequently occur in interstratified specimens with smectites (see Section 8.4.5).

The most frequent trioctahedral mica is biotite, a primary mineral that is found with diverse compositions, represented as $KMg_{0.6\text{-}1.3}Fe(II)_{2.4\text{-}1.2}(AlSi_3)O_{10}(OH)_2$; the dioctahedral layer has range of Mg:Fe(II) ratios. There are (very infrequent) members with Mg only (phlogopite: $Mg_3(AlSi_3)O_{10}(OH)_2$) and Fe only (annite: $Fe_3(AlSi_3)O_{10}(OH)_2$); thus, biotite can be considered as an intermediate of a composition sequence with end members phlogopite and annite, or as a solid solution of these two minerals; recently, the term "biotite" is being adopted as a series name rather than a specific mineral name. The presence of Fe(II) makes biotite less stable than muscovite and more susceptible to weathering; the oxidation of Fe(II) to Fe(III) results in less negative charge in the mineral layers and consequently release of K^+ and weakening of the interlayer attraction. The weathering of biotite may result in several secondary minerals, including kaolinite, vermiculite, gibbsite, and goethite (Thompson and Ukrainczyk 2002).

Micas have in general a low CEC, which is mainly due to near-edge zones depleted of K (Table 8.3); illite and glauconite have been reported to have a slightly higher capacity, probably due to smaller particle size. As with other phyllosilicates discussed so far, chemical reactivity is due mainly to terminal edge groups.

8.4.2 Vermiculites

Vermiculites are minerals similar to micas, but with lower permanent charge, which is compensated by hydrated cations in the interlayer. Figure 8.18 shows schematically the structure of vermiculites. The hydrated cations are easily exchangeable, due to the higher interlayer space; it is possible that the hydrated cations exchange with bulky organic cations expanding the mineral, or drying at high temperature after exchanged by K^+ (or other low hydration energy cations), which causes the collapse of the interlayer distance to typical mica values. Thus, these minerals have a high CEC and are important in regulating plant nutrient dynamics, and also are able to regulate metal contents by fixation of K^+ and NH_4^+ ions, and other metals by hydroxy interlayering, among other reactions (Malla 2002).

Vermiculites, as other phyllosilicates, exist in dioctahedral and trioctahedral variants, but the latter are more abundant and stable, whereas dioctahedrals are not found as macroscopic crystals, only in the clay-size fraction. In vermiculites, by definition, the layer charge ranges from −0.6 to −0.9 per formula unit, resulting from around 15 to 25% of tetrahedral sites substituted by Al(III). Thus, the ideal formula for trioctahedral vermiculite is (assuming divalent hydrated M cations at the interlayer) $(M.nH_2O)_{0.3\text{-}0.45}Mg_3(Al_{0.6\text{-}0.9}Si_{3.1\text{-}3.4})O_{10}(OH)_2$, whereas the dioctahedral has two Al atoms replacing the three Mg. Actual soil trioctahedral vermiculites may have Mg or Ca in the interlayer, often present Fe(III) substitution in the octahedral sheet and sometimes Fe(II) and Ti(III), also in the octahedral sheet. Dioctahedral vermiculites have been found with hydrated Na in the interlayer, both Fe(III) and Ti(III) plus Mn(II)

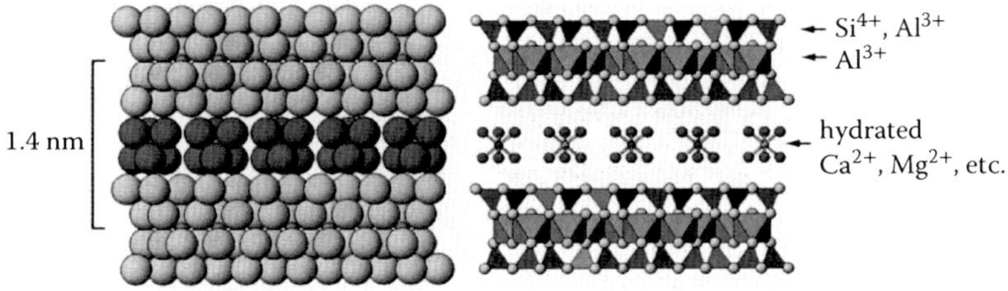

FIGURE 8.18 Lateral view of a scheme of two vermiculite layers, comprised of two Si tetrahedral sheets, with isomorphic Al(III) substitution and one Al dioctahedral sheet. The layers are bound together by hydrated cations, most commonly Ca^{2+} or Mg^{2+}, which compensate the negative charge of the tetrahedral sheets. Left: space-filling spheres representation; right: regular polyhedra representation. The Si atoms are indicated by black tetrahedra, Al atoms are indicated by gray octahedra, O atoms are light gray and OH groups are slightly darker than O atoms; the interlayer hydrated cations are medium gray and the water molecules are dark gray. (Adapted from Schulze, D. G., *Soil Mineralogy with Environmental Applications*, Soil Science Society of America, Madison, WI, 2002, with permission).

in the octahedral sheet, with some OH groups replacing O atoms in the anionic part. The presence of trivalent cations in the trioctahedral sheet lowers the charge, which is compensated by more Al substitution in the tetrahedral sheet; it should be noted that actually trivalent substitution in the octahedral sheet results in an intermediate mineral, between dioctahedral and trioctahedral (Malla 2002). Actual vermiculites consequently have a structure more or less distorted, compared with the ideal one shown in Figure 8.18. An important variant is the presence of hydroxy–Al species in the interlayer: the compensation of positive charge is partially achieved by the presence, along hydrated cations of blocks of positively charged Al–OH layers, where the stoichiometry is below that of aluminum hydroxide, 1:3, and where some sites are occupied by water molecules. In these cases, sometimes the mineral has been classified separately, as *hydroxy-interlayered vermiculite*. Hydroxy interlayering results in changes in the cation-exchange behavior, such as enhanced selectivity of K over Ca. Vermiculites can also incorporate water in the interlayer due to two mechanisms: one is the addition of up to two molecular layers (about 1.5 nm), a process akin to water crystallization in salts, and the other is the osmotic swelling, where the presence of certain ions in the interlayer (such as Li^+ or some organic cations) cause the incorporation of large quantities of water, similar to smectites (see Section 8.4.3).

As already discussed, the most important property of vermiculites is their relatively high permanent charge so as to acquire a considerable amount of cations to compensate, leading to an important CEC; for clay-size vermiculites, it is about 1.3 to 2.1 mol kg^{-1} (130–210 cmol kg^{-1}) (Table 8.3). Besides that, there are also OH edge groups that have acid–base behavior and contribute to the total particle charge; at high pH, these sites are negatively charged, thus the total CEC is higher than the permanent charge. At low pH, the OH edge groups are positive, thus giving some anion-exchange capacity to the mineral. These hydroxyl groups can also react with anions and other species to produce specific adsorption to some extent.

8.4.3 Smectites

Smectites are expansible clay minerals with a layer charge, by definition, between −0.2 and −0.6 per formula unit; because of the relatively lower charge, the structure expands and shrinks quite easily (compared with vermiculites) depending on the medium conditions. The interlayer contains, besides hydrated cations as in vermiculites, water molecules, relatively more free than those in the cation hydration layers, and thus the interlayer distance is both higher than that in vermiculites and readily variable; that is, smectites are easily expansible or *swelling minerals* with a wide interlayer distance (Figure 8.19). These minerals are highly sorbent, having important agricultural, engineering, and environmental applications. As other phyllosilicates, there are dioctahedral and trioctahedral variants. Trioctahedral smectites include montmorillonite (with substitution in the octahedral sheet) with the ideal formula (M being a monovalent cation) $M_{0.2\text{-}0.6}Al_{1.8\text{-}1.4}Mg_{0.2\text{-}0.6}Si_4O_{10}(OH)_2$; beidelite, substituted in the tetrahedral sheets: $M_{0.2\text{-}0.6}Al_2(Si_{3.8\text{-}3.4}Al_{0.2\text{-}0.6})O_{10}(OH)_2$; and nontronite: $M_{0.2\text{-}0.6}Fe(III)_2(Si_{3.8\text{-}3.4}Al_{0.2\text{-}0.6})O_{10}(OH)_2$. The ideal dioctahedral smectites include, for example, hectorite: $M_{0.2\text{-}0.6}Mg_{2.8\text{-}2.4}Li_{0.2\text{-}0.6}Si_4O_{10}(OH)_2$; saponite: $M_{0.2\text{-}0.6}Mg_3(Si_{3.8\text{-}3.4}Al_{0.2\text{-}0.6})O_{10}(OH)_2$; and sauconite: $M_{0.2\text{-}0.6}Zn_3(Si_{3.8\text{-}3.4}Al_{0.2\text{-}0.6})O_{10}(OH)_2$. Most actual soil smectites have different, more complex composition including, as in the

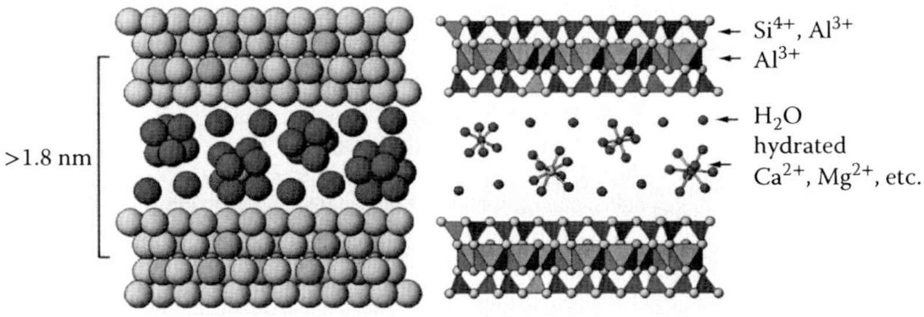

FIGURE 8.19 Lateral view of a scheme of two smectite layers, comprised of two Si tetrahedral sheets, with isomorphic Al(III) substitution, and one Al dioctahedral sheet. The layers are bound together by hydrated cations, most commonly Ca^{2+} or Mg^{2+}, which compensate the negative charge of the tetrahedral sheets; the interlayer is wider than vermiculites (Figure 8.18) containing more water molecules. Left: space-filling spheres representation; right: regular polyhedra representation. The Si atoms are indicated by black tetrahedra, Al atoms are indicated by gray octahedra, O atoms are light gray, and OH groups are slightly darker than O atoms; the interlayer hydrated cations are medium gray and the water molecules dark gray. (Adapted from Schulze, D. G., *Soil Mineralogy with Environmental Applications*, Soil Science Society of America, Madison, WI, 2002, with permission).

case of vermiculites, partially dioctahedral–trioctahedral sheets, and with substitution in both types of sheets. Nevertheless, many of them resemble beidelite and montmorillonite. Smectite minerals, if highly expanded, tend to dissociate in small particles (interlayer distances growing to the point of separation), thus they are mostly found in the fine clay fraction of soils and sediments, and consequently show high surface area, compared with other phyllosilicates (Table 8.3). Soil smectites are major contributors to soil CEC (as vermiculites); because of this, smectites can act as a reservoir for cationic macro and micronutrients. The easy swell-and-shrink behavior also has consequences in soil mechanics; soil dominated by smectites are known as expansible soils, and their movement is capable of great damage to man-built structures, producing landslides or soil creep (slow downward movement of soil horizons).

There are several possible origins for smectites in soils: inherited from parent materials, formed by weathering of other 2:1 phyllosilicates (especially micas), or precipitated from soil solutions after dissolution of other primary minerals.

8.4.4 Chlorites

Chlorites are 2:1 phyllosilicates with isomorphic substitution; the negative charge is compensated by a hydroxide sheet in the interlayer as seen in Figure 8.20 (Kohut and Warren 2002). This sheet has an approximately octahedral structure like gibbsite (e.g., $Al(OH)_3$); however, it is not a pure hydroxide that would be electrically neutral, but has some OH^- missing or replaced by water molecules, thus having a net positive charge (Velde and Meunier 2008). Because of the presence of this sheet, chlorites are often classified separately as 2:1:1 phyllosilicates (Olson, Thompson, and Wilson 1999). The hydroxide sheet usually contains Al(III) or Mg(II), but other cations such as Fe(III) can be found as well; it is sometimes termed gibbsite sheet or brucite sheet. This sheet also makes chlorites unstable in soil environments, weathering to

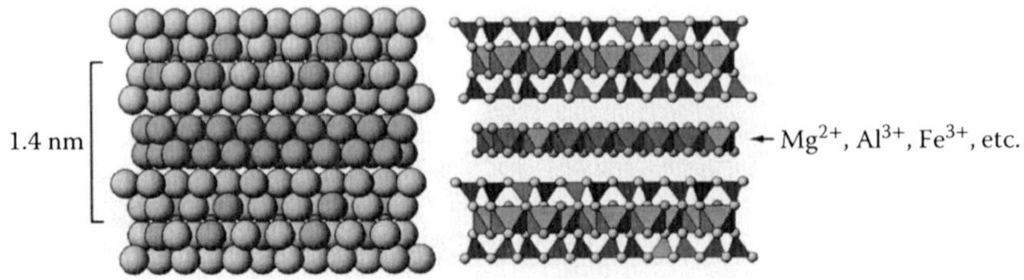

FIGURE 8.20 Lateral view of a scheme of two chlorite layers, comprised of two Si tetrahedral sheets, with isomorphic Al(III) substitution, and one Al dioctahedral sheet. The layers are bound together by a metal hydroxide sheet, usually of Al^{3+} or Mg^{2+}, which is not stoichiometric, having OH^- deficiency and thus compensating the negative charge of the tetrahedral sheets. Left: space-filling spheres representation; right: regular polyhedra representation. The Si atoms are indicated by black tetrahedra, Al atoms are indicated by gray octahedra, O atoms are light gray, and OH groups are slightly darker than O atoms; the interlayer hydroxide sheet is darker than the regular octahedral sheet. (Adapted from Schulze, D. G., *Soil Mineralogy with Environmental Applications*, Soil Science Society of America, Madison, WI, 2002, with permission).

smectites or vermiculites and releasing the interlayer cations in the environment; this may result in the release of significant amounts of Mg^{2+} in the soil.

The tetrahedral sheet of chlorites contains isomorphic substitution by Al(III), and sometimes Fe(III) or Cr(III) is found (Kohut and Warren 2002). The octahedral sheet normally contains Mg(II), Al(III), Fe(II), and Fe(III), but some substitution by Cr(III), Ni(II), Mn(II), or other metals occurs as well. Chlorites, in contrast with vermiculites and smectites, are hardly expandable, because the hydroxide cations are not easily exchangeable, nor is water easily incorporated to the intermediate octahedral sheet.

A general formula for chlorites can be written (Kohut and Warren 2002) in the roman numerals convention as $(M(II,III))_n{}^{VI}(Si_{4-x}Al_x)^{IV}O_{10}(OH)_2(M'(II,III))_m{}^{VI}(OH)_6$, where n and m may take the values 2 or 3 for dioctahedral or trioctahedral sheets, respectively; x ranges usually between 0.8 and 1.6; and M and M′ represent di- and trivalent metal cations, respectively. Note that the interlayer hydroxide sheet has been separated at the end, because it is often difficult to distinguish the cations between the two octahedral sheets, usually they are merged into the main formula. Because of the additional sheet, there are four variants for chlorites: full trioctahedral, full dioctahedral, di-trioctahedral (dioctahedral layers and trioctahedral intermediate sheet), and tri-dioctahedral (trioctahedral layers and trioctahedral intermediate sheet). The chlorites found in most soils are full trioctahedral. The trioctahedral sheets show replacement of Mg by Fe(II), forming a solid solution series without any changes in the structure; on the other hand, dioctahedral Al is replaced by Mg. Basic end member of trioctahedral chlorites is clinochlore, $Mg_5Al(Si_3Al)O_{10}(OH)_8$ (ideal formula); other common end members are chamosite (with Fe(II) instead of Mg), penninite (Mn instead of Mg), and nimite (Ni instead of Mg). Actual samples have intermediate compositions; the sample is named from the dominant octahedral cation. There are as well other naming schemes for chlorites (Bailey 1975).

Being almost non-expanding phyllosilicates, chlorites have relatively low surface area and CEC (Table 8.3), mostly due to edge OH groups, which increase with weathering (Kohut and Warren 2002). They have been shown to strongly adsorb several cations such as Ba(II).

8.4.5 Interstratified Phyllosilicates

Often, more than one kind of silicate layer is found in actual soil materials. In such cases, typically 2:1 layers with different charges are stacked and bound together either by dehydrated monovalent cations (typically K^+) or hydrated divalent metals, usually Ca^{2+}, depending on layer charges or the previous processes. Figure 8.21 shows a schematic representation of an interstratified illite-smectite mineral (Thompson and Ukrainczyk 2002); this is an example of an *irregular or random interstratification*; alternately, regular, predictable sequences can be found, usually alternating layers of two types, as in hydrobiotite, consisting of an alternating sequence of biotite and vermiculite layers. Similar to this, some systems present commonly *regular interstratification*, whereas others, such as illite-smectite, show usually irregular behavior. Interstratified minerals have been described in terms of crystallites composed of mixed layers (MacEwan 1956), and more recently as stacking of very small "pure" or "fundamental" particles of separate minerals (i.e., separate illite and smectite particles in Figure 8.21); these particles can be as thin as a couple of 2:1 layers bound together by K^+ ions—for example "illite, 2 nm" in Figure 8.21 (Nadeau et al. 1984). There are several arguments in favor of the interparticle model (Thompson and Ukrainczyk

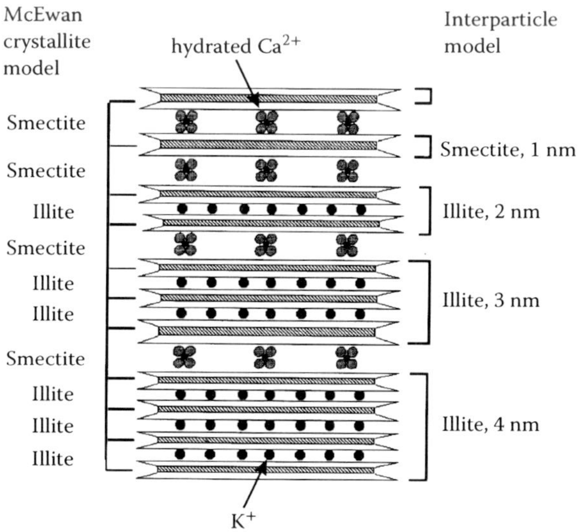

FIGURE 8.21 Schematic view of an interstratified illite–smectite mineral. At the left is the description in terms of the McEwan model and at right in the view of the interparticle model. (Adapted from Thompson, M.L. and Ukrainczyk, L., *Soil Mineralogy with Environmental Applications*, Soil Science Society of America, Madison, WI, 2002, with permission).

2002); that is, that the interstratified samples are simply aggregations of very small individual particles. More recently, theoretical arguments and some experimental evidence support precipitation of interstratified particles (both regular and irregular) directly from solution (Wang and Xu 2006). On the other hand, regular interstratified minerals such as hydrobiotite have been described as a result of weathering of biotite.

8.5 OTHER ALUMINOSILICATE MINERALS

8.5.1 Allophane and Imogolite

Allophane and imogolite are poorly crystalline aluminosilicate materials having high specific surface area, small particle size, and unique morphology, and may have permanent charge; these features lead to physicochemical properties that can noticeably affect the soil structure and properties, reacting with inorganic and organic species. They are actually transient phases in weathering, mainly of volcanic glass, albeit they have been associated with other environments where Si and Al are present in solution in enough concentration so as to precipitate (Harsh, Chorover, and Nizeyimana 2002). Often, they are termed as SRO (for short range order) aluminosilicates. These minerals are important components in soils derived from recently deposited volcanic materials.

The term allophane actually refers to a wide range of clay-size minerals with short range order containing silica, alumina, and water in several chemical structures. Its composition can be broadly described as $x\mathrm{SiO_2{\cdot}Al_2O_3{\cdot}}y\mathrm{H_2O}$, with $0.8 \leq x \leq 2$ and $y \geq 2.5$. Allophane has short- to medium-range crystalline order, as detected by X-ray diffraction and electron micrographs. *Short-range order* means that ordering according to some geometric structure (starting from an atom in the solid) spans only to a few near neighbors. It is comprised of small particles that aggregate forming highly porous flocs 3.5 to 5.5 nm in diameter, able to retain a high amount of hydration water. This behavior imparts low bulk densities to soils dominated by allophane. Also, very high specific surface areas result (Table 8.3).

Imogolite is a mineral with better defined composition, having the unit formula $\mathrm{Al_2SiO_3(OH)_4}$. It displays short-, medium-, and long-range order. Al is present only

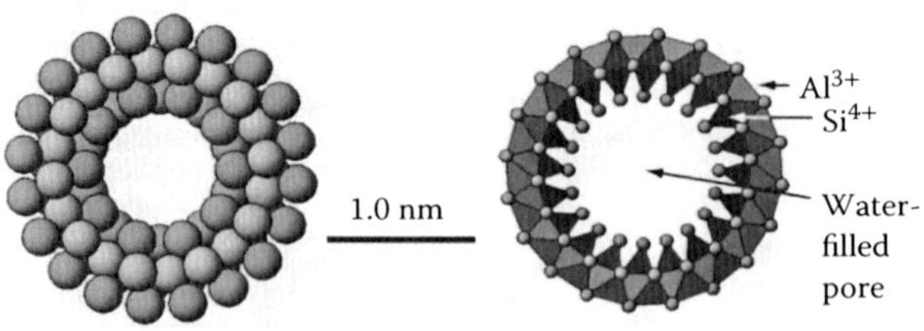

FIGURE 8.22 Representation of a cross section of an imogolite tube. The Si atoms are indicated by black tetrahedra, Al atoms are indicated by gray octahedra, O atoms are light gray, and OH groups are slightly darker than O atoms. Note that the tetrahedral "sheet" is highly distorted due to the tubular configuration, some "basal" O are shared with the octahedra and there are OH groups at the apical positions. (Adapted from Schulze, D. G., *Soil Mineralogy with Environmental Applications*, Soil Science Society of America, Madison, WI, 2002, with permission).

in octahedral configuration, whereas Si is in partially condensed tetrahedra. The structure is tubular, as shown in Figure 8.22, with an outer diameter of 2.3 to 2.7 nm; several tubes are usually arranged in bundles of 10 to 30 nm diameter and a few micrometer long.

These minerals are highly reactive and can adsorb specifically inorganic (especially heavy metals such as Co, Cu, Cd, Ni, Pb, etc.) and organic ions and compounds. This is due to the high surface area and the presence of aluminol groups at the outer surface and silanol groups at the inner surface. These groups also lead to a pH-dependent surface charge; permanent charge may be present by substitution of Si by Al, for Si-rich allophanes (because tetrahedral structure dominates). For Al-rich allophanes (known as proto-imogolite) and imogolites, the permanent charge is negligible.

8.5.2 Palygorskite and Sepiolite

These secondary minerals are considered phyllosilicates, even when their structure does not consist of true tetrahedral and octahedral sheets (Figure 8.23): the tetrahedral "sheet" consists of tetrahedra alternately pointing "up" and "down" (as seen in Figure 8.23, along the crystallographic *c* axis) with only the basal O atoms lying in the same plane (Singer 2002). The apical O atoms of adjacent Si sheets are facing each other and bound by octahedral sheet fragments as seen in the figure. This arrangement is continued back and forth from the plane of the figure, leaving channels of about 0.6 nm × 0.4 nm in palygorskite and 0.4 nm × 1 nm in sepiolite. The crystals have consequently a fibrous structure. In Figure 8.23, a palygorskite fragment is shown; the structure of sepiolite is very similar, the only difference is the width of the "ribbons," being one Si tetrahedron.

Palygorskite has the ideal composition $Mg_5Si_8O_{20}(OH)_2{\cdot}8H_2O$, but isomorphic substitution is almost always present; thus, in actual samples, there are considerable variations. Tetrahedral composition is generally of the form $Si_{8-x}Al_x$ with x ranging between 0.12 and 0.66. Octahedral cations replacing Mg are commonly Al and Fe, with an average of four cations in total, thus this mineral can be considered dioctahedral. There is some segregation of the octahedral cations, Al (smaller) tends to occupy internal positions, whereas Fe(II) and Mg tend to be found at edge positions (Singer 2002).

Sepiolite has the ideal formula $Mg_8Si_{12}O_{30}(OH)_4{\cdot}12H_2O$; substitution in both tetrahedral and octahedral positions is usually minor, and Mg fills not less than 90% of these positions; thus, this material is clearly trioctahedral. Other elements have been sometimes reported in the octahedral centers, such as Fe(II/III), Ni(II), or Mn(II).

Due to the particular structure, these minerals show a very high specific surface area (Table 8.3), and in fact the experimentally measured area is considerably smaller than the theoretical one, probably because the internal channels are not fully accessible to N_2 in the Brunauer, Emmet, and Teller (BET) method (see Section 7.6.4.2). The CEC (Table 8.3), on the other hand, is relatively low, given the specific surface area, due to low permanent charge. Palygorskite and sepiolite are usually found in arid and semiarid regions, and until recently they have not been studied thoroughly; little is known about the impact of these minerals on soil properties. They can be inherited from parent material or formed by weathering of primary aluminosilicates (Singer 2002). These materials have a high capacity for water sorption, and concentrated

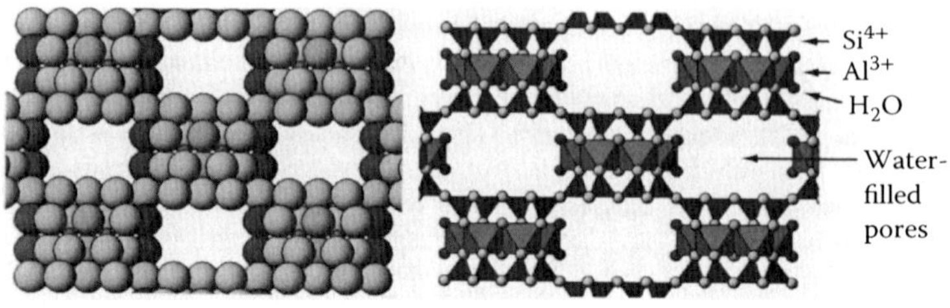

FIGURE 8.23 Structural model of a palygorskite crystal. The Si atoms are indicated by black tetrahedra, Al atoms are indicated by gray octahedra, O atoms are light gray, OH groups are slightly darker than O atoms, and coordinated water molecules are dark gray. Note that the basal O of the tetrahedra are all in the same plane, but the Si tetrahedral are alternately up and down, and the Al octahedral sheet is broken in segments, leaving channels. (Adapted from Schulze, D.G., *Soil Mineralogy with Environmental Applications*, Soil Science Society of America, Madison, WI, 2002, with permission).

clay suspensions are very stable. Surface reactivity is achieved mainly by silanol groups and Mg-coordinated water molecules forming hydrogen bonds, and small polar organic molecules are known to be sorbed in these minerals (Singer 2002).

8.5.3 Zeolites

Zeolites are minerals formed by Si tetrahedra in complex spatial arrangements, leaving pores or channels wider than those found in sepiolite. Some Si atoms are substituted by Al, thus the framework has negative charge. Figure 8.24 shows an example of a zeolite. The inner spaces are normally filled by water and exchangeable cations that compensate the negative charge. Due to the rigid structure, zeolites can undergo processes such as cation exchange, reversible dehydration, and hydration without suffering structural changes. This property leads to interesting applications, such as serving as ionic "sieves," because small cations can enter the pores and be exchanged, whereas larger ones are not allowed due to the fixed pore size (Boettinger and Ming 2002).

Zeolites are tectosilicates, this structure being responsible for the rigid mechanical characteristics. The void volume due to pores ranges between about 18 and 47% of total volume; this indicates that there are a number of different structures with a high range of pore sizes, normally filled by cations and water, which is often termed "zeolitic water."

The general formula for zeolites can be written as M_x^+, $M'^{2+}_y[Al_{x+2y}Si_{n-(x-2y)}O_{2n}]\cdot mH_2O$, where brackets enclose the framework structure. M and M′ are exchangeable cations and mH_2O is the zeolitic water. The atomic ratio Si/Al ranges between 5.0 and nearly 1 (Boettinger and Ming 2002), giving a theoretical CEC between about 2.3 and 5.7 mol kg^{-1}, based in the total charge; in practice, these values are not always reached, depending on the cations involved, indicating that some part of the pores are not accessible to all cations.

Zeolites are important components of sedimentary rocks, mainly of volcanic origin; also they are found in the cavities of igneous rocks such as basalt. Zeolite formation is in general not completely understood. In alkaline lake systems, zeolites precipitate at high pH, where Si and Al are most soluble; this process is dependent on

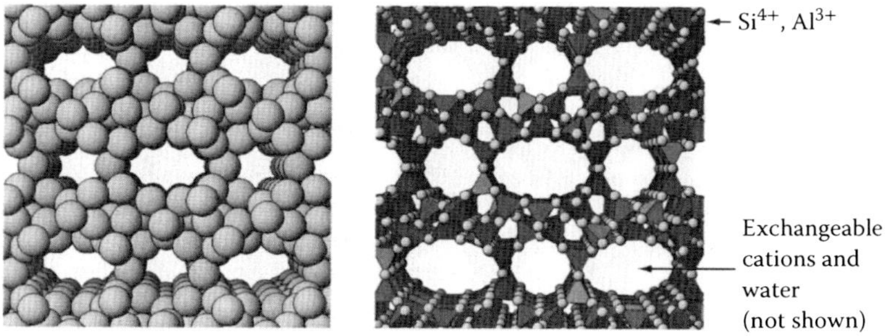

FIGURE 8.24 Structural model of a clinoptilolite crystal (a zeolite). The Si atoms are indicated by dark gray tetrahedra, Al atoms are indicated by gray tetrahedra, and O atoms are light gray. (Adapted from Schulze, D.G., *Soil Mineralogy with Environmental Applications*, Soil Science Society of America, Madison, WI, 2002, with permission).

pH, Si/Al ratio, salinity, cations present, and dissolution rate of volcanic glass. It has been difficult to obtain some zeolites under laboratory conditions.

8.6 CHARACTERIZATION OF SILICATE MINERALS

8.6.1 Identification and Crystalline Structure

As presented in Chapter 7, there are several types of methods that can be employed in the characterization of mineral substances (Amonette 2002). As with all minerals, x-ray diffraction (XRD) (Section 7.3.1) is the prime identification method, through coincidence of the lattice distances for the main diffractogram lines. In many cases, this leads directly to a clear assignment and sample identification. However, as described before, phyllosilicates present special problems due to the presence of interlayer cations that may be hydrated or not, variable interlayer space (smectites), and others. Often, the same mineral can have a mixture of interlayer cations, leading to broadened XRD peaks. Thus, a procedure has been devised to differentiate them properly, which is summarized in Figure 8.25. Briefly, the sample is homoionized with Mg^{2+} through ion exchange and equilibrated in an environment of controlled humidity. Next, it is placed in contact with glycerol. In the following step, the sample is now homoionized with K^+, and finally several heating steps reaching a final temperature of 550°C are conducted. The resulting effects on the XRD diagram are shown in Figure 8.25, and the effect on the minerals is schematized in Figure 8.26. Details of the procedure can be found in the literature (Whittig and Allardice 2010).

Besides XRD, other important studies are elemental analysis, either by chemical or physical methods, such as neutron activation analysis (NAA), x-ray fluorescence (XRF), or x-ray energy dispersive spectroscopy (X-EDS), for example (see Sections 7.6.1, 7.3.3, and 7.5.2, respectively); the advantage of these methods is that they are non destructive, as oppossed to wet chemical analysis. Additionally, IR spectroscopy can bring useful complementary information. Sometimes, the chemical composition is required along XRD analysis to fully identify a mineral. Also, thermal analysis (Section 7.6.5) is a useful tool in the qualitative and, sometimes, quantitative determination of clay minerals.

8.6.2 Ion Exchange

Other important characterization studies are ion-exchange capacity, especially cationic, and specific surface measurement. CEC is an important parameter in clay characterization and soil behavior, because it affects nutrient availability, among other soil properties. Some clay minerals (smectites and vermiculites) have a high

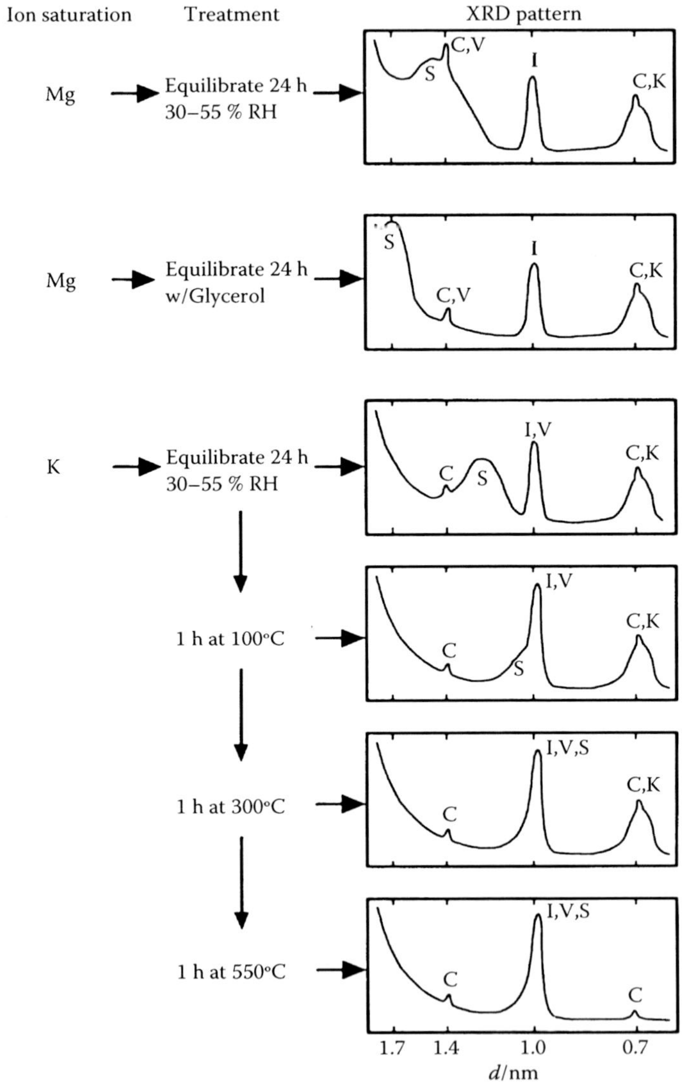

FIGURE 8.25 Simplified scheme of the procedure for XRD identification of clay minerals, showing the XRD patterns resulting for a hypothetical mixture of smectite (S), chlorite (C), vermiculite (V), illite (I), and kaolinite (K). (Reprinted from Amonette, J.E., *Soil Mineralogy with Environmental Applications*, Soil Science Society of America, Madison, WI, 2002, with permission).

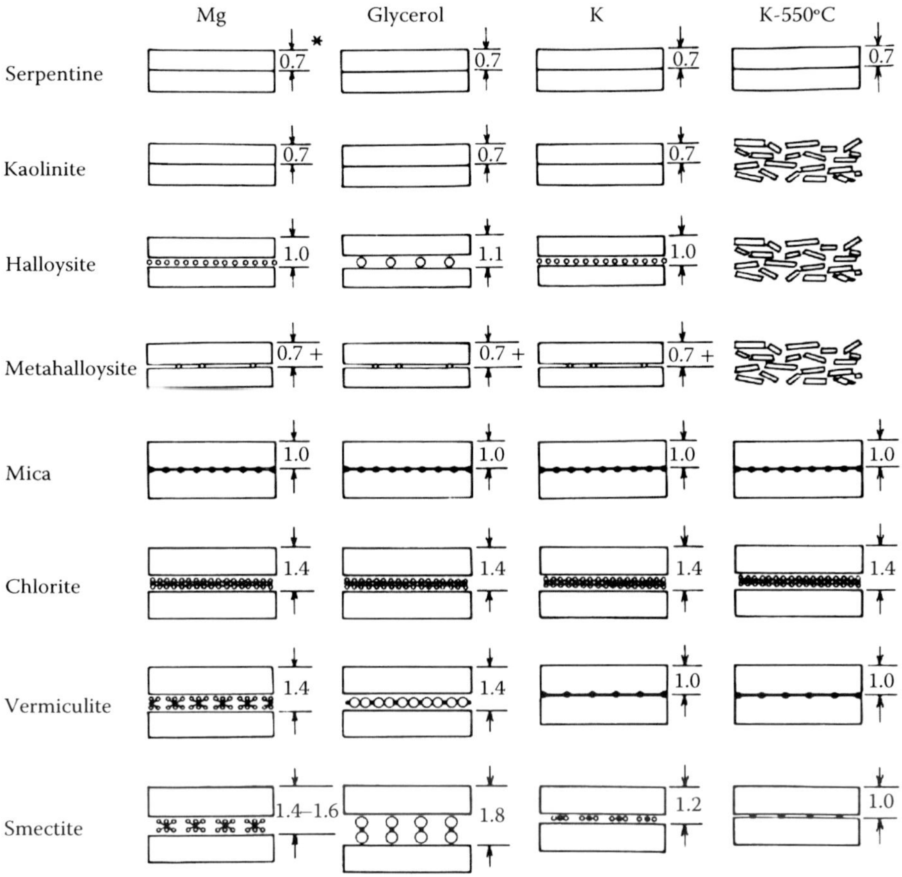

FIGURE 8.26 Schematic representation of the behavior of different phyllosilicates to the treatments indicated in Figure 8.25. The symbol (*) indicates approximate spacings in nanometer. (Adapted from Whittig, L.D. and Allardice, W.R., *Physical and Mineralogical Methods*, Soil Science Society of America, Madison, WI, 2010, with permission).

CEC, as exemplified in Table 8.3, due to permanent negative charge. On the other hand, anion-exchange capacity is generally low and present only at low pH values. The determination of CEC has been widely described in the literature (Sumner and Miller 1996; Zelazny, He, and Vanwormhoudt 1996 and references therein).

8.6.3 Specific Surface Area

Specific surface area determination is most commonly conducted by positive adsorption studies (Section 7.6.4.2), and here the election of the probe molecule is very important. The most "traditional" and widely used method is N_2 adsorption at 77 K, using the BET isotherm to evaluate monolayer coverage (Section 7.6.4.2); sometimes other inert gases, such as argon, are employed in the same conditions (Sposito 1984). In soil characterization, other substances are also used in the BET method at, or near to, room temperature (see Table 7.7); use of water in BET method was proposed by Orchiston (1953, 1954), and several polar organic molecules were also proposed (as discussed in Section 7.6.4.2), ethylene glycol monoethyl ether (EGME) being the most commonly employed (Carter, Mortland, and Kemper 2010). On the other hand, adsorption from solution is often employed, most usually using methylene blue, but several other substances have also been proposed. Table 8.4 presents results obtained with some clay minerals using N_2 alone, water, ethylene glycol, and EGME in gas phase, and methylene blue from solution. Table 8.4 lists some results for phyllosilicate mineral obtained with several methods. Even when there is no direct comparison with all methods applied to the same sample, it is seen that the area obtained is dependent on the probe molecule and is strongly dependent for expandable clay minerals such as montmorillonite. It is generally considered that N_2, being a non polar molecule, interacts only with low polarity surfaces; in particular, it is not able to enter the interlayer regions of clay minerals, thus the nitrogen area, A_{N_2}, is often considered as the *external area* of phyllosilicates. Polar molecules such as water or EGME are able to ingress in the interlayer region, thus the area obtained with EGME, A_{EGME} (and water or other polar substances), is considerably higher for vermiculites and smectites, whereas it is of the same order of magnitude for kaolinite. The differences in areas determined with polar molecules for different clay mineral groups suggest that the mineral properties determine to a greater extent the amount which can be sorbed in the interlayer. In the case of water, it was proposed (Sposito 1984) that water molecules covered the external

TABLE 8.4
Specific Surface Areas (m^2 g^{-1}) of Some Clay Minerals Determined by Adsorption of Different Probe Molecules[a]

Mineral	A_{N2}	A_{H2O}	A_{EG}	A_{EGME}	A_{MB}
Kaolinite (Peerless)[b]	18–19	12–23			
Kaolinite (Warren)[c]	21			28	
Kaolinite (Florida)[d]	33				35
Illite (Fifthian, Illinois)[c]	67			133	
Illite (Morris, Illinois)[e]			209	193	
Illite (Oswego)[d]	68				78
Ca-montmorillonite (Bidahochi)[c]	76			751	

[a] EG stands for ethylene glycol, EGME for ethylene glycol monoethyl ether, and MB for methylene blue.
[b] Keenan, Mooney, and Wood (1951).
[c] Chiou, Rutherford, and Manes (1993).
[d] Hang and Brindley (1970).
[e] Carter, Heilman, and Gonzalez (1965).
[f] Sposito (1984).
[g] Mooney, Keenan, and Wood (1952).

area and also the interlayer region with one monolayer; according to this view, the internal area would be *twice* the difference between A_{H_2O} and A_{N_2}, because every water molecule in the interlayer would be in contact with two "internal" surfaces (phyllosilicate layers). However, it was already found (Quirk 1955) that the amount of water sorbed by montmorillonite is strongly dependent on the interlayer cation, with a twofold increase in going from Cs^+ to Mg^{2+}. X-ray analysis shows that there are different numbers of cation-associated water molecules depending on the cation. This is true also for ethylene glycol (Hendricks and Dyal 1952), and likely also true for EGME. The assumption of a single molecular layer in the interlayer space, for polar molecules, seems to be valid only for vermiculites. On the other hand, ionic species adsorbing from solution, such as methylene blue, may exchange all the interlayer cations giving an interlayer adsorption driven by the permanent charge, but its ability to adsorb on the external surface is uncertain. One may ask which is the "true" or "absolute" specific surface; the answer is that there is no such thing; the "area" of interest for a particular phenomenon or process depends on the chemical species involved, and on how these species interact with the particle. In a general way, the BET-N_2 method is widely employed as a standard, which may not be very representative for 2:1 phyllosilicates, but nevertheless is a comparative parameter almost universally employed; in soil science, the BET-EGME and solution-MB methods are often employed, which again may not give exact results but are more representative than the nitrogen one, and have comparative significance as well. The specific surface areas shown in Table 8.3 are mostly obtained with the EGME method.

8.7 ION BINDING TO SILICATE-BASED MINERALS

The main ways for ions binding to minerals are basically two: electrostatic (ion exchange) and specific (surface complexation). In silicate minerals, the first is the most important and significant for soil properties. However, not all silicates, as described in the previous sections, have a noticeable permanent negative charge, but those (such as smectites and vermiculites) are capable of retaining greater amounts of cations. On the other hand, minerals with permanent positive charge are very rare.

Silanol and aluminol groups are present at layer edges and fractures, defects, and so on (whenever a normal layer bond is broken), which can be ionized depending on pH, and so can participate in ion exchange, but also bind ions more specifically, mainly transition metal cations. The type of interaction between a clay mineral and a metal cation may depend on medium conditions, such as ionic strength and pH, and can be revealed by XAS techniques (Section 7.3.2). In an equilibrium and XAFS study of Pb(II) adsorption on montmorillonite (Strawn and Sparks 1999), it was found that at low values of ionic strength and pH, adsorption is primarily outer-sphere and most likely occurs on the basal planes existing in the interlayer, regions of the montmorillonite, and the results suggest that there exists a layer of water molecules between the complexed Pb atoms and the mineral surface; as pH and ionic strength increase, inner-sphere adsorption becomes more important, being most likely that in the samples equilibrated at high ionic strength, and/or high pH, adsorption is occurring on the edges of the montmorillonite (where silanol and aluminol groups are present) and includes the formation of Pb polymers. At intermediate values, it is suggested that adsorbed Pb can exist as a mixture of inner- and outer-sphere complexes. In another study of Cu(II) adsorption on vermiculite, using polarized XANES and EXAFS spectroscopy (Furnare, Vailionis, and Strawn 2005), it was concluded that Cu forms inner-sphere and outer-sphere complexes on clay minerals, but does not form multinuclear surface precipitates that have been observed for Co, Zn, and Ni. The results obtained with polarized radiation suggest that Cu adsorption on vermiculite is not occurring in the interlayer, but rather Cu is adsorbing onto the edges of the mineral crystals. In the case of Cd and Pb adsorption on kaolinite (Gräfe, Singh, and Balasubramanian 2007), EXAFS experiments reveal that cadmium dominantly (>75%) formed outer-sphere complexes on kaolinite, whereas in contrast, Cd is adsorbed as an inner-sphere complex on gibbsite, suggesting that the Si tetrahedral sheet hindered Cd sorption to the Al octahedral sheet on kaolinite. Lead-formed polymeric complexes, which bonded to kaolinite via edge sharing with surface Al octahedra. Two distinct Pb–Al edge-sharing distances were observed on gibbsite, as opposed to one on kaolinite, suggesting a similar steric hindrance effect for the surface complexation of polymeric Pb complexes on kaolinite. The results of this study show that the Si tetrahedral sheet limited the surface complexation of Cd and Pb on kaolinite.

Anion adsorption on silicate minerals is usually low, and only present at low pH, where the aluminol groups are positively charged (Parfitt 1979). Phosphate adsorption on kaolinite was observed to occur on the edges, and mostly in exchangeable form; it has been suggested that some phosphate anions were bound in bidentate form, more specifically. On other phyllosilicates such as mica, the presence of iron hydroxyl groups, due to iron substitution, greatly enhanced phosphate adsorption; it is often observed that anion adsorption correlates positively with metal contents in sorbent (Kang et al. 2010). Sulfate showed a similar behavior, but with lower amounts adsorbed (Parfitt 1979; Rao and Sridharan 1984). Arsenate and arsenite adsorption on kaolinite and montmorillonite (Goldberg 2002) showed a more complex behavior, with a maximum in adsorption as a function of pH for kaolinite, at pH $\cong$ 5–6 for AsO_4^{3-} and pH $\cong$ 9 for AsO_2^-; on montmorillonite, arsenate also showed the same maximum, whereas arsenite had a more complex response, without a clear maximum. In general, most anions are retained by ion exchange on clay minerals; in some cases (phosphate, sulfate) specific adsorption by ligand exchange with aluminol groups has been proposed; in the case of brucite (a 2:1:1 phyllosilicate), a specific binding of arsenate to the intermediate hydroxide layer has been suggested (Beaulieu and Savage 2005).

The above examples show the general features of ion adsorption on phyllosilicates, which can be applied also to other silicate minerals: the involvement of edge groups, mainly aluminol, in the coordination with metal cations, and in the development of positive charges favoring anion binding. The quantitative treatment of these phenomena will be presented in Part III.

8.8 SUMMARY

Silicate minerals, in particular phyllosilicates, constitute the most important and abundant class of soil minerals. Here, the most important aluminosilicate minerals have been presented and its interaction with ionic species is described. The more significant of these minerals, phyllosilicates, has a unique layered structure that constitutes their landmark and is central to their properties. Heterovalent isomorphic substitution is present in many phyllosilicates, resulting in a net permanent negative charge, which is compensated by interlayer cations or positively charged layers. For high values of this charge (about −1 per four tetrahedral units), there is strong interaction with cations, this results in tightly bound cations that lose their hydration layer, and so the mineral has its layers closely bound, with small interlayer spacing; these constitute the mica group. For intermediate values of permanent charge, the interlayer is filled with hydrated cations compensating the mineral negative charge, thus having a more open interlayer structure and opening the way for exchange of interlayer cations; these minerals constitute the vermiculite group. Those phyllosilicates with relatively lower permanent charge have correspondingly weaker interactions, consequently not only have hydrated cations in the interlayer but also admit water and other substances with relative ease, thus being capable of swelling under appropriate conditions; these are the smectite group of minerals, which have, due to this special property, a high number of applications. The ability of the last two groups to easily exchange the interlayer cations results in a high CEC. Other silicate minerals have a more rigid structure and so they do not display such properties.

The ion-binding capacity of silicate minerals is mainly given by electrostatic interactions (i.e., ionic exchange), and there is relatively little specific binding, mainly through edge silanol and aluminol groups.

In the next chapters, other important families of soil components, oxide minerals and organic matter, will be described; the interaction between all of them will be treated in Chapter 14.

REFERENCES

Abd-Allah, S. M., O. M. El Hussaini, and R. M. Mahdy. 2007. Towards a more safe environment: (2) Characterization of some clay sediments in Egypt for safe environmental applications. *Australian Journal of Basic and Applied Sciences* 1, no. 4: 813–823.

Amonette, J. E. 2002. Methods for determination of mineralogy and environmental availability. In *Soil Mineralogy with Environmental Applications*, ed. J. B. Dixon and D. G. Schulze, 153–197. Soil Science Society of America Book Series 7. Madison, WI: Soil Science Society of America.

Bailey, S. W. 1975. Chlorites. In *Soil Components. Vol. 2: Inorganic Components*, ed. J. E. Gieseking, 191–263. New York: Springer-Verlag.

Beaulieu, B. T., and K. S. Savage. 2005. Arsenate adsorption structures on aluminum oxide and phyllosilicate mineral surfaces in smelter-impacted soils. *Environmental Science & Technology* 39, no. 10: 3571–3579. doi:10.1021/es048836f.

Boettinger, J. L., and D. W. Ming. 2002. Zeolites. In *Soil Mineralogy with Environmental Applications*, ed. J. B. Dixon and D. G. Schulze, 585–618. Soil Science Society of America Book Series 7. Madison, WI: Soil Science Society of America.

Carter, D. L., M. D. Heilman, and C. L. Gonzalez. 1965. Ethylene glycol monoethyl ether for determining surface area of silicate minerals. *Soil Science* 100, no. 5: 356–360.

Carter, D. L., M. M. Mortland, and W. D. Kemper. 2010. Specific surface. In *Methods of Soil Analysis. Part 1. Physical and Mineralogical Methods*, ed. A. Klute. Soil Science Society of America Book Series 5. Madison, WI: American Society of Agronomy-Soil Science Society of America.

Chiou, C. T., D. W. Rutherford, and M. Manes. 1993. Sorption of nitrogen and ethylene glycol monoethyl ether (EGME) vapors on some soils, clays, and mineral oxides and determination of sample surface areas by use of sorption data. *Environmental Science & Technology* 27, no. 8: 1587–1594. doi:10.1021/es00045a014.

Cope, C. B., W. H. Fuller, and S. L. Willetts. 1983. *The Scientific Management of Hazardous Wastes*. 1st ed. Cambridge, UK: Cambridge University Press.

Dixon, J. B., and D. G. Schulze, eds. 2002. *Soil Mineralogy with Environmental Applications*. Soil Science Society of America Book Series 7. Madison, WI: Soil Science Society of America.

Furnare, L. J., A. Vailionis, and D. G. Strawn. 2005. Polarized XANES and EXAFS spectroscopic investigation into copper(II) complexes on vermiculite. *Geochimica et Cosmochimica Acta* 69, no. 22: 5219–5231. doi:10.1016/j.gca.2005.06.020.

Goldberg, S. 2002. Competitive adsorption of arsenate and arsenite on oxides and clay minerals. *Soil Science Society of America Journal* 66, no. 2: 413–421. doi:10.2136/sssaj2002.4130.

Gräfe, M., B. Singh, and M. Balasubramanian. 2007. Surface speciation of Cd(II) and Pb(II) on kaolinite by XAFS spectroscopy. *Journal of Colloid and Interface Science* 315, no. 1: 21–32. doi:10.1016/j.jcis.2007.05.022.

Hang, P. T., and G. W. Brindley. 1970. Methylene blue absorption by clay minerals. determination of surface areas and cation exchange capacities (clay-organic studies XVIII). *Clays and Clay Minerals* 18, no. 4: 203–212. doi:10.1346/CCMN.1970.0180404.

Harsh, J., J. Chorover, and E. Nizeyimana. 2002. Allophane and imogolite. In *Soil Mineralogy with Environmental Applications*, ed. J. B. Dixon and D. G. Schulze, 291–322. Soil Science Society of America Book Series 7. Madison, WI: Soil Science Society of America.

Hendricks, S. B, and R. S Dyal. 1952. Formation of mixed layer minerals by potassium fixation in montmorillonite. *Soil Science Society of America Journal* 16, no. 1: 45–48. doi:10.2136/sssaj1952.03615995001600010014x.

Hofmann, U. 1955. Intracrystalline swelling, cation exchange, and anion exchange of minerals of the montmorillonite group and of kaolinite. *Clays and Clay Minerals* 4, no. 1: 273–287. doi:10.1346/CCMN.1955.0040132.

Kang, D. H, A. P. Schwab, C. T. Johnston, and M. K. Banks. 2010. Adsorption of iron cyanide complexes onto clay minerals, manganese oxide, and soil. *Journal of Environmental Science and Health. Part A: Toxic/Hazardous Substances & Environmental Engineering* 45, no. 11: 1391–1396. doi:10.1080/10934529.2010.500930.

Keenan, A. G., R. W. Mooney, and L. A. Wood. 1951. The relation between exchangeable ions and water adsorption on kaolinite. *Journal of Physical Chemistry* 55, no. 9: 1462–1474. doi:10.1021/j150492a006.

Kohut, C. K., and C. J. Warren. 2002. Chlorites. In *Soil Mineralogy with Environmental Applications*, ed. J. B. Dixon and D. G. Schulze, 531–553. Soil Science Society of America Book Series 7. Madison, WI: Soil Science Society of America.

Löwenstein, W. 1954. The distribution of aluminum in the tetrahedra of silicates and aluminates. *American Mineralogist* 39: 92–96.

MacEwan, D. M. C. 1956. Fourier transform methods for studying scattering from lamellar systems. *Kolloid-Zeitschrift [Journal of Colloids]* 149, no. 2–3: 96–108. doi:10.1007/BF01511475.

Malla, P. B. 2002. Vermiculites. In *Soil Mineralogy with Environmental Applications*, ed. J. B. Dixon and D. G. Schulze, 501–529. Soil Science Society of America Book Series 7. Madison, WI: Soil Science Society of America.

Mindat.org. 1993a. Mineralogy database—mineral collecting, localities, mineral photos and data. *Mineralogy Database*. http://www.mindat.org/.

Mindat.org. 1993b. Serpentine group. *Serpentine Group Mineral Information and Data*. http://www.mindat.org/min-11135.html.

Mineral gallery. 1995. Amethyst galleries' mineral gallery. *Mineral Gallery*. http://www.galleries.com/default.htm.

Minerals.net. 1997. The mineral and gemstone kingdom: Home. *The Mineral and Gemstone Kingdom*. http://www.minerals.net/Default.aspx.

Mooney, R. W., A. G. Keenan, and L. A. Wood. 1952. Adsorption of water vapor by montmorillonite. I. Heat of desorption and application of BET theory1. *Journal of American Chemical Society* 74, no. 6: 1367–1371. doi:10.1021/ja01126a001.

Nadeau, P. H., M. J. Wilson, W. J. Mchardy, and J. M. Tait. 1984. Interstratified clays as fundamental particles. *Science* 225, no. 4665: 923–925. doi:10.1126/science.225.4665.923.

Olson, G. C., M. L. Thompson, and M. A. Wilson. 1999. Phyllosilicates. In *Handbook of Soil Science*, ed. M. E. Sumner, F77–F123. 1st ed. Boca Raton, FL: CRC.

Orchiston, H. D. 1953. Adsorption of water vapor: I. Soils at 25 [degrees] c. *Soil Science* 76, no. 6: 453.

Orchiston, H. D. 1954. Adsorption of water vapor: II. Clays at 25 [degrees] C. *Soil Science* 78, no. 6: 463.

Parfitt, R. L. 1979. Anion adsorption by soils and soil materials. In *Advances in Agronomy*, ed. N. C. Brady, 1–50. San Diego, CA: Academic Press.

Quirk, J. P. 1955. Significance of surface areas calculated from water vapor sorption isotherms by use of the BET equation. *Soil Science* 80, no. 6: 423.

Rao, S. M., and A. Sridharan. 1984. Mechanism of sulfate adsorption by kaolinite. *Clays and Clay Minerals* 32, no. 5, 414–418.

Schulze, D. G. 2002. An introduction to soil mineralogy. In *Soil Mineralogy with Environmental Applications*, ed. J. B. Dixon and D. G. Schulze, 1–36. Soil Science Society of America Book Series 7. Madison, WI: Soil Science Society of America.

Singer, A. 2002. Palygorskite and sepiolite. In *Soil Mineralogy with Environmental Applications*, ed. J. B. Dixon and D. G. Schulze, 555–583. Soil Science Society of America Book Series 7. Madison, WI: Soil Science Society of America.

Sparks, D. L. 2002. *Environmental Soil Chemistry*, 2nd ed. San Diego, CA: Academic Press.

Sposito, G. 1984. *The Surface Chemistry of Soils*. New York: Oxford University Press.

Strawn, D. G., and D. L. Sparks. 1999. The use of XAFS to distinguish between Inner- and outer-sphere lead adsorption complexes on montmorillonite. *Journal of Colloid and Interface Science* 216, no. 2: 257–269. doi:10.1006/jcis.1999.6330.

Sumner, M. E., and W. P. Miller. 1996. Cation exchange capacity and exchange coefficients. In *Methods of Soil Analysis. Part 3. Chemical Methods*, ed. D. L. Sparks, 1201–1230. Soil Science Society of America Book Series 5. Madison, WI: American Society of Agronomy-Soil Science Society of America.

Thompson, M. L., and L. Ukrainczyk. 2002. Micas. In *Soil Mineralogy with Environmental Applications*, ed. J. B. Dixon and D. G. Schulze, 431–466. Soil Science Society of America Book Series 7. Madison, WI: Soil Science Society of America.

Tunega, D., M. H. Gerzabek, and H. Lischka. 2004. Ab initio molecular dynamics study of a monomolecular water layer on octahedral and tetrahedral kaolinite surfaces. *The Journal of Physical Chemistry B* 108, no. 19: 5930–5936. doi:10.1021/jp037121g.

Velde, B., and A. Meunier. 2008. *The Origin of Clay Minerals in Soils and Weathered Rocks*. Berlin: Springer.

Wang, Y., and H. Xu. 2006. Geochemical chaos: Periodic and nonperiodic growth of mixed-layer phyllosilicates. *Geochimica et Cosmochimica Acta* 70, no. 8: 1995–2005. doi:10.1016/j.gca.2006.01.002.

White, G. N., and J. B. Dixon. 2002. Kaolin-serpentine minerals. In *Soil Mineralogy with Environmental Applications*, ed. J. B. Dixon and D. G. Schulze, 389–414. Soil Science Society of America Book Series 7. Madison, WI: Soil Science Society of America.

Whittig, L. D., and W. R. Allardice. 2010. X-ray diffraction techniques. In *Methods of Soil Analysis. Part 1. Physical and Mineralogical Methods*, ed. A. Klute, 331–362. Soil Science Society of America Book Series 5. Madison, WI: American Society of Agronomy-Soil Science Society of America.

Yang, C. –S., J. M. Mora-Fonz, and C. R. A. Catlow. 2011. Stability and structures of aluminosilicate clusters. *Journal of Physical Chemistry C* 115, no. 49: 24102–24114. doi:10.1021/jp202394w.

Yukselen, Y., and A. Kaya. 2008. Suitability of the methylene blue test for surface area, cation exchange capacity and swell potential determination of clayey soils. *Engineering Geology* 102, no. 1–2: 38–45. doi:10.1016/j.enggeo.2008.07.002.

Zelazny, L. W., L. He, and A. M. Vanwormhoudt. 1996. Charge analyses of soils and anion exchange. In *Methods of Soil Analysis. Part 3. Chemical Methods*, ed. D. L. Sparks, 1231–1254. Soil Science Society of America Book Series 5. Madison, WI: American Society of Agronomy-Soil Science Society of America.

Zelazny, L. W., P. J. Thomas, and C. L. Lawrence. 2002. Pyrophyllite-talc minerals. In *Soil Mineralogy with Environmental Applications*, ed. J. B. Dixon and D. G. Schulze. Soil Science Society of America Book Series 7. Madison, WI: Soil Science Society of America.

9 Oxide Minerals

Oxides are the second most important family of minerals in soils, in some cases responsible for more than 50% mass and for important properties. Here, the most abundant classes of oxide minerals are described. More detailed discussions can be found in the literature (Kämpf, Scheinost, and Schulze 1999; Dixon and Schulze 2002).

9.1 SILICA MINERALS

The term silica refers to silicon dioxide (SiO_2), in its various forms, and silica minerals are the second most abundant group, behind silicate minerals discussed in Chapter 8 (Monger and Kelly 2002). The silica minerals include quartz (in several crystalline forms) and opals. All of them are tectosilicates, formed from strongly bound SiO_4 tetrahedra, which were discussed in Section 8.1.

Quartz is the most abundant and ubiquitous silica mineral, showing up in many soils over the world. Primary quartz, which appears as polymorphs crystallized from cooling magmas, is distinguished by the way the SiO_4 tetrahedra are arranged, as well as the presence of chemical impurities; there are also secondary quartz types, including coarsely crystalline and microcrystalline, which crystallize from supersaturated H_4SiO_4 solutions. Silica is thermodynamically stable enough so as to form both primary and secondary minerals and durable enough to persist in most environments.

Main SiO_2 polymorphs are α-quartz, α-trydimite, and α-cristobalite, all stable at low temperature; Figure 9.1 shows the structures of α-quartz and α-cristobalite. Other forms of quartz are summarized in Table 9.1, including high temperature and pressure polymorphs and some of the many varieties of secondary quartz minerals that include several gemstones. *Opals* are also included, which can be viewed as hydrated SiO_2 of variable crystallinity and often found in soil environments. They form in aqueous environments such as the soil solution, lakes, and so on. Opal-C and CT are composed of disordered spheres and are found in soils, specially the CT variety; opal-A forms from biological activity and is found in plants (phytoliths) and the skeletons of some organisms such as radiolaria, sponges, and so on.

SiO_2 is scarcely soluble, its solubility being dependent on crystallinity and impurities; moreover, its dissolution is a slow process. In soil, the average solubility of silica, as undissociated silicic acid (H_4SiO_4), is around 10^{-3} M (Lindsay and Elgawhary 1972) at pH below 8; at higher pH values, ionization causes a sharp increase in solubility. This increased solubility can lead to weathering of silica minerals, which is otherwise very slow.

Owing to the almost complete absence of heterovalent isomorphic substitution and low amounts of silanol groups, silica minerals have a very low adsorption capacity, either by cation exchange or by specific adsorption. However, cation-exchange capacity increases somewhat at high pH values. In general, these minerals have a small contribution to soil–ion binding.

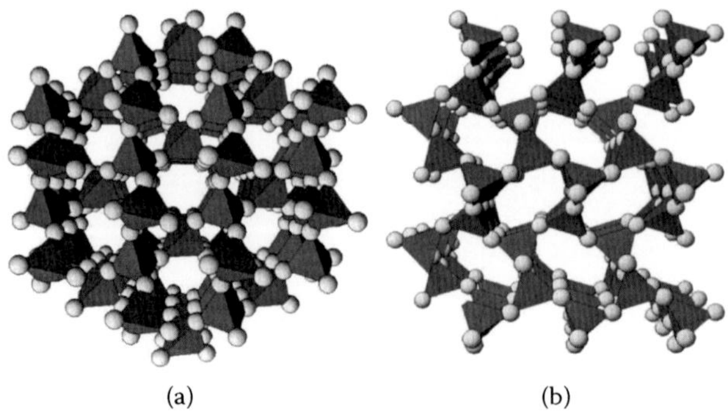

FIGURE 9.1 Crystalline structures of α-quartz (a) and α-cristobalite (b) showing different forms of linking SiO_4 tetrahedra (two out of nine). The cristobalite structure is more open, thus less dense and can contain more impurities than quartz. (Reprinted from Monger, H. C. and Kelly, E. F., *Soil Mineralogy with Environmental Applications*, Soil Science Society of America, Madison, WI, 611–636, 2002, with kind permission.)

9.2 ALUMINUM (HYDR)OXIDES

Aluminum is an abundant element that is present as a major element in all soils. However, it appears mainly in aluminosilicate minerals (Chapter 8), *gibbsite* being almost the only aluminum mineral appearing in an important extent in soils (Huang et al. 2002). There are also poorly crystalline aluminum hydroxide or oxohydroxide colloids that are highly reactive, which can strongly adsorb either nutrient or pollutant ionic species.

The main hydroxide mineral phases are gibbsite, *bayerite*, *norstrandite*, and *doyleite*, all sharing the chemical composition $Al(OH)_3$; of them, only gibbsite is frequently found in soils. Oxohydroxides (AlOOH) include *boehmite* and *diaspore*, and finally the only significant mineral oxide is *corundum* (Al_2O_3); none of these occur frequently in soils. All these minerals have octahedral conformation around Al^{3+} ions; Figure 9.2 shows models of the crystalline structures of an aluminum hydroxide, gibbsite, and an oxohydroxide, boehmite.

The aluminum speciation and precipitation equilibria play an important role both in its state in the soil solution (and soil acidity) and in the presence of aluminum minerals in soil environments; hence, we will devote some discussion to this point. Al(III) in aqueous solution undergo the following hydrolysis reactions involving aqueous mononuclear species:

$$Al^{3+} + H_2O \rightleftharpoons Al(OH)^{2+} + H^+ \qquad \log K_1 = -4.99 \qquad (9.1)$$

$$Al(OH)^{2+} + H_2O \rightleftharpoons Al(OH)_2^+ + H^+ \qquad \log K_2 = -5.30 \qquad (9.2)$$

$$Al(OH)_2^+ + H_2O \rightleftharpoons Al(OH)_3(aq) + H^+ \qquad \log K_3 = -6.40 \qquad (9.3)$$

$$Al(OH)_3(aq) + H_2O \rightleftharpoons Al(OH)_4^- + H^+ \qquad \log K_4 = -6.30 \qquad (9.4)$$

TABLE 9.1
Most Important Silica Minerals

Mineral	Stability	Density ($g \cdot cm^{-3}$)	Crystalline System
	SiO_2 polymorphs		
α-Quartz	Low temperature, low pressure, T<573°C	2.65	Trigonal
β-Quartz	High temperature, low pressure, 573°C<T<870°C	2.53	Hexagonal
α-Tridymite	High temperature, low pressure	2.26	Monoclinic
β-Tridymite	Less compact than α-tridymite, 870°C<T<1470°C	2.22	Hexagonal
α-Cristobalite	Highest temperature, low pressure	2.32	Tetragonal
β-Cristobalite	1470°C<T<1713°C (melting point)	2.20	Cubic
Coesite	High temperature and pressure, 450°C<T<800°C	3.01	Monoclinic
	Secondary quartz varieties		
Amethyst	Shades of violet from small amounts of Fe(III)		
Rose quartz	Rose-red or pink color due to small amounts of Ti(IV)		
Agate	Microcrystalline, alternating layers of fibrous quartz (chalcedony), in bands having different colors		
Onyx	Another microcrystalline banded mineral, having parallel layers		
Flint	Microcrystalline granular quartz, dark		
Jasper	Microcrystalline, granular, usually red due to occluded hematite		
	Opals ($SiO_2 \cdot nH_2O$)		
Opal-C	Opal composed of disordered cristobalite lepispheres		
Opal-CT	Similar to opal-C but with cristobalite and tridymite		
Opal-AG	Submicron silica spheres, displays opalescence, precious		
Opal-A	Biogenic opal, in phytoliths, diatoms, sponge spicules and radiolarian		

Source: Monger, H.C. and Kelly, E.F., *Soil Mineralogy with Environmental Applications*, Madison, WI, Soil Science Society of America, 611–636, 2002.

The K_i are the successive equilibrium constants. It should be noted that Al(III) in water is best represented as the hexaaquo complex $Al(H_2O)_6^{3+}$, and all the species in Equations 9.1 through 9.4 have been adjusted accordingly (e.g., Equation 9.1 becomes $Al(H_2O)_6^{3+} \rightleftharpoons Al(H_2O)_5(OH)^{2+} + H^+$); however, for simplicity, these reactions are usually written as above. The neutral compound $Al(OH)_3$(aq) is very scarcely soluble, but for the time being we will ignore the precipitation of solid phases. Besides the mononuclear species just presented, it is known that in the

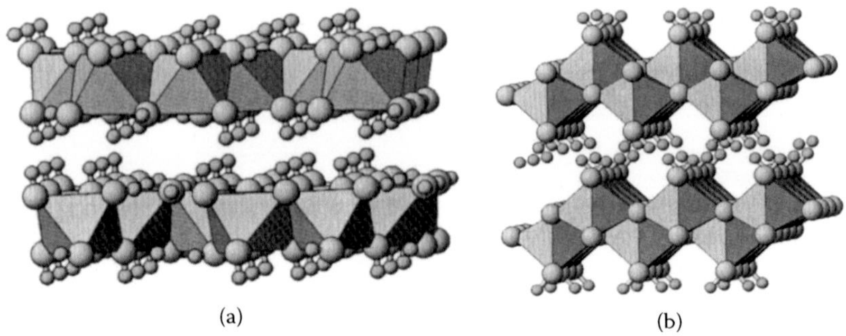

FIGURE 9.2 The crystalline structure of gibbsite (a) and boehmite (b). Gibbsite is the most common form of $Al(OH)_3$, formed by octahedral layers held together by hydrogen bonds; boehmite is an oxohydroxide: AlOOH organized as zigzag layers, where the "internal" vertices are occupied by O^{2-} species, whereas the external ones are OH^- species, linking other layers by hydrogen bonds. (Reprinted from Huang, P. M. et al., *Soil Mineralogy with Environmental Applications*, Soil Science Society of America, Madison, WI, 261–289, 2002, with kind permission.)

Al(III)–OH^- system, a number of "polymers" appear leading to neutral species and precipitation. The reactions leading to polynuclear Al(III) species are not well known in general; the only cases where equilibrium constants are known are $Al_2(OH)_2^{4+}$ and $Al_3(OH)_4^{5+}$, which can be written as follows:

$$2Al^{3+} + 2H_2O \rightleftharpoons Al_2(OH)_2^{4+} + 2H^+ \qquad \log K_{2,2} = -7.69 \quad (9.5)$$

$$3Al^{3+} + 4H_2O \rightleftharpoons Al_3(OH)_4^{5+} + 4H^+ \qquad \log K_{3,4} = -13.89 \quad (9.6)$$

Considering reactions 9.1 through 9.6, the Al(III) speciation in solution as a function of pH can be estimated. Figure 9.3 shows the speciation diagrams resulting for Al(III) total concentrations, c_T, of 1 mM and 10 μM; note that because polynuclear species are considered, the diagram is concentration dependent. At low concentrations, the polynuclear species are negligible, but as concentration increases, the proportion of such species becomes noticeable, as expected from reactions 9.5 and 9.6; later we will discuss its importance. Now, considering $Al(OH)_3$(aq), it is seen that its concentrations are predicted to be of the same order as c_T; however, this is not real as aluminum hydroxide is known to precipitate. In fact, several solid phases, either crystalline or amorphous, can be found; their solubilities can be evaluated from the respective *solubility product constants*, for example, for gibbsite in equilibrium with the free ions.

$$Al(OH)_3(\text{gibbsite}) \rightleftharpoons Al^{3+} + 3OH^- \qquad (9.7)$$

as

$$K_{SP}(\text{gibbsite}) = a_{Al^{3+}} \left(a_{OH^-}\right)^3 \qquad (9.8)$$

or, equivalently, as the *solubility constant*, K_{s0} for precipitation from the aqueous species $Al(OH)_3(aq)$ as follows, for several solid phases:

$$Al(OH)_3(aq) \rightleftharpoons Al(OH)_3(s, \text{amorphous}) \qquad \log K_{s0,\text{amorph}} = -5.9 \tag{9.9}$$

$$Al(OH)_3(aq) \rightleftharpoons Al(OH)_3(s, \text{soil}) \qquad \log K_{s0,\text{soil}} = -8.4 \tag{9.10}$$

$$Al(OH)_3(aq) \rightleftharpoons Al(OH)_3(s, \text{gibbsite}) \qquad \log K_{s0,\text{gibbsite}} = -8.95 \tag{9.11}$$

$$Al(OH)_3(aq) \rightleftharpoons AlOOH(s, \text{bohemite}) + H_2O \qquad \log K_{s0,\text{bohemite}} = -8.113 \tag{9.12}$$

$$Al(OH)_3(aq) \rightleftharpoons AlOOH(s, \text{diaspore}) + H_2O \qquad \log K_{s0,\text{diaspore}} = -9.818 \tag{9.13}$$

$$2Al(OH)_3(aq) \rightleftharpoons Al_2O_3(s, \text{corundum}) + 3H_2O \qquad \log K_{s0,\text{corundum}} = -13.73 \tag{9.14}$$

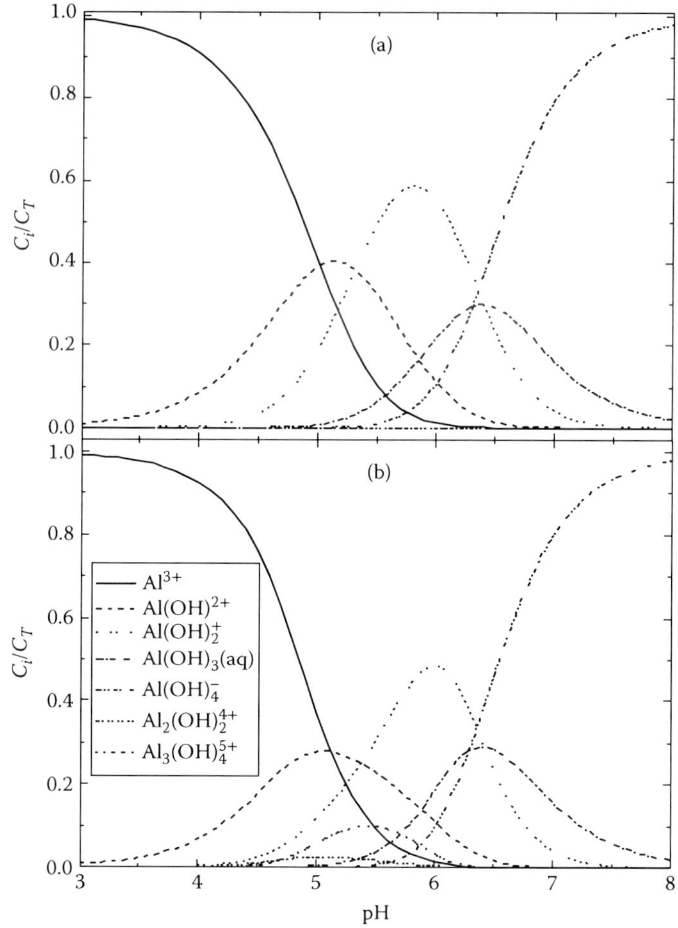

FIGURE 9.3 Speciation diagram of aluminum as a function of pH, according to reactions 9.1 through 9.6, neglecting precipitation; total aluminum concentration: (a) $c_T = 1$ mM; (b) 10 μM.

Gibbsite, boehmite, diaspore, and corundum are minerals found in soil; amorphous $Al(OH)_3$ is the form that commonly precipitates first when an aluminum solution is alkalinized; "soil" aluminum hydroxide refers to an average K_{s0} value introduced by Gustafsson (Gustafsson, Lumsdon, and Simonsson 1998). To evaluate the thermodynamic tendency to precipitate of any solid phase, it is convenient to compute the *saturation index*, *SI*, which can be defined as

$$SI = \log I_{SP} - \log K_{SP} \tag{9.15}$$

where I_{PS} is the "instantaneous" ionic activity product, calculated from the activities obtained; assumed absence of precipitation, a_i^0; I_{SP} reads, for reaction 9.7:

$$I_{SP} = a^0_{Al^{3+}} \left(a^0_{OH^-}\right)^3 \tag{9.16}$$

For reactions 9.9 through 9.13, we have simply

$$SI = \log a^0_{Al(OH)_3\,(aq)} - \log K_{s0} \tag{9.17}$$

where the value of K_{s0} obviously depends on the solid phase considered; for reaction 9.14 it reads

$$SI = 2\log a^0_{Al(OH)_3\,(aq)} - \log K_{s0}(\text{corundum}) \tag{9.18}$$

The simplicity of Equations 9.17 and 9.18 is an important reason to write reactions 9.9 through 9.14. Figure 9.4 shows the SI as a function of pH for the solid phases considered here. Of course, the shape of all curves resemble, in logarithmic form, that of $Al(OH)_3(aq)$ in Figure 9.3; the only difference is that of corundum, due to the coefficient 2 in Equation 9.14. Except for a small pH range between approximately 6 and 7, the most stable solid phase is diaspore, and the least one is (as expected) amorphous aluminum hydroxide. However, the most common aluminum oxide mineral is gibbsite, diaspore being only rarely found. This may appear surprising, but the kinetics of the process has also to be considered; as it is well known (Robarge 1998), the least stable phases are the first to precipitate, which afterward slowly evolve into more stable solids. Thus, as for other metals, the amorphous hydroxide precipitates first, which slowly evolves, eventually reaching the gibbsite crystal structure. The transformation of gibbsite to diaspore, albeit thermodynamically favored, is presumably extremely slow, because it involves the loss of water and the change from a monoclinic to an orthorhombic structure. Thus, in a soil environment, several solid Al(III) forms may coexist at a given time, from the highly reactive amorphous phase to the more crystalline but less reactive minerals such as gibbsite or diaspore. The "soil" K_{s0} introduced by Gustaffson represents an average value to account for global Al(III) solubility.

The species distribution are altered as a consequence of precipitation; Figure 9.5 shows the speciation diagram resulting from the presence of amorphous $Al(OH)_3$ solid in equilibrium; if other more stable solids are considered, the pH range of stability of aqueous species is reduced.

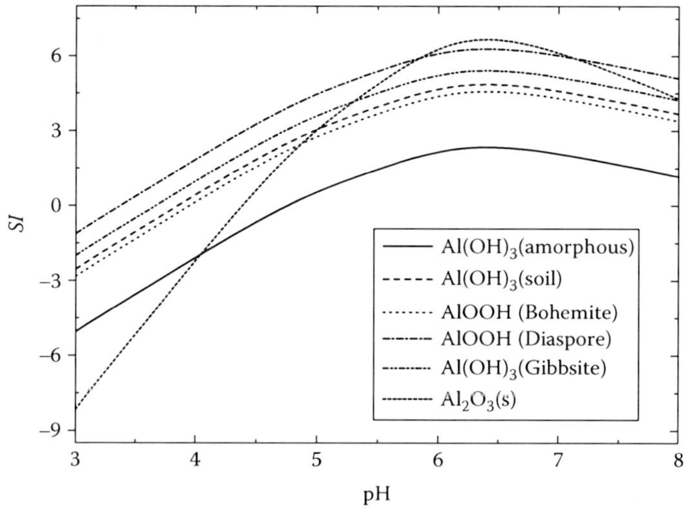

FIGURE 9.4 Saturation index (SI) as a function of pH for several aluminum solid phases; total aluminum concentration, $c_T = 1$ mM.

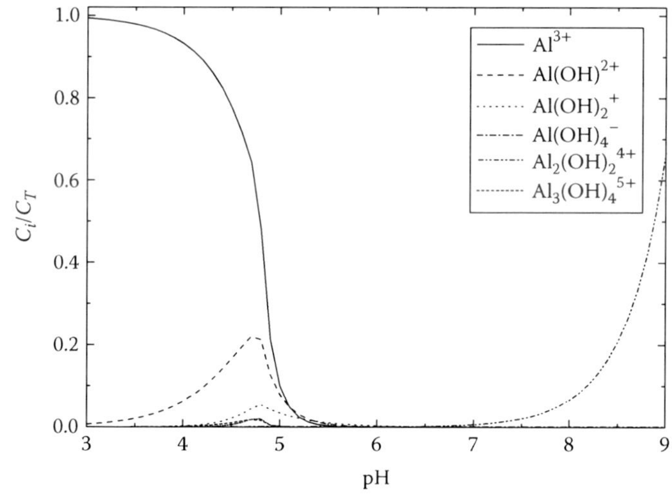

FIGURE 9.5 Speciation diagram of aluminum as a function of pH, according to reactions 9.1 through 9.6, considering precipitation of amorphous $Al(OH)_3$ according to reaction 9.9; total aluminum concentration, $c_T = 1$ mM.

The precipitation of aluminum hydroxides involves the polymerization of Al^{3+} and OH^- ions in solution. A first model of this process was proposed by Hsu and Bates (Hsu and Bates 1964), which is schematized in Figure 9.6. It is assumed that the species $Al(OH)^{2+}$ and $Al(OH)_2^+$ are not stable and bind, through OH^- ions, other Al^{3+} ions, which in turn bind more OH^- and so on. The figure schematizes the main stages of this process. These stages correspond approximately to reactions 9.1 through 9.3 but as a continuous process driven by the OH^-/Al(III) molar ratio in solution, R; as R increases, the polymerization progresses giving more extended structures

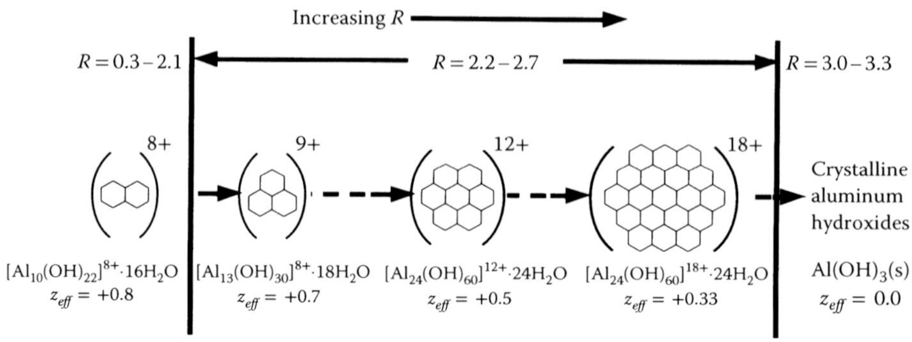

FIGURE 9.6 Scheme of Al(III) polymerization leading to precipitation according to Hsu and Bates. R is the molar ratio of OH^- to Al(III) in solution; z_{eff} is the effective charge per aluminum atom. (Reproduced with kind permission of the Mineralogical Society from a paper by Hsu and Bates [1964].)

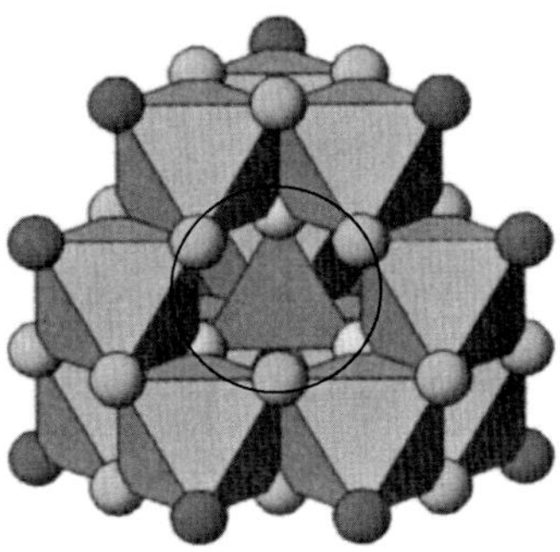

FIGURE 9.7 Model of an "Al_{13}" aluminum polymeric species; the circle indicates the central tetrahedral $Al(OH)_4^-$ group. (Reprinted from Huang, P. M. et al., *Soil Mineralogy with Environmental Applications*, Soil Science Society of America, Madison, WI, 261–289, 2002, with kind permission.)

with a decreasing effective average charge per Al atom, z_{eff}. The equilibrium constants K_1 to K_3 can thus be regarded as average values. The structures observed in Figure 9.6 clearly resemble octahedral sheets found in phyllosilicates and aluminum hydroxides such as gibbsite. More recently, a different structure, known as "Al_{13}" was identified in laboratory experiments (Anderson and Bertsch 1988; Parker and Bertsch 1992a, b); it is formed around an $Al(OH)_4^-$ tetrahedral anion, surrounded by 12 octahedral groups totalizing 13 Al(III) species as shown in Figure 9.7. The charge of such structure is theoretically +7, but values between +3 and +7 have been observed. It has been proposed that Al_{13} species can lead to the formation of nuclei of poorly crystalline phases, which would eventually transform to gibbsite or another crystalline solid. However, the existence in nature of Al_{13} cations has not been established as yet, and the mechanism of Al hydroxides precipitation is still an open question (Huang et al. 2002).

Al oxides and hydroxides present a high surface concentration of aluminol groups, which are highly reactive. In the first place, they show acid–base reactivity, leading to a strong dependence of the surface charge on pH, which can be determined by titration (see, e.g., Figure 5.19). Furthermore, these groups have the ability to specifically bind both anions and cations, thus being of high interest for our purposes here; this behavior will be addressed in Section 9.7.

9.3 IRON (HYDR)OXIDES

Iron oxide minerals (including hydroxides and oxohydroxides) are common and widely distributed in soils, having properties with an important impact on soil, such as their redox behavior, active surface chemistry, pigmenting power, and so on (Kämpf, Scheinost, and Schulze 1999; Bigham, Fitzpatrick, and Schulze 2002). Iron is an abundant element in Earth's crust, thus its minerals are widespread. The most important iron oxide minerals in soil are summarized in Table 9.2. Figure 9.8 shows the crystalline structure of some minerals. *Goethite* (α-FeOOH, Figure 9.8a) is the most common iron mineral and also the most stable thermodynamically. The Fe(III) ions have octahedral coordination, and the OH groups form hydrogen bonds with adjacent O atoms, resulting in a relatively open structure, which is reflected by the comparatively low density. *Hematite* (α-Fe_2O_3, Figure 9.8b), on the contrary, is the denser iron oxide, having a compact structure where the octahedra are arranged in a form very similar to phyllosilicate dioctahedral sheets, with one out of each three octahedral centers being vacant. *Magnetite* (Fe_3O_4, Figure 9.8c) is a mixed valence oxide (two Fe(III) for each Fe(II)), where the trivalent iron centers show a tetrahedral coordination, thus making it an inverse spinel (Deer, Howie, and Zussman 1996). It has a packed structure, showing a high density. *Maghemite* results from magnetite oxidation and thus has the same structure, albeit some positions are now occupied by Fe(III), whereas others are vacant; natural maghemites usually contain some amounts of Fe(II), due to incomplete oxidation. These two compounds show a strong ferrimagnetic behavior. *Lepidocrocite* (γ-FeOOH) is another common iron oxohydroxide, with a structure more compact than goethite, being similar to this one in other aspects. *Ferrihydrite* is a poorly crystalline mineral consisting of small spherical particles (2–6 nm), which are collected in large aggregates. There is no consensus about either its structure or its chemical composition (Eggleton and Fitzpatrick 1988; Drits et al. 1993; Manceau and Gates 1997; Jambor and Dutrizac 1998); the formula, besides that shown in Table 9.2, is often given as $Fe_2O_3 \cdot 1/2H_2O$, but it is usually considered as a hydrated (oxo)hydroxide. *Schwertmannite* is unique, being a mixed iron oxide-sulfate, having sulfate ions in the crystal structure, believed to fill tunnel cavities in a structure similar to that of akaganeite (see later in text); this is also a poorly crystalline mineral, thus being difficult to study in detail. Finally, the term *green rust* refers to a group of compounds of $Fe(OH)_2$ trioctahedral sheets where some Fe(II) ions are oxidized to Fe(III), the charge balanced by anions (such as Cl^-, SO_4^{2-}, and CO_3^{2-}) in the interlayer. Green rust phases are actually transient ones in the process of formation of more stable iron oxides.

Natural iron oxides show isomorphic substitution like phyllosilicates; however, the charge differences are usually compensated internally (e.g., by a change in the

TABLE 9.2
Most Important Iron Oxide and Hydroxide Minerals

Mineral	Formula	Crystalline System	Color	Density ($g \cdot cm^{-3}$)	Soil Environment Where It Is Found
Hematite	α-Fe_2O_3	Trigonal	Reddish gray, black, blackish red	5.3	Aerobic soils of tropics, subtropic, arid, semiarid regions, especially warm areas
Maghemite	γ-Fe_2O_3	Isometric	Dark brown, brick red	4.9	Highly weathered soils, tropics and subtropics, from magnetite
Magnetite	Fe_3O_4	Isometric (spinel)	Grayish black or iron black	5.15	Sand and silt fractions. Usually inherited from parent material
Goethite	α-FeOOH	Orthorhombic	Brown, reddish brown, yellowish brown, ocher yellow	3.8 (average)	All weathering regimes, especially cool, wet climates
Lepidocrocite	γ-FeOOH	Orthorhombic	Red, yellowish brown, blackish brown	4.0	Seasonally anaerobic soils, cool climates
Ferrihydrite	$Fe_5O_3(OH)_9$	Trigonal	Yellow brown, dark brown	3.8	Soils subject to rapid oxidation in the presence of organics
Fougerite (Green rust)	$(Fe(II), Mg)_6Fe(III)_2(OH)_{18} \cdot 4H_2O$	Trigonal	Bluish gray	2.5	Strongly hydromorphic soils
Schwertmannite	$Fe_8O_8(OH)_6SO_4 \cdot nH_2O$ (approximate)	Tetragonal	Brownish yellow	3.88	Acid sulfate soils, coastal and inland areas; also anthropogenic sites

Source: Bigham, J.M. et al., *Soil Mineralogy with Environmental Applications*, Soil Science Society of America, Madison, WI, 2002; Barthelmy, D., Mineralogy database, 2010, http://webmineral.com/.

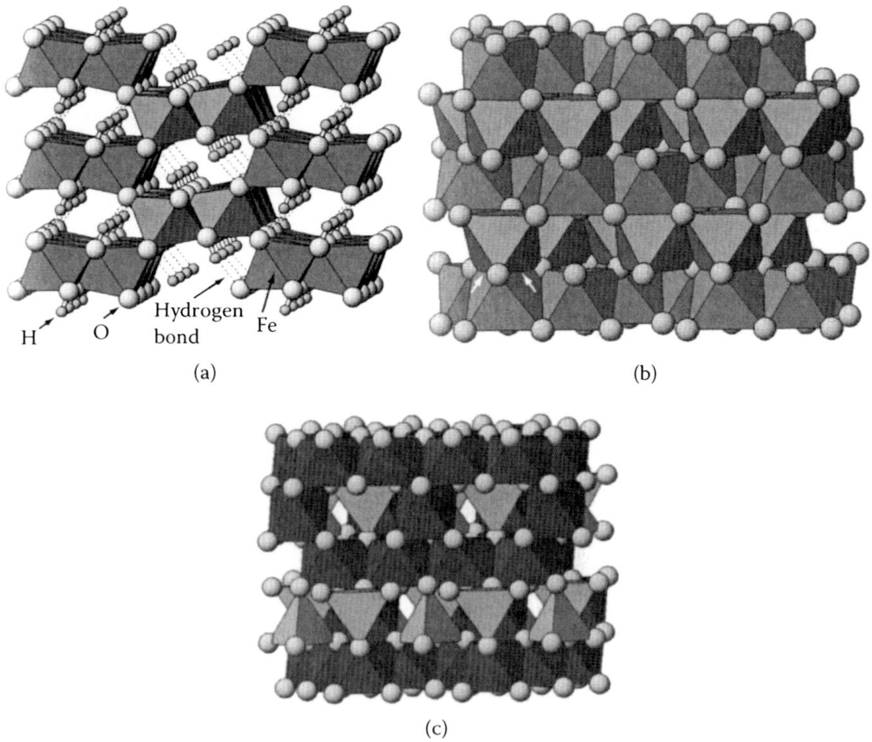

FIGURE 9.8 The crystalline structure of (a) goethite, (b) hematite, and (c) magnetite/maghemite. (Reprinted from Bigham, J. M. et al., *Soil Mineralogy with Environmental Applications*, Soil Science Society of America, Madison, WI, 323–366, 2002, with kind permission.)

number of vacancies); thus, these compounds do not present permanent charge. Owing to its abundance, Al(III) is the most common substituent found in Fe minerals; this substitution usually has a strong effect on crystal properties. Its extent is dependent on the crystal structure, being very low in lepidocrocite and reaching up to 33% in goethite. Another interesting example is that of *titanomagnetites*, where Ti(IV) replaces a proportional amount of Fe(II) and Fe(III), yielding a solid solution series of magnetite and *ulvospinel* (Fe_3O_4 and Fe_2TiO_4, respectively). These minerals are also discussed in Section 9.5.1.

The surface of iron oxides contain OH groups that are reactive, yielding marked acid–base behavior, thus a rather strong pH dependence of the surface charge; iron oxides are major components of variable charge soils, as discussed in Chapter 5. These surface groups react with ions, binding specifically both anions and cations. These features are discussed in Section 9.7.

All Fe oxides and oxohydroxides are scarcely soluble. The solubility usually follows the order ferrihydrite > maghemite > lepidocrocite > hematite > goethite (Schwertmann and Taylor 1989), which obviously is the order of the relative thermodynamic stability of these minerals. The reported solubility products of actual samples, however, show important variations due to differences in particle size and morphology, crystallinity, composition (presence of substitution), and so on. The solubility products, defined in Equation 9.8, range between 10^{-37} and 10^{-44} approximately (Schwertmann 1991). For schwertmannite, it has been proposed the following equilibrium:

$$Fe_8O_8(OH)_{4.8}(SO_4)_{1.6}(s) + 20.8H^+(aq) \rightleftharpoons 8Fe^{3+}(aq) + 1.6SO_4^{2-}(aq) + 12.8H_2O(aq) \quad (9.19)$$

with log $K_S = 18$ (Bigham et al. 1996).

Similar to Al oxides, the formation of Fe oxides start with the precipitation of less stable, poorly crystalline phases that slowly evolve toward the more stable minerals. Figure 9.9 shows schematically the most common processes; usually, it starts in soils and sediments from dissolved Fe(II), which can be produced by several reactions; one important source is complexation and reduction of Fe(III) minerals as part of the nutrition of both simple and higher organisms. In general, remobilization of iron as Fe(II) is achieved when the soil or sediment is depleted of O_2 due to microbial activity, and iron is reduced by organic matter: for example, as

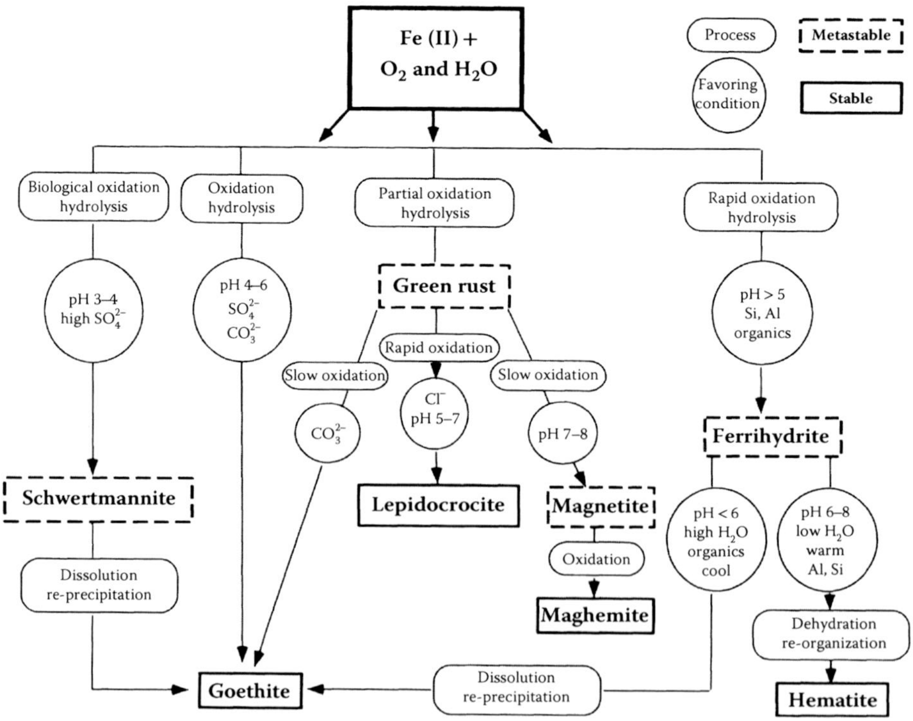

FIGURE 9.9 Diagram of the pathways in the formation and transformation of Fe oxides from Fe(II) in soil systems. (Reprinted from Bigham, J. et al., *Soil Mineralogy with Environmental Applications*, Soil Science Society of America, Madison, WI: 323–366, 2002, with kind permission.)

$$4FeOOH(s) + CH_2O(s) + 8H^+(aq) \rightleftharpoons 4Fe^{2+}(aq) + CO_2(g) + 7H_2O(l) \quad (9.20)$$

where CH_2O stands for organic fraction. Many microorganisms reduce iron as part of their metabolic activity, as a source of energy, but Fe(II) is not required, and so it is released. Also, reductive dissolution of Fe minerals may be promoted by photochemical reactions, especially in natural waters (Mcknight, Kimball, and Bencala 1988). Finally, weathering of iron sulfide minerals in soils and sediments usually proceeds according to

$$2FeS_2(s) + 15/2O_2(g) + 7H_2O(l) \rightleftharpoons 2Fe^{2+}(aq) + 4H^+(aq) + 4SO_4^{2-}(aq) \quad (9.21)$$

$$2Fe^{2+}(aq) + 3H_2O(l) + 1/2O_2(g) \rightleftharpoons 2FeOOH(s) + 4H_2(aq) \quad (9.22)$$

However, reaction 9.21 yields strongly acidic conditions; when the pH falls below 4, the kinetics of Fe(II) oxidation becomes very slow, thus it may remain for long times in reduced form.

As it is well known, goethite and hematite are the most stable iron oxides; in ambient conditions, goethite is the most stable, but at higher temperatures hematite becomes more stable. On the contrary, the particle size has influence: it was proposed that hematite is also more stable for particles smaller than 75 nm. However, goethite is widely found, even in smaller sizes, which do not convert easily to hematite; conversely, hematite is found in larger particles that do not transform into goethite. Moreover, other oxides such as lepidocrocite and maghemite are found to be quite stable; even magnetite and schwertmannite can be found to be persistent for long times. In general, the kinetics of conversions between these phases ranges from very to extremely slow, if any. The higher abundance of goethite can be explained, because there are four different routes that end in this mineral (Figure 9.9). The processes listed in this figure include chemical and biological oxidation of Fe(II) in different media. Besides these, an other pathway leading to secondary iron oxide minerals is the weathering of primary minerals, such as olivines, pyroxenes, amphiboles, and so on, or mixed Fe-Ti oxides, such as ilmenite. These minerals, formed under magma conditions, are often unstable in ambient conditions and dissolve slowly. The oxidation of Fe(II) ions in these minerals often accelerates the dissolution process, due to the charge imbalance created; examples are the oxidation of iron sulfides shown on Equations 9.21 and 9.22, often mediated by microorganisms, or the dissolution of minerals such as fayalite, an end member of the olivine series:

$$Fe_2SiO_4(s) + 1/2O_2(g) + 3H_2O(l) \rightleftharpoons 2FeOOH(s) + Si(OH)_4(aq) \quad (9.23)$$

The silicic acid formed may suffer other reactions, such as combination with Al(III) growing structures described in the previous section to form aluminosilicates.

9.4 MANGANESE OXIDES

Albeit it is not as abundant as aluminum or iron, manganese is a key element in soil chemistry, due to its redox behavior, and is an essential nutrient for animals and plants. Manganese occurs in nature mostly as oxides, which have environmental impact in several ways. These oxides adsorb heavy metals, some of them in a quite selective form (such as Ba and Pb), thus removing pollutants from the soil solution. Another important aspect is the oxidizing ability of oxides of Mn(III) and/or Mn(IV), which play a key role in the environment; in fact, these oxides usually mediate between the atmospheric oxygen and the reducing agents, such as Fe(II), organic matter, and so on. This oxidizing ability can be environmentally favorable—such as oxidizing the more toxic As(III) to the relatively less dangerous As(V)—and unfavorable, such as when Cr(III) is oxidized to the most toxic and mobile Cr(VI) (Dixon and White 2002).

There is a number of Mn minerals in soils, several of which are poorly known, because they tend to appear as small nodules of poor crystallinity or as coatings on other minerals, such as Fe oxides. Table 9.3 summarizes data for some relatively common Mn oxides; the last three are, in fact, less common but are included here for reference and comparison. Manganite and hausmannite are analogues of goethite and magnetite, respectively (Kämpf, Scheinost, and Schulze 1999; Dixon and White 2002).

Manganese in minerals exists mostly in tetravalent Mn(IV) form and in some cases in trivalent Mn(III); Mn(II) is usually soluble and found in the soil solution. The Mn in solids is normally found in octahedral coordination; in several cases, the octahedra share edges, forming octahedral sheets with randomly separated vacancies, resulting in a net negative charge. The sheets are separated by cations (along with water molecules) that counterbalance the charge; due to this layered structure, these minerals are termed *phyllomanganates*. *Birnessite* and *lithioporite* (Figure 9.10a–c) are phyllomanganates; the last one has positively charged Al-Li hydroxide sheets stacked between negative Mn sheets. In other cases, the octahedra share corners, forming a structure with channels, termed *tectomanganates* (Figure 9.10d).

It is worth here to discuss further the redox behavior of manganese oxides, as they have an important role in soil chemistry. Mn(II) is oxidized in soil by diverse routes. One prominent path is microbial oxidation to Mn(IV), and several bacterial species

TABLE 9.3
Important Manganese Oxides

Mineral	Formula	Crystalline System	Density ($g{\cdot}cm^{-3}$)
Birnessite	(Na, Ca) (Mn(III), Mn(IV))$_7$O$_{14}$ · 2 · 8H$_2$O	Monoclinic	3.0
Todorokite	(Mg$_{0.77}$Na$_{0.33}$) (Mg$_{0.18}$Mn(II)$_{0.60}$Mn(IV)$_{5.22}$) O$_{12}$ · 3 · 07H$_2$O	Monoclinic	3.65 (average)
Lithiophorite	Al$_2$LiMn(IV)$_2$Mn(III)O$_6$(OH)$_6$	Trigonal	3.25 (average)
Vernadite	δ-MnO$_2$	(Pseudo)tetragonal	3.0
Pyrolusite	α-MnO$_2$	Tetragonal	4.73 (average)
Manganite	γ-MnOOH	Monoclinic	4.34
Hausmannite	Mn$_3$O$_4$	Tetragonal	4.76

Source: Dixon, J.B. and White, G.N., *Soil Mineralogy with Environmental Applications*, Soil Science Society of America, Madison, WI, 2002.

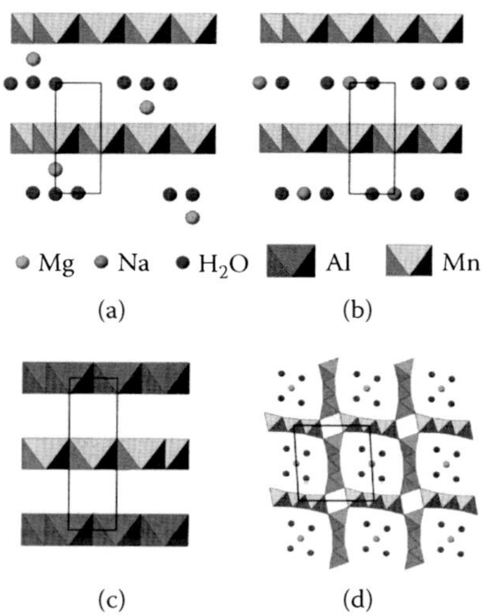

FIGURE 9.10 Polyhedral models of some Mn minerals: (a) birnessite with divalent (Mg) ions in the interlayer; (b) birnessite with monovalent (Na) ions in the interlayer—note that Na^+ ions are in the center, whereas Mg^{2+} ions in birnessite are closer to one of the sheets; (c) lithioporite, with an (Al, Li)-OH sheet stacked in the z direction; (d) todorokite, showing the channels containing cations surrounded by water molecules. (Reprinted from Dixon, J. B. and White, G. N., *Soil Mineralogy with Environmental Applications*, Soil Science Society of America, Madison, WI, 367–388, 2002, with kind permission.)

have been identified as capable of doing this (Tebo et al. 1997). Abiotic direct oxidation of Mn(II) by atmospheric O_2 has been demonstrated at pH > 8, yielding mixed Mn(III)-Mn(IV) oxides; in more acidic media, however, there is no conclusive evidence of the lack of abiotic oxidation paths, albeit in some reports the oxidation of Mn has been shown to occur in soils already containing Mn oxides (Wada et al. 1978; Ross and Bartlett 1981), thus suggesting an autocatalytic reaction. No further evidence, however, has been given to date.

Once formed, these Mn(III) and/or Mn(IV) oxides are strong oxidants. For Mn(IV), a generic standard potential is known:

$$MnO_2(s) + 4H^+(aq) + 2e^- \rightleftharpoons Mn^{2+}(aq) + 2H_2O(l) \quad E^o = 1.23\ V \tag{9.24}$$

Unfortunately, the actual values for different solid phases with the formula MnO_2, or other solids containing Mn(IV), are not known. For Mn(III) the following values are reported (James and Bartlett 1999):

$$Mn_3O_4 + 2e^- + 8H^+ \rightleftharpoons 3Mn^{2+} + 4H_2O \quad E^o = 1.82\ V \tag{9.25}$$

$$Mn_2O_3 + 2e^- + 6H^+ \rightleftharpoons 2Mn^{2+} + 3H_2O \quad E^o = 1.52\ V \tag{9.26}$$

$$\gamma\text{-}MnOOH + e^- + 3H^+ \rightleftharpoons Mn^{2+} + 2H_2O \quad E^o = 1.50\ V \tag{9.27}$$

The Mn^{3+} species is unstable and exists only in the presence of strong complexing agents; in soils it is almost always found in solid form.

All the aforementioned solid compounds of Mn(III) and Mn(IV) frequently form nodules or cover other minerals, thus generally have a relatively high area exposed (considering its mass). These surfaces are able to oxidize many inorganic and organic compounds (James and Bartlett 1999; Dixon and White 2002; Bartlett and Ross 2005). It is well known that Cr(III) is readily oxidized to Cr(VI) by Mn oxides; the oxidized product is a mobile pollutant and can be transported far from the oxidation site. Organic matter complexed Cr(III) is not so easily oxidized, so that it is used for tanning leather; however, in some cases, evidence of Cr oxidation in these conditions was reported. Natural Mn oxides have been shown to oxidize Co(II), As(III), and Se(IV). Co(III) is incorporated in the oxide structure, whereas As(V) and Se(VI) are released as anions. Oxidation of various organic compounds, both natural and pollutants, has been demonstrated, including hydroquinone and polyphenols. The mechanism usually involves sorption followed by electron transfer through surface–adsorbate oxygen bridges. The presence of intermediate Mn species (e.g., Mn[III]) remains uncertain; in some cases, it has been shown that Mn(III) sites in some Mn minerals are involved in the oxidation process.

It is worth noting that Mn(III) in solution is unstable, undergoing dismutation through

$$2Mn^{3+} + 2H_2O \rightleftharpoons Mn^{2+} + MnO_2 + 4H^+ \quad (9.28)$$

However, in the presence of complexing agents such as citrate, the reverse reaction occurs spontaneously; adsorption of organic compounds on MnO_2 surfaces favor the occurrence of this reaction; apparently, the organic compounds must be oxidizable by Mn(III) for the reaction to take place. Thus, Mn(III) is cycling in soil in the presence of organic matter, mostly coming from biological material (Bartlett and Ross 2005).

9.5 TITANIUM AND ZIRCONIUM OXIDE MINERALS

Both Ti and Zr form primary minerals in igneous environments, many of which persist in soils as dense minerals due to its resistance to weathering. Exceptions are titanium containing silicates, which dissolve, leaving soluble Ti species that result in secondary minerals. Ti and Zr minerals are often found in ancient sand dunes, being a source of these elements for commercial and industrial uses (Fitzpatrick and Chittleborough 2002).

9.5.1 Titanium Oxides

Titanium is the ninth-most-abundant element, and its oxides and other minerals are quite widespread. Both primary and secondary Ti minerals are found in soil (Table 9.4). There are several natural polymorphs of TiO_2, the most important being *rutile*, *anatase*, and *brookite*; others, much less common, have been described, including TiO_2 (B) and high pressure *columbite-*, *baddeleyite-*, *cotunnite-*, *pyrite-*, and *fluorite*-structured forms (Banfield, Veblen, and Smith 1991; Swamy, Gale, and Dubrovinsky 2001; Triebold et al. 2011). Rutile is the most common Ti mineral,

TABLE 9.4
Most Common Titanium Minerals

Mineral	Formula	Type and Habit	Crystalline System	Density (g·cm^{-3})
Rutile	TiO_2	Mostly primary, can also form from weathering; acicular, granular, or prismatic	Tetragonal	4.25
Anatase	TiO_2	Usually secondary; pyramidal or tabular	Tetragonal	3.9
Brookite	TiO_2	Low temperature rutile polymorph; platy or tabular	Orthorhombic	4.11
Ilmenite	$Fe(II)TiO_3$	Primary; lamellar or tabular	Trigonal-Rhombohedral	4.72
Ferrian ilmenite	$(FeTiO_3)_x(Fe_2O_3)_{1-x}$	Secondary, ilmenite-hematite solid solution	Trigonal-Rhombohedral	Varying
Pseudobrookite	$Fe(III)_2TiO_5$	Primary; crystalline fine	Orthorhombic	4.4
Pseudorutile	$Fe(III)_2Ti_3O_9$	Secondary, intermediate in ilmenite weathering; thin, irregular plates	Hexagonal	3.8
Ulvöspinel	$Fe(II)_2TiO_4$	Primary, usually on the surface of magnetite and ilmenite	Cubic	4.8 (calculated)

Source: Kämpf, N. et al., *Handbook of Soil Science*, CRC Press, Boca Raton, FL, 1999; Barthelmy, D., Mineralogy database, 2010, http://webmineral.com/.

composed of octahedral groups of Ti atoms surrounded by six O atoms, sharing the edges, forming chains in a structure similar to pyrolusite (Figure 9.11a). Anatase is a secondary TiO_2 mineral, considerably less abundant than rutile, where the octahedra share four O-O edges, two at one side and other two at right angles in the opposite side, outlining a three-dimensional network (Figure 9.11b). Brookite, a rather uncommon mineral, has a more complex structure, where the octahedra are deformed and share three O-O edges to form cross-linked chains (Figure 9.11c). These minerals often show substitution, typically Fe, Nb, and Ta in rutile and Fe(III) in anatase, with OH replacing O for charge balance. TiO_2 is also an important industrial supply, being used as a white pigment, as a photocatalyzer for applications such as effluent treatment, and so on. It is most commonly used, however, not in mineral form but as synthetic particles, obtained from rutile or, more usually, from ilmenite.

Besides the above-discussed TiO_2 forms, there are several Fe-Ti mixed oxides that deserve attention, included in Table 9.4. The most important is *ilmenite*, the second-most-abundant Ti mineral in soils; it has the ideal formula $Fe(II)TiO_3$ (52.65% TiO_2 and 47.35% w/w FeO), with a crystal structure similar to hematite (Figure 9.8b), with half of the Fe atoms replaced by Ti. The ideal composition is observed only in the absence of oxygen; in soils, ilmenite is found to contain a mix of FeO and Fe_2O_3,

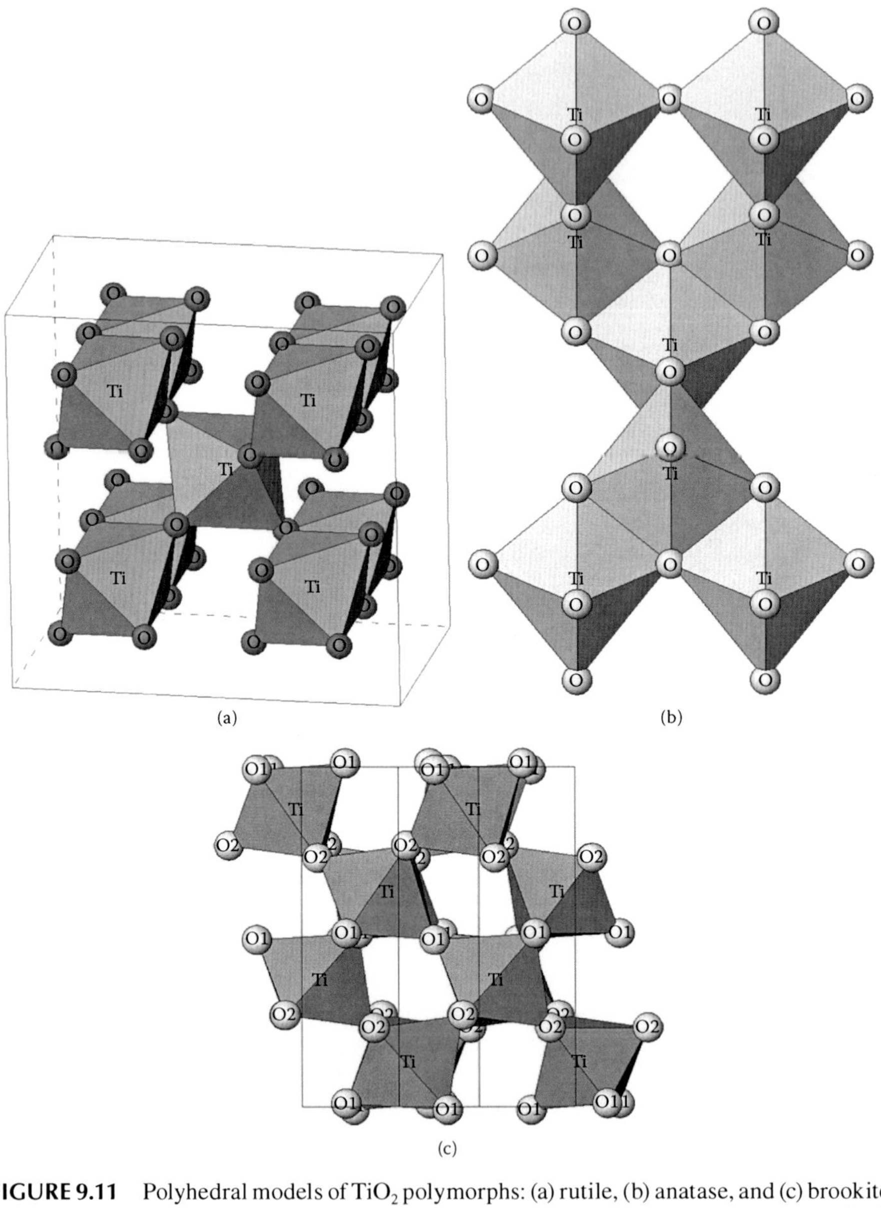

FIGURE 9.11 Polyhedral models of TiO_2 polymorphs: (a) rutile, (b) anatase, and (c) brookite. (Reprinted with kind permission from Dr. K. Nomura, Intermediate Temperature Solid Conductors—Crystal Structure Gallery—AIST, http://staff.aist.go.jp/nomura-k/english/itscgallary-e.htm.)

with the contents of TiO_2 increased up to 75% w/w due to oxidation and dissolution of Fe(II). *Ferrian ilmenites* $(FeTiO_3)_{1-x}$ $(Fe_2O_3)_x$ are considered to be the product of ilmenite oxidation at high temperatures (over 700°C), typically by fire. Ilmenite is often employed in the industrial synthesis of TiO_2. Pseudorutile is a disordered hydroxylated ferric titanate (ideally $Fe(III)_2Ti_3O_9$, in soils $Fe_2O_3 \cdot nTiO_2 \cdot mH_2O$ with

$3 \leq n \leq 5$ and $1 \leq m \leq 2$), consisting in intercalated goethite and rutile-like portions; it is primarily the result of ilmenite weathering. *Pseudobrookite* has two types of octahedral sites, strongly distorted, ideally occupied by Fe(III) and Ti(IV); it is believed to be formed by ilmenite oxidation at high temperatures, typically by fires. *Ulvöspinel* ($Fe(II)_2TiO_4$) is a common constituent of titaniferous magnetite iron ores; its structure is of an inverted spinel, as magnetite (Figure 9.8c), and forms solid solutions with this mineral, which are known as *titanomagnetites*. *Titanomaghemites* are oxidation products of titanomagnetites (Katsura and Kushiro 1961).

There are a number of transformations that Ti minerals may undergo, as shown schematically in Figure 9.12. Note that rutile, basically a primary mineral found in high-pressure, high-temperature igneous rocks, may also be formed by surface thermal alterations of secondary Ti minerals.

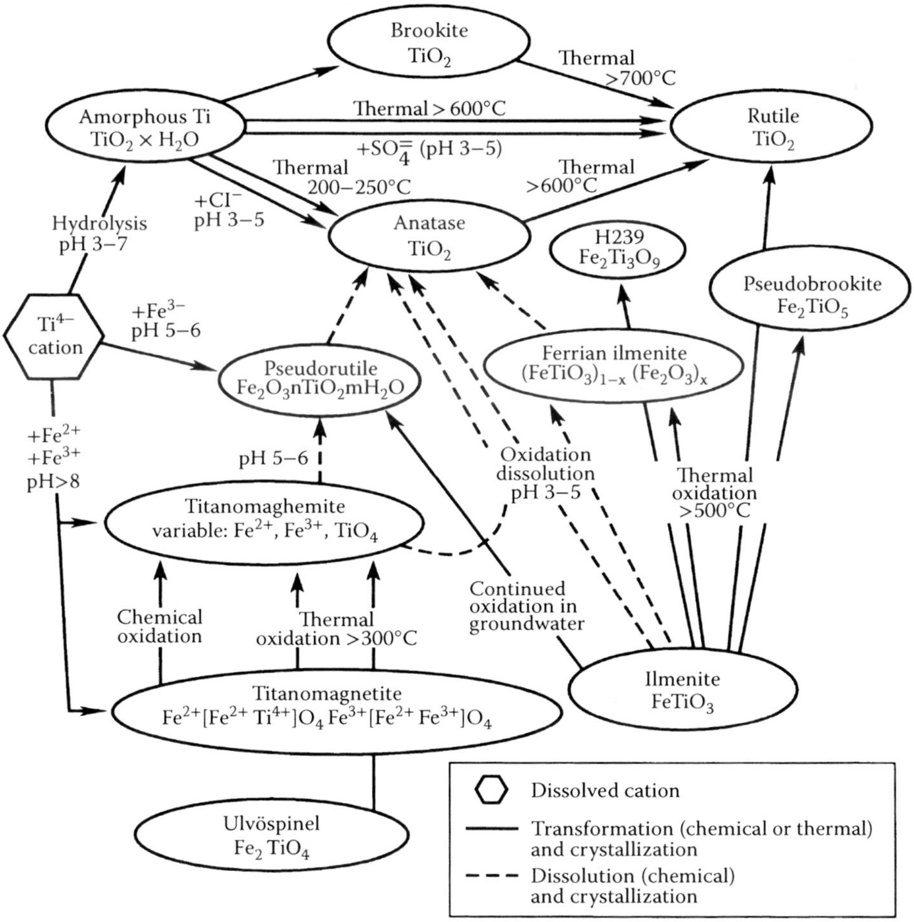

FIGURE 9.12 Transformation pathways of main titanium minerals upon weathering; thermal changes are caused by fire. (Reprinted from Fitzpatrick, R. W. and Chittleborough, D. J., *Soil Mineralogy with Environmental Applications*, Soil Science Society of America, Madison, WI, 667–690, 2002, with kind permission.)

9.5.2 Zirconium Minerals

The most abundant Zr mineral is *zircon*, $ZrSiO_4$, a tetragonal orthosilicate, frequently found in a prismatic, pyramidal habit, with Zr coordinated to eight oxygen atoms. A relatively common oxide is *baddeleyite*, ZrO_2, a monoclinic mineral found in prismatic habit. Other less common minerals are *calzirtite*, $CaZr_3TiO_9$ and *zirconolite*, $CaZr_3Ti_2O_7$, where zirconium appears associated with calcium and titanium. All these minerals can be found as primary or secondary. Hafnium is a common substituent for Zr, and rare earths such as yttrium are also found.

9.6 CHARACTERIZATION OF OXIDES

9.6.1 Identification and Crystalline Structure

As discussed in Chapter 7, there are a number of methods that can be employed in the characterization of mineral substances (Amonette 2002). As with all minerals, X-ray diffraction (XRD) (Section 7.3.1) is the prime identification method, through coincidence of the lattice distances for the main diagram lines, which usually leads to sample identification. An example for XRD patterns for several iron oxides is shown in Figure 9.13. The effect of crystallinity on the shape of the diffraction peaks, for example, in the case of the two goethite samples at the top left, or ferrihydrite and schwertmannite at the right, should be noted. On the contrary, comparing hematite, magnetite, and maghemite, some lines are coincident, due to the similar crystal structure, but others differ, allowing the proper identification. Because XRD analyses require a few percent contents for a particular mineral to be detectable, this method is more difficult to apply for less abundant elements, such as manganese; its minerals are rather rare, frequently found as nodules or covering other minerals. Thus, the experimental pattern quality decreases, as it can be observed for some Mn minerals in Figure 9.14. In such cases, differential XRD (DXRD) analysis may be useful; for instance, the lithiophorite pattern in Figure 9.14b was obtained as the difference between the original soil XRD pattern, and that obtained after dissolving Mn minerals with hydroxylamine (Golden, Dixon, and Kanehiro 1993).

9.6.2 Spectroscopic Studies

Spectroscopic characterization is also important in the study of oxide minerals. Among the many methods (Sections 7.3 through 7.7), it is important to mention X-ray fluorescence (XRF, Section 7.3.3), which allows a rapid elemental analysis of mineral samples. Electron microscopy allows observation of clay fraction particles shape and, coupled to X-ray spectroscopy, also elemental analysis (Section 7.5). Infrared spectroscopy (Section 7.4.3) is also frequently employed; here, difference spectra combined with selective extraction procedures (Hass and Fine 2010) can be used to identify and study minor components (Golden, Dixon, and Kanehiro 1993). Nuclear magnetic resonance (NMR, Section 7.4.4) is in some cases useful (with uneven spin nuclei), for example, to distinguish octahedral and tetrahedral Al centers (Bertsch and Parker 1996).

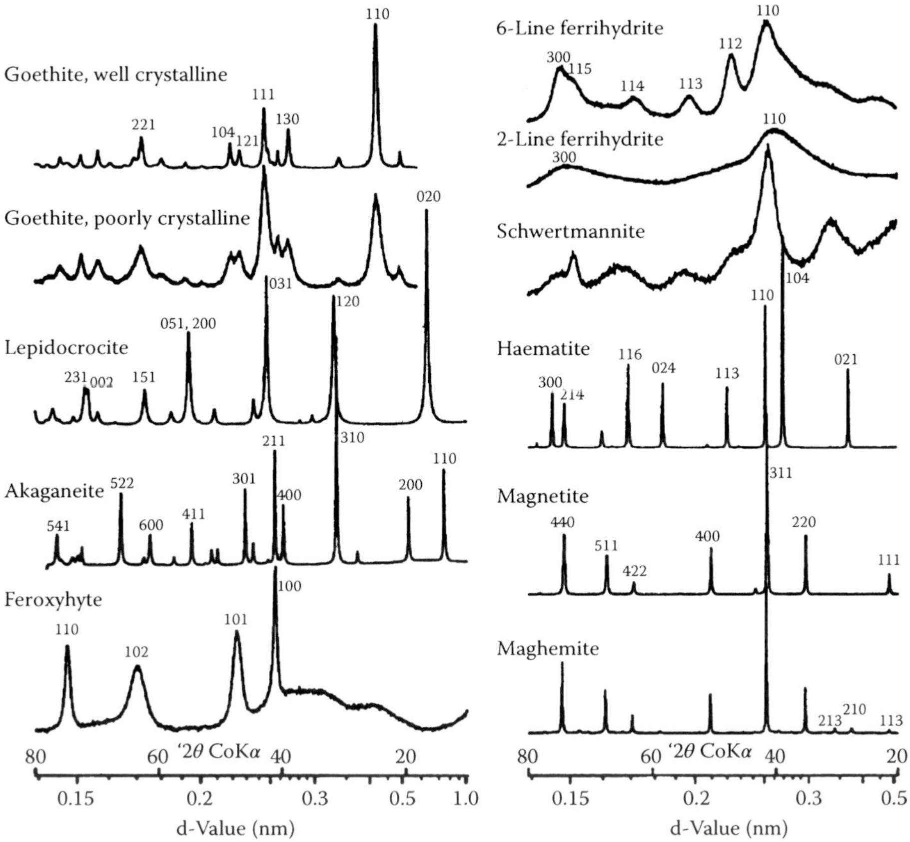

FIGURE 9.13 X-ray diffraction patterns for several common Fe oxides, including poorly crystalline compounds. (Cornell and Schwertmann: *The Iron Oxides: Structure, Properties, Reactions, Occurrences and Uses*. 2007. Copyright Wiley-VCH Verlag GmbH & Co. KGaA. Reproduced with permission.)

A particular type of spectroscopy, mainly applicable to iron minerals, is *Mössbauer spectroscopy*, which is based on the emission and absorption of γ rays by nuclei (Goldanskii and Herber 1969; Goodman 1980; Murad and Cashion 2004; Dickson and Berry 2005). Atomic nuclei undergo energy level transitions with absorption and emission of energy just as molecules, but the spacing between levels is higher, so that these transitions show up in the γ-ray region (Figure 7.10). These transitions show a hyperfine structure (i.e., very small changes of the transition energy) that is dependent on the electronic environment and applied electrical or magnetic fields, which can change or split the energy levels. These changes can provide information about the atom's state and local environment within a system and ought to be observed using resonance fluorescence. The basic setup would be to observe the interactions of the target atoms with γ-rays of an energy corresponding very precisely to a difference between two energy levels; this incident radiation should be generated by emission of the same element (resembling the atomic absorption spectrometry approach).

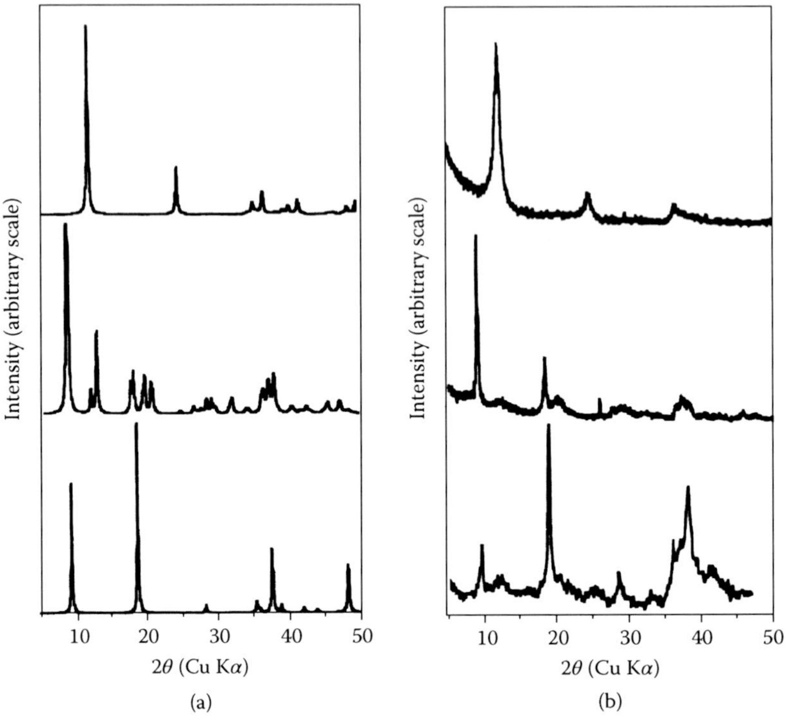

FIGURE 9.14 Comparison of calculated (a) and experimental (b) X-ray diffraction patterns for manganese minerals. Top patters: birnessite, calculated and natural from Quebec; middle: todorokite, calculated and natural from Cuba; bottom: lithioporite, calculated and from soil nodules. Note the broad peaks in the actual patterns, indicating small crystals. In the case of lithiophorite, the actual pattern was obtained using a subtraction procedure (differential X-ray diffraction). (Reprinted from Dixon, J. B. and White, G. N., *Soil Mineralogy with Environmental Applications*, Soil Science Society of America, Madison, WI, 367–388, 2002, with kind permission.)

There is a problem, however: when a photon is emitted from a particle (nucleus, molecules, etc.), there is a recoil effect, just as when a cannon fires a projectile. As a consequence, the photon energy is equal to the difference between levels minus the recoil energy, thus not matching exactly the transition; furthermore, because the recoil effect also affects the target nucleus upon photon absorption, its energy should be that of the transition *plus* the recoil energy. To complicate things further, Doppler effect causes an energy dispersion. What Rudolph Mössbauer discovered in 1957 is that in solid materials, under appropriate circumstances the recoil energy can be extremely low, because the whole solid recoils. Thus, if the emitting and absorbing nuclei are contained in solid matrices, resonance absorption is achieved. The Mössbauer effect is detected only in isotopes with very low lying excited states (low γ-ray energy); similarly the resolution is dependent on the lifetime of the excited state. These two factors limit the number of isotopes that can be used successfully for Mössbauer spectroscopy. The most used is ^{57}Fe; other isotopes commonly studied are ^{119}Sn, ^{121}Sb, ^{129}I, and ^{183}W. The vast majority of Mössbauer studies in soils is based on ^{57}Fe, as other soil-abundant elements do not show Mössbauer response. In situ applications have been reported (Feder et al. 2005), and it has also been employed in the study of Martian soils (Madsen et al. 2009).

9.6.3 Specific Surface Area and Surface Charge

The determination of specific surface area of oxide minerals can be performed by the Brunauer, Emmet, and Teller (BET) method with N_2 or by other techniques (Section 7.6.4); with the exception of some Mn minerals that have a layered structure, it is usually a straightforward measurement.

Surface charge determination is of high importance in the study of oxide minerals. These materials usually show a strong dependence of the charge on pH, as already discussed in Chapter 5. Most oxide minerals present amphoteric surface OH groups able to undergo acid–base reactions, which can be represented as

$$\text{M-OH} + \text{H}^+ \rightleftharpoons \text{M-OH}_2^+ \tag{9.29}$$

$$\text{M-OH} + \text{OH}^- \rightleftharpoons \text{M-O}^- + \text{H}_2\text{O} \tag{9.30}$$

These reactions determine the charge variability of most oxide minerals and consequently the pH dependence of the whole soil charge, for oxide-rich soils (Section 5.1.5).

The standard methods to measure the surface charge at different pHs are the measurement of electrophoretic mobilities (Section 5.4.1) and acid–base titration (Section 5.4.2) (see also Lyklema 1995; Goldberg, Lebron, and Suarez 1999; Kosmulski 2009a). The first method allows determination of the net particle charge, whereas titration brings the net protonic charge. Because oxide minerals usually do not have a measurable permanent charge, titration results directly give the intrinsic charge, and, in the absence of specific adsorption, the two methods should give coincident results. In the presence of specific adsorption, the net charge changes markedly (see Figure 5.25).

Typical charge versus pH (or mobility, or ζ potential, Section 5.4) curves can be seen in Figures 5.19 and 5.21. Aluminum, iron, and manganese oxides commonly show points of zero charge (pzc or pH_0, Table 5.3), roughly falling in the range 4–10. The most notable exception is quartz (and other silica minerals) that have very low pzc, because they usually do not develop negative charge easily (see the curves in Figure 5.22). Table 9.5 presents selected values of pH_0 for common oxide minerals, taken from the extensive compilation due to Kosmulski (Kosmulski 2009a, b, 2011). The pzc values of some soil mineral samples are compared with the approximate range observed in synthetic samples. It is well known that the pzc of a laboratory (or industrially) prepared solid sample shows variability, being dependent on the synthesis method, surface treatment, and so on; thus, it should be no surprise to find a high variability in natural samples, which can be affected by additional factors, such as the presence of impurities; for example, Al and Fe oxides may contain an appreciable percentage of SiO_2, which will lower the pzc value. It is not common to find experimentally determined the pH_0 of natural samples, due to difficulties

TABLE 9.5
Points of Zero Charge of Representative Oxide Minerals

Mineral	Conditions (Origin)	Points of Zero Charge
Corundum (Al_2O_3)	0.002 M KNO_3	9
Boehmite (γ-AlOOH)	Synthetic, various	7.7–9.5
Diaspore (α-AlOOH)	0.001 M KNO_3 (Great Wall)	5.7
	0.001 M KCl (Henan)	5.5
Gibbsite	(Steinbach)	4.8
	(New Caledonia)	5.2
	Synthetic, various	4.9–10.4
Magnetite (Fe_3O_4)	0.001–1 M KNO_3 (Quebec)	6.4
	0.01–1 M KCl (Serbia)	6.5
	0–0.01 M KNO_3 (Turkey)	5
	Synthetic, various	5–8
Hematite (α-Fe_2O_3)	0.01–0.1 M KNO_3 (Middleback, South Australia)	6.8
	0.0001 M KNO_3 (Labrador)	6.9
	0.001–1 M KNO_3, $NaClO_4$ (Quebec)	5.2
	0.0005 M $NaNO_3$ (Vesuvius)	6.1
	Synthetic, various	5.2–9.5
Goethite (α-FeOOH)	0.01 M $NaNO_3$ (Pikes peak batholith)	7.0
	$NaClO_4$ (concentration not specified) (Aue)	9.5
	Synthetic, various	6.7–10.4
Lepidocrocite (γ-FeOOH)	NaCl (concentration not specified) (Siegen Germany)	5.4
	Synthetic, various	5.8–8
Manganite (γ-MnOOH)	0–0.1 M $NaNO_3$ (Ilfeld, Germany; Pawling mine, New York)	5.4
	Synthetic, various	6.2–9.6
Pyrolusite (α-MnO_2)	0.001 M NaCl (UmBogma, Sinai)	6.4
	0.01 M $NaNO_3$ (Sitpar mine, India)	4.3
	0.01 M KNO_3 (Sitpar mine, India)	6.7
	Synthetic, various	3.3–7.3
Birnessite ($K_4Mn_{14}O_{27} \cdot 9H_2O$)	Synthetic	<2.0
Quartz (SiO_2)	0.001–0.1 M KCl (Argentina)	2.3
	0.0001 M NaCl (Brazil)	1–2
	0.001 M NaCl (Fontainebleau, France)	2.4 (electrophoresis)
	Synthetic, various	<1–4.7
Obsidian (70–75% SiO_2)	$NaClO_4$ (concentration not specified) (Iceland)	2
Rutile (TiO_2)	0.001–1 M KNO_3 (Mexico)	5.3
	0.1 M NaCl (North Carolina)	4.1
	Synthetic, various	3.5–7.2

Source: Kosmulski, M., *Surface Charging and Points of Zero Charge*. CRC Press, Boca Raton, FL, 2009a; *J. Colloid Interface Sci.*, 337, 2, 439–448, 2009b; *J. Colloid Interface Sci*. 353, 1, 1–15, 2011.

in proper sample isolation; for example, for boehmite, there is a lack of reliable values, because it is commonly found associated with other minerals in bauxite. It is interesting to note that for several oxides the pzc of natural samples tends to be lower than for synthetic samples; notably in the case of hematite, where soil samples often have $pH_0 < 7$, whereas the bulk of synthetic hematite pzc values reported fall between 8 and 9.

9.6.4 Miscellaneous Methods

Thermal methods of analysis are often useful in the characterization of minerals, as described in Section 7.6.5. Aluminum hydroxides such as gibbsite show a mass loss of 34.6% on dehydroxylation; thus, they show an important negative peak in DTA and a marked mass loss in TGA, and so these techniques are employed for both qualitative and quantitative characterization of these minerals. The same happens with iron hydroxides and oxohydroxides, such as goethite, lepidocrocite, and so on; also, the presence of OH groups in otherwise thermally inert minerals such as hematite can be detected.

Some minerals, most notably some iron oxides, have relatively marked magnetic properties that can affect soil properties and also are useful in its characterization. When a magnetically susceptible sample is subject to a magnetic field H of low intensity, it acquires a magnetization $M = \kappa H$, where κ is the volume susceptibility; the mass susceptibility is defined as $\chi = \kappa/\rho$, where ρ is the sample density. Ferromagnetic minerals are those that have high susceptibilities and also show a remnant magnetization after a strong field was applied and removed, whereas paramagnetic materials have low susceptibilities and do not show remnant magnetization (Butler 1992, Chapter 2; Buschow and Boer 2003; Evans and Heller 2003). Common ferromagnetic minerals include magnetite, maghemite, titanomagnetite, and titanomaghemite, which have values of χ of the order of 10^{-4} m^3 kg and are attracted by small hand-magnets revealing their presence in a sample. The magnetic susceptibility of a mixed sample is mainly determined by the concentration of ferromagnetic minerals and influenced by their size, composition, and shape. There are several methods to measure magnetic susceptibilities (Thompson and Oldfield 1986). Magnetism has important environmental effects (Thompson and Oldfield 1986; Evans and Heller 2003), and the magnetic response is used as a way to separate titanium minerals (Fitzpatrick and Chittleborough 2002).

9.7 ION BINDING TO OXIDE MINERALS

Oxide minerals, specially iron and aluminum (hydr)oxides, show an important activity toward ion binding. Unlike silicates, specific interactions play a paramount role here, because the surface hydroxyl groups are quite reactive. It should be noted that even "pure" oxides such as Al_2O_3 and Fe_2O_3 have surface hydroxyls because the "dangling bonds" of surface O atoms become OH groups. Albeit these minerals develop a surface charge through protonation/deprotonation reactions (such as Equations 9.29 and 9.30) and thus are able to bind both positive and negative ions electrostatically, this way of binding is less important.

It should be noted that minerals, and oxide minerals in particular, have different types of OH groups, depending on the coordination of the O atoms, as revealed by spectroscopic studies. Goethite (α-FeOOH) has four types of surface hydroxyls whose reactivities are a function of the coordination environment of the O in the FeOH group (Sposito 1984; Sparks 2002). The FeOH groups are A-, B-, or C-type sites, depending on whether the O is coordinated with 1, 3, or 2 adjacent Fe(III) ions. The fourth type of site is a Lewis acid-type site, which results from chemisorption of a water molecule on a bare Fe(III) ion. Only A-type sites are basic (can bind H^+), and, on the contrary, A-type and Lewis acid sites can release a proton. The B- and C-type sites are considered unreactive. Thus, A-type sites can be either a proton acceptor or a proton donor (i.e., they are amphoteric). Other spectroscopic studies have shown that boehmite (γ-AlOOH) and lepidocrocite (γ-FeOOH) have two types of OH, presumably associated with different crystal faces (Lewis and Farmer 1986).

Modern spectroscopic techniques, especially X-ray absorption methods (Section 7.3.2) have shed light on the interactions between oxide surfaces and ions, both cations and anions. In the following, a description of both cation and anion binding with representative examples is given.

9.7.1 Cation Binding

Cation binding to oxide minerals has been extensively investigated (Sposito 1984; McBride 1994; McBride 1999; Sparks 2002). Pb(II), in particular, has been the subject of much attention; here, as in other cases, X-ray absorption methods have played a key role in the study of Pb(II) adsorption on Al_2O_3 (Chisholm-Brause et al. 1990) and goethite (Roe et al. 1991). In the first case, analysis of radial structure functions yielded one O atom at 2.23 Å and two O at 2.46 Å. Further analysis for a relatively high metal concentration experiment yielded approximately one second-neighbor Pb atom at 3.45 Å and one second-neighbor Al atom at 3.72 Å, in addition to the three first-neighbor O ligands. These data led the authors to suggest that Pb(II) bonds directly to the γ-Al_2O_3 surface as an inner-sphere complex, discarding outer-sphere complexes, surface precipitates, or diffusion into the adsorbent; the presence of Pb atoms in the second coordination shell suggests that at least some Pb is sorbed as small multinuclear complex. On the basis of the presence of second-neighbor Al and the observed interatomic distances, the authors infer that the adsorption site is monodentate. In the case of Pb(II) onto goethite, the adsorption isotherms were successfully modeled assuming the formation of inner-sphere, monodentate, polymeric Pb(II) complex. X-ray absorption spectroscopy (XAS) studies have shown that the controlling sorption process depends on the relative and absolute concentrations of sorbate and sorbent. At low coverages, XAS results suggest that Pb forms inner-sphere surface complexes (Figure 9.15a); at higher coverages, the formation of surface polymers of Pb is suggested (Figure 9.15b). No evidence was obtained supporting the formation of a surface precipitate at high coverages. In a later study (Bargar, Brown, and Parks 1997), XAFS measurements were interpreted considering the formation of inner-sphere mononuclear bidentate complexes of Pb(II) species to edges of FeO_6 octahedra on both goethite and hematite, where each Pb(II) is bonded to the surface through two O-OH groups, retaining at least one OH ligand

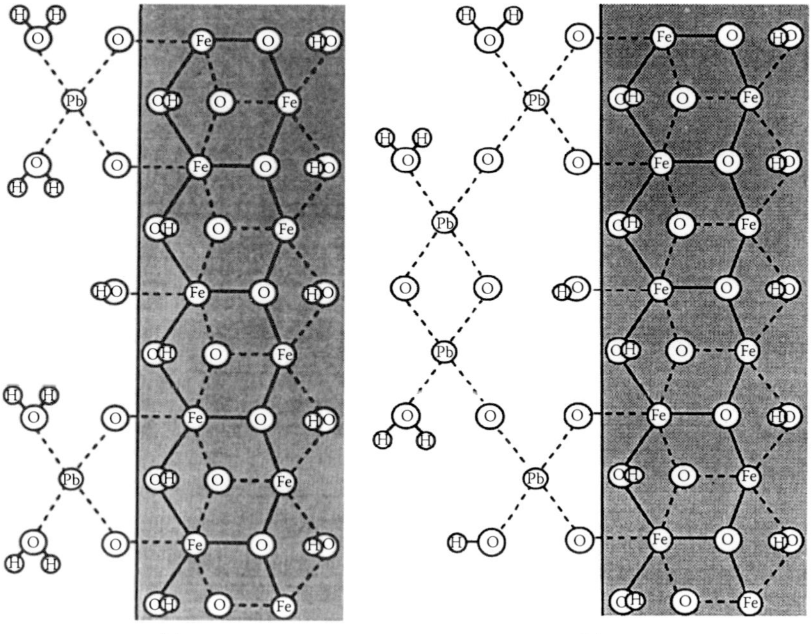

FIGURE 9.15 Schematic representation of structural models of Pb(II) adsorption onto goethite surfaces. (a) Inner-sphere surface complex; (b) surface polymerization anchored by some inner-sphere bound Pb(II) ions. (Reprinted with permission from Roe et al. 1991, 367–373. Copyright 1991 American Chemical Society.)

on the solution side. In another research, the binding of Pb(II) on pristine and anion-modified goethite and boehmite surfaces was studied by electrophoretic mobility and XAS (Weesner and Bleam 1998); as in previous studies, lead is found to bind as inner-sphere complex on goethite and as oxy-clusters on boehmite. Phosphate and sulfate anions increase the ability of the solids to sorb Pb(II); phosphate acts as a reactant to form lead phosphate surface phases that may be highly dispersed. The sulfate ion is more effectively adsorbed when lead is present and increases its adsorption on boehmite. More recently, Pb(II) sorption onto ferrihydrite was studied by traditional equilibrium methods complemented with XAS analyses. It was found that at pH > 5, edge-sharing bidentate complexes are mainly formed on the oxide surface (Figure 9.16a); at pH = 4.5; a mixture of monodentate and bidentate sorption complexes at the oxide surface is suggested (Figure 9.16b and 9.16c). It has also been studied the adsorption of Pb(II) on Mn oxides (Boonfueng et al. 2006; Zhao et al. 2010). In general, all studies show that Pb adsorbs onto oxides by inner-sphere complexes, that is, by specific adsorption; little or no evidence has been found of outer-sphere complexes.

Zn adsorption onto silica and gibbsite was also studied by XAS methods; here EXAFS spectroscopy was employed to observe both the sorbed Zn environment and the kinetics of the process (Roberts, Ford, and Sparks 2003). In the case of silica, EXAFS revealed that Zn was in octahedral coordination with first-shell

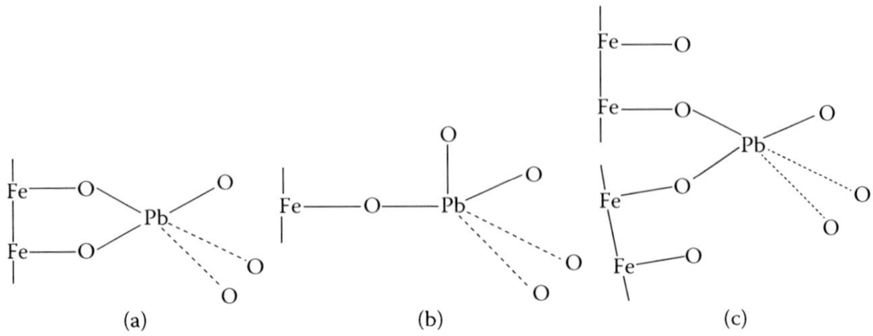

FIGURE 9.16 Schematic representations of predominant Pb(II)–ferrihydrite sorption complexes based on the interpretations of the X-ray absorption spectra analyses at (a) bidentate edge-sharing mononuclear; (b) monodentate mononuclear; and (c) bidentate corner-sharing mononuclear. At pH > 5.0, only configuration (a) is observed. At pH < 5.0, either a combination of configurations (a) and (b) or a combination of configurations (b) and (c) is expected. (Reprinted from *Journal of Colloid and Interface Science*, 205, Weesner and Bleam, Binding characteristics of Pb^{2+} on anion-modified and pristine hydrous oxide surfaces studied by electrophoretic mobility and X-ray absorption spectroscopy, 380–389. Copyright 1998 with permission from Elsevier.)

oxygen atoms up to a surface loading of approximately 1 μmol m^{-2}, changing to tetrahedral coordination as surface loading and pH increased. For the high-surface-area gibbsite system, the Zn–O first-shell distance was intermediate between values for tetrahedral and octahedral coordination over all loading levels. Zn formed inner-sphere adsorption complexes on both silica and high-surface-area gibbsite over all reaction conditions. For Zn sorption on low-surface-area gibbsite, formation of Zn–Al-layered double hydroxide occurred. The highest pH sample (7.51) in the Zn–amorphous silica system resulted in the formation of an amorphous $Zn(OH)_2$ precipitate with tetrahedral coordination between Zn and O. For both amorphous silica and high-surface-area gibbsite, Zn sorption kinetics was relatively rapid, completing within 24 hours. In contrast, Zn sorption on low-surface-area gibbsite was much slower, taking nearly 800 hours for a sorption plateau to be reached, due to the hydroxide formation. In another investigation, Zn(II) adsorption onto anatase was studied by EXAFS and XANES and modeled theoretically by density functional theory (He et al. 2011). A coordination number of average ~4.5 O atoms around a central Zn atom was obtained by EXAFS analysis. Figure 9.17 shows the structures considered, including the aqueous Zn(II) complex (Figure 9.17a). DFT calculations indicated that this coordination structure was consistent with the mixture of four-coordinated bidentate binuclear (Figure 9.17c) and five-coordinated bidentate mononuclear (Figure 9.17d) metastable equilibrium adsorption states. The monodentate mononuclear (Figure 9.17b) complex has six-coordinated Zn and was found to be less stable than the metastable equilibrium state.

Other X-ray spectroscopic studies include cadmium adsorption on aluminum oxides (Papelis et al. 1995), where it was found that Cd–O distances and coordination numbers (6 oxygens) on two aluminas agree with Cd–O distances and coordination in aqueous solutions. In low sorption density samples, absence of cadmium

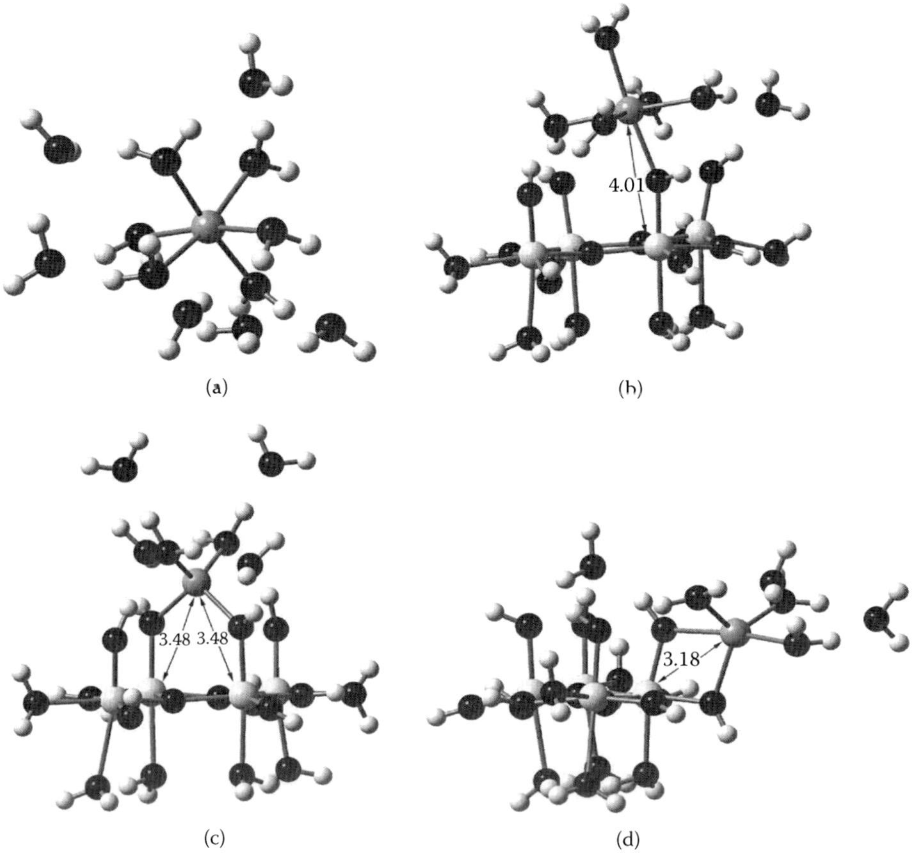

FIGURE 9.17 Calculated Zn(II)–TiO_2 surface complexes from density functional theory: (a) dissolved Zn(II) with six outer-sphere water molecules; (b) monodentate mononuclear; (c) bidentate binuclear (BB); (d) bidentate mononuclear (BM). Gray, black, large white, and small white circles denote Zn, O, Ti, and H atoms, respectively. (Reprinted with permission from He et al. 2011, 1873–1879. Copyright 2011 American Chemical Society.)

second neighbors suggests that the sorption complexes are mononuclear. In a sample prepared from a higher total cadmium concentration solution, in which the formation of a precipitate was observed during sample preparation, the presence of a Cd-Cd correlation suggests the formation of a precipitate or small multinuclear complexes. Absence of aluminum second neighbors indicates that either cadmium form outer-sphere complexes, or that aluminum second neighbors cannot be detected because of the low backscattering amplitude of aluminum or disorder effects. The ionic strength dependence of the adsorption suggests relatively strong, probably inner-sphere binding, so the latter explanation is more likely.

Of special importance is the adsorption of radionuclides onto soil components. The adsorption of Np(V) onto goethite has been studied by Combes et al. (Combes et al. 1992), finding that sorbed Np(V) is present as a mononuclear species similar to that of dissolved NpO_2^+ in water. The results do not rule out the possibility that the surface

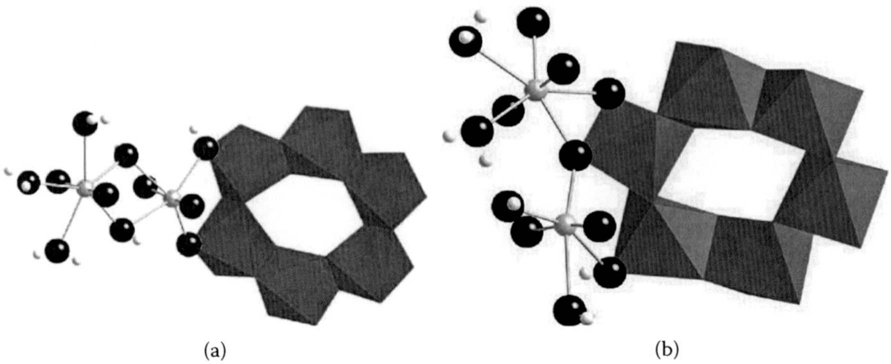

FIGURE 9.18 Optimized structures of uranyl-adsorbed species as proposed by Hattori et al. (a) Uranyl dimer in corner-sharing arrangement; (b) the corner- and the edge-sharing configuration next to each other. Black, gray, and white spheres represent O, U, and H atoms, respectively; gray octahedra represent $Al(OH)_6$ structures. (Reprinted from *Geochimica et Cosmochimica Acta*, 73, Hattori et al., The structure of monomeric and dimeric uranyl adsorption complexes on gibbsite: A combined DFT and EXAFS study, 5975–5988, Copyright 2009, with permission from Elsevier.)

complex may include hydroxide or carbonate ligands but provide permissive evidence for the formation of inner-sphere complexes of Np at the goethite/water interface, thus having relatively strong, specific binding. Uranyl adsorption onto oxides has been studied by several methods (Hattori et al. 2009; Galindo et al. 2010; Malin et al. 2011; Zhao et al. 2012). In a combined density functional and EXAFS study (Hattori et al. 2009) of uranyl adsorption onto gibbsite, adsorption to basal faces was discarded in favor of edge adsorption; furthermore, EXAFS results are consistent with dimeric adsorption. Two structural models are proposed: in the acidic region, a distortion of the uranyl dimer, to form both a corner-sharing and an edge-sharing linkage to neighboring Al octahedra, is proposed (Figure 9.18b). In the alkaline region, a corner-sharing uranyl dimer complex is the most favorable (Figure 9.18a).

These examples (by no means exhaustive) illustrate the general aspects of cation adsorption onto oxide minerals: inner-sphere complex formation, preferentially edge or corner binding as opposed to basal plane binding, eventually forming small polynuclear complexes or surface polymers but more rarely surface precipitates. All that is true of transition metal and p-block representative cations; alkaline metals normally bind only electrostatically, and alkali earth metals most often do the same.

9.7.2 Anion Binding

Much like cations, oxyanions adsorb strongly onto oxides, usually by a ligand exchange mechanism (Parfitt 1979; Sposito 1984; McBride 1999; Sparks 2002), which can be represented as follows

$$Me \cdots OH + PO_4^{3-} + H^+ \rightleftharpoons Me \cdots OPO_3^{2-} + H_2O \tag{9.31}$$

$$\begin{matrix} Me\cdots OH \\ Me\cdots OH \end{matrix} + PO_4^{3-} + 2H^+ \rightleftharpoons \begin{matrix} Me\cdots O\cdot \\ Me\cdots O\cdot \end{matrix} PO_2^- + 2H_2O \quad (9.32)$$

where Me represents a metal ion (such as Fe(III), Al(III), etc.), here exemplified for phosphate anions, but it is equally applicable to other oxyanions. Equation 9.31 represents a monodentate binding, whereas Equation 9.32 corresponds to bidentate binding. The protonation state is of course dependent on pH.

These processes have also been intensely investigated, especially for pollutants as As and Se anions. A classic case is phosphate, which has been investigated by diverse methods (Celi, Barberis, and Marsan 2000; Lefèvre 2004; Zhang et al. 2009; Castaldi et al. 2010 and references therein). Recently (Kim et al. 2011), the adsorption of PO_4^{3-} species onto iron oxohydroxides has been investigated using ^{31}P static spin-echo mapping NMR experiments. The results confirmed the presence of Fe–O–P covalent bonds, in turn confirming inner-sphere, specific adsorption. Very similar broad lineshapes are observed for dimethylphosphinic acid (DPA) versus PO_4^{3-} binding, but the ^{31}P hyperfine shifts and the linewidths for the phosphate spectra are essentially twice those observed for DPA binding, providing compelling evidence for bidentate binding (of one phosphate anion to two neighboring Fe^{3+} ions) in the case of the phosphate ion. Small ^{31}P shifts to lower frequency were, however, observed as the pH was increased, and they are tentatively ascribed to the presence of a small concentration of monodentate species at higher pH. Figure 9.19 shows the proposed binding sites.

Arsenic anion adsorption onto oxide minerals has been studied by numerous researchers (Waychunas et al. 1993; Manceau 1995; Manning, Fendorf, and Goldberg 1998; Foster, Brown, and Parks 2003; Wang et al. 2009). It has been concluded from EXAFS studies that the predominant adsorption mode of both As(V) and As(III) is bidentate binding (Figure 9.20a), mainly corner-sharing, but with some contribution of edge-sharing adsorption (Figure 9.20b), on representative Fe and Mn oxides. Very similar results are found for SeO_4^{2-} adsorption onto the same minerals and have also been proposed on TiO_2 (Svecova et al. 2011).

Chromate adsorption onto ferrihydrite was studied by attenuated total reflection IR spectroscopy (ATR-IR) and theoretical frequency calculations (Johnston and Chrysochoou 2012). The calculations were done for several model arrangements, shown in Figure 9.21. It was found that the results are consistent with the formation of monodentate and bidentate surface complexes. Monodentate complexes (Figure 9.21e) are dominant at low surface coverage and pH ≥ 6.5, and bidentate complexes (Figure 9.21f) form at high surface coverage and pH < 6.

The preceding examples consistently show that oxyanion adsorption on common oxides (of Al, Fe, Mn, Ti) leads to strong specific binding, commonly through bidentate covalent complexes and in some cases with monodentate binding as well.

Among simple, non-O-containing anions, fluoride is one of the most active and important, and the adsorption of F^- has been investigated due to its environmental importance (Fan, Parker, and Smith 2003; Mohapatra et al. 2009 and references therein). Fluoride has been found to enter the substrate crystalline lattice by solid diffusion and/or ion exchange mechanism (Fan, Parker, and Smith 2003).

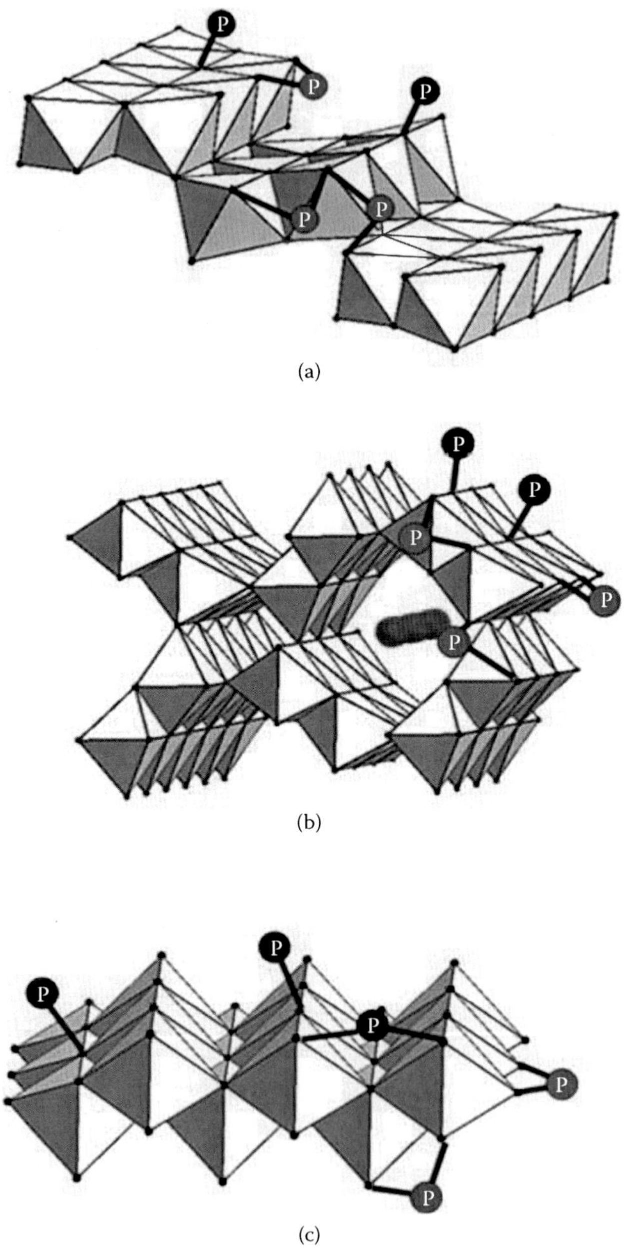

FIGURE 9.19 The crystal structures showing the possible adsorption sites of phosphate on the surface of (a) goethite (α-FeOOH), (b) akaganeite (β-FeOOH), and (c) lepidocrocite (γ-FeOOH). The octahedra represent FeO_6 groups. Hydrogen atoms are on split positions in (b), and (c) Cl^- is at the center of the tunnels in (b) The "P" labeled balls represent adsorbed phosphate. Black balls represent an unfavorable binding site, and gray balls represent a favorable site. For clarity the nonbonded oxygen atoms of the phosphate ions are omitted. (Kim et al. 2011. Reproduced by permission of the Royal Society of Chemistry.)

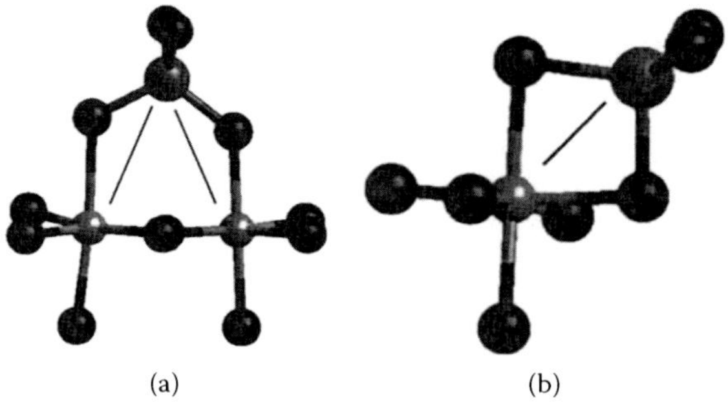

FIGURE 9.20 Idealized geometry of As(V) adsorption complexes on Mn oxides based on sorption complex geometries proposed for As(V) adsorption on hydrous Fe(III) oxides. The sorption complexes are illustrated for As(V), but for Se (IV) adsorption the geometries are almost identical. (a) Corner-sharing complex is generally agreed to be the predominant complex formed by sorbed As(V) and Fe(III) oxides; (b) edge-sharing complex is assumed to show minority sorption on the same substrates. The light gray spheres represent Mn atoms; whereas the large black spheres represent As; and the small, black spheres, O atoms. (Reprinted from *Geochimica et Cosmochimica Acta*, 67, Foster, Brown, and Parks, X-ray absorption fine structure study of As(V) and Se(IV) sorption complexes on hydrous Mn oxides, 1937–1953, Copyright 2003, with permission from Elsevier.)

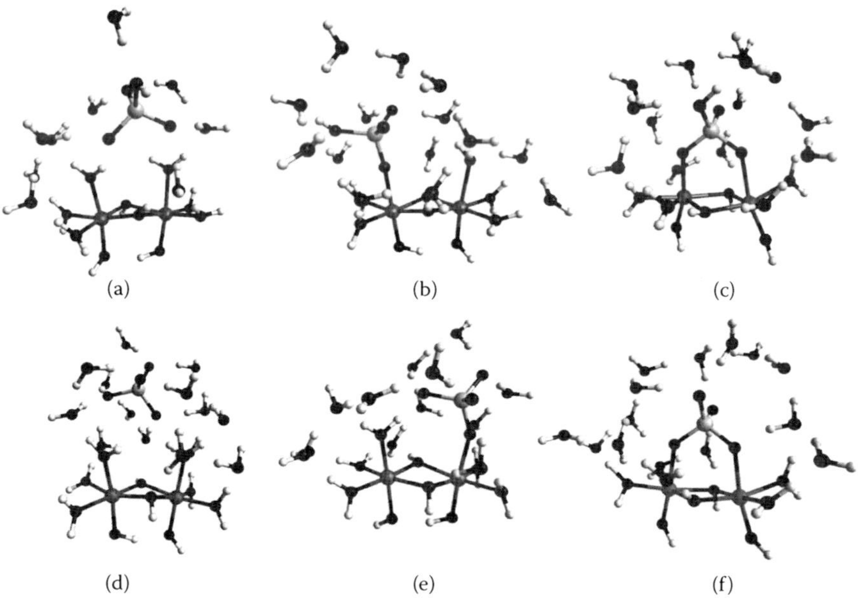

FIGURE 9.21 Molecular models employed in the theoretical analysis of the interaction of chromate anions with ferrihydrite. Protonated (top) and nonprotonated (bottom) cluster models used to calculate infrared frequencies of chromate coordinated to iron oxide: outer-sphere (a, d) monodentate (b, e) and bidentate (c, f). Some explicit water molecules were removed for clarity. Black, dark gray, light gray, and white spheres represent O, Fe, Cr, and H atoms, respectively. (Reprinted with permission from Johnston and Chrysochoou 2012, 5851–5858. Copyright 2011 American Chemical Society.)

9.8 SUMMARY

In this chapter, the structure of oxide minerals has been reviewed. With some exceptions (notably silica and some Si minerals), octahedral coordination around metal cations is the rule for the oxide structure. Oxide minerals (again with the exception of silica) have usually active surfaces with highly reactive hydroxyl groups. In particular, their surfaces are active ion adsorbents, mostly reacting to give specific adsorption, forming strong covalent bonds. Iron, aluminum, and manganese oxides are the most reactive.

REFERENCES

Amonette, J. E. 2002. Methods for determination of mineralogy and environmental availability. In *Soil Mineralogy with Environmental Applications*, ed. J. B. Dixon and D. G. Schulze, 153–197. Soil Science Society of America Book Series 7. Madison, WI: Soil Science Society of America.

Anderson, M. A., and P. M. Bertsch. 1988. Dynamics of aluminum complexation in multiple ligand systems. *Soil Science Society of America Journal* 52, no. 6: 1597–1602. doi:10.2136/sssaj1988.03615995005200060015x.

Banfield, J. F., D. R. Veblen, and D. J. Smith. 1991. The identification of naturally occurring TiO2 (B) by structure determination using high-resolution electron microscopy, image simulation, and distance-least-squares refinement. *American Mineralogist* 76, no. 3–4: 343–353.

Bargar, J. R., G. E. Brown Jr., and G. A. Parks. 1997. Surface complexation of Pb(II) at oxide-water interfaces: II. XAFS and bond-valence determination of mononuclear Pb(II) sorption products and surface functional groups on iron oxides. *Geochimica et Cosmochimica Acta* 61, no. 13: 2639–2652. doi:10.1016/S0016-7037(97)00125-7.

Barthelmy, D. 2010. Mineralogy database. http://webmineral.com/.

Bartlett, R. J., and D. S. Ross. 2005. Chemistry of redox processes in soils. In *Chemical Processes in Soils*, ed. M. A. Tabatabai and D. L. Sparks, 461–487. Madison, WI: Soil Science Society of America.

Bertsch, P. M., and D. R. Parker. 1996. Aqueous polynuclear aluminum species. In *The Environmental Chemistry of Aluminum*, ed. G. Sposito, 117–168. 2nd ed. Boca Raton, FL: CRC Press.

Bigham, J. M., R. W. Fitzpatrick, and D. G. Schulze. 2002. Iron oxides. In *Soil Mineralogy with Environmental Applications*, ed. J. B. Dixon and D. G. Schulze, 323–366. Soil Science Society of America Book Series 7. Madison, WI: Soil Science Society of America.

Bigham, J. M., U. Schwertmann, S. J. Traina, R. L. Winland, and M. Wolf. 1996. Schwertmannite and the chemical modeling of iron in acid sulfate waters. *Geochimica et Cosmochimica Acta* 60, no. 12: 2111–2121. doi:10.1016/0016-7037(96)00091-9.

Boonfueng, T., L. Axe, Y. Xu, and T. A. Tyson. 2006. Nickel and lead sequestration in manganese oxide-coated montmorillonite. *Journal of Colloid and Interface Science* 303, no. 1: 87–98. doi:10.1016/j.jcis.2006.07.056.

Buschow, K. H. J., and F. R. Boer. 2003. *Physics of Magnetism and Magnetic Materials*. New York: Kluwer Academic Publishers.

Butler, R. F. 1992. *Paleomagnetism: Magnetic Domains to Geologic Terranes*. Oxford, UK: Blackwell Scientific Publications. http://www.geo.arizona.edu/Paleomag/book/.

Castaldi, P., M. Silvetti, G. Garau, and S. Deiana. 2010. Influence of the pH on the accumulation of phosphate by red mud (a bauxite ore processing waste). *Journal of Hazardous Materials* 182, no. 1–3: 266–272. doi:10.1016/j.jhazmat.2010.06.025.

Celi, L., E. Barberis, and F. A. Marsan. 2000. Sorption of phosphate on goethite at high concentrations. *Soil Science* 165, no. 8: 657–664.

Chisholm-Brause, C. J., K. F. Hayes, A. L. Roe, G. E. Brown Jr., G. A. Parks, and J. O. Leckie. 1990. Spectroscopic investigation of Pb(II) complexes at the γ-Al2O3/water interface. *Geochimica et Cosmochimica Acta* 54, no. 7: 1897–1909. doi:10.1016/0016-7037(90)90260-R.

Combes, J. M., C. J. Chisholm-Brause, G. E. Brown, G. A. Parks, S. D. Conradson, P. G. Eller, I. R. Triay, D. E. Hobart, and A. Miejer. 1992. EXAFS spectroscopic study of neptunium(V) sorption at the alpha-iron hydroxide oxide (alpha-FeOOH)/water interface. *Environmental Science and Technology* 26, no. 2: 376–382. doi:10.1021/es00026a020.

Cornell, R. M., and U. Schwertmann. 2007. *The Iron Oxides: Structure, Properties, Reactions, Occurrences and Uses*. Weinheim, Germany: Wiley-VCH.

Deer, W. A., R. A. Howie, and J. Zussman. 1996. *An Introduction to the Rock-Forming Minerals*. 2nd ed. Upper Saddle River, NJ: Prentice Hall.

Dickson, D. P. E., and F. J. Berry, eds. 2005. *Mössbauer Spectroscopy*. New York: Cambridge University Press.

Dixon, J. B., and D. G. Schulze, eds. 2002. *Soil Mineralogy with Environmental Applications*. Soil Science Society of America Book Series 7. Madison, WI: Soil Science Society of America.

Dixon, J. B., and G. N. White. 2002. Manganese oxides. In *Soil Mineralogy with Environmental Applications*, ed. J. B. Dixon and D. G. Schulze, 367–388. Soil Science Society of America Book Series 7. Madison, WI: Soil Science Society of America.

Drits, V. A., B. A. Sakharov, A. L. Salyn, and A. Manceau. 1993. Structural model for ferrihydrite. *Clay Minerals* 28, no. 2: 185–207.

Eggleton, R. A., and R. W. Fitzpatrick. 1988. New data and a revised structural model for ferrihydrite. *Clays and Clay Minerals* 36, no. 2: 111–124. doi:10.1346/CCMN.1988.0360203.

Evans, M., and F. Heller. 2003. *Environmental Magnetism, Volume 86: Principles and Applications of Enviromagnetics*. 1st ed. San Diego, CA: Academic Press.

Fan, X., D. J. Parker, and M. D. Smith. 2003. Adsorption kinetics of fluoride on low cost materials. *Water Research* 37, no. 20: 4929–4937. doi:10.1016/j.watres.2003.08.014.

Feder, F., F. Trolard, G. Klingelhöfer, and G. Bourrié. 2005. In situ Mössbauer spectroscopy: Evidence for green rust (fougerite) in a gleysol and its mineralogical transformations with time and depth. *Geochimica et Cosmochimica Acta* 69, no. 18: 4463–4483. doi:10.1016/j.gca.2005.03.042.

Fitzpatrick, R. W., and D. J. Chittleborough. 2002. Titanium and zirconium minerals. In *Soil Mineralogy with Environmental Applications*, ed. J. B. Dixon and D. G. Schulze, 667–690. Soil Science Society of America Book Series 7. Madison, WI: Soil Science Society of America.

Foster, A. L., G. E. Brown Jr., and G. A. Parks. 2003. X-ray absorption fine structure study of As(V) and Se(IV) sorption complexes on hydrous Mn oxides. *Geochimica et Cosmochimica Acta* 67, no. 11: 1937–1953. doi:10.1016/S0016-7037(02)01301-7.

Galindo, C., M. D. Nero, R. Barillon, E. Halter, and B. Made. 2010. Mechanisms of uranyl and phosphate (co)sorption: Complexation and precipitation at α-Al2O3 surfaces. *Journal of Colloid and Interface Science* 347, no. 2: 282–289. doi:10.1016/j.jcis.2010.03.045.

Goldanskii, V. I., and R. H. Herber. 1969. *Chemical Applications of Mossbauer Spectroscopy*. San Diego, CA: Academic Press.

Goldberg, S., I. Lebron, and D. L. Suarez. 1999. Soil colloidal behavior. In *Handbook of Soil Science*, ed. M. E. Sumner, B195–B240. 1st ed. Boca Raton, FL: CRC Press.

Golden, D. C., J. B. Dixon, and Y. Kanehiro. 1993. The manganese oxide mineral, lithiophorite, in an oxisol From Hawaii. *Australian Journal of Soil Research* 31, no. 1: 51–66.

Goodman, B. A. 1980. Mossbauer spectroscopy. In *Advanced Chemical Methods for Soil and Clay Minerals Research: Proceedings of the Nato Advanced Study Institute Held at the University of Illinois, July 23-August 4, 1979*, ed. J. W. Stucki and W. L. Banwart. New York: Springer.

Gustafsson, J. P., D. G. Lumsdon, and M. Simonsson. 1998. Aluminium solubility characteristics of spodic B horizons containing imogolite-type materials. *Clay Minerals* 33, no. 1: 77–86.

Hass, A., and P. Fine. 2010. Sequential selective extraction procedures for the study of heavy metals in soils, sediments, and waste materials—A critical review. *Critical Reviews in Environmental Science and Technology* 40, no. 5: 365–399. doi:10.1080/10643380802377992.

Hattori, T., T. Saito, K. Ishida, A. C. Scheinost, T. Tsuneda, S. Nagasaki, and S. Tanaka. 2009. The structure of monomeric and dimeric uranyl adsorption complexes on gibbsite: A combined DFT and EXAFS study. *Geochimica et Cosmochimica Acta* 73, no. 20: 5975–5988. doi:10.1016/j.gca.2009.07.004.

He, G., G. Pan, M. Zhang, and G. A. Waychunas. 2011. Coordination structure of adsorbed Zn(II) at water–TiO2 interfaces. *Environmental Science and Technology* 45, no. 5: 1873–1879. doi:10.1021/es1035283.

Hsu, P. H., and T. F. Bates. 1964. Formation of X-ray amorphous and crystalline aluminium hydroxides1. *Mineralogical Magazine* 33, no. 264: 749–768. doi:10.1180/minmag.1964.033.264.04.

Huang, P. M., M. K. Wang, N. Kämpf, and D. G. Schulze. 2002. Aluminum hydroxides. In *Soil Mineralogy with Environmental Applications*, ed. J. B. Dixon and D. G. Schulze, 261–289. Soil Science Society of America Book Series 7. Madison, WI: Soil Science Society of America.

Jambor, J. L., and J. E. Dutrizac. 1998. Occurrence and constitution of natural and synthetic ferrihydrite, a widespread iron oxyhydroxide. *Chemical Review* 98, no. 7: 2549–2586. doi:10.1021/cr970105t.

James, B. R., and R. J. Bartlett. 1999. Redox phenomena. In *Handbook of Soil Science*, ed. M. E. Sumner, B169–B194. 1st ed. Boca Raton, FL: CRC Press.

Johnston, C. P., and M. Chrysochoou. 2012. Investigation of chromate coordination on ferrihydrite by in situ ATR-FTIR spectroscopy and theoretical frequency calculations. *Environmental Science and Technology* 46, no. 11: 5851–5858. doi:10.1021/es300660r.

Kämpf, N., A. C. Scheinost, and D. G. Schulze. 1999. Oxide minerals. In *Handbook of Soil Science*, ed. M. E. Sumner, F77–F123. 1st ed. Boca Raton, FL: CRC Press.

Katsura, T., and I. Kushiro. 1961. Titanomaghemite in igneous rocks. *American Mineralogist* 46, no. 1–2: 134–145.

Kim, J., W. Li, B. L. Philips, and C. P. Grey. 2011. Phosphate adsorption on the iron oxyhydroxides goethite (α-FeOOH), akaganeite (β-FeOOH), and lepidocrocite (γ-FeOOH): A 31P NMR Study. *Energy & Environmental Science* 4, no. 10: 4298. doi:10.1039/c1ee02093e.

Kosmulski, M. 2009a. *Surface Charging and Points of Zero Charge*. 1st ed. Surfactant Science Series 145. Boca Raton, FL: CRC Press.

Kosmulski, M. 2009b. pH-dependent surface charging and points of zero charge. IV. Update and new approach. *Journal of Colloid and Interface Science* 337, no. 2: 439–448. doi:10.1016/j.jcis.2009.04.072.

Kosmulski, M. 2011. The pH-dependent surface charging and points of zero charge: V. Update. *Journal of Colloid and Interface Science* 353, no. 1: 1–15. doi:10.1016/j.jcis.2010.08.023.

Lefèvre, G. 2004. In situ Fourier-transform infrared spectroscopy studies of inorganic ions adsorption on metal oxides and hydroxides. *Advances in Colloid and Interface Science* 107, no. 2–3: 109–123. doi:10.1016/j.cis.2003.11.002.

Lewis, D. G., and V. C. Farmer. 1986. Infrared absorption of surface hydroxyl groups and lattice vibrations in lepidocrocite (γ-FeOOH) and boehmite (γ-AlOOH). *Clay Minerals* 21, no. 1: 93–100. doi:10.1180/claymin.1986.021.1.08.

Lindsay, W. L., and S. M Elgawhary. 1972. Solubility of silica in soils. *Soil Science Society of America Journal* 36, no. 3: 439–442. doi:10.2136/sssaj1972.03615995003600030022x.

Lyklema, J. 1995. *Fundamentals of Interface and Colloid Science: Solid-Liquid Interfaces*. London, UK: Academic Press.

Madsen, M. B., W. Goetz, P. Bertelsen, C. S. Binau, F. Folkmann, H. P. Gunnlaugsson, J. í Hjøllum et al. 2009. Overview of the magnetic properties experiments on the Mars Exploration Rovers. *Journal of Geophysical Research* 114: 20. doi:200910.1029/2008JE003098.

Malin, J. N., J. G. Holland, S. A. Saslow, and F. M. Geiger. 2011. U(VI) adsorption and speciation at the acidic silica/water interface studied by resonant and nonresonant second harmonic generation. *The Journal of Physical Chemistry C* 115, no. 27: 13353–13360. doi:10.1021/jp203091x.

Manceau, A. 1995. The mechanism of anion adsorption on iron oxides: Evidence for the bonding of arsenate tetrahedra on free Fe(O, OH)6 edges. *Geochimica et Cosmochimica Acta* 59, no. 17: 3647–3653. doi:10.1016/0016-7037(95)00275-5.

Manceau, A., and W. P. Gates. 1997. Surface structural model for ferrihydrite. *Clays and Clay Minerals* 45, no. 3: 448–460. doi:10.1346/CCMN.1997.0450314.

Manning, B. A., S. E. Fendorf, and S. Goldberg. 1998. Surface structures and stability of Arsenic(III) on Goethite: Spectroscopic evidence for inner-sphere complexes. *Environmental Science and Technology* 32, no. 16: 2383–2388. doi:10.1021/es9802201.

McBride, M. B. 1994. *Environmental Chemistry of Soils*. New York: Oxford University Press, March 17.

McBride, M. B. 1999. Chemisorption and precipitation reactions. In *Handbook of Soil Science*, ed. M. E. Sumner, B265–B302. 1st ed. Boca Raton, FL: CRC Press.

Mcknight, D. M., B. A. Kimball, and K. E. Bencala. 1988. Iron photoreduction and oxidation in an acidic mountain stream. *Science* 240, no. 4852: 637–640. doi:10.1126/science.240.4852.637.

Mohapatra, M., S. Anand, B. K. Mishra, D. E. Giles, and P. Singh. 2009. Review of fluoride removal from drinking water. *Journal of Environmental Management* 91, no. 1: 67–77. doi:10.1016/j.jenvman.2009.08.015.

Monger, H. C., and E. F. Kelly. 2002. Silica minerals. In *Soil Mineralogy with Environmental Applications*, ed. J. B. Dixon and D. G. Schulze, 611–636. Soil Science Society of America Book Series 7. Madison, WI: Soil Science Society of America.

Murad, E., and J. Cashion. 2004. *Mössbauer Spectroscopy of Environmental Materials and Their Industrial Utilization*. Norwell, MA: Kluwer Academic Publishers.

Papelis, C., G. E. Brown, G. A. Parks, and J. O. Leckie. 1995. X-ray absorption spectroscopic studies of cadmium and selenite adsorption on aluminum oxides. *Langmuir* 11, no. 6: 2041–2048. doi:10.1021/la00006a033.

Parfitt, R. L. 1979. Anion adsorption by soils and soil materials. In *Advances in Agronomy*, ed. N. C. Brady, 1–50. San Diego, CA: Academic Press.

Parker, D. R., and P. M. Bertsch. 1992a. Formation of the "Al13" tridecameric aluminum polycation under diverse synthesis conditions. *Environmental Science and Technology* 26, no. 5: 914–921. doi:10.1021/es00029a007.

Parker, D. R., and P. M. Bertsch. 1992b. Identification and quantification of the "Al13" tridecameric aluminum polycation using ferron. *Environmental Science and Technology* 26, no. 5: 908–914. doi:10.1021/es00029a006.

Robarge, W. P. 1998. Precipitation/dissolution reactions in soils. In *Soil Physical Chemistry*, ed. D. L. Sparks, 193–238. 2nd ed. Boca Raton, FL: CRC Press.

Roberts, D. R., R. G. Ford, and D. L. Sparks. 2003. Kinetics and mechanisms of Zn complexation on metal oxides using EXAFS spectroscopy. *Journal of Colloid and Interface Science* 263, no. 2: 364–376. doi:10.1016/S0021-9797(03)00281-9.

Roe, A. L., K. F. Hayes, C. Chisholm-Brause, G. E. Brown, G. A. Parks, K. O. Hodgson, and J. O. Leckie. 1991. In situ X-ray absorption study of lead ion surface complexes at the goethite-water interface. *Langmuir* 7, no. 2: 367–373. doi:10.1021/la00050a029.

Ross, D. S., and R. J. Bartlett. 1981. Evidence for nonmicrobial oxidation of manganese in soil. *Soil Science* 132, no. 2: 153–160.

Schwertmann, U. 1991. Solubility and dissolution of iron oxides. *Plant and Soil* 130, no. 1: 1–25. doi:10.1007/BF00011851.

Schwertmann, U., and R. M. Taylor. 1989. Iron oxides. In *Minerals in Soil Environments*, ed. J. B. Dixon and S. B. Weed. 2nd ed. Madison, WI: Soil Science Society of America.

Sparks, D. L. 2002. *Environmental Soil Chemistry*. 2nd ed. San Diego, CA: Academic Press.

Sposito, G. 1984. *The Surface Chemistry of Soils*. New York: Oxford University Press.

Svecova, L., M. Dossot, S. Cremel, M. O. Simonnot, M. Sardin, B. Humbert, C. Den Auwer, and L. J. Michot. 2011. Sorption of selenium oxyanions on TiO2 (rutile) studied by batch or column experiments and spectroscopic methods. *Journal of Hazardous Materials* 189, no. 3: 764–772. doi:10.1016/j.jhazmat.2011.02.090.

Swamy, V., J. D. Gale, and L. S. Dubrovinsky. 2001. Atomistic simulation of the crystal structures and bulk moduli of TiO2 polymorphs. *Journal of Physics and Chemistry of Solids* 62, no. 5: 887–895. doi:10.1016/S0022-3697(00)00246-8.

Tebo, B. M., W. C. Ghiorse, L. G. V. Waasbergen, P. L. Siering, and R. Caspi. 1997. Bacterially-mediated mineral formation: Insights into Manganese(II) oxidation from molecular genetic and biochemical studies. In *Geomicrobiology: Interactions between Microbes & Minerals*, ed. J. F. Banfield and K. H. Nealson, 225–266. Chantilly, VA: Mineralogical Society of America.

Thompson, R., and F. Oldfield. 1986. *Environmental Magnetism*. 1st ed. London: Allen and Unwin.

Triebold, S., G. Luvizotto, R. Tolosana-Delgado, T. Zack, and H. V. Eynatten. 2011. Discrimination of TiO_2 polymorphs in sedimentary and metamorphic rocks. *Contributions to Mineralogy and Petrology* 161, no. 4: 581–596. doi:10.1007/s00410-010-0551-x.

Wada, H., A. Seirayosakol, M. Kimura, and Y. Takai. 1978. The process of manganese deposition in paddy soils. 1: A hypothesis and its verification. *Soil Science and Plant Nutrition* 24: 55–62.

Wang, Y., G. Morin, G. Ona-Nguema, F. Juillot, F. Guyot, G. Calas, and G. E. Brown. 2009. Evidence for different surface speciation of arsenite and arsenate on green rust: An EXAFS and XANES study. *Environmental Science and Technology* 44, no. 1: 109–115. doi:10.1021/es901627e.

Waychunas, G. A., B. A. Rea, C. C. Fuller, and J. A. Davis. 1993. Surface chemistry of ferrihydrite: Part 1. EXAFS studies of the geometry of coprecipitated and adsorbed arsenate. *Geochimica et Cosmochimica Acta* 57, no. 10: 2251–2269. doi:10.1016/0016-7037(93)90567-G.

Weesner, F. J., and W. F. Bleam. 1998. Binding characteristics of Pb2+on anion-modified and pristine hydrous oxide surfaces studied by electrophoretic mobility and X-ray absorption spectroscopy. *Journal of Colloid and Interface Science* 205, no. 2: 380–389. doi:10.1006/jcis.1998.5629.

Zhang, G., H. Liu, R. Liu, and J. Qu. 2009. Removal of phosphate from water by a Fe–Mn binary oxide adsorbent. *Journal of Colloid and Interface Science* 335, no. 2: 168–174. doi:10.1016/j.jcis.2009.03.019.

Zhao, W., Q. Wang, F. Liu, G. Qiu, W. Tan, and X. Feng. 2010. Pb2+ adsorption on birnessite affected by Zn2+ and Mn2+ pretreatments. *Journal of Soils and Sediments* 10, no. 5: 870–878. doi:10.1007/s11368-010-0219-1.

Zhao, D., X. Wang, S. Yang, Z. Guo, and G. Sheng. 2012. Impact of water quality parameters on the sorption of U(VI) onto hematite. *Journal of Environmental Radioactivity* 103, no. 1: 20–29. doi:10.1016/j.jenvrad.2011.08.010.

10 Humic Substances

10.1 NATURE AND CLASSIFICATION OF SOIL HUMIC SUBSTANCES

Soil humic substances (HS) compose the largest part of soil organic matter (SOM) and have a deep impact on soil properties such as buffering capacity, metal-binding ability, stability of aggregates of soil particles, water-holding capacity, and sorption of hydrophobic organic compounds, most of which in turn have a large impact on soil fertility (Senesi and Loffredo 1998; Baldock and Nelson 1999; Hayes and Malcolm 2001; González-Pérez et al. 2004; Clapp et al. 2005). It is a highly heterogeneous mixture (in chemical and physical sense) of organic compounds, many of colloidal and/or polyelectrolytic nature, having a medium to high molecular weight, yellow to black color, and a mixed aliphatic/aromatic composition. These substances are distinctive of soil systems, being the result of degradation processes of dead organisms (mainly vegetal, but also including animal) as a result of microbial activity, possibly followed for secondary synthesis/condensation reactions, in what is called the *humification* process (Clapp et al. 2005). It is interesting to note that even when their composition is location dependent (due to local vegetation, fauna, and climate), their properties are similar everywhere.

As introduced in Chapter 7, HS are usually classified into fractions in terms of solubility:

1. Fulvic acids (FAs): the fraction that is soluble in water at any pH value
2. Humic acids (HAs): the fraction that is soluble at pH > 1
3. Humin: the fraction insoluble in water at any pH

Figure 10.1 summarizes the conceptual separation scheme corresponding to that classification, including a further distinction between gray and brown HAs, the latter being stable in electrolyte media, thus being more polar; usually, what is obtained as "humic acid" is the gray fraction. It is important to emphasize that the classification is purely operational; it is not related to composition, structure, or properties of HS; of course, it results nevertheless in different composition and properties for each group.

HS are not alone in soils but often appear associated with mineral colloids, small ions, and other organic compounds; this point will be addressed in Chapter 14, but now we note that it implies a separation procedure to isolate HS for their study.

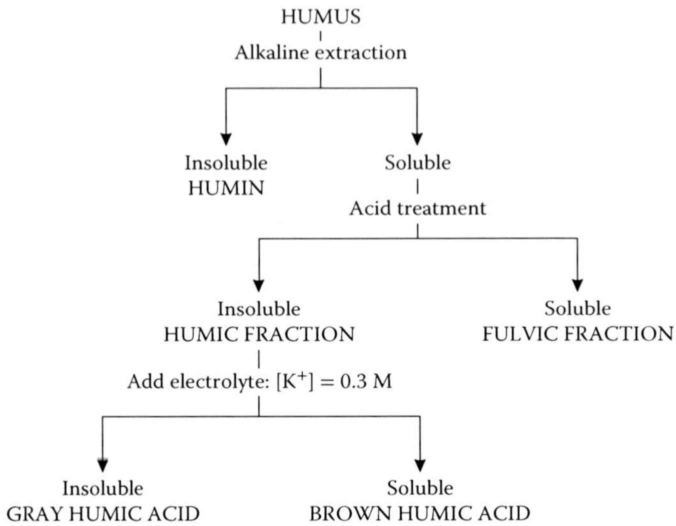

FIGURE 10.1 Fractionation scheme of humic substances, based on the IHSS procedure. (From Swift R. S., *Methods of Soil Analysis. Part 3. Chemical Methods*, Soil Science Society of America Book Series 5, Madison, WI: American Society of Agronomy-Soil Science Society of America, January 1, 1996.)

10.2 EXTRACTION OF HS

The isolation of HS is a difficult problem that has challenged researchers for many years. The ideal extraction method should meet the following objectives (Stevenson 1994):

1. The method leads to the isolation of unaltered material.
2. The extracted HS are free of inorganic contaminants, such as clay and polyvalent cations.
3. Extraction is complete, thereby ensuring the representation of fractions from the entire molecular-weight range.
4. The method is universally applicable to all soils.

At present, no method fulfills all these requirements. Furthermore, there is no clear limit between what is considered "true" HS and what fall in the category of organic "impurities." A variety of techniques have been employed, depending on the nature of the material to be examined or for the experiment purposes (Stevenson 1994; Senesi and Loffredo 1998; Hayes and Malcolm 2001). Usually, aqueous solvents are used for the isolation of HS from soils, but some organic solvents can have applications; thus, nonpolar compounds such as fats, waxes, resins, and others can be extracted with organic solvents such as hexane, ether, alcohol–benzene mixtures, and others. However, Hayes (1985) concluded that "good" organic solvents have relatively high values of dielectric constant and permittivity and are capable of accepting H bonding. Di-methylformamide and dimethylsulfoxide (DMSO) meet these requirements, but problems are presented for the recoveries of conventional fractions of HS from these solvents. Hydrolysis procedures have been used for isolating individual monomers, such as amino acids and sugars. The definition of HS given in the previous section, however, implies an alkaline extraction. Table 10.1 collects some extractants employed in the literature, which are discussed in the following sections.

10.2.1 Alkaline Extraction

The most used method is the alkaline extraction with NaOH, which has been proposed and endorsed by the International Humic Substances Society (IHSS) (Swift 1996; Hayes and Malcolm 2001). NaOH solution 0.1 to 0.5 M in water and a soil to extractant ratio of from 1:2 to 1:5 (g:ml) have been widely used for recovering organic matter; it is summarized in Figure 10.2. Leaching the soil with dilute HCl, which removes Ca and other polyvalent cations, increases the efficiency of extraction of organic matter with alkaline reagents. The original procedure employed subsequently a 0.1 M HCl/0.3 M HF treatment to remove finely divided inorganic soil colloids, thus providing materials with reasonably low ash contents (Swift 1996). The HCl/HF treatment is repeated to have ash contents <1%, followed by dialysis; these operations lead to considerable losses of the humic fractions. Clapp and Hayes (1999) found that filtration through partially clogged 0.45- or 0.2-μm filters very effectively lowers the ash contents of HS fractions, and of course this procedure avoids degradation of the humic materials.

However, there are some drawbacks in alkali extraction:

1. Alkali solutions dissolve silica from the mineral matter and this silica contaminates the organic fractions separated from the extract.
2. Alkali solutions dissolve protoplasmic and structural components from fresh organic tissues and these become mixed with the humified organic matter.

TABLE 10.1
Some Extractants for Humic Substances

Extractant	Organic Matter Yield
Strong alkaline extractants	
NaOH	80%
Na_2CO_3	30%
Ethylenediamine 2.5 M	About 63%
Mild extractans	
$Na_4P_2O_7$ and other	30%
Organic chelants: acetyloacetone, cupferron, hydroxyquinoline	30%
Organic solvents	
Formic acid (HCOOH)	55%
Ethylenediamine (anhydrous)	About 5%
Sulfur-containing (dimethylsulfoxide, etc.)	About 22%
Pyridine	About 36%

Source: Data from Stevenson, F. J., *Humus Chemistry: Genesis, Composition, Reactions*. 2nd ed., Wiley, Hoboken, NJ, 1994; Senesi, N., and E. Loffredo, *Soil Physical Chemistry*, 2nd ed., CRC Press, Boca Raton, FL, 1998.

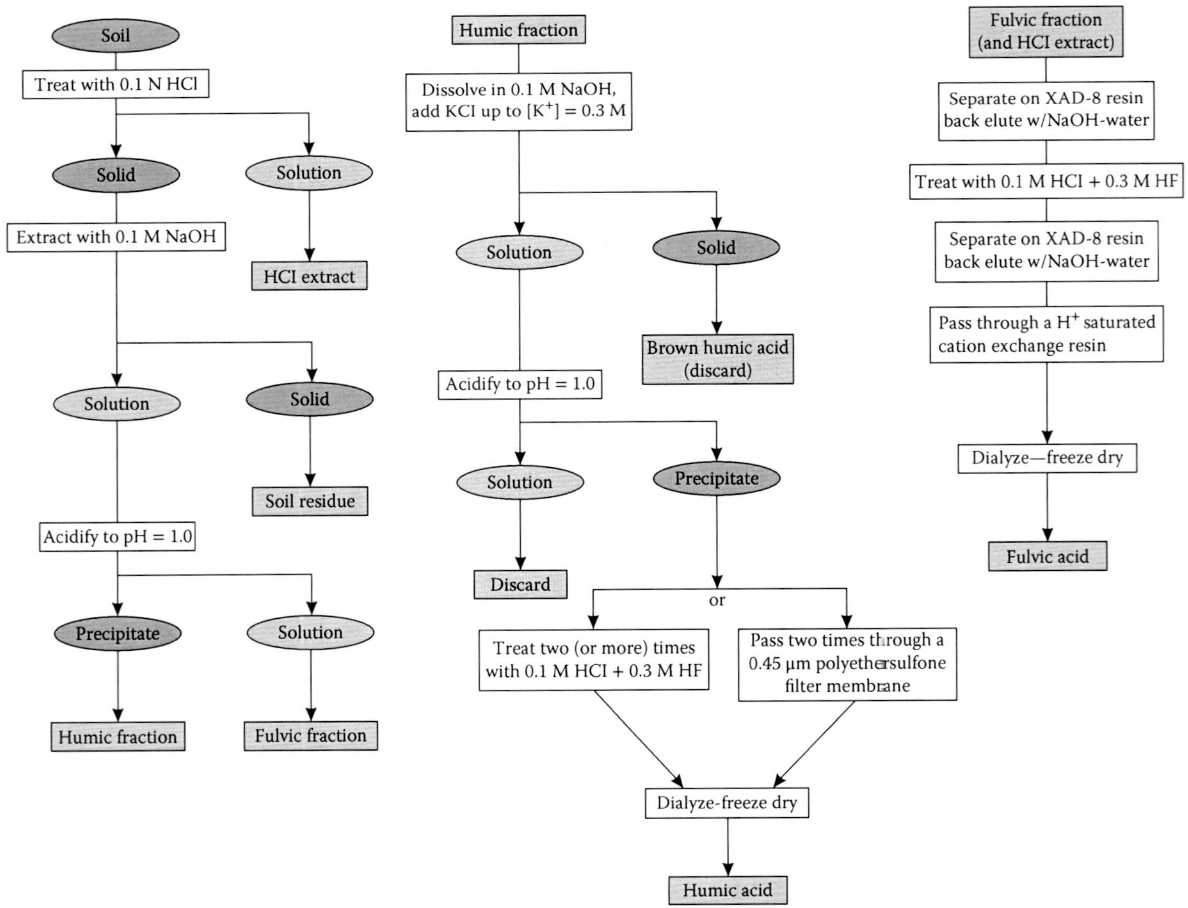

FIGURE 10.2 Separation and fractionation procedure according to the IHSS (Swift 1996) with some modifications from Hayes and coworkers. (From Clapp, C. E. and Hayes, M. H. B., *Soil Science*, 164, no. 12, 899–913, 1999; Hayes, M. H. B. and Malcolm, R. L., *Humic Substances and Chemical Contaminants*, Soil Science Society of America, Madison, WI, 2001.)

3. Under alkaline conditions, oxidation of some organic constituents occurs in contact with air, both during extraction and when the extracts are allowed to stand.
4. Other chemical changes can occur in alkaline solution such as condensation between amino acids and aldehydes or quinones. The more alkaline the solution and the longer the extraction period, the greater will be the chemical changes.

The oxidation can be prevented by working in an inert gas atmosphere, usually N_2. Other reactions can be reduced by minimizing the time that the HS are in strong alkaline medium.

10.2.2 Mild Extractants

Several milder and more selective extractants have been recommended in recent years as alternatives for the classical alkali extraction. Included are salts of complexing agents ($Na_4P_2O_7$ and ethylenediaminetetraacetic acid), organic complexing agents in aqueous media (acetylacetone), and organic solvents of various types. Although less alteration of organic matter may result, these extractants are much less effective than alkali hydroxides in removing organic matter (Table 10.1). In some cases, a mild extractant is definitely preferred; for others, a more complete extraction is achieved with alkali.

10.2.2.1 $Na_4P_2O_7$ and Other Neutral Salts

In most soils, polyvalent cations such as Ca(II), Fe(III), and Al(III) are responsible for maintaining organic matter associated with minerals and in a flocculated, insoluble condition. Accordingly, reagents that remove these cations by precipitation or complexation lead to solubilization of organic matter. Reagents such as ammonium oxalate, sodium pyrophosphate, and salts of weak organic acids have been used for this purpose. $Na_4P_2O_7$ has been the most widely used. However, the amount of organic matter recovered (<30%, Table 10.1) is considerably less than that with caustic alkali, but less alteration occurs.

10.2.2.2 Formic Acid (HCOOH)

Under certain circumstances, up to 55% of the organic matter in mineral can be extracted with formic acid containing LiF, LiBr, or HBF_4. An advantage of anhydrous formic acid for extraction of organic matter is that it is a polar compound that neither exhibits oxidizing nor hydrolytic properties. Furthermore, it is a good solvent for many compounds, including polysaccharides. Large quantities of Ca, Fe, Al, and other inorganic cations are dissolved from the soil along with the organic matter, and thus far it has not been possible to remove the inorganic material completely.

10.2.2.3 Organic Chelating Agents

Several organic compounds such as acetylacetone, cupferron, and hydroxyquinoline, which are capable of forming chelates with polyvalent metal ions, have been used for extracting illuvial organic matter from Spodosols but are rather ineffective for extracting organic matter from other soil types.

10.2.3 Sequential Extractions

Sequential extractions, that is successive extractions with different reagents, were performed by Hayes et al. (1996) who carried out exhaustive sequential extractions using water, $Na_4P_2O_7$ at pH 7, $Na_4P_2O_7$ at pH 10.6, and $Na_4P_2O_7$ + 0.1 M NaOH. Clapp and Hayes (1999) carried out a similar exhaustive sequential extraction process and followed that sequence with a DMSO/HCl treatment. The DMSO, a dipolar aprotic solvent, is a poor solvent for anions but a good solvent for cations. Thus, the metal cations that neutralize the charges on humics would be solvated, and the humic anionic groups (carboxylates and phenates) would not. However, DMSO will associate with the phenolic and carboxyl groups by hydrogen bonding, thus breaking the humic intra- and intermolecular H bonds. Also, the nonpolar part of DMSO can associate with the hydrophobic moieties in HS, thus making DMSO a very effective solvent for HS. Häusler and Hayes (1996) showed that dissolving HA in DMSO containing 1% HCl and passing the solution through an XAD 8 [(poly)methyl-methacrylate] resin column allows the HS to adsorb to the resin and the solvent to be eluted by passing excess 0.01 M HCl. This line has continued to be developed (Hayes 2006); recently, a sequential extractions approach has been reported to give yields around 90% (Hayes et al. 2010). Most importantly, this procedure seems to recover most of humin, which traditional methods discard.

In recent years, other fractionation methods have been proposed, such as ultrafiltration (Yu et al. 2008) or fractional precipitation by salting out (Zavarzina, Vanifatova, and Stepanov 2008). Size-exclusion chromatography (SEC, which is discussed in Section 10.3.2.3) in preparative form (low-pressure gel-permeation chromatography) has also been employed (De Nobili and Chen 1999; Perminova 1999; Clapp et al. 2005).

10.3 CHARACTERIZATION, COMPOSITION, AND STRUCTURE OF HS

10.3.1 Elemental Composition

HS, as almost all organic matter, is composed of C, H, O, and N; sometimes S and other elements are found; elemental analysis should be the first characterization step in the analysis of HS. It is usually performed by the classic method of full combustion followed by gas chromatography of the resulting products. Table 10.2 shows values found for HA and FA. It is observed that, besides C, O is the most important, whereas N contents are relatively low. Also, the O/C atomic ratio is higher for FA than for HA, which is consistent with their greater solubility, inclusive in acid media; this ratio is around 0.5 for HA and 0.7 for FA. The H/C ratio is around 1.0 and 1.4 for soil HA and FA, respectively (Table 10.2). As HS are actually mixtures of different substances, one cannot give a meaningful molecular formula; nevertheless, it is useful for several purposes, such as checking the efficiency of separation procedures, distinguishing classes of HS, and so on. In Table 10.2, it is seen that a clear dependence on climate is found for both HA and FA.

10.3.2 Molecular Weight and Size Distributions

There are several techniques available to study the molecular weight and size of HS; these are discussed briefly in the following paragraphs. It should be recalled that HS

TABLE 10.2
Elemental Composition (% Mass) of Humic and Fulvic Acids from Different Soils

Element	Soil: Arctic	Cool Acid	Subtropical	Tropical	Range	Average	Atomic X/C Ratio
			Fulvic acids				
C	47.7	47.6–49.9	42.2–44.3	42.8–50.6	40.7–50.6	45.7	1.00
H	5.4	4.1–4.7	5.9–7.0	3.8–5.3	3.8–7.0	5.4	1.4
N	1.1	0.9–1.3	3.1–3.2	2.0–3.3	0.9–3.3	2.1	0.04
S	1.6	0.1–0.5	2.5	1.3–3.6	0.1–3.6	1.9	0.016
O	44.2	43.6 –47.0	43.1–46.2	39.7–47.8	39.7–49.8	44.8	0.73
			Humic acids				
C	56.2	53.8–58.7	53.6–55.0	54.4–54.9	53.6–58.7	56.2	1.00
H	6.2	3.2–5.8	4.4–5.0	4.8–5.6	3.2–6.2	4.7	1.00
N	4.3	0.8–2.4	3.3–4.6	4.1–5.5	0.8–4.3	3.2	0.05
S	0.5	0.1–0.5	0.8–1.5	0.6–0.8	0.1–1.5	0.8	0.005
O	32.8	35.4–38.3	34.8–36.3	34.1–35.2	32.8–38.3	35.5	0.47

Source: Senesi, N., and E. Loffredo, *Soil Physical Chemistry*, 2nd ed., CRC Press, Boca Raton, FL, 1998.

are actually a mixture of molecules of different size, shape, and elemental composition; thus, they display *polydispersity* (a distribution of sizes and molecular weights) and are *polyelectrolytic* (i.e., they are charged with several elemental charges per molecule).

10.3.2.1 Colligative Properties

Colligative properties, mainly osmotic pressure and freezing point depression, have been traditionally employed to determine molecular weights (Levine 2008; Atkins and Paula 2009; Garland, Nibler, and Shoemaker 2009). These methods are useful for values of molecular weight, M, up to about 50,000. It has been often applied to HS (Aiken and Gillam 1990), despite significant problems: one is the deviation from ideal behavior (infinite dilution) due to the polydisperse and polyelectrolytic nature of HS, and the other, the partial dissociation of HS, due to their weak acid character. Thus, suitable corrections must be made (Senesi and Loffredo 1998), which are complicated resulting in M values of limited validity.

All the different methods to measure molecular properties yield average values for polydisperse samples. Colligative properties, being a property dependent on the number of solute molecules, result in the *number average* of M:

$$\overline{M}_n = \frac{\sum_i n_i M_i}{\sum_i n_i} \tag{10.1}$$

where M_i is the molecular weight of fraction i, and n_i is the number of molecules in that fraction.

10.3.2.2 Light Scattering

Macromolecular solutions and colloidal suspensions show light dispersion phenomena, which is found to be essentially dependent on the molecular weight of the dispersing substance, according to the Rayleigh–Gans–Debye approximation (Striegel et al. 2009):

$$\frac{K^*c}{R(\theta)} = \frac{1}{P(\theta)}\left(\frac{1}{M} + 2A_2c + 3A_3c + \cdots\right) \tag{10.2}$$

where c is the dispersing macromolecule concentration, K^* a constant essentially dependent on the solvent; the incident light wavelength, A_2, A_3, and so on are the virial coefficients; $P(\theta)$ is the particle scattering factor; and $R(\theta)$ is the Rayleigh *excess ratio*:

$$R(\theta) = \frac{I_\theta r^2}{I_0} \tag{10.3}$$

where I_0 is the intensity of the incident light, I_θ is the intensity of the scattered light at an angle θ, and r is the distance from the scattering particle to the detector. $P(\theta)$ is given by

$$P(\theta) = 1 + \frac{16\pi^2}{3\lambda^2} R_G^2 \sin^2\frac{\theta}{2} + \cdots \tag{10.4}$$

where R_G is the *radius of gyration*, given by

$$R_G^2 = \frac{\left\langle \sum_i w_i R_i^2 \right\rangle}{\sum_i w_i} \tag{10.5}$$

where the w_i are the masses of the atoms (or groups), the R_i are the distances to the molecule center of mass, and $\langle\ \rangle$ means an average over all orientations. R_G gives an approximation to the molecule size and can be determined using multiangle light scattering (Wyatt 1998; Striegel et al. 2009).

By measuring the excess ratio $R(\theta)$ at different concentrations, the molecular weight can be determined. Because in a humic sample (as with most macromolecular materials), a distribution of molecular weights is present, and because light dispersion is a property essentially dependent on M, the measurement yields the *weight average* molecular weight:

$$\overline{M}_w = \frac{\sum_i w_i M_i}{\sum_i w_i} = \frac{\sum_i n_i M_i^2}{\sum_i n_i M_i} \tag{10.6}$$

which, in general, will be higher than the number average M_n (Equation 10.1), unless the sample is strictly monodisperse (all molecules equal), where $M_w = M_n$. The *polydispersity*, a measure of the distribution width, is defined as

$$\rho_P = \frac{M_w}{M_n} \tag{10.7}$$

However, other parameters are required to fully characterize a distribution (Striegel et al. 2009).

10.3.2.3 Size-Exclusion Chromatography

In size-exclusion chromatography (SEC), the solutes are eluted through a column packed with porous particles; the larger molecules are unable to penetrate the pores, thus elute rapidly, whereas the smaller solute molecules enter the pores being retained longer times. As a consequence, in the chromatogram, the fractions appear in the order of decreasing molecular weight (Striegel et al. 2009). A calibration is needed, which requires compounds of known M of a more similar nature than the samples; for HS, polystyrene sulfonates (PSS) have been employed (Zhou, Cabaniss, and Maurice 2000). Figure 10.3a shows chromatograms for several PSS of known M (indicated in the graph) plus salicylic acid and acetone as low M standards; included is the IHSS Suwannee River standard fulvic acid. It is important to note that the chromatogram shows the molecular weight distribution, thus M_n and M_w can be evaluated as follows:

$$\overline{M}_n = \frac{\sum_i h_i M_i}{\sum_i h_i} \tag{10.8}$$

$$\overline{M}_w = \frac{\sum_i h_i M_i^2}{\sum_i h_i M_i} \tag{10.9}$$

where h_i is the height in the chromatogram for each M_i. Furthermore, other averages can be evaluated as well; the so-called z-average (the third moment of the distribution) can be determined in the same way:

$$\overline{M}_z = \frac{\sum_i w_i M_i^2}{\sum_i w_i M_i} = \frac{\sum_i n_i M_i^3}{\sum_i n_i M_i^2} = \frac{\sum_i h_i M_i^3}{\sum_i h_i M_i^2} \tag{10.10}$$

If multiangle light scattering detection is used with SEC, a distribution of R_G values can be obtained (Striegel et al. 2009). SEC is a powerful technique to characterize HS and has been applied by a number of researchers (Baigorri et al. 2007; Asakawa et al. 2008; Maia, Piccolo, and Mangrich 2008; Trubetskaya et al. 2008).

10.3.2.4. Ultracentrifugation

In analytical ultracentrifugation (Lebowitz, Lewis, and Schuck 2002; Scott, Harding, and Rowe 2006), the sample suspension is rotated at high speed with simultaneous

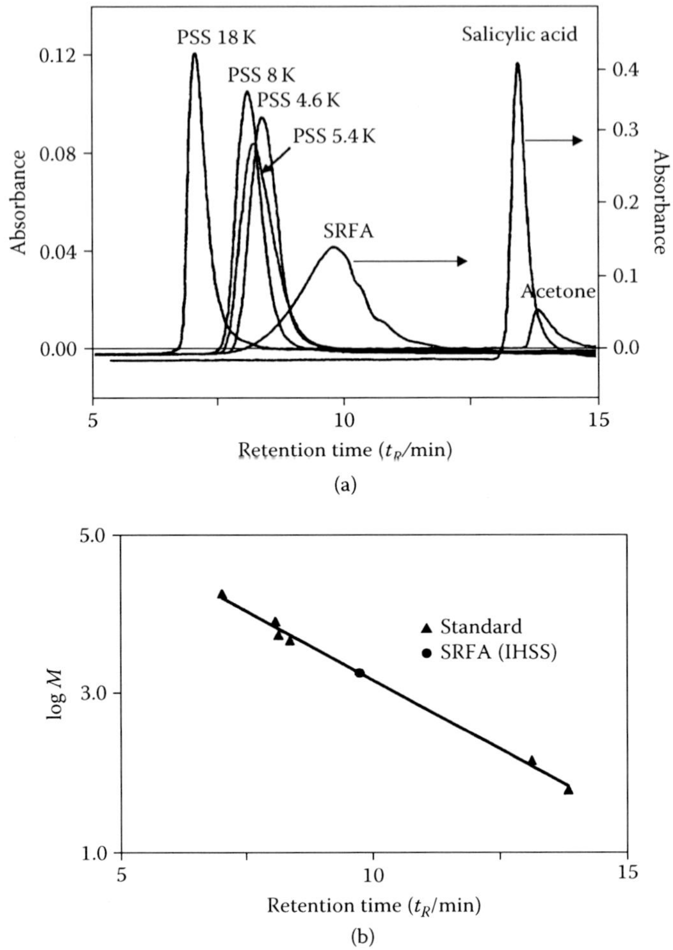

FIGURE 10.3 (a) Examples of high-pressure size-exclusion chromatograms for a variety of standards and the Suwannee River fulvic acid (IHSS). (b) Calibration function obtained by linear regression of the log M vs t_R plot for standards shown in (a). (Reprinted from *Water Research*, 34, no. 14, Zhou, Q. et al., Considerations in the use of high-pressure size exclusion chromatography (HPSEC) for determining molecular weights of aquatic humic substances, 3505–3514. Copyright 2000, with permission from Elsevier.)

monitoring of the concentration profile inside the tube by optical means (absorbance, turbidity, etc.). There are two forms of analysis: sedimentation velocity and sedimentation equilibrium.

In a sedimentation velocity experiment, the stationary sedimentation velocity u is measured and the *sedimentation coefficient*, s, is obtained by

$$s = \frac{u}{\omega^2 r} \tag{10.11}$$

where ω is the rotor angular speed and r the radial position of the sedimenting particle; it is independent of the operating conditions (Ralston 1992; Scott and Schuck 2006).

The molecular weight is related to s through the Svedberg equation (valid at infinite dilution):

$$M = \frac{\boldsymbol{R}Ts}{D(1 - v\rho_s)} \tag{10.12}$$

where D is the diffusion coefficient, v is the solute partial-specific volume (Levine 2008; Atkins and Paula 2009), and ρ_s is the solution density. However, this technique is not very appropriate for highly polydisperse materials such as HS, albeit it is faster than equilibrium measurements (Senesi and Loffredo 1998).

At centrifugal speeds lower than those generally used for velocity measurements, sedimentation is balanced by diffusional transport due to the concentration gradient established, and equilibrium is considered to be achieved when the net transport vanishes throughout the solution. It can be easily established by running the centrifuge until the concentration distribution does not show variations with time; this usually requires long times (up to 48 h) to attain equilibrium, which makes that a time-consuming technique. However, it is a powerful technique, allowing to find the different averages M_n, M_w, and M_z (Senesi and Loffredo 1998). The basic equation for molecular weight deductions is

$$M = \frac{\boldsymbol{R}T}{(1 - v\rho_S)\omega^2 rc}\frac{\partial c}{\partial r} \tag{10.13}$$

where c is the concentration (position-dependent).

10.3.2.5 Viscosimetry

Viscosity is a measure of the friction (resistance to mechanical movement) of a fluid. In the case of liquids, the viscosity of a solution is different from the pure solvent and is dependent on the nature and concentration of the solute. The following Staudinger equation (Tanford 1961; Clapp, Emerson, and Olness 1990; Stevenson 1994) allows to estimate molecular weights:

$$\eta = KM^{\alpha} \tag{10.14}$$

Here, η is the intrinsic viscosity, M is the molecular weight, and K and α are parameters dependent on the type of compound, usually found (in polymer science) by calibrating with known similar compounds. For an unsolvated sphere, $\alpha \approx 0$; for a thin disc, $\alpha \approx 1$; and for flexible polymers behaving hydrodynamically as spheres, $\alpha \approx 0.5$. For HS, values between 0.5 and 0.8 have been reported. The determination of molecular weights by viscosity is complicated, because HS are charged macromolecules, and their behavior is affected by factors such as pH and ionic strength, but it has been applied by some authors (Avena, Vermeer, and Koopal 1999; Kawahigashi, Sumida, and Yamamoto 2005). With viscosimetry, a different average is evaluated, namely the viscosity average $\overline{M}_v$:

$$\overline{M}_v = \left[\frac{\sum_i n_i M_i^{\alpha}}{\sum_i n_i}\right]^{1/\alpha} \tag{10.15}$$

where α is the exponent of Equation 10.14, which is dependent on the experimental conditions.

10.3.2.6 Microscopic Observation

In relatively recent times, microscopic observation of HS has been employed to study their size and shape. Albeit some research by electron microscopy has been reported (Rizzi et al. 2004; Chen et al. 2007), most studies have employed atomic force microscopy (AFM) of humic molecules adsorbed on surfaces (Plaschke et al. 1999; Balnois et al. 1999; Liu et al. 2000; Gerin and Dufrêne 2003; Assemi et al. 2004; Chen et al. 2007). AFM observations (Section 7.5.3) allow in principle the determination of the size of humic molecules. Figure 10.4 (Balnois et al. 1999) shows images of peat humic acid adsorbed on mica; it is seen that at high pH relatively small particles (about 2 nm) and at acidic pH larger particles, presumably aggregates, are observed. In other studies, small particles, of a few nanometers, have also been observed (see, for example, Figure 2.15). Figure 10.5 presents histograms of height measurements at different pH values; the presence of aggregates at pH = 3.2 is clearly observed, which is attributable to the lower negative charge due to proton binding to carboxylate groups. The images appear clear and well defined; however, there are a number of difficulties involved in these measurements (Balnois et al. 1999): (i) lateral dimensions are largely overestimated due to the geometry of the tip; (ii) the samples may be partially flattened or altered in some way due to drying before observation, which is ex-situ; and (iii) the force exerted by the AFM probe could be responsible for a partial flattening, or other lateral alterations of the particles on the surface, even when operating using the minimal forces of tapping mode.

Furthermore, height anomalies have been found in the measurement of the size of biomolecules using tapping mode AFM. When images are obtained in air, the adhesive and capillary forces, due to the presence of a water layer on the surface,

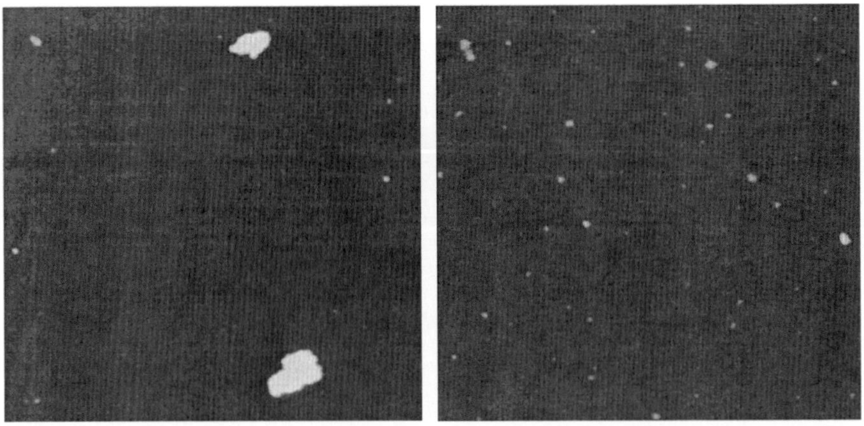

(a) (b)

FIGURE 10.4 Tapping mode AFM images of peat humic acid adsorbed on mica: 10 mg L^{-1} peat humic acid solution, 5 mM NaCl, and scan size is 2 μm × 2 μm; (a) pH 3.2, (b) pH 6.8. (Reprinted with permission from Balnois, E. et al., *Environmental Science and Technology* 33, no. 21, 3911–3917. Copyright 1999 American Chemical Society.)

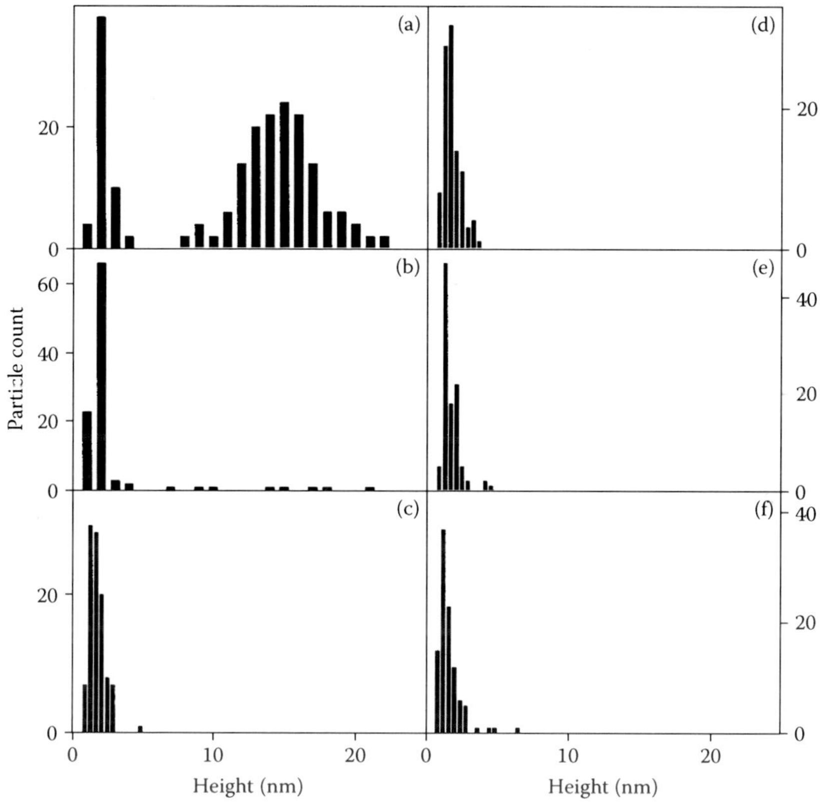

FIGURE 10.5 Height histograms of the peat humic acid (10 mg L^{-1}, 5 mM NaCl) as a function of pH: (a) pH 3.2, (b) pH 5.2, (c) pH 5.9, (d) pH 6.5, (e) pH 6.8, and (f) pH 7.2. (Reprinted with permission from Balnois, E. et al., *Environmental Science and Technology* 33, no. 21, 3911–3917. Copyright 1999 American Chemical Society.)

may dampen the cantilever amplitude, which can be interpreted as a height increase (Weisenhorn et al. 1989; Kühle et al. 1998).

10.3.2.7 Other Techniques

Several other techniques have been applied to the study of physical properties of HS. Field-flow fractionation (described in Section 7.6.3) has been applied to the study of size distribution of HS (Beckett, Jue, and Giddings 1987; Lead et al. 2000). Electrophoresis has been largely applied in the separation of charged components of HS, but was employed only a few times for the determination of molecular weight and size of HS (Senesi and Loffredo 1998). The retardation of macromolecules in a gel media under an applied electric field is related to the size of the pores and the migrating molecules. The retardation coefficient K_R can be related to the molecular weight by

$$M = a + bK_R \tag{10.16}$$

where *a* and *b* are constants (Hayes et al. 1990). A calibration with known standards of similar hydration, density, and shape as the sample should be performed, which poses difficulties for HS; notwithstanding this, it has been applied in the literature (Kasparov, Tikhomirov, and Fless 1981). Matrix-Assisted Laser Desorption Ionization—Time of Flight was also proposed to determine the molar mass distribution of HS (Shinozuka, Shibata, and Yamaguchi 2004). Finally, exchange methods on resins have been employed by several researchers (Clapp et al. 2005).

10.3.3 Spectroscopic Methods

Spectroscopic characterization is a very important tool in the study of HS; here, the more important techniques are reviewed.

10.3.3.1 Infrared Spectroscopy

Infrared (IR) spectroscopy has been, for a long time, a key tool in the structural analysis of organic molecules (Bellamy 1975; Johnston and Aochi 1996). However, for complex molecules like HS, the interpretation of IR spectra becomes difficult, because there is much band overlapping. Notwithstanding this, useful information can be extracted for IR spectra; most research has been conducted using mid-IR range, but the near-IR region has also been employed recently (Reeves III 2010; Bellon-Maurel and McBratney 2011).

The principles and general applications of IR spectroscopy have been discussed in Section 7.4.3; here, we will remark some aspects. The IR spectra allows to identify some characteristic functional groups through their vibrations appearing in definite spectral regions (see Figure 10.6): at higher wavenumbers, the higher energy vibrations appear, which are stretching of X–H bonds (X being mainly C, O, and N); these appear around 2900 cm^{-1} for C–H stretching vibrations and between 3000 and 3500 cm^{-1} for O–H and N–H vibrations, which are broadened in the presence of H bonding. In the region around 1700 cm^{-1}, the C=O vibrations corresponding to carboxylic acid groups are observed, whereas other carbonyl groups appear more close to 1600 cm^{-1}, and the carboxylate group (COO^-) has characteristic bands near 1600 and 1350 cm^{-1}. Aromatic rings show usually two bands around 1500–1400 cm^{-1}. At 1264 cm^{-1}, C–O and O–H deformations (i.e., lateral vibrations) are observed. More detailed descriptions can be found in the literature (Bellamy 1975; Pretsch, Bühlmann, and Badertscher 2009). In Figure 10.6, IR spectra at different pH values of a gray humic acid derived from a commercial sample are shown, where the above mentioned bands are observed. The main feature here is the marked decrease in the 1712 cm^{-1} carboxylate band due to ionization (Alvarez-Puebla and Garrido 2005).

10.3.3.2 Nuclear Magnetic Resonance

Nuclear Magnetic Resonance (NMR), which has been presented in Section 7.4.4, is one of the most important analytical tools in the study of natural organic matter (Steelink et al. 1990; Malcolm 1990; Simpson et al. 2001a; Cardoza et al. 2004; Baigorri et al. 2009; Thorn and Cox 2009). In an heteronuclear, multidimensional NMR study of a FA fraction isolated from the B horizon of an Irish Podzol, Simpson et al. (2001a) identified mono- and dicarboxylic acids, with an average chain length of about 10 carbon atoms

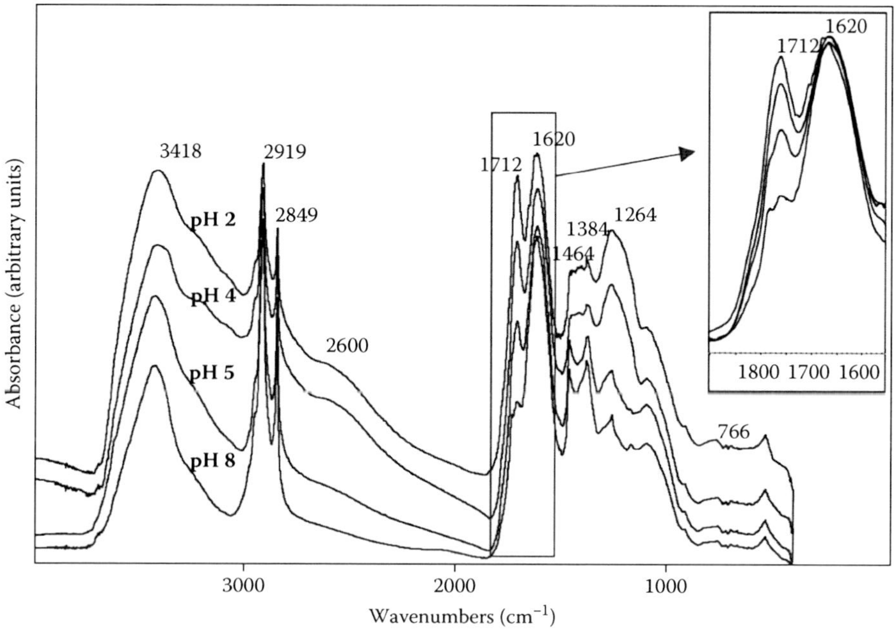

FIGURE 10.6 IR spectra of a humic acid at different pH values. Inset, detail of the carboxylic band region. (Reprinted from *Chemosphere*, 59, no. 5, Alvarez-Puebla, R. A. and Garrido, J. J., Effect of pH on the aggregation of a gray humic acid in colloidal and solid states, 659–667. Copyright 2005, with permission from Elsevier.)

(these are easily the major components); smaller amounts (about 10–20% of the acids) of esters and alcohols or ethers; some carbohydrate and amino acid residues; and very small amounts of 1,2-, 1,4-, and 1,3,4-substituted benzenes and of cinnamic acids as developed in Figures 10.7 through 10.10. Figure 10.7 shows the ^{1}H one-dimensional spectrum, where a higher proportion of aliphatic over aromatic groups is observed. The letters identify H atoms belonging to the groups indicated in Figure 10.10. Figure 10.8 shows the ^{13}C spectrum of the FA and a DEPT version of it (DEPT stands for Distortionless Enhancement by Polarization Transfer, a technique employed in ^{13}C-NMR to allow the number of hydrogens on a carbon to be determined. CH_3 and CH groups give negative peaks whereas CH_2 show positive peaks), where the assignments in the structures presented in Figure 10.10 are shown (the DEPT spectrum confirms some of the CH_n groups). Better resolution and further confirmation of assignments is obtained in two dimensions. The two-dimensional HSQC spectrum in Figure 10.9 (HSQC stands for Heteronuclear Single bond Quantum Correlation spectroscopy, a technique that detects H–C couplings over one bond and provides chemical shift data for both atoms in a CH unit) shows the same region, and the interpretation is relatively straightforward. For example, cross peaks a and b are consistent with chemical shifts (^{1}H and ^{13}C) found for CH_2 groups α and β to a carboxyl group. The other cross peaks in Figure 10.9 can be assigned similarly. Multidimensional NMR is a powerful technique to elucidate structures in complex samples such as NOM (Pommery et al. 1988; Steelink et al. 1990; Malcolm 1990; Otto, Robert Carper, and Larive 2001; Simpson et al. 2001a; Simpson,

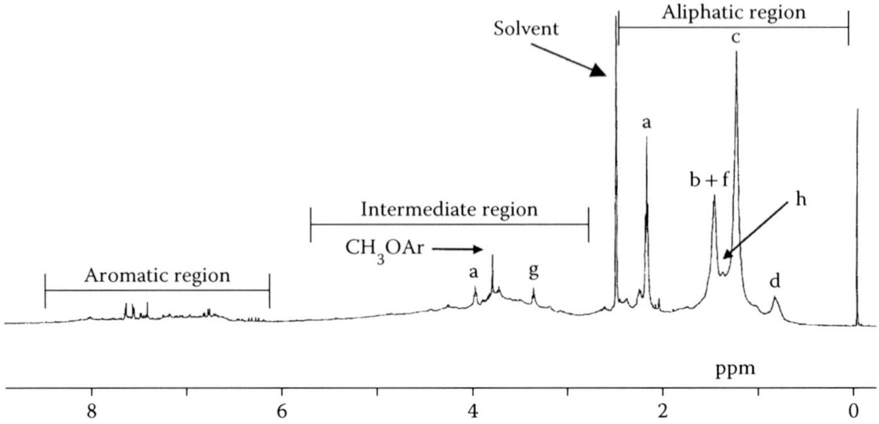

FIGURE 10.7 One-dimensional 500 MHz ^{1}H-NMR spectrum of fulvic acid. (Reprinted with permission from Simpson, A.J. et al., *Eur. J. Soil Sci.*, 52, no. 3, 495–509, 2001b. Copyright 2001 John Wiley & Sons.)

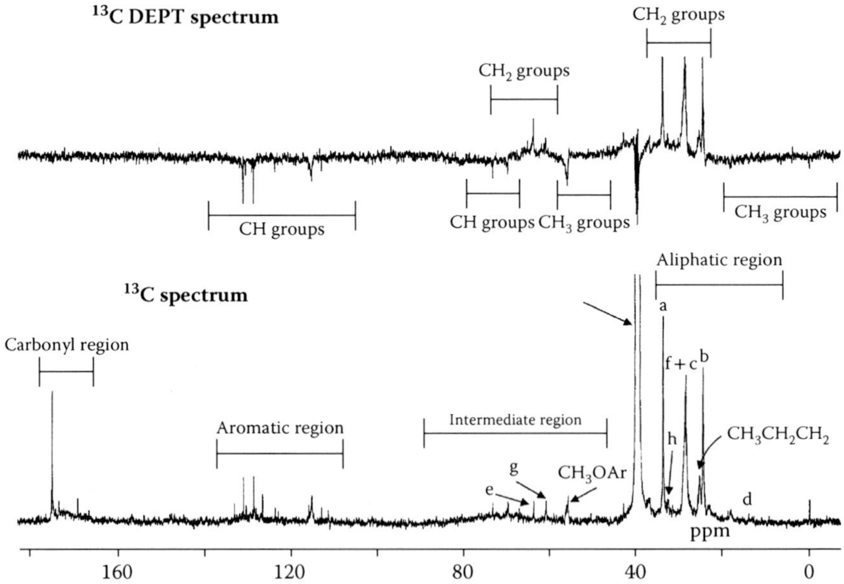

FIGURE 10.8 One-dimensional 500 MHz ^{13}C and DEPT ^{13}C spectra of fulvic acid. Reprinted from (Simpson et al. 2001b) with permission. Copyright (2001) John Wiley & Sons.

Kingery, and Hatcher 2003; Cardoza et al. 2004; Conte, Spaccini, and Piccolo 2006; Thorn and Cox 2009; Sanders, Washton, and Mueller 2010).

10.3.3.3 UV–Visible and Fluorescence Spectroscopy

Electronic spectroscopies of HS are the result of the absorption of incident radiation by a humic molecule, resulting in a loss of intensity in the transmitted beam

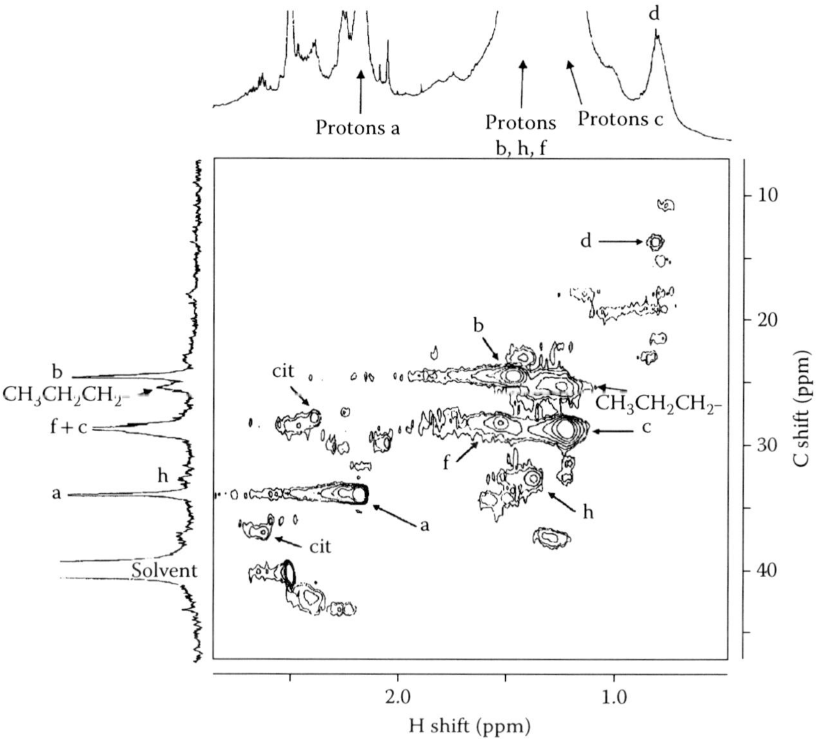

FIGURE 10.9 Two-dimensional HSQC spectrum, expanded view of the aliphatic region. The lettered cross peaks show one bond H-C coupling at the positions marked in Figure 10.10. (Reprinted with permission from Simpson, A.J. et al., *Eur. J. Soil Sci.*, 52, no. 3, 495–509, 2001b. Copyright 2001 John Wiley & Sons.)

I

$CH_3(CH_2)_nCH_2CH_2COOR$

d c b a

II

$ROOCCH_2CH_2(CH_2)_mCH_2CH_2COOR$

a b c b a

III

$CH_3(CH_2)_pCH_2CH_2OOCR$

d c f e

IV

$CH_3(CH_2)_qCH_2CH_2OR$

d c h g

FIGURE 10.10 Scheme of fulvic acid fragments according to the NMR spectra of Figure 10.7–10.9. (Reprinted with permission from Simpson, A.J. et al., *Eur. J. Soil Sci.*, 52, no. 3, 495–509, 2001b. Copyright 2001 John Wiley & Sons.)

and, usually, in an emission of fluorescence (Section 7.4.1). The measurement of absorption spectra results in relatively featureless spectra, as shown in Figure 10.11. However, the slope is dependent on the particular HS studied, thus the ratio

$$\frac{E_4}{E_6} = \frac{A(465\,\text{nm})}{A(665\,\text{nm})} \tag{10.17}$$

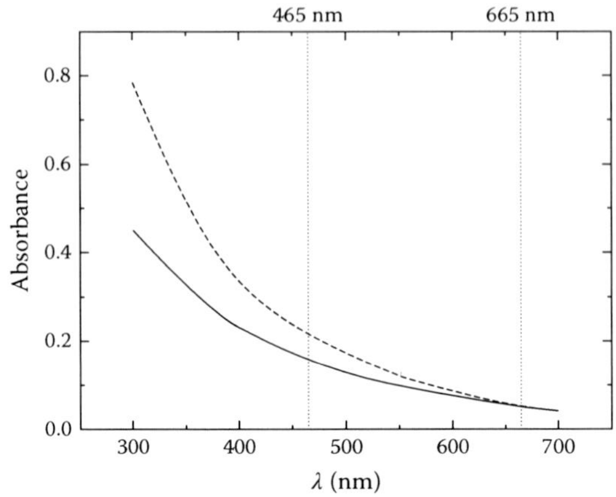

FIGURE 10.11 Typical absorbance spectra of humic substances. The vertical lines indicate the wavelengths where the absorbance ratio E_4/E_6 is measured.

where *A* is the absorbance, is frequently reported as a characterization parameter; it is reported to be $E_4/E_6 > 5$ for FA and <5 for HA. This ratio varies with pH and ionic strength, which has been attributed to changes in chromophores in highly conjugated polyaromatic structures (Baes and Bloom 1990). At a given pH and ionic strength, E_4/E_6 has been found to be dependent on the origin of the HS; this has been correlated by some authors with the degree of humification, which would be inversely related to E_4/E_6; however, others argued that the observed visible absorption could be due to scattering, thus the changes would be due to the different molecular weight of FA and HA and not related to the amount of condensated aromatic rings (Senesi and Loffredo 1998).

Most HS show a relatively weak but measurable fluorescent emission upon light absorption in the UV region and the blue end of the visible spectrum (up to ~450 nm). In a fluorescence measurement, the sample is irradiated with incident or *excitation beam* of wavelength λ_{ex}, and the emission is measured (usually at an angle of 90° relative to excitation, to minimize interferences) at a wavelength λ_{em}. The florescence spectra give useful information about the HS and their interaction with metal cations (Senesi 1990; Mobed et al. 1996; Bidoglio et al. 1997; Esteves da Silva et al. 1998; Patel-Sorrentino, Mounier, and Benaim 2002; Hernandez et al. 2007; Santín et al. 2009). There are several ways to analyze the fluorescence response: the most basic is a spectrum of emission intensity as a function of λ_{em} for a fixed λ_{ex}. Figure 10.12a shows typical emission spectra of a soil humic acid at two excitation wavelengths; the opposite way is an excitation spectrum, where the emission (measuring) wavelength is fixed, and the excitation one is scanned: Figure 10.12b shows examples of excitation spectra at different emission wavelengths of the same HA. A more complete measurement is the recording of emission spectra scanning the excitation wavelength, which is known as *excitation-emission matrix* (EEM), which is shown in Figure 10.12c, in landscape form, and in Figure 10.12d, as a contour plot; in both the representations, lighter pixels mean higher intensity of fluorescence emission. Another way of recording the fluorescence response is in the *synchronous spectrum*, where both λ_{ex} and λ_{em} are scanned at a fixed difference, $\Delta\lambda$; this is illustrated as follows (Sierra et al. 2005): in Figure 10.13a, the EEM map of an estuarine FA is shown, where the position of synchronous spectra at different $\Delta\lambda$ values are marked; it can be noted that these spectra correspond to diagonal cuts (with unity slope) on the EEM (also, emission and excitation spectra are horizontal and vertical cuts, respectively). Figure 10.13b shows the resulting spectra for the different $\Delta\lambda$ values. The peaks labeled α, α', and γ are assigned to different fluorophores; the two former are attributed to humic material, whereas the last is considered protein-like material, and have been related to tyrosine and/or tryptophan fluorescence (Sierra et al. 2005). Fluorescence measurements, especially of EEMs, have been found useful in the characterization of HS. In a comparative study of HS from diverse sources (Mobed et al. 1996), results show that EEMs may be used to discriminate between soil-derived and aquatic-derived IHSS HS and between HA and FA derived from the same source. Absorbance correction was shown to be essential for accurate representation and comparison of the EEMs of the HS at high concentrations.

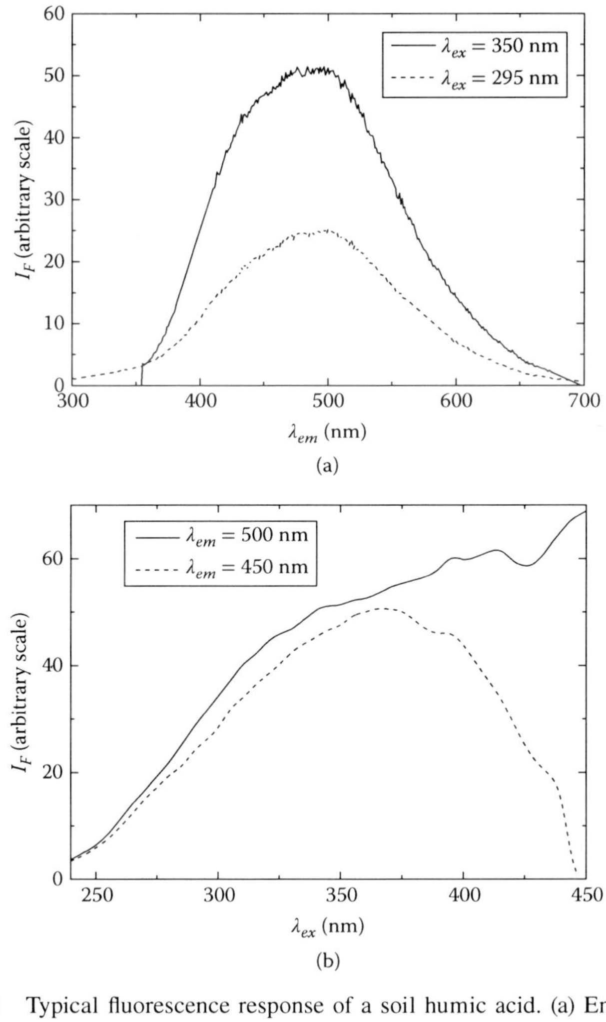

FIGURE 10.12 Typical fluorescence response of a soil humic acid. (a) Emission spectra; (b) excitation spectra.

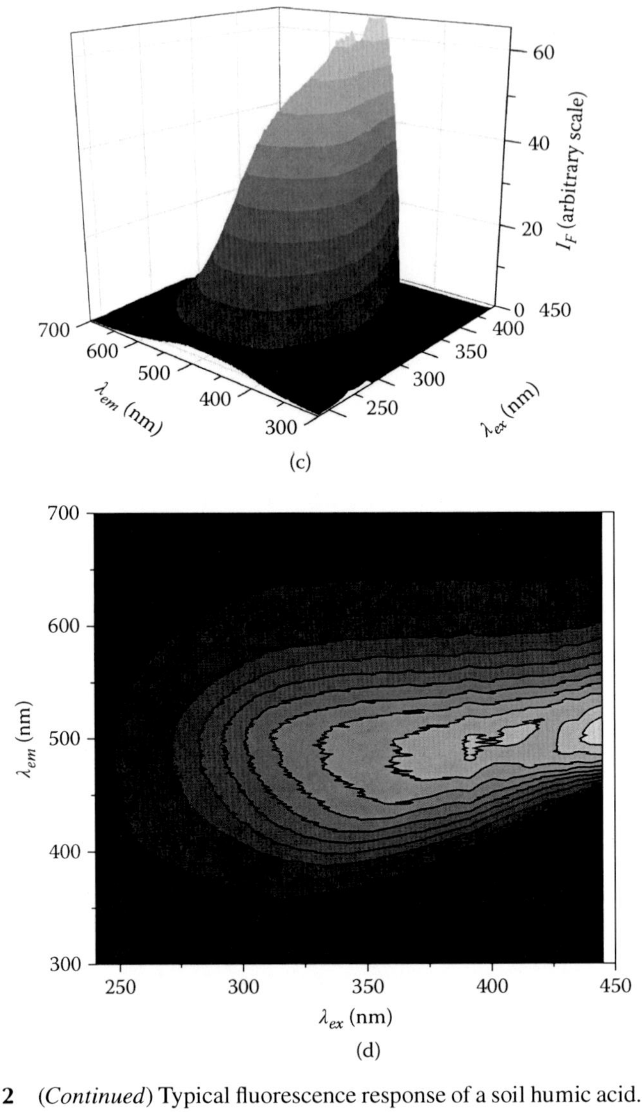

FIGURE 10.12 (*Continued*) Typical fluorescence response of a soil humic acid. (c) excitation-emission matrix (EEM) plot in landscape format; (d) EEM in contour plot form.

The fluorescent response has been shown to be important in the study of ion binding to HS. The interaction of UO_2^{2+} with three molecular weight fractions of a soil FA was studied by synchronous fluorescence (SyF) spectroscopy (Shin et al. 2001). Fluorescence quenching was observed to occur upon uranyl binding to the FA, and analysis of the SyF spectra showed that the fractions with molecules of larger sizes have a higher content of condensed aromatic compounds. From the analysis of the quenching profiles of the peaks, using a nonlinear method, the concentration of binding sites and the corresponding stability constants were calculated; these stability constants of the UO_2^{2+} –fulvate complexes indicated a stronger affinity for UO_2^{2+} in the

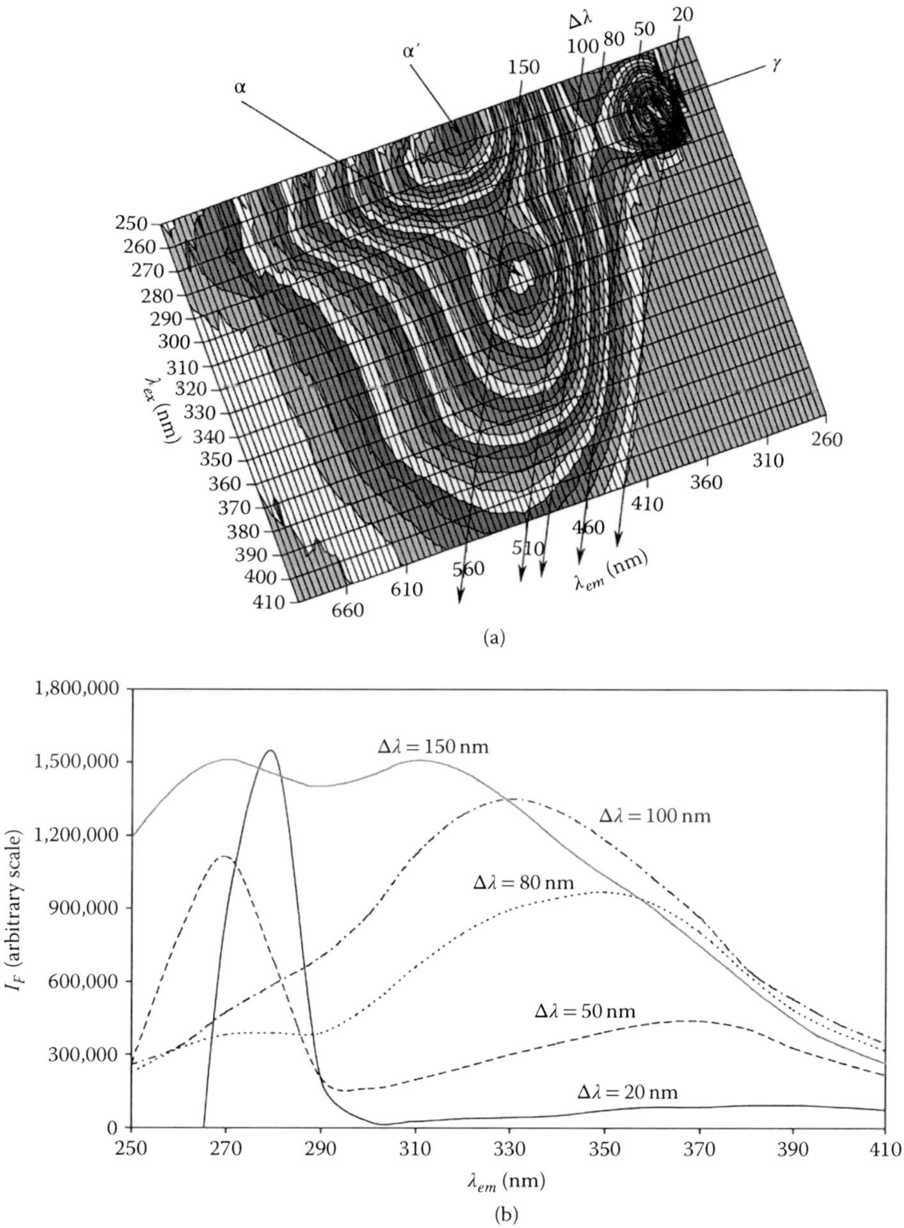

FIGURE 10.13 (a) EEM of an estuarine FA sample and (b) synchronous spectra of the same sample, at different $\Delta\lambda$ values, which are indicated in the EEM; it is seen that synchronous spectra correspond to diagonal cuts on the EEM. (Reprinted from *Chemosphere*, 58, no. 6, Sierra, M. M. D. et al., Fluorescence fingerprint of fulvic and humic acids from varied origins as viewed by single-scan and excitation/emission matrix techniques, 715–733. Copyright 2005, with permission from Elsevier.)

fraction with molecules of larger size. It is generally observed that cation binding causes quenching of fluorescence of HS, except a few cases like Al(III). Figure 10.14 (Hernandez et al. 2007) shows results for the interaction of Cu(II) and Zn(II) with FAs extracted from pig slurry (PS), from an unamended soil, and from PS-amended soils. Figure 10.14 shows EEM measurements of two amended soils, where the quenching due to Cu(II) adsorption is clearly seen, whereas Zn(II) produces only a small quenching. The data analysis indicates that the stability constants of Cu(II)–FA complexes and the Cu(II) complexing capacities of FAs isolated from PS, both unamended and PS-amended soils, are generally much larger than the corresponding values obtained for Zn(II). The PS–FA, which features a smaller content of acidic functional groups and a lower degree of aromatic polycondensation and humification than the unamended soil FA, is also characterized by smaller binding capacities and stability constants. The binding capacities and affinities of both PS-amended soil FAs are intermediate between those of native soil FA and PS–FA but closer to the former, thus suggesting a limited effect of PS-amendment at both rates onto native soil FA. Thus, fluorescence measurements are able to furnish useful information on cation binding to HS; there are many studies in the literature using fluorescence, specially with EEM, in that sense; chimiometric methods, mainly PARAFAC (Stedmon and Bro 2008), are of great help (Ohno, Amirbahman, and Bro 2008; Santín et al. 2009; He et al. 2010; Yu et al. 2010).

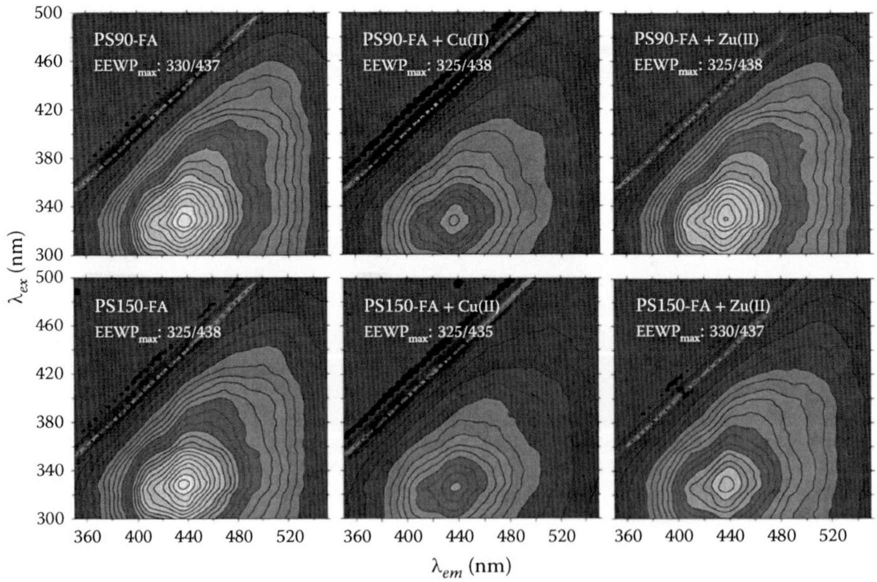

FIGURE 10.14 Fluorescence excitation-emission matrix (EEM) spectra of fulvic acids (FAs) isolated from soils amended with pig slurry (PS) at two different rates (PS90 and PS150, respectively) in the absence or presence of either Cu(II) or Zn(II). EEWPmax stands for excitation/emission wavelength pairs at the maximum fluorescence intensity. (Reprinted with permission from *European Journal of Soil Science*, 58, no. 4, Hernandez, D. et al., Fluorescence analysis of copper(II) and zinc(II) binding behaviour of fulvic acids from pig slurry and amended soils, 900–908. Copyright 2007, with permission from Elsevier.)

10.3.4 Functional Groups and Acid–Base Properties

10.3.4.1 Functional Group Analysis

The analysis of functional groups in HS is carried out by standard organic chemistry methods as follows (Swift 1996). Total acidity is usually found by titration of the HS (brought to H^+ form by use of a cation exchange resin) up to pH = 7.0 (which is an arbitrary end point). Alternatively, the HS are treated with a known excess of $Ba(OH)_2$, the precipitate is filtered and the remaining base is titrated with HCl to pH = 8.4. The amount of carboxylic groups is found by treatment with calcium acetate, and titration of the released protons with NaOH. The difference between the total acidity and the carboxylic contents is considered to be the phenolic (weaker acid groups) contents. On the other hand, total OH groups are determined by acetylation and posterior hydrolysis to acetic acid, which is titrated with standard base. The difference between total OH and phenolic OH, as determined above, is the amount of alcoholic groups. Finally, carbonyl groups are measured by reaction with an excess of hydroxylamine followed by titration of the unreacted hydroxylamine. Quinoid groups are determined by reduction with an excess amount of Fe(II) and amperometric titration of the remaining reductant; ketonic groups are found as difference between total C=O and the quinoidic ones. Table 10.3 presents results for FA and HA from different climates. It should be noted that, due to difficulties in precise quantification, these results should be taken with caution (Senesi and Loffredo 1998). Higher

TABLE 10.3
Main Functional Group Composition (mol kg^{-1}) of Humic and Fulvic Acids from Different Soils

	Soil					
Functional Group	**Arctic**	**Cool Acid**	**Subtropical**	**Tropical**	**Range**	**Average**
			Fulvic acids			
Total acidity	11.0	8.9–14.2	6.4–12.3	8.2–10.3	6.4–14.2	10.3
Carboxylate	8.8	6.1–8.5	5.2–9.6	7.2–11.2	5.2–11.2	8.2
Phenolic OH	2.2	2.8–5.7	1.2–2.7	0.3–2.5	0.3–5.7	3.0
Alcoholic OH	3.8	3.4–4.6	6.9–9.5	2.6–5.2	2.6–9.5	6.1
Quinoidic carbonyl	2.0	1.7–3.1	1.2–2.6	0.3–1.5	1.2–4.2	2.7
Ketonic carbonyl	2.0			1.6–2.7		
			Humic acids			
Total acidity	5.6	5.7–8.9	6.3–7.7	6.2–7.5	5.6–8.9	6.7
Carboxylate	3.2	1.5–5.7	4.2–5.2	3.8–4.5	1.5–5.7	3.6
Phenolic OH	2.4	3.2–5.7	2.1–2.5	2.2–3.0	2.1–5.7	3.9
Alcoholic OH	4.9	2.7–3.5	2.9	0.2–1.6	0.2–4.9	2.6
Quinoidic carbonyl	2.3	0.1–1.8	0.8–1.5	1.4–2.6	0.1–5.6	2.9
Ketonic carbonyl	1.7			0.3–1.4		

Source: Senesi, N., and E. Loffredo, *Soil Physical Chemistry*, 2nd ed., CRC Press, Boca Raton, FL, 1998.

acidity of FAs is clearly observed as compared with humics. The carbonyl contents are found to vary widely between different samples. Note that in some cases ketonic and quinoidic carbonyl are not differentiated, thus total carbonyl is reported.

The acid–base behavior of HS is complicated due to its complex composition. Figure 10.15 shows typical acid–base titration curves of 14 samples from the IHSS (Ritchie and Perdue 2003). These curves where obtained by titrating a known mass of HS, in 0.1 M NaCl supporting electrolyte, with standard NaOH, taking care to avoid problems such as HS oxidation in alkaline medium. The curves are plotted as the absolute value of the negative humic charge Q_A (which is actually the amount of dissociated protons), which is found from the charge balance:

$$Q_A/C_{HS} = [Na^+] + [H^+] - [Cl^-] - [OH^-] \tag{10.18}$$

where C_{HS} is the mass concentration (kg L^{-1}) of HS.

The shape of the curves do not resemble common weak acid titration curves; this is due to the high number of different acidic groups, mainly carboxylic and phenolic, with varying dissociation constants; the modeling of these curves will be treated in Chapter 13.

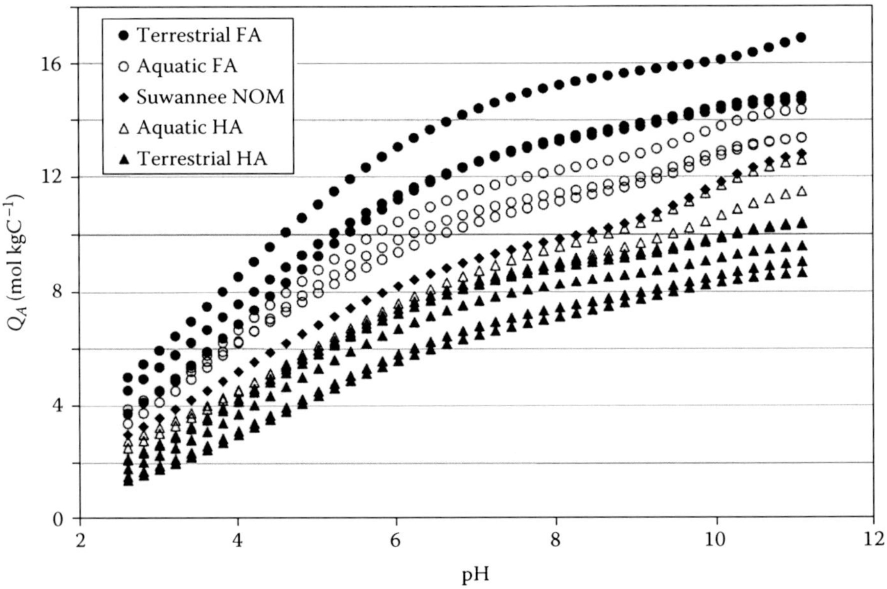

FIGURE 10.15 Acid–base titration curves of 14 humic substances, from IHSS standards: Terrestrial FA samples are standard Elliot soil, standard Pahokee Peat and reference Pahokee Peat; aquatic FA include standard Suwannee River, reference Suwannee River and reference Nordic Lake; NOM is Suwannee River; aquatic HA are standard Suwannee River and reference Nordic Lake; and terrestrial HA include standard Elliot soil, standard Pahokee Peat, standard Leonardite, reference Pahokee Peat and reference Summit Hill Soil. (Reprinted from *Geochimica et Cosmochimica Acta*, 67, no. 1, Ritchie, J. D and Perdue, E., Protonbinding study of standard and reference fulvic acids, humic acids, and natural organic matter, 85–93. Copyright 2003, with permission from Elsevier.)

10.3.4 Structural Studies by Chemical Degradative Methods

Chemical degradative methods applied to HS include acid or alkaline hydrolysis, reductive and oxidative degradation, chemical degradation with Na_2S or phenol; thermal degradation include thermogravimetry (TG), differential thermogravimetry (DTG), differential thermal analysis (DTA), and the more sophisticated pyrolysis methods (Senesi and Loffredo 1998; Clapp et al. 2005). Chemical degradation causes the breaking of chemical bonds with additions from the solution medium. Acid hydrolysis was shown to cause a weight loss around 42%; in other cases, NMR spectroscopy showed removal of sugar and amino acid groups, increase in aromaticity and loss of substituent carboxyl groups. Alkaline degradation in 5 M NaOH revealed the presence of up to 30 phenolic compounds by chromatography, accounting for 6% of the initial material at 250°C. After hydrolysis in 2 M NaOH and chromatographic separation, *n*-fatty acids, phenolic moieties, benzenecarboxylic acids, and some nitrogen compounds were identified by mass and micro-IR spectrometry. Alkaline hydrolysis was reported to be effective for C–O bonds cleavage (Senesi and Loffredo 1998 and references therein).

Reductive attack was conducted mainly with Na amalgam and Zn dust. Na amalgam relies in the production of nascent H, which causes the cleavage of ether linkages and liberation of phenol and phenolic acids from HS, thus being a relatively mild reactive. Labile OH groups are believed to be lost during amalgam attack. The products of degradation must be identified after the process; this task is most commonly performed using thin-layer chromatography. However, there is risk of dehydroxylation of phenolic moieties or reoxidation of some products, which may obscure the results. Zn dust, on the other hand, is a strong reducing agent, capable of breaking C–C and C–N bonds but not aromatic rings containing heteroatoms. It is applied by heating the sample with Zn at temperatures up to 530°C. As products, polycyclic aromatic hydrocarbons containing mainly two to five rings and also higher condensed products such as coronene have been found. It has been suggested, however, that this drastic method produces excessive bond breaking, molecular rearrangements, and recombination of fragments, leading to an artificial increment of condensed aromatic groups (Senesi and Loffredo 1998 and references therein). Oxidative degradation has been performed with several reactants; the most common are alkaline permangante, alkaline cupric oxide, and peroxoacetic acid. These procedures yield several degradation products that can be separated and identified by analytical methods, mainly different chromatographic techniques, including gas chromatography combined with mass spectrometry and micro-IR spectrometry. An important step is premethylation of HS prior to oxidation (or sometimes before products separation), to stabilize phenolic groups against degradation, increase of solubility on nonpolar solvents and increasing volatility. Alkaline potassium permanganate oxidation is a drastic procedure; nevertheless, it is considered the best, because it produces a relatively large number of identifiable degradation compounds. Major products identified (in both methylated and unmethylated HS) generally include benzenecarboxylic acids with 2-6 acid groups, phenolic acids containing 1-3 OH groups and 1-5 COOH groups, and aliphatic mono-, di-, and tri-carboxylic acids, usually found in the form of ethers and esters. These structures are thought by some authors to constitute the major "building blocks" of HS. The benzenecarboxylic acids are generally believed to derive from oxidation of aliphatic side chains associated with polycyclic aromatic compounds and/or aliphatic bridges between aromatic moieties, whereas the aliphatic carboxylic acids may arise from oxidation of straight-chain compounds or labile ring systems. However, these interpretations may change in view of recent results from NMR studies. Alkaline CuO oxidation is considered a relatively mild oxidative degradation procedure for HS, which is especially efficient for releasing phenolic structures, whereas aromatic structures bound through C–C bonds are not easily attacked. Degradation by alkaline CuO oxidation of a premethylated FA yielded about 30% products. More than half of the degradation products were identified. Phenolics, mainly phenolic acids, were around two thirds of the identified compounds, and benzenecarboxylic acids accounted for about 15%. Alkanes, fatty acid methyl esters, and aliphatic dicarboxylic acid esters were also detected. Peroxoacetic acid (CH_3COOOH) has been used as an oxidizing agent for HS with the purpose of minimizing the magnitude and extent of chemical alterations, and the formation of possible artifacts that may result in oxidations conducted under alkaline conditions. Products and yields obtained by peroxoacetic acid oxidation of HA and FA were phenolic acids, 4.3% and 4.1%, respectively, and benzenecarboxylic acids, 15.2% and 7.3%, respectively. In addition, small amounts of fatty acids and aliphatic dicarboxylic acids were detected (Senesi and Loffredo 1998; Clapp et al. 2005).

10.3.4.3 Structural Studies by Thermal Degradation

10.3.4.3.1 Thermal Analysis

Thermal analysis methods (introduced in Section 7.6.5) are useful in the characterization of SOM, and HS in particular. The TG and DTG show mass loss due to decomposition reactions. In DTG, soil HA generally present intense peaks at ~280 and 540°C, and a weak one near 100°C; for FA, there is a main peak at 420°C with two weak ones, below 100°C and about 270°C, respectively. The general interpretation of these experiments is difficult. It has been suggested that the main thermal reactions for HA are (1) dehydration up to 200°C; (2) elimination of functional groups (mainly carboxylates) between 200 and 250°C; and (3) further dehydration and dehydrogenation at higher temperatures; reactions between 500 and 600°C were attributed to breaking of aliphatic and alicyclic structures, and those above 600°C to decomposition of aromatic structures. For FA, the main reactions were dehydration and decarboxylation up to 400°C and further dehydration and deoxydation at higher temperatures (Senesi and Loffredo 1998 and references therein). DTA and differential scanning calorimetry give results similar to DTG. DTA curves usually show (1) a shallow endothermic peak near 100°C, due to loss of water; (2) an exothermic peak between 300 and 380°C, ascribed to decarboxylation and dehydration of phenol groups; and (3) a prominent exothermic peak between 440 and 490°C, attributed to decomposition of aromatic nuclei.

As an example, Figure 10.16 shows TG and DTA curves for three humic acids, from lignin (L), leonardite (Le), and peat (P). The features mentioned above appear for the three HA, but for lignin HA a sharp DTA peak at 699°C is found (Francioso et al. 2005).

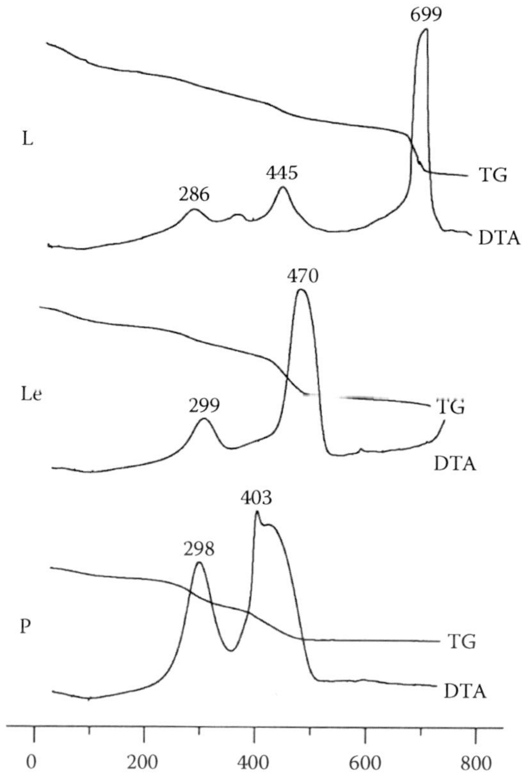

FIGURE 10.16 TG and DTA curves in air atmosphere of three humic acids extracted from peat (P), leonardite (Le), and lignite (L). (Reprinted with permission from Francioso, O. et al., *Appl. Geo.*, 20, no. 3, 537–544. Copyright 2005 Springer-Verlag.)

The first exothermic peak was considered to be the result of thermal combustion of polysaccharides, decarboxylation of acidic groups, and dehydration of hydroxylate aliphatic structures; the second exothermic peak was attributed to the combustion of aromatic structures and cleavage of C–C bonds. The high-temperature peak in the L sample was assumed to be related to high-temperature combustion accompanied by a polycondensation. In addition, the presence of long-chain hydrocarbons and N compounds can contribute to thermal reactions over 450°C.

10.3.4.3.2 Pyrolysis Methods

Pyrolysis is the decomposition of substances by application of thermal energy; of course, that is what happens in the techniques discussed in the previous paragraph, but in *analytical pyrolysis*, this process is combined with gas analytical methods to characterize the decomposition products. The reaction is produced in vacuum or in a rapid stream of a carrier gas to avoid as much as possible secondary reactions. The products are usually separated by gas chromatography and analyzed by mass spectrometry (in this technique, gaseous molecules and fragments are ionized, separated and detected according to their charge/mass ratio, allowing identification of the separated analytes [Hoffmann and Stroobant 2007]). Pyrolysis–gas chromatography–mass spectrometry (Py–GC–MS) is a powerful method, albeit it produces a huge amount of information that is of difficult interpretation (Irwin 1979; Howarth and Sinding-Larsen 1983; Saiz-Jimenez 1994; Moldoveanu 1998). Pyrolysates of HS contain a rich mixture of products that can be related to their constituent biopolymers and building blocks. A complete array of products identified in pyrolysates of HS and their parent biopolymers are listed in the literature (Senesi and Loffredo 1998; Clapp et al. 2005 and references therein). It appears that all HS give, in various proportions, products highly characteristic of polypeptide and polysaccharide moieties, lignin units, microbially synthesized polyphenols, substituted polycarboxylic acids, lipids, and other aliphatic constituents, amino sugars, and aromatic hydrocarbons. Broad differences were found in the proportions of pyrolysis products between FAs, HAs, and humins. Pyrolysates of HAs were enriched in polypeptide products, lignin derivates, and phenols, but contained little polysaccharide products, whereas FA pyrolysates contained less polypeptide products and higher levels of polysaccharides. Similar results were obtained in another study where carbohydrate and phenolic constituents were found to be more pronounced in pyrolysis products from FAs, whereas saturated and unsaturated hydrocarbons were the most relevant products obtained from HAs. In modern studies, Py–CG–MS is combined with other techniques, such as thermally assisted hydrolysis and methylation; in a recent study (Buurman et al. 2009), HAs from five reference soils called the Eurosoils, including a Vertic Cambisol (E1, Italy) and an Orthic Podzol (E5, Germany), were studied by Py–GC–MS and thermally assisted hydrolysis and methylation (THM) and subsequent analysis by GC–MS. Figure 10.17 shows the resulting GC chromatograms of the pyrolysates with assignments. The E5 sample showed the largest aliphaticity and the strongest degradation of aliphatics and lignin, whereas the E1 HA had abundant polyaromatics, which indicate the presence of charred material. Classification by "degradation state" based on Py–GC–MS largely coincided with acidity determined by titration, but FTIR data did not coincide. Py–GC–MS, with its much larger resolution, is a better tool to distinguish effects of vegetation, microbial input, and degradation.

10.3.5 Molecular Structure of HS

The determination of molecular structures of HS is a problem that has interested many researchers for a long time (Hayes et al. 1990; Stevenson 1994; Hayes and Malcolm 2001; Simpson et al. 2001a, 2002; Wershaw 2004; Sutton and Sposito 2005; Baigorri et al. 2009). The matter is complicated, because HA and FA are really fractions defined by their solubility properties, thus potentially encompassing a large number of organic compounds. Furthermore, it appears that some researchers attempted to search for a common general chemical structure for HS, which is not the case, as modern analytical techniques have revealed, but nevertheless it should have appeared unlikely in advance. It is interesting to review here some proposals that have been published over the years. In the 1980s, Stevenson (1994) on the basis of wet chemical methods plus 1-D NMR studies, proposed the structures shown in

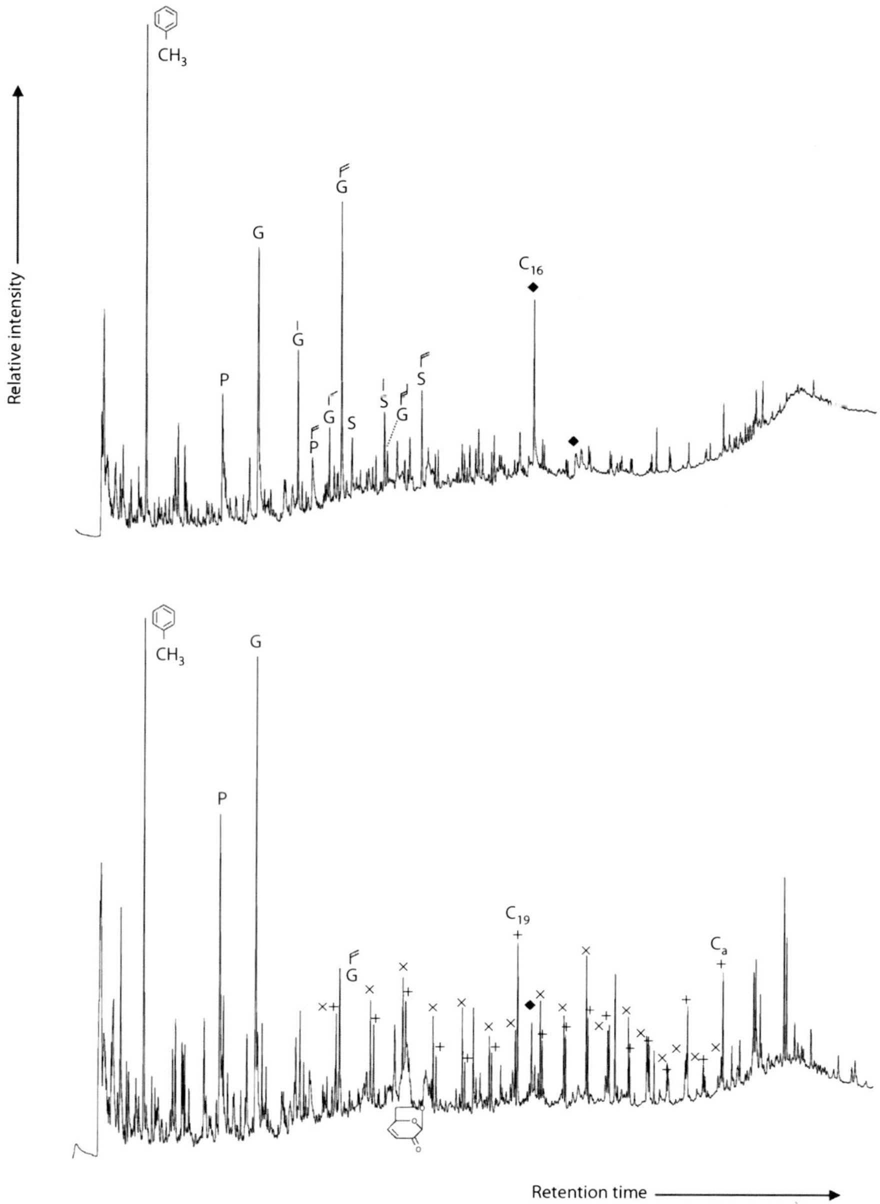

FIGURE 10.17 Gas chromatograms of the pyrolysates of Eurosoil humic acids. Legend: P = phenol; G = guaiacol (2-methoxyphenol); S = syringol (2,6-dimethoxyphenol); ◆ = fatty acid; × = *n*-alkene; + = *n*-alkane; C_n indicates chain length. (Reprinted from *Geoderma*, 150, no. 1–2, Buurman, P. et al., Analytical pyrolysis and thermally assisted hydrolysis and methylation of EUROSOIL humic acid samples—A key to their source, 10–22. Copyright 2009, with permission from Elsevier.)

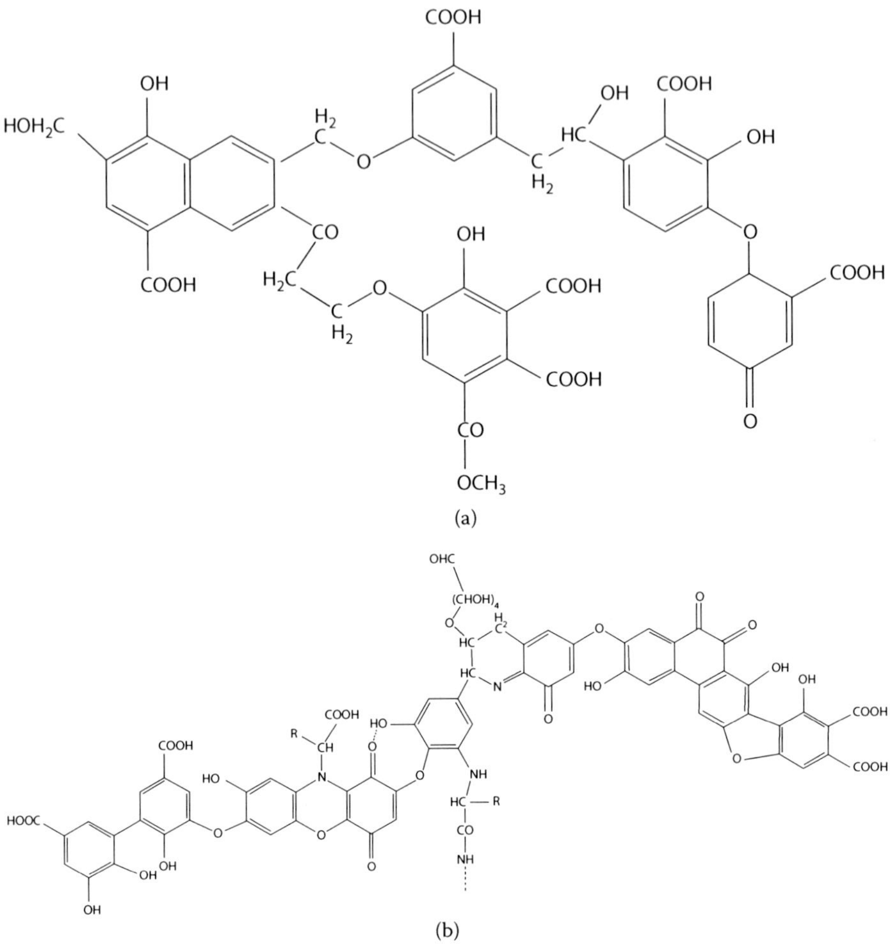

FIGURE 10.18 Humic structures proposed by Stevenson (1994): (a) fulvic acid; (b) humic acid.

Figure 10.18, where the main functional groups detailed above (Table 10.3) are present. In the humic structure, protein and carbohydrate residues are included, and a predominance of aromatic groups is observed. It has been a matter of debate whether HS were mainly aromatic or aliphatic (Farmer and Pisaniello 1985; Schulten et al. 1992; Hatcher and Minard 1995; Clapp et al. 2005); nowadays, NMR studies show that HS have usually a predominance of aliphatic residues, and also with the presence of aromatic moieties (see, for example, Figures 10.7 and 10.9); also, the features observed in the fluorescence of HS point strongly to condensed aromatic rings.

About a decade after, Schulten and Schnitzer (1993) published the HA molecular structure shown in Figure 10.19. Clearly, a macromolecular concept is implied here, which would require necessarily extensive coupling reactions of degradation fragments from plant residues. Another humic structure proposed was a polymeric one,

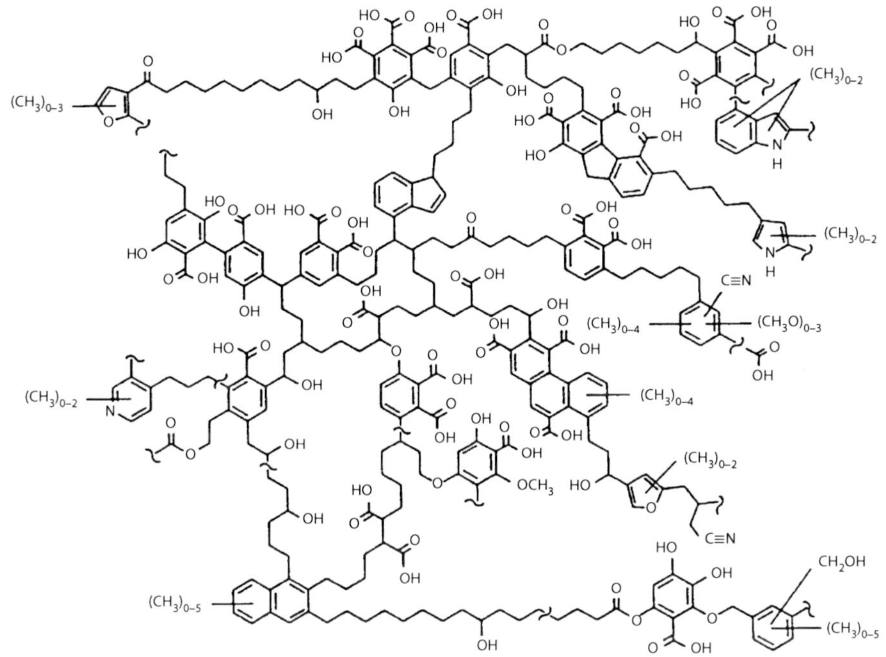

FIGURE 10.19 Humic acid structure proposed by (With kind permission from Springer Science+Business Media, *Naturwissenschaften*, Schulten, H.-R., and M. Schnitzer, 80, 1993, 29–30.)

based on the monomers presented in Figure 10.20 (Sein, Varnum, and Jansen 1999); these monomers were supposed to bind together by amide bonds. This is more a model enabling computational studies than a true representation of HS. Some authors have employed this model in theoretical studies (Sein, Varnum, and Jansen 1999; Alvarez-Puebla, Valenzuela-Calahorro, and Garrido 2006; Kalinichev and Kirkpatrick 2007); however, a polymer has properties arising from its regularity that are not present in HS.

Another important fact in the HS behavior is their capability to aggregate and disaggregate in solution, which has been observed by several authors (Tombacz and Meleg 1990; Engebretson and von Wandruszka 1998; Maurice and Namjesnik-Dejanovic 1999; Manning, Bennett, and Milton 2000; Pranzas et al. 2003). In particular, it is known that to dissolve an HA in a relatively acidic medium (pH about 4–5), the best way is to dissolve it first in alkali and then lower the pH to the desired value; at high pH, the HA is highly charged, thus favoring the disaggregation by repulsion between the molecules. After this process, AFM studies reveal relatively small molecular size (about 2–4 nm) (Balnois et al. 1999; Plaschke et al. 1999); see Figures 2.15 and 10.21. Analytical pyrolysis, in turn, has revealed a number of different fragments, including aromatics and aliphatics, carbohydrates, polypeptides, and so on. On the other hand, detailed 2-D NMR studies by Simpson et al (2002) have revealed the presence of molecular fragments of diverse origin (aliphatic, aromatic, polypeptides, polysaccharides, etc.), and proposed that HS in soil are essentially aggregates of fragments of different chemical nature, bound together by a combination of H bonding,

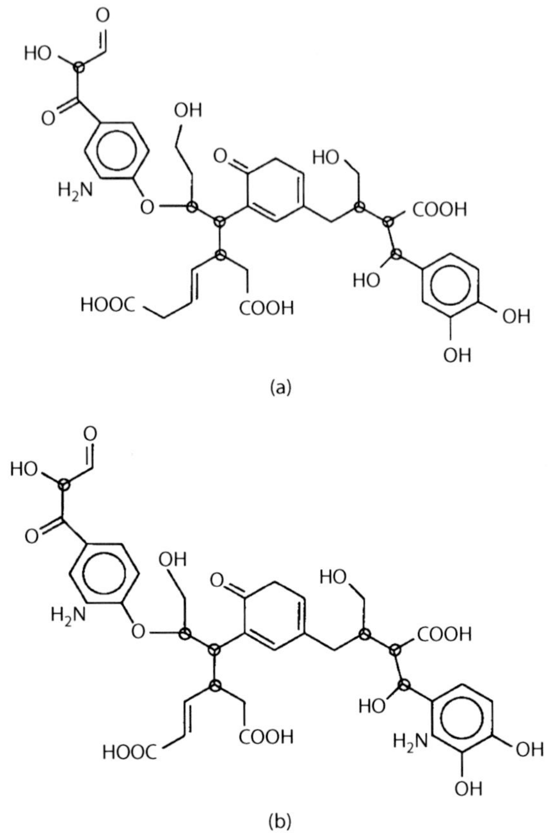

FIGURE 10.20 The Steelink (a) and TLB (b) monomer structures in the polymeric models of HS.

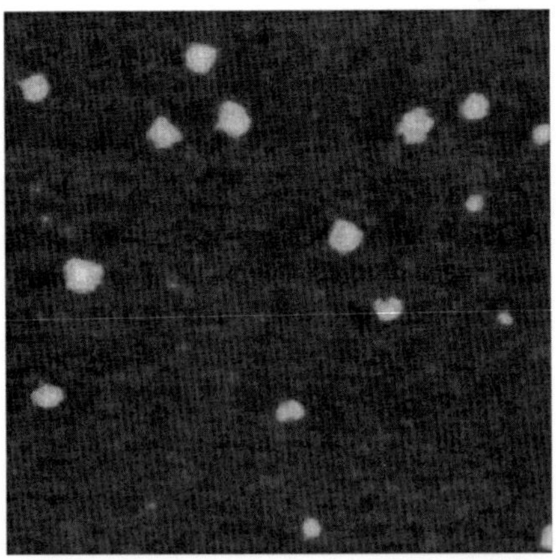

FIGURE 10.21 Typical TM-AFM image of isolated Suwannee River humic acid macromolecules adsorbed on mica. Scan size is 600 nm × 600 nm. (Reprinted with permission from Balnois, E. et al., *Environmental Science and Technology* 33, no. 21, 3911–3917. Copyright 1999 American Chemical Society.)

cation complexation, and van der Waals forces. Figure 10.22 shows the pictorial view of comparing with the polymer conception, whereas Figure 10.23 shows a molecular illustration of the concept. It is worth noting here that there is experimental evidence showing that HS suffer size changes with pH variations (Pinheiro et al. 1996; Avena, Vermeer, and Koopal 1999) and are permeable (Duval et al. 2005). This idea of the composition and state of HS in soils is consistent with all the currently available experimental information.

More recent spectroscopic studies have focused on the analysis of functional groups and their environment (Hay and Myneni 2007; Deshmukh et al. 2007), proposing a number of small structures that can be presented as isolated molecules or as part of more extended structures; they concluded that a significant portion of humic carboxyl groups exists predominantly in the form of unsubstituted and branched aliphatic and alicyclic structures. The NMR data also show that a portion of carboxyl groups in soils have electron-withdrawing substituents on neighboring carbon atoms; such groups not only increase the acidity of carboxyl groups but also improve their ability to bind to metals through chelation. Some samples also exhibit

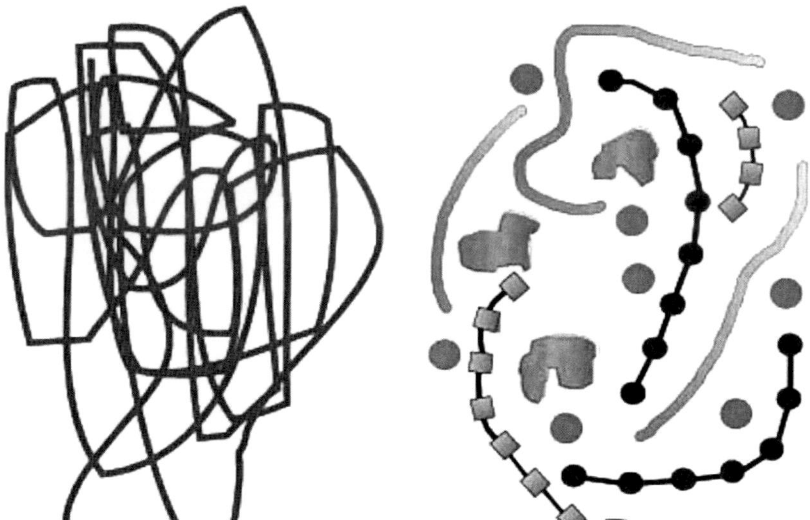

(a) (b)

FIGURE 10.22 Pictorial representation of: (a) the concept of humic substances (HS) as randomly coiled macromolecules; (b) the new concept that the major components found in alkaline extractable soil HS are in fact of relatively low molecular weight (<2,000 Da) and associate to form an aggregate. The symbols represent the isolated circles generic metal cations, the square chained units polypeptides, the chained circles polysaccharides, the coarse lines aliphatic chains, and the irregular blocks aromatic lignin fragments, respectively. (Reprinted with kind permission from Springer Science+Business Media: *Naturwissenschaften*, Simpson, A. J. et al., Molecular structures and associations of humic substances in the terrestrial environment, 89, 2002, 84–88.)

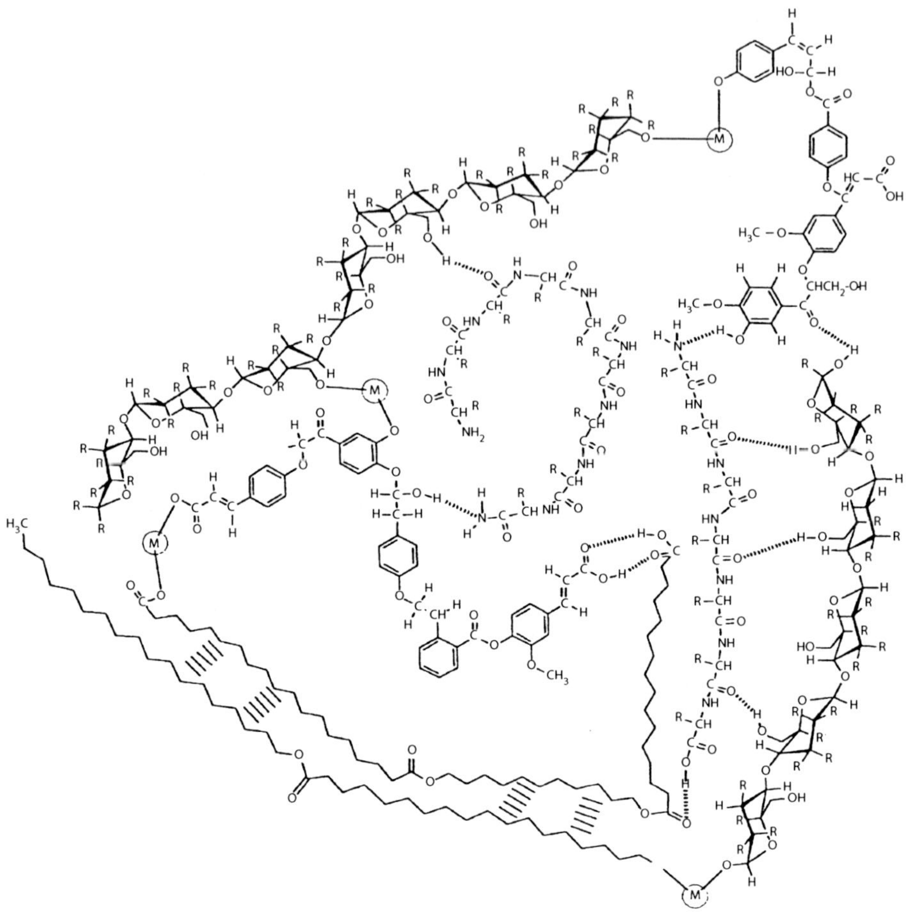

FIGURE 10.23 An illustration, after Clapp et al. (2005) of how components of soil HS associate in aqueous environments in the presence of metal cations to form aggregates. The "M" labeled spheres represent cations, the dashed lines indicate H bonds and the wide dashed lines represent hydrophobic associations, following Simpson et al. (2002). (Reprinted from Clapp et al., *Chemical Processes in Soils*, Soil Science Society of America, Madison, WI, 2005, with kind permission.)

distinct signals for aromatic acids, but these are in lower proportion; this is at a variance with other models that propose a high content of aromatic acid groups. These results suggest that aromatic carboxyls (particularly, substituted benzoic acids) may not necessarily be the dominant carboxyl structures in all humics.

10.4 ION BINDING TO HS

The interaction of metal cations with HS is a topic of great importance both from fundamental and environmental points of view and has attracted the attention of a high number of researchers (Clapp et al. 2001; Senesi and Loffredo 2005; Van Riemsdijk et al. 2006 and references therein). In a general approach, it is often useful to consider the reactivity of metal cations and ligand groups. Inorganic cations may be divided into three classes according to their reactivity with organic ligands: hard cations are those that interact mainly electrostatically forming weak outer-sphere complexes with hard oxygen ligands, such as Ca^{2+}, Mg^{2+}, and others; soft cations are those that have strong affinity for binding covalently to intermediate and soft ligands such as N or S, for example, Cd^{2+}, Pb^{2+}, or Hg^{2+}; and intermediate or "borderline" cations have an intermediate character forming relatively strong bonds with both types of ligands. On the other hand, hard ligands (Lewis bases) include carboxylate, ester, phenolic and alcoholic OH, carbonyl, and ether groups; intermediate ligands are mainly those containing nitrogen (amines, amides, etc.), and soft bases are mainly S-containing groups (Buffle 1988). Usually, it is expected that hard cations will react preferably with hard ligands and so on; however, exceptions are observed, for example Pb(II), a soft cation forms stable coordination complexes with carboxylates, a hard ligand.

As with mineral soil components, modern spectroscopic techniques have been of great help in the elucidation of the structures existing around humic-complexed metal cations. XAS studies (both XANES and EXAFS) have shed light about the coordination environment of metal cations, as reviewed by several authors (Bloom, Bleam, and Xia 2001; Senesi and Loffredo 2005). In these and more recent studies (Karlsson, Persson, and Skyllberg 2006; Gustafsson et al. 2007; Karlsson and Persson 2010), most evidence points to the existence of inner-sphere tetragonal complexes for cations like Cu(II) and Pb(II) and octahedral coordination for others, such as Fe(III), Ni(II), Co(II), or Zn(II). In all cases, the first coordination sphere is consistent with O or N ligand atoms, because in EXAFS it is not possible to distinguish between both; however, given the very different proportion of O over N in HS, it is generally accepted that the first shell is formed mostly by O atoms. In the second shell, two to four C atoms are generally found, indicating multiple coordination to one or two HS molecules. Figure 10.24 shows a proposed structure for the binding of Cu(II) to NOM, after examining by EXAFS four SOM samples and three DOM

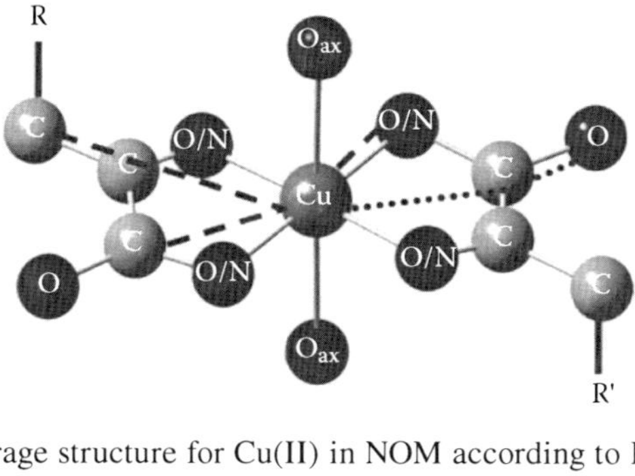

FIGURE 10.24 Average structure for Cu(II) in NOM according to Karlsson, Persson, and Skyllberg (2006). The dashed lines indicate the three single scattering (SS) paths, and the dotted line indicates the multiple scattering (MS) path, identified in the EXAFS analysis. The more distant axial oxygens (Oax) are not included in the final fits to the EXAFS data. R represents carbon chains. (Reprinted with permission from Karlsson, T. et al., *Environmental Science & Technology*, 40, no. 8, 2623–2628. Copyright 2006 American Chemical Society.)

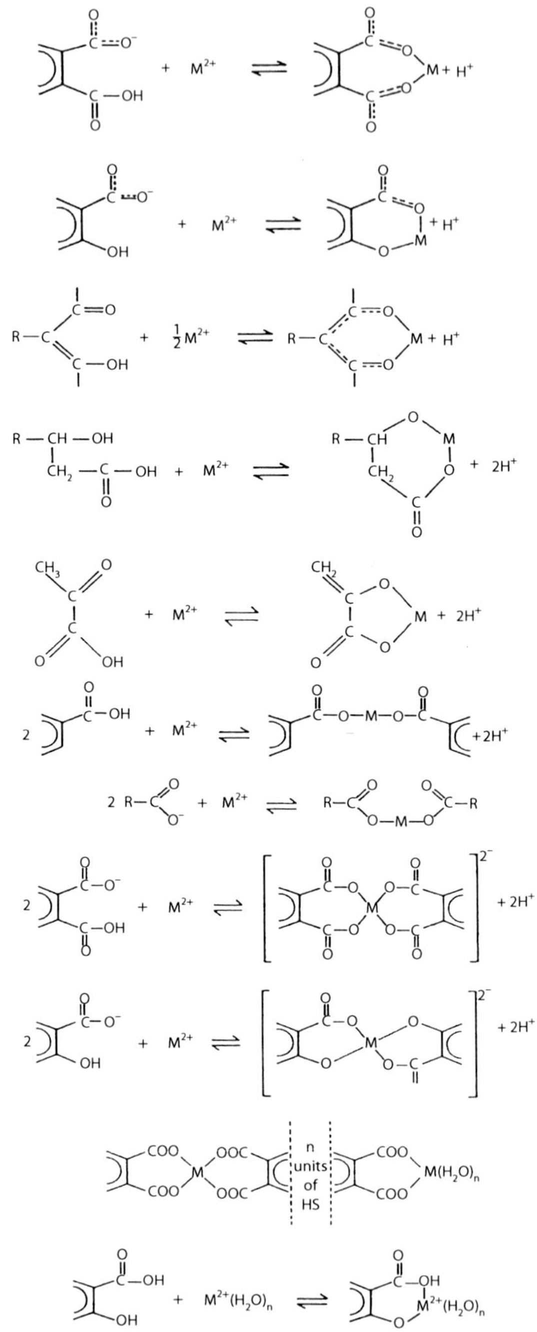

(1)
(2)
(3)
(4)
(5)
(6)
(7)
(8)
(9)
(10)
(11)

FIGURE 10.25 Different metal–humic complexation structures proposed by Senesi and Loffredo. (Reprinted from Senesi, N., and E. Loffredo, *Chemical Processes in Soils*, Soil Science Society of America, Madison, WI, 563–617, 2005, with kind permission.)

samples (Karlsson, Persson, and Skyllberg 2006). A similar behavior for Pb(II) is suggested by several studies (Xia, Bleam, and Helmke 1997; De La Rosa et al. 2005). In a detailed study on Fe(III) interaction with peat humic acid (Karlsson and Persson 2010), it was found that mononuclear complexes are formed exclusively at low Fe loadings, but at high loadings, a mixture of mononuclear complexes and polymeric Fe(III) (hydr)oxides, with an increasing amount of (hydr)oxides at higher pH, was found. Chelate structures were detected for the mononuclear complexes, and no appreciable reduction of Fe(III) was observed.

Other studies of metal–HS interactions include NMR experiments (Kingery et al. 2001; Senesi and Loffredo 2005 and references therein), which have been used to correlate metal-binding structures with metal-binding characteristics, to elucidate binding mechanisms and to determine complexation equilibrium constants. IR experiments are less helpful, but they generally confirm that the carboxylate groups are often the main ligand for complexed metals (Davies et al. 2001).

Senesi and Loffredo (2005) proposed, on the basis of the available experimental results, a number of possible metal–HS complexation structures, which are shown in Figure 10.25. It should be noted that chelate structures are predominant, and that in some cases, metal bridging of two different molecules is proposed.

10.5 SUMMARY

In this chapter, the structure and ion-binding behavior of HS were considered. After a long evolution of ideas, the current knowledge is that HS are composed of molecules of intermediate size (M < 2000 Da), including carbohydrates, polypeptides, and aliphatic and aromatic hydrocarbons, bound together in the solid state by H bonding, metal complexation, and hydrophobic interactions. Functional groups are mainly oxygenated ones, including carboxylic acids, aliphatic and phenolic OH, ketonic carbonyls, esters, and ethers. Carboxylates appear to dominate metal humic interactions, and complexation of transition metals tends to form chelate structures. This chapter ends the study of the main soil components and their ion-binding behavior. In the next Part, the modeling of ion binding to soil colloids will be treated.

REFERENCES

Aiken, G. R., and A. H. Gillam. 1990. Determination of molecular weights of humic substances by colligative property measurements. In *Humic Substances II: In Search of Structure*, ed. M. H. B. Hayes, P. MacCarthy, R. L. Malcolm, and R. S. Swift, chap. 18. 1st ed. New York: John Wiley & Sons.

Alvarez-Puebla, R. A., C. Valenzuela-Calahorro, and J. J. Garrido. 2006. Theoretical study on fulvic acid structure, conformation and aggregation a molecular modelling approach. *Science of the Total Environment* 358, no. 1–3: 243–254.

Alvarez-Puebla, R. A., and J. J. Garrido. 2005. Effect of pH on the aggregation of a gray humic acid in colloidal and solid states. *Chemosphere* 59, no. 5: 659–667. doi:10.1016/j.chemosphere.2004.10.021.

Asakawa, D., T. Kiyota, Y. Yanagi, and N. Fujitake. 2008. Optimization of conditions for high-performance size-exclusion chromatography of different soil humic acids. *Analytical Sciences* 24, no. 5: 607–613.

Assemi, S., P. G. Hartley, P. J. Scales, and R. Beckett. 2004. Investigation of adsorbed humic substances using atomic force microscopy. *Colloids and Surfaces A: Physicochemical and Engineering Aspects* 248, no. 1–3: 17–23. doi:10.1016/j.colsurfa.2004.08.035.

Atkins, P., and J. de Paula. 2009. *Physical Chemistry*. 9th ed. New York: W. H. Freeman.

Avena, M. J., A. W. P. Vermeer, and L. K. Koopal. 1999. Volume and structure of humic acids studied by viscometry pH and electrolyte concentration effects. *Colloids and Surfaces A: Physicochemical and Engineering Aspects* 151, no. 1–2: 213–224.

Baes, A. U., and P. R. Bloom. 1990. Fulvic acid ultraviolet-visible spectra: Influence of solvent and pH. *Soil Science Society of America Journal* 54, no. 5: 1248–1254. doi:10.2136/sssaj1990.03615995005400050008x.

Baigorri, R., M. Fuentes, G. González-Gaitano, and J. M. García-Mina. 2007. Analysis of molecular aggregation in humic substances in solution. *Colloids and Surfaces A: Physicochemical and Engineering Aspects* 302, no. 1–3: 301–306.

Baigorri, R., M. Fuentes, G. González-Gaitano, J. M. García-Mina, G. Almendros, and F. J. González-Vila. 2009. Complementary multianalytical approach to study the distinctive structural features of the main humic fractions in solution: Gray humic acid, brown humic acid, and fulvic acid. *Journal of Agricultural and Food Chemistry* 57, no. 8: 3266–3272. doi:10.1021/jf8035353.

Baldock, J. A., and P. N. Nelson. 1999. Soil organic matter. In *Handbook of Soil Science*, ed. M. L. Sumner, B75–B84. 1st ed. Boca Raton, FL: CRC.

Balnois, E., K. J. Wilkinson, J. R. Lead, and J. Buffle. 1999. Atomic force microscopy of humic substances: Effects of pH and ionic strength. *Environmental Science and Technology* 33, no. 21: 3911–3917.

Beckett, R., Z. Jue, and J. C. Giddings. 1987. Determination of molecular weight distributions of fulvic and humic acids using flow field-flow fractionation. *Environmental Science & Technology* 21, no. 3: 289–295. doi:10.1021/es00157a010.

Bellamy, L. J. 1975. *The Infra-Red Spectra of Complex Molecules*. 3rd ed. New York: Wiley.

Bellon-Maurel, V., and A. McBratney. 2011. Near-infrared (NIR) and mid-infrared (MIR) spectroscopic techniques for assessing the amount of carbon stock in soils—Critical review and research perspectives. *Soil Biology and Biochemistry* 43, no. 7: 1398–1410. doi:16/j.soilbio.2011.02.019.

Bidoglio, G., D. Ferrari, E. Selli, F. Sena, and G. Tamborini. 1997. Humic acid binding of trivalent Tl and Cr studied by synchronous and time-resolved fluorescence. *Environmental Science & Technology* 31, no. 12: 3536–3543. doi:10.1021/es970294x.

Bloom, P. R., W. F. Bleam, and K. Xia. 2001. X-ray spectroscopy applications for the study of humic substances. In *Humic Substances and Chemical Contaminants*, ed. C. E. Clapp, M. H. B. Hayes, N. Senesi, P. R. Bloom, and P. M. Jardine, 317–350. Madison, WI: Soil Science Society of America.

Buffle, J. 1988. *Complexation Reactions in Aquatic Systems: And Analytical Approach*. Chichester, UK: Ellis Horwood.

Buurman, P., K. G. J. Nierop, J. Kaal, and N. Senesi. 2009. Analytical pyrolysis and thermally assisted hydrolysis and methylation of EUROSOIL humic acid samples—A key to their source. *Geoderma* 150, no. 1–2: 10–22.

Cardoza, L. A., A. K. Korir, W. H. Otto, C. J. Wurrey, and C. K. Larive. 2004. Applications of NMR spectroscopy in environmental science. *Progress in Nuclear Magnetic Resonance Spectroscopy* 45, no. 3–4: 209–238. doi:10.1016/j.pnmrs.2004.06.002.

Chen, C., X. Wang, H. Jiang, and W. Hu. 2007. Direct observation of macromolecular structures of humic acid by AFM and SEM. *Colloids and Surfaces A: Physicochemical and Engineering Aspects* 302, no. 1–3: 121–125.

Clapp, C. E., W. W. Emerson, and A. E. Olness. 1990. Sizes and shapes of humic substances by viscosity measurements. In *Humic Substances II: In Search of Structure*, ed. M. H. B. Hayes, P. MacCarthy, R. L. Malcolm, and R. S. Swift, chap. 17. 1st ed. New York: John Wiley & Sons.

Clapp, C. E., and M. H. B. Hayes. 1999. Characterization of humic substances isolated from clay- and silt-sized fractions of a corn residue-amended agriculture soil. *Soil Science* 164, no. 12: 899–913.

Clapp, C. E., M. H. B. Hayes, N. Senesi, P. R. Bloom, and P. M. Jardine, eds. 2001. *Humic Substances and Chemical Contaminants*. Madison, WI: Soil Science Society of America.

Clapp, C. E., M. H. B. Hayes, A. J. Simpson, and W. L. Kingery. 2005. Chemistry of soil organic matter. In *Chemical Processes in Soils*, ed. M. A. Tabatabai and D. L. Sparks, 1–150. Madison, WI: Soil Science Society of America.

Conte, P., R. Spaccini, and A. Piccolo. 2006. Advanced CPMAS-13C NMR techniques for molecular characterization of size-separated fractions from a soil humic acid. *Analytical and Bioanalytical Chemistry* 386, no. 2: 382–390. doi:10.1007/s00216-006-0637-5.

De La Rosa, G., J. R. Peralta-Videa, J. G. Parsons, and J. L. Gardea-Torresdey. 2005. Using X-ray absorption spectroscopy to study the speciation and coordination of lead binding to humic materials. *Spectroscopy* 20, no. 7: 24–30.

Davies, G., E. A. Ghabbour, A. Cherkasskiy, and A. Fataftah. 2001. Tight metal binding by solid phase peat and soil humic acids. In *Humic Substances and Chemical Contaminants*, ed. C. E. Clapp, M. H. B. Hayes, N. Senesi, P. R. Bloom, and P. M. Jardine, 371–395. Madison, WI: Soil Science Society of America.

Deshmukh, A P., C. Pacheco, M. B. Hay, and S. C.B. Myneni. 2007. Structural environments of carboxyl groups in natural organic molecules from terrestrial systems. Part 2: 2D NMR spectroscopy. *Geochimica Et Cosmochimica Acta* 71, no. 14: 3533–3544. doi:10.1016/j.gca.2007.03.039.

Duval, J. F. L., K. J. Wilkinson, H. P. Van Leeuwen, and J. Buffle. 2005. Humic substances are soft and permeable: Evidence from their electrophoretic mobilities. *Environmental Science and Technology* 39, no. 17: 6435–6445.

Engebretson, R. R., and R. von Wandruszka. 1998. Kinetic aspects of cation-enhanced aggregation in aqueous humic acids. *Environmental Science & Technology* 32, no. 4 (February 1): 488–493. doi:10.1021/es970693s.

Esteves da Silva, J. C. G., A. A. S. C. Machado, C. J. S. Oliveira, and M. S. S. D. S. Pinto. 1998. Fluorescence quenching of anthropogenic fulvic acids by Cu(II), Fe(III) and UO22+. *Talanta* 45, no. 6: 1155–1165. doi:10.1016/S0039-9140(97)00224-5.

Farmer, V. C., and D. L. Pisaniello. 1985. Against an aromatic structure for soil fulvic acid. *Nature* 313, no. 6002: 474–475. doi:10.1038/313474a0.

Francioso, O., D. Montecchio, P. Gioacchini, and C. Ciavatta. 2005. Thermal analysis (TG–DTA) and isotopic characterization (13C–15N) of humic acids from different origins. *Applied Geochemistry* 20, no. 3: 537–544. doi:10.1016/j.apgeochem.2004.10.003.

Garland, C. W., J. W Nibler, and D. P. Shoemaker. 2009. *Experiments in Physical Chemistry*. 8th ed. Boston, MA: McGraw-Hill Higher Education.

Gerin, P. A., and Y. F. Dufrêne. 2003. Native surface structure of natural soil particles determined by combining atomic force microscopy and X-ray photoelectron spectroscopy. *Colloids and Surfaces B: Biointerfaces* 28, no. 4: 295–305. doi:10.1016/S0927-7765(02)00163-7.

González-Pérez, J. A., F. J. González-Vila, G. Almendros, and H. Knicker. 2004. The effect of fire on soil organic matter—a review. *Environment International* 30, no. 6: 855–870. doi:10.1016/j.envint.2004.02.003.

Gustafsson, J. P., I. Persson, D. B. Kleja, and J. W. J. van Schaiks. 2007. Binding of iron (III) to organic soils: EXAFS spectroscopy and chemical equilibrium modeling. *Environmental Science & Technology* 41, no. 4: 1232–1237.

Hatcher, P. G., and R. D. Minard. 1995. Comment on the origin of benzenecarboxylic acids in pyrolysis methylation studies. *Organic Geochemistry* 23, no. 10: 991–994. doi:10.1016/0146-6380(95)00071-2.

Häusler, M. J., and M. H. B. Hayes. 1996. Uses of the XAD-8 resin and acidified dimethylsulfoxide in studies of humic acids. In *Humic Substances and Organic Matter in Soil and Water Environments: Characterization, Transformations, and Interactions: Proceedings of the 7th International Conference of the International Humic Substances Society, University of the West Indies, St. Augustine, Trinidad and Tobago, 3-8 July 1994*, ed. C. E. Clapp, M. H. B. Hayes, N. Senesi, and S. M. Griffith, 25–32. St. Paul, MN: International Humic Substances Society.

Hays, M. B., and S. C. B. Myneni. 2007. Structural Environments of carboxyl groups in natural organic molecules from terrestrial systems. Part 1: Infrared spectroscopy. *Geochimica Et Cosmochimica Acta* 71, no. 14: 3518–3532.

Hayes, M. H. B., and R. L. Malcolm. 2001. Considerations of compositions and of aspects of the structures of humic substances. In *Humic Substances and Chemical Contaminants*, ed. C. E. Clapp, M. H. B. Hayes, N. Senesi, P. R. Bloom, and P. M. Jardine. Madison, WI: Soil Science Society of America.

Hayes, M. H. B. 1985. Extraction of humic substances from soil. In *Humic Substances in Soil, Sediment, and Water: Geochemistry, Isolation and Characterization.*, ed. G. R. Aiken, D. M. McKnight, R. L. Wershaw, and P. MacCarthy. New York: John Wiley & Sons. http://www.cabdirect.org/abstracts/19861902274.html;jsessionid=B9D267B7CCCA0DF0218051877C473F5E.

Hayes, M. H. B. 2006. Solvent systems for the isolation of organic components from soils. *Soil Science Society of America Journal* 70, no. 3: 986–994. doi:10.2136/sssaj2005.0107.

Hayes, M. H. B., P. MacCarthy, R. L. Malcolm, and R. S. Swift, eds. 1990. *Humic Substances II: In Search of Structure*. 1st ed. Hoboken, NJ: John Wiley & Sons.

Hayes, M. H. B., R. S. Swift, C. M. Byrne, and A. J. Simpson. 2010. The isolation and characterization of humic substances and humin from grey brown podzolic and grey grassland soils. In *Proceedings, Symposium 2.2.1*, ed. R. Gilkes and N. Prakongkep, 1:198–201. Brisbane, Australia: IUSS. http://www.iuss.org.

Hayes, T. M., M. H. B. Hayes, J. O. Skjemstad, R. S. Swift, and R. L. Malcolm. 1996. Isolation of humic substances from soil using aqueous extractants of different pH and XAD resins, and their characterization by 13C-NMR. In *Humic Substances and Organic Matter in Soil and Water Environments: Characterization, Transformations, and Interactions: Proceedings of the 7th International Conference of the International Humic Substances Society, University of the West Indies, St. Augustine, Trinidad and Tobago, 3-8 July 1994*, ed. C. E. Clapp, M. H. B. Hayes, N. Senesi, and S. M. Griffith, 13–24. St. Paul, MN: International Humic Substances Society.

He, Z., T. Ohno, D. C. Olk, and F. Wu. 2010. Capillary electrophoresis profiles and fluorophore components of humic acids in Nebraska corn and Philippine rice soils. *Geoderma* 156, no. 3–4: 143–151. doi:10.1016/j.geoderma.2010.02.011.

Hernandez, D., C. Plaza, N. Senesi, and A. Polo. 2007. Fluorescence analysis of copper(II) and zinc(II) binding behaviour of fulvic acids from pig slurry and amended soils. *European Journal of Soil Science* 58, no. 4: 900–908.

Hoffmann, E. D., and V. Stroobant. 2007. *Mass Spectrometry: Principles and Applications*. Chichester, UK: John Wiley & Sons.

Howarth, R. J., and R. Sinding-Larsen. 1983. Multivariate analysis. In *Statistics and Data Analysis in Geochemical Prospecting*, ed. R. J. Howarth, 207. Amsterdam: Elsevier Scientific Pub. Co.

Irwin, W. J. 1979. Analytical pyrolysis—an overview. *Journal of Analytical and Applied Pyrolysis* 1, no. 1: 3–25. doi:10.1016/0165-2370(79)80002-9.

Johnston, C. T., and Y. O. Aochi. 1996. Fourier Transform Infrared and Raman Spectroscopy. In *Methods of Soil Analysis. Part 3. Chemical Methods*, ed. D. L. Sparks, 269–321. Soil Science Society of America Book Series 5. Madison, WI: American Society of Agronomy-Soil Science Society of America.

Kalinichev, A. G., and R. J. Kirkpatrick. 2007. Molecular dynamics simulation of cationic complexation with natural organic matter. *European Journal of Soil Science* 58, no. 4: 909–917. doi:10.1111/j.1365-2389.2007.00929.x.

Karlsson, T., and P. Persson. 2010. Coordination chemistry and hydrolysis of Fe(III) in a peat humic acid studied by X-ray absorption spectroscopy. *Geochimica et Cosmochimica Acta* 74, no. 1: 30–40. doi:10.1016/j.gca.2009.09.023.

Karlsson, T., P. Persson, and U. Skyllberg. 2006. Complexation of copper(II) in organic soils and in dissolved organic matter—EXAFS evidence for chelate ring structures. *Environmental Science & Technology* 40, no. 8: 2623–2628. doi:10.1021/es052211f.

Kasparov, S. Y., F. A. Tikhomirov, and A. D. Fless. 1981. Use of disk electrophoresis to fractionate humic acids. *Soviet Soil Science* 36: 21.

Kawahigashi, M., H. Sumida, and K. Yamamoto. 2005. Size and shape of soil humic acids estimated by viscosity and molecular weight. *Journal of Colloid and Interface Science* 284, no. 2: 463–469. doi:10.1016/j.jcis.2004.10.023.

Kingery, W. L., A. J. Simpson, F. Han, and B. Xing. 2001. Nuclear magnetic resonance studies of metal interactions with humic substances. In *Humic Substances and Chemical Contaminants*, ed. C. E. Clapp, M. H. B. Hayes, N. Senesi, P. R. Bloom, and P. M. Jardine, 397–426. Madison, WI: Soil Science Society of America.

Kühle, A., A. H. Sorensen, J. B. Zandbergen, and J. Bohr. 1998. Contrast artifacts in tapping tip atomic force microscopy. *Applied Physics A: Materials Science & Processing* 66, no. 0: S329–S332. doi:10.1007/s003390051156.

Lead, J. R., K. J. Wilkinson, E. Balnois, B. J. Cutak, C. K. Larive, S. Assemi, and R. Beckett. 2000. Diffusion coefficients and polydispersities of the Suwannee River fulvic acid: comparison of fluorescence correlation spectroscopy, pulsed-field gradient nuclear magnetic resonance, and flow field-flow fractionation. *Environmental Science and Technology* 34, no. 16: 3508–3513.

Lebowitz, J., M. S. Lewis, and P. Schuck. 2002. Modern analytical ultracentrifugation in protein science: A tutorial review. *Protein Science: A Publication of the Protein Society* 11, no. 9: 2067–2079.

Levine, I. 2008. *Physical Chemistry*. 6th ed. New York: McGraw-Hill Science/Engineering/Math.

Liu, A., R. C. Wu, E. Eschenazi, and K. Papadopoulos. 2000. AFM on humic acid adsorption on mica. *Colloids and Surfaces A: Physicochemical and Engineering Aspects* 174, no. 1–2: 245–252.

Maia, C. M. B. F., A. Piccolo, and A. S. Mangrich. 2008. Molecular size distribution of compost-derived humates as a function of concentration and different counterions. *Chemosphere* 73, no. 8: 1162–1166.

Malcolm, R. L. 1990. Chapter 12: Application of solid state 13C NMR spectroscopy to geochemical studies of humic substances. In *Humic Substances II: In Search of Structure*, ed. M. H. B. Hayes, P. MacCarthy, R. L. Malcolm, and R. S. Swift. 1st ed. Hoboken, NJ: John Wiley & Sons.

Manning, T. J., T. Bennett, and D. Milton. 2000. Aggregation studies of humic acid using multiangle laser light scattering. *Science of the Total Environment* 257, no. 2–3: 171–176.

Maurice, P. A., and K. Namjesnik-Dejanovic. 1999. Aggregate structures of sorbed humic substances observed in aqueous solution. *Environmental Science & Technology* 33, no. 9: 1538–1541. doi:10.1021/es981113+.

Mobed, J. J., S. L. Hemmingsen, J. L. Autry, and L. B. McGown. 1996. Fluorescence characterization of IHSS humic substances: Total luminescence spectra with absorbance correction. *Environmental Science & Technology* 30, no. 10: 3061–3065. doi:10.1021/es960132l.

Moldoveanu, S. C. 1998. *Analytical Pyrolysis of Natural Organic Polymers*. Amsterdam: Elsevier.

De Nobili, M., and Y. Chen. 1999. Size exclusion chromatography of humic substances: Limits, perspectives and prospectives. *Soil Science* 164, no. 11: 825–833.

Ohno, T., A. Amirbahman, and R. Bro. 2008. Parallel factor analysis of excitation–emission matrix fluorescence spectra of water soluble soil organic matter as basis for the determination of conditional metal binding parameters. *Environmental Science & Technology* 42, no. 1: 186–192. doi:10.1021/es071855f.

Otto, W. H., W. Robert Carper, and C. K. Larive. 2001. Measurement of cadmium(II) and calcium(II) complexation by fulvic acids using 113Cd NMR. *Environmental Science and Technology* 35, no. 7: 1463–1468.

Patel-Sorrentino, N., S. Mounier, and J. Y. Benaim. 2002. Excitation-emission fluorescence matrix to study pH influence on organic matter fluorescence in the Amazon basin rivers. *Water Research* 36, no. 10: 2571–2581. doi:10.1016/S0043-1354(01)00469-9.

Perminova, I. V. 1999. Size exclusion chromatography of humic substances: Complexities of data interpretation attributable to non-size exclusion effects. *Soil Science* 164, no. 11: 834–840.

Pinheiro, J. P., A. M. Mota, J. M. R. d' Oliveira, and J. M. G. Martinho. 1996. "Dynamic properties of humic matter by dynamic light scattering and voltammetry." *Analytica Chimica Acta* 329, no. 1–2: 15–24. doi:10.1016/0003-2670(96)00097-9.

Plaschke, M., J. Römer, R. Klenze, and J. I. Kim. 1999. In situ AFM study of sorbed humic acid colloids at different pH. *Colloids and Surfaces A: Physicochemical and Engineering Aspects* 160, no. 3: 269–279. doi:10.1016/S0927-7757(99)00191-0.

Pommery, J., J. P. Ebenga, M. Imbenotte, G. Palavit, and F. Erb. 1988. Determination of the complexing ability of a standard humic acid towards cadmium ions. *Water Research* 22, no. 2: 185–189.

Pranzas, P., R. Willumeit, R. Gehrke, J. Thieme, and A. Knöchel. 2003. Characterisation of structure and aggregation processes of aquatic humic substances using small-angle scattering and X-ray microscopy. *Analytical and Bioanalytical Chemistry* 376, no. 5: 618–625. doi:10.1007/s00216-003-1970-6.

Pretsch, E., P. Bühlmann, and M. Badertscher. 2009. *Structure Determination of Organic Compounds: Tables of Spectral Data*. 4th ed. Berlin: Springer.

Ralston, G. 1992. Introduction to Analytical Ultracentrifugation. Beckman Coulter. https://www.beckmancoulter.com/wsrportal/bibliography?docname = 361847.pdf.

Reeves III, J. B. 2010. Near- versus mid-infrared diffuse reflectance spectroscopy for soil analysis emphasizing carbon and laboratory versus on-site analysis: Where are we and what needs to be done? *Geoderma* 158, no. 1–2: 3–14. doi:16/j.geoderma.2009.04.005.

Van Riemsdijk, W. H., L. K. Koopal, D. G. Kinniburgh, M. F. Benedetti, and L. Weng. 2006. Modeling the interactions between humics, ions, and mineral surfaces. *Environmental Science & Technology* 40, no. 24: 7473–7480.

Ritchie, J. D., and E. Michael Perdue. 2003. Proton-binding study of standard and reference fulvic acids, humic acids, and natural organic matter. *Geochimica et Cosmochimica Acta* 67, no. 1: 85–93.

Rizzi, F. R., S. Stoll, N. Senesi, and J. Buffle. 2004. A transmission electron microscopy study of the fractal properties and aggregation processes of humic acids. *Soil Science* 169, no. 11: 765–775.

Saiz-Jimenez, C. 1994. Analytical pyrolysis of humic substances: pitfalls, limitations, and possible solutions. *Environmental Science & Technology* 28, no. 11: 1773–1780. doi:10.1021/es00060a005.

Sanders, R. L., N. M. Washton, and K. T. Mueller. 2010. Measurement of the reactive surface area of clay minerals using solid-state NMR studies of a probe molecule. *Journal of Physical Chemistry C* 114, no. 12: 5491–5498. doi:10.1021/jp906132k.

Santín, C., Y. Yamashita, X. Otero, M. Álvarez, and R. Jaffé. 2009. Characterizing humic substances from estuarine soils and sediments by excitation-emission matrix spectroscopy and parallel factor analysis. *Biogeochemistry* 96, no. 1: 131–147. doi:10.1007/s10533-009-9349-1.

Schulten, H. -R., and M. Schnitzer. 1993. A state of the art structural concept for humic substances. *Naturwissenschaften* 80, no. 1: 29–30. doi:10.1007/BF01139754.

Schulten, H.-R., B. Plage, M. Schnitzer, J. W. De Leeuw, and P. G. Hatcher. 1992. A chemical structure for humic substances. *Naturwissenschaften* 79, no. 7: 330–331.

Scott, D. J., S. E. Harding, and A. J. Rowe, eds. 2006. *Analytical Ultracentrifugation: Techniques and Methods*. 1st ed. Cambridge, UK: Royal Society of Chemistry.

Scott, D. J., and P. Schuck. 2006. A brief introduction to the analytical ultracentrifugation of proteins for beginners. In *Analytical Ultracentrifugation: Techniques and Methods*, ed. D. J. Scott, S. E. Harding, and A. J. Rowe, 1–25. 1st ed. Cambridge, UK: Royal Society of Chemistry.

Sein, J., M. Varnum, and S. A. Jansen. 1999. Conformational modeling of a new building block of humic acid: Approaches to the lowest energy conformer. *Environmental Science & Technology* 33, no. 4: 546–552. doi:10.1021/es9805324.

Senesi, N. 1990. Molecular and quantitative aspects of the chemistry of fulvic acid and its interactions with metal ions and organic chemicals: Part II. The fluorescence spectroscopy approach. *Analytica Chimica Acta* 232, no. 1: 77–106.

Senesi, N., and E. Loffredo. 1998. The chemistry of soil organic matter. In *Soil Physical Chemistry*, ed. D. L. Sparks, 239–370. 2nd ed. Boca Raton, FL: CRC Press.

Senesi, N., and E. Loffredo. 2005. Metal ion complexation by soil humic substances. In *Chemical Processes in Soils*, ed. M. A. Tabatabai and D. L. Sparks, 563–617. SSSA Book Series 8. Madison, WI: Soil Science Society of America.

Shin, H.-S., K.-H. Hong, M.-H. Lee, Y.-H. Cho, and C.-W. Lee. 2001. Fluorescence quenching of three molecular weight fractions of a soil fulvic acid by UO2(II). *Talanta* 53, no. 4: 791–799. doi:10.1016/S0039-9140(00)00567-1.

Shinozuka, T., M. Shibata, and T. Yamaguchi. 2004. Molecular weight characterization of humic substances by MALDI-TOF-MS. *Journal of the Mass Spectrometry Society of Japan* 52, no. 1: 29–32.

Sierra, M. M. D., M. Giovanela, E. Parlanti, and E. J. Soriano-Sierra. 2005. Fluorescence fingerprint of fulvic and humic acids from varied origins as viewed by single-scan and excitation/emission matrix techniques. *Chemosphere* 58, no. 6: 715–733. doi:10.1016/j.chemosphere.2004.09.038.

Simpson, A. J., J. Burdon, C. L. Graham, M. H. B. Hayes, N. Spencer, and W. L. Kingery. 2001a. Interpretation of heteronuclear and multidimensional NMR spectroscopy of humic substances. *European Journal of Soil Science* 52, no. 3: 495–509. doi:10.1046/j.1365-2389.2001.00402.x.

Simpson, A. J., J. Burdon, C. L. Graham, M. H. B. Hayes, N. Spencer, and W. L. Kingery. 2001b. Interpretation of heteronuclear and multidimensional NMR spectroscopy of humic substances. *European Journal of Soil Science* 52, no. 3: 495–509. doi:10.1046/j.1365-2389.2001.00402.x.

Simpson, A. J., W. L. Kingery, M. H. Hayes, M. Spraul, E. Humpfer, P. Dvortsak, R. Kerssebaum, M. Godejohann, and M. Hofmann. 2002. Molecular structures and associations of humic substances in the terrestrial environment. *Naturwissenschaften* 89, no. 2: 84–88.

Simpson, A. J., W. L. Kingery, and P. G. Hatcher. 2003. The identification of plant derived structures in humic materials using three-dimensional NMR spectroscopy. *Environmental Science & Technology* 37, no. 2: 337–342. doi:10.1021/es025956j.

Stedmon, C. A., and R. Bro. 2008. Characterizing dissolved organic matter fluorescence with parallel factor analysis: A tutorial. *Limnology and Oceanography: Methods* 6, no. Nov: 572–579.

Steelink, C., R. L. Wershaw, K. A. Thorn, and M. A. Wilson. 1990. Chapter 10: Application of liquid-state NMR spectroscopy to humic substances. In *Humic Substances II: In Search of Structure*, ed. M. H. B. Hayes, R. L. Malcolm, P. MacCarthy, and R. S. Swift. 1st ed. Hoboken, NJ: John Wiley & Sons.

Stevenson, F. J. 1994. *Humus Chemistry: Genesis, Composition, Reactions*. 2nd ed. Hoboken, NJ: Wiley.

Striegel, A., W. W. Yau, J. J. Kirkland, and D. D. Bly. 2009. *Modern Size-Exclusion Liquid Chromatography: Practice of Gel Permeation and Gel Filtration Chromatography*. Hoboken, NJ: John Wiley & Sons.

Sutton, R., and G. Sposito. 2005. Molecular structure in soil humic substances: The new view. *Environmental Science & Technology* 39, no. 23: 9009–9015. doi:10.1021/es050778q.

Swift, R. S. 1996. Organic matter characterization. In *Methods of Soil Analysis. Part 3. Chemical Methods*, ed. D. L. Sparks. Soil Science Society of America Book Series 5. Madison, WI: American Society of Agronomy-Soil Science Society of America.

Tanford, C. 1961. *Physical Chemistry of Macromolecules*. New York: John Wiley & Sons Inc.

Thorn, K. A., and L. G. Cox. 2009. N-15 NMR spectra of naturally abundant nitrogen in soil and aquatic natural organic matter samples of the International Humic Substances Society. *Organic Geochemistry* 40, no. 4: 484–499. doi:10.1016/j.orggeochem.2009.01.007.

Tombacz, E., and E. Meleg. 1990. A theoretical explanation of the aggregation of humic substances as a function of pH and electrolyte concentration. *Organic Geochemistry* 15, no. 4: 375–381. doi:10.1016/0146-6380(90)90164-U.

Trubetskaya, O. E., O. A. Trubetskoi, B. A. Borisov, and N. F. Ganzhara. 2008. Electrophoresis and size-exclusion chromatography of humic substances extracted from detritus and soils of different geneses. *Eurasian Soil Science* 41, no. 2: 171–175.

Weisenhorn, A. L., P. K. Hansma, T. R. Albrecht, and C. F. Quate. 1989. Forces in atomic force microscopy in air and water. *Applied Physics Letters* 54, no. 26: 2651–2653. doi:10.1063/1.101024.

Wershaw, R. L. 2004. *Evaluation of conceptual models of natural organic matter (humus) from a consideration of the chemical and biochemical processes of humification*. Scientific Investigations Report. Denver, CO: USGS. http://pubs.er.usgs.gov/usgspubs/sir/sir20045121.

Wyatt, P. J. 1998. Submicrometer particle sizing by multiangle light scattering following fractionation. *Journal of Colloid and Interface Science* 197, no. 1: 9–20. doi:10.1006/jcis.1997.5215.

Xia, K., W. Bleam, and P. A. Helmke. 1997. Studies of the nature of Cu2+ and Pb2+ binding sites in soil humic substances using X-ray absorption spectroscopy. *Geochimica et Cosmochimica Acta* 61, no. 11: 2211–2221. doi:10.1016/S0016-7037(97)00079-3.

Yu, C.-H., C.-H. Wu, C.-H. Lin, C.-H. Hsiao, and C.-F. Lin. 2008. Hydrophobicity and molecular weight of humic substances on ultrafiltration fouling and resistance. *Separation and Purification Technology* 64, no. 2: 206–212.

Yu, G.-H., Y.-H. Luo, M.-J. Wu, Z. Tang, D.-Y. Liu, X.-M. Yang, and Q.-R. Shen. 2010. PARAFAC modeling of fluorescence excitation-emission spectra for rapid assessment of compost maturity. *Bioresource Technology* 101, no. 21: 8244–8251. doi:10.1016/j.biortech.2010.06.007.

Zavarzina, A. G., N. G. Vanifatova, and A. A. Stepanov. 2008. Fractionation of humic acids according to their hydrophobicity, size, and charge-dependent mobility by the salting-out method. *Eurasian Soil Science* 41, no. 12: 1294–1301.

Zhou, Q., S. E. Cabaniss, and P. A. Maurice. 2000. Considerations in the use of high-pressure size exclusion chromatography (HPSEC) for determining molecular weights of aquatic humic substances. *Water Research* 34, no. 14: 3505–3514. doi:10.1016/S0043-1354(00)00115-9.

Part III

Ion Binding to Soil Colloids

11 Modeling Ion Binding
General Concepts

11.1 GENERAL PROBLEM

In the preceding parts of this book, many different basic concepts have been reviewed: in Part I, Chapters 3 through 5, important principles of surface chemistry have been addressed—the equilibrium adsorption behavior of ions to soil colloids (and to colloidal particles in general), the necessary electrostatic considerations for an interface between two charge-bearing phases, and the concomitant development of surface charges; in Part II, Chapters 8 through 10, the chemical structure and surface properties of the most important types of soil colloids have been discussed, where it can be seen that, despite the different types of surfaces and surface groups, there are two main modes of ion–surface interaction—electrostatic attraction/repulsion between a charged surface and ions of opposite/same charge leading to outer sphere complexes in the case of attraction, and specific interaction resulting in chemically bound inner sphere complexes. A general treatment of ion binding should in principle take into account all these elements. As a detailed microscopic knowledge of the surface of soil colloids is lacking (not to speak of the multicomponent aggregates found in the actual soils, to be considered in Chapter 14), phenomenological modeling is the usual way to reach the ability to make quantitative predictions, which are useful in many aspects of soil science, particularly in geochemistry (Mattigod and Zachara 1996; Zachara and Westall 1998; Koretsky 2000; Perdue 2001; Goldberg 2005; Senesi and Loffredo 2005; Van Riemsdijk et al. 2006).

The basic statement of the problem is as follows: We start comparing solution phase reactions with surface adsorption processes. In the first case, we consider a coordination complex forming in solution between a metal ion and a ligand with a reaction like this:

$$Cu^{2+} + Cl^- \rightleftharpoons CuCl^+ \tag{11.1}$$

and the equilibrium of a weak acid, written as an association reaction:

$$A^- + H^+ \rightleftharpoons HA \tag{11.2}$$

The equilibrium condition is governed by the equilibrium constant K or, equivalently, the standard free Gibbs energy change ΔG^0, both related by the following equation (McQuarrie and Simon 1997; Berry, Rice, and Ross 2000; Atkins and Paula 2009):

$$\Delta G^0 = -\boldsymbol{R}T\ln K \tag{11.3}$$

In the process, ΔG^0 can be viewed as the reversible work performed in reactions such as Equations 11.1 and 11.2; in these reactions, the fact that ions are involved does not add nothing new to Equation 11.3; merely, the interaction between charges is considered in the activity coefficients, as discussed in Chapter 3. There are no electrostatic energies involved, because the solution is a homogeneous phase; thus, the electric potential is constant everywhere.

The problem is different when an interface is involved. Figure 11.1 shows schematically the situation: the ion, ideally approaching the surface from the solution bulk, faces a potential gradient near the surface. A positive ion thus encounters a repulsive gradient near the surface; it is still possible to reach the surface and become specifically adsorbed, provided the *affinity*—that is, the free energy change—is negative enough, for example, for the case of a transition cation (like Cu^{2+} or Fe^{3+}) adsorbing onto an oxide particle that may have net positive charge but still have OH groups able to bind the cation. On the contrary, outer sphere adsorption of ions on a surface of like charge is unfavorable. In the opposite case, ions of charge opposite to that of the surface are favored in their approach to the surface, for both specific and unspecific adsorption.

To account quantitatively for the electrostatic effect, the energy change (reversible work) involved in bringing a mole of ions of charge z from the solution bulk (where the electric potential is taken as zero) to the surface having a potential ψ_0 is equal to $z\boldsymbol{F}\psi_0$; the total standard Gibbs free energy change upon adsorption is thus

$$\Delta\tilde{G}^0 = \Delta G^0 + z\boldsymbol{F}\psi_0 \tag{11.4}$$

or, written more generally,

$$\Delta\tilde{G}^0 = \Delta G^0_{ch} + \Delta G^0_{el} \tag{11.5}$$

$\Delta\tilde{G}^0$ is the *electrochemical standard Gibbs free energy* change (or electrochemical free energy for short), $\Delta G^0{}_{ch}$ is the chemical contribution, and $\Delta G^0{}_{el}$ is the electrical

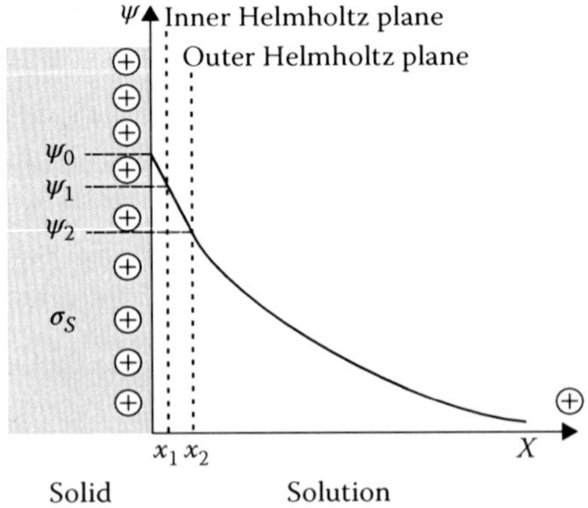

FIGURE 11.1 Schematic representation (based on Figure 3.13, Gouy–Chapman–Stern–Grahame model) of the approach of an ion (right) to a charged surface (left), where an electric potential gradient is present.

contribution. The thermodynamic equilibrium and spontaneity criteria now are based on $\Delta\tilde{G}^0$. The electrochemical potential introduced in Chapter 3 is related to $\tilde{G}$ as the chemical potential is related to G (Bard and Faulkner 2000; Levine 2008; Atkins and Paula 2009):

$$\tilde{\mu}_i = \left(\frac{\partial \tilde{G}}{\partial n_i}\right)_{T,P,n_j} \tag{11.6}$$

Thus, any approach has to consider in principle both contributions: the chemical, nonelectrostatic, or *intrinsic* contribution, and the electrostatic one. Now, to model ion sorption to soil colloidal particles, two main approaches exist: *chemical modeling* (should not be confused with the chemical contribution mentioned in the preceding discussion) and *adsorption modeling*; in the first case, the sorption is treated as a chemical reaction, with the free surface sites considered as a reactant, and a chemical equilibrium expression in terms of the bulk activities is written. In adsorption modeling, an appropriate adsorption isotherm is employed; however, it is not written in terms of the bulk activity a_i of the adsorbate, but instead it should be related to the activity at the outer Helmholtz plane $a_{i,2}$, that is, at the limit of the diffuse double layer, x_2 in Figure 11.1 (see also Section 3.3.3). Both are related, due to the condition of constancy of the electrochemical potential (Section 3.1.1), by

$$a_{i,2} = a_i e^{-\frac{z_i F \psi_2}{RT}} \tag{11.7}$$

Thus, adsorption modeling requires finding ψ_2 with a detailed knowledge of inner and outer sphere adsorbed species to find the surface potential and/or charge, that is, solving the electrostatic problem entirely, or in some approximate way. In chemical modeling, only the surface potential is required; thus, simplified electrostatic models can be employed; this approach is usually considered advantageous and has been widely employed (Mattigod and Zachara 1996; Zachara and Westall 1998; Perdue 2001; Goldberg 2005; Van Riemsdijk et al. 2006).

11.1.1 Chemical Modeling

In chemical modeling, the surface species are treated as aqueous solutes (considering the solid concentration and site density), and the problem is treated mathematically as homogeneous competitive equilibria (with mass and charge balances, etc.), but activity coefficient for surface species is not explicitly considered, being in fact included in the electrostatic term of the surface reactions. It has been stated that chemical modeling is superior to adsorption modeling, but this apparent superiority comes from the fact that no complete models for surface binding have been formulated in terms of adsorption isotherms; these are usually applied in a conditional way, as in many examples presented in Chapter 4. However, it is perfectly possible to couple adsorption isotherms to electrostatic modeling to formulate more rigorous models.

We introduce here chemical modeling that considers the proton adsorption reaction leading to the protonic surface charge on an oxide surface. Typical surface reactions can be written as

$$\text{X-OH} + \text{H}^+ \rightleftharpoons \text{X-OH}_2^+ \tag{11.8}$$

$$\text{X-OH} \rightleftharpoons \text{X-O}^- + \text{H}^+ \tag{11.9}$$

where X represents a cationic, Lewis acid, lattice species (Al(III), Fe(III), etc.). This process is different from ordinary acid–base equilibrium due to the presence of the charged surface. Thus, the equilibrium condition for reactions 11.8 and 11.9 should be written in terms of the electrochemical potentials

$$\tilde{\mu}_{s0} + \tilde{\mu}_{H^+} = \tilde{\mu}_{s+} \tag{11.10}$$

$$\tilde{\mu}_{s0} = \tilde{\mu}_{s-} + \tilde{\mu}_{H^+} \tag{11.11}$$

where $_s+$, $_s-$, and $_s0$ represent positively charged (protonated), negatively charged (dissociated) and uncharged sites, respectively. Recalling (Equation 3.3) the electrochemical potential for a species i in a phase α:

$$\tilde{\mu}_i^\alpha = \mu_i^{0,\alpha} + \boldsymbol{RT} \ln a_i^\alpha + z_i \boldsymbol{F} \psi^\alpha \tag{11.12}$$

we can write for Equation 11.10 (dropping the phase superscript for standard chemical potentials) as

$$\mu_{s0}^0 + \boldsymbol{RT} \ln a_{s0}^S + \mu_{H^+}^0 + \boldsymbol{RT} \ln a_{H^+} + \boldsymbol{F} \psi^B = \mu_{s+}^0 + \boldsymbol{RT} \ln a_{s+}^S + \boldsymbol{F} \psi^S \tag{11.13}$$

where the superscripts B and S refer to the solution bulk and surface, respectively. Rearranging, and recalling that it is taken $\psi^B = 0$, we get the following equilibrium expression:

$$-\left(\mu_{s+}^0 - \mu_{s0}^0 - \mu_{H^+}^0\right) - \boldsymbol{F} \psi^S = \boldsymbol{RT} \ln \frac{a_{s+}^S}{a_{s0}^S a_{H^+}^B} \tag{11.14}$$

The term in parenthesis in the left-hand side of Equation 11.14 is the Gibbs free energy change of reaction 11.8. The next term is the electrical contribution to the electrochemical Gibbs free energy change of reaction 11.8:

$$\Delta\tilde{G}^0 = \mu_{s+}^0 - \mu_{s0}^0 - \mu_{H^+}^0 + \boldsymbol{F} \psi^S \tag{11.15}$$

Now, the equilibrium constant for reaction 11.8 will be given by

$$K_{eq,1} = \boldsymbol{e}^{-\frac{\Delta G^0}{\boldsymbol{RT}}} \boldsymbol{e}^{-\frac{\boldsymbol{F}\psi^S}{\boldsymbol{RT}}} = K^{int}{}_{,1} e_f\left(\psi^S\right) = \frac{a_{s+}^S}{a_{s0}^S a_{H^+}^B} \tag{11.16}$$

where $K^{int}_{,1}$ is the intrinsic or "chemical" constant for Reaction 11.8, and $e_f(\psi) = \exp(-z_i\boldsymbol{F}\psi/\boldsymbol{RT})$ is the *electrostatic Boltzmann factor*. In the same way, for reaction 11.9 one finds

$$K_{eq,2} = e^{-\frac{\Delta G^0}{RT}} e^{-\frac{F\psi^S}{RT}} = K^{int}_{,2} e_f\left(\psi^S\right) = \frac{a^S_{s-} a^B_{H^+}}{a^S_{s0}} \tag{11.17}$$

Note that the electrostatic factor is the same for both reactions as long as the potential is the same; that is, the location of the surface charges is the same. This is obviously true for these cases, but not in others, as we will discuss in the next chapters.

There are several aspects that must be remarked at this point (Zachara and Westall 1998; McBride 1999; Sparks 2002; Sposito 2008). First, which is the "activity" of the surface sites? The only thermodynamically meaningful quantity is the surface excess (Chapter 2); however, frequently it is not accessible. Instead, the surface concentration, coverage degree, or the amount sorbed Q_i may be used; this last is done frequently. Recalling that they are treated like chemical species, this is akin, in a sense, to the treatment of a solid as a solid solution and expressing its composition in terms of the molar fractions of the "pure" components. Thus, the activities of negatively and positively charged sites and of the neutral ones could be expressed in terms of the respective mole fractions. Alternatively, one can express the activities of charged sites ("adsorbed species") in terms of coverage degrees: $a_i = \theta_i/\theta_0$ where θ_0 is the fraction of uncharged ("free") sites, but this involves extrathermodynamical assumptions. In any case, from a thermodynamic point of view, proper activity coefficients should also be included. Of course, changing the way to express activities will change the numerical value of ΔG^0 and consequently that of K^{int}. A second point is that the surface potential ψ^S cannot be experimentally measured, and consequently should be obtained from some suitable model. Thus, it has significance only in the context of the assumptions and conventions made in the applied model. In Chapter 3, it was discussed the calculation of electrostatic potential profiles, mainly based on the Poisson–Boltzmann (PB) equation was discussed. The results so obtained are valid in the context of the diffuse layer model and, strictly speaking, cannot be replaced in expressions such as Equation 11.13 to do further thermodynamic derivations. Relating to these considerations, in Equation 11.12, ψ^α is strictly the potential in the bulk of phase α; using this equation along the interfacial region, which is implicit in the application to surface species, is strictly nonthermodynamic. Finally, the distinction between inner- and outer sphere complexes and charges is also nonthermodynamic.

11.1.2 Modeling Bases

Overall, either in chemical or in adsorption modeling, there are two basic elements in the model: the intrinsic adsorption/equilibrium constants, which are the "chemical part" or submodel in the sense that electrostatics is not involved, and the electrostatic contribution or submodel, which should provide a description of the potential profile along the interface, which in turn should be able to predict potential values needed along the constants to find the amounts adsorbed. In fact, there is an interrelation between these factors, because ion adsorption modifies the surface charge and thus the surface potential that affects the extension of adsorption. In practice, these considerations usually lead to iterative numerical calculations.

The intrinsic constants can be a few, as usually happens with oxide minerals, or can be a large number, as it is the case with organic matter in general, and humic substance (HS) in particular, or with complex associations of soil components. In general, the set of equilibrium constants, more properly termed a *distribution of intrinsic constants*, is described as number of sites with the same K^{int} as a function of K^{int} or log K^{int}; the last form is preferred, because it is directly related to free energy (Equation 11.3). Thus, the description of the distribution is effectively of an intensity (number of sites) as a function of an energy, so that it is generally termed *affinity spectrum*. Affinity spectra may consist of a small number of discrete sites (just as in the case of a normal polyprotic acid such as phosphoric or citric acids) or a high number of these, which may be considered as forming a continuous distribution or spectrum. The affinity spectra will be considered in Section 11.2. The electrostatic submodel has had several approaches, which are introduced in Chapters 12 and 13. The expressions "affinity spectra" and "constant distribution" are commonly found and are used interchangeably here.

A general aspect to take into account in the model formulation is the number of adjustable parameters. Because the experimental information is often limited, many model parameters have to be found from data fitting. The more elaborated and complex the model, the higher the number of adjustable parameters; more of these usually led to a better fitting, but also to a higher probability of resulting physically unrealistic. It should be noted at this point that, as of this writing, the modeling is largely empiric. There have been several attempts to formulate models based on sound statistical mechanics foundations, but several aspects remain to be solved (Borkovec and Koper 1994b; Borkovec et al. 1996; Cernik, Borkovec, and Westall 1996; Garcés, Mas, and Puy 1999; Borkovec, Hamáček, and Piguet 2004; Garces, Mas, and Puy 2004; Borkovec, Koper, and Piguet 2006; Companys et al. 2007).

11.2 AFFINITY SPECTRA

The affinity spectra or constant distribution is a matter that has attracted several researchers, especially in the case of HS (Koopal and Vos 1993; Borkovec and Koper 1994a; Manunza et al. 1995; Borkovec et al. 1996; Lin and Rayson 1998; Avena, Koopal, and van Riemsdijk 1999; Lin, Drake, and Rayson 2002; Garces, Mas, and Puy 2004; Orsetti, Andrade, and Molina 2009; David et al. 2010). The proton affinity spectrum is treated separately from the spectra of other ions. The reason is that, as H^+ binding is almost always present, for all other ions (either metal cations or anions), there is always competition with protons; thus, competitive adsorption must be considered. Here, discrete and continuous affinity spectra are discussed, both for noncompetitive and for competitive cases, and in Section 11.3, methods to extract them from experimental data are presented.

11.2.1 Discrete Affinity Spectra

11.2.1.1 Proton-Binding Equilibrium

Discrete affinity spectra are composed of a few, well-characterized binding sites with known binding constants; this is the case of common polyprotic acids, such as phosphoric acid

$$PO_4^{3-} + H^+ \rightleftharpoons HPO_4^{2-} \qquad \log K_1 = pKa_3 = 12.67 \tag{11.18}$$

$$HPO_4^{2-} + H^+ \rightleftharpoons H_2PO_4^{-} \qquad \log K_2 = pKa_2 = 7.21 \tag{11.19}$$

$$H_2PO_4^{-} + H^+ \rightleftharpoons H_3PO_4 \qquad \log K_3 = pKa_1 = 2.12 \tag{11.20}$$

where K_i represents association (binding) equilibrium constants, and Ka_i is the commonly employed dissociation constant; in environmental chemistry, these reactions are often written in association form, as in Equations 11.18 through 11.20. Thus, PO_4^{3-} has an affinity spectrum with three delta function-type spikes of the same height; this is schematically shown in Figure 11.2a, plotted as fraction of total sites, f, against the logarithm of the association constant. Another example is an equimolar mixture of the monoprotic acetic acid ($\log K_{ac} = 4.76$) and the diprotic dipicolinic acid ($\log K_{pi1} = 2.16$, $\log K_{pi2} = 4.76$), which will show two spikes, one at 2.16 and the second one, of double height, at 4.76 (Figure 11.2b).

For proton binding to soil minerals, in most cases, reactions 11.8 and 11.9 are considered in the modeling, which is a spectrum consisting of two sites of equal height ($f = 0.5$ for each one); this is termed the 2-pK model. However, it is also employed as the so-called 1-pK model, where the protonation equilibrium is represented by

$$X\text{-}OH^{-1/2} + H^+ \rightleftharpoons X\text{-}OH_2^{+1/2} \qquad \log K = pH_{H,0} \tag{11.21}$$

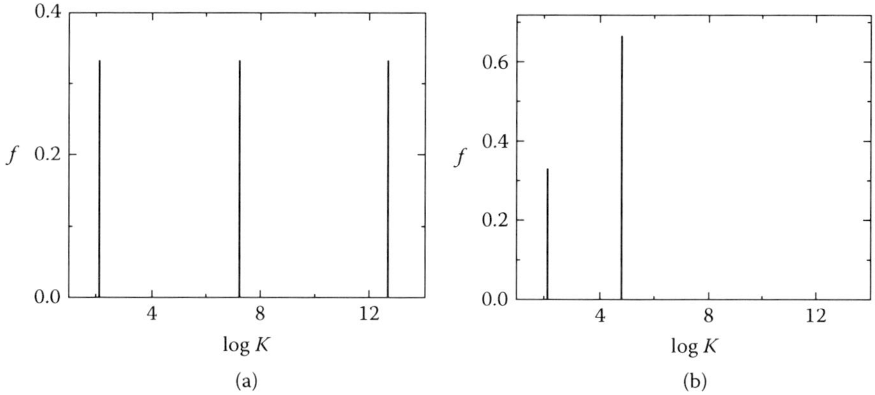

FIGURE 11.2 Affinity spectra (represented as fraction of sites against the logarithm of association constant) for (a) phosphoric acid; (b) an equimolar mixture of acetic and dipicolinic acids.

where log K is equal to the point of zero net proton charge (Chapter 5); in the following, we will drop the subscript "int" for intrinsic constants. It has been shown (Lützenkirchen 1998) that the two models are equivalent in predicting experimental titration data; the 1-pK model has one adjustable parameter less than the other model, thus it would be preferable. However, the 2-pK model is in widespread use (Zachara and Westall 1998; Goldberg 2005), because it appears physically more comfortable.

In some cases, two types of proton-binding sites are considered, such as in the case of goethite, where singly and triply bound surface OH have been distinguished by Rahnemaie, Hiemstra, and Van Riemsdijk (2006), using the 1-pK model for each, namely

$$\text{Fe-OH}^{-1/2} + \text{H}^+ \rightleftharpoons \text{Fe-OH}_2^{+1/2} \qquad K_1 \tag{11.22}$$

$$\text{Fe}_3\text{-O}^{-1/2} + \text{H}^+ \rightleftharpoons \text{Fe}_3\text{-OH}^{+1/2} \qquad K_3 \tag{11.23}$$

The total surface coverage for discrete affinity spectra is simply the weighted sum of the coverages of the diverse sites, θ_i, as in Equation 4.53:

$$\theta = \sum_i f_i \theta_i \tag{11.24}$$

where θ is the total coverage and f_i the fraction of sites i with constant K_i.

11.2.1.2 Competitive Binding

When a metal cation or an anion binds to the surface of a soil colloid, especially oxide particles, it always occurs in competition with proton-binding equilibrium. For transition cations such as Cu(II), Pb(II), and so on, both monodentate binding and bidentate binding have been proposed, that is,

$$\text{X-OH} + \text{M}^{z+} \rightleftharpoons \text{X-O-M}^{(z-1)+} + \text{H}^+ \tag{11.25}$$

$$2\text{X-OH} + \text{M}^{z+} \rightleftharpoons (\text{X-O})_2\text{-M}^{(z-2)+} + 2\text{H}^+ \tag{11.26}$$

Reactions 11.25 and 11.26 correspond to monodentate binding and bidentate binding, respectively; these structures are depicted in Figure 11.3, exemplifying with Cu for bidentate-binding case and Pb for monodentate-binding case. For bidentate, two proton sites are involved usually. These two binding modes appear as different spikes in the affinity spectra, because they have different K values. Often, they correspond to physically different sites on the surface, but it is in principle possible for the same sites to show bidentate complexation at low loadings and monodentate binding for high loadings, which introduces some complications in the treatment. Also, in some cases, other small ligands (mainly OH^-) are involved in ion binding such as (Gu, Evans, and Barabash 2010 and references therein)

$$\text{X-OH} + \text{M}^{z+} + \text{H}_2\text{O} \rightleftharpoons \text{X-O-MOH}^{(z-2)+} + 2\text{H}^+ \tag{11.27}$$

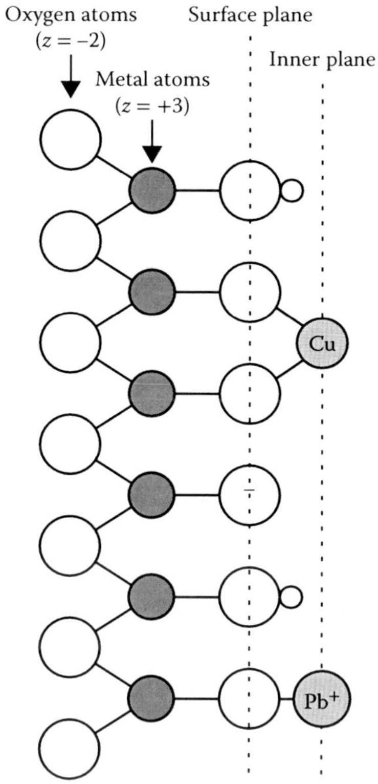

FIGURE 11.3 Schematic view of a mineral surface with monodentate (Pb(II))- and bidentate (Cu(II))-specific binding to surface sites.

The case of oxoanions like phosphate or sulfate is similar, presenting both monodentate and bidentate binding depending on the anion and the surface. For example, As(V) is known to bind to iron oxides almost exclusively in a bidentate way (Wang and Mulligan 2008):

$$2\text{X-OH} + H_2AsO_4^- + \rightleftharpoons (\text{X-O})_2\text{-}AsO_2^- + 2H_2O \tag{11.28}$$

whereas chromate at pH > 6.5 on ferrihydrite forms monodentate complexes (Johnston and Chrysochoou 2012):

$$\text{X-OH} + HCrO_4^- \rightleftharpoons \text{X-O-}CrO_3{}^- + H_2O \tag{11.29}$$

All these reactions involve the competition of H^+ ions for the same sites; with discrete complexation constants, competitive binding is treated like common competitive equilibria, even in the adsorption description.

11.2.2 Continuous Affinity Spectra

Soil organic matter (SOM), particularly HSs, presents many functional groups that bind protons and metal cations. The intrinsic affinity distribution is often assumed to be continuous, but in any case, two main types of groups or sites are included:

carboxylic sites, with pKa in the range 3–5, and phenolic ones, with pKa ranging between 8 and 10. One of the first models was the Gaussian humic model, where a bimodal Gaussian function, including the two types mentioned, without electrostatic contribution, was applied (Manunza et al. 1995). The most commonly employed is the Sips distribution, which was introduced in Section 4.4.7 and is quite similar to the Gaussian one.

11.2.2.1 Proton Binding

For noncompetitive proton binding, the proton coverage degree (fraction of sites that are protonated) is given, following Equation 4.69, as

$$\theta_H = \int_{-\infty}^{\infty} q_H(K_H, a_H) f(\log K_H) \mathrm{d}\log K_H = \int_{-\infty}^{\infty} \frac{K_H a_H}{1+K_H a_H} f(\log K_H) \mathrm{d}\log K_H \quad (11.30)$$

where q_H is the local isotherm, here (as usual) taken as the Langmuir expression; Equation 11.30 reads, in terms of amount sorbed, as

$$Q_H = Q_S \int_{-\infty}^{\infty} \frac{K_H a_H}{1+K_H a_H} f(\log K_H) \mathrm{d}\log K_H \quad (11.31)$$

Here, the site fraction f becomes a continuous function of the proton-binding parameter K_H. As mentioned in the preceding discussion, usually a two-peak distribution is assumed for f, one centered about log K_H = 3–5 and the other one about 8–10. As it was discussed in Chapter 4, introduction of the Sips distribution (Equation 4.70) leads to the Langmuir–Freundlich (LF) isotherm:

$$\theta_H = \frac{Q_H}{Q_S} = \frac{(\bar{K}_H a_H)^n}{1+(\bar{K}_H a_H)^n} \quad (11.32)$$

where $\bar{K}_H$ is the average proton-binding constant (center of the distribution), and n is the heterogeneity parameter. For two types of sites, as it is generally considered, Equation 11.32 results in

$$\theta_H = \frac{Q_H}{Q_S} = x_1 \frac{(\bar{K}_{H,1} a_H)^{n_1}}{1+(\bar{K}_{H,1} a_H)^{n_1}} + x_2 \frac{(\bar{K}_{H,2} a_H)^{n_2}}{1+(\bar{K}_{H,2} a_H)^{n_2}} \quad (11.33)$$

where the subscripts 1 and 2 refer to the two types, and x means the fraction of sites for each type. Equation 11.33 corresponds to a two-peak Sips distribution, given by

$$f(\log K_H) = \frac{\ln(10)}{2\pi} \times \left\{ x_1 \frac{\sin(\pi n_1)}{\cosh\left[n_1 \ln(10)\left(\log \bar{K}_{H,1} - \log K_H\right)\right] + \cos(\pi n_1)} + x_2 \frac{\sin(\pi n_2)}{\cosh\left[n_2 \ln(10)\left(\log \bar{K}_{H,2} - \log K_H\right)\right] + \cos(\pi n_2)} \right\} \quad (11.34)$$

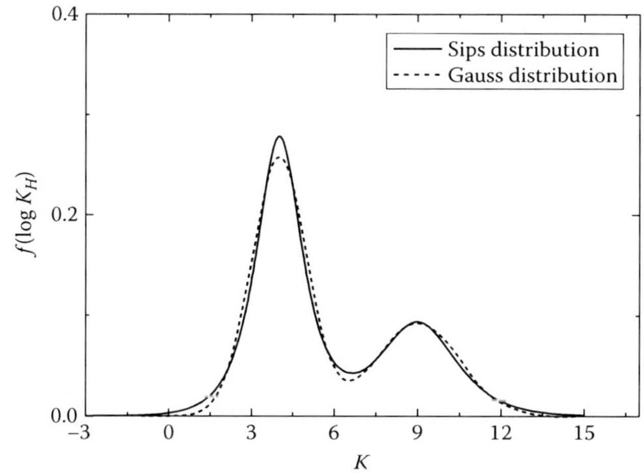

FIGURE 11.4 Typical Sips and Gauss distribution shapes for proton binding to humic substances. The Sips spectrum is given by Equation 11.34, with $\log \bar{K}_{H,1} = 4$, $\log \bar{K}_{H,2} = 9$, $x_1 = 0.65$, $x_2 = 0.35$, $n_1 = 0.55$, and $n_2 = 0.45$; the Gaussian spectrum has the same parameters, except $w_1 = 2$ and $w_2 = 4$.

For comparison, a dual peak Gaussian distribution reads

$$f(\log K_H) = \frac{1}{\sqrt{\pi/2}} \left[\frac{x_1}{w_1} e^{-\frac{2(\log \bar{K}_{H,1} - \log K_H)^2}{w_1^2}} + \frac{x_2}{w_2} e^{-\frac{2(\log \bar{K}_{H,2} - \log K_H)^2}{w_2^2}} \right] \quad (11.35)$$

Figure 11.4 compares dual peak Sips and Gaussian distributions with parameters typical of HS; it is seen, as already shown in Figure 4.27a, that the two are similar.

11.2.2.2 Competitive Metal Binding

Continuous affinity spectra are almost exclusively applied to SOM, which does little anion binding; thus, we will refer here to cation complexation only. In general, both M^{z+} and H^+ bind to the same sites, so that the competitive Langmuir equation can be used as local isotherm for M^{z+}. Assuming that each M^{z+} binds to a single site, where H^+ also binds, it can be written as

$$q_M = q_S \frac{K_M a_M}{1 + K_H a_H + K_M a_M} \quad (11.36)$$

where K_M is the metal binding constant, q_M is the local amount of metal bound, q_S is the saturation amount for the local isotherm, and a_M is the metal activity. Now for the heterogeneous case, Equation 11.36 should be integrated for a bidimensional distribution $h(\log K_H, \log K_M)$, which in the general case will include the correlations between H^+ and M^{z+} binding:

$$Q_M = Q_S \iint q_M(K_M, K_H, a_M, a_H) h(\log K_M, \log K_H) \, d\log K_M \, d\log K_H \quad (11.37)$$

The general problem of treating continuous affinity spectra for competitive binding is rather complex and has been addressed by several authors (Koopal et al. 1994; Borkovec et al. 1996; Cernik, Borkovec, and Westall 1996; Rusch et al. 1997; Garces, Mas, and Puy 2004; Orsetti, Andrade, and Molina 2009; Rey-Castro et al. 2009; David et al. 2010); we will not discuss all theoretical aspects at this point.

Koopal et al. (1994) deduced a LF type isotherm for competitive binding, resulting in

$$\theta_{i,k} = \frac{n_i}{n_0}\frac{\left(\bar{K}_{i,k}a_i\right)^{n_i}}{\sum_j\left(\bar{K}_{j,k}a_j\right)^{n_j}} \times \frac{\left[\sum_j\left(\bar{K}_{j,k}a_j\right)^{n_j}\right]^{p_k}}{1+\left[\sum_j\left(\bar{K}_{j,k}a_j\right)^{n_j}\right]^{p_k}} \tag{11.38}$$

where $\theta_{i,k}$ is the coverage (fraction of sites bound) of species i on sites type k; the sums run over all binding species. Here, p_k is the Sips distribution width parameter of site type k, representing the *intrinsic ion i heterogeneity* of the sorbing colloid, whereas n_i is the *nonideality parameter* of ion i; n_0 is a reference n parameter, taken as that belonging to protons, n_H. Equation 11.38 is known as the nonideal competitive adsorption (NICA) isotherm or equation, but when it was first proposed, the quotient n_i/n_H was not included; it was later added (Kinniburgh et al. 1999) to incorporate thermodynamic consistency. The total coverage of species i is the sum over all sites:

$$\theta_i = \frac{Q_i}{Q_S} = \sum_k \theta_{i,k} \tag{11.39}$$

It should be noted that, in Equation 11.38, the inclusion of both particle-specific (p_k) and ion-specific (n_i) parameters amounts to a separation of surface heterogeneity and ion nonideality effects, which both are in fact included in Equations 11.31 and 11.37; however, this is a model assumption and is not based on fundamental theory; in practice, they are obtained from fitting to experimental data (see Section 11.3 and Chapter 13). It is also interesting to note that the NICA equation is based in adsorption modeling, by the use of the LF isotherm. As it will be discussed in Chapter 13, the NICA isotherm has been widely employed to analyze metal complexation by SOM. Figure 11.5 shows NICA affinity spectra for binding of metal cations to humic acid, using the generic parameters reported by Milne et al. (2003) in Equation 11.34. Note the difference in the peak positions between weakly binding cations such as Ca(II) and strongly binding ones such as Cu(II) and Al(III).

Returning to Equation 11.37, as a way to simplify the problem, it can be assumed that the bidimensional spectrum h can be written as the product of two distributions corresponding to K_H, $g(\log K_H)$, and K_M, $f(\log K_M)$. This assumption implies in principle that these distributions are different and independent (uncorrelated). In practice, if g is the distribution deduced for H^+ binding in the absence of any interfering ions, the affinity distribution f will carry the H^+–M^{z+} correlations. Because

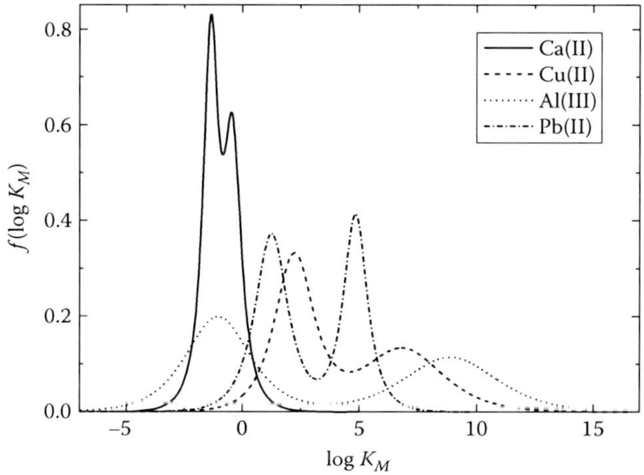

FIGURE 11.5 NICA isotherm plots of the generic affinity spectra for metal binding to humic acid. (Data from Milne, C.J. et al., *Environ. Sci. Technol.*, 37, 958–971, 2003.)

binding of metal cations in the full absence of protons is, at best, difficult to achieve experimentally, the "pure" distribution for M^{z+} binding is not accessible experimentally. Thus, the amount of metal bound can be written as

$$Q_M = Q_S \iint \frac{K_M a_M}{1 + K_H a_H + K_M a_M} g(\log K_H) f(\log K_M) \mathrm{d}\log K_H \, \mathrm{d}\log K_M \quad (11.40)$$

The assumption of single site binding for H^+ and M^{z+} is not of general validity, as there is spectroscopic evidence indicating that for many cations (such as Cu(II), Fe(II/III), etc.), a fraction of the bound metal is in bidentate or even tetradentate form (Christensen et al. 1998; Alvarez-Puebla, Valenzuela-Calahorro, and Garrido 2006; Orsetti, Andrade, and Molina 2009 and references therein). Because of the heterogeneous nature of HS, a mix of monodentate and polydentate complexes can be expected (due to the presence of both isolated and vicinal binding groups; see Chapter 10). Thus, because different metal fractions will be bound with different stoichiometries, we will introduce here an average stoichiometric coefficient r such as follows:

$$rA^- + M^{z+} \rightleftharpoons A_rM^{z-r} \quad (11.41)$$

In this case, considering a Langmuir-type equilibrium for the local isotherm leads to the following equations:

$$K_H a_H = \frac{q_H}{q_S - q_H - r q_M} \quad (11.42)$$

$$K_M a_M = \frac{q_M}{(q_S - q_H - r q_M)^r} \quad (11.43)$$

Except for integer r, Equations 11.42 and 11.43 cannot be solved for q_M and q_H, but should be solved numerically and inserted in Equation 11.40 (Orsetti, Andrade, and Molina 2009).

An average number of sites per cation can be estimated as the H^+/M^{z+} exchange ratio (Kinniburgh et al. 1999):

$$r_{ex} = -\frac{\frac{\partial Q_H}{\partial a_H}}{\frac{\partial Q_M}{\partial a_H}} = -\left(\frac{\partial \ln a_H}{\partial \ln a_M}\right)_{Q_M} \tag{11.44}$$

The rightmost part of Equation 11.44 allows determining r_{ex} from experimental data by measuring the horizontal distance between isotherms at different pH for the same Q_M. In the NICA equation, the factor n_M/n_0 appears as a stoichiometric factor, and in fact, it can be shown that r_{ex} is mainly determined by n_M/n_0 (Kinniburgh et al. 1999), but the n_i are also the ion nonideality parameters, thus they are playing different roles. In principle, it is considered that substrate heterogeneity and adsorbate–adsorbate interactions are difficult to separate, at best.

In the general case, multisite binding of metals would have to be considered in Equation 11.40. Note also that r_{ex} is not necessarily equal to r defined in Equations 11.42.

11.3 EXTRACTION OF AFFINITY SPECTRA

Besides the application of specific models, an important amount of work has been devoted to extract affinity distributions directly from experimental data, attempting to invert Equations 11.31, 11.37, and/or 11.40 or similar. The extraction of affinity distribution from experimental data is an ill-posed problem (Provencher 1982a; Cernik, Borkovec, and Westall 1995; Borkovec et al. 1996), because with experimental data one actually has, in the case of Equation 11.31,

$$Q_H = Q_S \int_{-\infty}^{\infty} \frac{K_H a_H}{1 + K_H a_H} f(\log K_H) \mathrm{d}\log K_H + \varepsilon \tag{11.45}$$

where ε is the experimental error. Now, it can be proved that for $\varepsilon > 0$, if $f(\lambda)$ is a solution of Equation 11.45, then for an arbitrary A there exists an ω such as $f(\lambda) + A\sin(\omega\lambda)$ is also a solution of 11.45; that is, there exists a large (in principle, infinite) number of solutions to Equation 11.45 within experimental error, most of them oscillating. Thus, common least squares methods would lead most probably to a very good fit but with any of a nearly infinite number of oscillating solutions. To overcome this problem, several approaches have been employed:

1. Semianalytical approaches, which involve introducing approximations to invert Equation 11.31 analytically
2. Regularization methods, to invert Equation 11.31 numerically by a modified least-squares scheme
3. The maximum entropy (MaxEnt) approach, a different numerical method which searches for the "most probable" constant distribution

In the following subsections, these approaches are reviewed. It is important to remark here that the deduction of affinity spectra from binding (titration) curves is a hard numerical problem, where the accuracy of the resulting spectrum depends critically on the quality of the input data; it is well known that "exact" distributions can only be obtained with accordingly exact data; even with 1% random deviations, variations in the results are found. Besides, titration curves are sensitive only to the main moments of the distribution (peak positions and width); thus in the absence of additional information, further details cannot be obtained (Merz 1980; Borkovec et al. 1996; Orsetti, Andrade, and Molina 2009).

11.3.1 Intrinsic and Conditional Equilibrium Constants

In Equations 11.30, 11.31, and 11.37, the intrinsic binding constants K_i are true thermodynamic quantities, related to the free energy by Equation 11.3. However, when aqueous ion activities are replaced by concentrations and the ionic strength (and/or pH and/or other concentrations) is kept constant, or when (more frequently) surface electrostatic interactions are neglected, results for the binding constants can be found, but which are valid only in the same experimental conditions as the original data; these are called *conditional affinity spectra* or *conditional constants*, a concept widely used in analytical chemistry (Garces, Mas, and Puy 2006; Orsetti, Andrade, and Molina 2009). Numerical methods for affinity spectrum extraction almost always yield conditional spectra.

11.3.2 Semianalytical Methods

Semianalytical methods attempt to solve analytically Equation 11.31 by introducing diverse approximations; this methodology has attracted several researchers (Gamble, Underdown, and Langford 1980; Gamble and Langford 1988; Buffle et al. 1990; Nederlof, Van Riemsdijk, and Koopal 1992, 1994), but currently numerical techniques are preferred. The main approaches are the *differential equilibrium function* (DEF) and the *local isotherm approximations* (LIA).

The DEF approach starts with the application of the mass action law to the overall ligand system. In this case, the metal binding to the substrate is represented as (dropping charges for simplicity)

$$\mathrm{M} + \mathrm{L} \rightleftharpoons \mathrm{ML} \tag{11.46}$$

where L represents a free ligand site. For each M concentration value, an effective mass action law is written as

$$\tilde{K} = \frac{[\mathrm{LM}]}{[\mathrm{L}][\mathrm{M}]} \tag{11.47}$$

where the square brackets indicate concentration, and $\tilde{K}$ is the *weighted average equilibrium function* (Gamble, Underdown, and Langford 1980), which is not a constant but a function of [M] or the metal loading of the substrate. A plot of $\tilde{K}$ versus, for instance, [LM] provides the course of this weighted average affinity as a function of the metal complexation. Dividing by [S], the total site concentration, one finds a Langmuir-type expression:

$$\tilde{K} = \frac{\theta}{[\mathrm{M}](1-\theta)} \tag{11.48}$$

Thus, $\tilde{K}$ represents a special, weighted, average affinity, but its calculation has some disadvantages (Gamble, Underdown, and Langford 1980; Gamble and Langford 1988); thus, the DEF was introduced as

$$\tilde{K}_{DE} = \frac{\mathrm{d}\{\tilde{K}(1-\theta\}}{\mathrm{d}\{1-\theta\}} = -\frac{\mathrm{d}\{[\mathrm{LM}]/[\mathrm{M}]\}}{\mathrm{d}[\mathrm{LM}]} \tag{11.49}$$

A plot of log K_{DE} (differential equilibrium function) has been accepted as characterization of heterogeneity by some authors (Gamble and Langford 1988), but later an estimation of the true cumulative distribution function was proposed as (Buffle et al. 1990) follows:

$$f_{DE}(\log K_{DE}) = -\frac{\mathrm{d}\theta}{\mathrm{d}\log(K_{DE})} \tag{11.50}$$

It can be easily shown (Nederlof, Van Riemsdijk, and Koopal 1992) that the DEF distribution implies a stepwise sorption–desorption process; that is, the site types fill or empty sequentially.

In the LIA approach, different approximations are introduced as the integral kernel (local isotherm) function q_H in Equation 11.30. The simplest form is the condensation approximation (Harris 1968), CA, where a step function is introduced as the local isotherm:

$$q_H = \begin{cases} 0 & \text{for } [\mathrm{M}] < 1/K \\ 1 & \text{for } [\mathrm{M}] \geq 1/K \end{cases} \tag{11.51}$$

Equation 11.51 can be substituted into Equation 11.30 and the affinity spectrum found as

$$f_{CA}(\log K) = \frac{\mathrm{d}\theta}{\mathrm{d}\log[\mathrm{M}]} \tag{11.52}$$

$$\log K = -\log[\mathrm{M}] \tag{11.53}$$

Another LIA approach is the asymptotically correct approximation (ACA), as a first improvement over the CA (Cerofolini 1974), defined as

$$q_H = \begin{cases} K[\mathrm{M}] & \text{for } [\mathrm{M}] < 1/K \\ 1 & \text{for } [\mathrm{M}] \geq 1/K \end{cases} \tag{11.54}$$

and the corresponding distribution function is

$$f_{ACA}(\log K) = \frac{\mathrm{d}\theta}{\mathrm{d}\log[\mathrm{M}]} - \frac{1}{\ln 10}\frac{\mathrm{d}^2\theta}{\mathrm{d}\log[\mathrm{M}]^2} \tag{11.55}$$

with Equation 11.53 giving K as a function of [M]. This approximation results in an asymmetrical distribution, which is unrealistic. Besides, both the CA and ACA approaches led to poor agreement, being crude approximations to the Langmuir function. A better approximation introduced by Nederlof, Van Riemsdijk, and Koopal (1990) is the logarithmic symmetrical approximation (LSA), defined by

$$q_H = \begin{cases} \alpha(K[\mathrm{M}])^{\beta} & \text{for} \quad [\mathrm{M}] < 1/K \\ 1-\alpha(K[\mathrm{M}])^{-\beta} & \text{for} \quad [\mathrm{M}] \geq 1/K \end{cases} \tag{11.56}$$

where α and β are parameters that have to be optimized for fitting to the Langmuir isotherm; it is found (Nederlof, Van Riemsdijk, and Koopal 1992) that $\alpha = 0.5$ and $\beta = 0.7$ reproduce well the Langmuir isotherm (albeit $\beta = 1$ gives a reasonable approximation also), and the distribution function results in

$$f_{LSA}(\log K) = \frac{\mathrm{d}\theta}{\mathrm{d}\log[\mathrm{M}]} - \frac{1}{(\beta \ln 10)^2}\frac{\mathrm{d}^3\theta}{\mathrm{d}\log[\mathrm{M}]^3} \tag{11.57}$$

with Equation 11.53 again defining K.

These different methods were evaluated and compared (Nederlof, Van Riemsdijk, and Koopal 1992, 1994); it is found that LSA gives the better approximation, whereas the DEF function is perfect only for homogeneous substrates. Analysis with cases with known affinity spectra show that LSA returns good approximations but flattened, whereas the DEF method gives distorted spectra but well detects small, thin peaks.

11.3.3 Modified Least Squares Methods

At the beginning of Section 11.3, the calculation of constant distributions from experimental data is an ill-posed problem that, if approached by classical solution schemes, leads to serious numerical problems and instabilities. To circumvent these difficulties, least squares minimization procedures are modified by the introduction of constrains, regularization, or both of them, to provide a substantial stabilization of the resulting distributions; these approaches have been widely applied (Provencher 1982a; Cernik, Borkovec, and Westall 1995; Borkovec et al. 1996; Rusch et al. 1997; Bersillon et al. 2001). In general terms, all modifications are applied over the common least squares problem, where the function to be minimized is the variance:

$$\chi^2 = \sum_j \left(Q_{j,\mathrm{exp}} - Q_{j,\mathrm{ctd}}\right)^2 \tag{11.58}$$

where $Q_{j,\mathrm{exp}}$ are the experimental points and $Q_{j,\mathrm{ctd}}$ are those calculated with Equation 11.31 or similar. The constrain most used by far is the *nonnegativity*, where function

11.58 is minimized with the condition that $f(\log K) \geq 0$ for all K. This helps greatly in eliminating many oscillating solutions. However, in many cases, there exists still variability in the solution, so regularization has to be applied. In a regularization procedure, the objective function to be minimized, Equation 11.58, is modified by a regularizor term, which penalizes the unwanted solutions:

$$O(\alpha) = \chi^2 + \alpha^2 R \tag{11.59}$$

where O is the objective function to minimize, R is the regularizor, and α is the regularization parameter, its value controlling the weight of the regularization function R. For these procedures, the integral in Equation 11.31 is discretized in M parts:

$$Q_H = Q_S \sum_{i=1}^{M} \frac{K_{H,i} a_H}{1 + K_{H,i} a_H} f_i \tag{11.60}$$

with f_i as the fraction of sites with constant $K_{H,i}$. There are in principle two types of regularization possible for these problems (Cernik, Borkovec, and Westall 1995): minimum number of peaks or maximum smoothness. In the first case, the regularizor is

$$R = \sum_{i=1}^{M} \sum_{k=1}^{i-1} f_i f_k \tag{11.61}$$

which increases as the number of nonzero f_i values increases; this usually leads to a discrete affinity spectrum. For maximum smoothness, the regularizor is

$$R = \int_{\Delta \log K} \left(\frac{\partial^2 f}{\partial \log K^2} \right)^2 \mathrm{d}\log K \approx \sum_{i=2}^{M-1} \left(f_{i-1} - 2 f_i + f_{i+1} \right)^2 \tag{11.62}$$

which penalizes solutions showing rapidly changing behavior, favoring instead smooth, continuous solutions, with few peaks. Note that this regularizor approaches the second derivative of the solution. Even when it is difficult to extract true, exact distributions from titration data (Borkovec et al. 1996), at least without additional knowledge, nevertheless useful information can be obtained with such procedures. Regularized numerical procedures have been extensively applied to many problems (Provencher 1982a; Cernik, Borkovec, and Westall 1995; Mao, Tang, and Swanson 2000; Ravikovitch et al. 2000; Ruf, Gould, and Haase 2000; Lin, Drake, and Rayson 2002; Eberhardt, Elder, and Labbé 2007; Orsetti, Andrade, and Molina 2009). Provencher (1982a,b) has developed a regularization method, implemented in the CONTIN program code, which has been often employed (Cao et al. 1997; Nomula and Cooper 1997; Morris et al. 1999; Puziy et al. 1999; Rasteiro, Lemos, and Vasquez 2008; Orsetti, Andrade, and Molina 2009). In the application of this and other numerical methods, conditional affinity spectra are commonly obtained, valid in given conditions, usually fixed ionic strength and, in the case of metal binding, also fixed pH (Garces, Mas, and Puy 1998, 2006; Puy et al. 2008).

Cernik, Borkovec, and Westall (1995) studied the application of these methods to simulated and real data, finding that very good fitting could be obtained with both discrete and continuous spectra, both being almost indistinguishable. Figure 11.6 shows the sorption of dodecylpiridinium cation in 0.01 M NaCl on an Environmental Protection Agency (EPA) soil sample; optimal regulation for a small number of sites results in four sites as indicated in Figure 11.6. Regularization for smoothness yields three peaks, qualitatively coincident with the discrete spectrum. It is worth noting that either mode results in very good fitting and is almost indistinguishable from the other. As it has been discussed by several authors, this should not be surprising regarding generic data fitting, since only the main moments are important for the description of the titration (binding) curve. Thus, the question of discrete versus continuous affinity spectra cannot be answered from the fitting of experimental curves; instead it should be addressed from the nature of the substrate. For mineral particles, a few discrete sites are generally expected, thus a discrete distribution is in order; on the contrary, in SOM, the main proton- and metal-binding sites are carboxylic and phenolic groups with potentially many molecular environments (Chapter 10), which as it is well known affects the respective binding constants; therefore, a continuous distribution appears physically more reasonable.

Orsetti, Andrade, and Molina (2009) employed the CONTIN package to analyze proton- and cation-binding data to HS, including humic and fulvic acids, obtaining conditional affinity distributions. For metal binding, separation of the proton affinity spectrum according to Equation 11.40 was employed; as the log K_H distribution is determined previously from a separate data set, Equation 11.40 can be solved with CONTIN by grouping as

$$Q_M = \int_{-\infty}^{\infty}\left[\int_{-\infty}^{\infty}\frac{K_M a_M}{1+K_H a_H + K_M a_M} g'(\log K_H)\mathrm{d}\log K_H\right] f(\log K_M)\mathrm{d}\log K_M \qquad (11.63)$$

with the factor in brackets as the kernel. The results were analyzed in terms of relevant spectroscopic studies. Figure 11.7a compares the conditional affinity distributions found at different ionic strengths for purified peat humic acid (PPHA) (Milne et al. 1995). The general shape of the distributions found is similar to that described in the literature, interpreted in terms of carboxylic sites (the lower proton affinity, the higher area peak) and phenolic ones (high proton affinity, the broad peak near log K_H = 9). Thus, one can attempt to deconvolute the peaks accordingly. Figure 11.7b shows the deconvolution of the affinity distribution at the higher ionic strength (where minimum electrostatic effects are expected) in terms of Gaussian functions. It is found that three Gaussian peaks are needed to obtain a satisfactory fit, with a small peak in the middle of the two main ones. The middle peak found in this and several distributions (Orsetti, Andrade, and Molina 2009) is centered about pKa = 6–7, which lies in the range of the pKa_2 values for adjacent carboxylic groups (such as phthalic or oxalic acids); a similar result was found by fluorescence measurements (Silva and Tauler 2006). Thus, this finding would indicate that some carboxylic groups are vicinal; the subject, however, requires further clarification.

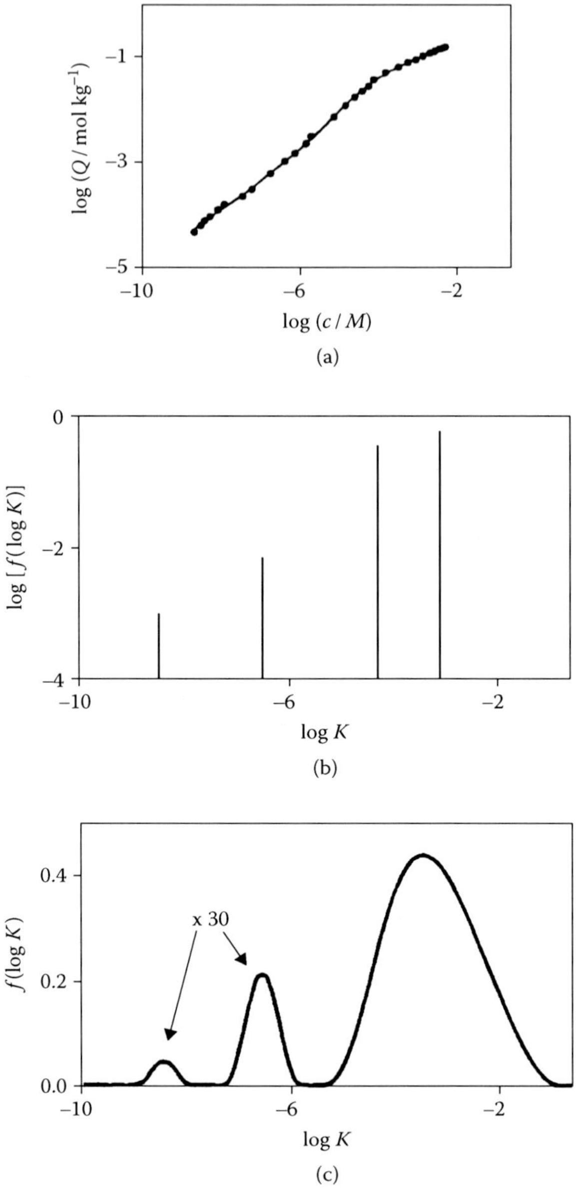

FIGURE 11.6 Sorption isotherms of dodecylpiridinium on a soil material (EPA-12). (a) Points: experimental data; line: fitted isotherms (almost indistinguishable) with discrete and continuous affinity spectra; (b) discrete affinity spectrum, found using regularization for a small number of sites; (c) continuous affinity spectrum, after regularizing for smoothness; note that the smaller peaks are magnified by a factor of 30. (Reprinted with permission from Cernik, M. et al., *Environ. Sci. Technol.,* 29, 2, 413–425. Copyright 1995 American Chemical Society.)

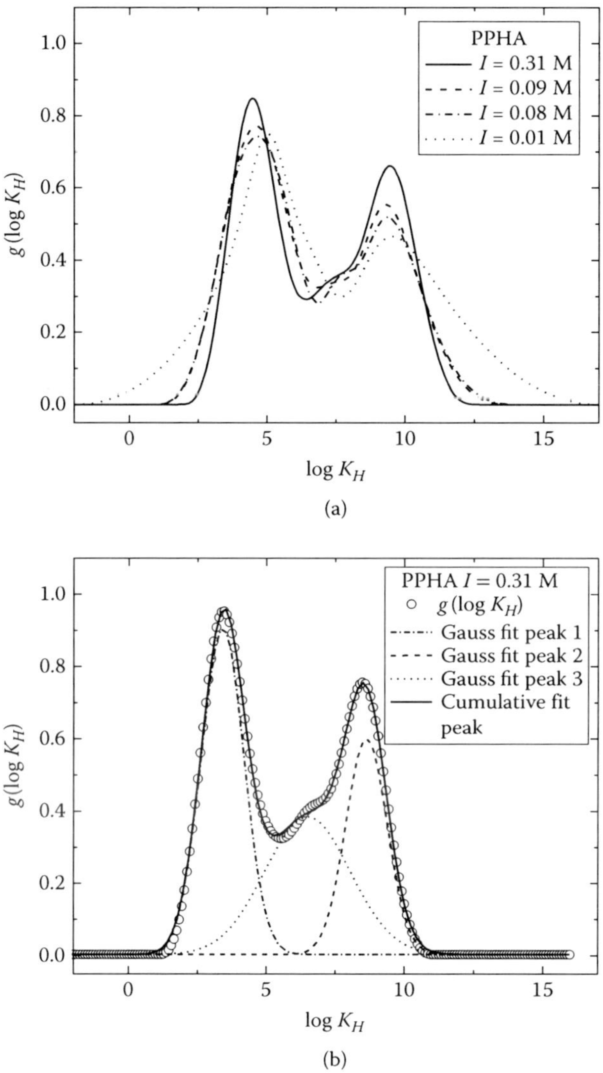

FIGURE 11.7 (a) Affinity distributions obtained at different ionic strengths from experimental data for purified peat humic acid (PPHA); (b) deconvolution of the distribution obtained for PPHA with Gaussian functions. (Data from Milne, C.J. et al., *Environ. Sci. Technol.*, 37, 958–971, 2003; Reprinted from *J. Colloid Interface Sci.*, 336, Orsetti, S. et al., Application of a constrained regularization method to extraction of affinity distributions: Proton and metal binding to humic substances, 377–387. Copyright 2009, with permission from Elsevier.)

Figure 11.8 shows the conditional affinity distributions found for Cd(II) and Pb(II) binding to purified peat humic acid, using data published by Milne et al. (2003) from previous work (Benedetti et al. 1995; Christl et al. 2001). As it is observed, the results are quite similar, showing relatively narrow peaks, the most prominent being that at low log K_M values, with one or two smaller peaks for higher stability binding sites.

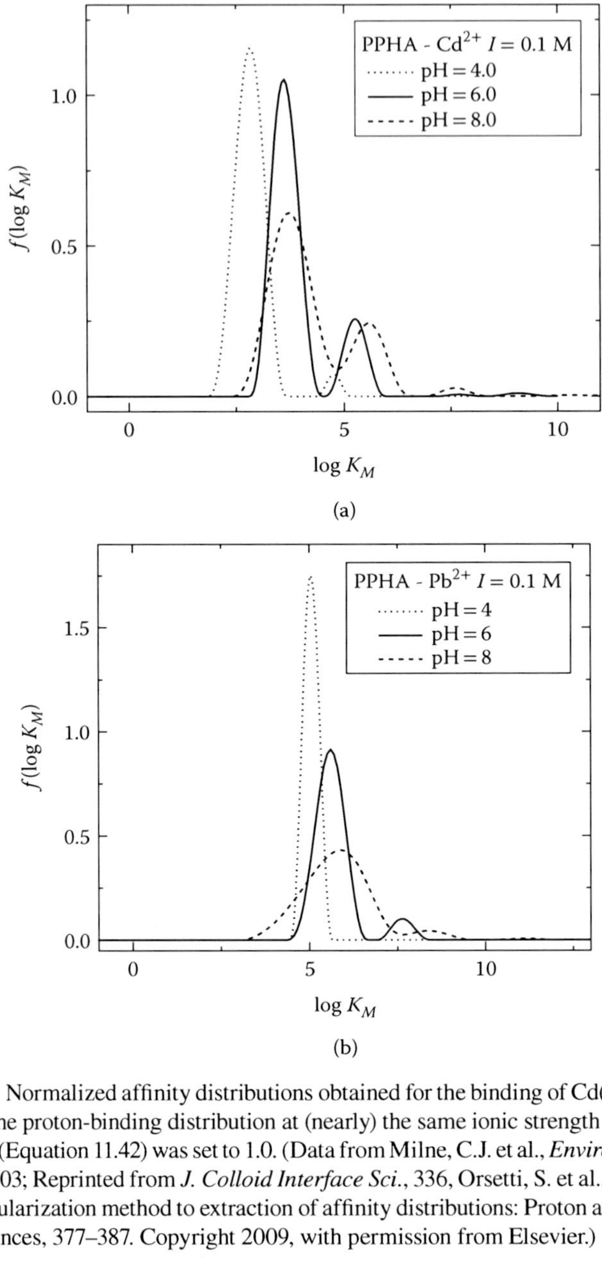

FIGURE 11.8 Normalized affinity distributions obtained for the binding of Cd(II) (a) and Pb(II) (b) to PPHA. The proton-binding distribution at (nearly) the same ionic strength (see Figure 11.7) was used, and *r* (Equation 11.42) was set to 1.0. (Data from Milne, C.J. et al., *Environ. Sci. Technol.*, 37, 958–971, 2003; Reprinted from *J. Colloid Interface Sci.*, 336, Orsetti, S. et al., Application of a constrained regularization method to extraction of affinity distributions: Proton and metal binding to humic substances, 377–387. Copyright 2009, with permission from Elsevier.)

There is a pH dependence of the distributions, shifting to higher values as the pH increases. The displacement observed is consistent with an ionic strength effect: as the electrostatic effect becomes more pronounced, the affinity curves shift to higher log K values. Based on spectroscopic evidence (Orsetti, Andrade, and Molina 2009 and references therein), the results of Figure 11.8 could, in principle, be interpreted in terms of two type of sites: a weaker, monodentate carboxylic, more abundant binding site with log K_{Cd} about 2–4 and a stronger, bidentate carboxylic, less-abundant site with log K_{Cd} near 6; however, more evidence is required to confirm this interpretation. In the case of Pb(II), the curves are similar, albeit only one peak is found at pH = 4; this fact can be attributed to the second type of sites having higher proton affinity, thus displacing completely Pb(II) ions at pH = 4. An interpretation similar to Cd(II) could in principle be given for lead: the higher peak at low log K_{Pb} values attributed to monodentate binding to carboxylic groups, and the peak at higher values to bidentate binding.

11.3.4 Maximum Entropy Formalism

The MaxEnt formalism seeks to find the most probable distribution compatible with all known constraints of a given problem; it was introduced by Jaynes (1957, 1982) and since then applied to many different statistical problems (Christakos 1998; Cheeseman and Stutz 2005; Komnitsas and Modis 2006; Vakarin and Badiali 2006; Orton and Lark 2007). Garces, Mas, and Puy (1998) proposed its application to the extraction of affinity spectra as an alternative to regularization methods as follows. The MaxEnt formalism attempts to find the most probable distribution $\{f_i\}$ of sites, compatible with the set of the M experimental data points obtained, where

$$f_i = \frac{n_i}{N}, \qquad 1 \le i \le m \tag{11.64}$$

is the fraction of sites having the same binding constant K_i, N being the total number of sites, and m the number of different constants. The probability of a given distribution $\{f_i\}$ of the N sites among all the possible sets is given by

$$P\left[\{f_i\}\right] = \frac{N!}{\prod_{j=1}^{m} n_j} \prod_{l=1}^{m} q_l^{n_l} \tag{11.65}$$

where q_i indicates the a priori probability of a given type i of sites. Extending the MaxEnt formalism to continuous distribution functions defined in a domain $[a, b]$ and applying the Shannon entropy $S\left[\{f_i\}\right] = \ln P\left[\{f_i\}\right] / N$, the sought distribution is that which maximizes

$$S\left[f(x)\right] = -\int_a^b f(x) \ln \frac{f(x)}{q(x)} \mathrm{d}x \tag{11.66}$$

where it has been substituted $x = \log K$. The distribution results to be

$$f(x) = \frac{q(x)}{Z} \exp\left(-\sum_{i=1}^{M} \frac{\lambda_i}{1 + 10^{-x-\mu_i}}\right) \tag{11.67}$$

where $\mu_i = \log c_i$, the experimental data points being $\{c_i, \theta_i\}$, $1 \le i \le M$. The λ_i are Lagrange multipliers associated with the experimental points, which have to be determined; note that the denominator is the Langmuir isotherm. Z is the partition function, defined by

$$Z = \int_a^b q(x)\exp\left(-\sum_{j=1}^{M} \frac{\lambda_j}{1+10^{-x-\mu_j}}\right)dx \tag{11.68}$$

The Lagrange multipliers λ_i are found from the constrains, which merely are the experimental data; that is, the affinity spectrum should fit the experimental data through Equation 11.30:

$$\theta_j = \int_a^b \frac{q(x)}{Z}\exp\left(-\sum_{i=1}^{M} \frac{\lambda_i}{1+10^{-x-\mu_i}}\right)\left(\frac{1}{1+10^{-x-\mu_j}}\right)dx \qquad 1 \le j \le M \tag{11.69}$$

Equations 11.69 form a nonlinear system of M equations with M unknowns; its solution yields the λ_i, which, inserted in Equation 11.67, results in the sought distribution. The a priori probability $q(x)$ should be appropriately chosen. Garces, Mas, and Puy (1998) chose the equiprobable distribution $q(x) = 1/(a - b)$ so that the final distribution is as close as possible to flatness. Figure 11.9 presents an example with simulated data for a discrete affinity spectrum corresponding to the following intrinsic stability constants and respective weights: $\log K_1 = 5.55, f_1 = 0.32$; $\log K_2 = 4.92$, $f_2 = 0.22$; $\log K_3 = 4.14, f_3 = 0.16$; $\log K_4 = 3.20, f_4 = 0.30$. Figure 11.9a presents the simulated binding data, and Figure 11.9b shows the affinity spectrum recovered, compared with the original employed in data simulation: the four peaks of the spectrum appear clearly defined and clearly tend to Dirac delta functions. The resolution of the spectrum is also remarkable, despite the peaks being separated by less than 1 log K unit. Garces, Mas, and Puy (1998) concluded that when accurate experimental data are available, the MaxEnt approach can be used to obtain the affinity distribution function that describes the heterogeneous adsorption onto a surface or the heterogeneous complexation of a small complexing agent to a macromolecule. However, it should be kept in mind that this method poses a stiff numerical problem: Equation 11.69 forms a system of strongly nonlinear equations, with as many equations and unknowns as data points; as the number of points increases (which is always desirable), the computational difficulty increases and consequently the accuracy of the results decreases; this situation imposes strong requirements on the numerical procedure employed. Perhaps because of this, the MaxEnt approach has had little application to date.

11.4 KINETIC ASPECTS

Throughout this book, the binding of ions in soils is treated from the point of view of chemical equilibrium, as it is done usually. It should be borne in mind, however, the implications of such approach. A reaction (or any other process that can be reversed) is in equilibrium when the direct and inverse reactions rates are equal, so that there are no changes in the concentrations of reactants and products unless the system is

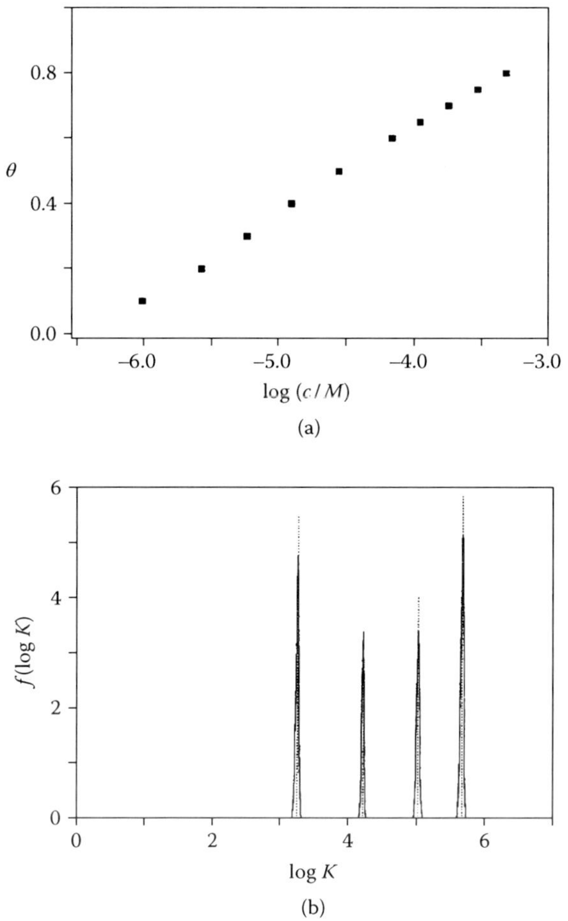

FIGURE 11.9 Extraction of a discrete affinity spectrum using the MaxEnt formalism. (a) Simulated mean coverage data versus the logarithm of the free complexing agent concentration. (b) $f(\log K)$ versus $\log K$; dotted lines indicate the exact K values with a height proportional to the probability of each kind of site; continuous line is the spectrum found. (Reprinted with permission Garces, J.L. et al., *J. Chem. Phys.*, 124, 4, 044710–044714. Copyright 1998 American Chemical Society.)

perturbed in some way. In other words, enough time has lapsed so as to reach a point where no changes are observed; the key words here are "enough time," because that is dependent on the timescale of the processes. Figure 11.10 shows the timescales needed to attain equilibrium for different types of soil reactions (Amacher 1991). Some of them (ion association, hydrolysis) are homogeneous reactions, whereas most (including sorption processes) are heterogeneous, which implies that, besides the rate of the chemical reaction itself, transport to the interface plays a relevant role (Aharoni and Sparks 1991; Sparks 1998; Leij and Genutchen 1999). For a complex set of chemical reactions, mixed homogeneous and heterogeneous (as it is always the case in soils) to be considered in equilibrium, each of them must either have reached

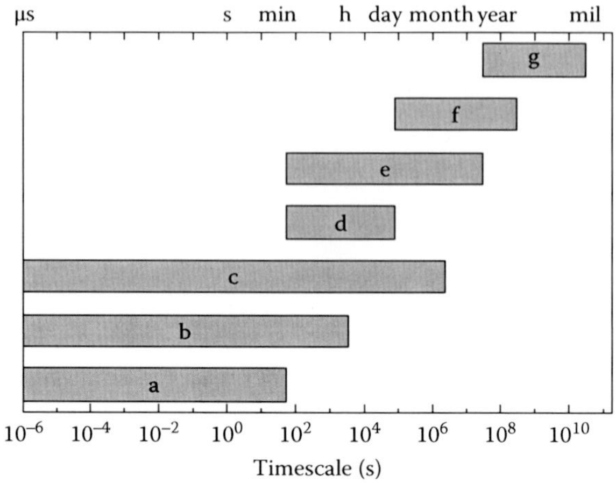

FIGURE 11.10 Timescales required to attain equilibrium of typical soil reactions classes: (a) ion association, (b) ion exchange and adsorption, (c) sorption processes with transport, (d) gas–water reactions, (e) multivalent ion hydrolysis, (f) mineral–solution reactions, and (g) mineral weathering and crystallization. (Redrawn after Amacher, M.C., *Rates of Soil Chemical Processes*, Soil Science Society of America, Madison, WI, 1991.)

the equilibrium, or to be comparatively so slow that it can be considered as frozen. This is usually the case for reactions involving small ion and molecules, so that the equilibrium treatment is sensible, but cases may well be found where it is not entirely applicable, mostly due to transport issues.

11.5 SUMMARY

In this chapter, the bases of the phenomenological models for ion binding to soil colloids, as equilibrium processes, have been stated. There are two main aspects to be considered: the sorption process itself (either specific or not) and the electrostatic problem associated with the reaction of a charged species at an interface. The sorption process can be modeled either through adsorption isotherms (adsorption modeling) or as chemical reactions (chemical modeling). The electrostatic models will be introduced in Chapters 12 and 13. The set of intrinsic binding (complexation) constants, known as affinity spectrum or constant distribution, may be found or estimated by semianalytical or special regression methods. In those chapters, these concepts will be applied to the main types of soil colloids: minerals and organic matter.

REFERENCES

Aharoni, C., and D. L. Sparks. 1991. Kinetics of soil chemical reactions—A theoretical treatment. In *Rates of Soil Chemical Processes*, eds. D. L. Sparks and D. L. Suarez, 19–59. SSSA Special Publications 27. Madison, WI: Soil Science Society of America. https://dl.sciencesocieties.org/publications/books/pdfs/sssaspecialpubl/ratesofsoilchem/19.

Alvarez-Puebla, R. A., C. Valenzuela-Calahorro, and J. J. Garrido. 2006. Theoretical study on fulvic acid structure, conformation and aggregation. A molecular modelling approach. *The Science of the Total Environment* 358, no. 1–3: 243–254.

Amacher, M. C. 1991. Methods of obtaining and analyzing kinetic data. In *Rates of Soil Chemical Processes*, eds. D. L. Sparks and D. L. Suarez, 19–59. SSSA Special Publications 27. Madison, WI: Soil Science Society of America. https://dl.sciencesocieties.org/publications/books/pdfs/sssaspecialpubl/ratesofsoilchem/19.

Atkins, P., and J. de Paula. 2009. *Physical Chemistry*. 9th ed. New York: W. H. Freeman.

Avena, M. J., L. K. Koopal, and W. H. van Riemsdijk. 1999. Proton binding to humic acids: Electrostatic and intrinsic interactions. *Journal of Colloid and Interface Science* 217, no. 1: 37–48. doi:10.1006/jcis.1999.6317.

Bard, A. J., and L. R. Faulkner. 2000. *Electrochemical Methods: Fundamentals and Applications*. 2nd ed. New York: John Wiley & Sons.

Benedetti, M. F., C. J. Milne, D. G. Kinniburgh, W. H. Van Riemsdijk, and L. K. Koopal. 1995. Metal ion binding to humic substances: Application of the non-ideal competitive adsorption model. *Environmental Science & Technology* 29: 446–457. doi:10.1021/es00002a022.

Berry, R. S., S. A. Rice, and J. Ross. 2000. *Physical Chemistry*. 2nd ed. New York: Oxford University Press.

Bersillon, J. -L., F. Villieras, F. Bardot, T. Gorner, and J. -M. Cases. 2001. Use of the Gaussian distribution function as a tool to estimate continuous heterogeneity in adsorbing systems. *Journal of Colloid and Interface Science* 240, no. 2: 400–411. doi:10.1006/jcis.2001.7657.

Borkovec, M., J. Hamáček, and C. Piguet. 2004. Statistical mechanical approach to competitive binding of metal ions to multi-center receptors. *Dalton Transactions*, no. 24: 4096. doi:10.1039/b413603a.

Borkovec, M., and G. J. M. Koper. 1994a. Affinity distributions of polyampholytes with interacting acid-base groups. *Langmuir* 10, no. 9: 2863–2865. doi:10.1021/la00021a004.

Borkovec, M., and G. J. M. Koper. 1994b. Ising models of polyprotic acids and bases. *The Journal of Physical Chemistry* 98, no. 23: 6038–6045. doi:10.1021/j100074a034.

Borkovec, M., G. J. M. Koper, and C. Piguet. 2006. Ion binding to polyelectrolytes. *Current Opinion in Colloid and Interface Science* 11, no. 5: 280–289.

Borkovec, M., U. Rusch, M. Cernik, G. J. M. Koper, and J. C. Westall. 1996. Affinity distributions and acid-base properties of homogeneous and heterogeneous sorbents: Exact results versus experimental data inversion. *Colloids and Surfaces A: Physicochemical and Engineering Aspects* 107: 285–296.

Buffle, J., R. S. Altmann, M. Filella, and A. Tessier. 1990. Complexation by natural heterogeneous compounds: Site occupation distribution functions, a normalized description of metal complexation. *Geochimica et Cosmochimica Acta* 54, no. 6: 1535–1553. doi:10.1016/0016-7037(90)90389-3.

Cao, H., G. H. Dai, J.-P. Yuan, and Y. C. Jean. 1997. Reliability of ortho-positronium lifetime distribution analysis in polymers by using CONTIN program. *Materials Science Forum* 255–257: 238–242.

Cernik, M., M. Borkovec, and J. C. Westall. 1995. Regularized least-squares methods for the calculation of discrete and continuous affinity distributions for heterogeneous sorbents. *Environmental Science and Technology* 29, no. 2: 413–425.

Cernik, M., M. Borkovec, and J. C. Westall. 1996. Affinity distribution description of competitive ion binding to heterogeneous materials. *Langmuir* 12, no. 25: 6127–6137. doi:10.1021/la960008f.

Cerofolini, G. F. 1974. Localized adsorption on heterogeneous surfaces. *Thin Solid Films* 23, no. 2: 129–152. doi:10.1016/0040-6090(74)90235-1.

Cheeseman, P., and J. Stutz. 2005. Generalized maximum entropy. *AIP Conference Proceedings* 803, no. 1: 374–381. doi:10.1063/1.2149816.

Christakos, G. 1998. Spatiotemporal information systems in soil and environmental sciences. *Geoderma* 85, no. 2–3: 141–179.

Christensen, J. B., E. Tipping, D. G. Kinniburgh, C. Grøn, and T. H. Christensen. 1998. Proton binding by groundwater fulvic acids of different age, origins, and structure modeled with the model V and NICA-Donnan model. *Environmental Science and Technology* 32, no. 21: 3346–3355.

Christl, I., C. J. Milne, D. G. Kinniburgh, and R. Kretzschmar. 2001. Relating ion binding by fulvic and humic acids to chemical composition and molecular size. 2. Metal binding. *Environmental Science & Technology* 35, no. 12 (June 1): 2512–2517. doi:10.1021/es0002520.

Companys, E., J. L. Garcés, J. Salvador, J. Galceran, J. Puy, and F. Mas. 2007. Electrostatic and specific binding to macromolecular ligands: A general analytical expression for the Donnan volume. *Colloids and Surfaces A: Physicochemical and Engineering Aspects* 306, no. 1–3: 2–13. doi:10.1016/j.colsurfa.2007.01.016.

David, C., S. Mongin, C. Rey-Castro, J. Galceran, E. Companys, J. L. Garcés, J. Salvador et al. 2010. Competition effects in cation binding to humic acid: Conditional affinity spectra for fixed total metal concentration conditions. *Geochimica et Cosmochimica Acta* 74, no. 18: 5216–5227. doi:10.1016/j.gca.2010.06.023.

Eberhardt, T. L., T. Elder, and N. Labbé. 2007. Analysis of ethanol-soluble extractives in southern pine wood by low-field proton NMR. *Journal of Wood Chemistry and Technology* 27, no. 1: 35–47.

Esteves Da Silva, J. C. G., and R. Tauler. 2006. Multivariate curve resolution of synchronous fluorescence spectra matrices of fulvic acids obtained as a function of pH. *Applied Spectroscopy* 60, no. 11: 1315–1321.

Gamble, D. S., and C. H. Langford. 1988. Complexing equilibria in mixed ligand systems: Tests of theory with computer simulations. *Environmental Science & Technology* 22, no. 11: 1325–1336. doi:10.1021/es00176a013.

Gamble, D. S., A. W. Underdown, and C. H. Langford. 1980. Copper(II) titration of fulvic acid ligand sites with theoretical, potentiometric, and spectrophotometric analysis. *Analytical Chemistry* 52, no. 12: 1901–1908. doi:10.1021/ac50062a029.

Garces, J. L., F. Mas, and J. Puy. 1998. Application of maximum entropy formalism in the determination of the affinity spectrum in macromolecular complexation. *Environmental Science & Technology* 32, no. 4: 539–548. doi:10.1021/es970468a.

Garcés, J. L., F. Mas, and J. Puy. 1999. Complexation to macromolecules with a large number of sites. *The Journal of Chemical Physics* 111: 2818.

Garces, J. L., F. Mas, and J. Puy. 2004. Affinity distribution functions in multicomponent heterogeneous adsorption. Analytical inversion of isotherms to obtain affinity spectra. *The Journal of Chemical Physics* 120, no. 19: 9266–9276. doi:10.1063/1.1710857.

Garces, J. L., F. Mas, and J. Puy. 2006. Conditional equilibrium constants in multicomponent heterogeneous adsorption: The conditional affinity spectrum. *The Journal of Chemical Physics* 124, no. 4: 044710–044714. doi:10.1063/1.2162876.

Goldberg, S. 2005. Equations and models describing adsorption processes in soils. In *Chemical Processes in Soils*, eds. M. A. Tabatabai and D. L. Sparks, 489–518. Madison, WI: Soil Science Society of America.

Gu, X., L. J. Evans, and S. J. Barabash. 2010. Modeling the adsorption of Cd (II), Cu (II), Ni (II), Pb (II) and Zn (II) onto montmorillonite. *Geochimica et Cosmochimica Acta* 74, no. 20: 5718–5728. doi:10.1016/j.gca.2010.07.016.

Harris, L. B. 1968. Adsorption on a patchwise heterogeneous surface; mathematical analysis of the step-function approximation to the local isotherm. *Surface Science* 10, no. 2: 129–145. doi:10.1016/0039-6028(68)90015-0.

Jaynes, E. T. 1957. Information theory and statistical mechanics. *Physical Review* 106, no. 4: 620. doi:10.1103/PhysRev.106.620.

Jaynes, E. T. 1982. On the rationale of maximum-entropy methods. *Proceedings of the IEEE* 70, no. 9: 939–952.

Johnston, C. P., and M. Chrysochoou. 2012. Investigation of chromate coordination on ferrihydrite by in situ ATR-FTIR spectroscopy and theoretical frequency calculations. *Environmental Science and Technology* 46, no. 11: 5851–5858. doi:10.1021/es300660r.

Kinniburgh, D. G., W. H. Van Riemsdijk, L. K. Koopal, M. Borkovec, M. F. Benedetti, and M. J. Avena. 1999. Ion binding to natural organic matter: Competition, heterogeneity, stoichiometry and thermodynamic consistency. *Colloids and Surfaces A: Physicochemical and Engineering Aspects* 151, no. 1–2: 147–166.

Komnitsas, K., and K. Modis. 2006. Soil risk assessment of As and Zn contamination in a coal mining region using geostatisretics. *The Science of the Total Environment* 371, no. 1–3: 190–196.

Koopal, L. K., W. H. van Riemsdijk, J. C. M. de Wit, and M. F. Benedetti. 1994. Analytical isotherm equations for multicomponent adsorption to heterogeneous surfaces. *Journal of Colloid and Interface Science* 166, no. 1 (August): 51–60. doi:10.1006/jcis.1994.1270.

Koopal, L. K., and C. H. W. Vos. 1993. Adsorption on heterogeneous surfaces. Calculation of the adsorption energy distribution function or the affinity spectrum. *Langmuir* 9, no. 10: 2593–2605. doi:10.1021/la00034a019.

Koretsky, C. 2000. The significance of surface complexation reactions in hydrologic systems: A geochemist's perspective. *Journal of Hydrology* 230, no. 3–4: 127–171. doi:10.1016/S0022-1694(00)00215-8.

Leij, F. J., and M. T. van Genutchen. 1999. Solute transport. In *Handbook of Soil Science*, ed. M. E. Sumner, A183–A228. 1st ed. Boca Raton, FL: CRC.

Levine, I. 2008. *Physical Chemistry*. 6th ed. New York: McGraw-Hill Science/Engineering/Math.

Lin, S., L. R. Drake, and G. D. Rayson. 2002. Affinity distributions of lead ion binding to an immobilized biomaterial derived from cultured cells of Datura innoxia. *Advances in Environmental Research* 6, no. 4: 523–532. doi:10.1016 /S1093-0191(01)00078-8.

Lin, S., and G. D. Rayson. 1998. Impact of surface modification on binding affinity distributions of Datura innoxia biomass to metal ions. *Environmental Science & Technology* 32, no. 10: 1488–1493. doi:10.1021/es970843z.

Lützenkirchen, J. 1998. Comparison of 1-pK and 2-pK versions of surface complexation theory by the goodness of fit in describing surface charge data of (Hydr)oxides. *Environmental Science & Technology* 32, no. 20: 3149–3154. doi:10.1021/es980125s.

Manunza, B., S. Deiana, V. Maddau, C. Gessa, and R. Seeber. 1995. Stability constants of metal-humate complexes: Titration data analyzed by bimodal Gaussian distribution. *Soil Science Society of America Journal* 59, no. 6: 1570–1574.

Mao, R., J. Tang, and B. G. Swanson. 2000. Relaxation time spectrum of hydrogels by CONTIN analysis. *Journal of Food Science* 65, no. 3: 374–381.

Mattigod, S. V., and J. M. Zachara. 1996. Equilibrium modeling in soil chemistry. In *Methods of Soil Analysis. Part 3. Chemical Methods*, ed. D. L. Sparks, 1309–1358. Soil Science Society of America Book Series 5. Madison, WI: American Society of Agronomy-Soil Science Society of America.

McBride, M. B. 1999. Chemisorption and precipitation reactions. In *Handbook of Soil Science*, ed. M. E. Sumner, B265–B302. 1st ed. Boca Raton, FL: CRC Press.

McQuarrie, D. A., and J. D. Simon. 1997. *Physical Chemistry: A Molecular Approach*. Herndon, VA: University Science Books.

Merz, P. H. 1980. Determination of adsorption energy distribution by regularization and a characterization of certain adsorption isotherms. *Journal of Computational Physics* 38, no. 1: 64–85. doi:10.1016/0021-9991(80)90012-1.

Milne, C. J., D. G. Kinniburgh, J. C. M. de Wit, W. H. van Riemsdijk, and L. K. Koopal. 1995. Analysis of metal-ion binding by a peat humic acid using a simple electrostatic model. *Journal of Colloid and Interface Science* 175, no. 2: 448–460.

Milne, C. J., D. G. Kinniburgh, W. H. van Riemsdijk, and E. Tipping. 2003. Generic NICA–Donnan model parameters for metal-ion binding by humic substances. *Environmental Science & Technology* 37, no. 5: 958–971. doi:10.1021/es0258879.

Morris, K. F., B. J. Cutak, A. M. Dixon, and C. K. Larive. 1999. Analysis of diffusion coefficient distributions in humic and fulvic acids by means of diffusion ordered NMR spectroscopy. *Analytical Chemistry* 71, no. 23: 5315–5321.

Nederlof, M. M., W. H. van Riemsdijk, and L. K. Koopal. 1990. Determination of adsorption affinity distributions: A general framework for methods related to local isotherm approximations. *Journal of Colloid and Interface Science* 135, no. 2: 410–426. doi:10.1016/0021-9797(90)90011-C.

Nederlof, M. M., W. H. van Riemsdijk, and L. K. Koopal. 1992. Comparison of semianalytical methods to analyze complexation with heterogeneous ligands. *Environmental Science & Technology* 26, no. 4: 763–771. doi:10.1021/es00028a015.

Nederlof, M. M., W. H. van Riemsdijk, and L. K. Koopal. 1994. Heterogeneity analysis for binding data using an adapted smoothing spline technique. *Environmental Science & Technology* 28, no. 6: 1037–1047. doi:10.1021/es00055a012.

Nomula, S., and S. L. Cooper. 1997. Structural studies of polyurethane ionomer solutions. 1. Dynamic light scattering. *Macromolecules* 30, no. 5: 1355–1362.

Orsetti, S., E. M. Andrade, and F. V. Molina. 2009. Application of a constrained regularization method to extraction of affinity distributions: Proton and metal binding to humic substances. *Journal of Colloid and Interface Science* 336, no. 2: 377–387.

Orton, T. G., and R. M. Lark. 2007. Estimating the local mean for Bayesian maximum entropy by generalized least squares and maximum likelihood, and an application to the spatial analysis of a censored soil variable. *European Journal of Soil Science* 58, no. 1: 60–73.

Perdue, E. M. 2001. Modeling concepts in metal-humic complexation. In *Humic Substances and Chemical Contaminants*, eds. C. E. Clapp, M. H. B. Hayes, N. Senesi, P. R. Bloom, and P. M. Jardine, 305–316. Madison, WI: Soil Science Society of America.

Provencher, S. W. 1982a. A constrained regularization method for inverting data represented by linear algebraic or integral equations. *Computer Physics Communications* 27, no. 3: 213–227.

Provencher, S. W. 1982b. CONTIN: A general purpose constrained regularization program for inverting noisy linear algebraic and integral equations. *Computer Physics Communications* 27, no. 3: 229–242.

Puy, J., J. Galceran, C. Huidobro, E. Companys, N. Samper, J. L. Garcés, and F. Mas. 2008. Conditional affinity spectra of Pb2+-humic acid complexation from data obtained with AGNES. *Environmental Science and Technology* 42, no. 24: 9289–9295.

Puziy, A. M., T. Matynia, B. Gawdzik, and O. I. Poddubnaya. 1999. Use of CONTIN for calculation of adsorption energy distribution. *Langmuir* 15, no. 18: 6016–6025.

Rahnemaie, R., T. Hiemstra, and W. H. van Riemsdijk. 2006. Inner- and outer-sphere complexation of ions at the goethite-solution interface. *Journal of Colloid and Interface Science* 297, no. 2: 379–388. doi:10.1016/j.jcis.2005.11.003.

Rasteiro, M. G., C. C. Lemos, and A. Vasquez. 2008. Nanoparticle characterization by PCS: The analysis of bimodal distributions. *Particulate Science and Technology* 26, no. 5: 413–437.

Ravikovitch, P. I., A. Vishnyakov, R. Russo, and A. V. Neimark. 2000. Unified approach to pore size characterization of microporous carbonaceous materials from N_2, Ar, and CO_2 adsorption isotherms. *Langmuir* 16, no. 5: 2311–2320. doi:10.1021/la991011c.

Rey-Castro, C., S. Mongin, C. Huidobro, C. David, J. Salvador, J. L. Garcés, J. Galceran, F. Mas, and J. Puy. 2009. Effective affinity distribution for the binding of metal ions to a generic fulvic acid in natural waters. *Environmental Science & Technology* 43, no. 19: 7184–7191. doi:10.1021/es803006p.

Ruf, H., B. J. Gould, and W. Haase. 2000. Effect of nonrandom errors on the results from regularized inversions of dynamic light scattering data. *Langmuir* 16, no. 2: 471–480.

Rusch, U., M. Borkovec, J. Daicic, and W. H. Van Riemsdijk. 1997. Interpretation of competitive adsorption isotherms in terms of affinity distributions. *Journal of Colloid and Interface Science* 191, no. 1: 247–255.

Senesi, N., and E. Loffredo. 2005. Metal ion complexation by soil humic substances. In *Chemical Processes in Soils*, eds. M. A. Tabatabai and D. L. Sparks, 563–617. *SSSA Book Series 8*. Madison, WI: Soil Science Society of America.

Sparks, D. L. 1998. Kinetics and mechanisms of chemical reactions at the soil mineral/water interface. In *Soil Physical Chemistry*, ed. D. L. Sparks, 135–191. 2nd ed. Boca Raton, FL: CRC.

Sparks, D. L. 2002. *Environmental Soil Chemistry*. 2nd ed. San Diego, CA: Academic Press.

Sposito, G. 2008. *The Chemistry of Soils*. 2nd ed. New York: Oxford University Press, April 30.

Vakarin, E. V., and J. P. Badiali. 2006. Maximum entropy approach to characterization of random media. *AIP Conference Proceedings* 872: 411–418. doi:10.1063/1.2423301.

Van Riemsdijk, W. H., L. K. Koopal, D. G. Kinniburgh, M. F. Benedetti, and L. Weng. 2006. Modeling the interactions between humics, ions, and mineral surfaces. *Environmental Science & Technology* 40, no. 24: 7473–7480.

Wang, S., and C. N. Mulligan. 2008. Speciation and surface structure of inorganic arsenic in solid phases: A review. *Environment International* 34, no. 6: 867–879. doi:16/j.envint.2007.11.005.

Zachara, J. M., and J. C. Westall. 1998. Chemical modeling of ion adsorption in soils. In *Soil Physical Chemistry*, ed. D. L. Sparks, 47–96. 2nd ed. Boca Raton, FL: CRC Press.

12 Ion Binding to Minerals
Discrete Site Modeling

12.1 ION BINDING TO MINERAL COMPONENTS

In previous chapters, the main modes of interaction between ions and soil mineral colloids have been discussed: in Chapters 4 and 5, the principles of ion–surface interactions have been laid down, and in Chapters 8 and 9, the main features of adsorption onto silicate and oxide minerals have been reviewed. As introduced in Chapter 11, two main contributions should be considered, namely, the nonelectrostatic forces composed of physical (van der Waals forces) and/or chemical (specific bonding) interactions and the electrostatic forces arising from the charged nature of both the adsorbate and the surface; this is reflected in two contributions (considered independent) to the Gibbs free energy, as in Equation 11.5:

$$\Delta\tilde{G}^0 = \Delta G^0_{ch} + \Delta G^0_{el} \tag{12.1}$$

The "chemical" part ΔG^0_{ch} corresponds in most cases to specific adsorption on reactive surface hydroxyl groups, whereas the electrical part ΔG^0_{el} is due to the coulombic interaction. In the first part, it arises quite naturally the concept of "binding site" as the place on the surface where the adsorbate binds; these sites can be associated with individual hydroxyl groups, but not necessarily, because when bidentate adsorption takes place, as in the case of phosphate or arsenate adsorption onto iron oxides, two hydroxyls may be considered as forming a single site.

These problems are almost always treated by chemical modeling, as described in Chapter 11, and reviewed by numerous authors (Mattigod and Zachara 1996; Zachara and Westall 1998; Goldberg 2005; Merdy, Koopal, and Huclier 2006; Van Riemsdijk et al. 2006; Gu, Evans, and Barabash 2010 and references therein). Albeit in most cases electrostatic interactions are considered, there are some authors proposing nonelectrostatic modeling for these phenomena (Pagnanelli et al. 2006).

When cation-specific adsorption takes place on soil minerals, especially oxides, the increasing amount of bound metal (sometimes termed "loading") because of increased solution activity can lead to surface clustering, at first, and eventually to "surface precipitation" (Sposito 1987) of metal hydroxide or salt, in a process similar to condensation after gas adsorption, as described in Chapter 4 in connection with type II isotherms (Section 4.2.1). An example is presented in Figure 9.15, where surface polymerization (equivalent to clustering) is proposed to explain the behavior of Pb(II) adsorption onto goethite. Figure 12.1 shows a kaolinite crystal with surface precipitation of $Al(OH)_3$. Surface precipitation has been studied and modeled by many authors (Sposito 1987; Westall 1987; Lützenkirchen and Behra 1995;

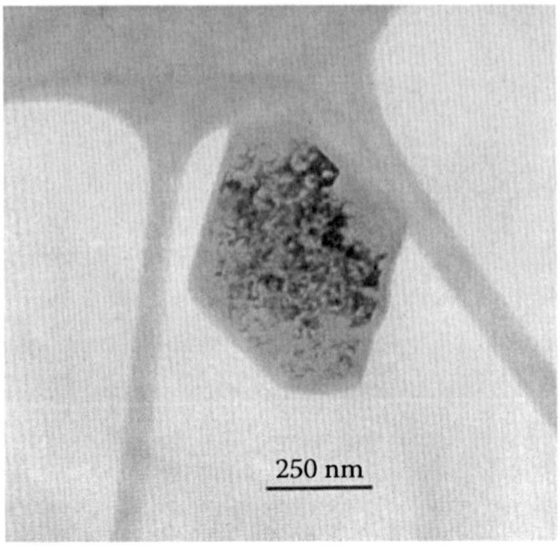

FIGURE 12.1 Transmission electron microscopy image of kaolinite particle showing surface precipitation of $Al(OH)_3$. Mottled appearance results from precipitation of the $Al(OH)_3$ phase on basal surface. (Reprinted from *Geochimica et Cosmochimica Acta*, 63, Thompson et al., Dynamic interactions of dissolution, surface adsorption, and precipitation in an aging cobalt(II)-clay-water system, 1767–1779. Copyright 1999, with permission from Elsevier.)

Karthikeyan, Elliott, and Chorover 1999; Li and Stanforth 2000) and will not be covered here.

12.2 GENERIC SURFACE DESCRIPTION: CLASSICAL ELECTROSTATIC MODELS

To treat the electrostatic effects in ion adsorption, a model of the interface is required, so as to relate the surface charge and potential with the system composition and conditions. Several models have been proposed in the literature, as we will develop in the following section.

12.2.1 Triple-Layer Model

12.2.1.1 Model Statement

The triple-layer model (TLM) was proposed by Davis, James, and Leckie (1978) and subsequently applied by many researchers (Westall 1987; Hayes et al. 1991; Mattigod and Zachara 1996; Sahai and Sverjensky 1997; Zachara and Westall 1998; Leroy and Revil 2004; Goldberg 2005; Tournassat et al. 2009; Kanematsu et al. 2010). The assumed interface structure is very similar to the Gouy–Chapman–Stern–Grahame (GCSG) double-layer model (Section 3.3) and is depicted in Figure 12.2 (compare with Figure 3.13). The plane "0" corresponds to the surface and is the place where the surface charge resides; this charge is composed of the intrinsic charge (protonic and permanent, if any) in addition to the inner-sphere complex ions. The plane "1" (also often termed "β") corresponds to the location of outer sphere complex ions, and the plane "2" or "d" is the start of the diffuse layer. Planes 1 and 2 resemble the inner and

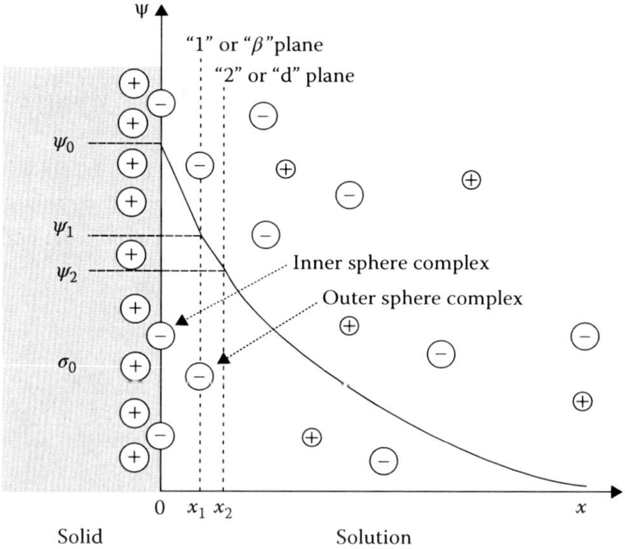

FIGURE 12.2 Scheme of the triple-layer model. The "0" plane is where the surface charge resides (including protonic, inner-sphere complexes and, if any, permanent charge); the "1" or β plane is where outer sphere complex ions are located, and the "2" or "d" plane is the start of the diffuse layer.

outer Helmholtz planes in the GCSG model, except for the location of adsorbed ions. The chemical part includes at first reactions for acid–base surface reactions, in either a 2-pK scheme

$$\text{X-OH} + \text{H}^+ \rightleftharpoons \text{X-OH}_2^+ \tag{12.2}$$

$$\text{X-OH} \rightleftharpoons \text{X-O}^- + \text{H}^+ \tag{12.3}$$

or 1-pK-type scheme:

$$\text{X-OH}^{-1/2} + \text{H}^+ \rightleftharpoons \text{X-OH}_2^{+1/2} \tag{12.4}$$

Then, the specific (inner-sphere) binding may be represented by diverse reactions; for example, single-site complexation of a cation M^{z+}

$$\text{X-OH} + \text{M}^{z+} \rightleftharpoons \text{X-O-M}^{(z-1)+} + \text{H}^+ \tag{12.5}$$

or two-site anion binding

$$2\text{X-OH} + \text{H}_2\text{YO}_4^- \rightleftharpoons (\text{X-O})_2\text{-YO}_2^- + 2\text{H}_2\text{O} \tag{12.6}$$

where Y may be P or As, for example; other anions such as SeO_4^{2-} give similar reactions.

In addition, the formation of outer sphere complexes (strictly, ion pairing) with ions that are either active (those adsorbing specifically) or from indifferent electrolyte (such as KNO_3) is also modeled as chemical reactions:

$$\text{X-OH}_2^+ + \text{YO}_4^{z-} \rightleftharpoons \text{X-OH}_2^+\text{YO}_4^{z-} \tag{12.7}$$

$$\text{X-O}^- + \text{M}^{z+} \rightleftharpoons \text{X-O}^-\text{M}^{z+} \tag{12.8}$$

$$\text{X-OH}_2^+ + \text{NO}_3^- \rightleftharpoons \text{X-OH}_2^+\text{NO}_3^- \tag{12.9}$$

$$\text{X-O}^- + \text{K}^+ \rightleftharpoons \text{X-O}^-\text{K}^+ \tag{12.10}$$

All these reactions are treated as chemical equilibria, such as (for Reaction 12.2 as example)

$$K_{\text{H}^+} = \frac{a_{\text{X-OH}_2^+}}{a_{\text{X-OH}}a_{\text{H}^+}} \approx \frac{[\text{X-OH}_2^+]}{[\text{X-OH}]a_{\text{H}^+}} \approx \frac{[\text{X-OH}_2^+]}{[\text{X-OH}][\text{H}^+]} \tag{12.11}$$

In Equation 12.11, the thermodynamically sound variables are the activities; the surface species X-OH_2^+ and X-OH are treated as solution reactant and product species (of course, with concentration given by $[\text{X-OH}_2^+] = \Gamma_{\text{X-OH}_2^+} A_S c_s$ where Γ is the surface concentration [mol·m^{-2}]; A_S is the specific surface area [m^2·g^{-1}]; and c_S is the solid concentration [g·dm^{-3}]), but the calculation of activity coefficients is troublesome, because the Debye–Hückel theory cannot be applied to surface adsorbates; thus, the thermodynamically correct expression in activities in the second member is approximated as shown in the following member, with the surface species expressed as concentrations; often, the whole expression is written in concentrations, as in the rightmost member of Equation 12.11. In doing so, the parameters deduced from fitting to experimental data are, strictly speaking, conditional, valid in the experimental conditions employed to obtain them. However, usually it is difficult to have sufficiently accurate experimental results, so as to be able to discriminate conditional parameters found under different conditions; thus, the use of concentrations appears sometimes acceptable.

The equilibrium constant in Equation 12.11 (and also in all surface reactions) is, as shown in Equation 11.16, an intrinsic chemical constant times an electrostatic Boltzmann factor:

$$K = K^{int} e_f(\psi) = K^{int} \exp\left(-\frac{zF\psi}{RT}\right) \tag{12.12}$$

The electrostatic potential ψ should be that corresponding to the plane where the appropriate charge is located. Thus, for Reactions 12.2 through 12.6, the potential is that at the surface plane, ψ_0, whereas for Reactions 12.7 and 12.8, the relevant potential is ψ_1 at the inner plane. These potentials are related to the model structure as follows: the spaces between planes 0 and 1, and between 1 and 2, are treated as capacitors with capacitances C_1 and C_2, respectively; thus, the following electrostatic relations hold:

$$\sigma_0 = C_1\left(\psi_0 - \psi_1\right) \tag{12.13}$$

$$\sigma_0 + \sigma_1 = C_2\left(\psi_1 - \psi_2\right) \tag{12.14}$$

where σ_0 and σ_1 are the net surface charges at the 0 and 1 planes, respectively, and C_1 and C_2 are the respective surface capacitances (F·m^{-2}). The charge balance at the interface is

$$\sigma_0 + \sigma_1 + \sigma_d = 0 \tag{12.15}$$

where σ_d is the diffuse charge given, for a z:z electrolyte, by Equation 3.43

$$\sigma_d = -\left(8c^{\infty}\varepsilon \boldsymbol{RT}\right)^{1/2}\sinh\left(\frac{z\boldsymbol{F}\psi_2}{2\boldsymbol{RT}}\right) \tag{12.16}$$

where c^{∞} and z are the electrolyte bulk concentration and ionic charge, respectively, and ε is the dielectric constant ($=\varepsilon_0\varepsilon_r$). The net surface charge is given by the surface charge balance:

$$\begin{aligned}\sigma_0 = \frac{\boldsymbol{F}}{A_S c_S}\Big\{&\left[\text{X-OH}_2^+\right]+\left[\text{X-OH}_2^+\text{YO}_4^{z-}\right]+(z-1)\left[\text{X-O-M}^{(z-1)+}\right]\\ &-\left[\text{X-O}^-\right]-\left[(\text{X-O})_2\text{-YO}_2^-\right]-\left[\text{X-O}^-\text{M}^{z+}\right]\Big\}\end{aligned} \tag{12.17}$$

Here, the surface species presented in Equations 12.2 through 12.8 are included, but of course the adsorbed species are dependent on the system at hand; it should be also noted that the outer sphere complexes are included by the "surface protonic charge" they are compensating; the outer sphere complexes form σ_1 (see below). The initial factor converts concentrations to surface charge: A_S is the surface specific area and c_S is the solid mass concentration. The net charge at the inner plane is given by the following:

$$\sigma_1 = \frac{\boldsymbol{F}}{A_S c_S}\left\{z_+\left[\text{X-O}^-\text{M}^{z+}\right]-z_-\left[\text{X-OH}_2^+\text{YO}_4^{z-}\right]\right\} \tag{12.18}$$

Finally, the mass balance at the surface should take into account the fact that there is a fixed number of surface sites

$$\begin{aligned}\Gamma_S A_S c_S =& \left[\text{X-OH}_2^+\right]+\left[\text{X-O}^-\right]+\left[\text{X-OH}_2^+\text{YO}_4^{z-}\right]\\ &+\left[\text{X-O}^-\text{M}^{z+}\right]+\left[\text{X-O-M}^{(z-1)+}\right]+2\left[(\text{X-O})_2\text{-YO}_2^-\right]\end{aligned} \tag{12.19}$$

where Γ_S is the number of sites per unit area. The complete model has as adjustable parameters all the intrinsic equilibrium constants, Γ_S, C_1, and C_2, albeit this last parameter is frequently taken as 0.2 F·m^{-2} (Hayes et al. 1991; Sahai and Sverjensky 1997), a value actually dating back to Grahame (1947, 1954). The full nonlinear equation system can be solved by computer numerical methods, as proposed originally by Westall and coworkers (Westall 1987; Zachara and Westall 1998), and currently implemented in several computer applications, such as the original Westall's FITEQL (Herbelin and Westall 1999), the U.S. Environmental Protection Agency (EPA) MINTEQA2 (EPA Center for Exposure Assessment Modeling 2006), PHREEQC (Parkhurst et al. 2008) and Visual MINTEQ (Gustafsson 2011; also check here for other software sites)

12.2.1.2 Applications of the Triple-Layer Model

The TLM has been widely applied to sorption analysis in environmental studies (Westall 1987; Mattigod and Zachara 1996; Sahai and Sverjensky 1997; Zachara and Westall 1998; Leroy and Revil 2004; Goldberg 2005; Sverjensky and Fukushi 2006; Goldberg and Kabengi 2010; Kanematsu et al. 2010; Nagata and Fukushi 2010; Mitchell, Goldberg, and Al-Abadleh 2011). It is a rather general model, in the sense that it includes both inner and outer sphere complexation, and conforms to the accepted GCSG double-layer model. On the other hand, it results in a relatively high number of adjustable parameters, which raises caution regarding the physical reality of the quantitative parameter results. Nevertheless, it has been shown to describe well numerous adsorption processes. Originally, it was developed with nonspecific, outer sphere surface complexation in mind. In an investigation of chromate adsorption on subsurface soil horizons (Zachara et al. 1989), the TLM was found to produce good qualitative predictions (Figure 12.3) on the effects of adsorbate concentration, pH, solids concentration, ionic strength, and competing ions on CrO_4^{2-} adsorption in soil, using outer sphere complexation constants from pure-phase Fe oxides (goethite, noncrystalline Fe_2O_3-H_2O [amorphous]) and a site density fitted from the clay fractions. Chromate adsorption on both the oxidic and montmorillonitic soils could also be simulated using CrO_4^{2-} adsorption constants on kaolinite and a site density that was 10 times the crystallographic estimate.

The TLM model has been applied not only to inorganic anions but also to organic species. In a study of 2,4-dichlorophenol, 2,4,6-trichlorophenol, and pentachlorophenol sorption in an allophanic soil (Cea et al. 2010), the isotherm experiments were well described (as shown in Figure 12.4) by the TLM in which monodentate outer and inner sphere complexes were considered to form between deprotonated organic molecules and active sites on the variable-charge soil. The calculated thermodynamic parameters suggest that chlorophenol sorption is a spontaneous ($\Delta G < 0$), endothermic ($\Delta H > 0$), and entropy-driven reaction ($\Delta S > 0$).

Both types of adsorption should not necessarily be considered in the TLM. In a study of As-organic compounds (Mitchell, Goldberg, and Al-Abadleh 2011), the triple-layer surface complexation model was applied to adsorption isotherm and pH envelope data for dimethylarsinic acid (DMA) and *p*-arsanilic acid (*p*-AsA) on

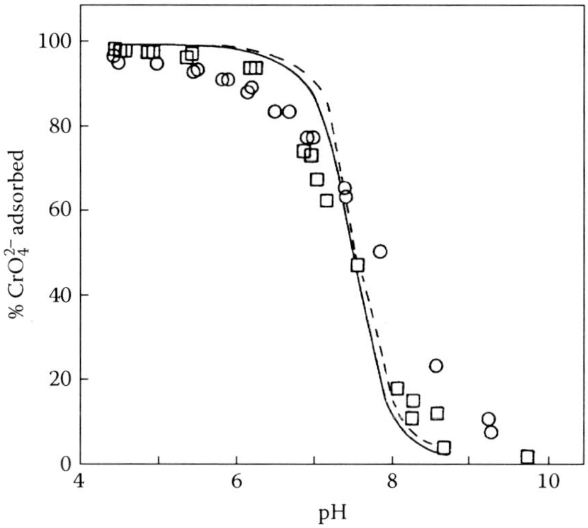

FIGURE 12.3 Chromate adsorption on two B horizons of North American soils. Experimental data (symbols) and the triple-layer model predicted results (lines) for (□, ——) a Tennessee Btx sandy horizon and (○, - - -) a North Carolina silty Be horizon. (Reprinted from Zachara, J. M. et al. *Soil Science Society of America Journal*, 53, 418–428, 1989. With kind permission from the Soil Science Society of America.)

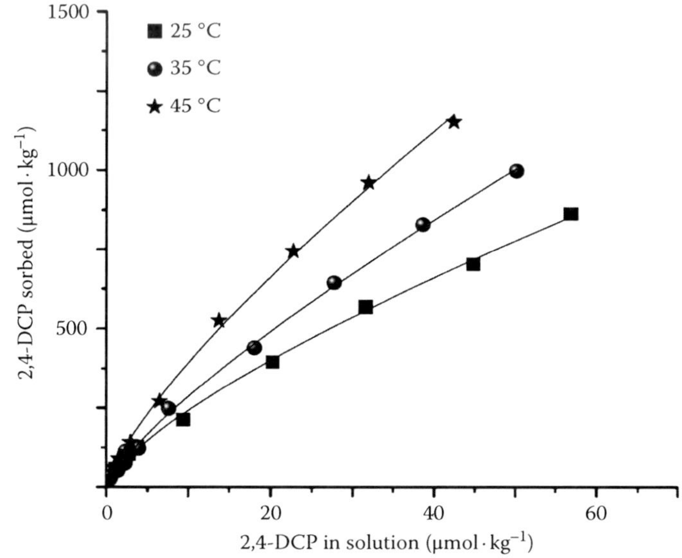

FIGURE 12.4 2,4-Dichlorophenol isotherms in allophanic soil, at pH 4.5, in 0.1 mol·L^{-1} KCl background electrolyte at temperatures of 25°C, 35°C, and 45°C. (Reprinted from *Chemosphere*, 78, cea, M. et al., Kinetic and thermodynamic study of chlorophenol sorption in an allophanic soil, 86–91. Copyright 2010, with permission from Elsevier.)

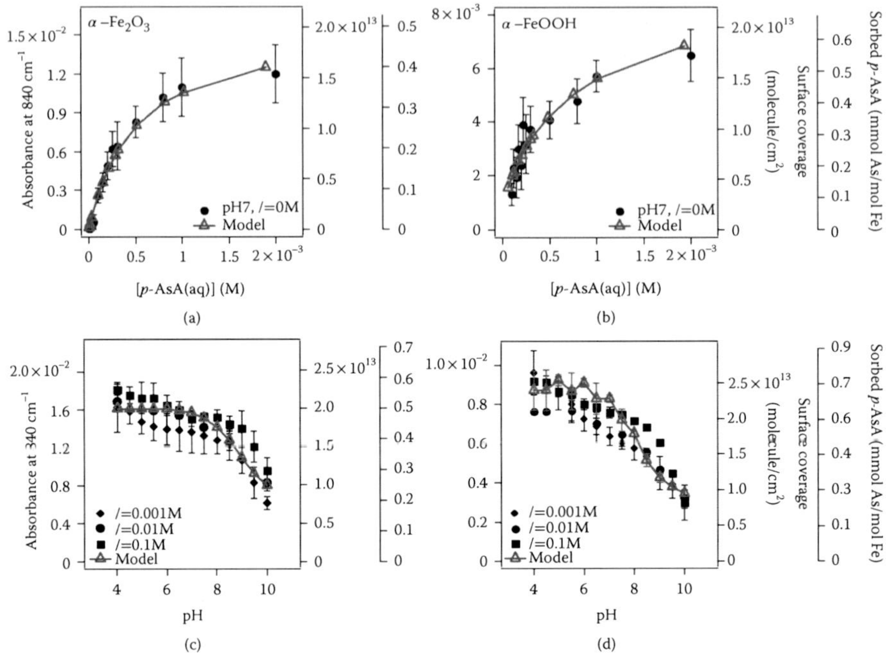

FIGURE 12.5 Adsorption isotherms (measured by the absorbance of the As = O stretching vibration at 840 cm^{-1}) of *p*-arsanilic acid (*p*-AsA) on hematite (a and c) and goethite (b and d) at 298 K as a function of concentration at pH = 7 and $I = 0.01$ M (upper panel), and as a function of pH and I using 1 mM *p*-AsA solution (lower panel). Closed symbols are experimental data (Depalma et al. 2008; Chabot, Hoang, and Al-Abadleh 2009). Empty triangles represent the fit to the triple-layer model. (Reprinted from *Journal of Colloid and Interface Science*, 358, Mitchell, W. et al., In situ ATR–FTIR and surface complexation modeling studies on the adsorption of dimethylarsinic acid and *p*-arsanilic acid on iron-(oxyhydr)oxides, 534–540. Copyright 2011, with permission from Elsevier.)

hematite and goethite. Kinetic and equilibrium adsorption measurements were done by attenuated total reflection–Fourier transform infrared (ATR–FTIR) spectroscopy; from spectra analysis, ligand exchange reactions were proposed. The best fit to the DMA adsorption data was obtained using outer sphere complex formation, whereas that for *p*-AsA was obtained using two monodentate inner-sphere surface complexes (Figure 12.5); it is noteworthy that inclusion of both complex types did not yield good fitting.

12.2.1.3 Further Advances: Extended Triple-Layer Model

Recently, the extended triple layer model (ETLM) was introduced by Sverjensky and coworkers (Sverjensky 2005; Sverjensky and Fukushi 2006), incorporating prediction of intrinsic equilibrium constants based on crystal chemical, electrostatic, and thermodynamic theory, allowing the reduction of adjustable parameters. In the ETLM, proton sorption Reactions 12.2 and 12.3 are written as associations:

$$\text{X-OH} + H^+ \rightleftharpoons \text{X-OH}_2^+ \quad K_{H,1} \tag{12.20}$$

$$X\text{-}O^- + H^+ \rightleftharpoons X\text{-}OH \quad K_{H,2} \tag{12.21}$$

The thermodynamic equilibrium constants are related to the point of zero net proton charge (pznpc) (Section 5.5) by

$$\log K_{H,1} = pH_{H,0} - \frac{\log K_n}{2} \tag{12.22}$$

$$\log K_{H,2} = pH_{H,0} + \frac{\log K_n}{2} \tag{12.23}$$

where $\log K_n$ is a measure of the separation between constants; $pH_{H,0}$ and K_n can be determined experimentally and predicted as follows. The Gibbs free energy change in Reactions 12.20 and 12.21 is considered to have three contributions:

$$\Delta\tilde{G}^0 = \Delta G_S^0 + \Delta G_{p,el}^0 + \Delta G_{p,i}^0 \tag{12.24}$$

where ΔG_S^0 is a Born solvation contribution, $\Delta G_{p,el}^0$ is an electrostatic proton interaction term, and $\Delta G_{p,i}^0$ is an intrinsic proton interaction term. It can be shown that

$$pH_{H,0} = 2\left[-\frac{\Delta\Omega_{H,0}}{\ln 10 \boldsymbol{RT}\varepsilon} - B_{H,0}\left(\frac{s}{r_{H^+}}\right) + \log K''_{H,0}\right] \tag{12.25}$$

$$\log K_n = -B_n\left(\frac{s}{r_{H^+}}\right) + \log K''_n \tag{12.26}$$

where $\Delta\Omega_{H,0}$ represents a Born solvation coefficient for the proton-binding reaction; ε, s, and r_{H^+} are the solid dielectric constant, the Pauling bond strength, and the bond distance, respectively. $B_{H,0}$, B_n, $K''_{H,0}$, and K''_n are coefficients to be determined by fitting to experimental data. For outer sphere ion binding, the treatment is similar. For specific binding, the electrostatic work involved in the removal of water dipoles is considered. For example, for bidentate As(V) binding, the reaction

$$2X\text{-}OH + H_2AsO_4^- \rightleftharpoons (X\text{-}O)_2\text{-}AsO_2^- + 2H_2O \tag{12.27}$$

is split in two stages:

$$2X\text{-}OH + H_2AsO_4^- + 2H^+ (\text{to plane } 0) \rightleftharpoons (X\text{-}OH_2^+)_2AsO_4^{3-} + 2H_2O + 2H^+ (\text{from plane } 1) \tag{12.28}$$

$$(X\text{-}OH_2^+)_2AsO_4^{3-} \rightleftharpoons (X\text{-}O)_2AsO_2^- + 2H_2O \tag{12.29}$$

In the first stage, Reaction 12.28, the surface is protonated, and the anion adsorbs in the 1 plane as outer sphere complex, losing two protons; the protons cancel out but are retained here to remark the interface planes involved, which define the contributions to the electrostatic work. In Reaction 12.29, the anion binds specifically, and two water molecules are removed from the 0 plane; because water has a dipolar moment, there is an electrostatic work involved, which is not considered in the original TLM. The electrostatic factor involved in Reaction 12.28 results from $H_2AsO_4^-$ adsorbing at the 1 plane, two protons adsorbing at the 0 plane, and two other protons desorbing from the 1 plane:

$$\Delta\psi_H = 2\psi_0 - \psi_1 - 2\psi_1 \tag{12.30}$$

In the case of Reaction 12.29, the molar electrostatic work released by the desorption of n water molecules from the 0 plane is given, according to Sverjensky and Fukushi (2006), by

$$W_{el} \approx -n\boldsymbol{F}\left(\psi_0 - \psi_1\right) \tag{12.31}$$

Thus, the water desorption electrostatic factor is

$$\Delta\psi_W = -2(\psi_0 - \psi_1) \tag{12.32}$$

The total electrostatic factor is

$$\Delta\psi = \Delta\psi_H + \Delta\psi_W = 2\psi_0 - 3\psi_1 - 2(\psi_0 - \psi_1) = -\psi_1 \tag{12.33}$$

which differs from the traditional TLM assumption $\Delta\psi = -\psi_0$. In general, the electrostatic predictions of the ETLM are at variance with the "traditional" TLM; this extended model has been applied in a number of recent studies (Sverjensky and Fukushi 2006; Kanematsu et al. 2010; Nagata and Fukushi 2010). Figure 12.6 shows some results of an application of the ETLM to arsenate and phosphate adsorption on a goethite-based granular porous adsorbent (Kanematsu et al. 2010); the adsorption was modeled considering surface protonation reactions such as Reactions 12.20 and 12.21, outer sphere complexation of supporting electrolyte ions such as Reactions 12.9 and 12.10, and three surface species of As and P: $(X\text{-}O)_2\text{-}AsO_2^-$, $(X\text{-}O)_2\text{-}AsO_2H$, and $X\text{-}O\text{-}AsO_3^{2-}$. The ETLM parameters were deduced from one set of curves (Figure 12.6a for As) and used to successfully predict results under different experimental conditions (such as Figure 12.6b).

12.2.2 Simplified Modeling: Basic Stern and Constant Capacitance Models

In the basic Stern model (BSM), also known as diffuse layer model (DLM), the "1" or β plane of the TLM is suppressed, as shown in Figure 12.7 (Hayes et al. 1991; Mattigod and Zachara 1996; Zachara and Westall 1998). Thus, outer-sphere complexation is not considered, and two adjustable parameters are eliminated (both capacitances C_1 and C_2). This simplification is reasonable when there is strong inner-sphere adsorption

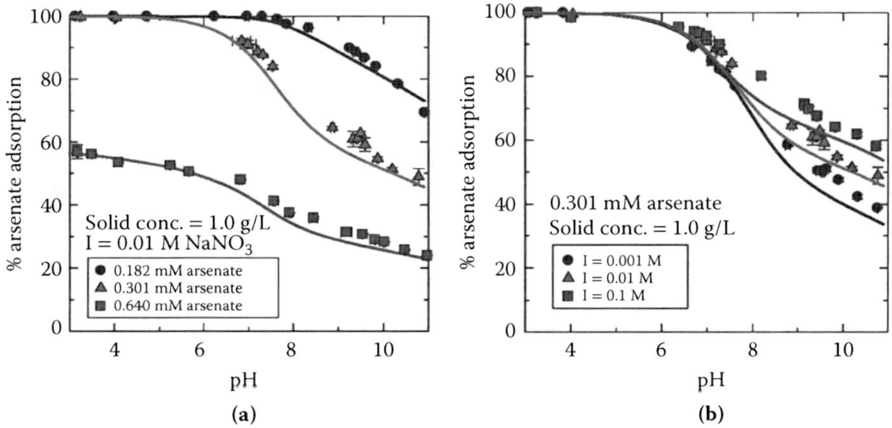

FIGURE 12.6 Arsenate adsorption edge results at the solid concentration of 1.0 g·L^{-1}: (a) at three different arsenate concentrations and (b) arsenate adsorption edge results at ionic strength of 0.001, 0.01, and 0.1 M. The experimental results shown in panel a were used to estimate the extended triple-layer model parameters used to calculate the curves in panels a and b. (Reprinted with permission from Kanematsu, M. et al., *Environ. Sci. Technol.*, 44, 2010. Copyright 2010 American Chemical Society.)

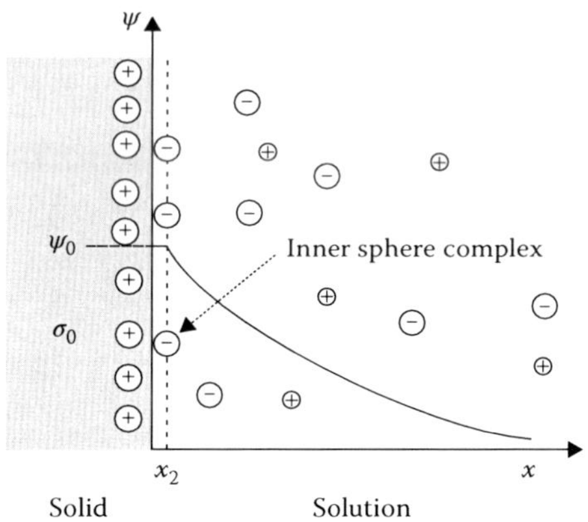

FIGURE 12.7 Scheme of the basic Stern or diffuse layer model. The "0" plane is where the surface charge resides (protonic and, if any, permanent) and the "2" or "d" plane is the start of the diffuse layer. Notice that the "1" plane of the triple-layer model (compare with Figure 12.2) is not considered here.

and outer sphere binding is not appreciable. In this model, the equations considered include surface acid–base equilibria (Reactions 12.2 and 12.3), inner-sphere complexation Reactions such as 12.5 and 12.6, and the diffuse layer Equations 12.15 and 12.16, with $\sigma_1 = 0$ and $\psi_2 = \psi_0$. Alternatively, ψ_2 may be considered different from ψ_0, both related through Equation 12.13, with ψ_2 instead of ψ_1 and retaining C_1. The surface mass balance (Equation 12.19) also holds, excluding the ion pair (outer sphere) species. The BSM has been applied to many problems where specific adsorption is the dominant process (Rytwo 2004; Hizal and Apak 2006; Villaseñor Nano and Strathmann 2006; Schaller et al. 2009; Akafia, Reich, and Koretsky 2011).

The constant capacitance model (CCM) is a further simplification, where the interfacial capacitance is assumed constant, related to the surface potential by

$$\sigma_0 = C_1 \psi_0 \tag{12.34}$$

The picture of the interface as assumed in the CCM is that shown in Figure 12.8, with the exponential decay of Figure 12.7 replaced by the potential drop across a capacitor. This implies that the diffuse layer is confined to a short distance from the interface; in turn, the model requires relatively high and constant ionic strength, which are the experimental conditions for this model to be applicable. Furthermore, different ionic strengths require different C_1 values. Notwithstanding that, the CCM model has been quite frequently applied to ion-sorption experiments (Goldberg 2005). Goldberg (2002) studied the adsorption of As species on soil colloids and applied the CCM for data analysis. Arsenate adsorption on oxides and clays was maximal at low pH and decreased with increasing pH above 9 for Al oxide, 7 for Fe oxide, and 5 for clays. Arsenite adsorption exhibited parabolic behavior with an adsorption maximum around pH 8.5 for all materials. The CCM was able to fit the arsenate and arsenite adsorption envelopes to obtain values of the intrinsic As surface complexation constants. Figure 12.9 shows results for As(III) and As(V) sorption to amorphous aluminum oxide and kaolinite; the fitting is good for Al oxide but only fair for kaolinite; notice the very different extent of adsorption in each case. In general, the CCM has been shown to be able to provide reasonable fit on cases where strongly specific adsorption is the main process.

Figure 12.10 shows a typical application of the BSM (DLM) to As adsorption on soil components also, here iron oxyhydroxides: hydrated ferric oxide (HFO) and goethite (Dixit and Hering 2003). A rather good agreement is observed for both As(V) and As(III) adsorption onto HFO, goethite, and magnetite. In general, good agreement is found for specific adsorption problems, somewhat better than the CCM.

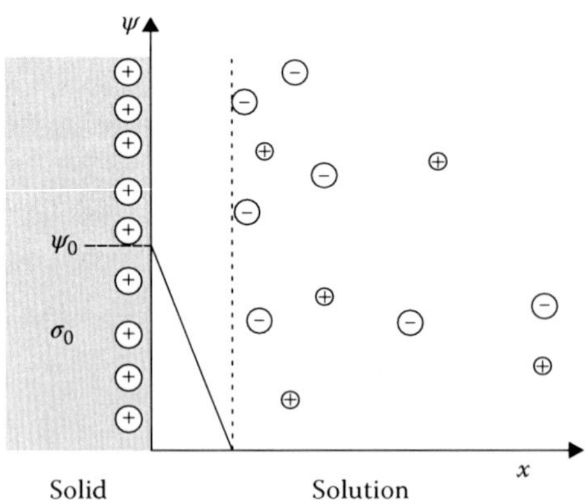

FIGURE 12.8 Scheme of the constant capacitance model. Note that the diffuse layer region is replaced by a constant capacitance, where the full potential drop is confined.

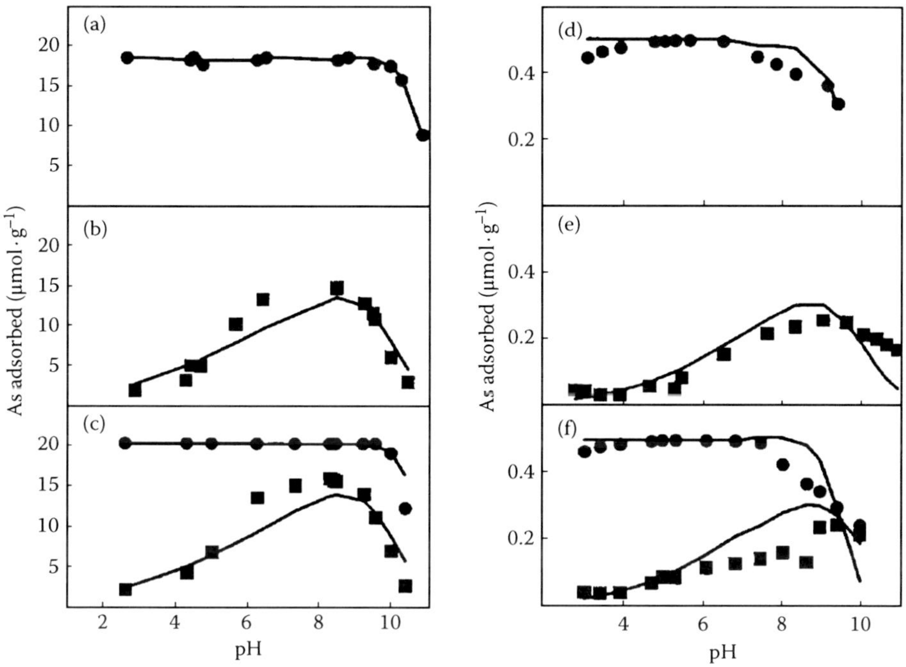

FIGURE 12.9 Adsorption of arsenic on (a–c) amorphous aluminum oxide and (d–f) kaolinite as a function of pH. (a) and (d) As(V) alone; (b) and (e) As(III) alone; (c) and (f) As(V) and As(III) together. Squares: As(III) experimental points; circles: As(V) data points; lines: predicted by the constant capacitance model. (Reprinted from Goldberg, S., *Soil Science Society of America Journal*, 66, 413–421, 2002. With kind permission from the Soil Science Society of America.)

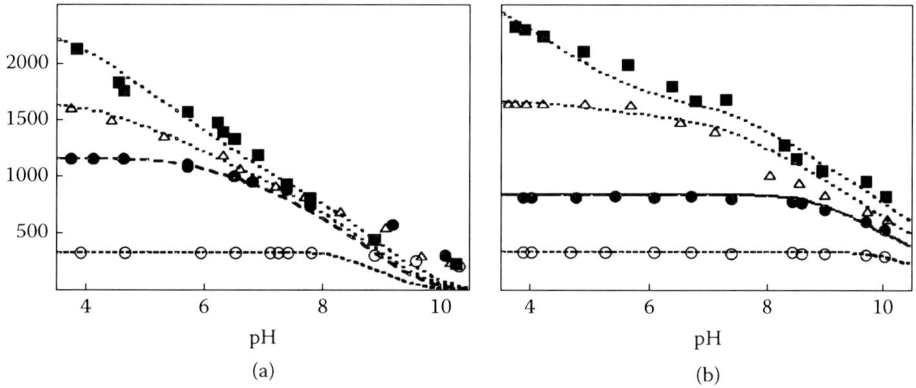

FIGURE 12.10 Arsenate sorption onto (a) hydrated ferric oxide (HFO) and (b) goethite for total As(V) concentrations (in μm) of (■) 100, (Δ) 50, and (●) 35 for HFO and 25 for goethite, (O) 10. Experimental conditions: 0.01 M $NaClO_4$; 0.03 g L^{-1} HFO or 0.5 g L^{-1} goethite. The lines represent fits to the diffuse layer surface complexation model. (Reprinted with permission from Dixit, S. and J. G. Hering, *Environ. Sci. Technol.,* 37, 2003. Copyright 2010 American Chemical Society.)

12.2.3 Stern Extended or Quad Layer Model

This model, also known as variable surface charge–variable surface potential model (Goldberg 2005) was introduced in Australia by Bowden, Barrow, and coworkers (Bowden, Posner, and Quirk 1977; Barrow et al. 1980; Bowden et al. 1980; Barrow 1983) but has not been applied much by other authors. As shown in Figure 12.11, a fourth layer is introduced (comparing with Figure 12.2): the "a" plane where the charge centers of specifically (inner-sphere) adsorbed ions are located; at the surface (0) plane, only protons are adsorbed. Further, a surface site is considered to have the structure shown in Figure 12.12, composed of a surface OH in addition to an adsorbed water molecule, which can acquire positive or negative charge as a whole through the reactions shown in Figure 12.12; note that the deprotonation is treated as OH^- adsorption.

Further, this model is based on adsorption modeling (Section 11.1); thus, the binding of ions at each plane is treated as Langmuir competitive adsorption, with consideration for electrostatic interactions in the adsorption equilibrium constants as in Equations 11.16 and 11.17. Thus, the charge at the 0 plane is given by the balance between adsorbed protons and "adsorbed" hydroxide ions

$$\sigma_0 = \frac{\Gamma_S\left[K_H a_{\mathrm{H^+}} \exp(-\boldsymbol{F}\psi_0/\boldsymbol{RT}) - K_{OH} a_{\mathrm{OH^-}} \exp(\boldsymbol{F}\psi_0/\boldsymbol{RT})\right]}{1 + K_H a_{\mathrm{H^+}} \exp(-\boldsymbol{F}\psi_0/\boldsymbol{RT}) + K_{OH} a_{\mathrm{OH^-}} \exp(\boldsymbol{F}\psi_0/\boldsymbol{RT})} \tag{12.35}$$

where K_H and K_{OH} are the proton and hydroxyl intrinsic adsorption constants and Γ_S is the total number of sites per unit area. Likewise, the charge at the "a" plane is given by

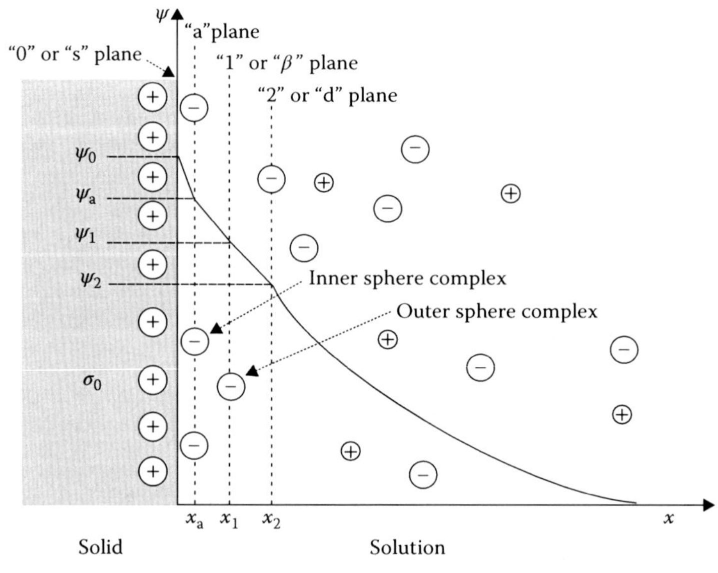

FIGURE 12.11 Scheme of the quad layer model. The "0" plane is where the surface charge resides (protonic and permanent charge, if present), the "a" plane corresponds to inner-sphere complexes, the "1" or β plane is where outer sphere complex ions are located, and the "2" or "d" plane is the start of the diffuse layer.

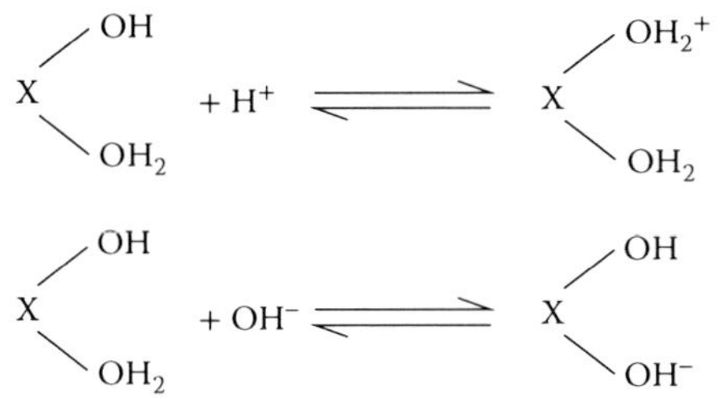

FIGURE 12.12 Scheme of surface protonation and deprotonation reactions according to the quad layer model. (From Bowden, J. W. et al., *Soil Res.*, 15, 2, 121–136, 1977.)

$$\sigma_a = \frac{\Gamma_T \sum_i z_i K^{int} a_i \exp(-z_i F\psi_a/RT)}{1+\sum_i K^{int} a_i \exp(-z_i F\psi_a/RT)} \tag{12.36}$$

where Γ_T is the number of specific adsorption sites per unit area (not necessarily equal to Γ_S), K^{int} is the intrinsic adsorption constant, and z_i is the charges of the specifically adsorbed ions i. For the "1" plane, we have the same type of expression, that is,

$$\sigma_1 = \frac{\Gamma_S \sum_j z_j K_j a_j \exp(-z_j F\psi_1/RT)}{1+\sum_i K_j a_j \exp(-z_j F\psi_1/RT)} \tag{12.37}$$

where j denotes outer sphere complex ions. Finally, the diffuse charge is given by Equation 12.16 (for 1:1 electrolytes). Obviously, the charge balance holds:

$$\sigma_0 + \sigma_a + \sigma_1 + \sigma_d = 0 \tag{12.38}$$

Between the planes, fixed surface capacitances are assumed; thus, the following charge–potential relationships are included:

$$\sigma_0 = C_1(\psi_0 - \psi_a) \tag{12.39}$$

$$\sigma_0 + \sigma_a = C_a(\psi_a - \psi_1) \tag{12.40}$$

$$\sigma_d = -C_2(\psi_1 - \psi_2) \tag{12.41}$$

The model parameters are Γ_S, Γ_T, the intrinsic constants K_i, K_j, and the capacitances, which are obtained from fitting to experimental data. The model was further extended (Barrow 1983, 1986; Barrow and Ellis 1986a) to include adsorption kinetics, diffusive penetration of adsorbates into the solid, and surface heterogeneity, with Gaussian affinity spectra.

Due in part to the large number of adjustable parameters, this model provides in general good fitting to experimental data. Figure 12.13 shows an example for zinc adsorption onto an Australian soil, with good fitting (Barrow and Ellis 1986b).

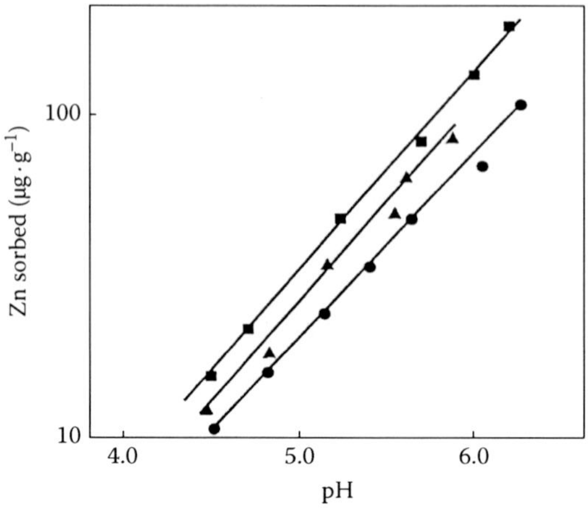

FIGURE 12.13 Zinc adsorption on an Australian soil as a function of pH for different supporting electrolytes: (■) water; (▲) 1.0 M $NaNO_3$; (●) 1.0 M NaCl. Lines: fit to the quad layer model. (Reprinted with permission from Barrow, N. J. and Ellis, A. S., *J.Soil Sci.*, 37, 2, 303–310, 1986b. Copyright 1986, John Wiley and Sons.)

12.3 DETAILED SURFACE DESCRIPTION: CD-MUSIC MODEL

12.3.1 Model Formulation

In the charge distribution—multisite complexation (CD-MUSIC) model of Hiemstra and Van Riemsdijk (Hiemstra and Van Riemsdijk 1996; Venema, Hiemstra, and van Riemsdijk 1996, 1997), a different conception of the interface structure is introduced, based on detailed spectroscopic studies of the solid surface and adsorbate binding sites, along with the concept of charge distribution between different O ligands of an adsorbed species. The spectroscopic analysis allows identification of the structure of inner-sphere complexes, which leads to a precise formulation of the surface reactions. The assumed structure is presented in Figure 12.14a; it resembles the TLM, but differs in the treatment of inner-sphere complexes: for adsorbed anions such as phosphate, the four O atoms bound to the P atom are divided in two parts: those bound to the solid are considered to reside in the 0 plane, whereas the other two are located in the 1 plane; these two types of O ligands are assumed to have, in general, different formal charges as a result of their locations. That leads to the "charge distribution" concept, shown in Figure 12.14b: here, two surface OH sites of a sixfold coordinate trivalent metal oxide (where Me may be, for example, Fe(III) or Al(III)) have reacted with a $Si(OH)_4$ molecule by a ligand exchange reaction

$$2MeOH + Si(OH)_4 \rightleftharpoons (MeO)_2Si(OH)_2 + 2H_2O \tag{12.42}$$

releasing two water molecules and giving the surface species shown in the figure. The O atoms bridging the Me and Si atoms are part of the 0 plane, whereas the OH groups to the right are located in the 1 plane. The model applies Pauling rules to the solid (Sposito 2008), in particular the concept of formal bond valence; thus, an atom *j*

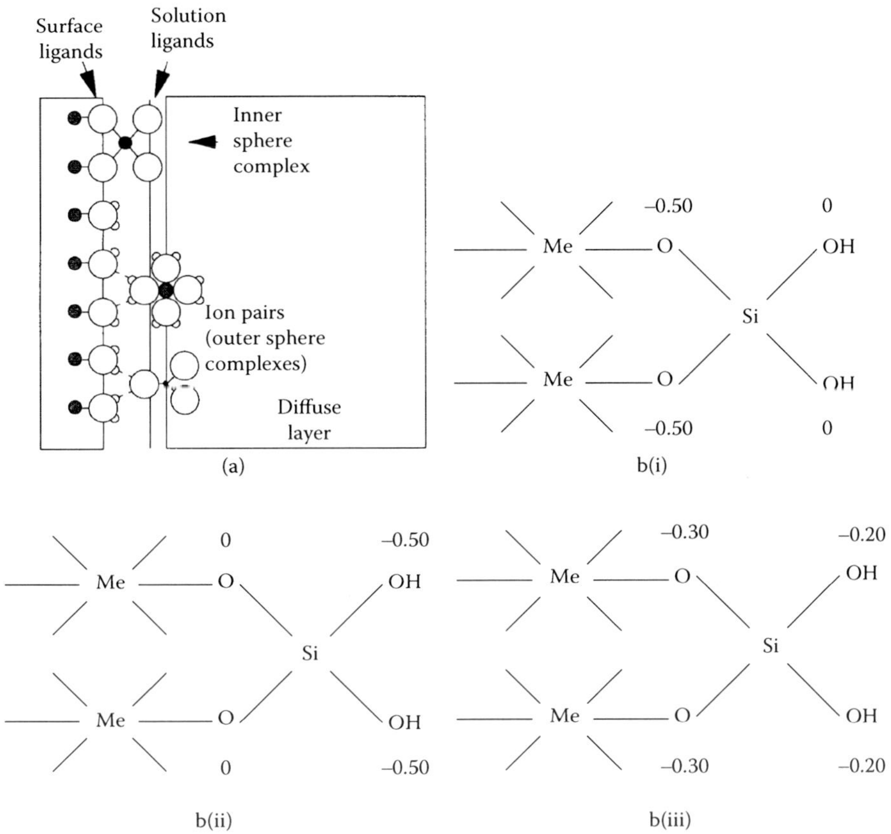

FIGURE 12.14 (a) Scheme of the interface as assumed in the CD-MUSIC model. Note that the planes do not pass through the center of inner-layer complexed ions (such as the P atom in adsorbed PO_4^{3-}) but through the O atoms; for example, in an adsorbed PO_4^{3-} anions, the two O atoms closer to the solid are considered as part of the surface plane, whereas the other two are assumed as lying in the 1 plane. The outer sphere complex ions (here termed ion pairs) are assumed to lie in the outer (2) plane. (b) A schematic representation of a bidentate silicon oxo complex bound to a Me hydroxide with trivalent Me ions in sixfold coordination ($\nu_{Me} = 3/6 = 0.5$), as modeled in the CD-MUSIC approach. The charge of the various ligands is given for various choices of the value of the charge distribution factor f: (i) equal distribution of the charge of the central ion over the surrounding ligands; (ii) an asymmetric distribution with neutral surface/solid oxygen; (iii) the actual distribution with $f = 0.6$. (Reprinted from *Journal of Colloid and Interface Science*, 179, Hiemstra, T. and Van Riemsdijk, W. H., A surface structural approach to ion adsorption: The charge distribution (CD) model, 488–508. Copyright 1996, with permission from Elsevier.)

in a crystal with a charge z_j surrounded by Z ligands leads to a formal bond valence ν_j given by

$$\nu_j = \frac{z_j}{Z} \tag{12.43}$$

For a trivalent Me(III) cation with sixfold coordination, $\nu_{Me} = +3/6 = +0.5$, thus each ligand has to contribute with a charge of −0.5 for electroneutrality. Applying this to the surface, a singly coordinate surface OH group (as in the example shown in Figure 12.14b) will have a net charge of −0.5 (furthermore, in the cases of doubly or triply coordinated OH groups, the net charge would be 0.0 and +0.5, respectively), as in the 1-pK model for acid–base behavior (Equation 11.21). For the adsorbed silicate group, applying Equation 12.43 to the Si(IV) cation results in $\nu_{Si} = +1$; now, if this average value is applied equally to the surrounding O atoms, it would result in a net charge of −0.5 for the inner atoms and 0 for the outer OH groups, as shown in Figure 12.14b-i. Here, the model introduces the assumption that given the asymmetry around the Si atom, its charge is not equally distributed between the 0 and 1 planes, but a fraction f of the charge of the central atom will be attributed to the surface, whereas $1 - f$ will belong to the outer ligands; thus, the inner ligands will receive a charge $fz_{Si}/2$ and the outer ones $(1 - f)\, z_{Si}/2$. The case $f = 0.5$ is the symmetric distribution (Figure 12.14b-i), which appears in conflict with the concept of local neutralization of charge; a value $f = 0.75$ would result in a full neutralization of the inner ligands, but now the outer OH groups carry a charge of −0.5 (Figure 12.14b-ii). A more balanced distribution comes with $f = 0.60$, resulting in the distribution shown in Figure 12.14b-iii. Actually, f is treated as an adjustable parameter, and for goethite, it effectively results to be close to 0.60 (Hiemstra and Van Riemsdijk 1996). With these concepts in mind, the surface binding reactions are written, taking as an example monodentate and bidentate phosphate binding to an iron oxide, as

$$\mathrm{FeOH^{-0.5}(su) + PO_4^{3-} + H^+ \rightleftharpoons FeO^{p}PO_3^{q}(su) + H_2O} \tag{12.44}$$

$$\mathrm{2FeOH^{-0.5}(su) + PO_4^{3-} + 2H^+ \rightleftharpoons (FeO^{p})_2\, PO_2^{q}(su) + H_2O} \tag{12.45}$$

where su means surface species, with p and q being the charges of the surface- and solution-oriented P ligands, respectively. According to the discussion above, $p = fz_{su}$ and $q = (1 - f)z_{su}$, where z_{su} is the resultant charge of the surface species, which is −2.5 for Reaction 12.44 and −2.0 for Reaction 12.45.

In the CD-MUSIC model, the activity of the surface species are expressed as coverage θ_i (fraction of surface sites occupied by species i, see Chapter 4), and the electrostatic contribution (Equation 12.1) is given by

$$\Delta G^{\circ}_{el} = \sum_i \frac{\Delta z_i \boldsymbol{F} \psi_i}{\boldsymbol{RT}} \tag{12.46}$$

where the sum runs over all positions (planes) i, and Δz_i is the charge change at position i on adsorption and ψ_i is the potential at this position. For the specific adsorption of a given species, the changes of charge at the different planes should be considered in the electrostatic contribution. For the 0 plane, it can be written as

$$\Delta z_0 = \Delta n_H z_H + f z_{Me} \tag{12.47}$$

where Δn_H is the change in the number of protons involved in the surface reaction (which can be positive, negative, or zero) and $z_H = +1$ its charge. For the 1 plane, it results

$$\Delta z_1 = (1-f)z_{\mathrm{Me}} + \sum_l n_l z_l \tag{12.48}$$

where n_l is the number of ligands with charge z_l positioned in the 1 plane. For example, in the bidentate phosphate adsorption (Reaction 12.45) for $f = 0.5$, it is readily found $\Delta z_0 = 0.5$ and $\Delta z_1 = -1.5$, thus the equilibrium expression for Reaction 12.45 results to be

$$K = K^{int} \exp\left(-\frac{0.5\boldsymbol{F}\psi_0}{\boldsymbol{RT}}\right)\exp\left(\frac{1.5\boldsymbol{F}\psi_1}{\boldsymbol{RT}}\right) = \frac{\theta_{(\mathrm{FeO})_2\mathrm{PO}_2}}{\theta^2_{\mathrm{FeOH}}a_{\mathrm{PO}_4^{3-}}a^2_{\mathrm{H}^+}} \tag{12.49}$$

and similarly for other reactions. The charges of surface species have been omitted for clarity.

12.3.2 Application of the CD-MUSIC Model

As in the other surface complexation models, the "chemical" part consists in several adsorption reactions. In the application of the CD-MUSIC model, the choice of these reactions has been carefully based on spectroscopic evidence; of course, that can be done also in other models such as TLM, BSM, and so on, but here it is rather essential for proper elucidation of the charge distribution. The original paper (Hiemstra and Van Riemsdijk 1996) applied the model to phosphate adsorption on goethite. Based on spectroscopic studies, they proposed the existence of monodentate-bound species (Reaction 12.44), an unprotonated bidentate-bound species (Reaction 12.45), and a protonated bidentate species, following:

$$2\mathrm{FeOH}^{-0.5}(\mathrm{su}) + \mathrm{PO}_4^{3-} + 3\mathrm{H}^+ \rightleftharpoons \left(\mathrm{FeO}^p\right)_2 \mathrm{PO}_2^q\mathrm{H}(\mathrm{su}) + \mathrm{H_2O} \tag{12.50}$$

The additional proton is bound to an "external" ligand; thus, it does not enter in the calculation of Δz_0 (Equation 12.47). Finally, two site types are considered regarding surface protonation: singly and triply coordinated surface OH groups.

The model parameters are the capacitances C_1 and C_2 (as in the TLM), the intrinsic protonation and complexation constants, and f. The model has shown good predictive capabilities, as shown in Figure 12.15, where data from different authors (Hiemstra and Van Riemsdijk 1996; Bowden et al. 1980) are seen to fit well by the CD-MUSIC model with the same parameter values.

Venema, Hiemstra, and van Riemsdijk (1996) studied the adsorption of Cd on goethite applying this model. Here, studies found that it is important to consider the crystal faces; goethite particles usually have a high proportion of 110 surface (about 90%) and 021 (the remaining ~10%). There are two possible Cd adsorption modes on each face, as presented in Figure 12.16, for a total of four adsorbed species. The model predicts well the experimental behavior, as shown for example in Figure 12.17; other experimental results are also well predicted, and the results are consistent with spectroscopic evidence (Venema, Hiemstra, and van Riemsdijk 1996, 1997).

In spite of the successful results, because of the requirement of detailed knowledge of surface species, the application of the CD-MUSIC model has been until present

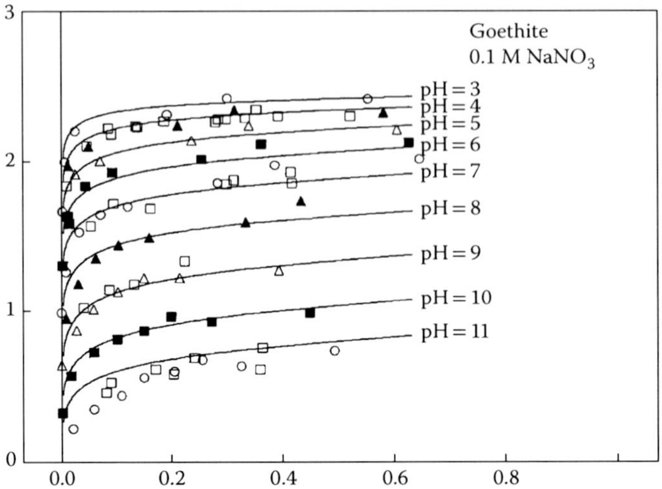

FIGURE 12.15 The adsorption of PO_4^{3-} by monodomainic goethite used by Hiemstra and Van Riemsdijk (1996) and that of Bowden et al. (1980). The lines are calculated with the charge distribution model and are consistent with the surface species observed by cylindrical internal reflection–Fourier transform infrared spectroscopy. (Reprinted from *Journal of Colloid and Interface Science*, 179, Hiemstra, T. and Van Riemsdijk, W. H. 488–508. Copyright 1996, with permission from Elsevier.)

somewhat restricted; it has been mainly applied to goethite and other iron oxides (Hiemstra and Van Riemsdijk 1999; Rahnemaie, Hiemstra, and van Riemsdijk 2006; Ponthieu et al. 2006; Antelo et al. 2010; Goli et al. 2011). Recently, some studies on TiO_2 (Svecova et al. 2011) and on calcite (Sø et al. 2012) have been published.

12.4 COMPARISON BETWEEN MODELS

Comparison between the different models requires application of those models to the same set of experimental data, which has not been much covered in the literature; we recall some important examples here. Hayes et al. (1991) compared the "classic" CCM, DLM (BSM), and TLM models using acid–base titration data for goethite, TiO_2, and α-Al_2O_3. Of all the three models evaluated, the TLM gave the best fitting to the titration data over the widest range in ionic strength conditions, but it requires the greatest effort to estimate the necessary parameters from titration data. At constant ionic strength, the CCM and TLM fitted the titration data quite well for all data sets. The DLM was not as successful at constant ionic strength conditions. Even though the DLM was not able to fit the titration as well as the TLM or CCM, the simplicity of the model and the ability to arrive at a unique set of parameter values without the need for arbitrary assignment of parameters makes it an attractive alternative to the other two models. Increasing the model input value for the total number of surface sites caused a decrease in the best-fit log K values of the surface protolysis constants. In the case of the CCM, the best-fit surface protolysis constants were relatively insensitive to changes in the value of the capacitance fitting parameter.

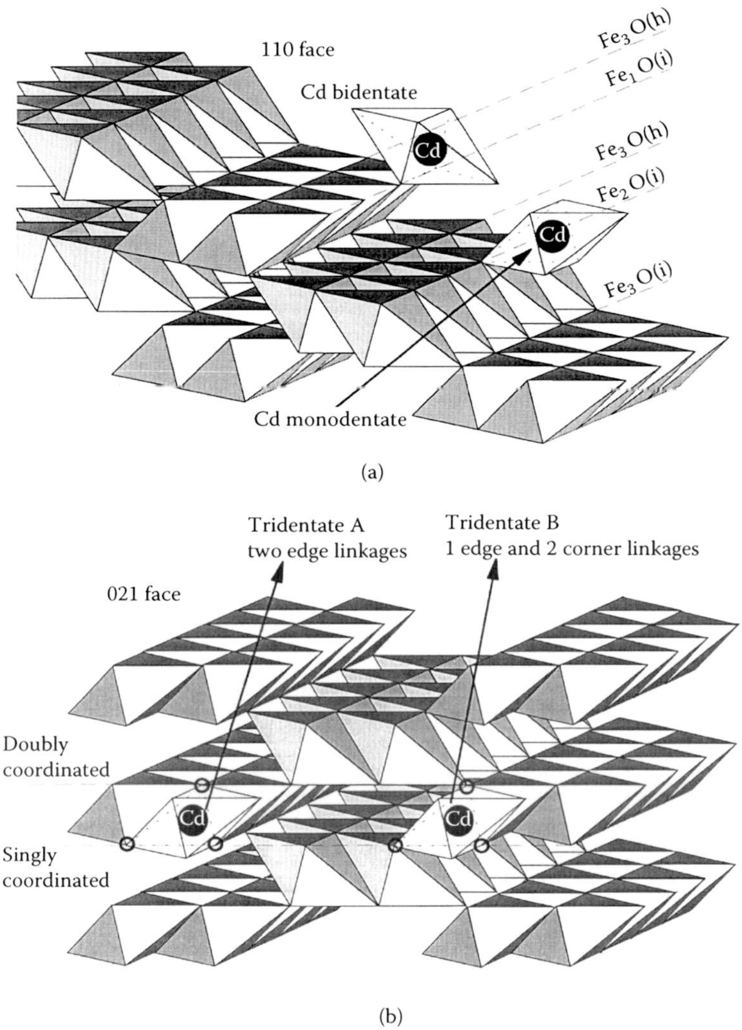

FIGURE 12.16 Cd adsorbed species on goethite. (a) on the 110 crystal face of goethite, where an adsorbing cadmium ion can form two different surface complexes in a growth position, a bidentate with the singly coordinated surface groups and a monodentate with the doubly coordinated surface groups; (b) on the 021 face both complexes are tridentate complexes with two singly coordinated and one doubly coordinated surface group. (Reprinted from *Journal of Colloid and Interface Science*, 183, Venema, P. et al., Multisite adsorption of cadmium on goethite, 515–527. Copyright 1996, with permission from Elsevier.)

Similarly, the best-fit values of TLM surface electrolyte binding constants were less influenced by changes in the value of C_1 for $C_1 > 1.2$ F m^{-2}. For a given C_1 value, the best-fit TLM values of the electrolyte binding constants were sensitive to changes in ΔpK, for $\Delta pK \leq 3$.

Wen, Du, and Tang (1998) compared the same models using experimental data of the potentiometric titration and the metal adsorption isotherms for Cu and Cd using

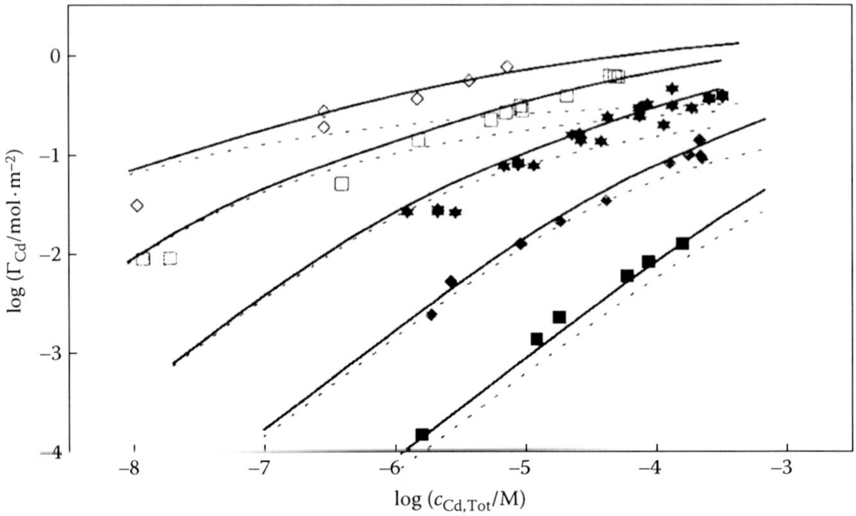

FIGURE 12.17 Cd adsorption isotherms on goethite as a function of total Cd concentration in 0.1 M $NaNO_3$ at different pH values: (■), pH = 5; (◆), pH = 6; (★), pH = 7; (□), pH = 8; and (◇), pH = 9. (Reprinted from *Journal of Colloid and Interface Science*, 183, Venema, P. et al., Multisite adsorption of cadmium on goethite, no. 2, 515–527, Copyright (1996), with permission from Elsevier.)

Lean River (China) sediment. The experimental data of the adsorption edges were used to verify the performance of the models. This work indicated that all three models can simulate the experimental results very well. In predicting the adsorption behavior of the sediment sample, the relative errors of these three models were quite close.

Gustafsson (2003) modeled the adsorption of molybdate and tungsate on ferrihydrite applying the CCM and the CD-MUSIC models, using literature-suggested reactions and parameters. The adsorption edge experiments (pH dependence) were well modeled by both approaches considering two monodentate complexes, but modeling of competition with PO_4^{3-} was less satisfactory; this was attributed to uncertainty in the PO_4^{3-} complexation constants, in turn due to the lack of spectroscopic studies on this system.

Tournassat et al. (2009) compared the BSM and TLM models with molecular dynamics simulations of a montmorillonite/water interface at the pore scale in 0.1 M NaCl. Simulation-derived values were compared with macroscopic model results obtained from the classical models. Although the Na concentration profile is well reproduced in the diffuse layer, anion exclusion is overestimated by the BSM and TLM theories under the experimental conditions employed; the agreement between molecular dynamics simulated and modeled diffuse-layer composition is less accurate with TLM than with BSM. However, the potentials at the three planes of interest are accurately reproduced. It was also showed that molecular dynamics simulations can be used to constrain BSM parameters or, in combination with zeta potential measurements, TLM parameters, by providing suitable values for the capacitance parameters.

In a study of Br^- adsorption by amorphous Al and Fe oxides, montmorillonite, kaolinite, and temperate and tropical soils (Goldberg and Kabengi 2010), the CCM and TLM models were used to analyze the results, fitting only the intrinsic complexation constants for bromide (outer sphere in the case of TLM); all the other parameters were taken from the literature. Albeit both models fitted the experimental data, the CCM gave a better prediction of the adsorption edge experiments; only bromide adsorption on amorphous Fe oxide as a function of pH and ionic strength was well described using the TLM.

Finally, Sø et al. (2012) studied the competitive adsorption of arsenate and phosphate onto calcite in batch experiments and modeled the results using the CCM and the CD-MUSIC model. It was found that both models predicted well the separate adsorption of both anions, but the CCM underpredicted the competition when combining the models for single sorbate systems; instead the CD-MUSIC model predicted satisfactorily the competitive adsorption of phosphate and arsenate onto calcite in the binary system by combining the parameters of the separate systems.

From the above-referred investigations, it should be clear that in many cases different models provide similar prediction of experimental results, especially when comparing the classical CCM, BSM, and TLM approaches. The CD-MUSIC model, albeit more complex, appears to give better results when predicting competitive adsorption using parameters derived from noncompetitive experiments. The question to be answered next to find a satisfactory fitting is that of the physical reality of the parameter values found. Here, the emphasis on applying modern spectroscopic techniques (such as EXAFS, other x-ray absorption techniques, or advanced IR studies) to elucidate the surface species, introduced with the CD-MUSIC model but equally applicable to other models, is a welcome addition; it allows to select the surface adsorption reactions with sound foundations. Apart from this, the choice of electrostatic model should take into account the physical nature of the problem; a typical case is when strong, specific adsorption is the dominant mode, thus a simplified model such as BSM or even CCM may work well if it is able to handle ionic strength changes (which is often a problem in the CCM). The CD-MUSIC model is a more detailed and complete model; however, it depends on a precise knowledge of the surface crystallography, which may be a difficult problem in natural samples, which often are heterogeneous and/or have many crystal imperfections.

12.5 APPLICATION TO CHARGE REGULATION BETWEEN PARTICLES

In Chapter 6, the interaction of colloidal particles by charge regulation was introduced, but its treatment requires use of the concepts developed in Chapter 11 and in this chapter. We will show here the results of a rather recent study (Chan et al. 2006).

Figure 12.18 shows schematically two plane surfaces of dissimilar particles, such as oxides, which are in equilibrium with the surrounding solution. The surfaces are modeled following the GCSG theory (Chapter 3, Section 3.3). Proton charging is assumed to define the intrinsic charge. "Ion pairing" is considered to occur on charged sites following Yates, Levine, and Healy (1974), which can be regarded, in terms of the description given in Chapter 5 (Section 5.1), as inner-sphere complexes; however, no distinction is made here between inner and outer spheres. As it can be

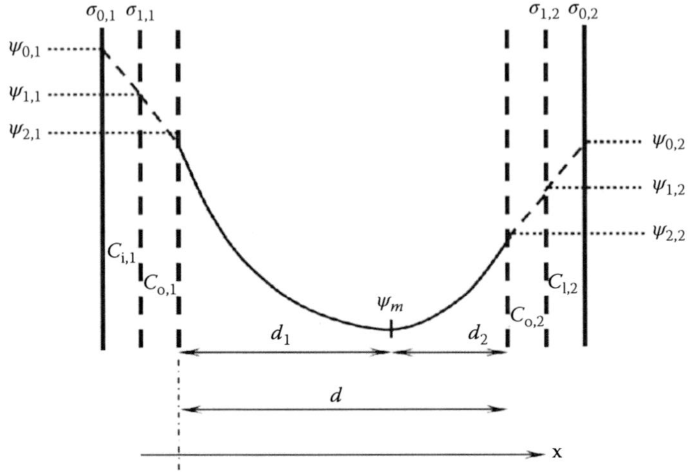

FIGURE 12.18 Schematic representation of the model for two parallel plane surfaces in the charge regulation approach, assumed to be oxide surfaces in equilibrium with the solution. (Adapted from *Journal of Colloid and Interface Science*, 296, Chan, D. Y. C. et al., Electrical double layer interactions between dissimilar oxide surfaces with charge regulation and Stern-Grahame layers, no. 1 (April 1), 150–158, Copyright (2006), with permission from Elsevier.)

seen in Figure 12.18, the electrolyte ions forming pairs are assumed to be located in the inner Helmholtz plane (plane 1), and diffuse ions are present at the outer plane (plane 2) and beyond. The surface is assumed to be in equilibrium with the solution pH through the following reactions, typical of oxide surfaces (here, S stands for a surface site):

$$SOH_2^+ \rightleftharpoons SOH + H^+ \tag{12.51}$$

$$SOH \rightleftharpoons SO^- + H^+ \tag{12.52}$$

with the equilibrium constants, respectively, given for surface j by (see Section 11.1)

$$K_{+,j} = \frac{\Gamma_{SOH,j}a_{H^+}}{\Gamma_{SOH_2^+,j}}e^{-F\psi_{0,j}/RT} \tag{12.53}$$

$$K_{-,j} = \frac{\Gamma_{SO^-,j}a_{H^+}}{\Gamma_{SOH,j}}e^{-F\psi_{0,j}/RT} \tag{12.54}$$

where $\Gamma_{i,j}$ is surface concentration of species i on surface j and a_i is the bulk solution activity of species i. These are related to the point of zero charge, pH_0, by

$$pH_{0,j} = \frac{pK_{+,j} + pK_{-,j}}{2} \tag{12.55}$$

Also, it is defined

$$\Delta pK_j = pK_{-,j} - pK_{+,j} \tag{12.56}$$

Ion pairing (considering a 1-1 MX electrolyte) occurs through the following reactions (written as dissociations):

$$SOH_2^+ X^- \rightleftharpoons SOH_2^+ + X^- \tag{12.57}$$

$$SO^- M^+ \rightleftharpoons SO^- + M^+ \tag{12.58}$$

the respective constants being

$$K_{X,j} = \frac{\Gamma_{SOH_2^+,j} a_{X^-}}{\Gamma_{SOH_2^+X^-,j}} e^{-F\psi_{0,j}/RT} \tag{12.59}$$

$$K_{M,j} = \frac{\Gamma_{SO^-,j} a_{M^+}}{\Gamma_{SO^-M^+,j}} e^{-F\psi_{0,j}/RT} \tag{12.60}$$

In this work, the activities are approximated by concentrations, so that $a_{H^+} = c_{H^+}$ and $a_{X^-} = a_{M^+} = c$, where c is the electrolyte concentration. The potentials at the respective planes are related through the integral capacitances of the inner part of the double layer, $C_{I,j}$ and that of the outer part, $C_{O,j}$ by

$$\sigma_{0,j} = C_{I,j}\left(\psi_{0,j} - \psi_{1,j}\right) \tag{12.61}$$

and

$$\sigma_{1,j} = -C_{I,j}\left(\psi_{0,j} - \psi_{1,j}\right) + C_{O,j}\left(\psi_{1,j} - \psi_{2,j}\right) \tag{12.62}$$

Also, the respective charges are related to the surface concentrations by

$$\sigma_{0,j} = F\Gamma_{s,j} \frac{\Gamma_{SOH_2^+,j} - \Gamma_{SO^-,j} + \Gamma_{SOH_2^+X^-,j} - \Gamma_{SO^-M^+,j}}{\Gamma_{SOH,j} + \Gamma_{SOH_2^+,j} + \Gamma_{SO^-,j} + \Gamma_{SO^-M^+,j} + \Gamma_{SOH_2^+X^-,j}} \tag{12.63}$$

$$\sigma_{1,j} = F\Gamma_{s,j} \frac{\Gamma_{SO^-M^+,j} - \Gamma_{SOH_2^+X^-,j}}{\Gamma_{SOH,j} + \Gamma_{SOH_2^+,j} + \Gamma_{SO^-,j} + \Gamma_{SO^-M^+,j} + \Gamma_{SOH_2^+X^-,j}} \tag{12.64}$$

where $\Gamma_{s,j}$ is the number of total ionizable sites on surface j. The potential in the electrolyte region between the two outer planes is given by the PB equation for the Gouy–Chapman model (Equation 6.37) with the boundary conditions

$$\left.\frac{d\psi}{dx}\right)_{x_j} = -\frac{\kappa\varepsilon RT}{F}\left[4\sinh^2\left(\frac{F\psi_{2,j}}{2RT}\right) - \frac{P}{c^\infty RT}\right]^{1/2} \tag{12.65}$$

calculated at $x_1 = 0$ for $\psi_{2,1}$ and $x_2 = d$ for $\psi_{2,2}$; here, P is the interaction pressure and c^∞ is the bulk electrolyte concentration. Also,

$$\left.\frac{d\psi}{dx}\right)_{x=0} = -\frac{1}{\varepsilon}\left(\sigma_{0,1} + \sigma_{1,1}\right) \quad \text{at} \quad x = 0 \tag{12.66}$$

and

$$\left.\frac{d\psi}{dx}\right)_{x=d} = \frac{1}{\varepsilon}\left(\sigma_{0,2} + \sigma_{1,2}\right) \quad \text{at} \quad x = d \tag{12.67}$$

Note that, because the PB equation is valid beyond the outer planes, for the diffuse layers whose potential values at the limits are the $\Psi_{2,j}$, the whole set of Equations 12.53 through 12.56 and 12.59 through 12.67 form the boundary conditions. We shall not go into the resolution details here; they are described in the original paper (Chan et al. 2006); here we will describe the main results. Figure 12.19 shows results for this model (curve 3) compared with those obtained under constant charge and constant potential conditions; the three curves were obtained setting the two diffuse layer potentials $\Psi_{2,j}$ at infinite separation to the same values. It can be seen there that, for the parameters indicated, the charge regulation model predicts a behavior similar to that of constant charge, being repulsive at all distances, albeit with smaller force values. Given that the surfaces have pH_0 values of 9 and 7, and the medium pH is 6, both surfaces have at large separations charges of equal sign but dissimilar value: under constant potential conditions, the interaction is initially repulsive but becomes attractive at short separations (as discussed in Chapter 6), whereas in the other cases, it is always repulsive. It is interesting to compare these curves with those shown in Figure 6.10, in particular with the linear superposition approximation: this can approximate the charge regulation results at relatively large separations, but fails

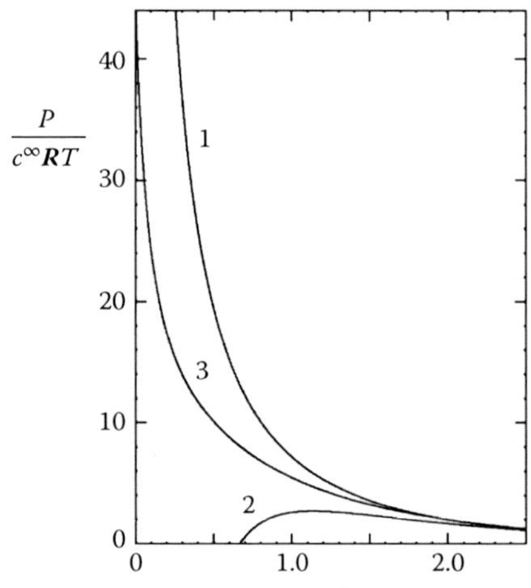

FIGURE 12.19 Interaction pressure as a function of surface separation as predicted for a model system as in Figure 12.18 at pH = 6, c = 1 mM with $pH_{0,1} = 9$, $\Delta pH_1 = 3$, $pH_{0,2} = 7$, $\Delta pH_2 = 2$, $C_{I,j}$ = 140 μF cm^{-2}, C_{Oj} = 20 μF cm^{-2}, and all ion pair intrinsic constants set to 1. Curve 1 is under constant charge conditions, curve 2 at constant potential, and curve 3 under charge regulation. (Adapted from *Journal of Colloid and Interface Science*, 296, Chan, D. Y. C. et al., Electrical double layer interactions between dissimilar oxide surfaces with charge regulation and Stern-Grahame layers, no. 1 (April 1), 150–158, Copyright (2006), with permission from Elsevier.)

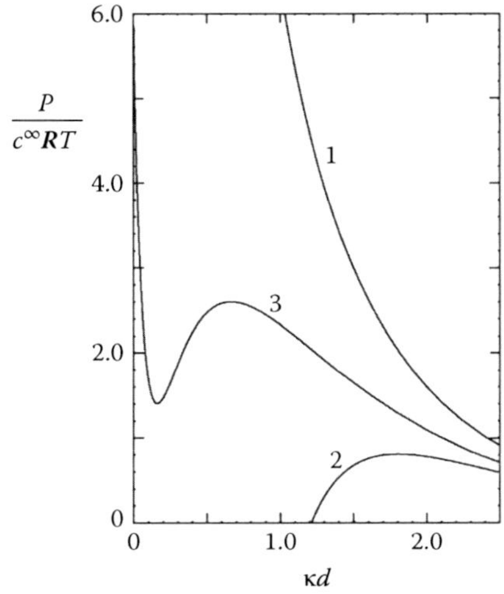

FIGURE 12.20 Interaction pressure as a function of surface separation as predicted for a model system as in Figure 12.18. All parameters as in Figure 12.19, except $pH_{0,2} = 6.55$. Curve 1 is under constant charge conditions, curve 2 at constant potential, and curve 3 under charge regulation. (Adapted from *Journal of Colloid and Interface Science*, 296, Chan, D. Y. C. et al., Electrical double layer interactions between dissimilar oxide surfaces with charge regulation and Stern-Grahame layers, no. 1 (April 1), 150–158, Copyright (2006), with permission from Elsevier.)

at shorter distances. The charge regulation model, moreover, does not necessarily predict monotonic curves; in Figure 12.20, the interactions for the same system, only changing $pH_{0,2}$ from 7 to 6.55, are plotted. Now, because the second surface is closer to zero charge conditions, the predicted behavior is no longer monotonic but shows a local minimum at relatively short distances. Note that this implies that the interaction energy becomes "attractive" in a range of distances. This behavior is correlated with nonmonotonic changes of potentials and ionic concentrations for the surface with "higher" charge, here surface 1 (not shown), see Chan et al. (2006) for details.

12.6 SUMMARY

This chapter has been devoted to the modeling of ion binding (mainly specific) to mineral soil colloids. Here, the chemical contribution to the interaction is generally modeled with discrete sites which, in modern studies, are postulated based on advanced spectroscopic techniques such as x-ray absorption methods. Usually, a few ligand-exchange reactions have to be considered, including monodentate and bidentate binding, and for anion adsorption, some protonated species. Several theories have been proposed for the electrostatic contribution, all based in some way or another on the GCSG model of the double layer, presented in Chapter 3. The most detailed approach, the CD-MUSIC model, considers a charge distribution for the adsorbed species unsymmetrical, at variance with solution species, because of the asymmetry created by the adsorption reaction. These models are generally able to fit quite well the experimental data, but the CD-MUSIC model has been shown to be capable of predicting competitive adsorption from fitting to noncompetitive experimental data. Its application, however, turns difficult by the need of a detailed knowledge of the surface crystallography.

REFERENCES

Akafia, M. M., T. J. Reich, and C. M. Koretsky. 2011. Assessing Cd, Co, Cu, Ni, and Pb Sorption on montmorillonite using surface complexation models. *Applied Geochemistry* 26, Supplement: S154–S157. doi:10.1016/j.apgeochem.2011.03.091.

Antelo, J., S. Fiol, C. Pérez, S. Mariño, F. Arce, D. Gondar, and R. López. 2010. Analysis of phosphate adsorption onto ferrihydrite using the CD-MUSIC model. *Journal of Colloid and Interface Science* 347, no. 1: 112–119. doi:10.1016/j.jcis.2010.03.020.

Barrow, N. J. 1983. A mechanistic model for describing the sorption and desorption of phosphate by soil. *Journal of Soil Science* 34, no. 4: 733–750. doi:10.1111/j.1365-2389.1983.tb01068.x.

Barrow, N. J. 1986. Testing a mechanistic model. I. The effects of time and temperature on the reaction of fluoride and molybdate with a soil. *Journal of Soil Science* 37, no. 2: 267–275. doi:10.1111/j.1365-2389.1986.tb00028.x.

Barrow, N. J., J. W. Bowden, A. M. Posner, and J. P. Quirk. 1980. An objective method for fitting models of ion adsorption on variable charge surfaces. *Soil Research* 18, no. 1: 37–47.

Barrow, N. J., and A. S. Ellis. 1986a. Testing a mechanistic model. III. The effects of pH on fluoride retention by a soil. *Journal of Soil Science* 37, no. 2: 287–293. doi:10.1111/j.1365-2389.1986.tb00030.x.

Barrow, N. J., and A. S. Ellis. 1986b. Testing a mechanistic model. V. The points of zero salt effect for phosphate retention, for zinc retention and for acid/alkali titration of a soil. *Journal of Soil Science* 37, no. 2: 303–310. doi:10.1111/j.1365-2389.1986.tb00032.x.

Bowden, J. W., S. Nagarajah, N. J. Barrow, A. M. Posner, and J. P. Quirk. 1980. Describing the adsorption of phosphate, citrate and selenite on a variable-charge mineral surface. *Soil Research* 18, no. 1: 49–60.

Bowden, J. W., A. M. Posner, and J. P. Quirk. 1977. Ionic adsorption on variable charge mineral surfaces. Theoretical charge development and titration curves. *Soil Research* 15, no. 2: 121–136.

Cea, M., J. C. Seaman, A. Jara, M. L. Mora, and M. C. Diez. 2010. Kinetic and thermodynamic study of chlorophenol sorption in an allophanic soil. *Chemosphere* 78, no. 2: 86–91. doi:10.1016/j.chemosphere.2009.10.040.

Chabot, M., T. Hoang, and H. A. Al-Abadleh. 2009. ATR-FTIR studies on the nature of surface complexes and desorption efficiency of p-arsanilic acid on iron (oxyhydr)oxides. *Environmental Science & Technology* 43, no. 9: 3142–3147. doi:10.1021/es803178f.

Chan, D. Y. C., T. W. Healy, T. Supasiti, and S. Usui. 2006. Electrical double layer interactions between dissimilar oxide surfaces with charge regulation and Stern-Grahame layers. *Journal of Colloid and Interface Science* 296, no. 1: 150–158. doi:10.1016/j.jcis.2005.09.003.

Davis, J. A., R. O. James, and J. O. Leckie. 1978. Surface ionization and complexation at the oxide/water interface: I. Computation of electrical double layer properties in simple electrolytes. *Journal of Colloid and Interface Science* 63, no. 3: 480–499. doi:10.1016/S0021-9797(78)80009-5.

Depalma, S., S. Cowen, T. Hoang, and H. A. Al-Abadleh. 2008. Adsorption thermodynamics of p-arsanilic acid on iron (oxyhydr)oxides: In-situ ATR-FTIR studies. *Environmental Science & Technology* 42, no. 6: 1922–1927. doi:10.1021/es071752x.

Dixit, S., and J. G. Hering. 2003. Comparison of Arsenic(V) and Arsenic(III) sorption onto iron oxide minerals: Implications for arsenic mobility. *Environmental Science & Technology* 37, no. 18: 4182–4189. doi:10.1021/es030309t.

EPA Center for Exposure Assessment Modeling. 2006. *MINTEQA2*. Washington, DC: US EPA. http://www.epa.gov/ceampubl/mmedia/minteq/.

Goldberg, S, 2002. Competitive adsorption of arsenate and arsenite on oxides and clay minerals. *Soil Science Society of America Journal* 66, no. 2: 413–421. doi:10.2136/sssaj2002.4130.

Goldberg, S. 2005. Equations and models describing adsorption processes in soils. In *Chemical Processes in Soils*, ed. M. A. Tabatabai and D. L. Sparks, 489–518. Madison, WI: Soil Science Society of America.

Goldberg, S., and N. J. Kabengi. 2010. Bromide adsorption by reference minerals and soils. *Vadose Zone Journal* 9, no. 3: 780. doi:10.2136/vzj2010.0028.

Goli, E., R. Rahnemaie, T. Hiemstra, and M. J. Malakouti. 2011. The interaction of boron with goethite: Experiments and CD–MUSIC modeling. *Chemosphere* 82, no. 10: 1475–1481. doi:10.1016/j.chemosphere.2010.11.034.

Grahame, D. C. 1947. The electrical double layer and the theory of electrocapillarity. *Chemical Reviews* 41, no. 3: 441–501. doi:10.1021/cr60130a002.

Grahame, D. C. 1954. Differential capacity of mercury in aqueous sodium fluoride solutions. I. Effect of concentration at 25°. *Journal of the American Chemical Society* 76, no. 19: 4819–4823. doi:10.1021/ja01648a014.

Gu, X., L. J. Evans, and S. J. Barabash. 2010. Modeling the adsorption of Cd (II), Cu (II), Ni (II), Pb (II) and Zn (II) onto montmorillonite. *Geochimica et Cosmochimica Acta* 74, no. 20: 5718–5728. doi:10.1016/j.gca.2010.07.016.

Gustafsson, J. P. 2003. Modelling molybdate and tungstate adsorption to ferrihydrite. *Chemical Geology* 200, no. 1–2: 105–115. doi:10.1016/S0009-2541(03)00161-X.

Gustafsson, J. P. 2011. *Visual Minteq*. Stockholm, Sweden: KTH, Department of Land and Water Resources Engineering.

Hayes, K. F., G. Redden, W. Ela, and J. O. Leckie. 1991. Surface complexation models: An evaluation of model parameter estimation using FITEQL and oxide mineral titration data. *Journal of Colloid and Interface Science* 142, no. 2: 448–469. doi: 10.1016/0021-9797(91)90075-J.

Herbelin, A. L., and J. C. Westall. 1999. *FITEQL*. Oregon: Department of Chemistry, Oregon State University.

Hiemstra, T., and W. H. Van Riemsdijk. 1996. A surface structural approach to ion adsorption: The charge distribution (CD) model. *Journal of Colloid and Interface Science* 179, no. 2: 488–508. doi:10.1006/jcis.1996.0242.

Hiemstra, T., and W. H. Van Riemsdijk. 1999. Surface structural ion adsorption modeling of competitive binding of oxyanions by metal (hydr)oxides. *Journal of Colloid and Interface Science* 210, no. 1: 182–193. doi:10.1006/jcis.1998.5904.

Hizal, J., and R. Apak. 2006. Modeling of copper(II) and lead(II) adsorption on kaolinite-based clay minerals individually and in the presence of humic acid. *Journal of Colloid and Interface Science* 295, no. 1: 1–13. doi:10.1016/j.jcis.2005.08.005.

Kanematsu, M., T. M. Young, K. Fukushi, P. G. Green, and J. L. Darby. 2010. Extended triple layer modeling of arsenate and phosphate adsorption on a goethite-based granular porous adsorbent. *Environmental Science & Technology* 44, no. 9: 3388–3394. doi:10.1021/es903658h.

Karthikeyan, K. G., H. A. Elliott, and J. Chorover. 1999. Role of surface precipitation in copper sorption by the hydrous oxides of iron and aluminum. *Journal of Colloid and Interface Science* 209, no. 1: 72–78. doi:10.1006/jcis.1998.5893.

Leroy, P., and A. Revil. 2004. A triple-layer model of the surface electrochemical properties of clay minerals. *Journal of Colloid and Interface Science* 270, no. 2: 371–380. doi:10.1016/j.jcis.2003.08.007.

Li, L., and R. Stanforth. 2000. Distinguishing adsorption and surface precipitation of phosphate on goethite (α-FeOOH). *Journal of Colloid and Interface Science* 230, no. 1: 12–21. doi:10.1006/jcis.2000.7072.

Lützenkirchen, J., and Ph. Behra. 1995. On the surface precipitation model for cation sorption at the (hydr)oxide water interface. *Aquatic Geochemistry* 1, no. 4: 375–397. doi:10.1007/BF00702740.

Mattigod, S. V., and J. M. Zachara. 1996. Equilibrium modeling in soil chemistry. In *Methods of Soil Analysis. Part 3: Chemical Methods*, ed. D. L. Sparks, 1309–1358. Soil Science Society of America Book Series 5. Madison, WI: American Society of Agronomy-Soil Science Society of America.

Merdy, P., L. K. Koopal, and S. Huclier. 2006. Modeling metal–particle interactions with an emphasis on natural organic matter. *Environmental Science & Technology* 40, no. 24: 7459–7466. doi:10.1021/es0628203.

Mitchell, W., S. Goldberg, and H. A. Al-Abadleh. 2011. In situ ATR–FTIR and surface complexation modeling studies on the adsorption of dimethylarsinic acid and p-arsanilic acid on iron-(oxyhydr)oxides. *Journal of Colloid and Interface Science* 358, no. 2: 534–540. doi:10.1016/j.jcis.2011.02.040.

Nagata, T., and K. Fukushi. 2010. Prediction of iodate adsorption and surface speciation on oxides by surface complexation modeling. *Geochimica et Cosmochimica Acta* 74, no. 21: 6000–6013. doi:10.1016/j.gca.2010.08.002.

Pagnanelli, F., L. Bornoroni, E. Moscardini, and L. Toro. 2006. Non-electrostatic surface complexation models for protons and lead(II) sorption onto single minerals and their mixture. *Chemosphere* 63, no. 7: 1063–1073. doi:10.1016/ j.chemosphere.2005.09.017.

Parkhurst, D., S. Charlton, A. Riggs, and R. Webb. 2008. *PHREEQC*. Washington, DC: United States Geological Survey.

Ponthieu, M., F. Juillot, T. Hiemstra, W. H. van Riemsdijk, and M. F. Benedetti. 2006. Metal ion binding to iron oxides. *Geochimica et Cosmochimica Acta* 70, no. 11: 2679–2698. doi:10.1016/j.gca.2006.02.021.

Rahnemaie, R., T. Hiemstra, and W. H. van Riemsdijk. 2006. Inner- and outer-sphere complexation of ions at the goethite-solution interface. *Journal of Colloid and Interface Science* 297, no. 2: 379–388. doi:10.1016/j.jcis.2005.11.003.

Rytwo, G. 2004. Applying a Gouy-Chapman-Stern model for adsorption of organic cations to soils. *Applied Clay Science* 24, no. 3–4: 137–147. doi:10.1016/ j.clay.2003.01.001.

Sahai, N., and D. A. Sverjensky. 1997. Evaluation of internally consistent parameters for the triple-layer model by the systematic analysis of oxide surface titration data. *Geochimica et Cosmochimica Acta* 61, no. 14: 2801–2826.

Schaller, M. S., C. M. Koretsky, T. J. Lund, and C. J. Landry. 2009. Surface complexation modeling of Cd(II) adsorption on mixtures of hydrous ferric oxide, quartz and kaolinite. *Journal of Colloid and Interface Science* 339, no. 2: 302–309. doi:10.1016/j.jcis.2009.07.053.

Sø, H. U., D. Postma, R. Jakobsen, and F. Larsen. 2012. Competitive adsorption of arsenate and phosphate onto calcite; experimental results and modeling with CCM and CD-MUSIC. *Geochimica et Cosmochimica Acta* 93: 1–13. doi:10.1016/ j.gca.2012.06.021.

Sposito, G. 1987. Distinguishing adsorption from surface precipitation. In *Geochemical Processes at Mineral Surfaces*, ed. J. A. Davis and K. F. Hayes, 217–228. ACS Symposium Series 323. Washington, DC: American Chemical Society.

Sposito, G. 2008. *The Chemistry of Soils*. 2nd ed. New York: Oxford University Press.

Svecova, L., M. Dossot, S. Cremel, M.-O. Simonnot, M. Sardin, B. Humbert, C. D. Auwer, and L. J. Michot. 2011. Sorption of selenium oxyanions on TiO2 (rutile) studied by batch or column experiments and spectroscopic methods. *Journal of Hazardous Materials* 189, no. 3: 764–772. doi:10.1016/j.jhazmat.2011.02.090.

Sverjensky, D. A, 2005. Prediction of surface charge on oxides in salt solutions: Revisions for 1:1 (M+L−) electrolytes. *Geochimica et Cosmochimica Acta* 69, no. 2: 225–257. doi:10.1016/j.gca.2004.05.040.

Sverjensky, D. A. and K. Fukushi. 2006. A predictive model (ETLM) for As(III) adsorption and surface speciation on oxides consistent with spectroscopic data. *Geochimica et Cosmochimica Acta* 70, no. 15: 3778–3802. doi:10.1016/j.gca.2006.05.012.

Thompson, H. A., G. A. Parks, and G. E. Brown Jr. 1999. Dynamic interactions of dissolution, surface adsorption, and precipitation in an aging cobalt(II)-clay-water system. *Geochimica et Cosmochimica Acta* 63, no. 11–12: 1767–1779. doi:10.1016/S0016-7037(99)00125-8.

Tournassat, C., Y. Chapron, P. Leroy, M. Bizi, and F. Boulahya. 2009. Comparison of molecular dynamics simulations with triple layer and modified Gouy-Chapman models in a 0.1 M NaCl-montmorillonite system. *Journal of Colloid and Interface Science* 339, no. 2: 533–541. doi:10.1016/j.jcis.2009.06.051.

Van Riemsdijk, W. H., L. K. Koopal, D. G. Kinniburgh, M. F. Benedetti, and L. Weng. 2006. Modeling the interactions between humics, ions, and mineral surfaces. *Environmental Science & Technology* 40, no. 24: 7473–7480.

Venema, P., T. Hiemstra, and W. H. van Riemsdijk. 1996. Multisite adsorption of cadmium on goethite. *Journal of Colloid and Interface Science* 183, no. 2: 515–527. doi:10.1006/jcis.1996.0575.

Venema, P., T. Hiemstra, and W. H. van Riemsdijk. 1997. Interaction of cadmium with phosphate on goethite. *Journal of Colloid and Interface Science* 192, no. 1: 94–103. doi:10.1006/jcis.1997.4988.

Villaseñor, N., G., and T. J. Strathmann. 2006. Ferrous iron sorption by hydrous metal oxides. *Journal of Colloid and Interface Science* 297, no. 2: 443–454. doi:10.1016/j.jcis.2005.11.030.

Wen, X., Q. Du, and H. Tang. 1998. Surface complexation model for the heavy metal adsorption on natural sediment. *Environmental Science & Technology* 32, no. 7: 870–875. doi:10.1021/es970098q.

Westall, J. C. 1987. Reactions at the oxide-solution interface: Chemical and electrostatic models. In *Geochemical Processes at Mineral Surfaces*, ed. J. A. Davis and K. F Hayes, 54–78. ACS Symposium Series 323. Washington, DC: American Chemical Society.

Yates, D. E., S. Levine, and T. W. Healy. 1974. Site-binding model of the electrical double layer at the oxide/water interface. *Journal of the Chemical Society, Faraday Transactions 1* 70: 1807. doi:10.1039/f19747001807.

Zachara, J. M., C. C. Ainsworth, C. E. Cowan, and C. T. Resch. 1989. Adsorption of chromate by subsurface soil horizons. *Soil Science Society of America Journal* 53, no. 2: 418–428. doi:10.2136/sssaj1989.03615995005300020018x.

Zachara, J. M., and J. C. Westall. 1998. Chemical modeling of Ion adsorption in soils. In *Soil Physical Chemistry*, ed. D. L. Sparks, 47–96. 2nd ed. Boca Raton, FL: CRC Press.

13 Ion Binding to Humic Substances

Distributed Site Modeling

13.1 ION BINDING TO SOIL ORGANIC MATTER

The binding of ions to soil organic matter (SOM), and particularly soil humic substances (HSs), is, at first sight, akin to metal ion complexation to small organic molecules, where cations such as Cu(II), Cd(II), and Fe(II/III) are coordinated to Lewis basic groups (electron-pair donors), such as $-COO^-$ and $-NH_2$. However, HSs are relatively large molecules (or molecular associations, see Section 10.3.5) with colloidal properties; consequently they also retain cations electrostatically, showing cation exchange properties. As discussed in Chapter 10, HSs, being anionic, show little interaction with other anions. In HSs, groups able to complexate (i.e., to bind specifically) are mostly oxygen-containing ones, given the typical composition of HSs; most of such groups are carboxylates; some others are also considered usually, mostly phenolic moieties (Section 10.3.5). Despite the exact nature of these groups, the approach is basically that introduced in Chapter 11, considering both specific and electrostatic interactions. At variance with mineral colloids, the range of binding groups and binding constants is higher and is also poorly known (see Section 10.4), so that detailed treatments, such as the CD-MUSIC model discussed in Section 12.3, are not possible. Thus, the assortment of different coordinating groups (which can be regarded as adsorption sites) is generally modeled through some sort of distribution of interaction free energies (or, equivalently, of binding equilibrium constants), that is, an affinity spectrum, as often termed (Section 11.2). For the electrostatic contribution, diverse approaches can be found; some of them borrow concepts from models developed for mineral particles, presented in Chapter 12. However, as HSs are not really solid, hard particles but softer molecular aggregates (Duval et al. 2005), several more specific treatments have been developed.

13.2 AFFINITY SPECTRA OF HUMIC SUBSTANCES

In the modeling of ion binding to HSs, both discrete and continuous types of affinity spectra have been used. It should be noted that here we are not using the term "discrete" in the same sense of Chapter 12, where each binding constant is expected to correspond to an actual adsorption reaction, but instead they are separate representative values of all the binding constants in the HSs. It should be recalled here that, from a mathematical point of view, these two types are equivalent for describing titration curves, as discussed in Section 1.3.3 (see also Borkovec et al. 1996).

13.2.1 Discrete Affinity Spectra

Discrete distributions are assumed in several speciation schemes, such as Tipping's Windermere humic aqueous model (WHAM) V, VI, VI-FD, and VII (Tipping 1994, 1998; Smith et al. 2004; Tipping, Lofts, and Sonke 2011); and Gustafsson's Stockholm humic model (SHM) (Gustafsson 2001). Essentially, it consists of two sets of four constant values (sites) each; one set (labeled A) corresponds to carboxylic-type coordinating groups, whereas the other one (B) is associated with "high affinity" or "phenolic" type chemical groups; here, the proton reactions are written as dissociation (there are eight overall, four type A and four type B):

$$\mathrm{R-OH} \rightleftharpoons \mathrm{R-O^- + H^+} \tag{13.1}$$

To reduce the number of adjustable parameters, the A- and B-type proton dissociation constants are given, respectively, by

$$\mathrm{p}K_{H,i} = \mathrm{p}K_A + \frac{2i-5}{6}\Delta\mathrm{p}K_A, \quad 1 \le i \le 4 \tag{13.2}$$

$$\mathrm{p}K_{H,i} = \mathrm{p}K_B + \frac{2i-13}{6}\Delta\mathrm{p}K_B, \quad 5 \le i \le 8 \tag{13.3}$$

where $\mathrm{p}K_A$, $\Delta\mathrm{p}K_A$, $\mathrm{p}K_B$, and $\Delta\mathrm{p}K_B$ are adjustable parameters. To complete the description, the distribution of the total number of sites between these eight values must be specified: in the WHAM V and VI, it is assumed that two-thirds of the sites correspond to the set A and one-third to set B, and within each set, the four constants have equal weights. In the SHM, a slightly different approach is taken: for fulvic acids (FAs), sets A and B get 70% and 30% of the sites, respectively; whereas for humic acids (HAs), all the eight constant values have the same weight. Figure 13.1 shows schematically the proton affinity spectra discussed, for $\mathrm{p}K_A = 4.13$, $\Delta\mathrm{p}K_A = 3.0$, $\mathrm{p}K_B = 9.0$, and $\Delta\mathrm{p}K_B = 3.0$.

For metal ion binding, there are different approaches: monodentate metal complexation is treated as exchange with proton, such as

$$\mathrm{RH + M^{z+}} \rightleftharpoons \mathrm{RM + H^+} \tag{13.4}$$

where R stands from the humic molecule, its charge has been dropped (in the SHM, Reaction 13.4 is written oppositely, i.e., as dissociation). In WHAM V, Reaction 13.4 has the unique equilibrium constant value K_{HMA} for sites A and K_{HMB} for sites B, thus the reaction

$$\mathrm{R + M^{z+}} \rightleftharpoons \mathrm{RM} \tag{13.5}$$

has the constant values for each site type given by

$$\log K_{Mm,i} = \mathrm{p}K_{HA} + \frac{2i-5}{6}\Delta\mathrm{p}K_A + \log K_{HMA}, \quad 1 \le i \le 4 \tag{13.6}$$

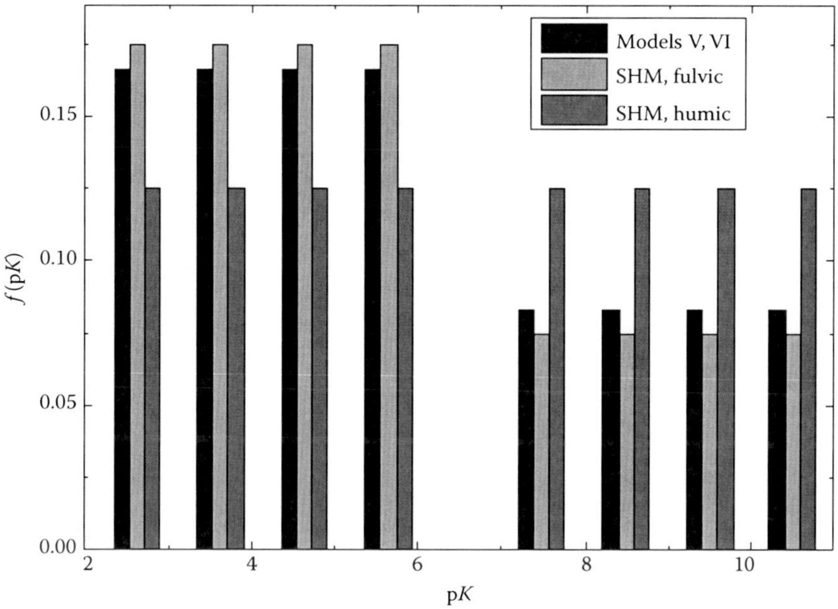

FIGURE 13.1 Scheme of the proton affinity spectra in Tipping Models V and VI and in Gustaffson Stockholm humic model.

$$\log K_{Mm,i} = \mathrm{p}K_{HB} + \frac{2i-5}{6}\Delta \mathrm{p}K_B + \log K_{HMB}, \quad 5 \le i \le 8 \tag{13.7}$$

where m indicates monodentate binding. It is clearly seen that with these definitions there is a strong correlation between metal and proton binding. In Model VI, Equations 13.6 and 13.7 are replaced with the introduction of new parameters K_{MA}, K_{MB}, ΔLK_A, and ΔLK_B by

$$\log K_{Mm,i} = \log K_{MA} + \frac{2i-5}{6}\Delta LK_A, \quad 1 \le i \le 4 \tag{13.8}$$

$$\log K_{Mm,i} = \log K_{MB} + \frac{2i-5}{6}\Delta LK_B, \quad 5 \le i \le 8 \tag{13.9}$$

Most frequently, when there are less data available, ΔLK_A and ΔLK_B are both set equal to a single parameter ΔLK_1 (Tipping 1998). In the SHM, a simpler approach is taken, assuming a single-base metal binding constant for the A proton sites, no monodentate metal binding to B sites, and heterogeneity is accounted for by a parameter ΔLK_1:

$$\log K_{Mm,i} = \log K_{Mm} + x_i \Delta LK_1, \quad 1 \le i \le 3 \tag{13.10}$$

where the value of x_1 is set to 0 for 90.1% of the A sites, x_2 to 1 for 9%, and x_3 to 2 for 0.9%. This implies that only carboxylic groups are assumed to bind metal cations in monodentate form and that a small fraction of these has higher affinity; it is assumed

(Gustafsson 2001) that in this way a few higher affinity sites (N groups, thiols) are accounted for. Discrete affinity spectra models allow for multidentate binding. Here, the approaches are more different; Model V considers that, when a cation M^{z+} binds simultaneously to sites *i* and *j*, the respective constant is given by (*b* standing for bidentate)

$$\log K_{Mb,i,j} = \log K_{Mm,i} + \log K_{Mm,j} \tag{13.11}$$

where $\log K_{Mm,i}$ and *j* are given by Equations 13.6 and 13.7. In Model VI, both bidentate and tridentate sites are considered, with fractions f_b and f_t of the total sites that can make bidentate and tridentate binding, respectively; their ranges of binding strength are increased by additional terms. Thus, for bidentate sites,

$$\log K_{Mb,i,j} = \log K_{Mm,i} + \log K_{Mm,j} + x\Delta LK_2 \tag{13.12}$$

where x is a weight parameter with the same meaning as x_i in Equation 13.10. Likewise, tridentate sites are also included with

$$\log K_{Mt,i,j,k} = \log K_{Mm,i} + \log K_{Mm,j} + \log K_{Mm,k} + y\Delta LK_2 \tag{13.13}$$

where y is another weight parameter; the values of x are chosen as above; for tridentate sites, y is set to 0, 1.5, and 3, again for 90.1%, 9%, and 0.9%, respectively. In principle, each metal has a characteristic ΔLK_2. In the SHM, only bidentate sites are included, with constants given by

$$\log K_{Mb,i} = \log K_{Mb} + 2x_i\Delta LK_1, \quad 1 \le i \le 3 \tag{13.14}$$

Based on empirical correlations, in Model VI, the parameter ΔLK_2 was set in terms of the first formation constant for the complexation of the metal M^{z+} with NH_3:

$$M^{z+} + NH_3 \rightleftharpoons (MNH_3)^{z+} \quad K_{MNH_3} \tag{13.15}$$

in the form

$$\log \Delta LK_2 = 0.55 \log K_{MNH_3} \tag{13.16}$$

Also, a correlation between K_{MA} and K_{MB} (Equations 13.8 and 13.9) was found in the form

$$\log K_{MB} = 3.39 \log K_{MA} - 1.15 \tag{13.17}$$

This correlation (which in fact is a linear relationship between the free energy changes in the binding of M to sites A and B) was later found to be more general (Carbonaro and Di Toro 2007). Furthermore, not every combination of monodentate sites into bidentate and tridentate is allowed: in Model VI, eight bidentate and 16 tridentate sites are included (Tipping 1998), given by diverse combinations of carboxylic- and phenolic-type groups; in the SHM, only six bidentate sites are allowed (Gustafsson 2001), phenolic–phenolic bidentate binding is not allowed. Figure 13.2 shows schematically these metal binding spectra for $\log K_M = 5$ and $\Delta LK_1 = 1$.

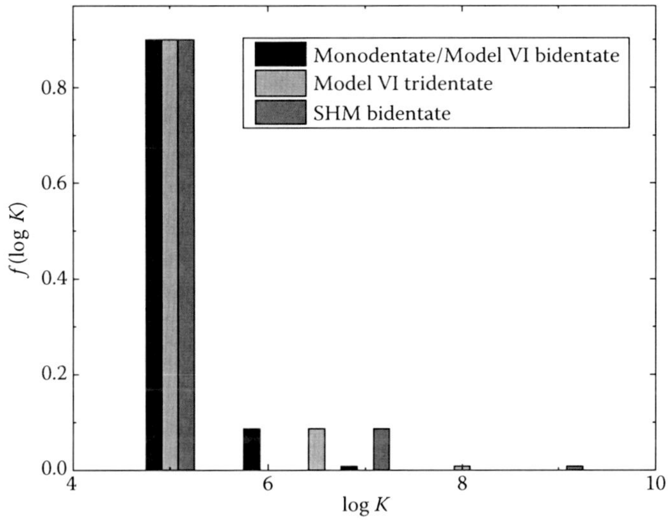

FIGURE 13.2 Schemes of metal cation affinity spectra according to Tipping's Model VI and Gustafsson's Stockholm humic model.

More recently, Tipping and coworkers (Lofts and Tipping 2011; Stockdale, Bryan, and Lofts 2011; Tipping, Lofts, and Sonke 2011) improved the treatment, introducing Model VII, where the main changes from Model VI affinity spectra are the following:

1. The replacement of Equation 13.17 by

$$\log K_{MB} = \log K_{MA}\left(\frac{\mathrm{p}K_{HB}}{\mathrm{p}K_{HA}}\right) \tag{13.18}$$

2. The parameters ΔLK_{A} and ΔLK_{B} are dropped altogether, as the fitting was generally not sensitive to its value.
3. The number of bidentate site combinations was reduced from eight to six and that of tridentate sites was reduced to eight.

These somewhat different proposals are descriptions of proton and metal ion binding based on, in principle, arbitrary constant distributions. They have, of course, been chosen with the overall HS composition in mind but are not directly related either to spectroscopic data or other experimental evidence. The proton affinity spectrum, arranged as two sets of sites, is reasonable in terms of the known facts about the humic structure (Chapter 10) and is consistent with affinity spectra found by numerical inversion methods (Section 11.3.3). The metal binding constants include multidentate binding, which is expected to occur (Section 10.4); the definitions chosen, however, came from a trade-off between inclusion of diverse complexation modes and preventing the number of adjustable parameters from getting too high. In that sense, and taking into account the chemistry of metal complex formation, the exclusion of monodentate binding to phenolic groups, which is generally not observed with small molecules, appears adequate.

13.2.2 Continuous Affinity Spectra

As presented in Section 11.2.2, there are essentially two forms of continuous affinity spectra proposed: Gaussian (Manunza et al. 1995) and Sips (Sips 1948; Koopal et al. 1994) functions, the latter underlying the well-known nonideal competitive adsorption (NICA) isotherm (Benedetti et al. 1995). Although the two are very similar (see Figure 11.4), the NICA equation is very often used, because it leads to an analytical expression for the adsorption isotherm, which is lacking in the Gaussian case.

A general Gaussian affinity spectrum has the form

$$f(\log K) = \sum_{i=1}^{n_p} \frac{y_i}{w_i\sqrt{\pi/2}} e^{-2\left((\log K - \log K_i)/w_i\right)^2} \tag{13.19}$$

where n_p is the number of peaks, y_i the fraction of total sites corresponding to sites of type i, K_i the location of the Gaussian function center (mean binding equilibrium constant), and w_i the width (heterogeneity) parameter. It has been used mainly for conditional affinity spectra (Manunza et al. 1995; Bersillon et al. 2001; Orsetti, Andrade, and Molina 2009; Orsetti, Andrade, and Molina 2010) because it does not support, at least now, competitive binding. The Sips affinity spectrum is given by

$$f(\log K) = \frac{\ln 10}{2\pi} \sum_{i=1}^{n_p} \frac{y_i \sin(\pi n_i)}{\left\{\cosh\left[n_i \ln 10(\log K - \log K_i)\right] + 2\cos(\pi n_i)\right\}} \tag{13.20}$$

where n_i is the width parameter and K_i has the same meaning as in Equation 13.19. The Sips function leads to the Langmuir–Freundlich isotherm (Section 4.4.7.2) and, considering competitive binding, to the NICA isotherm (Section 11.2.2) given by

$$Q_i = \sum_{k=1}^{m} \frac{n_{i,k}}{n_{1,k}} Q_{\max,k} \theta_{i,k}(a_1,\ldots,a_n) = Q_{\max} \sum_{k=1}^{m} \frac{n_{i,k}}{n_{1,k}} y_k \theta_{i,k}(a_1,\ldots,a_n) \tag{13.21}$$

with

$$\theta_{i,k}(a_1,\ldots,a_n) = \frac{\left(K_{i,k}a_i\right)^{n_{i,k}}}{\sum_{j=1}^{N}\left(K_{j,k}a_j\right)^{n_{j,k}}} \times \frac{\left[\sum_{j=1}^{N}\left(K_{j,k}a_j\right)^{n_{j,k}}\right]^{p_k}}{1+\left[\sum_{j=1}^{N}\left(K_{j,k}a_j\right)^{n_{j,k}}\right]^{p_k}} \tag{13.22}$$

where $\theta_{i,k}$ is the coverage (fraction of sites bound) of species i on sites type k; p_k is the Sips distribution width parameter of site type k, representing the "intrinsic heterogeneity" of the sorbing colloid, whereas n_i is the nonideality parameter of ion i; $n_{1,k}$ is a reference n parameter, normally taken as that belonging to protons n_H; and N is the number of competitively binding species. Usually, only two site types are included, thus $m = 2$ in Equation 13.21. For proton-only binding ($N = 1$), Equation 13.22 reduces to

$$Q_H = Q_{\max,1} \frac{(K_{H,1}a_H)^{m_1}}{1+(K_{H,1}a_H)^{m_1}} + Q_{\max,2} \frac{(K_{H,2}a_H)^{m_2}}{1+(K_{H,2}a_H)^{m_2}} \quad (13.23)$$

where $m_i = n_{H,i}p_i$; note that this is simply a two-site Langmuir–Freundlich isotherm. The NICA isotherm constitutes the intrinsic part of the well-known NICA–Donnan (ND) model (Kinniburgh et al. 1996, 1999; Milne, Kinniburgh, and Tipping 2001; Milne et al. 2003). Figures 11.4 and 11.5 show examples of NICA isotherms found by fitting of titration data.

13.2.3 Structure-Based Affinity Spectra

There is ongoing work looking for affinity spectrum models more closely related to the HS molecular structure. Atalay, Carbonaro, and Di Toro (2009) used the system performs automated reasoning in chemistry (SPARC) computational method (Hilal, Karickhoff, and Carreira 2003; Hilal et al. 2007) to predict proton-binding constants from molecular structure. A series of molecular structures that have been suggested for both HA and FA, with different structural arrangements of acidic and other functional groups, were used as an equimolar mixture to represent HAs and FAs. For aqueous HAs and FAs, the results are comparable to the WHAM distribution. For soil HA, the WHAM probability distribution is less acidic than the SPARC predictions for the carboxylic sites but similar to that of the phenolic sites. Matynia et al. (2010) estimated semiempirically average proton-binding constants ($K_{H,i}$) for published structural models of HA and FA (Figure 13.3) by breaking down the macromolecules into reactive structural units (RSUs) and calculating $K_{H,i}$ values of the RSUs using linear free energy relationships. The predicted constants for phenolic-type sites are generally higher than those derived from potentiometric titrations, but the difference may not be significant in view of the considerable uncertainty of the acidity constants determined from acid–base measurements at high pH. The predicted constants for carboxylic-type sites agree well with titration data analyzed with the WHAM VI, the Impermeable Sphere model (Avena, Koopal, and Van Riemsdijk 1999), and the SHM but differ by about one log unit from those obtained with the ND model (Milne, Kinniburgh, and Tipping 2001) and used to derive recommended generic values. Recent efforts mainly use linear free energy relationships to seek prediction of HA and FA binding constants (Carbonaro, Atalay, and Di Toro 2011; Sundararajan, Rajaraman, and Ghosh 2011).

On the other hand, Cabaniss (2008, 2009, 2011) proposed the prediction of metal–natural organic matter (NOM) complexation constants using a methodology termed "quantitative structure–property relationships," which is essentially based on linear free energy relationships, seeking to predict binding constant values through calibration against reported complex formation values in the NIST databases. Now, NOM conditional binding constants have been predicted for several cations at pH = 7.0 and I = 0.1 M and only for 1:1 complexes. Some predictions of this model are consistent with a variety of experimental data from the literature, but others await validation by molecular-level analysis.

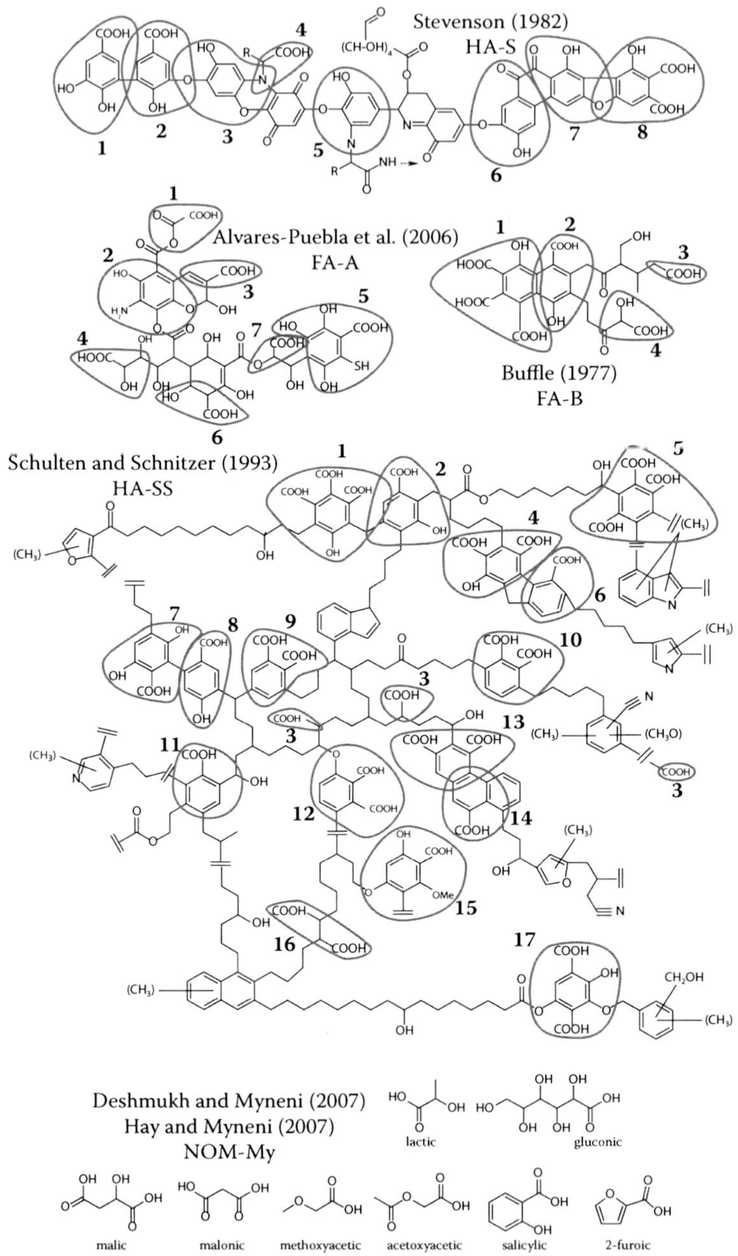

FIGURE 13.3 Structure models for humic and fulvic acids proposed in the literature and used to obtain average proton-binding constants according to Matynia et al. (2010). Circled areas delimit the reactive structural units. (Reprinted from *Geochimica et Cosmochimica Acta*, 74, Matynia, A. et al., 2010. Semiempirical proton binding constants for natural organic matter, 1836–1851. Copyright 2010, with permission from Elsevier.)

13.3 ELECTROSTATIC MODELING OF ION BINDING TO HUMIC PARTICLES

The electrostatic contribution modeling has also followed two basic lines (Saito et al. 2005): one treated the HSs as rigid particles (Bartschat, Cabaniss, and Morel 1992; de Wit, van Riemsdijk, and Koopal 1993; Milne et al. 1995; Gustafsson 2001; Companys et al. 2007), applying standard colloid chemistry models, whereas other authors applied, although sometimes in a restricted and approximate way, the concept of Donnan equilibrium (Benedetti, Van Riemsdijk, and Koopal 1996; Tipping 1998; Companys et al. 2007).

13.3.1 Particle Models

There are several possible forms for modeling the HSs as a particle (Bartschat, Cabaniss, and Morel 1992; Avena, Koopal, and van Riemsdijk 1999; Ohshima 2006; Orsetti, Andrade, and Molina 2010). HSs usually are modeled as spherical particles, which can be considered as either "hard or soft particles" (Figure 13.4). A hard particle has a fixed size, independent of medium conditions such as pH and ionic strength that affect, among other things, the humic charge. A soft particle undergoes size changes as conditions change, most prominently its charge; typically, the radius grows as the charge increases, because of increased repulsion. On the other hand, there are two further possibilities regarding its morphology: the "impermeable" and "permeable" spheres (Figure 13.5); in the first case, the charge is assumed to be located on the surface and counterions are located outside, just as with a mineral particle (Figure 13.5a), whereas in permeable particles, the negative humic charge is spread over the particle volume, and ions are allowed to penetrate the particle (Figure 13.5b), compensating, at least partially, the charge. Note the potential curves indicated in the figure. Impermeable particles behave as any solid particle discussed so far. The potential profiles start at the surface with a given potential ψ_0 (negative in Figure 13.1) and decay toward the bulk solution; the internal particle potential is in principle not known and is irrelevant. In a permeable sphere, there is a continuous potential profile outside and inside the particle, which is schematically shown in Figure 13.5b (bottom). It is shown here as a smooth line because, generally, the charge is considered averaged over the neighbor sites; however, because there is actually a discrete charge distribution, the potential should show local variations around

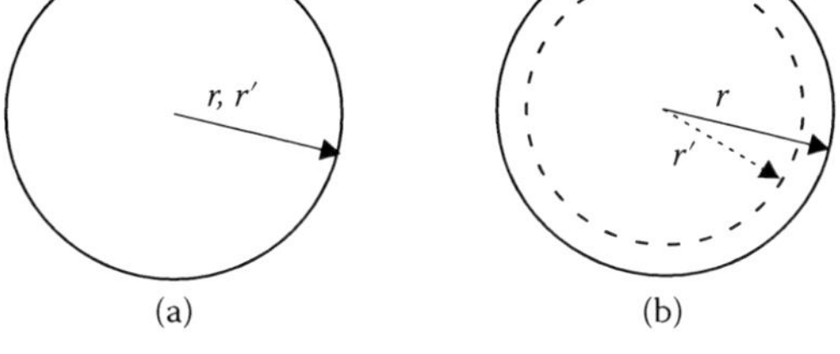

FIGURE 13.4 (a) Hard and (b) soft sphere models; r and r' are the sphere radii for higher and lower humic charges, respectively.

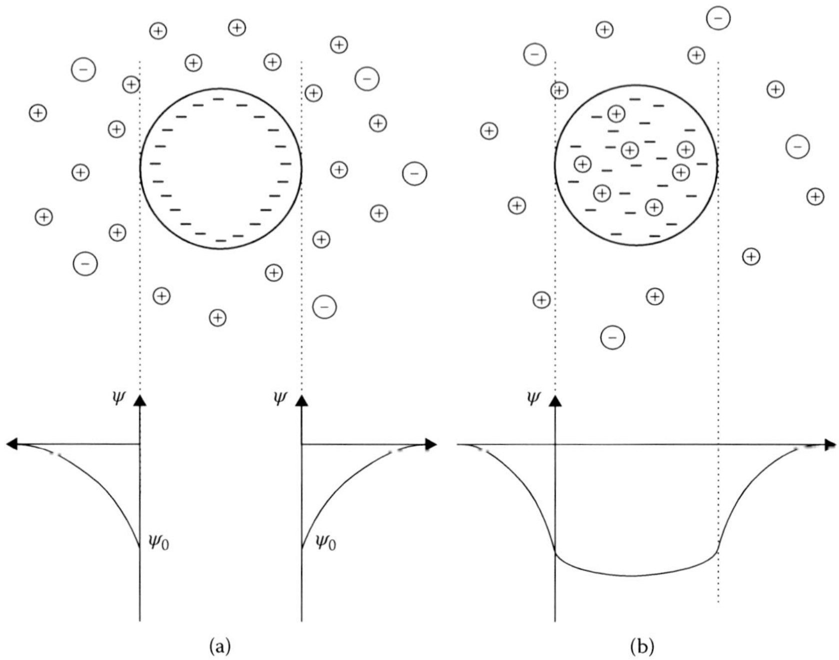

FIGURE 13.5 *Top*: impermeable (a) and permeable (b) sphere models. *Bottom*: schematic electric potential profiles for each case.

the ions and ionized groups, locations, leading to an irregularly oscillating profile; this profile, however, is extremely difficult to solve.

In principle, the different assumptions shown in Figures 13.4 and 13.5 can be combined in all ways, leading to four possibilities. In practice, only two models are usually considered. In the permeable sphere model (PSM), the humic particle is considered a soft separate phase or gel, where the humic molecules are penetrated—swelled—by water and ions, and surrounded by a diffuse layer region. In the impenetrable sphere model (ISM), the humic molecules are treated as hard spheres, similarly to solid colloidal particles (such as oxides), having a surface charge and a diffuse layer. Saito et al. (2005) analyzed several electrostatic models, separating them in permeable and impenetrable sphere types, and used the master curve approach to evaluate the models considered. They found that, under particular assumptions, both types were applicable. Ohshima (2006) introduced a PSM model with a hard, impenetrable core that could be either charged or uncharged (but considered uncharged in the model developed there), termed "soft particle model."

Many recent studies on HSs show that humic molecules are associated and/or entangled, forming aggregates in most of the pH range (Simpson et al. 2002; Rizzi et al. 2004; Sutton and Sposito 2005; Kawahigashi, Sumida, and Yamamoto 2005; Baigorri et al. 2007a,b; Kucerík et al. 2007), and behave as soft, permeable particles (Duval et al. 2005). Thus, the PSM appears to be a more realistic view of humic particles.

13.3.2 Donnan Phase Models

In Donnan phase models (Marinsky 1985; Marinsky, Baldwin, and Reddy 1985; Benedetti, Van Riemsdijk, and Koopal 1996; Miyajima, Mori, and Ishiguro 1997), a Donnan equilibrium is assumed between the bulk solution and ions contained in a permeable particle or in a separate phase (Figure 13.6) separated by a virtual "membrane." In Figure 13.6, the dashed line represents the virtual membrane and the grayed area the internal Donnan phase; effectively, this amounts to treat the diffuse layer as a separate phase. The Donnan equilibrium, in a general form, is established as follows (Stell and Joslin 1986; Chang 2005; Atkins and Paula 2011): a solution containing macromolecules with a total (expressed as mole number) negative charge $z_h\left(z_h < 0\right)$ and an electrolyte constituted by small cations and anions of charge z_+ and z_- is enclosed by a membrane (as schematized in Figure 13.6) in the presence, outside the membrane, of a solution of the same electrolyte ions in concentrations c_+ and c_-, respectively, which can cross the membrane. Inside the membrane, the solution has volume V_{D} (the Donnan volume), and the electrolyte ions have concentrations c_+^{D} and c_-^{D}. In the bulk, the electroneutrality condition obviously imposes that

$$z_+c_+ + z_-c_- = 0 \tag{13.24}$$

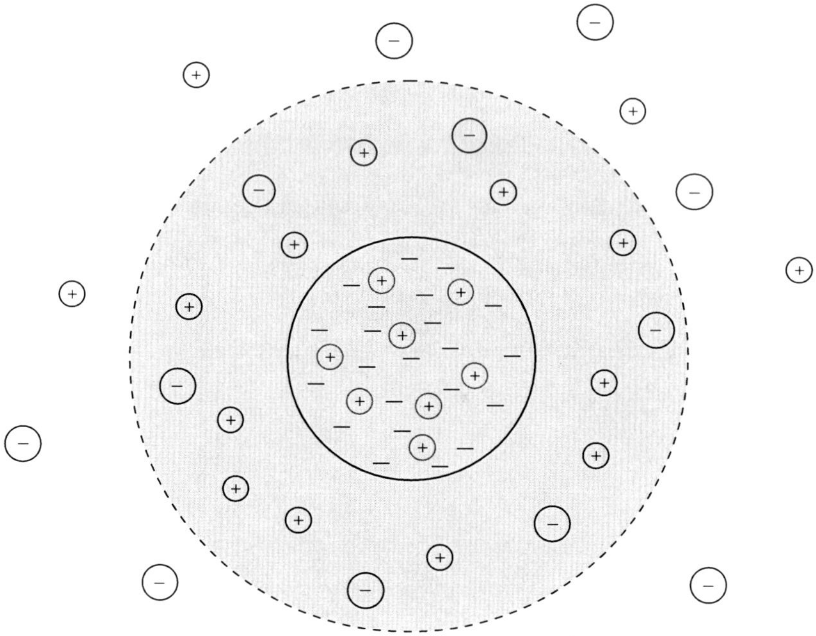

FIGURE 13.6 A scheme of a virtual Donnan phase: a portion of the solution around the humic particle, effectively containing the diffuse layer (grayed area) is considered as a different phase separated from the bulk solution, in electrochemical equilibrium.

Inside the Donnan volume, Equation 13.24 no longer holds because of the presence of the negatively charged macromolecule; instead, the electroneutrality condition is now (assuming that the Donnan volume is large enough as to neglect border effects)

$$\frac{z_h}{V_D} + z_+ c_+^D + z_- c_-^D = 0 \tag{13.25}$$

Clearly, the concentrations inside V_D cannot be equal to the bulk solution values: because of the macromolecule charge, the cation concentration should be higher than the bulk, whereas the anion concentration should be lower. The ions able to cross the membrane must be in electrochemical (osmotic) equilibrium between the two solutions; that is, in Equations 3.1 through 3.3, the electrochemical potentials must be equal,

$$\mu_i^o + \boldsymbol{R}T \ln\left(\frac{c_i \gamma_i}{m^\upsilon}\right) + z_i \boldsymbol{F} \psi = \mu_i^o + \boldsymbol{R}T \ln\left(\frac{c_i^D \gamma_i^D}{m^\upsilon}\right) + z_i \boldsymbol{F} \psi_D \tag{13.26}$$

where the left-hand side corresponds to the bulk solution and the right one to the Donnan volume; ψ_D is the potential inside the membrane, the Donnan potential. The potential in the bulk is normally set to zero, thus the Donnan potential is given by the ratio of activities:

$$\frac{a_i}{a_i^D} = \frac{c_i \gamma_i}{c_i^D \gamma_i^D} = \boldsymbol{e}^{z_i \boldsymbol{F} \psi_D / \boldsymbol{R}\mathrm{T}} = \chi_D \tag{13.27}$$

If the activity coefficients are assumed equal at both sides of the membrane, the Donnan potential for a 1:1 electrolyte of concentration c results to be

$$\psi_D = \frac{\boldsymbol{R}T}{\boldsymbol{F}} \operatorname{arcsinh}\left(\frac{z_h}{2cV_D}\right) \tag{13.28}$$

In principle, the Donnan potential and the concentrations inside the Donnan volume can be found if z_h is known and the activity coefficients can be obtained. However, in several models, as discussed later, ψ_D is not explicitly calculated, and only Equations 13.25 and 13.27 are used to close the system.

In Tipping's models WHAM V and VI (Tipping and Hurley 1992; Tipping 1998), a Donnan phase is assumed, surrounding the humic particles and containing only positively charged counterions, whereas the electrostatic effect on binding is accounted for by an empirical Boltzmann factor; here, the humic particles are assumed to be of fixed size (one value for FAs and another one for HA). In the Model VI-FD (Smith et al. 2004), the Donnan volume is fixed and contains agglomerates of humic molecules (Figure 13.7). In the ND model (Kinniburgh et al. 1999; Milne, Kinniburgh, and Tipping 2001), and a model based on the Henderson–Hasselbach equation (Companys et al. 2007), a "Donnan volume" of solution around the humic particle is assumed to contain enough ions—the whole system of particle plus solution— so as to be electrically neutral, the volume being adjustable as a function of ionic strength. Both approaches lead to good fit with experimental data (Christensen et al. 1998); the Donnan volume model, however, has been criticized, as it leads in some situations to unreasonably larger volume values (Avena, Koopal, and Van Riemsdijk 1999).

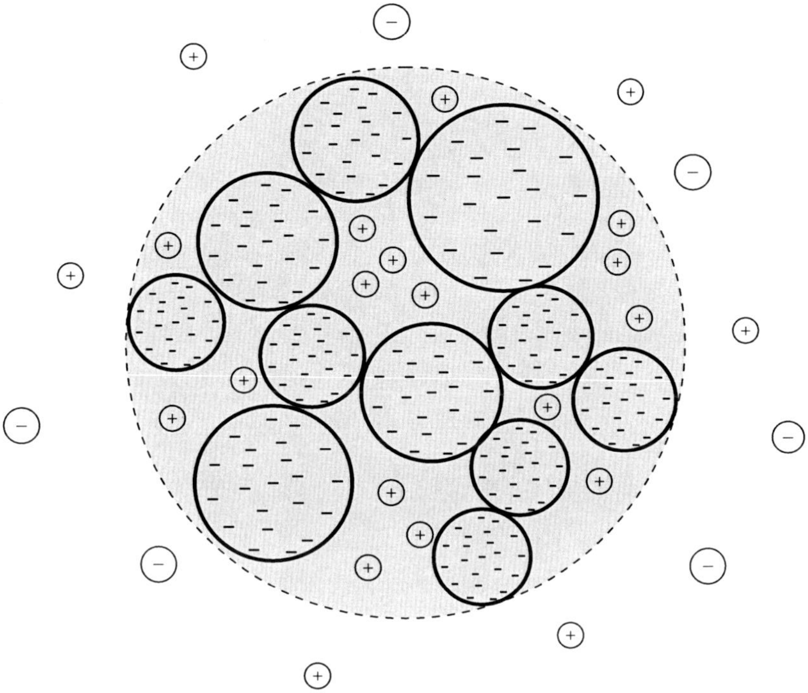

FIGURE 13.7 A scheme of a Donnan phase surrounding an aggregate of humic molecules, as assumed in WHAM VI-FD. (From Smith, E. J. et al., *European Journal of Soil Science*, 55, no. 3, 433–447, 2004.)

A recent model, based on the Donnan equilibrium concept, but with a different approach, is the elastic polyelectrolyte network (EPN) model. In the EPN model (Orsetti, Andrade, and Molina 2010), the HS particles are considered as divided in to two fractions: an inner fraction g_f, which behaves as a gel in Donnan equilibrium with the bulk solution, and an external fraction $1-g_f$, which is in equilibrium with the bulk (Figure 13.8); thus, it is assumed that a fraction g_f of the total number of sites resides inside the gel phase, whereas a fraction $1-g_f$ resides outside the gel; given an affinity spectrum, it applies to the external sites in equilibrium with the bulk solutions, with activity a_i for ion i, whereas for the internal medium, it is applied the internal activity a_i^D, in Donnan equilibrium with the bulk:

$$a_j^D = a_j e^{-F\psi_D/RT} \tag{13.29}$$

The gel is treated as an elastic-charged polymeric network that swells and shrinks as its charge changes; the Donnan potential (for a 1:1 electrolyte) is given by

$$\psi_D = \frac{RT}{F}\operatorname{arcsinh}\left(\frac{g_f \vartheta_A \delta \varphi_2}{2I(1-\varphi_2)}\right) \tag{13.30}$$

where ϑ_A is the net negative charge per unit of mass of the humic particle, δ is the density of the dry HS, $\delta = 1.5$ g cm^{-3} (Dinar, Mentel, and Rudich 2006), φ_2 is the

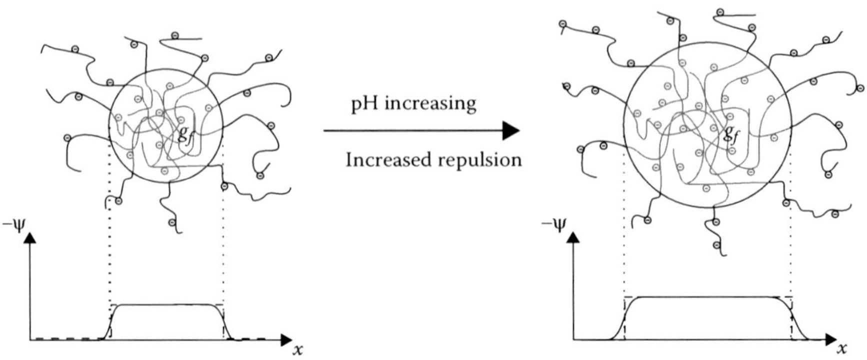

FIGURE 13.8 Schematic representation of the humic electrostatic model: a fraction g_f of the total humic particle is modeled as a gel-like region in Donnan equilibrium with the bulk solution, defining a Donnan potential ψ_D, which affects the ion activity inside the gel. The gel behaves as a soft, penetrable particle, expanding as the humic charge increases. The continuous lines in the potential plots at the bottom represent the actual potential profiles expected; the dashed lines show the model assumed profile. The small circles represent negatively charged sites. Ion charges are not shown for clarity. (Reprinted with permission from Orsetti, S. et al., *Langmuir*, 26, no. 5, 3134–3144. Copyright 2010 American Chemical Society.)

volume fraction of the HS in the gel, and I is the ionic strength; φ_2 is obtained from the swelling equilibrium condition given below:

$$\frac{1}{v_1}\left[\chi\varphi_2^2+\varphi_2+\ln(1-\varphi_2)\right]+\frac{1}{v_2}\left(\varphi_2^{1/3}-\frac{2\varphi_2}{f_P}\right)+2I\left[1-\sqrt{1+\left(\frac{g_f\vartheta_A\delta\varphi_2}{2I(1-\varphi_2)}\right)^2}\right]=0 \qquad (13.31)$$

where v_1 is the water molar volume, v_2 is the average molar volume of a humic equivalent chain, χ is a Van Laar heat of mixing parameter, and f_P is the Flory functionality parameter that is set equal to 3.

13.4 MODELS FOR ION BINDING TO HUMIC SUBSTANCES

13.4.1 Windermere Humic Models

As already discussed, the WHAM family postulated a discrete affinity spectra, as in Equations 13.2 and 13.3, for proton exchange (reactions written as dissociations in this case); for metal binding, there are differences between the model versions. In Model V (Tipping and Hurley 1992), a strong correlation between proton and metal binding was assumed (Equations 13.6 and 13.7) that was not satisfactory; in Model VI (Tipping 1998), this correlation was relaxed (Equations 13.8 and 13.9) improving the fitting. Multidentate binding is allowed, with a simple relationship to monodentate constants in Model V (Equation 13.11), whereas in Model VI, bi- and tridentate binding is considered (Equations 13.12, 13.13, and 13.15 through 13.17). In Model VII, finally (Tipping, Lofts, and Sonke 2011), the number of possible bidentate and tridentate sites is reduced, some parameters are dropped, and Equation 13.17 is replaced by Equation 13.18.

The electrostatic part of these models comprises a Boltzmann-type factor affecting the equilibrium constants, as in chemical models for minerals (Equation 11.16):

$$K_i = K_i^{int} e_f(\psi_S) \tag{13.32}$$

Here, the electrostatic factor is a function of humic charge,

$$K_i = K_i^{int} \exp(2wz_h) \tag{13.33}$$

where z_h is the average number of charges in a humic molecule. The parameter w is given by

$$w = P \log I \tag{13.34}$$

where I is the ionic strength (molar) and P is an adjustable parameter; Equation 13.34 implies that w is zero for $I = 1$ M, thus it is valid at lower values of I. Although Equation 13.33 resembles 13.32, a Donnan potential is not calculated, and the exponential concentration profiles expected in the diffuse layer (Chapter 3) are replaced by a region of fixed volume, a Donnan volume region around the particle, where cation accumulation takes place. The Donnan volume is assumed to be a spherical casket around the particle of thickness κ^{-},

$$V_D = \frac{N_A}{M_r} \frac{4\pi}{3} \left[\left(r + \frac{1}{\kappa} \right)^3 - r^3 \right] \tag{13.35}$$

where M_r is the molecular weight of the HS, r is its radius (assumed as a spherical particle), and κ is the Debye inverse length, given by Equation 3.15:

$$\kappa = \sqrt{\frac{2\mathbf{F}^2 I}{\varepsilon \mathbf{R} T}} \tag{13.36}$$

In WHAM, M_r and r are estimated from literature (see Table 13.1). It is assumed in WHAM that the humic charge is completely counterbalanced by cations present in the Donnan volume, whereas anions are excluded. The concentration of a cation i inside this volume is given by

$$c_i^D = c_i K_{sel,i} R^{|z_i|} \tag{13.37}$$

Here, R is the ratio required for the sum of counterion charges to balance the humic charge. Equation 13.37 implies that, for low ionic strengths, the Donnan phase may have a mixture of different cations, whereas for high ionic strengths, the dominant cation will predominate in the Donnan volume.

Table 13.1 presents the parameters of Models VI–VII. It is seen that the number of adjustable parameters is relatively high, as it is the case with most models currently

TABLE 13.1
Summary of Parameters in WHAM VI and VII

Parameter	Description	Origin
Q_A	Abundance of type A sites	Fitted from experiment
Q_B	Abundance of type B sites	0.5 Q_A
pK_A	Intrinsic proton dissociation constant for type A sites (Equation 13.2)	Fitted from experiment
pK_B	Intrinsic proton dissociation constant for type B sites (Equation 13.3)	Fitted from experiment
ΔpK_A	Distribution term that modifies pK_A (Equation 13.2)	Fitted from experiment
ΔpK_B	Distribution term that modifies pK_B (Equation 13.3)	Fitted from experiment
$\log K_{MA}$	Intrinsic equilibrium constant for metal binding at type A sites (Equation 13.8)	Fitted from experiment
$\log K_{MB}$	Intrinsic equilibrium constant for metal binding at type B sites (Equation 13.9)	Fitted from experiment, or estimated by correlation with $\log K_{MA}$ (Equation 13.17)
ΔLK_{MA} ΔLK_{MB} ΔLK_1	Distribution term(s) that modify $\log K_{MA}$ (Equations 13.8 and 13.9)	Fitted from experiment, or fixed at an universal value; dropped in Model VII
ΔLK_2	Distribution term that modifies the strengths of bidentate and tridentate sites (Equations 13.12 and 13.13)	Fitted from experiment, or estimated by correlation with $\log K_{MNH3}$ (Equation 13.16)
P	Electrostatic parameter (Equation 13.34)	Fitted from experiment
K_{sel}	Selectivity coefficient for counterion accumulation (Equation 13.37)	Fitted from experiment, or usually set to unity
f_b	Fraction of proton sites that can make bidentate sites	Calculated from geometry
f_t	Fraction of proton sites that can make tridentate sites	Calculated from geometry
M	Molecular weight: 1500 (FA), 15000 (HA)	Estimated from literature
r	Molecular radius 0.8 nm (FA), 1.72 nm (HA)	Estimated from literature

Source: Adapted from Tipping, E., *Aquat. Geochem.*, 4, 3–48, 1998; Tipping, E. et al., *Environ. Chem.*, 8, 225–235, 2011.

used; even adopting values from literature or known correlations, there are six for proton binding and, at least, one more for each metal cation. Good fitting to experimental data is found with Model VI and VII (Tipping 1998; Tipping, Lofts, and Sonke 2011). Figure 13.9 shows an example for proton binding to the purified peat humic acid (PPHA) of Milne et al. (1995), and Figure 13.10 shows the prediction of Cd binding by the same acid, data from Benedetti et al. (1995).

13.4.2 Stockholm Humic Model

The affinity part of the SHM proposed by Gustafsson (2001) is similar to the WHAM case, but simpler, with less parameters. The proton-binding part is given by Equations 13.2 and 13.3, whereas the monodentate metal binding is allowed only

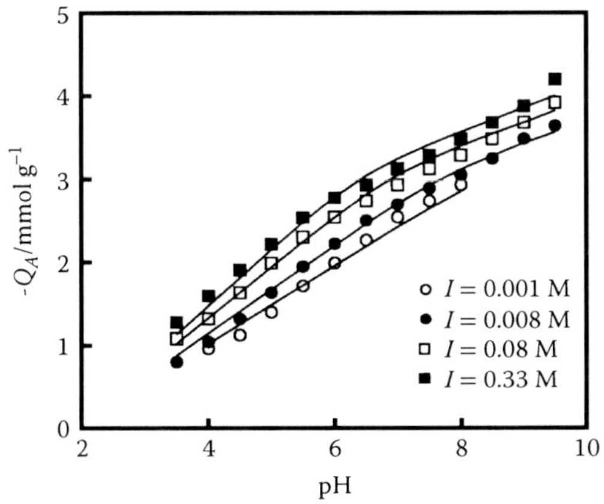

FIGURE 13.9 Proton dissociation fitting to model for HA, using data for purified peat humic acid (PPHA), dataset HH-09 in Milne compilation (Milne et al. 2003), originally form Milne et al. (1995). The lines are model VI fits. (Reprinted with kind permission from Springer Science + Business Media: *Aquatic Geochemistry*, Tipping E., Humic ion-binding model VI: An improved description of the interactions of protons and metal ions with humic substances, 4, 1998, 3–48, Copyright 1998.)

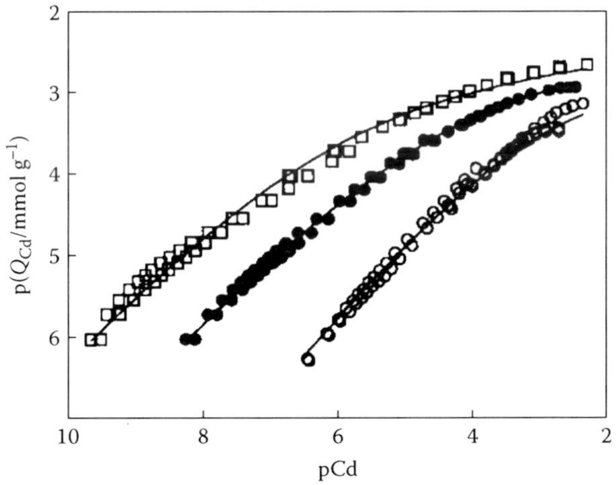

FIGURE 13.10 Cadmium binding by HA. The data are from Benedetti et al. (1995). Lines are fits with WHAM VI (4 adjustable parameters). (○) pH = 4; (●) pH = 6; (□) pH = 8. (Reprinted with kind permission from Springer Science + Business Media: *Aquatic Geochemistry*, Tipping E., Humic ion-binding model VI: An improved description of the interactions of protons and metal ions with humic substances, 4, 1998, 3–48, Copyright 1998.).

for type A sites (Equation 13.10). Bidentate binding is allowed between selected A and B sites (also similar to modelWHAM) with a simpler spread definition (Equation 13.14). In addition, in some cases, the monodentate adsorption of MOH^+ complexes is included, using the standard aqueous hydrolysis constant for M^{2+} and OH^- complexation.

In the electrostatic part, instead, the SHM uses an impermeable sphere model, applying the basic Stern model (BSM) (Section 12.2.2) with some concepts borrowed from the CD-MUSIC model (Section 12.3). In the SHM, the bulk of the HSs is considered to form gels, modeled using the BSM. However, depending on their configuration and size, some humic molecules may have certain groups outside the gel that behave like normal monomeric molecules. Electrostatic interactions within this part of the HSs are considered to be negligible. This has some similarities with the EPN discussed above (Section 13.3.2), but here, the gel is actually an impermeable sphere.

In the BSM, the surface charge is related to the surface potential ψ_0 and the diffuse layer potential ψ_2 by Equation 12.13:

$$\sigma_0 = C(\psi_0 - \psi_2) \tag{13.38}$$

The charge–potential relationship for the spherical geometry assumed is given approximately for a 1:1 electrolyte of concentration c by (Ohshima, Healy, and White 1982)

$$\sigma_d = A\sqrt{8\boldsymbol{RT}c\varepsilon_0\varepsilon_r}\,\sinh\left(\frac{\boldsymbol{F}\psi_2}{2\boldsymbol{RT}}\right) - \sigma_0 \tag{13.39}$$

where ε_0 is the vacuum permittivity, ε_r the water dielectric constant, and

$$A = \sqrt{1 + \left[\frac{2}{\kappa r \cosh^2\left(\dfrac{\boldsymbol{F}\psi_2}{4\boldsymbol{RT}}\right)}\right] + \left[\frac{8\ln\cosh\left(\dfrac{\boldsymbol{F}\psi_2}{4\boldsymbol{RT}}\right)}{(\kappa r)^2 \sinh\left(\dfrac{\boldsymbol{F}\psi_2}{4\boldsymbol{RT}}\right)}\right]} \tag{13.40}$$

where κ is the Debye inverse length and r is the particle radius. The surface charge σ_0 is estimated, in the absence of specific adsorption, by

$$\sigma_0 = \frac{\boldsymbol{F}[\mathrm{R}^-]}{A_S c_{HA}} \tag{13.41}$$

where $[\mathrm{R}^-]$ is the molar concentration of dissociated sites, A_S the surface area, and c_{HA} the mass concentration of the HSs.

The equilibrium constants in Equations 13.2, 13.3, 13.10, and 13.14 contain an electrostatic factor for charged species; here, an approach similar to CD-MUSIC model is taken, dividing the charge of the specifically bound ion between the surface plane (potential ψ_0) and the Stern plane (ψ_2); the charge distribution factor f (Section 12.3.1) is assumed to be 0.25 for monodentate binding and 0.60 or 0.62 for bidentate binding. Thus, for modentate binding of a cation M^{2+} giving RM^+, the electrostatic factor is (taking into account that only a fraction g_F of the HSs is affected by electrostatic interactions)

$$e_f = -0.5g_f \exp\left(\frac{\boldsymbol{F}\psi_0}{\boldsymbol{RT}}\right) + 1.5g_f \exp\left(\frac{\boldsymbol{F}\psi_2}{\boldsymbol{RT}}\right) \tag{13.42}$$

For bidentate binding giving R_2M, the electrostatic factor is

$$e_f = -0.75 g_f \exp\left(\frac{F\psi_0}{RT}\right) + 0.75 g_f \exp\left(\frac{F\psi_2}{RT}\right) \tag{13.43}$$

Finally, nonspecific binding is considered, where electrolyte monovalent cations are allowed to adsorb at the Stern plane to compensate the gel charge $[X^-] = g_f[R^-]$,

$$XC \rightleftharpoons X^- + C^+ \tag{13.44}$$

where C^+ is a monovalent cation with a constant K_C.

See the original paper (Gustafsson 2001) for further details. In the SHM, there are thus several parameters, summarized in Table 13.2; some of them are calculated or estimated from literature as shown in Table 13.2, and after that there are still six parameters for proton binding plus three more for metal binding.

The SHM is somewhat simpler than the WHAM and shows good fitting of single-metal binding; but for competitive metal binding, it is not so satisfactory

TABLE 13.2
Summary of Parameters Used in SHM

Parameter	Description	Origin
Q_{max}	Amount of proton-dissociating sites (= $Q_{max,A} + Q_{max,B}$)	Fitted from experiment
$Q_{max,B}$	Amount of type B sites	0.5 $Q_{max,A}$ (HA); 0.3 $Q_{max,A}$ (FA)
$\log K_A$	Intrinsic proton dissociation constant for type A sites	Fitted from experiment
$\log K_B$	Intrinsic proton dissociation constant for type B sites	Fitted from experiment
ΔpK_A	Distribution term that modifies $\log K_A$	Fitted from experiment
ΔpK_B	Distribution term that modifies $\log K_B$	Fitted from experiment
$\log K_{Mm}$	Instrinsic equilibrium constant for monodentate complexation of metal M	Fitted from experiment
$\log K_{Mb}$	Instrinsic equilibrium constant for bidentate complexation of metal M	Fitted from experiment
ΔLK_2	Distribution term that modifies the strengths of complexation sites	Fitted from experiment
r	Radius of HA or FA	Fixed at 0.75 nm (FA); 1.8 nm (HA)
C	Sterm layer capacitance	Fixed at 2 F m^{-2}
Γ_S	Site density of HS functional groups	Fixed at 1.2 sites nm^{-2}
A_S	Specific surface area of HS	Calculated from geometry using r and Γ_S
g_f	Gel fraction parameter	Fitted from experiment
K_C	Instrinsic equilibrium constant for the accumulation of screening counterions	Fixed at universal value: $10^{0.8}$

Source: Adapted from Gustafsson, J. P., *J. Colloid Interf. Sci.*, *244*, 102–112, 2001. Table 13.3.

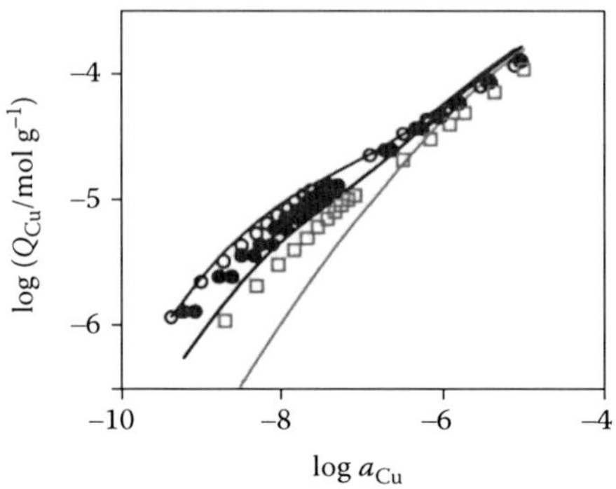

FIGURE 13.11 Copper complexation to 1 g/L Pennines (N. England) HA at pH 4 in the presence of Fe(III) at different ratios to HA (in μmol Fe/g HA): (○) 3.5; (●) 14; (□) 36. Points are observations of Tipping et al. (2002), and lines are model fits to the SHM. (Reprinted from *Environ Sci Technol, 41*, Gustafsson, J. P. et al., Binding of iron (III) to organic soils: EXAFS spectroscopy and chemical equilibrium modeling, 1232–1237. Copyright 2007, with permission from Elsevier.)

(see Figure 13.11). This model has been used mainly by Gustafsson and coworkers (Gustafsson and Van Schaik 2003; Gustafsson, Pechova, and Berggren 2003; Gustafsson and Kleja 2005; Gustafsson et al. 2007).

13.4.3 NICA–Donnan Model

The ND model, introduced by the Wageningen group (Benedetti et al. 1995; Benedetti, Van Riemsdijk, and Koopal 1996; Kinniburgh et al. 1996, 1999; Milne, Kinniburgh, and Tipping 2001; Milne et al. 2003), is composed of the NICA isotherm for intrinsic interaction (Equations 13.21 through 13.23) and a simple Donnan phase for ion accumulation, a generalized form of Equation 13.25 is

$$\frac{z_h}{V_D} + \sum_{i=1}^{N} z_i c_i^D = 0 \tag{13.45}$$

where the sum runs over all ions present in the system. The concentrations in the Donnan phase, c_i^D, are related to the bulk concentrations, c_i, by a Boltzmann factor (Equation 13.27)

$$c_i^D = c_i \chi_D^{z_i} \tag{13.46}$$

with $\chi_D = \exp(-\boldsymbol{F}\psi_D / \boldsymbol{RT})$. The Donnan volume was found to vary linearly with the logarithm of ionic strength (Benedetti, Van Riemsdijk, and Koopal 1996), thus the following empirical relation was adopted,

$$\log V_D = b(1 - \log I) - 1 \tag{13.47}$$

b being an adjustable parameter, characteristic of the HSs and independent of experimental conditions. This is the only adjustable parameter for the electrostatic contribution, and that essentially completes the model as commonly used.

Table 13.3 summarizes the ND parameters. There are five parameters for proton binding, but when including metal complexation, four new parameters are included for each metal, plus two more, p_1 and p_2, which cannot be distinguished analyzing proton binding alone (see Equation 13.23), but require simultaneous fitting of proton- and metal-binding data. Unlike the earlier models, where some parameters can be estimated from independent experiments or previous knowledge, in principle, all these parameters should be found from fitting. In some cases, analyzing a large number of literature datasets, Milne et al. (2003) adopted values when dealing with small datasets, but that was often done on the basis of correlation with previous fitting results. Perhaps because of the number of parameters, the ND model has been shown to provide very good fitting and good predicting capability for metal binding to HSs. Figure 13.12 shows an example, with very good fitting for Cu interaction with peat HA at different pHs and ionic strengths (Vidali, Remoundaki, and Tsezos 2011). This model has been often used by several authors (Gustafsson and Kleja 2005; Plaza et al. 2005a; Fernández et al. 2007; Khai et al. 2008; Drosos, Jerzykiewicz, and Deligiannakis 2009; Puy et al. 2009; Groenenberg, Koopmans, and Comans 2010; van Schaik, Kleja, and Gustafsson 2010).

TABLE 13.3
Summary of Parameters in the ND Model

Parameter	Description	
$Q_{max,1}$	Abundance of type 1 sites (assumed carboxylic)	Fitted from experiment
$Q_{max,2}$	Abundance of type 2 sites (assumed phenolic)	Fitted from experiment
$\log K_{H,1}$	Average value (distribution center) of proton affinity constant for sites of type 1	Fitted from experiment
$\log K_{H,2}$	Idem for type 2	Fitted from experiment
$n_{H,1}$	Nonideality parameter for proton binding to type 1 sites	Fitted from experiment
$n_{H,2}$	Idem for type 2	Fitted from experiment
p_1	Intrinsic humic heterogeneity parameter for type 1 sites	Fitted from experiment or assumed universal value
p_2	Idem for type 2	Fitted from experiment or assumed universal value
$\log K_{M,1}$	Average value (distribution center) of cation M^{z+} affinity constant for sites of type 1	Fitted from experiment
$\log K_{M,2}$	Idem for type 2	Fitted from experiment or estimated from literature
$n_{M,1}$	Nonideality parameter for metal binding to type 1 sites	Fitted from experiment
$n_{M,2}$	Idem for type 2	Fitted from experiment or estimated
b	Donnan volume parameter	Fitted from experiment

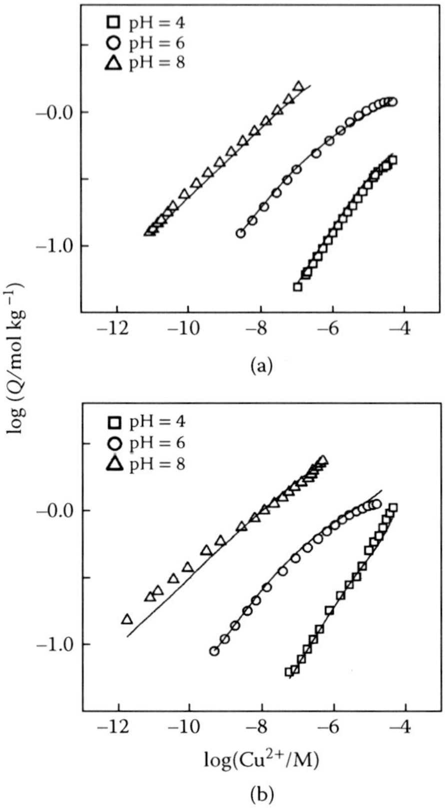

FIGURE 13.12 Copper binding to PPHA at concentration 100 mg L^{-1} (a) and 20 mg (b) at pH 4, 6 and 8 plotted against the free Cu ion concentration [Cu^{2+}] (M). I = 0.1 M $NaNO_3$ (upper panel) and I = 0.01 M $NaNO_3$ (lower panel). Symbols are data from Cu titrations. The curves are obtained by fitting to the ND model. (Reprinted with kind permission from Springer Science + Business Media: *Water, Air, & Soil Pollution*, Vidali R. et al., An experimental and modeling study of Cu_{2+} binding on humic acids at various solution conditions. Application of the NICA-Donnan model, 218, 2011, 487–497, Copyright 2011.)

13.4.4 NICA–EPN Model

The recently proposed EPN model (Section 13.3.2) is an electrostatic contribution scheme (or submodel) that attempts to predict the shrinking and swelling of the HS particles concomitant with cation binding. It assumes that the humic particle is composed of internal gel fraction, which is a soft, permeable particle, and an external part which behaves as normal aqueous weak acids; the internal medium is in Donnan equilibrium with the bulk solution (Equations 13.29 through 13.31). In principle, this scheme can be combined either with discrete or continuous affinity spectra binding models. Now, it has been combined with the NICA isotherm and Gaussian distributions (Orsetti, Andrade, and Molina 2010). In the NICA–EPN model, the total amount bound of an ion i is given by (recalling Equations 13.21 and 13.22)

$$Q_i = (1 - g_f)\sum_{k=1}^{m} \frac{n_{i,k}}{n_{1,k}} Q_{\max,k}\theta_{i,k}(a_1,\ldots,a_N) + g_f \sum_{k=1}^{m} \frac{n_{i,k}}{n_{1,k}} Q_{\max,k}\theta_{i,k}(a_1^D,\ldots,a_n^D) + Q_{i,exc} \tag{13.48}$$

In Equation 13.48, the first term in the right-hand side is the fraction outside the gel, where the bound ions are in equilibrium with the bulk solution; the next term is the binding inside the gel, in equilibrium with the activities in the Donnan phase, and the last term is the accumulation inside the gel phase, given by

$$Q_{i,exc} = \left(c_i^D - c_i\right)\frac{g_f}{\delta\varphi_2} \tag{13.49}$$

The binding inside the gel is expected to be relatively larger than outside because of the effect of the negative Donnan potential. The solid volume fraction φ_2 which is dependent on the net humic charge, and thus on the extension of binding, allows the prediction of the swelling/shrinking of the HS as its negative charge increases/decreases respectively. The adjustable parameters are all the NICA isotherm parameters (Table 13.3, with the exception of b) plus g_f, χ and v_2, thus the number of parameters is higher than in the ND model; however, the increase of parameters should in principle be related to the electrostatic contribution. On the other hand, it is possible to incorporate data for size variations of HSs with pH, although such data are scarce and difficult to obtain; in the original paper (Orsetti, Andrade, and Molina 2010), qualitative comparisons with literature data were made with encouraging results. As of this writing, the NICA–EPN model has been applied only to proton binding, with good results (Orsetti, Andrade, and Molina 2010). Figure 13.13 shows the very good model fitting to the well-known data for PPHA of Milne et al., originally published in 1995 (Milne et al. 1995) but made available later in 2003 (Milne et al. 2003).

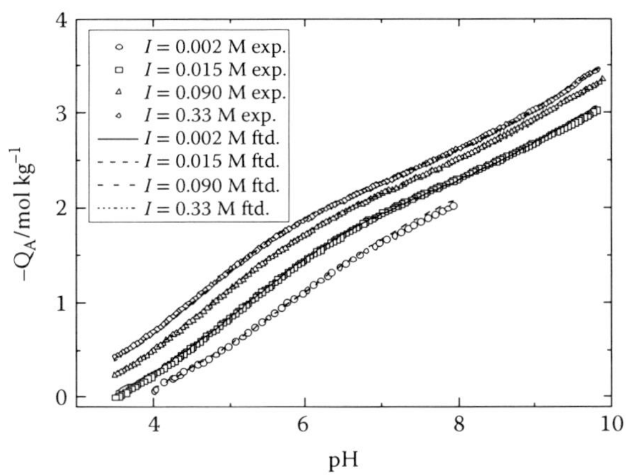

FIGURE 13.13 Fitting of data for proton binding to PPHA, dataset HH-09 of Milne et al. (From Milne, Kinniburgh, and Tipping 2001; Milne et al. 2003) to the NICA–EPN model. (Reprinted with permission from Orsetti, S. et al., 2010, *Langmuir*, 26, no. 5, 3134–3144. Copyright 2010 American Chemical Society.)

13.4.5 Comparison of the Different Models

In the humic speciation models used to date, as discussed in the earlier section, the affinity spectra proposed are only loosely based on knowledge of the HS structure, just in the distinction between carboxylic and phenolic-type sites; moreover, this distinction is sound for proton binding but becomes less reasonable regarding metal complexation, because phenolic groups usually are not observed to bind metal cations appreciably. In addition, recent spectroscopic (Deshmukh et al. 2007) and nonaqueous titration (Kirishima et al. 2009) studies indicate that phenolic groups are largely unsubstituted; thus, their participation in metal complexation with high affinity appears questionable. In the affinity spectra presented here, metal binding behavior is mainly assumed to be similar to proton binding with some allowance, in the discrete spectrum models, for a few high-affinity metal complexation constants. It is worth noting that in the SHM, binding to phenolic groups exclusively is not allowed (Gustafsson 2001), which appears to be a sensible restriction. Another aspect to note is that some studies indicate that not all carboxylic and phenolic groups are detected in acid–base titrations (Janos, Krízenecká, and Madronová 2008; Ritchie and Perdue 2008), but those not contributing to acid–base equilibrium could have a role on metal binding. On the other hand, in the NICA isotherm, the structure of Equations 13.21 and 13.22 implies that only monodentate binding is considered, and that high-affinity metal binding occurs on high-affinity proton sites ("phenolic sites") that, as discussed above, appears to be not very realistic; coordination chemistry concepts strongly suggest that high-affinity complexation should occur on multidentate sites (see also Section 10.4); besides, the presence of "nonideality" parameters $n_{i,k}$ both as isotherm exponents and as a sort of stoichiometric coefficients in Equations 13.21 and 13.22 (required by thermodynamic consistency) feels uncomfortable from a microscopic point of view.

Overall, the discrete affinity spectra models appear to be more close to what is well known in complexation by small ligands; on the other hand, the lack of detailed knowledge about complexing groups in SOM forces the definition of generic sites and the adoption of schemes such as in Equations 13.2 and 13.3 to reduce the number of adjustable parameters, which is equivalent to the parameters in a continuous function, either Gaussian or Sipsian. It is worth noting here that using discrete sites is a chemical modeling scheme, formally identical to those discussed in Chapter 12, whereas the use of continuous affinity spectra, in the context of Equations 11.30 and 11.37, is a form of adsorption modeling.

In the electrostatic part, the Donnan phase model has found quite widespread application, possibly because of its simplicity—at least in the basic way applied in the ND and WHAM VI models—and resulting in good fitting capability; however, in the ND model, the V_D values may become unrealistically high under certain conditions (Avena, Koopal, and Van Riemsdijk 1999). On the other hand, the ISM has been used by some authors (Tipping and Hurley 1992; Avena, Koopal, and van Riemsdijk 1999; Gustafsson 2001) but with less success; only the SHM has shown good fitting/prediction ability.

Of the models above presented, the ND has found wider acceptance and application because of its good predicting capability, including competence between metal cations (Kinniburgh et al. 1999). However, recent discrete site models (WHAM VI and VII) also are able to predict competence fairly well; in fact, at low loadings, any model predicting well single-cation binding should expected to be adequate for competence between cations, of course, also considering proton competition.

The proton binding is the more extensively studied case, and many studies have proposed average/generic values for the mean binding constants of both types of proton sites, carboxylic and phenolic. Table 13.4 summarizes many reported values for mean binding constants and site fractions; the second column specifies the origin of these values: model fitting (with the electrostatic submodel in parenthesis) or estimation from other sources. Most of them are averages resulting from fitting of the above discussed models to experimental data, whereas a few are proposed values

TABLE 13.4
Comparison of Literature Proton-Binding Constants and Site Fractions for Different Humic Models[a]

Reference	Model/Analysis (Electrostatics)	HS	log K_1	log K_2	y_1	y_2
Tipping 1998	WHAM VI (Donnan phase)	FA	3.20	9.40	0.67[b]	0.33[b]
		HA	4.10	8.80	0.67[b]	0.33[b]
Avena, Koopal, and Van Riemsdijk 1999	Numerical analysis of master curve (ISM)	HA	3.5–4.5	7.5–8.5	n.d.	n.d.
Milne, Kinniburgh, and Tipping 2001	ND (Donnan phase)	FA	2.34	8.60	0.69	0.31
		HA	2.93	8.00	0.54	0.46
Gustafsson 2001	Stockholm humic model (ISM, for a fraction of HS)	FA	3.50	8.75	0.77[b]	0.23[b]
		HA	4.10	8.95	0.67[b]	0.33[b]
Ritchie and Perdue 2003	Conditional analysis with modified Henderson–Hasselbalch equation, no electrostatics	FA	3.80	9.78	0.92	0.08
		HA	4.38	9.72	0.88	0.12
Plaza et al. 2005b	ND (Donnan phase)	FA	2.83	7.02	0.65	0.35
		HA	3.74	7.68	0.72	0.28
Hay and Myneni 2007	IR measurements	HS	3.3–3.5	—	—	—
Drosos, Jerzykiewicz, and Deligiannakis 2009	ND (Donnan phase)	HA	3.62	8.54	0.76	0.24
Orsetti, Andrade, and Molina 2010	NICA–EPN (PSM for a fraction of HS)	FA	4.02[c]	8.30[c]	0.63[c]	0.37[c]
		HA	4.22	9.46	0.62	0.38
Matynia et al. 2010	Semiempirical theoretical estimations	FA	3.80	9.87	—	—
		HA	3.73	9.83	9.83	—

[a] Largely based on data collected by Matynia et al. (2010). Fitted values unless otherwise noted.
[b] Assumed, not obtained from fitting.
[c] Average of only three values.

from analysis of other data (such as infrared [IR]) or semiempirical calculations. In general, the fitted values are based on data corresponding to different HSs, but several authors (Tipping 1998; Gustafsson 2001; Milne, Kinniburgh, and Tipping 2001; Orsetti, Andrade, and Molina 2010) have used equal or similar datasets from the compilation of Milne, Kinniburgh, and Tipping (2001) that improves comparison. It is apparent that the ND model results are generally lower than the other fitted values. Considering the carboxylic-type constant, the ND model gives for FA log K_1 values of 2.3–2.8, whereas all other values fall in the range 3.2–4.0; for HA, the values are 2.9–3.7 for ND and 4.1–4.4 for the rest. The non ND results compare reasonably well with model independent results: a range 3.5–4.5 range of Avena (Avena, Koopal, and Van Riemsdijk 1999), the IR deduced result of 3.3–3.5 (Hay and Myneni 2007), or the semiempirical calculations of Matynia (Matynia et al. 2010), giving 3.8 for FA and 3.7 for HA. Thus, the ND results, especially those of Milne et al., appear to be significantly lower. For the phenolic-type sites, there is less dispersion between different models: for FA, the log K_2 fitted values range is 7.0–9.8, and for HA is 7.7–9.7, with the ND results being in the lower part of the range. The semiempirical calculations produce results of about 9.8 for both types of HSs, whereas Avena et al. (1999) deduced a range of 7.5–8.5; thus, for these sites the differences between models are not so marked. Overall, the observation is that different models fit the data equally well with different values for parameters that would have expected to be comparable. This is not only due to the high number of adjustable parameters in the models but also due to the fact that only titrations curves, which are neither very detailed nor sensitive to parameter values, are considered; additional, independent data are needed to advance further.

Observing the average fractions y_1 and y_2 of carboxylic and phenolic types, in general, a higher proportion of carboxylic types is found in fitted results (except WHAM and SHM, where they are fixed), which is in agreement with chemical analysis methods (Senesi and Loffredo 1998). The exception is again the ND model, particularly the results of Milne et al. for HA; from this point of view also, questions arise about how realistic is this model.

13.5 SUMMARY

In this chapter, the modeling of ion binding to HSs has been presented. As in earlier chapters, the models are comprised of two parts (or submodels): the "chemical" contribution to the electrochemical Gibbs free energy change on adsorption and the electrostatic contribution arising from the presence of an electric potential gradient in the interface. Because of the complex and as yet not fully established structure of HSs, well-defined, discrete site models (such as those discussed in Chapter 12) cannot be applied. Instead, a distribution of sites, or affinity spectrum, is assumed, based on the concept that there are two main types of sites: carboxylic and phenolic groups. Affinity spectra can have either discrete or continuous form; the discrete affinity spectra proposed in this case are composed of generic sites assumed to be representative of the actual composition, but bearing no direct relation to the HS structure. Continuous affinity spectra have a Gaussian or quasi-Gaussian (Sips) form; the Sips function leads to the Langmuir–Freundlich isotherm, which is at the heart of the

NICA isotherm for competitive binding. Diverse electrostatic models are considered, mainly based on one out of two concepts: the impermeable sphere model (similar to a mineral particle) or the Donnan phase model, where a certain volume is assumed to be in Donnan equilibrium with the bulk. Considering these contributions, several models are proposed in the end showing similar fitting quality to titration curves because these curves have poor sensitivity to the many adjustable parameters; additional, independent data are needed to improve modeling of cation binding to HSs.

REFERENCES

Atalay, Y. B., R. F. Carbonaro, and D. M. Di Toro. 2009. Distribution of proton dissociation constants for model humic and fulvic acid molecules. *Environmental Science & Technology* 43, no. 10: 3626–3631.

Atkins, P., and J. de Paula. 2011. *Physical Chemistry for the Life Sciences*. 2nd ed. New York: W. H. Freeman & Company.

Avena, M. J., L. K. Koopal, and W. H. van Riemsdijk. 1999. Proton binding to humic acids: Electrostatic and intrinsic interactions. *Journal of Colloid and Interface Science* 217, no. 1: 37–48. doi:10.1006/jcis.1999.6317.

Baigorri, R., M. Fuentes, G. González-Gaitano, and J. M. García-Mina. 2007a. Analysis of molecular aggregation in humic substances in solution. *Colloids and Surfaces A: Physicochemical and Engineering Aspects* 302, no. 1–3: 301–306.

Baigorri, R., M. Fuentes, G. Gonzalez-Gaitano, and J. M. Garcia-Mina. 2007b. Simultaneous presence of diverse molecular patterns in humic substances in solution. *The Journal of Physical Chemistry B* 111, no. 35: 10577–10582. doi:10.1021/jp0738154.

Bartschat, B. M., S. E. Cabaniss, and F. M. M. Morel. 1992. Oligoelectrolyte model for cation binding by humic substances. *Environmental Science & Technology* 26, no. 2: 284–294. doi:10.1021/es00026a007.

Benedetti, M. F., C. J. Milne, D. G. Kinniburgh, W. H. Van Riemsdijk, and L. K. Koopal. 1995. Metal ion binding to humic substances: Application of the non-ideal competitive adsorption model. *Environmental Science & Technology* 29, no. 2: 446–457. doi:10.1021/es00002a022.

Benedetti, M. F., W. H. Van Riemsdijk, and L. K. Koopal. 1996. Humic substances considered as a heterogeneous Donnan gel phase. *Environmental Science & Technology* 30, no. 6: 1805–1813. doi:10.1021/es950012y.

Bersillon, J. -L., F. Villieras, F. Bardot, T. Gorner, and J. -M. Cases. 2001. Use of the Gaussian distribution function as a tool to estimate continuous heterogeneity in adsorbing systems. *Journal of Colloid and Interface Science* 240, no. 2: 400–411. doi:10.1006/jcis.2001.7657.

Borkovec, M., U. Rusch, M. Cernik, G. J. M. Koper, and J. C. Westall. 1996. Affinity distributions and acid-base properties of homogeneous and heterogeneous sorbents: Exact results versus experimental data inversion. *Colloids and Surfaces A: Physicochemical and Engineering Aspects* 107: 285–296.

Cabaniss, S. E. 2008. Quantitative structure—property relationships for predicting metal binding by organic ligands. *Environmental Science & Technology* 42, no. 14: 5210–5216. doi:10.1021/es7022219.

Cabaniss, S. E. 2009. Forward modeling of metal complexation by NOM: I. A priori prediction of conditional constants and speciation. *Environmental Science & Technology* 43, no. 8: 2838–2844. doi:10.1021/es8015793.

Cabaniss, S. E. 2011. Forward modeling of metal complexation by NOM: II. Prediction of binding site properties. *Environmental Science & Technology* 45, no. 8: 3202–3209. doi:10.1021/es102408w.

Carbonaro, R. F., Y. B. Atalay, and D. M. Di Toro. 2011. Linear free energy relationships for metal–ligand complexation: Bidentate binding to negatively-charged oxygen donor atoms. *Geochimica et Cosmochimica Acta* 75, no. 9: 2499–2511. doi:10.1016/j.gca.2011.02.027.

Carbonaro, R. F., and D. M. Di Toro. 2007. Linear free energy relationships for metal-ligand complexation: Monodentate binding to negatively-charged oxygen donor atoms. *Geochimica et Cosmochimica Acta* 71, no. 16: 3958–3968.

Chang, R. 2005. *Physical Chemistry for the Biosciences*. 1st ed. Sausalito, CA: University Science Books.

Christensen, J. B., E. Tipping, D. G. Kinniburgh, C. Grøn, and T. H. Christensen. 1998. Proton binding by groundwater fulvic acids of different age, origins, and structure modeled with the model V and NICA-Donnan model. *Environmental Science & Technology* 32, no. 21: 3346–3355.

Companys, E., J. L. Garcés, J. Salvador, J. Galceran, J. Puy, and F. Mas. 2007. Electrostatic and specific binding to macromolecular ligands: A general analytical expression for the Donnan volume. *Colloids and Surfaces A: Physicochemical and Engineering Aspects* 306, no. 1–3: 2–13. doi:10.1016/j.colsurfa.2007.01.016.

Deshmukh, A. P., C. Pacheco, M. B. Hay, and S. C. B. Myneni. 2007. Structural environments of carboxyl groups in natural organic molecules from terrestrial systems. Part 2: 2D NMR spectroscopy. *Geochimica et Cosmochimica Acta* 71, no. 14: 3533–3544. doi:10.1016/j.gca.2007.03.039.

de Wit, J. C. M., W. H. van Riemsdijk, and L. K. Koopal. 1993. Proton binding to humic substances. 2. Chemical heterogeneity and adsorption models. *Environmental Science & Technology* 27, no. 10: 2015–2022. doi:10.1021/es00047a005.

Dinar, E., T. F. Mentel, and Y. Rudich. 2006. The density of humic acids and humic like substances (HULIS) from fresh and aged wood burning and pollution aerosol particles. *Atmospheric Chemistry and Physics* 6, no. 12: 5213–5224.

Drosos, M., M. Jerzykiewicz, and Y. Deligiannakis. 2009. H-binding groups in lignite vs. soil humic acids: NICA-Donnan and spectroscopic parameters. *Journal of Colloid and Interface Science* 332, no. 1: 78–84. doi:10.1016/j.jcis.2008.12.023.

Duval, J. F. L., K. J. Wilkinson, H. P. Van Leeuwen, and J. Buffle. 2005. Humic substances are soft and permeable: Evidence from their electrophoretic mobilities. *Environmental Science & Technology* 39: 6435–6445.

Fernández, J. M., C. Plaza, N. Senesi, and A. Polo. 2007. Acid-base properties of humic substances from composted and thermally-dried sewage sludges and amended soils as determined by potentiometric titration and the NICA-Donnan model. *Chemosphere* 69, no. 4: 630–635. doi:10.1016/j.chemosphere.2007.02.063.

Groenenberg, J. E., G. F. Koopmans, and R. N. J. Comans. 2010. Uncertainty analysis of the nonideal competitive adsorption—Donnan model: Effects of dissolved organic matter variability on predicted metal speciation in soil solution. *Environmental Science & Technology* 44, no. 4: 1340–1346.

Gustafsson, J. P. 2001. Modeling the acid-base properties and metal complexation of humic substances with the Stockholm humic model. *Journal of Colloid and Interface Science* 244, no. 1: 102–112.

Gustafsson, J. P., and D. B. Kleja. 2005. Modeling salt-dependent proton binding by organic soils with the NICA-Donnan and Stockholm humic models. *Environmental Science & Technology* 39, no. 14: 5372–5377.

Gustafsson, J. P., P. Pechova, and D. Berggren. 2003. Modeling metal binding to soils: The role of natural organic matter. *Environmental Science & Technology* 37, no. 12: 2767–2774.

Gustafsson, J. P., I. Persson, D. B. Kleja, and J. W. J. van Schaiks. 2007. Binding of iron (III) to organic soils: EXAFS spectroscopy and chemical equilibrium modeling. *Environmental Science & Technology* 41, no. 4: 1232–1237.

Gustafsson, J. P., and J. W. J. Van Schaik. 2003. Cation binding in a mor layer: Batch experiments and modelling. *European Journal of Soil Science* 54, no. 2: 295–310.

Hay, M. B., and S. C. B. Myneni. 2007. Structural environments of carboxyl groups in natural organic molecules from terrestrial systems. Part 1: Infrared spectroscopy. *Geochimica et Cosmochimica Acta* 71, no. 14: 3518–3532. doi:10.1016/j.gca.2007.03.038.

Hilal, S. H., S. W. Karickhoff, and L. A. Carreira. 2003. *Prediction of Chemical Reactivity Parameters and Physical Properties of Organic Compounds from Molecular Structure using SPARC*. Research Triangle Park, NC: National Exposure Research Laboratory, Office of Research and Development, US Environmental Protection Agency. http://www.epa.gov/ATHENS/publications/reports/EPA_600_R03_030.pdf.

Hilal, S. H., A. N. Saravanaraj, T. Whiteside, and L. A. Carreira. 2007. Calculating physical properties of organic compounds for environmental modeling from molecular structure. *Journal of Computer-Aided Molecular Design* 21, no. 12: 693–708. doi:10.1007/s10822-007-9134-y.

Janos, P., S. Krízenecká, and L. Madronová. 2008. Acid-base titration curves of solid humic acids. *Reactive and Functional Polymers* 68, no. 1: 242–247. doi:10.1016/j.reactfunctpolym.2007.09.005.

Kawahigashi, M., H. Sumida, and K. Yamamoto. 2005. Size and shape of soil humic acids estimated by viscosity and molecular weight. *Journal of Colloid and Interface Science* 284, no. 2: 463–469. doi:10.1016/j.jcis.2004.10.023.

Khai, N. M., I. Öborn, S. Hillier, and J. P. Gustafsson. 2008. Modeling of metal binding in tropical Fluvisols and Acrisols treated with biosolids and wastewater. *Chemosphere* 70, no. 8: 1338–1346.

Kinniburgh, D. G., C. J. Milne, M. F. Benedetti, J. P. Pinheiro, J. Filius, L. K. Koopal, and W. H. Van Riemsdijk. 1996. Metal ion binding by humic acid: Application of the NICA-Donnan model. *Environmental Science & Technology* 30, no. 5: 1687–1698. doi:10.1021/es950695h.

Kinniburgh, D. G., W. H. Van Riemsdijk, L. K. Koopal, M. Borkovec, M. F. Benedetti, and M. J. Avena. 1999. Ion binding to natural organic matter: Competition, heterogeneity, stoichiometry and thermodynamic consistency. *Colloids and Surfaces A: Physicochemical and Engineering Aspects* 151, no. 1–2: 147–166.

Kirishima, A., T. Ohnishi, N. Sato, and O. Tochiyama. 2009. Determination of the phenolic-group capacities of humic substances by non-aqueous titration technique. *Talanta* 79, no. 2: 446–453. doi:10.1016/j.talanta.2009.04.008.

Koopal, L. K., W. H. van Riemsdijk, J. C. M. de Wit, and M. F. Benedetti. 1994. Analytical isotherm equations for multicomponent adsorption to heterogeneous surfaces. *Journal of Colloid and Interface Science* 166, no. 1: 51–60. doi:10.1006/jcis.1994.1270.

Kucerík, J., D. Šmejkalová, H. Cechlovská, and M. Pekar. 2007. New insights into aggregation and conformational behaviour of humic substances: Application of high resolution ultrasonic spectroscopy. *Organic Geochemistry* 38, no. 12: 2098–2110.

Lofts, S., and E. Tipping. 2011. Assessing WHAM/Model VII against field measurements of free metal ion concentrations: Model performance and the role of uncertainty in parameters and inputs. *Environmental Chemistry* 8, no. 5: 501–516.

Manunza, B., S. Deiana, V. Maddau, C. Gessa, and R. Seeber. 1995. Stability constants of metal-humate complexes: Titration data analyzed by bimodal Gaussian distribution. *Soil Science Society of America Journal* 59, no. 6: 1570–1574.

Marinsky, J. A. 1985. An interpretation of the sensitivity of weakly acidic (basic) polyelectrolyte (cross-linked and linear) equilibria to excess neutral salt. *The Journal of Physical Chemistry* 89, no. 24: 5294–5302. doi:10.1021/j100270a035.

Marinsky, J. A., R. Baldwin, and M. M. Reddy. 1985. Interpretation with a Donnan-based concept of the influence of simple salt concentration on the apparent binding of divalentions to the polyelectrolytes polystyrenesulfonate and dextran sulfate. *The Journal of Physical Chemistry* 89, no. 24: 5303–5307. doi:10.1021/j100270a036.

Matynia, A., T. Lenoir, B. Causse, L. Spadini, T. Jacquet, and A. Manceau. 2010. Semi-empirical proton binding constants for natural organic matter. *Geochimica et Cosmochimica Acta* 74, no. 6: 1836–1851.

Milne, C. J., D. G. Kinniburgh, J. C. M. De Wit, W. H. Van Riemsdijk, and L. K. Koopal. 1995. Analysis of proton binding by a peat humic acid using a simple electrostatic model. *Geochimica et Cosmochimica Acta* 59, no. 6: 1101–1112.

Milne, C. J., D. G. Kinniburgh, and E. Tipping. 2001. Generic NICA-Donnan model parameters for proton binding by humic substances. *Environmental Science & Technology* 35, no. 10: 2049–2059. doi:10.1021/es000123j.

Milne, C. J., D. G. Kinniburgh, W. H. van Riemsdijk, and E. Tipping. 2003. Generic NICA–Donnan model parameters for metal-ion binding by humic substances. *Environmental Science & Technology* 37, no. 5: 958–971. doi:10.1021/es0258879.

Miyajima, T., M. Mori, and S. Ishiguro. 1997. Analysis of complexation equilibria of polyacrylic acid by a Donnan-based concept. *Journal of Colloid and Interface Science* 187, no. 1: 259–266. doi:10.1006/jcis.1996.4694.

Ohshima, H. 2006. Chapter 2. Potential distribution around a soft particle. In *Theory of Colloid and Interfacial Electric Phenomena*, ed. h. Ohshima, 12:39–55. Interface Science and Technology. Amsterdam: Academic Press.

Ohshima, H., T. W. Healy, and L. R. White. 1982. Accurate analytic expressions for the surface charge density/surface potential relationship and double-layer potential distribution for a spherical colloidal particle. *Journal of Colloid and Interface Science* 90, no. 1: 17–26. doi:10.1016/0021-9797(82)90393-9.

Orsetti, S., E. M. Andrade, and F. V. Molina. 2009. Application of a constrained regularization method to extraction of affinity distributions: Proton and metal binding to humic substances. *Journal of Colloid and Interface Science* 336, no. 2: 377–387.

Orsetti, S., E. M. Andrade, and F. V. Molina. 2010. Modeling ion binding to humic substances: Elastic polyelectrolyte network model. *Langmuir* 26, no. 5: 3134–3144. doi:10.1021/la903086s.

Plaza, C., G. Brunetti, N. Senesi, and A. Polo. 2005a. Proton binding to humic acids from organic amendments and amended soils by the NICA-Donnan model. *Environmental Science & Technology* 39, no. 17: 6692–6697. doi:10.1021/es0503377.

Plaza, C., N. Senesi, A. Polo, and G. Brunetti. 2005b. Acid-base properties of humic and fulvic acids formed during composting. *Environmental Science & Technology* 39, no. 18: 7141–7146.

Puy, J., C. Huidobro, C. David, C. Rey-Castro, J. Salvador, E. Companys, J. L. Garcés, J. Galceran, J. Cecília, and F. Mas. 2009. Conditional affinity spectra underlying NICA isotherm. *Colloids and Surfaces A: Physicochemical and Engineering Aspects* 347, no. 1–3: 156–166. doi:10.1016/j.colsurfa.2009.03.053.

Ritchie, J. D., and E. M. Perdue. 2003. Proton-binding study of standard and reference fulvic acids, humic acids, and natural organic matter. *Geochimica et Cosmochimica Acta* 67, no. 1: 85–93.

Ritchie, J. D., and E. M. Perdue. 2008. Analytical constraints on acidic functional groups in humic substances. *Organic Geochemistry* 39, no. 6: 783–799. doi:10.1016/j.orggeochem.2008.03.003.

Rizzi, F. R., S. Stoll, N. Senesi, and J. Buffle. 2004. A transmission electron microscopy study of the fractal properties and aggregation processes of humic acids. *Soil Science* 169, no. 11: 765–775.

Saito, T., S. Nagasaki, S. Tanaka, and L. K. Koopal. 2005. Electrostatic interaction models for ion binding to humic substances. *Colloids and Surfaces A: Physicochemical and Engineering Aspects* 265, no. 1–3: 104–113.

Senesi, N., and E. Loffredo. 1998. The chemistry of soil organic matter. In *Soil Physical Chemistry*, ed. D. L. Sparks, 239–370. 2nd ed. Boca Raton, FL: CRC Press.

Simpson, A. J., W. L. Kingery, M. H. Hayes, M. Spraul, E. Humpfer, P. Dvortsak, R. Kerssebaum, M. Godejohann, and M. Hofmann. 2002. Molecular structures and associations of humic substances in the terrestrial environment. *Naturwissenschaften* 89, no. 2: 84–88.

Sips, R. 1948. On the structure of a catalyst surface. *The Journal of Chemical Physics* 16, no. 5: 490–495. doi:10.1063/1.1746922.

Smith, E. J., C. Rey-Castro, H. Longworth, S. Lofts, A. J. Lawlor, and E. Tipping. 2004. Cation binding by acid-washed peat, interpreted with humic ion-binding model VI-FD. *European Journal of Soil Science* 55, no. 3: 433–447. doi:10.1111/j.1365-2389.2004.00607.x.

Stell, G., and C. G. Joslin. 1986. The Donnan equilibrium. *Biophysical Journal* 50, no. 5: 855–859.

Stockdale, A., N. D. Bryan, and S. Lofts. 2011. Estimation of model VII humic binding constants for Pd^{2+}, Sn^{2+}, U^{4+}, NpO_2^{2+}, Pu^{4+} and PuO_2^{2+}. *Journal of Environmental Monitoring* 13, no. 10: 2946–2950. doi:10.1039/C1EM10407A.

Sundararajan, M., G. Rajaraman, and S. K. Ghosh. 2011. Speciation of uranyl ions in fulvic acid and humic acid: A DFT exploration. *Physical Chemistry Chemical Physics* 13, no. 40: 18038–18046. doi:10.1039/C1CP21238A.

Sutton, R., and G. Sposito. 2005. Molecular structure in soil humic substances: The new view. *Environmental Science & Technology* 39, no. 23: 9009–9015. doi:10.1021/es050778q.

Tipping, E. 1994. WHAMC-A chemical equilibrium model and computer code for waters, sediments, and soils incorporating a discrete site/electrostatic model of ion-binding by humic substances. *Computers and Geosciences* 20, no. 6: 973–1023.

Tipping, E. 1998. Humic ion-binding model VI: An improved description of the interactions of protons and metal ions with humic substances. *Aquatic Geochemistry* 4, no. 1: 3–48.

Tipping, E., and M. A. Hurley. 1992. A unifying model of cation binding by humic substances. *Geochimica et Cosmochimica Acta* 56, no. 10: 3627–3641.

Tipping, E., S. Lofts, and J. E. Sonke. 2011. Humic ion-binding model VII: A revised parameterisation of cation-binding by humic substances. *Environmental Chemistry* 8, no. 3: 225–235.

Tipping, E., C. Rey-Castro, S. E. Bryan, and J. Hamilton-Taylor. 2002. Al(III) and Fe(III) binding by humic substances in freshwaters, and implications for trace metal speciation. *Geochimica et Cosmochimica Acta* 66, no. 18: 3211–3224. doi:10.1016/S0016-7037(02)00930-4.

Van Schaik, J. W. J., D. B. Kleja, and J. P. Gustafsson. 2010. Acid-base and copper-binding properties of three organic matter fractions isolated from a forest floor soil solution. *Geochimica et Cosmochimica Acta* 74, no. 4: 1391–1406. doi:10.1016/j.gca.2009.11.007.

Vidali, R., E. Remoundaki, and M. Tsezos. 2011. An experimental and modelling study of Cu^{2+} binding on humic acids at various solution conditions. Application of the NICA-Donnan model. *Water, Air, & Soil Pollution* 218, no. 1: 487–497. doi:10.1007/s11270-010-0662-z.

14 Ion Binding to Soils

In the preceding chapters, we revisited the interaction of ions with soil components, mainly minerals, and humic substances. As it should be apparent, there are a high number of studies on sorption of ions to soil components, and in many aspects, it is a field fairly well known. All these are laboratory studies, under well-controlled conditions. Now, the question arising is, How do we apply all these results to a natural soil, where the components are interacting between themselves, and the conditions are changing due to climate, human, and animal activity? There is no easy answer to this question, and in fact, nowadays, we are rather far away from satisfactory results in this respect. In this chapter, we will try to reflect the state-of-the-art response to this problem.

14.1 STATE OF SOIL COMPONENTS IN NATURAL SOILS

Mineral oxides, clay minerals, and, especially, soil organic matter (SOM) form aggregates that not only have an impact on soil physical properties but also affect their interaction with ions and small molecules. The formation of humic–mineral complexes is well known (Schnitzer 1986; Hayes and Himes 1986), and the interaction of HS with minerals, especially oxides, has received attention in the literature by several authors (Kretzschmar, Sticher, and Hesterberg 1997; Avena and Koopal 1998; Vermeer, van Riemsdijk, and Koopal 1998; Saito et al. 2004; Tombácz et al. 2004; Gasser, Mohsen, and Aly 2008; Yang, Lin, and Xing 2009). Organic matter in soil adsorbs onto minerals forming aggregates. As discussed in Chapters 10 and 13, humic substances are anionic polyelectrolytes; as such, they strongly bind to mineral surfaces by several forms of interaction, the most important being specific adsorption by ligand exchange with hydroxyl groups. Figure 14.1 shows a scanning electron microscopic image of soil minerals partially or totally covered by organic matter, a feature that is usual in topsoil horizons with an important proportion of SOM. Organic matter plays an important role in the aggregation of SOM, as visualized in Figure 14.2. As a consequence, SOM is often distributed into fractions as follows: within clay plus silt particles, within microaggregates (i.e., intra-microaggregate), within macroaggregates but external to microaggregates (i.e., includes light fraction, macroorganic matter, and microbial biomass), and free macroorganic matter (Carter 2002). These different locations for the SOM imply differences in stability, in properties regarding soil stability and farming, and in interactions with ions; free organic matter will likely behave close to laboratory studies, whereas aggregated SOM is expected to show differences.

Mineral components are modified not only by adsorption of organic matter but also due to binding or precipitation of inorganic solids, mainly due to weathering.

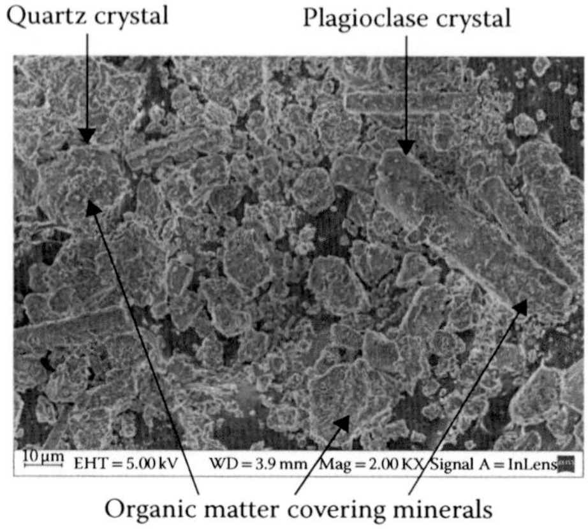

FIGURE 14.1 Scanning electron microscopic image of a soil sample from Buenos Aires province, Argentina. The presence of organic matter covering minerals is observed as an irregular surface appearance. (From G. V. Ferreyroa, R. S. Lavado, and F. V. Molina, unpublished data.)

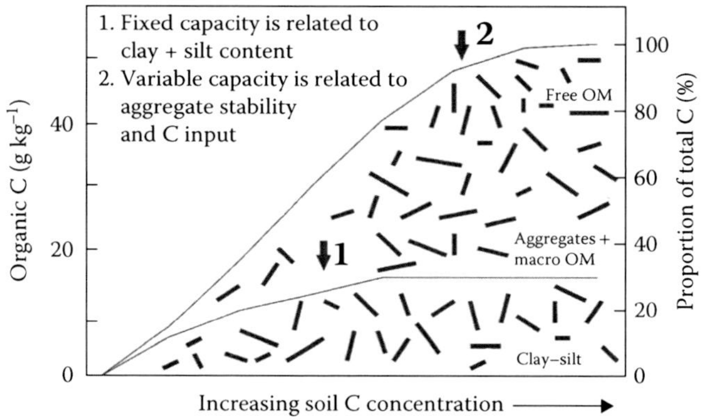

FIGURE 14.2 Conceptual model of soil organic matter (SOM) content and proportion in soil particles and aggregates of a Charlottetown fine sandy loam illustrating different measurable capacity levels. (Reprinted from Carter, M. R., 2002, with kind permission of the American Society of Agronomy.)

Figure 14.3 shows schematically possible cases (Sposito 1984), including hydroxide polymer precipitation among phyllosilicate layers, organic matter adsorption, metal oxide covering either the clay surface or the organic layer, and so on.

14.2 ION BINDING TO SEVERAL SOIL COMPONENTS

14.2.1 Additivity Rule

If the interaction of a given ion with diverse mineral and organic soil components is known and can be successfully predicted by suitable models, how can the amount adsorbed to a combination of two or more components can be predicted? The "ideal"

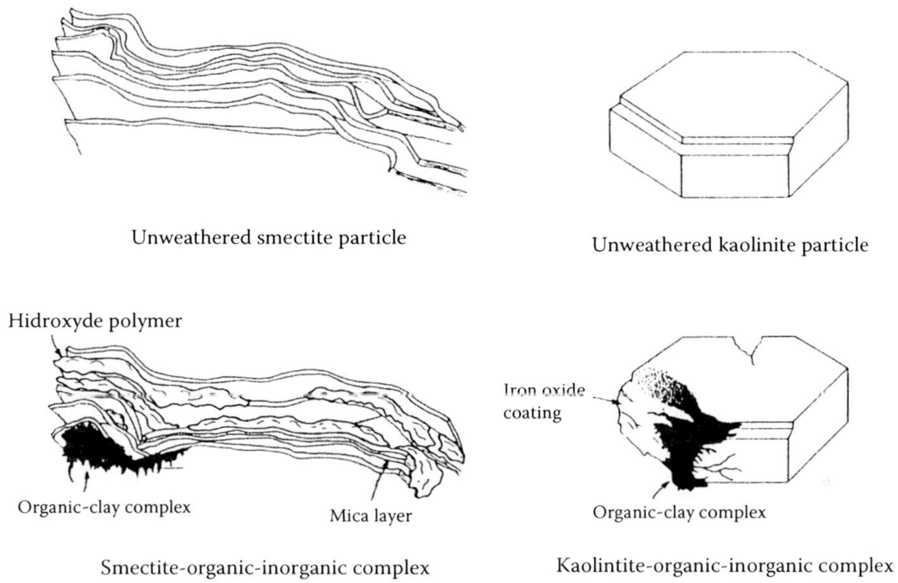

FIGURE 14.3 The alteration of minerals upon weathering, resulting in the formation of complexes between minerals, inorganic coatings, and organic matter. (Reproduced from Sposito, G., *The surface chemistry of soils*, 1984, by permission of Oxford University Press.)

answer is the simple sum of the amounts predicted for the individual components separately. This is known as the *additivity rule* or *linear additivity model*, first proposed by Zachara, Resch, and Smith (1994). It has been examined by several authors for different systems, as discussed next.

14.2.2 Sorption to Binary and Ternary Component Systems

The sorption of ions in laboratory systems formed by a few (mainly two or three) soil components has received attention, albeit not very extensive, from several research groups (note that here "binary" and "ternary" refer to the number of soil components; in the literature, the term "ternary" often refers to two components plus one adsorbing ion). We will summarize here some important results.

Christl and Kretzschmar (2001) studied the interaction of copper with fulvic acid and hematite, and the results were compared with model calculations based on the linear additivity assumption. The sorption data for the ion-single sorbent systems were modeled with a basic Stern model (BSM) (Section 12.2.2) for hematite and the NICA-Donnan model for fulvic acid. In the second step, pH-dependent sorption of Cu and fulvic acid in systems containing Cu, hematite, and fulvic acid was investigated in batch sorption experiments. Comparison of the experimental data with model calculations shows that Cu sorption in binary hematite–fulvic acid systems is systematically underestimated by up to 30% by using the linear additivity assumption, as shown in Figure 14.4.

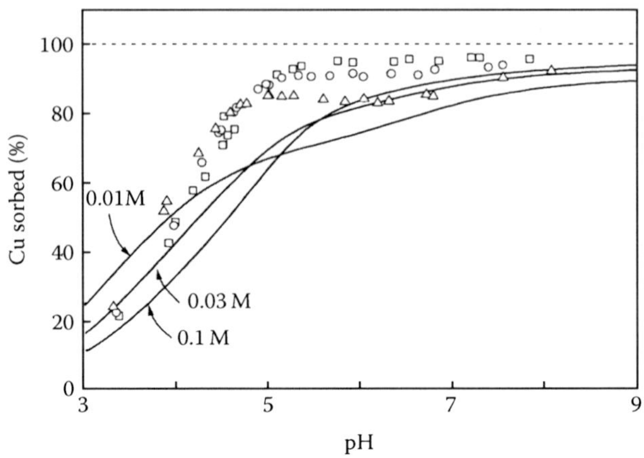

FIGURE 14.4 Cu sorption to the solid phase in binary systems hematite–fulvic acid at three different ionic strengths. Experimental data: (□) 0.1 M; (○) 0.03 M; (Δ) 0.01 M, and corresponding model calculations (lines) based on the additivity assumption. (Reprinted from *Geochimica et Cosmochimica Acta*, 65, Christl, I. and Kretzschmar, R., Interaction of copper and fulvic acid at the hematite-water interface, 3435–3442. Copyright 2001, with permission from Elsevier.)

Alcacio et al. (2001) performed extended X-ray absorption fine structure (EXAFS) spectroscopy analyses on aqueous pastes containing Cu(II) and goethite with different amounts of humic acid adsorbed. Three types of metal complexes were considered (Figure 14.5): metal bound to the mineral surface as an inner-sphere bidentate complex, Cu ions bridging the goethite surface and humic molecules (type A complex), and Cu bound to the humic substance in a multidentate complex (type B). The proposed type A complex involves Cu(II) bonded with goethite 001 surfaces as an inner-sphere complex, with humic acid involving carboxylic or phenolic acid functional groups. The type B complex involves Cu(II) bound to humic acid, which in turn is bound to goethite. Analysis of the x-ray absorption near-edge structure (XANES) and EXAFS data suggested that Cu(II) was bonded on average to both inorganic and organic functional groups as a type A complex at lower levels of adsorbed humate, and to organic groups as a type B complex at higher levels. The higher levels of adsorbed humic acid apparently made the goethite surface sites less accessible to Cu(II).

Saito et al. (2005) investigated the amount of Cu(II) bound in the binary system formed by purified Aldrich humic acid (PAHA) and goethite compared with the separate components. The experiments revealed that Cu^{2+} binding in the binary system is enhanced with respect to the sum of Cu^{2+} binding in the corresponding single sorbent systems (Figure 14.6). From the analysis of the charging behavior of the adsorbed PAHA as well as the smeared-out potential profile near the PAHA/goethite interface, the increase of Cu^{2+} binding to the complex was mainly attributed to the decrease of proton competition to the functional groups of the adsorbed PAHA and the change of the electrostatic potential in the vicinity of the goethite surface. These conclusions, however, appear to be in conflict with the EXAFS studies mentioned above.

In a similar study (Orsetti, Quiroga, and Andrade 2006), the binding of Pb(II) to the humic acid–goethite binary system and the individual components was

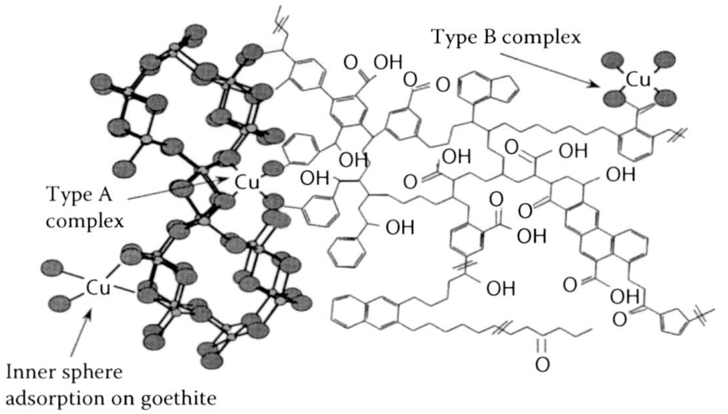

FIGURE 14.5 Illustration of some possible bonding configurations of Cu(II) on goethite–humate complexes: inner-sphere complexation of Cu(II) at goethite sites, and type A and B ternary complexes. (Reprinted from *Geochimica et Cosmochimica Acta*, 65, Alcacio, T. E. et al., Molecular scale characteristics of Cu (II) bonding in goethite–humate complexes, 1355–1366. Copyright 2001, with permission from Elsevier.)

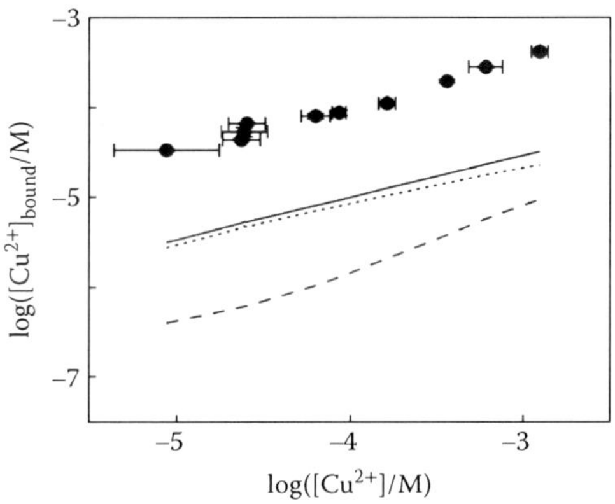

FIGURE 14.6 Comparison of the measured Cu^{2+} binding isotherms with the PAHA/goethite complex at 0.088 g of PAHA (total)/g of goethite with the calculated isotherms to PAHA (dotted curves, amount equivalent to the adsorbed amount) and bare goethite (dashed curves) and the sums of the PAHA and goethite isotherms (solid curves, additivity rule) for pH = 4.0, 0.1 M KNO_3. (Reprinted with permission from Saito et al., *Environ. Sci. Technol.*, 39, 13, 4886–4893, 2005. Copyright 2005 American Chemical Society.)

investigated. Comparison of the total amount bound in the binary system reveals that the system presents positive deviations; that is, the amount sorbed is about an order of magnitude higher than that predicted by the additivity rule (Figure 14.7). In situ and ex situ IR results suggest that the presence of Pb(II) increases the amount of humic acid bound to the oxide. It is proposed that proton displacement due to the interaction between humic substances and the oxide, along with the formation of ternary complexes with the Pb(II) cation bridging the oxide and the humic substances (type A complexes), cause the deviation from additivity.

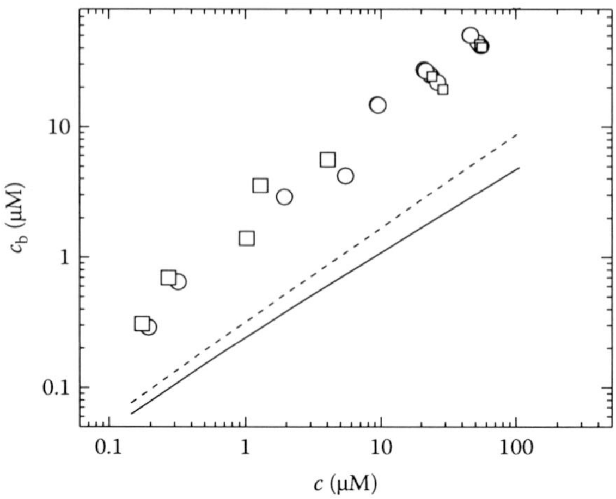

FIGURE 14.7 Amount of Pb(II) bound as a function of Pb(II) in solution for the system humic acid–goethite in 0.02 M $NaClO_4$ at pH 4.0. Experimental data: (○) 8 mg L^{-1} HA + 80 mg L^{-1} goethite; (□) 16 mg L^{-1} HA + 80 mg L^{-1} goethite. Additivity prediction: (—) 8 mg L^{-1} HA + 80 mg L^{-1} goethite; (- - -) 16 mg L^{-1} HA + 80 mg L^{-1} goethite. (Reprinted from *Chemosphere*, 65, Orsetti, S. et al., Binding of Pb(II) in the system humic acid/goethite at acidic pH, 2313–2321. Copyright 2006, with permission from Elsevier.)

Koretsky and coworkers studied sorption of Cu(II) and Co(II) cations in binary (hydrous ferric oxide [HFO] and kaolinite) and ternary (HFO, quartz, and kaolinite) mineral systems, respectively (Lund et al. 2008; Landry et al. 2009). The experimental studies consisted in adsorption edge experiments; the cation sorption to the corresponding separate minerals was modeled applying the diffuse layer model (DLM) (Section 12.2.2) with a varying number of binding sites; satisfactory fitting was found, except for quartz, where the slope could not be adequately predicted. A linear additivity approach was used to predict the sorption to binary and ternary mineral sorbent systems of both Cu (Figure 14.8) and Co (Figure 14.9); discrepancies between predictions and measurements were quantitatively similar to those observed for the pure mineral systems. It was concluded that a simple component additivity approach provides useful predictions of metal sorption in the mixed solid systems.

In an another study related to the additivity rule (Alessi and Fein 2010), the sorption of Cd(II) in mixtures of kaolinite, *Bacillus subtilis* bacterial cells, and iron oxyhydroxide (HFO) was performed in the absence and presence of an organic ligand, acetate. The results indicate that for systems containing *B. subtilis*, HFO, and kaolinite, the component additivity approach is a reasonable predictor of metal distribution, with the accuracy limited by the accuracy of the stability constants of the important surface complexes. However, in systems including acetate, the additivity rule predictions significantly underestimate the extent of adsorption above pH 5, likely due to the formation of ternary Cd–acetate surface complexes on each surface.

From the studies discussed above, albeit further research may be necessary to reach firm conclusions, an emerging picture is that when considering mixtures of mineral components, the additivity rule could predict reasonably well the extension of ion binding, but in the presence of organic matter, this is no longer valid.

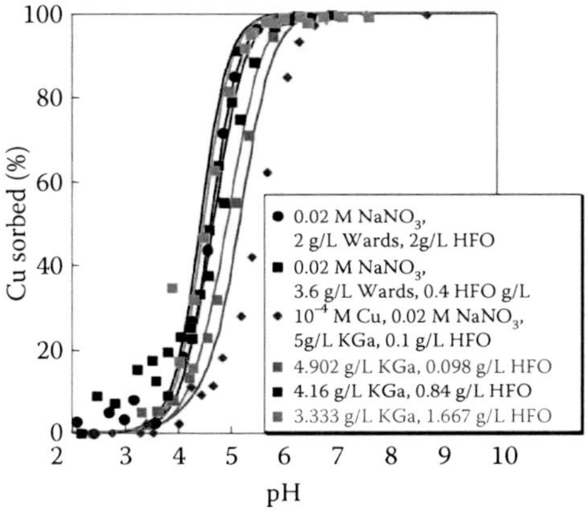

FIGURE 14.8 Predicted Cu adsorption as a function of pH on mixtures of kaolinite and HFO. Lines indicate fits using the 1-site HFO model and 1-site kaolinite model (Sverjensky and Sahai 1996) with formation of a monodentate Cu complex on a variable-charge site in the presence of 10^{-5} M Cu, 0.01 M $NaNO_3$. (Reprinted from Lund, T. et al., *Geochem. Trans.*, 9, 1, 9, 2008.)

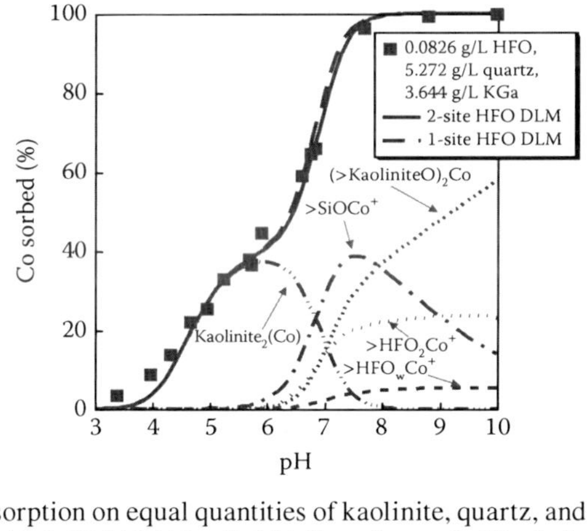

FIGURE 14.9 Co sorption on equal quantities of kaolinite, quartz, and HFO by surface area as a function of pH with 0.01M $NaNO_3$, 10^{-5} M Co and a total of 7 g L^{-1} solid. Solid lines indicate predictions of total Co sorbed using a two-site HFO model, a one-site quartz model with formation of a monodentate surface complex and a two-site kaolinite model with formation of bivalent Co complexes on variable charge and ion-exchange sites. Dashed lines show predictions of total Co sorbed using the same kaolinite and quartz models, but with a one-site HFO model. Speciation is shown for calculations using a two-site HFO model. (Reprinted from *Geochimica et Cosmochimica Acta*, 73, Landry, C. J. et al., Surface complexation modeling of Co(II) adsorption on mixtures of hydrous ferric oxide, quartz and kaolinite, 3723–3737. Copyright 2009, with permission from Elsevier.)

14.2.3 Sorption in Complex Systems and Whole Soils

Studies of ion sorption in soils have been conducted in many cases; often, these studies are analyzed in terms of basic adsorption isotherms, as in the examples given in Chapter 4 for the different types of them. It is perfectly possible to fit data from an experiment to an adsorption isotherm (such as Langmuir and Langmuir–Freundlich), possibly allowing for a small number of site types. Of course, the equation parameters (such as adsorption equilibrium constants) found in that way are almost always valid only in the experimental conditions used to obtain them; changing the conditions (such as pH and ionic strength) implies new parameter values, thus the need for comprehensive modeling leading to the variety of models discussed in Chapters 11 to 13. The question now is if these models can be straightforwardly applied to whole soils; there is no clear answer to date, but the discussion of the previous section suggests that the answer, at least in some cases, is no. We will now present some results regarding this point.

When ion binding to a given soil sample can be modeled (even empirically, giving conditional constants) with a few sites, often this is due to the soil behavior being dominated by a few components, each showing up as one or two sites in the whole. Calvet, Barriuso, and Dubus (2007) modeled the sorption data for three weak organic acids [2,4-D, 2,4-dichloro-phenoxy acetic acid; salicylic acid, 2-hydroxybenzoic acid; and clofencet, 2,4-(chlorophenyl)-3-ethyl-2,5-dihydro-5-oxopyridazine-4-carboxylic acid] in three Cambisols (soils with low oxide contents) and two Ferralsols (also oxisols, soil with high contents of Fe and Al oxides). These are relatively weak acids, with pK_a values between 2.6 and 2.8; because most measurements were done at pH > 4.0, these acids are mainly in anionic form in the experiments. The models applied were the constant capacitance model (Section 12.2.2) and the quad layer model (or variable charge–variable potential, Section 12.2.3). An additive approach was taken, where the five soils were considered to behave as mixtures of goethite and gibbsite, with values taken from the literature assuming monodentate complexes formation. Variations given by the two models in adsorbed amounts as a result of pH changes were well described for clofencet and 2,4-D but were less successful for salicylic acid (Figure 14.10). Simulations of sorption isotherms at various pHs ranging from 5 to 8 matched the experimental data closely for clofencet. The poorer fit observed for salicylic acid was attributed to the formation of bidendate complexes.

In assessing the relative importance of minerals and organic matter in ion binding and retention, there are varying reports in the literature. In an extensive study of Cd, Cr, Cu, Ni, Pb, and Zn sorption in acid soils, Covelo, Vega, and Andrade (2007a, 2007b) found that Pb and Cu were sorbed and retained to a greater extent than Cd, Ni, or Zn, which had low retention. Pb was sorbed more than any other metal. Cr was generally sorbed only slightly more than Cd, Ni, or Zn, but was strongly retained. The sorption of Pb and Cu correlated with organic matter content, while the retention of these and the other metals considered appeared to depend on clay minerals, especially kaolinite, gibbsite, and vermiculite. Overall capacity for sorption of heavy metals was positively related to organic matter, kaolinite, and Fe oxides contents and negatively related to cation exchange capacity (CEC) and to vermiculite and hematite contents. Overall capacity for retention of heavy metals was positively related to organic matter and kaolinite contents and negatively related to CEC and vermiculite content.

In a study of copper sorption to ternary systems composed of organic matter (leaf compost), ferrihydrite and montmorillonite (Martínez-Villegas and Martínez 2008), and its time evolution over 8 months, it was found that copper sorption followed the order organic matter > silicate clays > iron oxides (Figure 14.11a). Within each solid

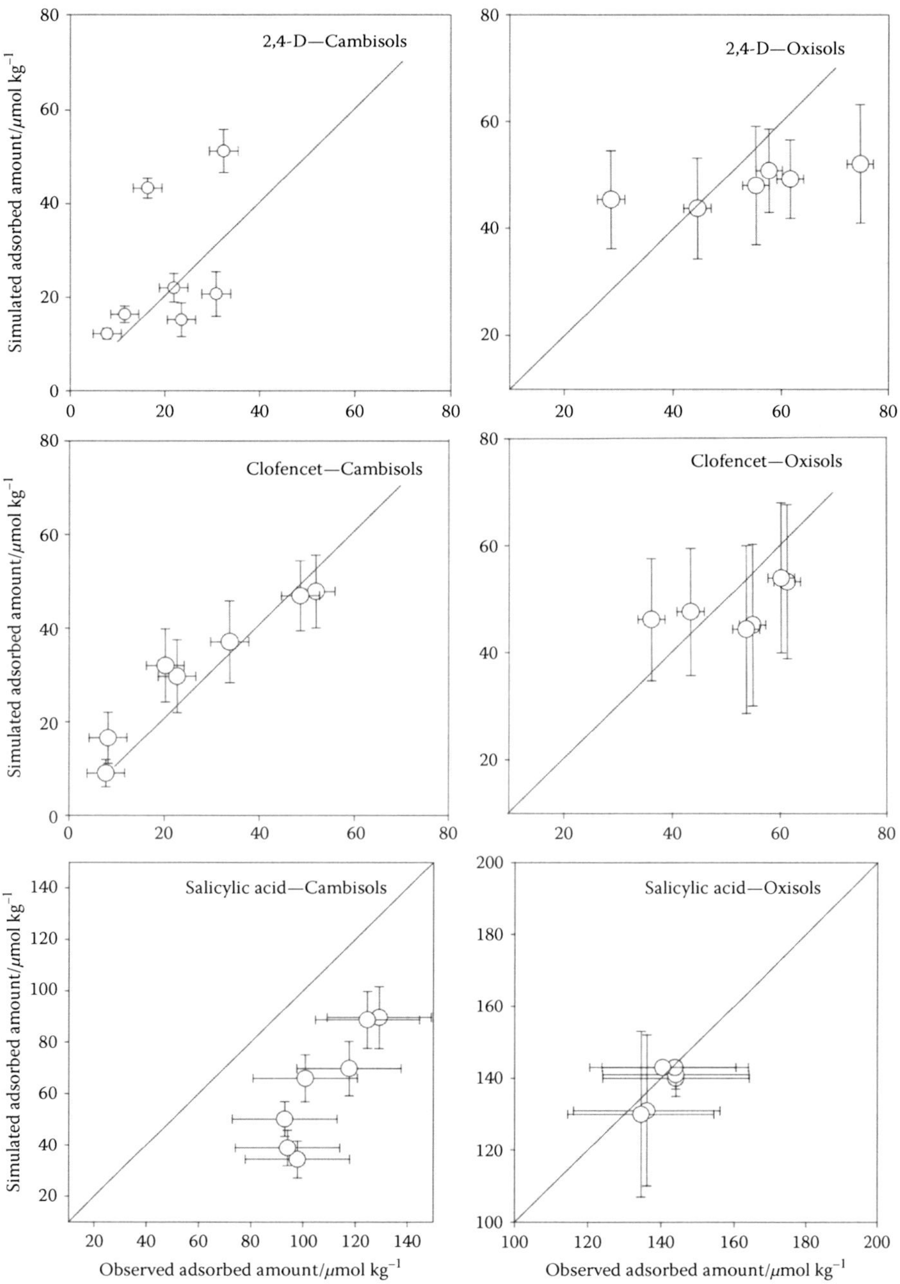

FIGURE 14.10 Comparison between observed and calculated adsorbed amounts of three organic acids (2,4-d, 2–4-dichloro-phenoxy acetic acid; salicylic acid, and clofencet, 2–4-(chlorophenyl)-3-ethyl-2,5-dihydro-5-oxopyridazine-4-carboxylic acid) on several Cambisols and Ferralsols at their native soil pH. The bars indicate the confidence limits (0.05). (Reprinted from Calvet, R., *Eur. J. Soil Sci.*, 58, 3, 609-624. Copyright 2007, with permission from John Wiley and Sons.)

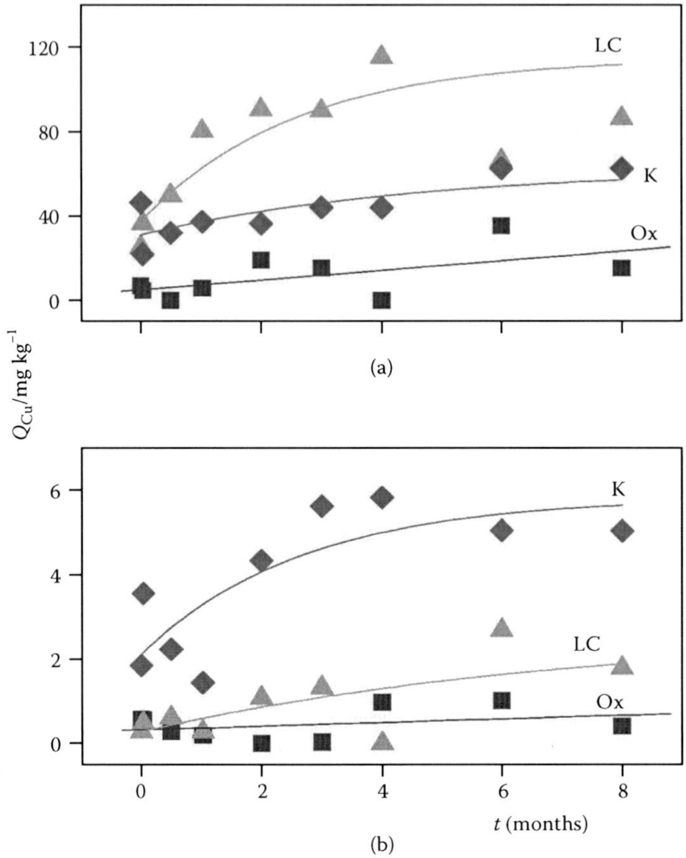

FIGURE 14.11 Total Cu (a) and exchangeable Cu (b) associated with different solid phases as a function of time in a multicomponent system. K: K^+-saturated Wyoming montmorillonite; LC: leaf compost; Ox: two-line ferrihydrite. (Reprinted from Martínez-Villegas, N. and Martínez, C. E., *Environ. Sci. Technol.*, 42, 8, 2833–2838, 2008. Copyright 2008 American Chemical Society.)

phase, exchangeable Cu was less than 10% of the total Cu sorbed (Figure 14.11b) suggesting that most of the metal is held by stronger interactions such as complexation and chemisorption. Dissolved organic carbon (DOC) reached steady state after 4 months and seemed to control Cu solubility and sorption behavior by the formation of soluble Cu–DOC complexes and by sorbing onto the mineral phases. DOC sorption onto ferrihydrite prevented Cu retention by this solid phase. Thus, organic matter showed a relevant role in the retention and lability of copper in this study.

Investigating the sorption of Pb, Cu, and Zn to samples of several horizons of forest soils (Sipos et al. 2008, 2009), Cu and Pb were found to have higher and stronger sorption on the studied samples than Zn. Only the former two metals showed significant differences in their immobilized metal amounts on the studied samples and soil mineral particles. Copper and zinc sorbed mostly on soil mineral constituents, whereas lead was associated mainly to SOM. Highest metal amounts were sorbed on the swelling clay mineral particles (smectites and vermiculites). Alkaline conditions due to the carbonate content of soils resulted both in increased sorption on the mineral particles for Cu and in enhanced role of precipitation for all the studied metals. On the other hand, the intimate association of phyllosilicates and iron resulted in significant increase in metal sorption capacity of the given particle. Note that this means that the resulting amount of metal sorbed by two components may depend on the association mechanism between them.

Vega, Covelo, and Andrade (2009) studied the competitive sorption of Cd, Cu, and Pb on samples of 20 soil horizons, evaluating the irreversibility of the adsorption by analyzing the hysteresis of the sorption–desorption curves. They concluded that the sorption of Cd, Cu, and Pb is most irreversible in the more basic soils with high Mn oxides content, large CEC, and, except in the case of Cd, high organic matter content. The irreversibility of the sorption of all three metals is also favored by clay (vermiculite being especially important for Cu and both vermiculite and chlorite for Pb); the irreversibility of Cd sorption is also favored by Fe oxides and exchangeable K, Ca, and Mg; and that of Cu sorption by Al oxides.

The effect of organic matter can also be appreciated when it is added as an amendment to, for instance, remediate contaminated soils. In a recent study, Garrido, Mendoza, and Arriagada (2012) investigated the effect of mine tailings containing high amounts of copper on Chilean soils, and the results of the addition of sewage sludge as amendment. An example of the results is shown in Figure 14.12; it is found that in soils contaminated with mine tailings, the sorption capacity and affinity of the substrate for Cu decreased substantially compared to uncontaminated soil. This effect is probably due to the decrease in the binding sites contributed by the soil and to the lower pH resulting from the soil-tailings mixture. On the other hand, the sewage sludge constitutes a good source of organic matter that contributes new sites for the sorption of metals, particularly Cu. These materials applied to soils contaminated with mine tailings improve Cu retention.

The preceding cases show that the additivity rule can be applied to ion binding to soils but, as discussed in the previous section, in many cases, it may not be fulfilled,

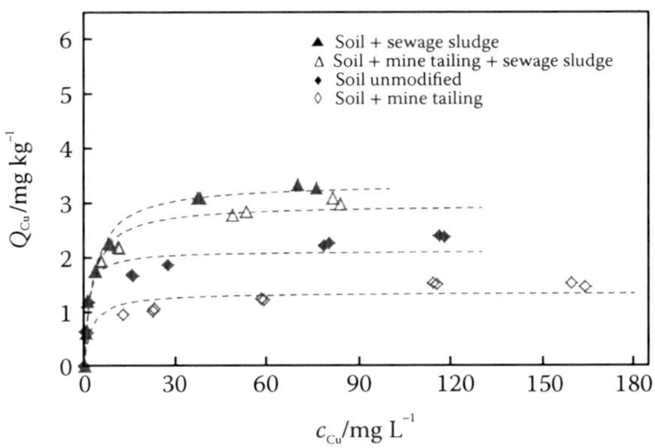

FIGURE 14.12 Sorption isotherms of Cu on a Chilean soil treated with mine tailings containing a high amount of Cu and with sewage sludge. Symbols represent experimental data and the dotted line represents the fit of data to the Langmuir model. (Reprinted from *Journal of Environmental Sciences*, 24, Garrido, T. et al., Changes in the sorption, desorption, distribution, and availability of copper, induced by application of sewage sludge on Chilean soils contaminated by mine tailings, 912–918. Copyright 2009, with permission from Elsevier.)

mainly due to the presence of organic matter, which can interact strongly with minerals, forming aggregates that modify the binding behavior of the separate components.

14.3 MODELS FOR ION SORPTION TO SOILS

Having a reasonably good model of a soil system, giving the capacity to predict with at least fair accuracy, the state and speciation of ionic species would be a welcome ability but implies a difficult task. In the last decades, several efforts have been conducted, with partial success. Some of these models are presented in the following sections.

14.3.1 Assemblage Models

An *assemblage* or *multisurface model* is a description of the soil solid phases as a set of surfaces where ions are bound by different mechanisms (i.e., ion exchange, surface complexation); the total amount bound of a species i is simply the sum over all the representative surfaces, such as

$$Q_i = y^{OM} Q_i^{OM} + y^{FeOx} Q_i^{FeOx} + y^{phsil} Q_i^{phsil} + \cdots = \sum_{\alpha} y^{\alpha} Q_i^{\alpha} \quad (14.1)$$

where OM is organic matter; $FeOx$, iron oxides; and phsil, phyllosilicates; each term in turn can be separated into several contributions—for example, organic matter would be composed of carboxylic and phenolic type sites, and others; all the contributions are included in the sum of the rightmost side of Equation 14.1 and are assumed to be independent; that is, interaction between components are not considered in ion-binding modeling; thus, this is effectively an application of the additivity rule discussed in the previous section. Q_i^{α} is the amount sorbed per unity of mass of phase α, and y^{α} is the respective mass fraction. The first proposal of an assemblage model was made by Lofts and Tipping (1998), known as SCAMP (surface chemistry assemblage model for particles), where the organic part was modeled using Windermere humic aqueous model (WHAM) V (Section 13.4.1), adsorption by oxides with a surface complexation model (SCM) that allows for site heterogeneity and an idealized cation exchanger. The SCM incorporates surface reactions as those presented in Section 12.2 (i.e., Equations 12.2 through 12.6) but with an electrostatic contribution consistent with model V; for example, for a proton dissociation reaction as

$$\text{X-OH}_2^+ \rightleftharpoons \text{X-OH} + \text{H}^+ \quad (14.2)$$

the equilibrium expression is given by

$$\frac{[\text{X-OH}]\, a_{\text{H}^+}}{[\text{X-OH}_2^+]} = K_{H,1} \exp(2wz_S) \quad (14.3)$$

where z_S is the surface charge and w is given by (Section 13.4.1)

$$w = P \log I \quad (14.4)$$

where I is the ionic strength.

Also, a Donnan volume occupied by counterions (Equation 13.35 and following) is included; see the original paper (Lofts and Tipping 1998) for further details. The fractions y^{α} are initially set in terms of the known soil composition but adjusted to fit the experimental data. This model was first applied to two samples of aquatic suspended particulate matter, one riverine, the other estuarine, taking into account the proportions of the different binding phases. Figure 14.13 shows the net proton charge of the particulate matter, as found by standard acid–base titration (Section 5.4.2). The lines correspond to different assemblage compositions (i.e., a different set of f^{α} values). Composition R1 was determined based on sample analysis and contained iron and manganese oxides, quartz, and aluminosilicates. Organic matter was set considering fulvic acid, humic acid, and inert fractions, with $f^{FA} = f^{HA} = 0.013$ and $f^{inert} = 0.214$, respectively; in composition R2, all the organic C was assumed to be HS, thus $f^{FA} = f^{HA} = 0.12$ and $f^{inert} = 0.0$; and in R3 all organics were assumed to be FA, thus $f^{FA} = 0.24$; the remaining were 0. The R3 composition is the one best fitting the experiment, but at the cost of the somewhat forced assumption of all organic matter being FA. Metal binding was calculated from constants found for the particulate components separately and found to give fair fitting; Figure 14.14 shows examples for Cd and Co; note that Cd seems to be better predicted with R3 than R2, whereas the opposite is true for Co. It should also be noted that suspended colloidal particles in a water body or stream have less interaction between components than aggregate soil colloids, and thus that system should be expected to be better described by a multi-surface model.

Gustafsson and coworkers (Khai et al. 2008) applied a similar approach in a study of metal binding in selected Vietnamese soils. The multisurface models considered metal binding to three types of surfaces: iron (oxo)hydroxides (modeled using the DLM, Section 12.2.2), organic matter (using either the Stockholm Humic model or

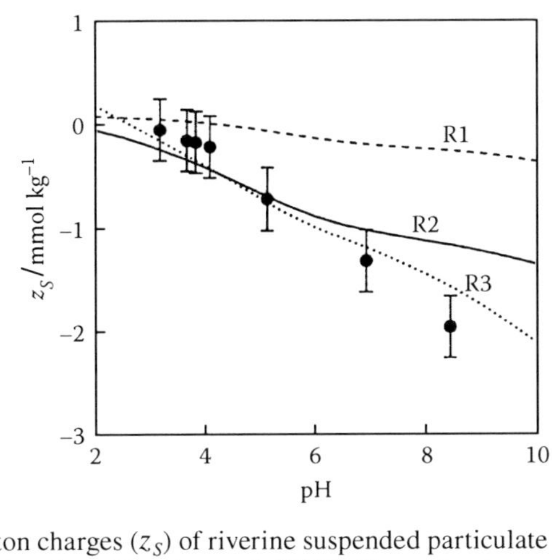

FIGURE 14.13 Proton charges (z_S) of riverine suspended particulate matter in 0.1 M NaCl. Points: experimental data from Ferreira et al. (1997); lines, simulations of the SCAMP models for different assemblage compositions. (Reprinted from *Geochimica et Cosmochimica Acta*, 62, Lofts, S. and Tipping, E., An assemblage model for cation binding by natural particulate matter, 2609–2625. Copyright 1998, with permission from Elsevier.)

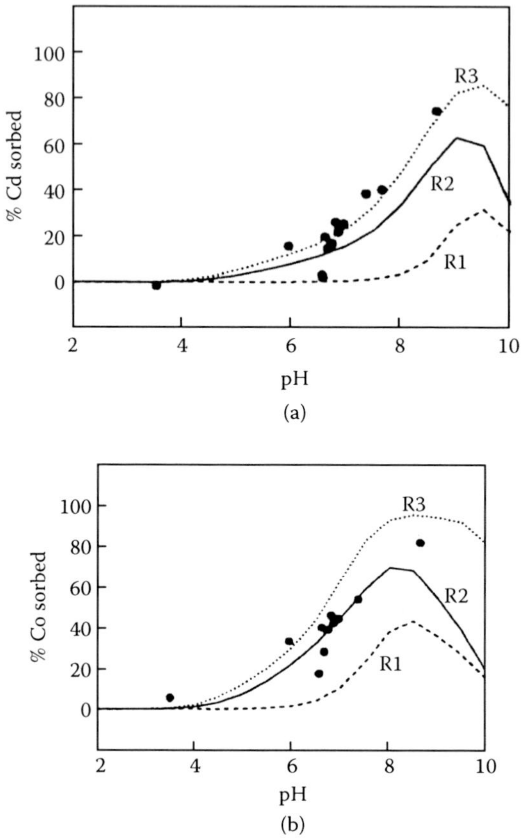

FIGURE 14.14 Metal adsorption by riverine particulate matter in 0.1 M NaCl: data for Cd (a) and Co (b). The points are measured values (Ferreira et al., 1997), the lines are predictions with SCAMP for different assemblage compositions. (Reprinted from *Geochimica et Cosmochimica Acta*, 62, Lofts, S. and Tipping, E., An assemblage model for cation binding by natural particulate matter, 2609–2625. Copyright 1998, with permission from Elsevier.)

NICA–Donnan model, Sections 13.4.2 and 13.4.3), and phyllosilicate clay (modeled as ion exchanger through the Gaines–Thomas equation, Section 5.3.2.2), as well as complexation to dissolved organic and inorganic ligands in the solution phase. Model parameters where taken from the literature: for oxide binding from Dzombak and Morel (1990) and Dijkstra, Meeussen, and Comans (2004); for humic proton binding from Gustafsson and Kleja (2005); for humic metal binding from Linde, Öborn, and Gustafsson (2007) and Milne et al. (2003); for ion exchange, all selectivity coefficients were taken as unity. It was found that for total dissolved Cd, Cu, and Zn, the two multi-surface models being tested provided good model fits for all soils; the results suggest that organic matter is an important sorbent for many metals in these soils. On the other hand, poor fits were obtained for Cr(III), Mn, and Pb for all soils. The study also suggests that the pH is the main factor that controls the solubility of metals in this type of soils.

In less-complex systems, modeling can be made simple; for example, in clay mineral dominated soils and sediments, ion exchange is expected to be the dominant soil–ion interaction and modeled as such. In a study of Zn(II) sorption onto clayey sediments (Tertre et al. 2009), a multisite ion-exchange model was employed. The experimental isotherms of Zn(II) on two different sediments at different pH and ionic strengths were interpreted assuming that clay minerals are the main ion-exchanging phases, and fitted using the previously obtained database on montmorillonite. Whatever the physicochemical conditions tested, a relatively good agreement is observed between experimental results and the predicted sorption behavior.

On a different approach, non-electrostatic modeling has been applied to describe the sorption of Pb and Cd in single- and binary-metal systems to three Spanish topsoils (Serrano et al. 2009). Both surface complexation and ion-exchange reactions were included in the model. Soil heterogeneity was modeled with functional groups, representing oxohydroxide minerals and phyllosilicate clay mineral edge sites, and two ion-exchange sites representing clay mineral exchange. An optimization process was carried out using the entire experimental sorption dataset to determine the binding constants for Pb and Cd surface complexation and ion-exchange reactions. The experimental data were well described by the model; however, as the authors describe, it is suitable for low organic contents soils.

An important aspect is the effect of soil dynamics on ion behavior. Albeit in this book soil dynamics is not addressed, it is worth mentioning a recent study (Bonten et al. 2011) where SCMs were combined with the soil dynamic model, Simulation Model for Acidification's Regional Trends (SMART; de Vries, Posch, and Kämäri 1989), which includes the effects of natural soil acidification resulting from the dissociation of CO_2 and the impact of the deposition of NO_x and NH_3. Parameters for SCMs were taken from generic datasets and not calibrated; modeling results for soil and soil solution concentrations for major elements in the period 1970–2010 matched the observations well. Furthermore, trace metals were included in the model, also using the existing framework of SCMs; the predicted sorption for most trace elements was satisfactory.

14.3.2 Incorporation of HS–Mineral Interaction: LCD Model

Albeit it can give reasonable results in some cases, the simple additivity of the amount bound to individual components cannot be expected to be of general application; component interaction must be taken into account. Many studies suggest that organic–mineral interaction has an important impact on ion binding, among other soil properties (see Section 14.1). Van Riemsdijk and coworkers introduced a model for such interactions, termed the *ligand and charge distribution* (LCD) model (Weng et al. 2006; Weng, Van Riemsdijk, and Hiemstra 2007), which was applied to the adsorption of phosphate ion on goethite (Weng, Van Riemsdijk, and Hiemstra 2008) and in soil samples (Weng, Vega, and Van Riemsdijk 2011). The LCD model combines the NICA-Donnan humic model (Section 13.4.3) with the CD-MUSIC model (Section 12.3), plus free-energy considerations to describe humic adsorption to mineral (especially oxide) surfaces.

In the adsorption of a humic molecule to a mineral surface, several reactions may occur simultaneously: formation of inner-sphere complexes between carboxylate groups and the mineral through ligand-exchange reactions, formation of outer-sphere complexes with carboxylate ($RCOO^-$) and hydroxyl (RCOH) groups, and protonation–deprotonation reactions. The overall adsorption reaction for an HS—for example a fulvic acid molecule—is written as (Filius et al. 2001, 2003)

$$(a+b)\mathrm{FeOH}^{-1/2} + (a+b+c)\mathrm{H}^+ + \mathrm{FA}^{z_h} \rightleftharpoons \mathrm{Fe}_a\left(\mathrm{FeOH}_2\right)_b \mathrm{FAH}_c^{(a+b)/2+c+z_h} + a\mathrm{H}_2\mathrm{O} \quad (14.5)$$

where FeOH represents surface hydroxyl groups, either singly or triply coordinated to Fe atoms; a is the number of surface groups forming inner-sphere complexes; b is the number of surface groups that form outer-sphere complexes; c is the number of protons forming non-coordinated RCOOH and RCOH groups; and z_h represents the charge of the fully deprotonated FA molecule. Reaction 14.5 has an overall effective equilibrium constant given by

$$\begin{aligned}\log K_{ov} &= a\log \overline{K}_{in} + b_1 \log \overline{K}_{out,1} + b_2 \log \overline{K}_{out,2} + \left(a+b_1+b_{2,1}\right)\log K_H \\ &\quad + c_1 \log \overline{K}_{H,1} + \left(b_{2,2}+c_2\right)\log \overline{K}_{H,2}\end{aligned} \quad (14.6)$$

where the subscript i in $\overline{K}_{H,i}$, b_i, $b_{i,j}$, and c_i distinguishes carboxylic (1) and phenolic (2) groups; j refers to the proton location: (1) is the case where proton is closest to the reactive site of the goethite surface, whereas (2) refers to where the proton is closest to the hydroxylic group; note that these coefficients are not constant but depend on experimental conditions. The first term in the right-hand side corresponds to inner-sphere complex formation, the next two correspond to outer-sphere complexes, the fourth term corresponds to protonation reaction of the metal oxide surface groups, and the last two terms correspond to protonation of the free groups of the FA molecule. The overbar indicates average constants of the NICA-type isotherm. The adsorption occurs at an electrified interface as schematized in Figure 14.15; thus, the reactions are affected by the potentials ψ_0 and ψ_1 at the respective planes, which is treated with the CD-MUSIC model (see Filius et al. 2003 for details).

The LCD model has been recently applied to phosphate adsorption on goethite in the presence of HS (Weng, Van Riemsdijk, and Hiemstra 2008) and to the sorption of phosphate in soils (Weng, Vega, and Van Riemsdijk 2011). In this work, goethite was used as a model of mineral oxides, which are the most important phosphate sorbents in soils, and FA was used as model of SOM. Thus, there are little changes in the modeling compared with goethite-FA laboratory studies. From six soils studied, five were well described by the model in the pH range 5–8. However, these soils have high Fe contents, so that using goethite as a mineral model is a rather logical step in this case. However, in soils with low oxide contents, this approach may not be appropriate; further studies are needed to test this model, at least in its present form.

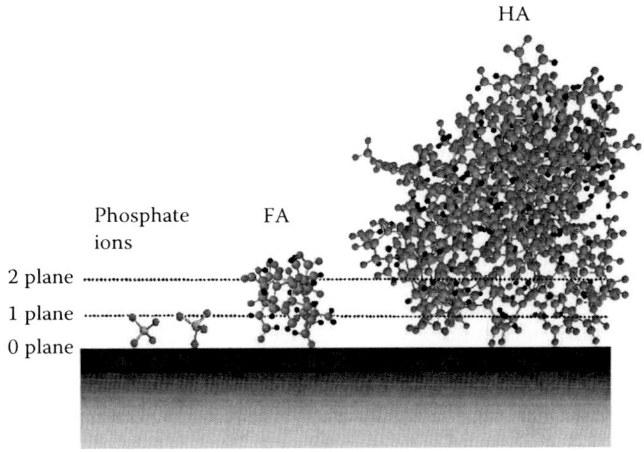

FIGURE 14.15 Schematic representation of phosphate, FA, and HA at goethite surface in which the extended Stern model is used to describe the structure of the electric double layer. (Reprinted with permission from Weng, L. et al., *Environ. Sci. Technol.*, 42, 23, 8747–8752, 2008. Copyright 2008 American Chemical Society.)

14.4 SUMMARY AND PERSPECTIVES

In this chapter, the problem of the analysis and modeling of ion binding in soils has been presented and briefly reviewed, trying to put forward the most important aspects. The sorption and retention of ions in soils cannot, in general, be regarded as the simple superposition of the adsorption to the individual components, because these interact among themselves (especially when SOM is present) altering their sorption capacity, among other properties. Notwithstanding that, in some studies, models based in additivity approaches have been applied with relative success, mainly for low-SOM soils. On the other hand, a model based in the mineral–humic interaction have been applied in some cases with good results, but more studies with different soil types are needed to ascertain its validity and applicability.

We have devoted a greater part of this book to the experimental features of ion sorption in complex systems as soils are and, as introduced in Chapter 1, particularly most of this third part is dedicated to its physicochemical modeling. The interest, as of this writing, is mostly oriented to models capable of predicting the state of chemical species with reasonable accuracy, either natural or pollutants, in a given soil under determinate conditions. Thus, to date most models are phenomenological, primarily based in experimental correlations; even when theories based on fundamental statistical mechanics principles may be desirable, this appears to be not feasible for the time being.

Modeling of ion sorption can be done in a simple but conditional way assigning some adsorption isotherm (Chapter 4), or in a more integrated way, so as to consider other variables such as pH, ionic strength, and so on. This can be done quite well for separate components, such as minerals either for ionic exchange (Chapter 5) or for specific adsorption (Chapter 12), and humic substances and SOM, in general

(Chapter 13). In the case of minerals, generic models such as the TLM, or more specific ones such as the CD-MUSIC model can be used with reasonable confidence; for HS, the NICA-Donnan, WHAM, or SHM models are rather equivalent regarding fitting and predictive capabilities. A difficulty with these humic models is that they appear to be overparametrized; in fact, the number of different parameters is to be expected from the point of view of the physical nature of the system, but the problem resides in that they can be calibrated only with equilibrium interaction curves (i.e., titration curves), which are difficult to measure with high accuracy. The Wageningen group introduced the exchange ratio as a separate measure (see Section 11.2.2.2), but because it involves differentiation of experimental data, it has not been used in the model calibration as yet. More experimental data are needed to improve in this respect.

For the whole soils, the problem of accurate modeling is far from solved. There are attempts based on a simple additivity rule, which are not expected to work in the presence of organic matter. Also, there is some recent application of the LCD model, which considers humic–mineral interaction and so appears more promising. However, this model is based on the CD-MUSIC one, which requires detailed knowledge of the mineral surface sites, which has to be obtained for many minerals and, besides, its applicability to natural samples is to be tested, due to their inherent variability (presence of surface defects, substitution, etc.).

The overall conclusion is that a lot has been done but a lot more remains to be achieved. A key factor is, as always, to obtain new experimental information, both qualitatively and quantitatively. The development of new and improved experimental methods is needed; great hopes should be put in X-ray absorption techniques, which provide key information about the chemical environment of a given atom, thus shedding light, among other things, on the chemical structure of sorbed species.

REFERENCES

Alcacio, T. E., D. Hesterberg, J. W. Chou, J. D. Martin, S. Beauchemin, and D. E. Sayers. 2001. Molecular scale characteristics of Cu (II) bonding in goethite–humate complexes. *Geochimica et Cosmochimica Acta* 65, no. 9: 1355–1366.

Alessi, D. S., and J. B. Fein. 2010. Cadmium adsorption to mixtures of soil components: Testing the component additivity approach. *Chemical Geology* 270, no. 1–4: 186–195. doi:10.1016/j.chemgeo.2009.11.016.

Avena, M. J., and L. K. Koopal. 1998. Desorption of humic acids from an iron oxide surface. *Environmental Science & Technology* 32, no. 17: 2572–2577. doi:10.1021/es980112e.

Bonten, L. T. C., J. E. Groenenberg, H. Meesenburg, and W. de Vries. 2011. Using advanced surface complexation models for modelling soil chemistry under forests: Solling forest, Germany. *Environmental Pollution* 159, no. 10: 2831–2839. doi:10.1016/ j.envpol.2011.05.002.

Calvet, R., E. Barriuso, and I. G. Dubus. 2007. Application of two surface complexation models to the adsorption of weak organic acids by soil: An additive approach. *European Journal of Soil Science* 58, no. 3: 609–624. doi:10.1111/j.1365-2389.2006.00846.x.

Carter, M. R. 2002. Soil quality for sustainable land management. *Agronomy Journal* 94, no. 1: 38–47. doi:10.2134/agronj2002.3800.

Christl, I., and R. Kretzschmar. 2001. Interaction of copper and fulvic acid at the hematite-water interface. *Geochimica et Cosmochimica Acta* 65, no. 20: 3435–3442.

Covelo, E. F., F. A. Vega, and M. L. Andrade. 2007a. Simultaneous sorption and desorption of Cd, Cr, Cu, Ni, Pb, and Zn in acid soils. I. Selectivity sequences. *Journal of Hazardous Materials* 147, no. 3: 852–861. doi:10.1016/j.jhazmat.2007.01.123.

Covelo, E. F., F. A. Vega, and M. L. Andrade. 2007b. Simultaneous sorption and desorption of Cd, Cr, Cu, Ni, Pb, and Zn in acid soils. II. Soil ranking and influence of soil characteristics. *Journal of Hazardous Materials* 147, no. 3: 862–870. doi:10.1016/j.jhazmat.2007.01.108.

Dijkstra, J. J., J. C. L. Meeussen, and R. N. J. Comans. 2004. Leaching of heavy metals from contaminated soils: An experimental and modeling study. *Environmental Science & Technology* 38, no. 16: 4390–4395. doi:10.1021/es049885v.

Dzombak, D. A., and F. M. M. Morel. 1990. *Surface Complexation Modeling: Hydrous Ferric Oxide*. 1st ed. Hoboken, NJ: Wiley-Interscience.

Ferreira, J. R., A. J. Lawlor, J. M. Bates, K. J. Clarke, and E. Tipping. 1997. Chemistry of riverine and estuarine suspended particles from the Ouse-Trent system, UK. *Colloids and Surfaces A: Physicochemical and Engineering Aspects* 120, no. 1–3. 183–198. doi:10.1016/S0927-7757(96)03721-1.

Filius, J. D., J. C. L. Meeussen, T. Hiemstra, and W. H. van Riemsdijk. 2001. Modeling the binding of benzenecarboxylates by goethite: The ligand and charge distribution model. *Journal of Colloid and Interface Science* 244, no. 1: 31–42. doi:10.1006/jcis.2001.7927.

Filius, J. D., J. C. L. Meeussen, D. G. Lumsdon, T. Hiemstra, and W. H. van Riemsdijk. 2003. Modeling the binding of fulvic acid by goethite: The speciation of adsorbed FA molecules. *Geochimica et Cosmochimica Acta* 67, no. 8: 1463–1474. doi:10.1016/S0016-7037(02)01042-6.

Garrido, T., J. Mendoza, and F. Arriagada. 2012. Changes in the sorption, desorption, distribution, and availability of copper, induced by application of sewage sludge on Chilean soils contaminated by mine tailings. *Journal of Environmental Sciences* 24, no. 5: 912–918. doi:10.1016/S1001-0742(11)60876-0.

Gasser, M. S., H. T. Mohsen, and H. F. Aly. 2008. Humic acid adsorption onto Mg/Fe layered double hydroxide. *Colloids and Surfaces A: Physicochemical and Engineering Aspects* 331, no. 3: 195–201.

Gustafsson, J. P., and D. B. Kleja. 2005. Modeling salt-dependent proton binding by organic soils with the NICA-Donnan and Stockholm Humic models. *Environmental Science and Technology* 39, no. 14: 5372–5377.

Hayes, M. H. B., and F. L. Himes. 1986. Nature and properties of humus–mineral complexes. In *Interactions of Soil Minerals with Natural Organics and Microbes*, ed P. M. Huang and M. Schnitzer, 103–158. SSSA Special Publications 17. Madison, WI: Soil Science Society of America.

Khai, N. M., I. Öborn, S. Hillier, and J. P. Gustafsson. 2008. Modeling of metal binding in tropical Fluvisols and Acrisols treated with biosolids and wastewater. *Chemosphere* 70, no. 8: 1338–1346.

Kretzschmar, R., H. Sticher, and D. Hesterberg. 1997. Effects of adsorbed humic acid on surface charge and flocculation of kaolinite. *Soil Science Society of America Journal* 61, no. 1: 101–108.

Landry, C. J., C. M. Koretsky, T. J. Lund, M. Schaller, and S. Das. 2009. Surface complexation modeling of Co(II) adsorption on mixtures of hydrous ferric oxide, quartz and kaolinite. *Geochimica et Cosmochimica Acta* 73, no. 13: 3723–3737. doi:10.1016/j.gca.2009.03.028.

Linde, M., I. Öborn, and J. P. Gustafsson. 2007. Effects of changed soil conditions on the mobility of trace metals in moderately contaminated urban soils. *Water, Air, & Soil Pollution* 183, no. 1: 69–83.

Lofts, S., and E. Tipping. 1998. An assemblage model for cation binding by natural particulate matter. *Geochimica et Cosmochimica Acta* 62, no. 15: 2609–2625.

Lund, T., C. Koretsky, C. Landry, M. Schaller, and S. Das. 2008. Surface complexation modeling of Cu(II) adsorption on mixtures of hydrous ferric oxide and kaolinite. *Geochemical Transactions* 9, no. 1: 9. doi:10.1186/1467-4866-9-9.

Martínez-Villegas, N., and C. E. Martínez. 2008. Solid- and solution-phase organics dictate copper distribution and speciation in multicomponent systems containing ferrihydrite, organic matter, and montmorillonite. *Environmental Science & Technology* 42, no. 8: 2833–2838. doi:10.1021/es072012r.

Milne, C. J., D. G. Kinniburgh, W. H. van Riemsdijk, and E. Tipping. 2003. Generic NICA–Donnan model parameters for metal-ion binding by humic substances. *Environmental Science & Technology* 37, no. 5: 958–971. doi:10.1021/es0258879.

Orsetti, S., M. M. Quiroga, and E. M. Andrade. 2006. Binding of Pb(II) in the system humic acid/goethite at acidic pH. *Chemosphere* 65, no. 11: 2313–2321.

Saito, T., L. K. Koopal, S. Nagasaki, and S. Tanaka. 2005. Analysis of copper binding in the ternary system Cu^{2+}/humic acid/goethite at neutral to acidic pH. *Environmental Science & Technology* 39, no. 13: 4886–4893. doi:10.1021/es0500308.

Saito, T., L. K. Koopal, W. H. Van Riemsdijk, S. Nagasaki, and S. Tanaka. 2004. Adsorption of humic acid on goethite: Isotherms, charge adjustments, and potential profiles. *Langmuir* 20, no. 3: 689–700.

Schnitzer, M. 1986. Binding of humic substances by soil mineral colloids. In *Interactions of Soil Minerals with Natural Organics and Microbes*, ed. P. M. Huang and M. Schnitzer, 77–101. SSSA Special Publications 17. Madison, WI: Soil Science Society of America.

Serrano, S., P. A. O'Day, D. Vlassopoulos, M. T. García-González, and F. Garrido. 2009. A surface complexation and ion exchange model of Pb and Cd competitive sorption on natural soils. *Geochimica et Cosmochimica Acta* 73, no. 3: 543–558. doi:10.1016/j.gca.2008.11.018.

Sipos, P., T. Németh, V. K. Kis, and I. Mohai. 2008. Sorption of copper, zinc and lead on soil mineral phases. *Chemosphere* 73, no. 4: 461–469. doi:10.1016/j.chemosphere.2008.06.046.

Sipos, P., T. Németh, V. K. Kis, and I. Mohai. 2009. Association of individual soil mineral constituents and heavy metals as studied by sorption experiments and analytical electron microscopy analyses. *Journal of Hazardous Materials* 168, no. 2–3: 1512–1520. doi:10.1016/j.jhazmat.2009.03.033.

Sposito, G. 1984. *The surface chemistry of soils*. New York: Oxford University Press.

Sverjensky, D. A., and N. Sahai. 1996. Theoretical prediction of single-site surface-protonation equilibrium constants for oxides and silicates in water. *Geochimica et Cosmochimica Acta* 60, no. 20: 3773–3797. doi:10.1016/0016-7037(96)00207-4.

Tertre, E., C. Beaucaire, N. Coreau, and A. Juery. 2009. Modelling Zn(II) sorption onto clayey sediments using a multi-site ion-exchange model. *Applied Geochemistry* 24, no. 10: 1852–1861. doi:10.1016/j.apgeochem.2009.06.006.

Tombácz, E., Z. Libor, E. Illés, A. Majzik, and E. Klumpp. 2004. The role of reactive surface sites and complexation by humic acids in the interaction of clay mineral and iron oxide particles. *Organic Geochemistry* 35, no. 3: 257–267. doi:10.1016/j.orggeochem.2003.11.002.

Vega, F. A., E. F. Covelo, and M. L. Andrade. 2009. Hysteresis in the individual and competitive sorption of cadmium, copper, and lead by various soil horizons. *Journal of Colloid and Interface Science* 331, no. 2: 312–317. doi:10.1016/j.jcis.2008.11.047.

Vermeer, A. W. P., W. H. van Riemsdijk, and L. K. Koopal. 1998. Adsorption of humic acid to mineral particles. 1. Specific and electrostatic interactions. *Langmuir* 14, no. 10: 2810–2819. doi:10.1021/la970624r.

de Vries, W., M. Posch, and J. Kämäri. 1989. Simulation of the long-term soil response to acid deposition in various buffer ranges. *Water, Air, and Soil Pollution* 48, no. 3–4: 349–390. doi:10.1007/BF00283336.

Weng, L., W. H. Van Riemsdijk, and T. Hiemstra. 2007. Adsorption of humic acids onto goethite: Effects of molar mass, pH and ionic strength. *Journal of Colloid and Interface Science* 314, no. 1: 107–118. doi:10.1016/j.jcis.2007.05.039.

Weng, L., W. H. Van Riemsdijk, and T. Hiemstra. 2008. Humic nanoparticles at the oxide–water interface: Interactions with phosphate ion adsorption. *Environmental Science & Technology* 42, no. 23: 8747–8752. doi:10.1021/es801631d.

Weng, L., W. H. Van Riemsdijk, L. K. Koopal, and T. Hiemstra. 2006. Ligand and Charge Distribution (LCD) model for the description of fulvic acid adsorption to goethite. *Journal of Colloid and Interface Science* 302, no. 2: 442–457. doi:10.1016/j.jcis.2006.07.005.

Weng, L., F. A. Vega, and W. H. Van Riemsdijk. 2011. Competitive and synergistic effects in pH dependent phosphate adsorption in soils: LCD modeling. *Environmental Science & Technology* 45, no. 19: 8420–8428. doi:10.1021/es201844d.

Yang, K., D. Lin, and B. Xing. 2009. Interactions of humic acid with nanosized inorganic oxides. *Langmuir* 25, no. 6: 3571–3576. doi:10.1021/la803701b.

Zachara, J. M., C. T. Resch, and S. C. Smith. 1994. Influence of humic substances on Co^{2+} sorption by a subsurface mineral separate and its mineralogic components. *Geochimica et Cosmochimica Acta* 58, no. 2: 553–566. doi:10.1016/0016-7037(94)90488-X.

Index

D

E

F

G

N

O

P